biology evolution and adaptation to the environment

Mahlon G. Kelly
University of Virginia

John C. McGrath

Houghton Mifflin Company

Boston Atlanta Dallas Geneva, Illinois
Hopewell, New Jersey Palo Alto London

biology evolution and adaptation to the environment

Illustrated by B. J. and F. W. Taylor

Cover photo: Ernst Hass, Magnum Photos, Inc.

p. i: Ira Kirschenbaum; pp. ii–iii: Frank Siteman from Stock, Boston; pp. iv, v: Ira Kirschenbaum; p. vi: © Jacques Jangoux; p. ix: Ira Kirschenbaum; p. x: Charlie Ott from National Audubon Society; pp. xiv–1: Nancy Lockspeiser from Stock, Boston; pp. 14–15, 253: Ira Kirschenbaum; pp. 472–3: Van Bucher from Photo Researchers, Inc.

Printed in the United States of America

preface

The publication of an introductory textbook in biology at the present is not an unusual event. Indeed, it is almost a commonplace one. Thus, it is necessary to consider for a moment the reasons for writing and publishing this book. It has been frequently and well said that the problems that will face mankind in this century and probably well into the twenty-first are essentially biological or have important biological components. It can also be said that compared to our understanding of other sciences, particularly the physical sciences of chemistry and physics, our understanding of biological sciences is relatively undeveloped. The questions concerning the growth of the human population, feeding a growing number of people, water and air pollution, and energy use and misuse are basically biological questions or have profound and significant biological implications. These are but a few of the problems that the readers of this book will have to face each day for most of the rest of their lives. They are not likely to be easily and quickly solved. Above all, they are not likely to be solved at all without a continuing expansion of biological knowledge on the part of scientists and nonscientists. One reason, then, for this book, is to examine some of the larger issues that face us and to attempt to convey not only what we know about the issues, but also to indicate those areas in which we lack sufficient knowledge to attempt solutions.

It is not enough for us to deal simply with the larger issues, since each of them is composed of a number of smaller components. An understanding of these subunits allows us to appreciate the complexities of life (in its broadest sense) in the late twentieth century. We are bombarded each day with news about decisions concerning world food supply, energy use, water pollution control programs, and so forth. The news media, whether printed or televised, usually cannot explore such decisions in depth because of self-imposed constraints of space and time. Nevertheless, we are expected to evaluate such decisions, and to do so with intelligence and honesty we must evaluate them on the basis of information that we can supply ourselves. Thus, another reason for this book is to supply some of the information, when available, that will allow the reader to function intelligently in today's world. Again, it is equally important to indicate those areas in which our knowledge is insufficient. It will certainly be necessary to make some decisions on the basis of insufficient data, but we must recognize that such decisions are subject to serious reconsideration in the light of new data that may become available.

Yet another reason for publishing this book follows from a trend in other books for the same course. In an attempt to please every reviewer's tastes, recent manuscripts have been reshaped into books of a thousand or more pages (''mega-tomes,'' as a friend of one of the authors has called them). In spite of its insufficiencies, our knowledge of biology has grown rapidly over the past twenty-five years, and so have biology textbooks. We anticipated that the students who would be reading this book, however, would not take any further courses in biology, and we were thus faced with having to make some choices about what should and should not be included here. Our first choice, a purely arbitrary and subjective one, was to attempt to write a book less than 600 pages long. Scientific progress does not take place evenly; rather one field moves ahead while another lags. Thus, textbooks tend to seem fragmentary in their scope if they attempt to cover all of biology. Since this unevenness of coverage has a tendency sometimes to frustrate the nonscience major, we choose to emphasize certain topics that, taken together, would lend a more narrative quality to the book. We hope that in so choosing our topics we have avoided the obstacle to learning that the frustration mentioned above can create.

An extremely important point in our considerations was the fact that in many recent books, evolution by natural selection, *the* unifying theme of all biology, has been explored in depth only at the end of the book. Yet it is evolution that is at the core of modern biology—it is evolution that has produced the dynamically balanced ecosystems of the planet upon which we live. We have chosen to reverse the more common approach and begin with the process of evolution. By doing so, we hope to be able to avoid producing the restlessness of students striving to see why a particular topic is treated at a particular time and how that topic relates to the others that have been examined. We think that our approach to the study of biology will enable the student to participate in the process of fitting each piece of knowledge into the whole of biology. A warning is in order, however: It will not always be perfectly and immediately apparent how each topic relates to every other one, and in some cases the student will have to suspend judgment temporarily as to the value of examining certain specific subjects at specific times. Well before the final two chapters of the book are reached, however, all should be clear.

acknowledgements

A number of persons have been of great assistance to the authors throughout the entire process of writing this book. We owe to them our deepest thanks for the many times that they helped us to avoid errors great and small. For such errors as may remain, we are wholly responsible.

Particular thanks are extended to Professors David Evans of the University of Miami, Floyd Grimm of Harford Community College, and Michael Soule of the University of California at San Diego for their careful and perceptive comments on various sections of the text. We are most deeply indebted to Gerard O'Donovan of Texas A & M University for his detailed and enlightening critique of the entire manuscript. We also owe an unpayable debt to our patient and forgiving wives and children: Gretchen, Jo, Clay, Scott, and Chrissy, all of whom paid the price of sacrificed weekends and evenings for longer than we like to recall.

We dedicate this book to three groups of people: our parents in the personal sense, the parents of our knowledge, and, as importantly, to the students who will read this book and will in turn use the knowledge that we have been able to impart.

contents

biology evolution and adaptation to the environment

part one introduction

A simple tree in a field is only the most obvious part of a community of hundreds of organisms, each with evolved biochemical and physiological adaptations that allow it to survive in the habitat of the community. *(Courtesy of Nicholas Sapieha from Stock, Boston.)*

the organization of diversity

chapter one It takes no special knowledge to recognize that the planet we live on is inhabited by a marvelous diversity of other living things. Life is everywhere. Even as this is being written, a proverbially desolate and inhospitable environment—Death Valley, California—is carpeted with a spring bloom of wildflowers. Unusually heavy rains have brought forth a profusion of life that has lain dormant for a long period of time. Nor is a blooming desert the only evidence of the universality of living forms. If we were to look for life anywhere on the surface of the earth, or in the depths of the ocean, we would almost certainly find it. Near the summit of Mount Everest man has found minute spiders, and in hot springs, where the water temperature is near the boiling point, we have found bacteria. In deep ocean trenches we have found fantastic species of fish, and were we to skim a quart of water from the surface of the ocean, we would find thousands of microscopic plants and hundreds of very small animals. Moreover, the composition of the atmosphere and the oceans, the makeup of many deposits of rock, and even the climates of some portions of the earth can be shown to be profoundly affected by the activities of plants, animals, and microorganisms. Very nearly all of the energy that man uses to maintain his civilization and economic well-being is the product of living or once-living things. Thus, life is not only present everywhere, but as a process can be shown to have shaped man's past, and to play a large role in his future. We say this not only because man is counted among the living things of the world but, more importantly, because man is completely dependent upon other living things.

Man and the Universality of Living Things

From the moment life begins for a single human being, he is in constant contact with other forms of life. Even a solitary man in a space capsule harbors, on his skin and in his intestines, a usually benign population of bacteria. The oxygen inhaled by a human being has been cycled through plants; the carbon dioxide that a human being exhales will, sooner or later, also be cycled through plants. Other chemical elements, such as hydrogen, nitrogen, phosphorus, sulfur, iron, and copper, which man incorporates into his body and is dependent upon for survival, are present in the earth or atmosphere and are taken up by plants through roots and cell membranes. In the plant, such elements are converted into complex organic chemicals that control life's processes. These organic compounds enter the human population as man eats the plants or eats the animals that have eaten the plants.

At the basis of this movement of materials from the earth to an organism, and from one organism to another, is energy. Chemical reactions occurring in the sun release vast amounts of energy. The greatest proportion of this energy never reaches the earth, and of that which does arrive here, only a small proportion is used in living systems. The solar energy of greatest concern to students of living systems is light energy. When the light energy of the sun reaches the earth, it is converted by plants into chemical energy through a process called *photosynthesis*. Photosynthetically produced chemical energy is the primary source of energy for all forms of plant and animal life.

Clearly, then, we cannot take a narrowly human-centered point of view toward life. Man is but one of an estimated 1,200,000 species of animals that depend upon 500,000 species of plants. Indeed, the catalogue of plant and animal species is expanded each year by an estimated 15,000 never-before-described species. Man is not merely one of many; he is a relative newcomer to the scene. Life is estimated to have arisen on this planet at least 3 billion years ago. A liberal estimate of the age of modern man is 500,000 years. Thus, for slightly more than 99.98% of the time that life has been present here, modern man has been absent. Yet, in the short time that we have been present, we have achieved a position of dominance over many other species. We have also often acted in ways that have disrupted the lives of other species. Since we depend upon other living things as thoroughly as we do, and since we also directly or indirectly intervene in the lives of the species with which we share this planet, it is incumbent upon us to try to understand not only the biology of man but the biology of other species as well.

Indeed, it is surprising that we understand so little about the effects of man on the species with which he interacts. And it is even more surprising that we so rarely apply our scant understanding when we contemplate some action. Until recently, when we built a city or a dam, disposed of our wastes, or cultivated a field of grain, we gave little thought to the impact of such activities on other forms of life—on our own life-support systems. As the physical and biological systems that support us have grown increasingly endangered, and as we have grown increasingly aware of our dependence upon these systems, we have begun to demand a more rigorous environmental accounting before undertaking the building of a highway or a pipeline. Among other things that have become painfully clear from such accountings is how little we know.

In attempting to understand man and his relationships with the living world, we must appreciate the unique and almost paradoxical position of the human species. The juxtaposition of two quotations from the Old Testament points up the paradox. The first, from Genesis, claims that man shall have

dominion over the fish of the sea, and over the fowl of the air, and over the cattle, and over all the earth, and over every creeping thing that creepeth upon the earth.

But from Ecclesiastes we hear

that which befalleth the sons of men befalleth beasts; even one thing befalleth them: as the one dieth, so dieth the other; yea, they have all one breath; so that a man hath no preeminence above a beast: for all his vanity.

Like all species of animals, man is a product of the process of evolution. As an animal species, we are related to, and interact with, other forms of life. Our relatedness and interactions are sometimes remote and indirect, but we are nevertheless a part of a vast, complex, and widely distributed system. Certain of our inherited abilities have enabled us to survive, and such abilities are firmly rooted in our biological past. At the same time, however, we have evolved to a state that is unique in the biological world: we are capable of abstract thought. This, like all other human abilities, is basically a biological phenomenon. Indeed, our closest relatives in the animal world have also been shown to be capable of grasping some abstract relationships. The difference lies in the *extent* to which the ability has evolved in us. We can predict the future because of our ability to think abstractly. We can imagine the existence of things that do not exist in nature, and we can turn imagined things into real things. Such artifacts, as we call them, have enabled us to control our environment, and we are thus able to live anywhere on the earth, at least for some period of time. Our capacity for abstract thought, or to give credit where it is due, our ancestor's capacity for abstract thought, has led to the development of culture. Through culture, we have been allowed to transmit information from one generation to the next through channels not available to other animal species. The latter are limited to a strictly biological information-transfer process—a constraint that we have been able to transcend through the evolution of language. Our unique ability to think abstractly is expressed in our tool-making, whether the tool be a crude chopper made of chipped stone or a space capsule, a book or a machine-driven combine for harvesting corn. We have even developed the ability to alter plant and animal species so that they may more suitably serve our purposes. We have, then, reached the stage at which other living things may be considered to be among our tools! Through culture and technology we have achieved an impressive degree of dominion over the earth and the things that live upon it. We can control our own environment, and we can control the environments of other species. But our control is not—we emphasize—it is *not* complete. In the very long term we are utterly dependent on the natural world. Because of our dependency, an attempt to understand the relationships between species of organisms, including man, and the relationships between organisms and their environments, may be among the most important endeavors we can undertake.

To understand the interactions of man and the rest of the living world, we must first look at some of the processes of life, such as evolution, adaptation, and the genetics of organisms and populations. We must then examine how organisms interact with each other and with their chemical and physical environments. With an understanding of such processes, we may then consider the evolution of man and examine how our evolved abilities have allowed us to survive in the living world. However, the living world is a vast, complex system, and in order to gain some understanding of the processes that take place in it, we must have some sort of structural framework upon which to build. Life itself suggests at least one such framework.

The Hierarchy of Life

One way of examining the processes of life would be to study units of varying complexity. Beginning at the simplest level, we could first consider atoms and then go on to study molecules, cells, tissues, organs, organisms, populations, species, communities, and ultimately the biosphere. (See box 1.1.) As we progressed from the simpler to the more complex levels we could consider the problems of sustaining life at each level.

As an alternative to such an approach, we could consider certain processes shared by all organisms, and examine how these processes are carried out at each level of complexity. Thus, for example, we can consider inheritance and look first at the genetics of cells, then of organisms, and then of populations. Indeed, that is the approach we shall take in this book, although we will not adhere to it with perfect rigidity. We shall review certain processes of life, and in dealing with them, we shall examine the implications of the process at each level.

One may ask why, in a book that claims to emphasize the biology of the environment, it is necessary to treat cellular and subcellular phenomena at all. If we are going to understand our own role in the biosphere, it is necessary for us to understand life at each of the levels on which it exists. For example, to understand what it means to pollute a lake, we must understand the chemistry of photosynthesis, as well as the multitude of chemical reactions that make up the metabolism of a cell. To understand man's effects on communities of organisms, we must understand the genetics and evolution of species and how these aspects of life are affected by the chemistry of the gene and the mathematics of population growth.

One of the first steps that we must take in studying living systems is to identify some basic unit of life. It has often been stated that the cell is such a unit. Certainly this is true if we are concerned primarily with how an organism functions internally, for it is the cell which reproduces and processes energy and materials. As we can see from bacteria and other single celled organisms, all the problems of living are solved within the limits of a single cell. On the other hand, some have claimed that the carbon atom is the basic unit, for it is the carbon atom that is the building block of all large biological molecules. Of course, it is impossible to define a basic unit of life that would satisfy everyone, since such a unit depends upon the interest of the observer. If we are interested in the development of an embryo or the chemistry of the gene, then the molecule and the cell are the basic units of life. If we are interested in the behavior of organisms or the interactions of one species with another, then the species is the primary unit with which we must deal, and we must therefore identify the species as the basic unit of life.

For our purposes, there are two reasons for using the species as the basic unit of life. First, it is the species with which man most often interacts. Man—especially the human populations we too condescendingly call "primitive man"—recognizes the many different species of animals and plants around him. He also recognizes how he interacts with them. Above all, he recognizes that they are distinctly different from each other. Later we will consider just what a species is, but for the present we can define a *species* as *a group of similar and related organisms that reacts to, and interacts with, its physical and biological environment in a characteristically unique way*. (This is essentially an ecological definition of the species; in Chapter 8 we will refine the definition in genetic terms.) We are referring to species when we refer to pine trees or cypresses, grass or weeds, carnivores (meat eaters) or herbivores (plant eaters). We also recognize our interactions with the species when we classify plants as edible or poisonous, or animals as benign or dangerous. In this context it is interesting to note the relative impoverishment of knowledge exhibited by a "civilized" American or European when compared to a so-called "primitive" man. A modern, civilized

From Atom to Biosphere

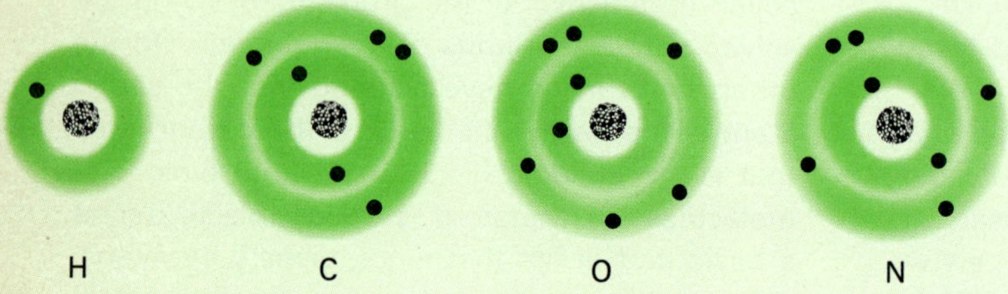

Atoms

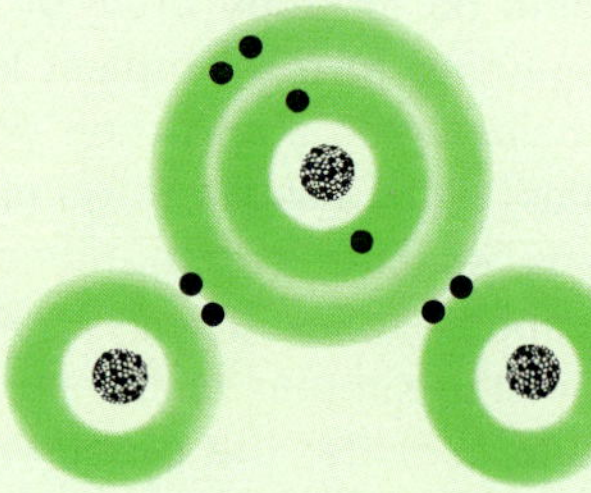

Water Molecule

(Keith R. Porter.)

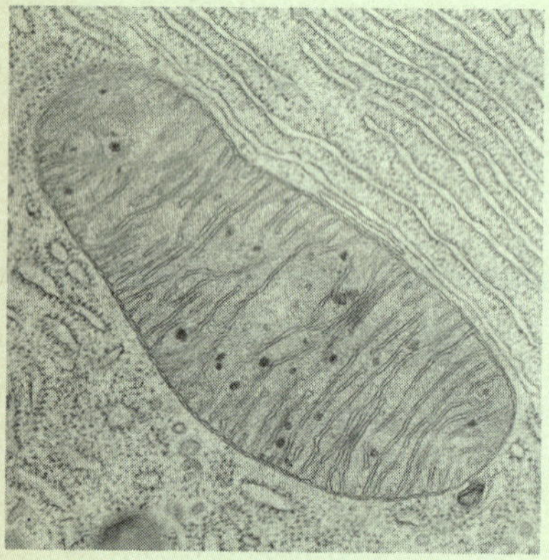

Mitochondrion

(Albert E. Vatter.)

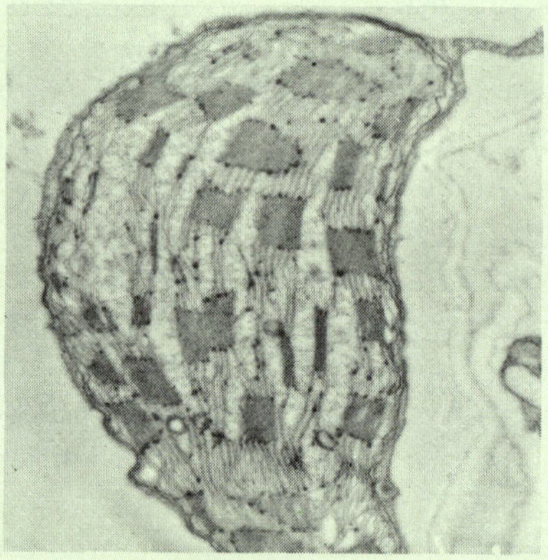

Chloroplast

Organelles

(Keith R. Porter.)

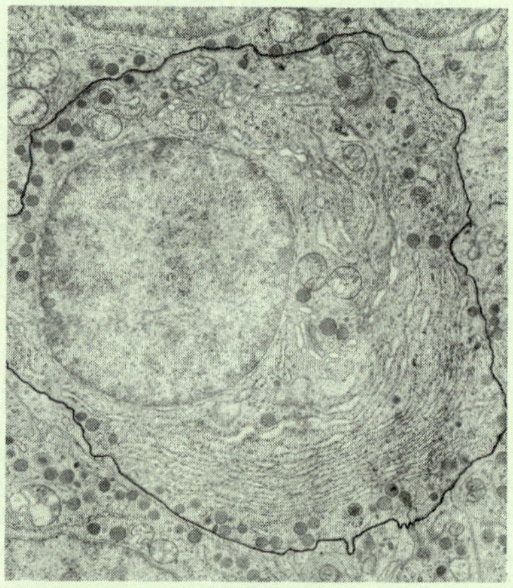

Cell

(Grant Heilman.)

Plant

(Russ Kinne, Photo Researchers, Inc.)

Animal

Organisms

BOX 1.1

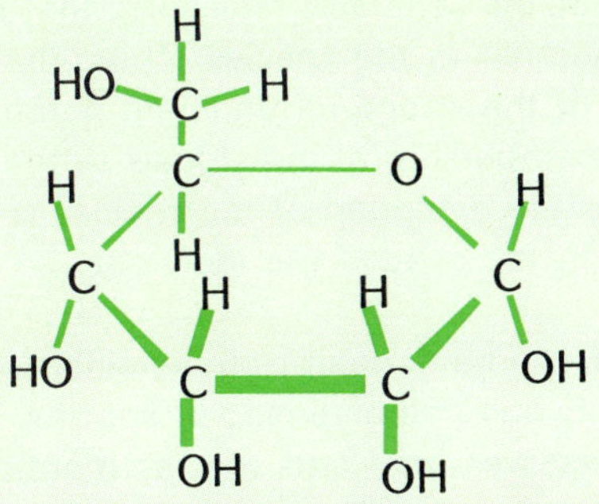

Glucose Molecule

(M. H. F. Wilkins.)

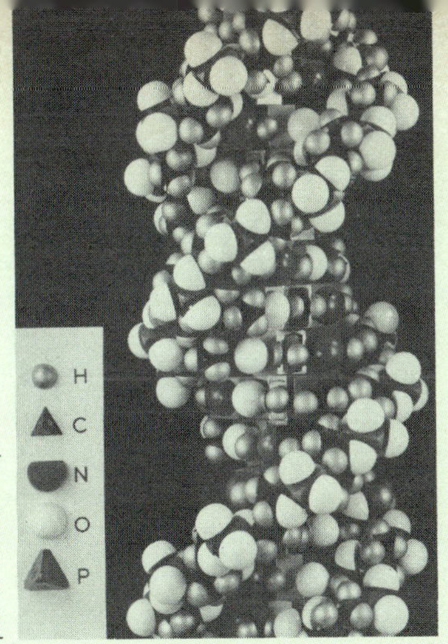

DNA Molecule

(Runk / Schoenberger from Grant Heilman.)

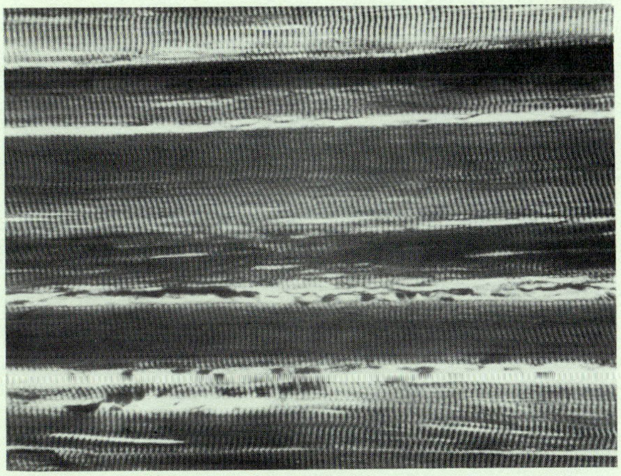

Tissue

(Camera M. D. Studios, Inc.)

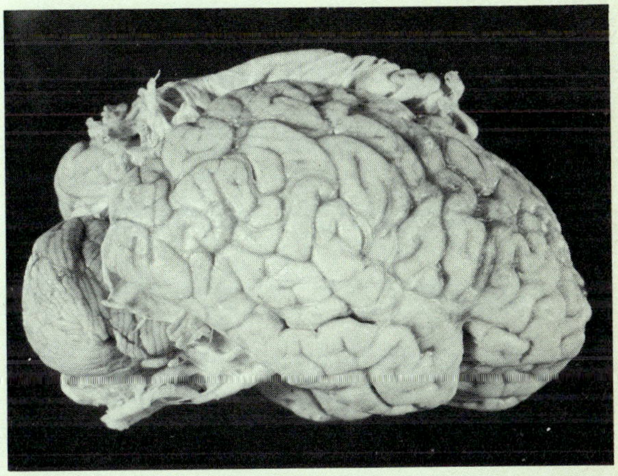

Organ

Population

(Grant Heilman.)

Community

(NASA.)

Biosphere

man recognizes very few species as resources of any sort. Were we to list only those species Americans regularly eat, we might arrive at a total of around fifty animals and plants, including such regional resources as abalone, cod, and okra. Australian aborigines recognize over eighty-five edible species of plants and animals, and they live in a desert! As we will see later, it is the species which interacts with other species to form communities, and the sum total of these interactions produce one form of what we call the "balance of nature." When we interact with communities, when we remove one species or promote the growth of another, when we introduce some element into the physical environment or remove some element from it, it is the species which is affected. We shall see frequent examples of such interactions throughout this book.

The second reason for considering the species as the basic unit of life is that it is the species which undergoes change through evolution. Over time, species arise from other species (*speciation*), and species become extinct. As we have said, and as we will later show, all members of a species are related to one another, since they are all descended from a common, ancestral stock. If the species reproduces sexually, all of its members are "cousins many times removed." As mating takes place among the members of the species, genes are exchanged over and over again. Such genetic exchange maintains the characteristic identity of the species that allows us to recognize individual organisms as members of some particular species. At the same time, as certain genes combine and change in ways that impair the ability of some individuals to survive and reproduce, such genes or gene combinations are eliminated from the species. As mating continues, as generations arise and die, and are replaced by new generations, the elimination of certain genes means that the species itself changes. We will devote most of the first part of this book to the details of this process as it is expressed at the levels of the cell, the organism, and the population. For as the characteristics of the species change, so too do the characteristics of the cells and molecules that make up the individuals in the species. Thus, the characteristics of the structural units are determined by the membership of individuals in the species, and the evolution of the species is expressed in the cells of the individual.

Although we have dealt in generalities up to this point, one characteristic of species, the *niche*, deserves brief and specific mention here. We can understand the niche in an intuitive way if we think of it as the role of a species in the community, as its "way of making a living." Within a particular community, a species may feed upon certain organisms and be fed upon by still other organisms; it may require certain minimal amounts of a nonbiological resource such as water; and it may only be able to tolerate certain ranges of temperature. All of these factors constitute a portion of a species' niche, which as we will see, may be very difficult to measure. We shall return to, and expand upon, the concept of the niche later, and we shall see that it is the niche which evolves and enables a species to survive, and it is the niche, therefore, which determines the interactions of species within communities.

As we said earlier, it is impossible to define a basic unit of life. Any such definition must be arbitrary and somewhat artificial. Species do not exist or evolve independently of the cells that make up the individuals, nor do cells and molecules exist and evolve independently of the species. However, we can identify, describe, and study the hierarchical levels of life separately to some degree. Studying them independently, though, involves the danger of losing sight of their interrelationships. Before beginning our study of the processes of life, then, it is desirable to examine a bit more closely the different levels of complexity that we will encounter and loosely tie together the interrelationships between them.

FIGURE 1.1 Mosaic of natural and domestic communities in Alaska. At least twelve separate communities are recognizable. *(Courtesy of Pro Pix from Monkmeyer Press.)*

The Living Landscape

No species consists of a single organism. Rather, in any area we were to observe, we would find *populations* of individuals. Because all of the individuals in the population are members of the same species, they interact with each other. They also interact, as a population, with populations of other species. The maple trees in a forest, for example, may constitute a single population that interacts with other populations of plants in the same forest. Exactly what constitutes a population is difficult to say. All of the individuals in a species may be thought of as making up a single population. We frequently refer to the "human population," meaning every human being on the planet. But we may also refer to the "Greenland Eskimo population," referring to a much smaller group—indeed, a relative handful of humans. What the precise limits might be in identifying some number of individuals as a population depends to some extent on what we are interested in learning about them.

As we have implied, no population lives alone—at least not for very long. If we were to spend a few hours driving—or better yet, walking—through almost any part of this country, we would encounter a mosaic of fields, meadows, clumps of trees, forests, and bodies of water. (See fig. 1.1.) Even in a sizable urban center, such as New York City or San Francisco, we could find more limited, but nonetheless revealing, examples of this mosaic quality of the environment. We would see communities, each consisting of several obvious, and many less obvious, populations. A *community* is *a definable array of populations which interact with each other*. As with our definition of a population, it is difficult to sharply limit most communities. Usually one community gradually gives way to another, although occasionally the boundaries between communities are easily identified. Within a large field, for example, we might encounter a sizable clump of trees which could be considered a community. The clump might contain only one species of tree, a small population of small mammals, such as mice, and some larger populations of insects, living either in or on the soil. The community would also undoubtedly contain populations of single-celled organisms, such as bacteria or protozoa. In defining the populations of this community, some species would pose a problem. For example, there might be some birds nesting in the trees, but because they are so highly mobile, could we consider them as a part of the community? Each of the other species we mentioned might spend all or most of its life in the clump and therefore be specific to the community. The birds might nest in the clump and interact in other ways with the other species there, but they would also probably spend a significant proportion of their lives elsewhere. The question is a rhetorical one, of course, and we use it to point out that a definition of a community also depends upon what we are interested in learning.

Although the populations in a community interact with each other, they also interact with their physical environment. Taken together, *the community and the nonliving environment make up the ecosystem*. The ecosystem is the basic unit of ecology, for it is at the level of the ecosystem that most ecological interactions take place. It is the ecosystem within which individuals survive and reproduce, and it is the ecosystem that determines the evolution of the species. Our example of the clump of trees as a community becomes an example of an ecosystem when we take into consideration the soil, light, water availability, and temperature in and around the clump. Another example of an ecosystem is a pond. Within the pond are a number of populations that constitute the pond community, and we could identify subcommunities in the pond. Thus, in the shallow water near the shoreline, we would find several species of plants, insects, worms, and bacteria not generally distributed throughout the pond. Each community in the pond, and the whole pond community, would interact with the water, of course, as well as with the light, temperature, and dissolved chemicals in the water. Indeed, the pond ecosystem would probably be a part of a larger ecosystem, which would include all of the populations and communities and physical conditions in the watershed of which the pond is a part.

For the species or populations to survive, of course, the individuals must survive. There is usually no difficulty involved in identifying an individual organism, although in some cases it is not as straightforward as you might think. Certain simple plants, for example, called *Volvox,* are made up of a number of cells enclosed in a jellylike capsule. (See fig. 1.2.) The cells of *Volvox* are not very different from one another, and if they are separated, each will develop into a new colony of cells. The question, of course, is whether to consider each cell an individual, or whether to designate the whole colony as an individual. Although certain organisms may present a problem of this sort, for the most part there is no difficulty in identifying individuals. Most of the organisms we see around us are relatively complex, multicellular ones, made up of highly specialized cells which function together and thus sustain life.

The cell is the smallest unit in which life can be observed, and it is at the cellular level that all the basic processes of life are carried out. But, the cell is not an undifferentiated thing in which life inheres. It is made up of a number of subcellular structures called *organelles* (''little organs''), the most important of which, for our purposes, are the mitochondrion, the chloroplast, and the nucleus. We shall examine the structure and function of chloroplasts and mitochondria in later chapters, but a few words are appropriate here. The *chloroplast* is the site in the plant cell where the light energy of the sun is converted into chemical energy that can be used by the plant or even stored elsewhere in the cell for future use. The *mitochondrion* is the structure in the cell where chemical energy is processed for use by the organism. Since plants are the only organisms which can convert solar energy into chemical energy, and since they also process the chemical energy for use, plant cells contain both chloroplasts and mitochondria. Nonphotosynthetic organisms (animals) are, obviously, made up of cells which contain no chloroplasts, but do contain mitochondria. Cells contain other complex organelles, such as ribosomes, which we shall discuss later, and still others as shown in figure 1.3. The cell is capable of reproduction, and when reproduction takes place, two usually identical cells are produced. Since the cell is a highly structured entity, its reproduction must be carefully controlled, and the center for such control is the *nucleus*. The nucleus is the organelle in which the genes, the units of inheritance, are located, and the genes control the precise organization of the cell. All of the functions of the cell take place within a *membrane*, the structure of which is not yet certainly known in chemical terms. It is through this permeable membrane that the materials necessary to sustain life must enter and the waste products of cellular metabolism must exit. The membrane cannot be thought of as a "sieve," however, since it appears to exercise almost absolute selectivity over the entry and exit of various materials. The cell membrane, then, is said to be *selectively permeable*.

Although life can be sustained at the level of a single cell, cells have evolved in certain ways that have given rise to multicellular organisms. A variety of hypotheses have been developed to explain the rise of multicellularity, and we shall examine some of them in forthcoming chapters. In general, however, it would seem that cells became specialized for particular functions and organized into *tissue* which performed such functions. Thus, in a multicellular organism, we find tissue that is specialized to produce move-

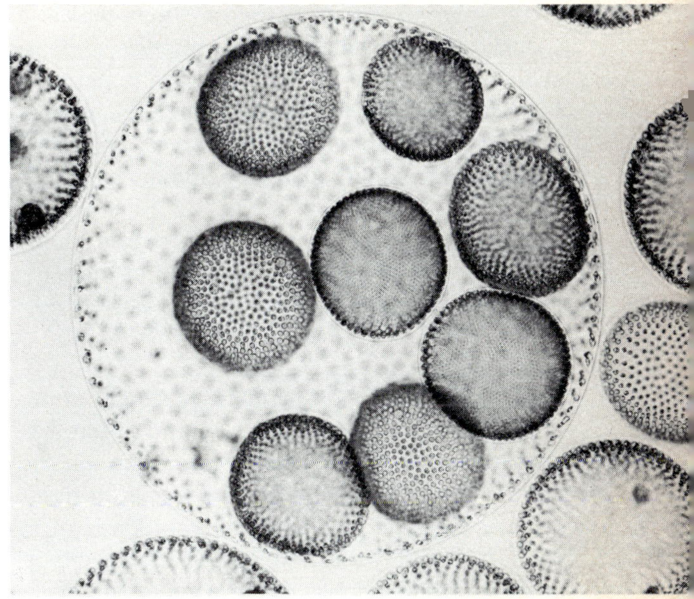

FIGURE 1.2 *Volvox.* A large colony with daughter colonies inside is surrounded by smaller nonreproducing colonies. *(Courtesy of Carolina Biological Supply Company.)*

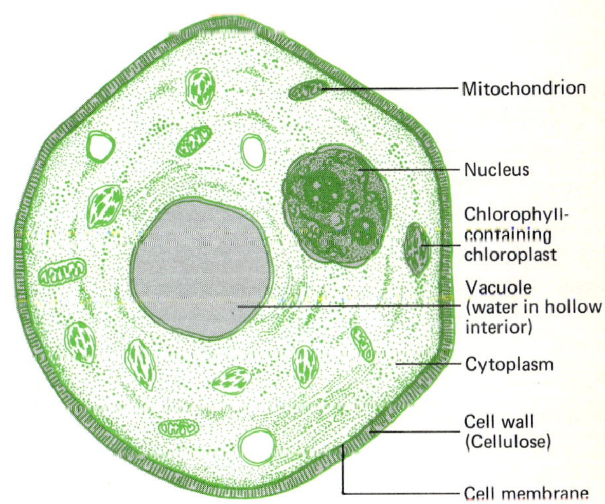

FIGURE 1.3 Generalized plant cell.

ment, such as muscle tissue, or tissue that is specialized for the delivery of materials to cells at a distance from the source of the material, such as blood. Still other tissue may be specialized for the reception and delivery of environmental information, such as nervous tissue.

Obviously, if tissue is specialized, then a multicellular organism must be made up of more than one sort of tissue. But clearly, an individual is not simply a mass of randomly distributed tissues. Just as a cell contains structures that perform specific functions, the body of an individual is made up of organs and organ systems which are specialized to carry out specific functions and are made up of specialized tissue. Usually, an organ contains different kinds of tissue. The brain, for example, is made up almost entirely of nervous tissue but contains other forms of tissue to supply the organ with blood.

Thus, we have species of individuals made up of cells, tissues, and organs that enable them to survive and to fit a niche. These individuals, in turn, survive within populations, communities, and ecosystems. Although we have emphasized the diversity of living things and the diverse ways in which we may look at them, the process that we call life is common to all of them, and thus we might expect that they all share certain problems associated with sustaining life. As we shall see in greater detail later, among the more important requirements shared by all living things are the needs to transfer energy and to cycle materials.

Let us first consider energy. As we have said, solar energy is converted by plants into chemical energy that can be stored or used by the plant. We shall see later that the storage and use of energy depends upon the structures of certain biological molecules. Thus, the capture and transfer of energy is a subcellular event. In a large, multicellular plant such as a palm tree, many cells do not participate in the conversion of solar energy, so energy must be transferred within the plant from cell to cell. As the plant grows, energy is used for the growth of cells and for the production of new cells. At some time in its life, the plant may be eaten by an animal, and then the chemical energy of the plant will be transferred to the animal. Thus the energy captured in a population of plants in a community is transferred to populations of animals. As animals in one population eat those in other populations, the energy is further transferred through the community along a ''food chain'' that begins with the plant. For reasons that we shall examine later, one characteristic of energy is that as it is transferred from plant to animal and then to other animals, less and less of it is available to each consumer. Thus, energy can be thought of as *flowing through* the biosphere in one direction, always diminishing, and thus always requiring replenishment from the sun.

In contrast to the one-way flow of energy in the biosphere, materials tend to cycle through the system. Like energy, materials are transferred through food chains from plants to animals, but unlike energy, the materials do not diminish in quantity as they pass from population to population. Ultimately, the materials taken up from the earth or the atmosphere by plants are returned to the earth and the atmosphere to be taken up again by other plants. Among the important materials that we will consider later are carbon, hydrogen, oxygen, nitrogen, and phosphorus.

Both the flow of energy and the cycling of materials are processes that are basic to man's relationships with the living world. Indeed, they are critical to our dependence on other living things and are therefore critical to our survival. If it is clear that biological systems are of immense importance to man, then it is also clear that we must understand something of the processes of life if we are to make intelligent, constructive decisions about our activities and the ways they affect other living things. In other words, everyone must understand the biological implications of human activities. To understand such implications it is not sufficient to know only how organisms function internally. We must concern ourselves with higher levels of organization, because it is at these levels that our activities have the most immediate and powerful impact. How then do we proceed toward an understanding of living systems and the implications of our activities?

Assuming that understanding the past is a valid means of understanding the present, we will first look at the process and products of evolution, since it was by this process that we arrived at the present state of affairs. We will then examine how organisms and cells function, for the processes of the biosphere work through the cells of the organism. Cells are, so to speak, the grist of the evolutionary mill. We will then look at the mill itself, at the processes of inheritance and genetics, and at how these processes are expressed in populations, which are the grinding wheels of the mill. Next we will take up the subjects of energy transfer and the cycling of materials through communities, for these processes are the ones that man depends upon and most strongly affects by his activities. Then we will consider how energy transfer and material cycling affect the growth of populations and how populations interact within communities. We will include a discussion of human population growth—what it means and how it happens—as well as a description of man's interaction with the communities of the biosphere. Finally, all of these topics must be put into the perspective of human evolution and the development of culture. Hopefully, proceeding in this way will produce a whole, although complex, portrait of the present and will allow us to make some tentative statements about the future.

A Yale University ecologist, G. E. Hutchinson, entitled one of his books, *The Ecological Theatre and the Evolutionary Play*. Man has been on stage but a moment or two, and yet in that time he has redesigned the set and rewritten the roles of many of the actors. This book is an attempt to provide a very, very brief synopsis of the plot and a description of the setting, so that we may try to become less destructive players while improvising on the original scenario.

part two
evolution and adaptation in the environment

The notched shell of one species of Galapagos Islands tortoise enables it to browse on high vegetation (a), whereas another species (b) does not have the notch and feeds closer to the ground. The evolution and speciation of such organisms on these islands were important factors in the formation of Darwin's ideas on the origin of species. *(Courtesy of Ira Kirschenbaum.)*

the mechanism of evolution

chapter two Man has speculated on nature, the source of life, and life's diversity for thousands of years. Indeed, the literature of any culture includes myths explaining the diverse forms of life. A number of explanations were in circulation in the nineteenth century, including the idea that the different forms of life had evolved from one another. Until the publication in 1859 of Charles Darwin's *On the Origin of Species by Means of Natural Selection,* however, the idea of evolution was considered somewhat unscientific and was entertained by proportionally more laymen than scientists. In fact, in the thirty years preceding the publication of Darwin's book, many prominent natural historians had written essays attacking the idea of evolution or supporting the idea of "special creation" in opposition to evolution. According to this theory, each species of animal or plant could be seen in the present as an exact representation of what it had been in the past; each had always been distinct and unique and would always remain distinct and unique. Even Darwin's closest associates and colleagues adhered to special creationism, and during the twenty or so years in which Darwin developed his ideas about evolution he refrained from discussing them at length with other scientists. (See box 2.1.)

In 1858, scarcely two years after he had started to work on the manuscript of *On the Origin of Species,* Darwin received a paper from Alfred Russel Wallace, a naturalist studying in southeast Asia. Wallace's ideas were so similar to Darwin's that the latter sent the paper, as well as one of his own, to a scientific society to be read. A year later Darwin's book was published, and the first printing was entirely sold out on the day of publication.

The effect of the book, both on the study of biology and on modern thought, was revolutionary. The impact of the Darwinian theory of evolution on modern thought has been of enduring importance, particularly in the areas of man's thought about himself. We shall briefly examine this aspect of evolution later. For the biological sciences, the book provided a conceptual framework that brought order to immense masses of accumulated observations about the anatomy, physiology, behavior, and the environmental interactions of plants and animals. It changed the concept of the static species to a concept of the species as a dynamically changing population of varying individuals.

The immediate effect of the publication of *On the Origin of Species* was to alter the thinking of most biologists about the operation of living systems. In part, this abrupt about-face can be attributed to the skill with which Darwin presented his argument. It is also partially explained by the immense mass of data that Darwin brought to bear on the topic. Just as important, perhaps, is the basic simplicity of the idea itself. Once Darwin articulated the process of evolution by natural selection, the logic of the concept became immediately apparent to almost anyone who was willing to look. T. H. Huxley, one of the foremost anatomists of the time, said after reading Darwin's book, "How extremely stupid not to have thought of that." The fact is that in a sense, the process had

Darwin and the Beagle

The undergraduate career of Charles Darwin (1809–1882) gave little hint that he would become one of the world's leading thinkers on any subject. He had withdrawn from the study of medicine at Edinburgh to pursue the study of theology at Cambridge, and in his autobiography he tells of having taken some "incredibly dull" courses in science. He had been an avid collector from his youth, and although his interest in collecting beetles became a passion of sorts in his undergraduate days, he did not dissect many of the specimens he collected, nor did he even bother to take any measurements of them. On the whole, it appeared to Darwin's father that Charles might become "an idle sporting man," more interested in autumnal shooting trips to Scotland than the pursuit of a profession like medicine or the clergy.

Darwin as a young man. *(Courtesy of Medical Illustration Unit, The Royal College of Surgeons of England.)*

In September, 1831, nine months after his graduation from Cambridge, Darwin learned of the availability of a post as naturalist aboard HMS *Beagle.* The ship was to leave soon on a voyage around the world to chart harbors frequented by English vessels, and the captain, Robert Fitz-Roy, was willing to share his cabin with "any young man who would volunteer to go with him without pay. . . ." It seemed that Fitz-Roy wanted a naturalist along to collect evidence to support the Biblical account of creation. With the help of his uncle, Darwin convinced his father to let him take the position, and soon after Christmas, 1831, the *Beagle* sailed, not to return to England until early October, 1836.

Although Darwin began the voyage with some familiarity with theories of evolution, he was a believer in special creation. As the voyage wore on, despite bouts of seasickness and an extremely heavy work schedule, Darwin read Lyell's *Principles of Geology.* Lyell's work and his own observations slowly raised questions in Darwin's mind regarding the validity of a literal interpretation of the Biblical story of creation and the age of the earth. By the time that the *Beagle* reached the Galapagos Islands, in 1835, Darwin may already have found his earlier position untenable. If not, his observations of the flora and fauna of the Galapagos marked the beginning of his conversion to the theory of evolution. Ultimately, his observations in these "enchanted isles," as Herman Melville called them, led to the formulation of the theory of evolution by natural selection.

BOX 2.1

been a part of man's cultural equipment all the time. Among the most persuasive pieces of evidence for natural selection was the process of "artificial" selection, an activity that had engaged man's efforts for thousands of years.

Artificial and Natural Selection

As a human activity, artificial selection predates written history. The artificial selection of species of grasses has provided man with the greatest bulk of his foodstuffs throughout history and indeed is in part the basis for the high levels of agricultural productivity that man has achieved in the recent past. Some of the side effects of the activity, as we shall see, have had extremely important environmental consequences. For our present purposes, however, it is only necessary to discuss the basic features of artificial selection.

All breeders of plants or animals operate in essentially the same way. Knowing that individuals within any population differ from each other, and that offspring tend to resemble their parents, the breeder merely selects those individuals that possess characteristics that he deems desirable. He encourages reproduction among the individuals he values and discourages reproduction among the individuals he does not value. Thus, in the next generation the breeder's population will contain a greater proportion of desirable individuals than in the preceding generation. Since any pair of parents can produce more than two offspring, the breeder is assured that within some limits, and given some time, he can produce a population in which very nearly all of the individuals may have a particular desirable characteristic.

If the characteristics of a population of animals or plants can be radically altered through the activities of man, Darwin argued, then there is *no reason to believe that a similar process cannot occur in nature without the intervention of*

The Beagle *at Sydney Harbour, 1841. (Courtesy of The National Maritime Museum, London.)*

man. The essential difference between the processes of artificial and natural selection is that, in the latter, the selection involves *no element of human choice.* In artificial selection the will of man is involved; in natural selection, survival and reproduction involve no conscious will. Just as the breeder prevents reproduction among those individuals he does not value, conditions of the environment prevent the survival and reproduction of certain individuals. Thus, survival within the framework of an ecosystem produces those organisms best adapted to the ecosystem. The lack of a conscious will, of course, was a factor that disturbed many of Darwin's opponents.

Evolution by natural selection is basically a two-stage process: (1) the production of variation, and (2) the selection of variants. For our purposes, there are four observable steps in the process.

First, is *variation.* It can easily be observed that any population of organisms comprises individuals that differ from one another. We are most aware of such variation in the organisms with which we are most familiar, such as the variations among humans, puppies in a litter, or house cats. Variations are less evident, but no less frequent, in populations of mosquitoes, sequoias, or cockroaches.

Second, we need no understanding of the details of genetics to appreciate that *offspring inherit the characteristics of their parents.* Although Darwin's ideas concerning genetics were about as wrong as they could be, they did not prevent him from formulating an essentially valid theory. An understanding of modern genetics greatly enhances our understanding of the basic process of evolution, but is not a prerequisite.

Third, it can also easily be observed that a pair of sexually reproducing parents can produce more than two offspring. This *reproductive potential* may itself vary from species to species. Thus, a human female may give birth to only one offspring at a time, but over her entire reproductive life she may have twelve or more children; a female mosquito may lay several hundred eggs at a time; a plant may produce several hundred or thousand seeds; and a parent pair of oysters may produce an average of one million offspring each time they reproduce.

Each of these observations was apparent to Darwin and to all of his contemporaries. Darwin, however, also saw the interrelationships between the observations. The key observation involves the reproductive potential of organisms. Since it can also be observed that the populations of organisms do not usually differ greatly from one generation to the next, then something must act to limit population growth. In a population under the control of a plant or animal breeder, the growth is limited by the activities of the breeder. By selecting only certain organisms to reproduce, the breeder promotes the survival of some and selects against the survival of others. Thus, within the breeder's population there is *differential survival*; that is, not all organisms survive equally well. The qualities that determine survival in the breeder's population are the variations that confer some advantage on the breeder. Could it not also be the case, under natural conditions, that the variations that confer some advantage on the organism itself may determine survival?

Thus, the *fourth* step is *differential survival*, or natural selection. Darwin defined natural selection as the "preservation of favourable variations and the rejection of injurious variations." Survival itself, of course, is not sufficient for evolution to take place; the organism must also reproduce. When surviving organisms reproduce, the variations that they possess are passed on to the next generation. Because the variations of the nonsurviving individuals are not passed on, the characteristics in the next generation are those that have been contributed by the survivors.

Understanding the concept of differential survival is of key importance in understanding the living world, for it implies that species become well adapted to their environments. Thus, the "balance of nature" is the result of the survival and reproduction of organisms most suited or "fit" to survive within the system. Throughout the rest of this book, we will see that adaptations for survival are of basic importance to the functioning of communities and ecosystems.

We must point out two things here. First, the only thing that is "required" of a species is that it survive and reproduce. Second, we can not ascribe a "purpose" to life, since the process goes on without a conscious plan or "higher purpose." That there may be no higher purpose to life is often a difficult concept for us to grasp, let alone live with, because we are used to perceiving things which have been obviously planned or designed. The irony of saying that there is no purpose, of course, is that the statement is not subject to logical proof; we cannot *prove* that a purpose does not exist. Nevertheless, starting from this paradoxical position, modern science has made enormous strides in understanding the natural world. The reason is simple: if no higher plan exists, then the answer to a question about nature can only be found *in* nature, not elsewhere.

Since variation, inheritance, reproductive potential, and differential survival are basically important to the evolutionary process, let us consider them in some more detail.

The Production of Variation

Variation

If all individuals in a species or population were identical, then selection would produce changes in the numbers of individuals but could not affect any of the characteristics of the individuals. Selection acts upon the small differences that exist among the individuals in any population of organisms. We should note that the variations that occur in a population are not limited to the outward appearance or behavior of an individual but may be expressed in numerous less visible ways as well. For example, on the average, the size of the brain of a contemporary human being is about 1350 cubic centimeters, and yet the brain of the Russian novelist Turgenev was measured at 2012 cc. and the brain of the French essayist and novelist Anatole France at barely half that, or 1017 cc. Physiological activities such as digestion, metabolism, and the rate of muscle contraction, among any number of other things, are also subject to variations within any population. The *physically manifested characteristics* of an individual are called the *phenotype* (Gr. *phaincin* to show + *typos* type). The phenotype of an organism is determined in part by its inheritance but also by its environment, and it *directly determines the ability of the individual to survive.* The inherited portion of an organism's phenotype is the result of its *genotype*; or *an individual's complete and unique genetic constitution.*

In nature it is often very difficult to determine how much phenotypic variation is due to genetic and how much due to environmental influences. Obviously, there are some characteristics shared by all members of a species or population which can easily be attributed to the heritable portion of the phenotype. All humans usually have five fingers on each of two hands and two eyes, the color of which is determined by heredity and appears never to change under changing environmental conditions. Less visibly, each human has one of only a few types of blood—A, B, O, or AB—and under varying conditions of the environment the blood type of an individual does not change. In contrast, certain physically manifested characters change readily as the conditions of the environment change. In a white human population, the amount of exposure to direct sunlight can be seen to alter the pigmentation of the skin, and the amount and quality of food eaten by any human can affect the size of his body. The amount and nature of the pigmentation in the skin and the shape and growth rate of bones are characteristics that are to some degree under the control of the genotype. As we shall see in Chapter 7, sorting out the relationships between the genotype and the phenotype is not a simple matter. For the present, we can say that the genotype contains the basic, inherited "plan" of an individual and determines how that individual develops and reacts in response to the environment; that is, the genotype indicates the upper and lower limits of the individual's body size, weight, metabolism, rate of growth, and age. The environment (diet, disease, exercise, and other factors) acts within the genotypically established limits to produce the phenotype. The implications of this relationship, to which we shall return later, is that the genotype contains more potential variability than is expressed in the phenotype.

While the survival of the individual is determined by the phenotype, it is the survival of the genotype which is of importance in evolution, since it is the genotype that is passed on from parents to their offspring. Let us briefly consider inheritance, then.

Inheritance

Just as we shall later return to the problem of the relationships between the phenotype and the genotype, so shall we also return to the subject of heredity in Chapter 6. What follows here is an oversimplified and superficial account of inheritance, but one which nevertheless will allow us to more fully appreciate the concept of evolution by natural selection.

Almost all organisms are the product of the fusion of two germ cells, or *gametes* (Gr. *gamēte* wife, *gamētēs* husband) produced by the parent organisms. Such fused gametes form the *zygote* (Gr. *zygotos* yoked) from which the individual grows and develops. Thus the gametes are direct, physical links between generations. Each gamete contains *chromosomes,* which in turn contain very large molecules of deoxyribonucleic acid (*DNA*), one of the two important nucleic acids we shall consider. Biological information is passed from one generation to the next by the *genes*—many thousands of them, arranged in a very precise order on the chromosomes in the nuclei of sperms and eggs. Each gene contains, coded upon it, the information for the production, or *synthesis*, of a specific protein. This biological information is stored in the sequence of the *bases* which make up the molecules of DNA. The genetic information of each gene is *transcribed* (in the same language, the language of nucleic acids) via a messenger, namely ribonucleic acid *(RNA),* and a specific protein is synthesized, or *translated* (into a different language, the language of proteins) from the message encoded upon such messenger-RNA.

Proteins, in turn, control most of the functions of the cells, tissues, and organs of the individual. Thus DNA dictates how the organism develops, how its form and function are arrived at. *Mutations* are changes in the structure of DNA, or even whole chromosomes. Such changes give rise to variations in the genotype and, via the proteins, produce changes in the phenotype. If the variations so produced favorably affect the survival of the individuals which are heir to them, they will be passed on to succeeding generations. In a sense, then, we can understand natural selection as the survival of specific structures of molecules; that is, the structure of the DNA in the chromosomes of a particular individual is the eventual product of natural selection.

In summary, we may consider *genetics* to be *the mechanism by which biological information is transmitted from generation to generation.* As we shall see, genetics is *much* more complex than we have presented it here, and indeed, it is also one of the means by which information is transmitted from cell to cell within the body of a developing organism.

The implication of this mechanism of biological communication is that *only those variations contained in the genotype can be transmitted from parent to offspring.* One of the theories of evolution that preceded Darwin's stated that the characteristics acquired by an organism during its lifetime would be transmitted to its offspring. Darwin did not wholly accept this interpretation of evolution, but lacking a knowledge of the mechanisms of genetics, he was hard put to refute it. In fact, he admitted the possibility that some acquired characteristics might be transmitted to the offspring.

There is another very important implication to be drawn from an understanding of the mechanisms of genetics: *the variations which are selected exist prior to the actual operation of selection.* Selection does not produce variations which then survive or fail to survive. The variations arise first, and then selection determines which shall and which shall not survive and be communicated to succeeding generations. Variations that are selected are called *adaptations.* In a sense this is a poor term, since it means something slightly different in ordinary usage than we wish it to mean here. In common usage "adaptation" refers to an adjustment or accommodation made in response to certain circumstances. In the biological sense of the word, however, an adaptation is a characteristic of an organism that arises *without any reference to environmental circumstances*, and then is "tested," by the environment to determine its value for the survival of the organism. A characteristic may be *adaptive* and thus contribute to, or enhance, the probability of survival; but it may become *maladaptive* and contribute to the failure to survive or reproduce.

Variations in the phenotypes of individuals are then selected by factors in the environment. We shall now examine briefly the next two steps in the process.

The Selection of Variations

Reproductive Potential

Obviously, any species or population must reproduce at least as many offspring in each generation as there are parents in the preceding generation if the population is not to become extinct. In fact, some organisms produce immense numbers of offspring and rely on a high *rate of production of new individuals,* or a high *natality,* to insure that sufficient numbers survive. Among such organisms the *death rate,* or *mortality,* is usually also very high, but in the long run enough offspring survive to insure the success of the species. We mentioned that a pair of oysters may produce as many as one million offspring each time they reproduce. The reason that the seas have not been filled with oysters is that, on the average, only two individuals out of the million offspring survive to reproduce. It is clear that the reproductive potential of one species may vary greatly from that of other species. It is also clear that in spite of the huge differences that may exist, all organisms can produce more than two offspring per pair of parents. Any population in which there are, on the average, fewer than two surviving offspring per parental pair is headed for extinction.

As we shall see in greater detail in Chapter 12, it can be shown that populations increase geometrically. A common analogy demonstrates this quite readily. If \$1000.00 is left on deposit in a bank to collect interest at the rate of 5% per year compounded annually, in the first year the interest will amount to \$50.00. In the next year, the interest will draw interest, and so the amount upon which the rate is computed is \$1050.00. The interest on that amount is \$52.50, and the deposit grows to \$1102.50. At the end of ten years, the \$1000.00 will have accumulated interest in the amount of \$795.85. Like money in the bank, individuals added to a population in turn add individuals to the population. Imagine, for example, what would occur in a population of insects if one pair of parents produced 400 offspring (which is not a very large brood among insects). There would be 200 new pairs of parents. If each new pair of parents also produced 400 offspring, then the population in the next generation would be 80,000. If these insects then paired off, they could produce 16 million offspring. Thus, in four generations the population could rise from 2 individuals to 16 million individuals, or more if preceding generations were still alive. In not too many generations there would be little room on the planet for anything but such insects! Obviously something must severely limit the growth of populations. According to the theory of evolution, such growth is limited by natural selection.

Differential Survival

The process of differential survival means that only *some* individuals will survive to reproduce, and thus only these survivors will contribute their genes to the gene pool of the next generation. Some characteristic of the environment exerts *selective pressure* on the population and determines which variations are favorable and which are unfavorable to the survival of the species. Such selective pressure may be *biotic*—that is, it may result from the activities or characteristics of other *living* things—or it may be *abiotic* and result from some *nonliving* aspect of the environment. Imagine, as an example of biotic selective pressure, that there are enough mice in a particular area to feed 1000 hawks *or* 1000 owls. If in the aggregate there are more than 1000 hawks and owls in that area, then there will be competition for the food (mice) and some of the birds will have to move out or will fail to survive. Now, suppose that the mice feed on grass and that there is a shortage of grass in the area because the soil lacks certain chemicals necessary for plant growth. Then, the mice will be subject to biotic selective pressures (from predators and from the supply of grass), and the grass will be subject to abiotic pressure (from the lack of chemicals).

Among the more important biotic pressures are competition, predation, disease, and parasitism. Important abiotic pressures include temperature extremes, soil type, rainfall, and light.

By differentially influencing the survival of individual organisms, selective pressure, since it is a characteristic of the environment, produces selection, the preservation or elimination of variations in a population. Since selection thus determines the adaptations of organisms, selective pressure can be understood as having produced the fangs of carnivores, the fast legs of deer, and the coloration and behavior of birds that allows them to recognize each other and select mating partners. Indeed, every adaptation of every organism is the product of natural selection and is therefore the product of the interactions of organisms with their biotic and abiotic environments.

All functions of an organism, then, have evolved through natural selection in such a way as to insure that the organism can reproduce. Its adaptations serve to fit it to its environment. Thus the environment of a species or population is inextricably bound up with its evolution.

Our understanding of evolution is based upon the accumulated evidence of over one hundred years of experimentation, observation, and reasoning. At times in the past, some aspects of the theory have come under criticism from scientists, but in nearly every case it has been found that the basic outlines of the process, as articulated by Darwin, are essentially valid. Among the strongest early criticisms were those directed against the theory on the grounds of genetics. The concepts of genetics that were current at the time did not support the theory of evolution by natural selection. Basically, inheritance was thought to blend the characteristics of the parents in the offspring. In a system of *blending inheritance,* the variations in a population would gradually disappear as time elapsed, and thus natural selection would have nothing to work upon. Either the theory of blending inheritance or Darwin's theory was incorrect. As we shall see, the blending theory of inheritance was invalid.

Although Darwin's reasoning was basically accurate, reasoning was not the only tool that he used in developing his theory. Indeed, one of the factors that contributed to the rapid acceptance of the theory was the vast amount of evidence that Darwin brought forth. Examining some of this evidence will provide us not only

Some Ideas That Influenced Darwin

As a concept to explain the biological world, evolution did not spring newly formed from the mind of Charles Darwin. Such ideas, more or less resembling that of evolution, can be traced back to antiquity. Predarwinian statements of evolutionary theory range from the casually speculative, and occasionally wildly fanciful, to serious, careful attempts to establish evolution as a theory capable of explaining the diverse phenomena of living systems. In a sense, when Darwin elucidated his ideas, the world was ready for them, the way having been paved by earlier thinkers. Two thinkers who preceded Darwin and exerted some influence on him were Jean Baptiste de Lamarck, an eighteenth century French naturalist and evolutionist, and Thomas Malthus, an eighteenth century Englishman and practitioner of the "dismal science," economics.

Lamarckism is a theory of evolution that holds that characteristics acquired during the lifetime of an individual can be inherited by its offspring. An example used by Lamarck to make his case has become a classical one, since it lends itself so well to an explanation of natural selection. According to Lamarck, the necks of present day giraffes are as long as they are because their ancestors stretched their necks to reach the upper branches of trees. Thus the length of the neck of an offspring would be determined by how much its parent was able to stretch. Explained in terms of natural selection, a few giraffes in some ancestral population probably had slightly longer necks than other giraffes. Such longer-necked individuals were able to exploit greater food resources, and thus were enabled to survive and reproduce more than their shorter-necked contemporaries. As time passed, the offspring of the longer-necked individuals made up a larger proportion of giraffe populations. In the first edition of *On the Origin of Species* Darwin treats Lamarckism with some skepticism, although criticism of his book and his own inability to explain inheritance caused him to soften his view somewhat in later editions.

Darwin had probably begun thinking about evolution before the end of the voyage of the *Beagle,* but it was not until a few years after his return that he read Malthus' "Essay on the Principles of Populations." Malthus' thesis was based on the observation that human populations tend to grow geometrically while the food supply increases arithmetically. As a consequence of the difference in these growth rates, Malthus concluded, man's population would soon exceed his capacity to provide enough food for itself unless something checked its growth. The checks to growth that he identified were "misery and vice." This unappealing vision was largely dismissed by other economists and has continued to be dismissed by most economists since, an attitude that we shall comment upon later. Darwin saw a parallel to natural populations in Malthusian economics, with natural selection serving as a check on their potential for growth. Indeed, Darwin gives a great deal of credit to the influence of Malthus' ideas on his formulation of the theory of evolution by natural selection. As he reports in his autobiography, after reading Malthus "for amusement," he had "at last got a theory by which to work" out "how selection could be applied to organisms living in a state of nature."

BOX 2.2

with indications of the validity of the theory but also with some examples of how selection works under natural conditions.

Evidence and Examples of Natural Selection

Among the books that played an important role in shaping Darwin's thinking about evolution (see box 2.2) was Lyell's *Principles of Geology.* Geological studies had begun to indicate that the earth was far, far older than had been assumed. Before the evidence of geology had been examined, the generally accepted age of the earth was about 6000 years. One cleric, calculating from the ages of the Patriarchs as reported in the Bible, had determined that the earth was created at 9:00 A.M., October 23, 4004 B.C. Geologists, however, were drastically revising such estimates, concluding that the earth was hundreds, perhaps thousands of millions of years old. It was also clear to geologists that major changes had taken place on the surface of the earth as a consequence of very small forces operating over very long periods of time. If it was conceivable that immense changes could be made in the face of the earth through apparently small forces, it seemed at least possible that correspondingly large changes could be made in living things in similar time spans. Among other prevalent geological concerns were *fossils,* or the remains of once-living organisms found in the strata of geological deposits.

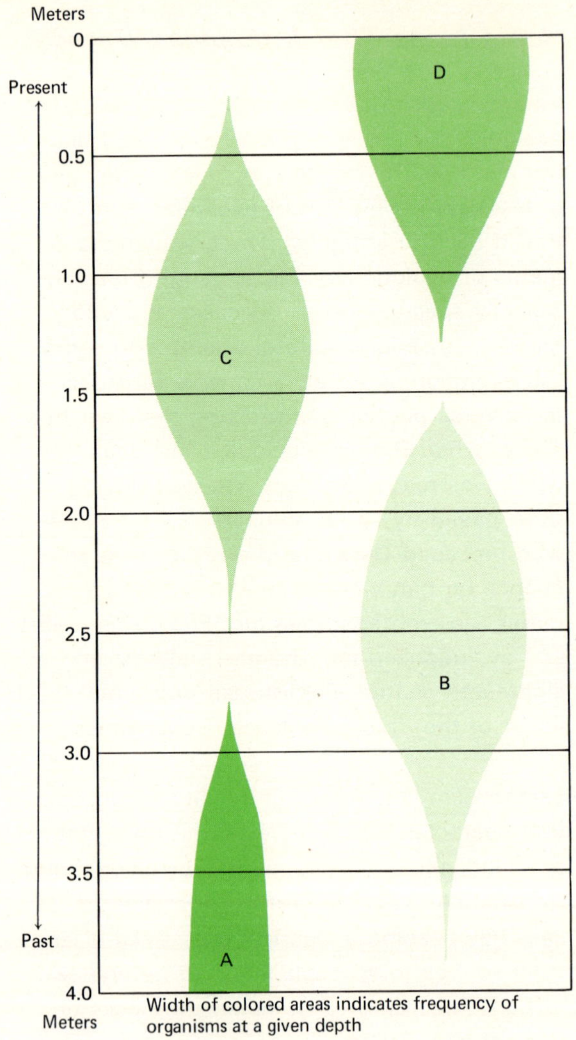

FIGURE 2.1 Change in fossil frequency from one stratum to another. A, B, C, and D represent groups of organisms in the strata shown.

The Fossil Record

The evidence upon which a great deal of the case for evolution rested, at least until the development of modern genetics, was the *fossil record.* The fossil record is the result of the laying down of a covering of soil on the remains of organisms, stratum by stratum, until such remains are many feet beneath the earth's surface. Fortunately, most species have "cooperated" with the development of the record by not burying their dead—a practice that man is thought to have begun only 75,000 years ago. In rare instances, chemical changes take place that partially preserve the remains, but usually the only parts that are preserved are bones, teeth, or shells. The softer parts of the body deteriorate rapidly, and evidence permitting the reconstruction of the probable appearance of an entire organism is extremely rare and largely indirect. Most frequently, all traces of any life are obliterated by the action of the soil, rock formation, and erosion. Here and there, however, deposits of rock are found that contain fossils, and from such scattered fragments of bone, teeth, and shells a great deal is deduced. We shall examine the fossil record of man's evolutionary history in Chapter 16, where we shall discuss some of the ways in which such deductions are made. The ages of fossils can be determined to some degree of accuracy by knowing the rate at which the soil cover was laid down. More recent methods, which yield more accurate dating of fossils, are based on the chemical and physical changes that are known to take place in rocks and soil and the rates at which such changes take place. The different strata of geological deposits can often be shown to be associated with significant environmental changes, such as periods of glaciation or important volcanic activity. One characteristic of the fossil record that was apparent, and that served to support Darwin's theory, was that the relative proportions of the variations between fossils of the same species in one stratum differed in the next higher stratum. From one stratum to another, one variation or range of variations could be seen to become more frequent, while another gradually disappeared. (See fig. 2.1.) This strongly suggested that species descended from one another and that in instances in which there seemed to be very abrupt changes in the frequencies of certain variations, the abruptness could be accounted for in part by gaps in the record. In fact, Darwin predicted that some transitional fossil forms would be found as the record became more thoroughly known, and indeed some of his predictions in this respect have been shown to be valid.

Comparative Anatomy

Among the other evidence that Darwin cited was the existence of anatomical similarities between species. Such similarities were often seen in the structures of the skeletons of animals, and parts of the body that were used for very different purposes among different species were nevertheless structured similarly. Such similarities suggested that a relationship existed between species. The similarities could be explained if at some past time some population of a species had been subjected to environmental conditions that were different from those affecting the other populations of the species. If the isolated population survived, then depending upon its environment, such structures would change according to the different selective pressures brought to bear. As can be seen in figure 2.2, the bones in the wing of a bat resemble the bones in the hand of a man and the bones in the wing of a pigeon. Human beings and bats bear other resemblances to each other, particularly in their reproductive structures, as will be seen later.

FIGURE 2.2 The bones of an ancestor may evolve to many different uses. The shaded colors show the evolutionary relationships. *(Redrawn from* Biology *by Karl Von Frisch. Modified from Linder. (Harper & Row, 1964.))*

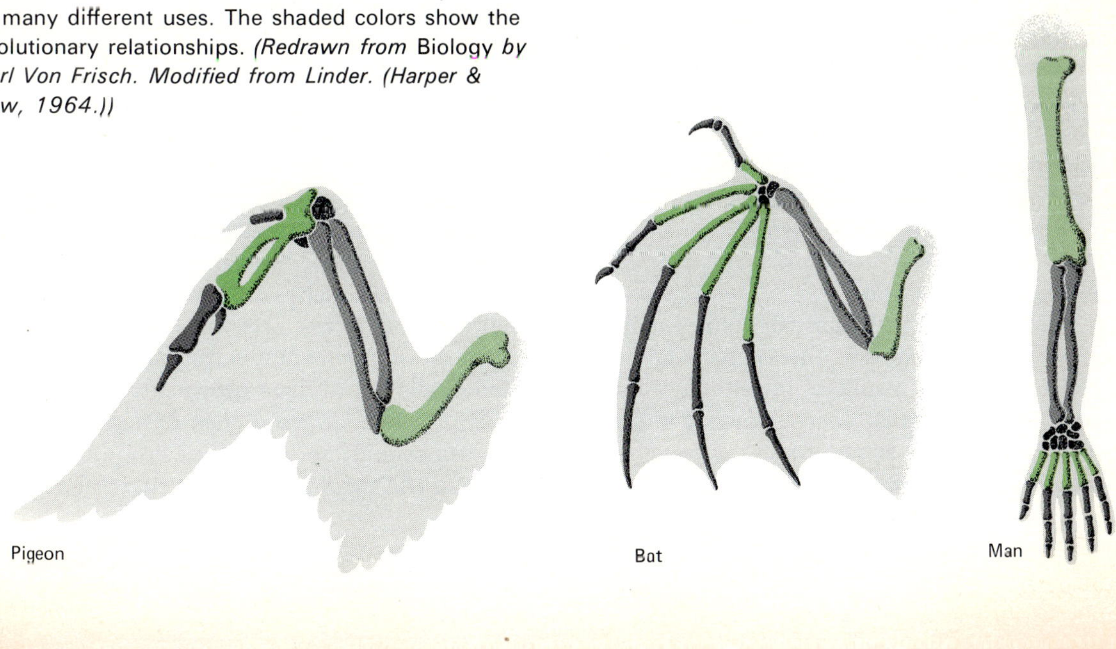

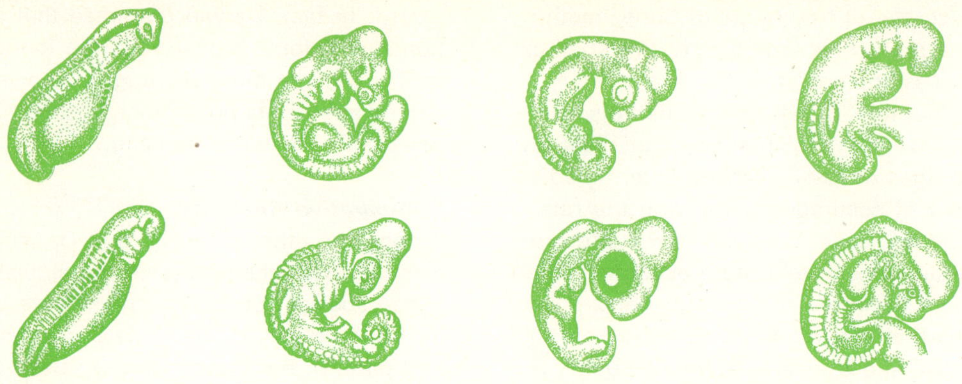

FIGURE 2.3 Embryos of man and other species showing the resemblances during their development. *(Redrawn from BSCS, Blue version.)*

Comparative Embryology

Prior to the publication of *On the Origin of Species,* certain observations had been made concerning embryonic specimens from a variety of species. One characteristic of embryos was shown to be that as they developed they bore strong resemblances to each other, although as adults they were extremely different. Figure 2.3 shows some embryos from widely different species, including man. Darwin hypothesized that because selection usually worked on the adult forms of the animals, and because biological structures usually appeared in the life of an organism at about the same time they appeared in the life of its parents, the resemblances indicated that some ancestral relationships existed between species.

Although the evidence of evolution makes itself clear to us in the *results* rather than in the actual working of the process, certain experiments have been performed that provide us with additional evidence for evolution.

Human Induced Evolution

In a laboratory an investigator can control many of the variables, change the environment at will, and observe the results of his alteration of environmental conditions within relatively short periods of time. Since some organisms possess what we would consider to be extremely short life spans, they may produce tens or hundreds of generations in a matter of hours or days, allowing the experimenter to simulate some evolutionary processes. To give an idea of how rapidly some organisms can reproduce, consider the common human intestinal bacterium, *Escherichia coli.* This bacterium can divide every 20 minutes under ideal conditions, and at the end of 24 hours would be in its seventy-first generation. At the end of 8 hours under such conditions, one bacterium could give rise to nearly 17 million bacteria, and at the end of 12 hours, nearly 69 billion!

A classic experiment showing adaptation and selection was performed in the late 1940s by the Nobel Prize winning biologist Joshua Lederberg and his wife Esther. The Lederbergs chose to investigate bacterial resistance to the antibiotic drug streptomycin. They first cultured a large bacterial population in a broth of organic substances that promoted rapid growth of the bacteria. They then transferred samples of the broth to culture dishes. Within 24 hours or so, small colonies of bacteria were seen to be growing on the media in the culture dishes, each colony believed to have grown from a single bacterium. The *replica-plating* of such colonies onto dishes with an antibiotic, streptomycin, added to the medium revealed that although some colonies failed to grow, a few survived. (See fig. 2.4.) The question was whether the ability to survive in a hostile environment had existed prior to the transfer or had resulted from exposure to the streptomycin. To answer this question the Lederbergs returned to the original culture dishes containing colonies known to be resistant and transferred samples of the presumed resistant strains to entirely new dishes. The media in the new culture dishes contained no streptomycin, and the bacteria in them grew until they covered the entire dish. Samples from

FIGURE 2.4 Replica-plating. The method used in steps 1 through 6 identifies potentially resistant bacteria. To determine whether the resistance is genetically controlled, samples from the original colonies are transferred to a new culture dish containing streptomycin. These samples have not been in contact with the antibiotic, and if they grow, they are probably genetically resistant to streptomycin.

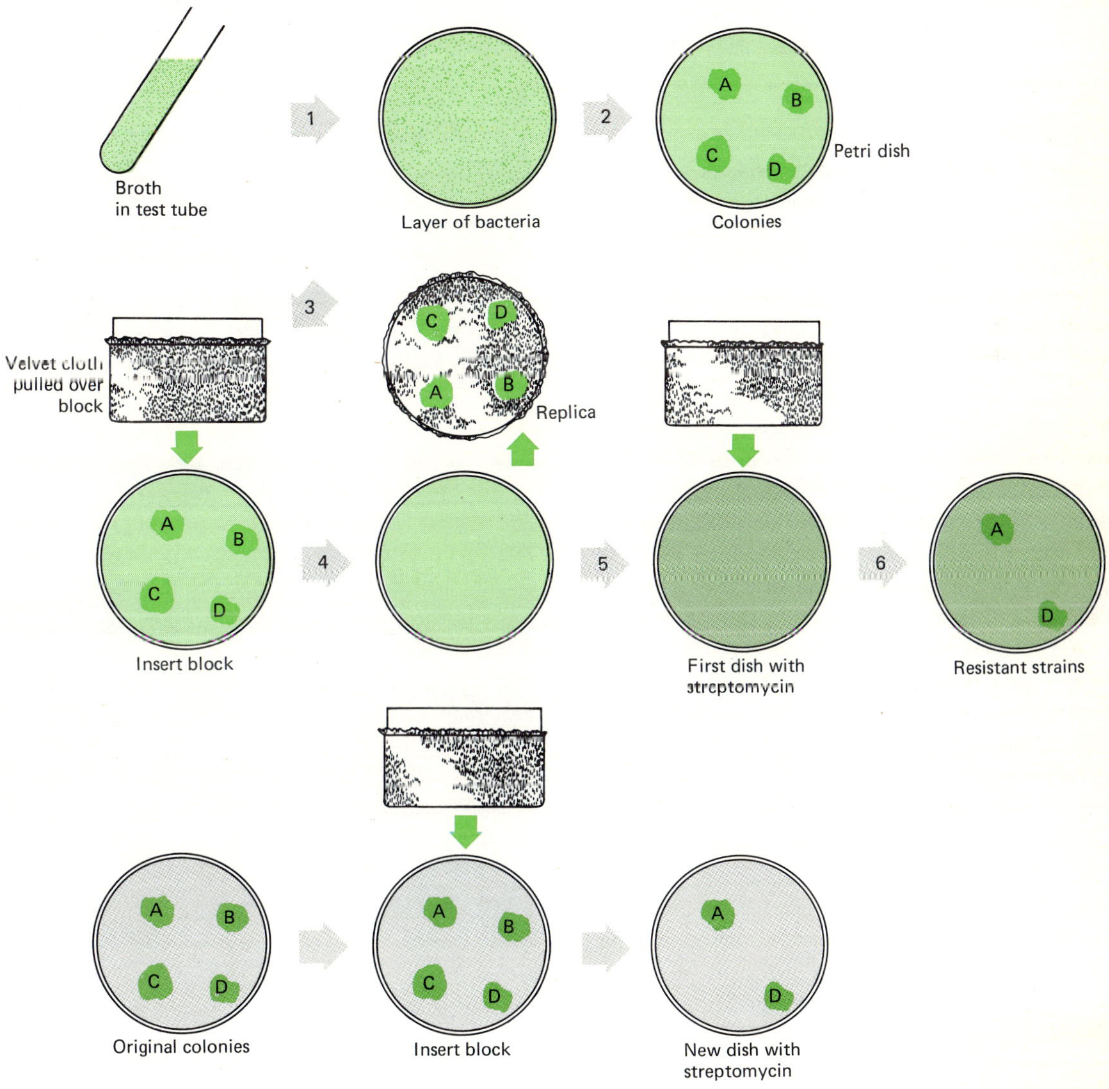

these dishes were then transferred to culture dishes containing streptomycin, and they were able to grow well in the hostile environment.

The conclusion drawn from the Lederberg experiment was that the resistant bacteria possessed the characteristic before they encountered streptomycin. As the resistant colonies were selected experimentally, they transmitted the streptomycin resistance to their offspring. An adaptation had been selected that permitted survival in a changed environment.

In another example of evolution operating in a very short time span, a human activity produced selective pressures of an entirely unintended nature.

Industrial Melanism

The peppered moth, *Biston betularia,* is a greyish moth with an irregular pattern of black markings on its body and wings. Around 1850, near Manchester, England, a new form of the moth was found, and because it was all black, it was given the name *B. betularia carbonaria,* the second name denoting the color of the new form. Such black, or *melanic,* forms are regular, although rare, occurrences in many species of moths. The trees and rocks in the area of Manchester, as in other parts of England, support growths of lichens—light-colored, greenish-grey plants. (See fig. 2.5.) When a peppered moth is resting on a lichen-covered tree or rock, it is very difficult to distinguish it from the background. Lichens, as it happens, are particularly sensitive to air pollution, and as the city of Manchester grew into an industrial center the factories in the area produced increasing amounts of black, sooty smoke. Ultimately, air pollution levels rose sufficiently to kill the lichens around the city, and soot deposits on the trees made them increasingly darker. In 1850, it was estimated that fewer than 1% of the peppered moth population near Manchester was made up of melanic forms. By the turn of the century, however, melanics had become so common that they were estimated to amount to 99% of the population. How had such a drastic

(a) (b)

FIGURE 2.5 *B. betularia.* (a) Light and dark varieties on a lichen-covered tree trunk. (b) The same on a dark, lichen-free tree trunk. *(Courtesy of M. F. W. Tweedie from National Audubon Society.)*

change taken place in the population in so short a time?

In 1937, one investigator hypothesized that the grey, or typical, peppered moth was preyed upon by visually hunting predators and that the disappearance of the lichens from the area had made the light-colored moth more visible and thus more susceptible to predation as it rested on the tree trunks. At the same time, the environmental change had made the black moths less visible and therefore less susceptible to predation. The spread of industrialism into other parts of England and the accompanying change in the proportion of black moths in other *B. betularia* populations supported the hypothesis. Actual proof, however, was not recorded for another twenty years.

In 1957, two investigators tested the hypothesis experimentally. After marking both grey and black moths in a way that was not visible to a predator, the investigators released a known number of moths of each kind both in areas that were free from lichens and in areas where lichens were growing. In the lichen-free area, they were able to recapture 40% of the black moths but only 19% of the grey moths, suggesting that black moths were preyed upon less in such environments. In the areas where lichens grew, they recaptured 12.5% of the grey moths but only 6% of the melanics. In each environment, then, the experimenters were able to recapture nearly twice as many "invisible" as visible forms of moths.

Still, no one had ever seen any organism preying upon the moths. To that end the experimenters set up a blind in each area and equipped them with motion picture cameras. They then placed both forms of the moth on nearby trees and were able to film birds in the act of eating the moths. In fact, at least seven different species of birds were identified as major predators on the moths. In the unpolluted environment, the birds ate 164 melanic moths and only 26 grey ones, while in the polluted environment, they ate 15 melanic and 43 grey moths. The motion pictures even showed birds searching the tree trunks for moths and capturing conspicuous ones, while apparently overlooking inconspicuous ones nearby. We should point out that the birds capture and eat the moths so quickly that it is not surprising that no one had witnessed it before.

The case of the peppered moths demonstrates how subtle, complex, and often indirect the workings of evolution can be. If we contrast the Lederberg experiment with the peppered moth case, we find that in the laboratory there were only two organisms involved, while in the peppered moth case there were five. The selective pressure that operated on the moth population was not simply a biotic one; an abiotic factor selected against the lichens. Moreover, the selective pressure produced by birds, if it were expressed as a quantitative value, would be seen to change over time, as one segment of the population slowly experienced a reduction in the pressure while another segment experienced an increase. Although the time factor was short when measured in human terms, we must remember that a moth generation is considerably shorter than a human one. In fact, such color changes have been reported in other populations of organisms. Undoubtedly others have occurred without our knowledge, either because the changes have been too subtle to detect readily or simply because we have not examined the situations.

We can draw two generalizations about selection from the Lederberg experiment and the peppered moth example. First, selection is often *directional.* If the environment changes, or if a population is introduced into a new environment, selective pressure produces changes in the characteristics of the population that adapt the population to the new conditions. If the population is already very specialized for life in the environment in which it lives, a change in the environment may cause extinction. If directional selection occurs, it may be followed by *stabilizing* selection; that is, as the population becomes adapted to the environment, selective pressure continues to operate, but it does so on the least well-adapted individuals in the population. Thus stabilizing selection serves to inhibit change in the population rather than promote it.

One example of both directional and stabilizing selection can be found in an examination of the species of finches found in the Galapagos Islands, 600 miles off the coast of Ecuador.

Darwin's Finches

When the *Beagle* put in at the Galapagos Islands in 1835, Darwin began a month-long study of the flora and fauna of these fantastic islands. The Galapagos is a volcanic archipelago that is thought to have risen from the sea a little more than one million years ago. Although the islands are relatively close to South America, the environments of the islands differ considerably from the coastal regions of the continent. Of the numerous plant and animal species in the Galapagos, many are found only on these islands—that is, they are *endemic* to the archipelago. Moreover, most of the endemic species appear to be related to South American species. Of particular interest are the thirteen species of finches that Darwin, and more recently several others, have observed. (See fig. 2.6.)

Six of the species are ground finches, six are tree finches, and the thirteenth appears not to be a finch at all, but a warbler. An internal examination of this thirteenth species reveals, however, that it is a finch. Of the ground finches, four species are found on all of the islands in the archipelago, but two other species are found only on the outlying islands. The four common species include one species that eats fruit and has a long pointed beak that adapts it to that food resource. The other three are seed eaters, but they are adapted by the sizes of their beaks to eating seeds of different sizes. Of the tree finches, one is a fruit eater, and the remainder are insect eaters; the latter differ in beak size and thus in the size of the insects they may eat. One of the insect eaters, the woodpecker finch, exhibits an amazing adapta-

FIGURE 2.6 Many different feeding habits and beak shapes have evolved in the Galapagos finches. The common ancestor and some of the adaptations are shown here.

tion. This remarkable bird lacks the long, prying tongue of the woodpecker, and although it has a woodpeckerlike beak, it cannot retrieve insects from the bark of trees with its tongue as a woodpecker does. Instead, it digs the insects out of the bark with a cactus spine—a tool!

As the evolutionary history of these finches has been reconstructed, it is thought that a small population of South American finches, probably ground finches, were blown across the ocean to the islands during an unusually severe storm. (Since finches are not good fliers, it is unlikely that they flew there without some aid. If it sounds implausible that they could have been transported to the islands in this manner, consider that each year North American birds are found on the coasts of Ireland and England at a distance two to three times as great as that separating South America and the Galapagos.) Once on the islands, probably only on one, the founding population of finches was faced with a new environment. As time passed, some finches may have flown or been blown to other islands. If the food resources on the other islands were different from those of the original island, then directional selection could have occurred that adapted the population to the resources of the new islands. Subsequently, as the different populations became separate species, and as the alternative resources dwindled because they were being used, stabilizing selection slowly replaced directional selection.

The Galapagos finches also demonstrate a process about which we shall learn more in the next two chapters—the process of *adaptive radiation.* Encountering a new environment, one in which opportunities existed that had gone unexploited, the founding population of finches underwent a series of evolutionary changes as the succeeding generations expanded, or radiated, into the available niches. As the environments of the earth have changed over the more than 3 billion years that life is believed to have existed on this planet, such radiation has produced a richly diverse biosphere. Indeed, the process continues today, although at such a slow pace that it is barely perceptible, if in fact it is perceptible at all.

Divergence and Convergence

When one species arises from another during the process of radiation, the two are said to have *diverged. Convergent* evolution, on the other hand, means that due to living in the same environment or under very similar environmental conditions, species that are only very remotely related may become very similar in some respects. Some of the clearest examples of convergence are seen in the *morphology,* or *body form,* of certain animals that live in the sea. The shark, the porpoise, and the ichthyosaurus are related only through some very distant ancestors, and yet they bear striking resemblances to each other as shown in figure 2.7. As we shall see in the next chapter, such

FIGURE 2.7 Convergent evolution of aquatic species. The ichthyosaurus was a reptile; the shark is a fish; and the porpoise, a mammal.

resemblances can be found among many almost unrelated species and are the product of adaptations to similar conditions of life.

Evolution and the Niche

The long-term result of the process of evolution is the production of a balanced ecological community. We shall return to this idea in greater detail later, although in a sense we shall never leave it. Implicit in the process of evolution is the determination of a species' niche, or the way it "makes a living." Through directional selection, a species is adapted to changing environmental conditions, and thus with time its niche changes. When environmental conditions cease changing, stabilizing selection maintains the niche as constant. The ecological niche of a species is therefore determined by the selection of the genetic characteristics that adapt the species to living in a particular ecological situation; that is, evolution produces the adaptations that define a species' way of making a living and thus fits the species into a biological community.

If two species in the same biological community are closely related and have very similar or identical ecological requirements, their niches will overlap or coincide, and the species will compete with each other. As we shall see, this usually means that one of the species will be eliminated from the community. However, there are instances in which recently evolved species can live side by side and appear to be identical or to have identical niches. When such *sibling species* are carefully examined or observed, they will reveal a number of minor morphological and ecological differences. For example, it was once thought that all European populations of the malarial mosquito, *Anopheles maculipennis,* were the same species. Although the populations exhibited certain differences in terms of their habitats and the animals they fed upon, judging from morphological evidence there appeared to be no differences at all between mosquitoes from one part of Europe and those from another. Only after the discovery that the markings and color of eggs laid by some populations differed from those laid by other populations was it apparent that there were at least six different sibling species of *Anopheles* in Europe. Indeed, after thirty years of further examination, the most reliable means of differentiating between the species is still by the eggs. The discovery that there were so many species explained why the young mosquitoes in some portions of Europe developed in brackish waters, while the young mosquitoes elsewhere on the continent developed in fresh water. It also explained why in some parts of Europe where the mosquito was abundant, the incidence of malaria was low, whereas equally abundant populations in other parts of Europe were accompanied by high rates of malarial infection.

Thus we can see how very subtle and specialized a niche can be. Determined by evolution, the niche is a delicately tuned mechanism that permits the species to survive and reproduce under a given set of environmental conditions. It cannot be too strongly emphasized that it is the niche of a species, not its appearance or even its "identifiability," that is the ecologically important factor in evolution.

We have only hinted thus far at some of the mechanisms by which species arise or descend from other species—that is, how *speciation* occurs. We shall postpone discussing this topic, however, until we have had an opportunity to examine the genetics of populations. It is obvious that speciation has occurred; the world around us is full of the evidence. Extended over 3 to 4 billion years, selection has produced men and cockroaches, elephants and jellyfish, grass and redwood trees—all from the same basic stuff, all possibly remotely related through some common, unpromising ancestor that arose on the primitive earth. Studies of the fossil record, anatomy, embryology, physiology, and behavior of closely and distantly related species of organisms allow us to understand the origin and evolution, or *phylogeny,* of organisms, which can aid in our understanding of the basic process of life. But before continuing with the biological aspects of evolution, let us turn very briefly to some of the implications of Darwin's theories for man.

The Impact of Evolution

Theories of man's nature abounded before Darwin defined the process of natural selection, just as they abound today, but basically, predarwinian concepts of man and nature were formulated around the idea of special creation. Man, the natural world, and the universe were thought to be the product of the design of God. Studies of the physical world as developed by Newton supported the idea that there was order and harmony in the universe and thus reinforced the concept of a designed universe. Since the predarwinian theories of evolution failed to provide a workable mechanism by which to explain the similarities and differences that could be observed in the natural world, they represented no real challenge to the idea of special creation. For the most part, before the publication of *On the Origin of Species,* man was placed nearly at the apex of God's creation, apart from, and dominant over, the rest of the natural world. The concept of natural selection, however, introduced randomness, as well as a process of trial and error, into the idea of nature, and this, of course, was at odds with the concept of design. One opponent of the idea of natural selection went so far as to declare that the Creator, in order to test man's faith, had designed the world in such a way as to make it appear that natural selection had occurred! Indeed, a great deal of resistance to the idea of evolution by natural selection has been based on a literal interpretation of Genesis. The idea that the same process that gave rise to an amoeba or an insect grub also has given rise to man is, to many persons, a repugnant thought. How could the beauty of nature and man be the product of an unthinking process of trial and error? How could human compassion, love, and reason be nothing more than the outcome of a series of random events that allowed the species to survive to reproduce? We can only pose such questions here, not consider what the possible answers might be, for that would take us too far from the point of this book.

Man's view of himself and of his institutions had begun to change before Darwin published *On the Origin of Species.* Differing schools of thought had asserted that man was little more than a series of chemical reactions, even going so far as to claim that thought was a product of the brain in the same way that urine was a product of the kidneys. Man having been reduced to a mere mechanism, then, man's relations with other men could be formulated into a series of laws that would completely predict his behavior. By removing man from a unique category and placing him within the process of natural selection, Darwin's theory seemed to lend credence to such ideas, and thus many who held to them adopted the idea of natural selection to their own uses.

The least responsible uses to which the idea of natural selection was put involved economics, politics, and race relations. It was argued that society was like nature and that survival in nature could be equated with power or wealth in society. It followed, then, that if a man or group of men were powerful or wealthy they had been "selected" and deserved their power and wealth because they were more fit to survive than others. The less powerful were not worthy of concern, for they were clearly less fit and therefore would not survive. Thus were justified some of the most ruthless and sometimes violent competitive methods on record.

Of all the pseudo-Darwinian justifications for a repulsive idea, perhaps none is worse than the nearly immediate adoption of natural selection as an underpinning for racial bigotry. Before the publication of *On the Origin of Species,* a

French writer, de Gobineau, published a work in which he claimed that a distinct hierarchy of human races existed, at the top of which were the blue-eyed, blonde people inhabiting northern England, France, and Belgium. Although de Gobineau wrote before Darwin, the ideas of natural selection were immediately adopted by the former's adherents, and thus such people became the ''fittest'' and were destined to survive and even to rule the world. Such ideas were later expanded by an Englishman, Chamberlain, who claimed that the modern-day Germans were the only representatives of the ''pure'' Aryan race and were the fullest expression of evolution. Chamberlain's thesis was later eagerly adopted by the German National Socialist, or Nazi, party as justification for the extermination of millions of Jews, Slavs, and other less ''fit'' human beings. We will not dwell here on the fallacy of such reasoning, for we shall be able to do so more thoroughly after we have gained an understanding of population genetics.

Perhaps the most significant social effect of Darwin's ideas was the removal of man from a central position in the universe, the end of anthropocentrism. Man being seen as a part of the process of evolution, the idea that one species is ''better'' or ''worse'' than another is meaningless. By definition, all surviving species are fit, and it is impossible to say objectively that a man is better than a bacterium. Placing man and bacteria on an equal footing compels us to view our relationship with nature in a more humble way than that to which we have been accustomed. Evolution also allows us to view our own activity and, indeed, our ''fate'' in a more objective way, for it leads us away from appeals to some abstract and absolute morality. The question that we must then ask about human actions is ''What meaning have they for survival?'' The idea of natural selection is by no means an argument against morality; it merely says that value judgments lie outside the realm of biology. As we shall see, a great deal of human behavior is culturally induced. Although we may find the roots of culture in the evolutionary history of man, culture itself is not a part of biology, in spite of the profound impact that it may have upon biology.

The objective view of life that Darwinian evolution allows us to take has led to more objective views in other areas of thought. Human behavior, for example, may be thought of as adaptive or maladaptive, and with this in mind anthropologists have come to view the cultures of man more objectively. Indeed, a great deal of modern anthropology is based upon the concept of evolution and is decidedly biological in tone and emphasis. In the area of philosophy, the idea of evolution has had profound implications. Many philosophers have come to the conclusion that the existence of man was not a preordained or inevitable outcome of the process of evolution, but rather a matter of chance. The implications of this conclusion have provided a rich source of exploratory thinking, and not a little controversy, about the ''meaning'' of life. We could argue that other fields of human thought have yet to learn the value of thinking in terms of there being no design to the universe or higher purpose to life. Seen from the viewpoint of evolution and natural selection, and their implications for man's place in nature, the self-righteousness of many political, racial, and national groups seems at least ridiculous, if not dangerous.

As we shall see in a later chapter, evolution has much to tell us about the equality of races and the ''places'' of men and women in society, as well as about the ''superiority'' and ''inferiority'' of various cultures. It is our opinion that the humility and objectivity of the ideas of Darwinian evolution can teach us lessons of immense importance to the welfare of individuals and humanity as a whole and can do so within the context of a morally acceptable ethic.

Selected Readings

*Darwin, Charles R., *Autobiography.* Edited by Nora Barlow. New York: W. W. Norton & Co., Inc., 1969. Darwin did not write his autobiography for publication, but rather at the request of his children, to tell them what his life was like.

*Darwin, Charles R., *On the Origin of Species.* New York: Atheneum Publishers, 1967. A facsimile of the First Edition, with an introduction by Ernst Mayr. The importance of the First Edition can be appreciated only by reading it. This is Darwin before he tried to please his critics, exhibiting the full power of an extraordinary mind.

*Darwin, Charles R., *The Voyage of the* Beagle. Garden City, N.Y.: Doubleday & Company, Inc., 1962. Darwin's fascinating and often charming account of his "voyage of discovery."

*Eiseley, Loren. *Darwin's Century.* Garden City, N.Y.: Doubleday & Company, Inc., 1958. A survey of the intellectual climate in which Darwin's ideas were formed and upon which they made their impact.

Hotton, Nicholas, III. *The Evidence of Evolution.* New York: American Heritage Publishing Company, Inc., 1968. An attractively illustrated book for the general reader, placing particular emphasis on the fossil evidence.

*Huxley, Julian S. *Evolution: The Modern Synthesis.* New York: Harper & Row Publishers, 1942. A classic overview of modern evolutionary thought.

*Irvine, William. *Apes, Angels, and Victorians.* New York: McGraw-Hill Book Company, 1972. An excellently conceived and written account of Darwin, his chief contemporary proponent, T. H. Huxley, and the furor raised by Darwin's theory.

*Moorehead, Alan. *Darwin and the* Beagle. New York: Harper & Row Publishers, 1969. A fine book, beautifully illustrated, dealing with the story of the voyage, and especially the relationship between Darwin and Fitz-Roy.

The following *Scientific American* articles, as well as those referenced in future chapters, are available as reprints, and augment or expand upon topics in the text.

Eiseley, Loren. "Charles Darwin." *Sci. Am.,* Vol. 194 (February, 1956), pp. 62–70. Offprint number 108.

Gilliard, E. Thomas. "The Evolution of Bowerbirds." *Sci. Am.,* Vol. 209 (December, 1963), pp. 38–46. Offprint number 1098.

Lack, David. "Darwin's Finches." *Sci. Am.,* Vol. 188 (April, 1953), pp 66–8. Offprint number 22.

* Paperback edition available.

The environment of the coral reef illustrates many of the adaptations of marine organisms. In this case the coral itself is composed of numerous animals of the most primitive kind, while the fish and crabs represent the most highly evolved of the animals that live in the sea. *(Courtesy of Jen and Des Bartlett from Photo Researchers, Inc.)*

adaptations to environments—water

chapter three Charles Darwin never presented rigorous proofs for his ideas about evolution; rather, he submitted masses of circumstantial evidence combined with tightly reasoned arguments. In the time that has passed since the publication of *On the Origin of Species*, Darwin's arguments have held up well under what may be the closest scrutiny ever given to an idea in the biological sciences. Today, nearly all biologists accept Darwin's arguments as "reasonable" and the idea of evolution by natural selection and the inheritance of selected characteristics as the equivalent of a law of nature.

We have briefly examined the process of evolution and have presented a few examples of the process. It should be clear that natural selection has led to the adaptations whereby organisms are able to gain needed energy and materials as well as to behave in such ways as are necessary to survive and reproduce. Clearly too, such adaptations exist within the context of environments. It is the purpose of this chapter to look at some adaptations that have allowed organisms to survive and reproduce, to see how the environment has operated to produce such adaptations, and to trace the evolutionary history of these adaptations.

Darwin was above all else a superb naturalist, as were so many of the men who "stood on his shoulders" to see the major evolutionary pathways leading to our present understanding of phylogeny and adaptation. In this chapter, the insight that we can hope to gain into the natural history of marine and fresh-water organisms is at best superficial. However, such insight should provide us with not only an appreciation of the mechanisms of adaptations to environments but also a background for understanding their human implications.

The Origin of Life

The question of how life arose on this planet has provided a source of continuing controversy since ancient times. Even among biologists there are many areas of disagreement. Nevertheless, there is general agreement that life probably arose in an aquatic environment and that the general environment was radically different from what it is today. There is also substantial agreement that the present composition of the atmosphere and oceans, the character of the soil, the climatic patterns, and even the composition of many of the rocks of the earth are the results of the effects of living organisms. That is, the evolution of life produced very great changes on the surface of the earth. We shall look first at how the early stages of life are thought to have arisen and then at how they are thought to have given rise to the organisms surviving in current environments.

Precisely how the earth was formed is still a controversial subject, but it is generally thought to have been formed around 4.5 billion years ago. (See table 3.1.) Subsequent to the formation of the planet, the atmosphere was formed, and it is generally agreed to have been radically different from the present atmosphere. It was rich in hydrogen, although the hydrogen was not in the form of free hydrogen—that is, not in the form of unattached hydrogen atoms. Rather, the atmosphere was made up largely of compounds of hydrogen and other elements such as carbon, oxygen, and nitrogen. The three compounds thought to have been most common were methane (CH_4), ammonia (NH_3) and water (H_2O). Several hundred million years after the formation of the earth, the surface cooled sufficiently to allow the condensation of atmospheric water vapor, thus beginning the formation of the oceans.

Since the origin of life was an unobserved and unrecorded event, our understanding of it must be based on an experimental re-creation of the conditions under which life is thought to have arisen. In the 1920s, a Russian biologist, A. I. Oparin, postulated that the abiological production of organic molecules was more likely to have occurred under the sorts of conditions that probably existed on the primitive earth than under present conditions. The hypothesis was also made by an English biologist, J. B. S. Haldane, who was at the time unfamiliar with Oparin's work. The idea was not tested until the 1950s. At that time the first modern experiment on the origin of life was performed. The experimenters circulated a mixture of methane, ammonia, and hydrogen through an aqueous solution and subjected the mixture to a continuous electrical impulse of around 60,000 volts. The electricity represented an energy source that simulated lightning on the primitive earth. After a time, it was noted that the water was turning brownish yellow, and after removal its composition was analyzed. It was found to contain a mixture of organic compounds and, significantly, a relative abundance of simple amino acids, the building blocks of proteins and thus of all life. Later experiments substituted ultraviolet light as an energy source and produced similar results. It appeared, from such experiments, that some of the basic compounds necessary to life as we know it could be synthesized from nonbiological molecules. Moreover, other experiments indicated that if the conditions within the experimental apparatus were altered to represent conditions of the environment as it is today, no amino acids were formed, thus suggesting that the conditions of the primitive earth were not only conducive, but even necessary, to the prebiological synthesis of organic molecules. Additional experiments have indicated that many of the chemical compounds utilized in living systems could also have arisen under primitive conditions.

TABLE 3.1 Timescale of Geological Eras and Evolution

Years Before Present	*Name of Era*	*Event*
100,000,000	Cenozoic	
225,000,000	Mesozoic	The transition to land
500,000,000	Paleozoic	The rise of vertebrates
1,000,000,000	Cambrian	Evolution of multicellular forms
2,000,000,000	Precambrian	Evolution of eucaryotes
3,000,000,000		Evolution of photosynthesis
		The origin of life on earth
4,000,000,000		Chemical evolution and the prebiological synthesis of organic molecules
4,500,000,000		Formation of the earth

Some problems are inherent in the explanation. Although the synthesis of biological molecules from abiological compounds in primitive environments can be accomplished under experimental conditions, the molecules found in living systems have a tendency to break down rapidly when subjected to ultraviolet radiation (see box 9.2). Moreover, those synthesized in the experiments were almost invariably small molecules, whereas the molecules found in living systems are almost always large molecules made up of a great number of simpler, similar small molecules. These problems are associated. At the present there is a layer of ozone (O_3) 40 kilometers (about 25 miles) from the surface of the earth, which was not present in the early stages of the formation of the planet. This ozone layer serves to filter out most of the ultraviolet radiation from the sun. If the ozone layer was not present when the basic molecules of living systems were being synthesized, it would seem to suggest that some other means of shielding them must have existed. Water provides just such a shield, since it absorbs ultraviolet radiations rather efficiently. In pure water all ultraviolet radiations are absorbed at a fairly shallow depth, and as organic compounds are formed in the water, they too serve to absorb ultraviolet radiation. Thus, at a depth of only a few centimeters in water containing organic compounds, there are no dangerous amounts of ultraviolet radiation. It seems therefore plausible to think that the syntheses necessary to produce the molecules on which living systems are based probably occurred in bodies of water.

However, the syntheses necessary to assemble large molecules from smaller constituents do not proceed readily in an aqueous solution, since they require the removal of a molecule of water from the two reacting molecules. Moreover, some important biological molecules also have a tendency to break down in aqueous solutions. Certain other properties of the primitive environment that offer a possible partial explanation of how the necessary large molecules could have been synthesized can be considered. Certain compounds, called *condensation agents*, are readily produced under primitive conditions, and these condensation agents could have served to promote the synthesis of large molecules in aqueous solutions. It may also be that the basic molecules were adsorbed on particles of clay or other mineral substrates and thus increased sufficiently in concentration to induce the necessary reactions. The temperature of the water may also have played a part in promoting the necessary reactions. Chemical reactions take place more readily in warm solutions than in cool ones.

The synthesis of some building block molecules, and even their assembly into appropriately large molecules, is not the origin of life. One of the basic attributes that we generally ascribe to a living thing is the ability to reproduce itself. Moreover, a living thing must also be capable of reproducing any changes in its genetically determined structure; it must be capable of mutating and of passing on those mutations to its offspring. The intricate system by which this is accomplished is not likely to have sprung into existence quickly and by chance; it probably required hundreds of millions of years to evolve to a crude level. As we understand it today, the proteins that make up the cell are assembled from amino acids according to instructions contained in the structure of the nucleic acid, DNA, but such syntheses are carried out through the action of a number of intermediary molecules called *enzymes.* A great deal of speculation, but *very* little evidence, is available to explain how such a system might have evolved. One line of speculation holds that in the absence of the required enzymes, weak chemical reactions could have taken place between certain nucleic acids and amino acids. If so, then such molecules would have had a clear selective advantage over others, since the others could come into existence only by chance. Another possibility is that life arose on a number of occasions in the primeval oceans and that all but one form was eliminated by competition. According to this speculation, all living things today, from the simplest bacterium to a human being, are descendants of that original successful creature.

At some time during the phase of chemical evolution, a living cell somehow evolved. At exactly what time and in exactly what manner this took place will probably never be known. Under certain conditions it has been found that synthetic proteins organize themselves into minute spheres that resemble some forms of small bacteria. These spheres can not by any means be considered cells, but they do offer a very tentative clue as to how the ancestors of cells might have arisen. (See fig. 3.1.) It is speculated that the first true cell, the "protobiont" as it has been called, resembled a modern spherical bacterium. It is also speculated that it was a *heterotroph* (*heteros* other + *trophein* to feed). Thus, the first form of life on the planet was incapable of synthesizing its own supply of energy and nutrients and must have fed on organic molecules taken from its environment. The alternative, an *autotroph* (*auto* self), seems unlikely to have arisen first. Autotrophs are capable of synthesizing organic compounds from inorganic ones, and like some modern bacteria and all plants, they can utilize the energy of the sun by converting it into chemical energy. In order to effect such syntheses, however, autotrophs generally employ complex and intricate series of chemical reactions that seem unlikely to have arisen very early in the evolution of life. Assuming that the first forms of life were heterotrophic, then, their appearance on the earth was probably soon followed by a population explosion, since they were probably immersed in a rich supply of organic molecules on which to feed (a "molecular Garden of Eden," as someone has called it). As the supply of nutrient molecules dwindled, selection must have begun to operate, and any individual with an ability to synthesize any molecule in short supply was conferred with a clear competitive advantage over those individuals unable to perform such syntheses. If, for example, molecule A grew scarce and an individual arose that could synthesize A from B, then that individual would survive and reproduce while those incapable of the synthesis would be selected against and die as compound A disappeared. If B then grew scarce, an individual capable of synthesizing B from C would have a competitive advantage. It has been suggested that the complex reactions necessary for sustaining life in contemporary organisms might have evolved in this fashion. In any case, if the original population of organisms was heterotrophic, and if all the descendants remained heterotrophic, they would have ultimately depleted the entire supply of nutrients in their environment, and the whole process would have to have begun anew. However, autotrophs are thought to have somehow evolved from an original heterotrophic population. Certain molecules produced nonbiologically probably came to be used by some heterotrophic organisms and then became incorporated into the synthesizing apparatus of these organisms in the manner described above. These molecules allowed the organisms which utilized them to begin the process of converting the energy of light into chemical energy. However it occurred, the evolution of autotrophs is still one of the great problems facing evolutionary biologists.

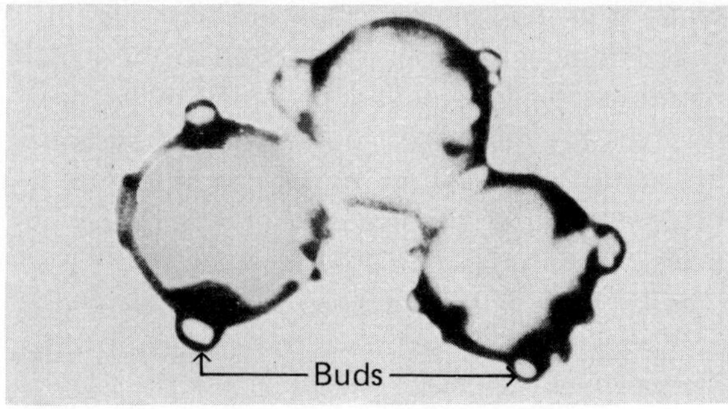

FIGURE 3.1 Proteinoid microspheres produced in the laboratory. The microspheres actually bud off new microspheres. Each large unit is about fifteen thousandths of a millimeter (0.0006 inches) in diameter. *(Courtesy of Sidney W. Fox, University of Miami.)*

The most widespread means by which autotrophs sustain themselves is photosynthesis, and it is probable that the ability to photosynthesize evolved in the original autotrophs. Once this process had begun the earth would never be the same again. The earliest fossil evidence of life indicates that the early autotrophs bore some resemblance to bacteria and blue-green algae. In some rocks of great antiquity there are traces of two molecules, *pristane* and *phytane*, which are known to be products of the break-

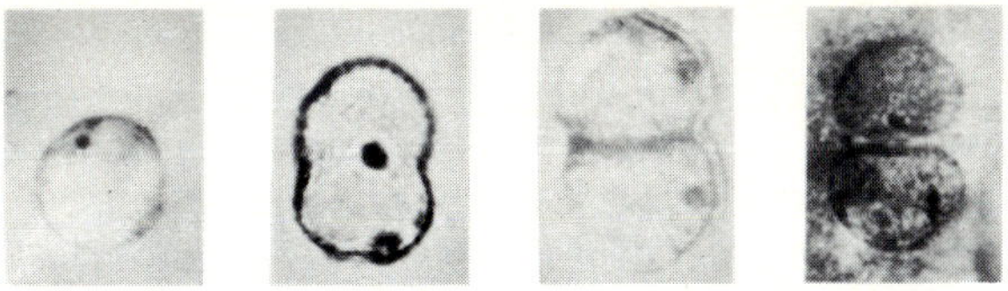

FIGURE 3.2 Photomicrographs of specimens of the alga *Glenobotrydion* taken from rocks that are one billion years old. It is highly unusual for such fossils to be preserved and most of the other evidence of the evolution of cells is much more indirect. These samples indicate that eucaryotes had arisen at least a billion years ago and probably much earlier. A cell division sequence is shown. The dark spots may be nuclei. *(Courtesy of J. William Schopf, University of California at Los Angeles.)*

down of chlorophyll, a key chemical in the photosynthetic process. Bacteria and blue-green algae are simple cells that reproduce by dividing into two cells and that contain no true nuclei. The genetic material of such cells, which are called *procaryotic* (L. *pro* before + Gr. *karyon* kernel) cells, is distributed unevenly in a nuclear region but is not bounded within a nuclear membrane. More recent rock specimens from a period of about two billion years ago contain traces of what appear to be cells with nuclei, while specimens of a more recent origin—about one billion years ago—contain a number of traces of what are clearly cells with nuclei, or *eucaryotic cells* (Gr. *eu* true) (See fig. 3.2.).

The distinction between procaryotic and eucaryotic cells is of considerable evolutionary significance for reasons that will be clear when we treat genetics in Chapter 6. Of all the organisms now populating the earth, only bacteria and blue-green algae are procaryotes. (See figs. 3.3 and 3.4.) All other cells, whether single-celled *Paramecia* or the cells in the body of a multicellular organism, are eucaryotes and generally possess membrane-bounded nuclei. (Human red blood cells, which live about four months, contain no nuclei when they mature, but they do contain nuclei for a short time when they are developing in the body.) Thus,

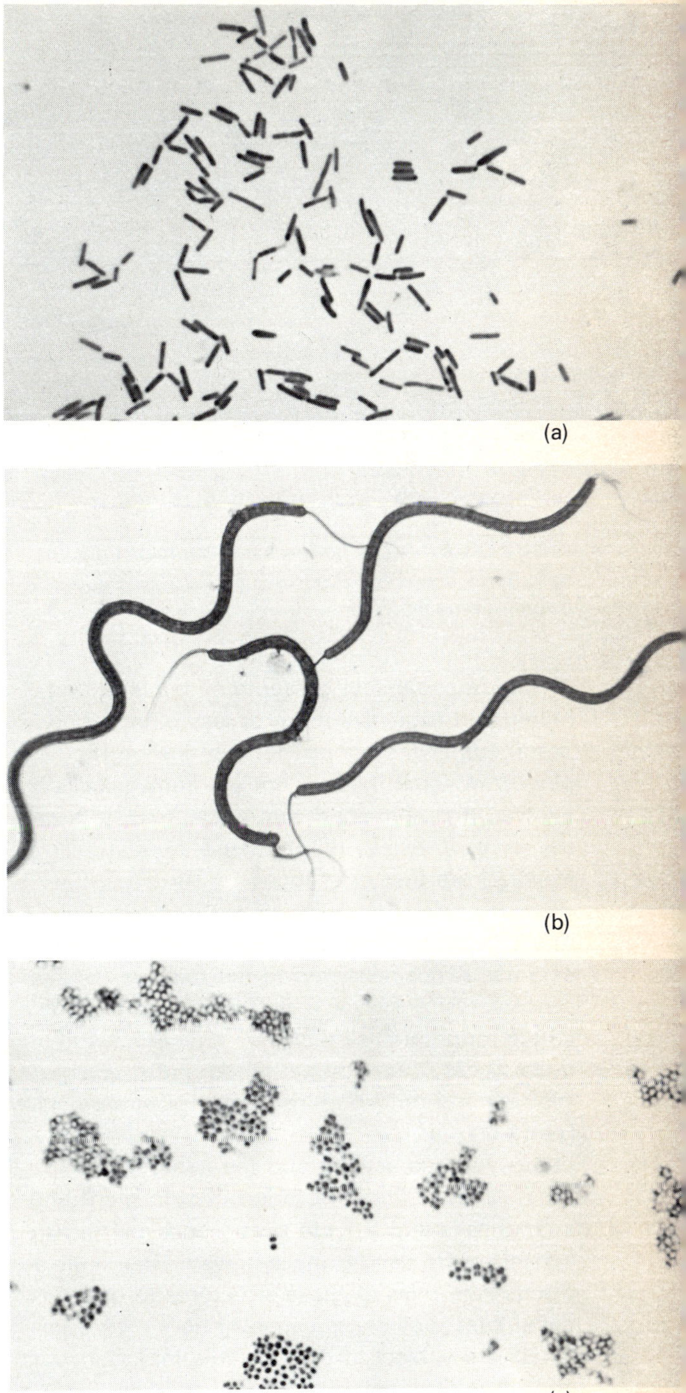

(a)

(b)

(c)

FIGURE 3.3 Species of bacteria showing (a) rod, (b) spiral, and (c) spherical forms. *(Courtesy of Carolina Biological Supply Company.)*

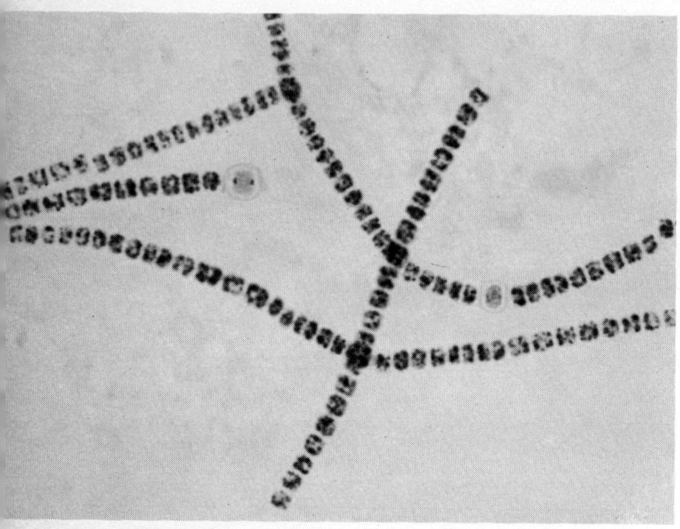

FIGURE 3.4 A blue-green alga (*Nostoc*). Cells in rows form filaments. *(Courtesy of Carolina Biological Supply Company.)*

it would appear that present-day bacteria and blue-green algae are more or less direct descendants of the earliest form of life on the planet. In some manner not yet known, cells evolved that contained true nuclei and, it is believed, gave rise to the immensely varying forms of life that exist today.

The difference between procaryotes and eucaryotes is not restricted to the nucleus. Unlike procaryotes, eucaryotes contain two important structural organelles that play critical roles in the process of energy gathering and use: chloroplasts and mitochondria. Chloroplasts are found only in plant cells and are the sites of photosynthetic activity and the capture of energy. Mitochondria are found in both plant and animal cells and are the sites where energy is converted for use by the organism. We shall discuss the roles of these structures in greater detail later, but the question of how they originated is a subject of considerable importance and of no little controversy.

One school of thought maintains that these structures were once free living organisms that came to live in *symbiosis* (''living together'') with procaryotic cells. Briefly stated, the argument for this theory goes as follows: The primitive atmosphere contained little or no free oxygen. The rise of photosynthesizing cells began an accumulation of free oxygen in the primitive environment. The original photosynthesizers and heterotrophs are thought to have been *anaerobic*—that is, organisms capable of living in the absence of molecular oxygen and to which, in fact, oxygen was a deadly poison. After some time, however, blue-green algae evolved and were capable of living in the presence of oxygen; that is, they were *aerobic*. Their appearance accelerated the accumulation of free oxygen in the primitive environment, and this posed a threat to the survival of the anaerobes. Some anaerobes came to live in oxygen-free environments, such as mud in oceans and lakes, while some others probably became extinct. Still others, it can be assumed, evolved in a way that allowed them to utilize the oxygen in their environment. An aerobic organism releases the energy stored in organic molecules by combining them with oxygen. Some form of aerobic, free-living organism is thought to have been the evolutionary forerunner of mitochondria, since the release of energy in cells with mitochondria occurs in those organelles by reactions with oxygen. The forerunners of mitochondria then somehow entered into a symbiotic relationship with anaerobes and thus conferred on their anaerobic partners an ability to survive in the presence of oxygen. Later, other free-living organisms became associated with the anaerobe/mitochondrion ''partnership.'' These new additions were similar to *spirochetes,* which are highly mobile, free-living bacteria. Their addition conferred an ability to move with some control and thus may have made the search for food more efficient. These spirochete-like partners may have become *flagella* (L. *flagellum* whip). (See fig. 3.5.) This three part association, according to the symbiotic theory, became the ancestral forerunner of animals. Then, it is postulated, blue-green algae may also have come to live symbiotically with the other three to give rise to higher plants. (See fig. 3.6.)

The theory of the symbiotic origin of mitochondria and chloroplasts has been disputed on several grounds. One of the theories proposed in response to the symbiotic theory maintains that the mitochondrion might have originated from an infolding of the cell membrane. (See fig. 3.7.) It has been pointed out that some procaryotes possess complex internal structures that could also have given rise to mitochondria. Basically, by this theory the eucaryotic cell arose in a direct line from an advanced procaryote. Other evidence regarding the origin of mitochondria and chloroplasts will be reviewed in Chapter 9, after we have had an opportunity to examine more thoroughly the processes of respiration and photosynthesis and the roles of mitochondria and chloroplasts in the life of the cell.

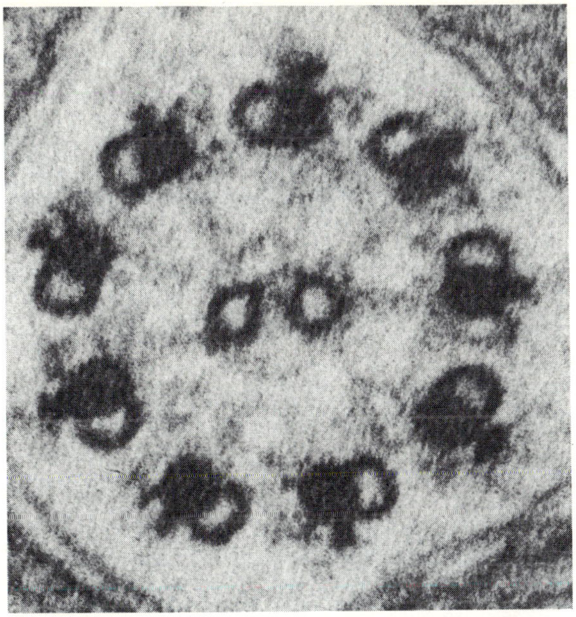

FIGURE 3.5 Cross section of a bird sperm flagellum. The "9 + 2" arrangement of contractile fibrils is characteristic of all higher animals. *(Courtesy of Don W. Fawcett, Harvard University Medical School.)*

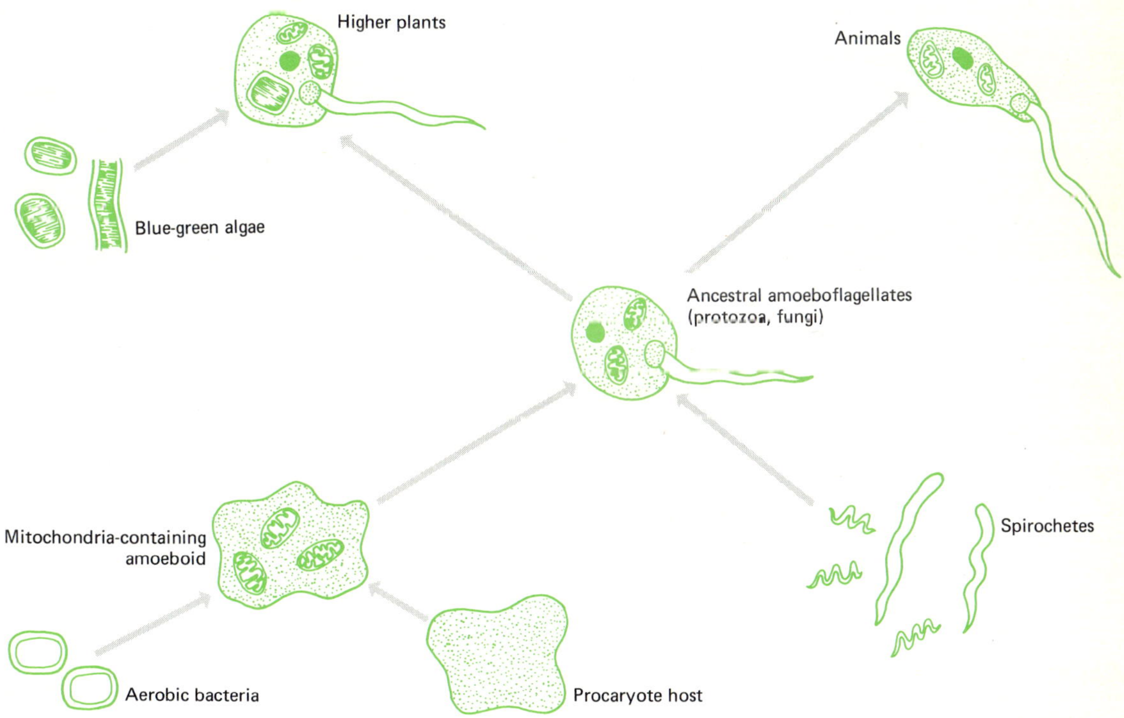

FIGURE 3.6 A proposed scheme to show the evolution of higher plants and animals from lower forms. *(Redrawn from "Symbiosis and Evolution" by Lynn Margulis. Copyright © August 1971 by Scientific American, Inc. All rights reserved.)*

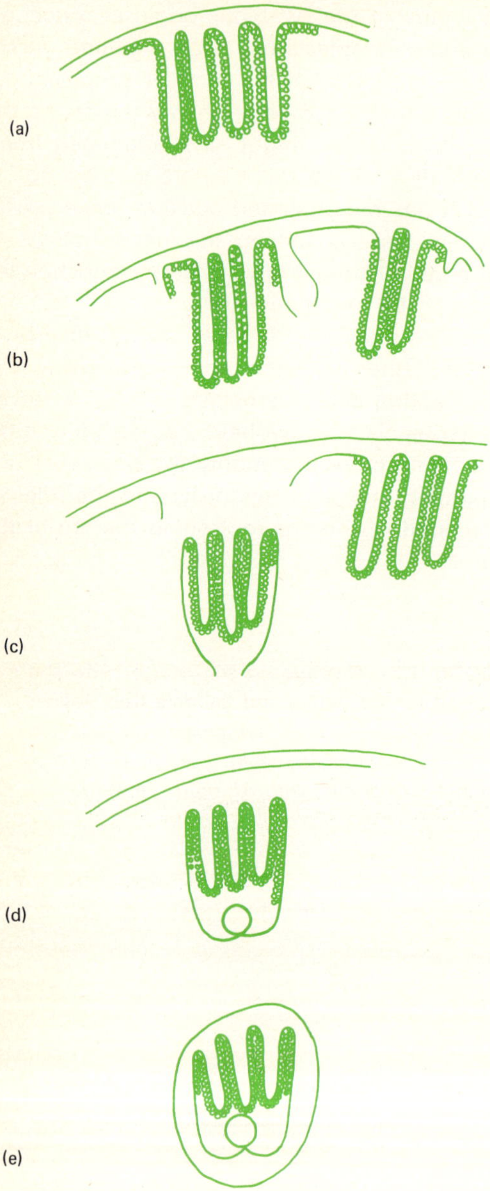

FIGURE 3.7 Hypothetical explanation of the origin of mitochondria. (a) Section of respiratory membrane, (b) part of membrane breaking free, (c) organelle forming by extension of membrane, (d) membrane completed, and (e) mitochondrion formed with a new outer membrane. *(Adapted from "The Nonsymbiotic Origin of Mitochondria" by R. A. Raff and H. R. Mahler. In* Science, *Vol. 177, pp. 575–582, Fig. 2, 18 August 1972; copyright 1972 by the American Association for the Advancement of Science.)*

The appearance of eucaryotic cells provided a line of demarcation in the living world that is still recognized as a key boundary in the evolution of complex multicellular organisms. Once cells with nuclei had arisen, the potential for cellular specialization that resulted in multicellular organisms could be exploited. These more complex forms were able to evolve specialized organs that conferred on them the capacity to survive in a variety of environments and thus increase the diversity of the biosphere.

Before concluding this section, we should turn briefly to the problems the biologist faces in organizing and sorting out this nearly bewildering array of diverse organisms. At one time in the development of biology as a science, such sorting appeared to be relatively simple. It appeared that one needed only to look around to see that there were basically two sorts of organisms—animals and plants. The invention of the microscope changed all that, of course, although some students of life remained steadfast, and arbitrarily insisted that bacteria were plants. Until recently, many biologists subscribed to a classification scheme consisting of four kingdoms: Monera, or procaryotic organisms; Protista, single-celled eucaryotes; Plantae, the higher plants; and Animalia, the higher animals. (We should note here, by the way, that the term "higher" is used not to denote superiority, but to indicate that the organism is of more recent evolutionary origin or of greater complexity.) The four kingdom classification is somewhat arbitrary, however, since the fungi and slime molds are included as plants. These two types of organisms, although they possess cells that are structured like plant cells and reproduce as plants, share certain characteristics of animals. (See fig. 3.8.) The cells of fungi and slime molds do not contain chlorophyll, and thus the organisms are not capable of photosynthesis. Instead, they derive energy by consuming organic compounds. Some forms live primarily on dead tissue and thus perform the ecologically important function of decomposition. Some other forms are parasitic; that is, they live on live tissue. Athlete's foot, for example, is

(a)

(b)

(c)

FIGURE 3.8 Fungi and slime mold. (a) A "bracket fungus" *(Polyporus versicolor)* (b) an edible morel mushroom *(Morchella esculenta)* (c) reproductive body of a slime mold *(Fuligo sp.)*. *((a) Courtesy of Grant Heilman; (b) courtesy of Hal Harrison from Grant Heilman; (c) courtesy of Hugh Spencer from National Audubon Society.)*

caused by a parasitic fungus, as are a number of agriculturally important blights. Taking into consideration the means by which higher organisms—that is, multicellular or multinucleate organisms—gather energy has provided the basis for a system of classification based on five kingdoms: Monera, Protista, Fungi, Plantae, and Animalia.

We shall examine a portion of this scheme of classification in more detail, for it provides us with a convenient summary of the characteristics of many organisms that we shall encounter in the chapters to come. The first kingdom, Monera, is made up of groups of organisms that are characterized by the absence of nuclear membranes, mitochondria, and chloroplasts. These organisms are usually solitary, unicellular forms, but there are some species which consist of unicellular individuals organized into colonies (see fig. 3.9). For the most part, Monera derive energy from organic sources, although an important group, the blue-green algae, are photosynthetic, and some photosynthetic bacteria utilize inorganic compounds like hydrogen sulfide (H_2S) in lieu of water. Reproduction in the kingdom Monera is predominantly by asexual fission. Some forms of Monera do not move about at all—that is, are nonmotile—but there are some forms that "glide" in an aquatic environment or propel themselves with simple flagella (see fig. 3.10). The kingdom Protista consists of eucaryotic unicellular organisms. No single means of gathering energy can be said to predominate among the protists. There are important groups that are photosynthetic, but there are other equally important groups that ingest food as a source of energy. There are yet others that digest food outside their bodies and absorb the products of digestion. There are even some forms that par-

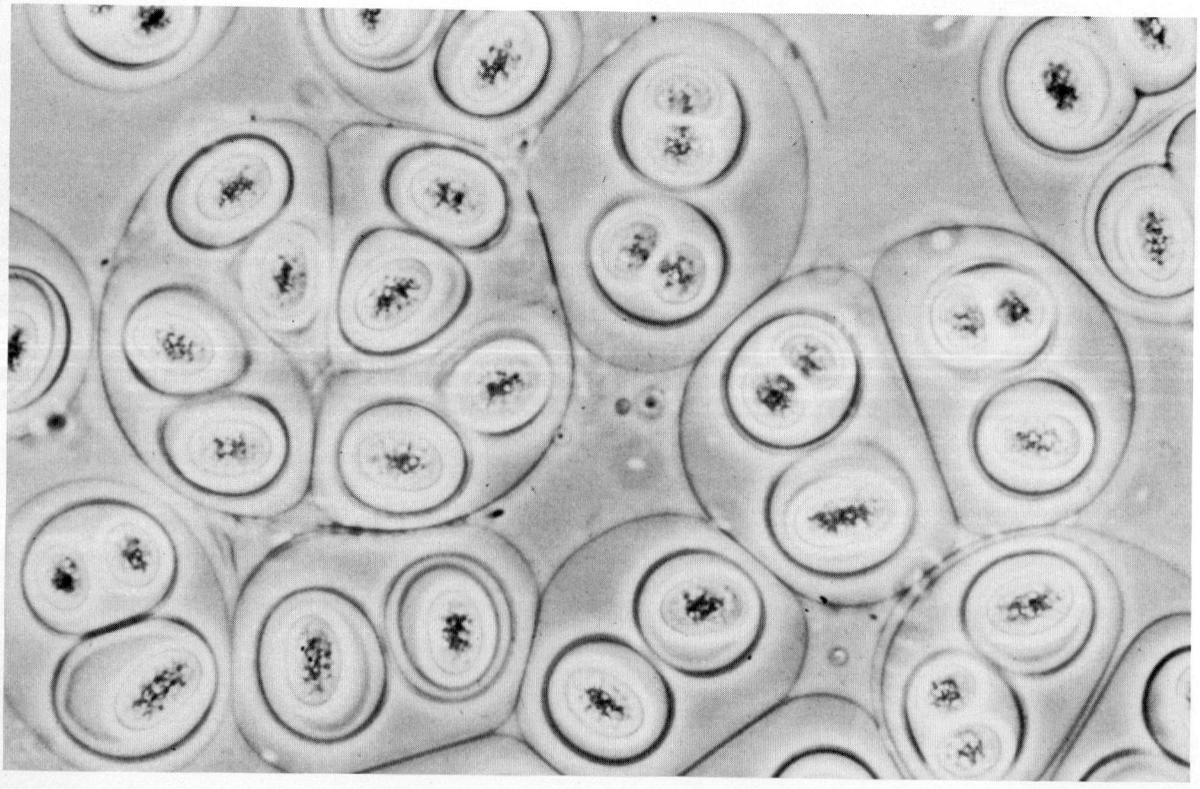

FIGURE 3.9 *Gloeocapsa*, a colonial blue-green alga. *(Courtesy of Ward's Natural Science Establishment, Inc., Rochester, New York.)*

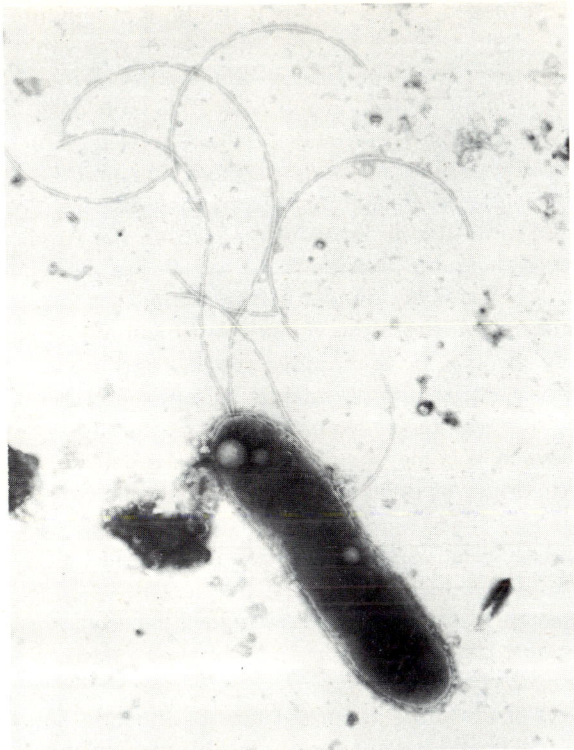

FIGURE 3.10 Many bacteria, such as this *Ectothiorhodospira,* have flagella for locomotion. Two may be seen attached to the cell, with other fragments present. This species, by the way, is one of the bacteria that is capable of photosynthesis. *(Courtesy of Stanley C. Holt, University of Massachusetts, Amherst.)*

take of more than one mode of energy gathering. Reproduction is more diverse among protists, with both sexual and asexual forms. Although some protists are nonmotile, a large proportion possess complex flagella or cilia that enable them to propel themselves.

The effect that the populations of lower organisms had on the earth itself must have been immense. The process of photosynthesis, carried on over hundreds of millions of years, produced a radical change in the atmosphere. To some extent oxygen can be dissolved in water, but only up to a certain concentration. After that the oxygen escapes from the water and enters the atmosphere. Our present atmosphere contains approximately 21% free oxygen, and the protective layer of ozone is the product of the reaction between sunlight and the free oxygen produced by photosynthesis. Over geological time, as many organisms died, the mineral parts of their bodies were laid down in deposits such as the famed White Cliffs of Dover in England, which are composed of the remains of single-celled organisms. We shall review some of the other geological effects of living systems in Chapter 11.

In summary then, two main types of organisms arose early in the evolutionary past: autotrophs, which produce chemical energy from sunlight and chemicals in the environment, and heterotrophs, which use the chemical energy to perform the activities necessary for survival. We shall see in forthcoming chapters that the so-called "biogeochemical" cycles of carbon, nitrogen, oxygen, and other elements are driven by the life processes of autotrophs and heterotrophs. The modification of the environment of the earth took place early, and the seas teemed with life billions of years ago. But life in the ocean changed ever so gradually as the organisms there adapted, radiating to fill the available niches and ultimately giving rise to the forms of today. Eventually plants and animals capable of surviving in the very rigorous, fluctuating terrestrial environment arose.

Adaptation in the Aquatic Environment

We can readily recognize that aquatic organisms are fundamentally different from terrestrial ones, but beyond noticing certain superficial distinctions, we rarely consider why they differ. The characteristics of water produced these differences.

By way of illustration, let us consider a shrimp living at a depth of 1000 meters (about 0.6 miles) in the ocean. Except for occasional flashes of light from luminescent organisms, the shrimp lives in perpetual darkness. The temperature of the water is about 8°C (45°F) and never varies more than a few degrees. At a depth of 1000 m. the shrimp is subjected to a pressure of about 1500 pounds per square inch, approximately 100 times greater than the pressure at the surface. The shrimp has never rested upon or even contacted any surface and, during its life, probably never will. From a human point of view such conditions may seem extreme, but in many ways they are not. Buoyed up and transported by water currents, the shrimp need not expend a great deal of energy on locomotion. It needs few adaptations to withstand temperature fluctuations, and compared to a terrestrial organism, it needs fewer adaptations to withstand drying. Because of the darkness, the shrimp is aided little by vision, but it is surrounded by all manner of dissolved chemicals that may provide it with clues about its environment. The great pressure is not a problem for the shrimp, since it is adapted to such pressures. The offspring of the shrimp are dispersed by water currents, and thus the reproductive mechanisms of the shrimp can be relatively simple. Currents also transport food to the shrimp and rapidly and easily transport waste products away.

On the whole, life is simpler in the oceans. Fewer complex adaptations are necessary, or perhaps even possible, to insure survival, and thus there exist in the ocean forms of life that are older, more diverse, and more primitive; forms of life that have never made the transition to land. Whereas the land is dominated by a few forms of life, the ocean contains a much greater variety of organisms.

Considering the ocean more closely, we can see that two basic environmental conditions can be described. The first is the water itself, and those forms of life that are found free in the water are called *pelagic* (Gr. *pelagos* sea). The second condition is the ocean bottom, and the forms of life found there are called *benthic* (Gr. *benthos* depth). Each of these environmental conditions has produced different life forms. The ocean bottom provides a surface to support organisms, and benthic organisms are usually stationary with respect to the water. If the bottom is hard, the organisms are usually attached; if it is soft, they are usually burrowers. Since most of the ocean bottom beyond the coasts is a soft ooze, there are many burrowers. Where coral reefs or rocky coasts exist, there are more attached forms of life. Pelagic organisms may also be divided into two groups: *plankton* (from the Greek word meaning "wandering"), which are more or less passively transported by water currents; and *nekton* (from the Greek word meaning "swimming"), which move actively and are capable of migrating against currents. Measured by live weight, or *biomass*, most of the life in the ocean is planktonic. By far, the greatest proportion of plankton are minute, single-celled plants called *phytoplankton*. Thus, although we shall look at some benthic forms because they are of evolutionary interest, they are not as important to the economy of life in the sea as are pelagic organisms.

One of the most important factors affecting life in the ocean is light for photosynthesis. In the open ocean at depths of greater than 30 meters (about 100 feet), there is not enough light for photosynthesis, and in more turbid coastal waters there is rarely enough light below 10 meters. Although some food enters the oceans by draining from coastal areas, nearly all of the food energy in the oceans is produced near the surface of the open sea and in coastal waters. As on land, the actual amounts of food produced depend on fertility, and wherever there is a good supply of nutrients there will be greater production. Thus, marine biologists think of the surface waters of the oceans as a teeming broth of life.

Water is also an excellent solvent, and thus the oxygen needed by animals and the carbon dioxide needed by plants can be transported in solution to the organisms that require them. As a solvent, water also carries dissolved nutrient material to plants and removes and dissolves the waste products of animals. Some of the dissolved organic compounds in water may be sensed by some organisms, just as we smell

TABLE 3.2 Composition of Sea Water

Element	*Tons per Cubic Mile*
Oxygen	4,037,000,000
Hydrogen	509,000,000
Chlorine	89,500,000
Sodium	49,500,000
Magnesium	6,125,000
Sulfur	4,240,000
Calcium	1,880,000
Potassium	1,790,000
Bromine	306,000
Carbon	132,000
Strontium	37,700
Boron	22,600
Silicon	14,130
Fluorine	6,125
Argon	2,825
Nitrogen	2,350

chemicals suspended in the atmosphere, and thus the solvent properties of water may affect the behavior of some organisms. Moreover, the solubility of some materials, including pollutants, may do much to determine where in the ocean which forms can live.

One important result of the solvent property of water is the *salinity*, or saltiness, of the ocean. Sea water contains very large and relatively constant quantities of chlorine, sodium, magnesium, sulfur, calcium, carbon, and at least fifty other elements, many of which are not used by organisms. (See table 3.2.) The salinity of the ocean is important in that it may make the ocean a drying environment. *Water has a tendency to move, by a process called osmosis, from areas of greater to areas of lesser concentration.* (See box 3.1.) Since the body fluids of all aquatic organisms are also saline solutions, an osmotic pressure may be set up that has a tendency to move water into or out of the organism. If the body fluids of the organism are less saline than the water of the external environment, water will be drawn out of the organism into the more saline sea water. Thus, the organism must possess some physiological mechanism for pumping water back into its body to prevent dehydration. Similarly, in fresh water the body fluids are more saline than the water in the environment, and thus the organism must have some sort of mechanism for pumping water out. It is because of this property of water that most freshwater species cannot live in sea water, and most marine species cannot live in fresh water. Certain species, such as some eels and salmon, live part of their lives in fresh water and part in salt water and possess elaborate physiological mechanisms to adjust to the differences in their environments. In areas where fresh and salt waters mix, as in estuaries where rivers empty into the sea, the fluctuations in salinity may produce conditions that place organisms under considerable osmotic stress.

The marine environment may well provide many advantages for plants and animals, since it is much less ''rigorous'' than the terrestrial environment. The adaptations required are very different, and the evolution of species in the ocean can be very instructive about the evolutionary process.

Animals in the Sea

We cannot consider here all of the important characteristics of marine organisms, for to do so would require far more space than is available. What we will do is concentrate on what we consider to be the two most important activities of animals—eating and reproducing. Nearly all of the abilities and activities of an organism are adaptations to these two ends, and since it is necessary to survive in order to reproduce, let us consider first the problem of getting the food and nutrients that are necessary for survival.

One of the unique characteristics of water is that it transports potential food for organisms. The food may be in the form of living phytoplankton or *detritus*—that is, dead particles of debris from once-living organisms. Thus, most marine animals are adapted to utilizing this

Osmosis, Diffusion, and Gills

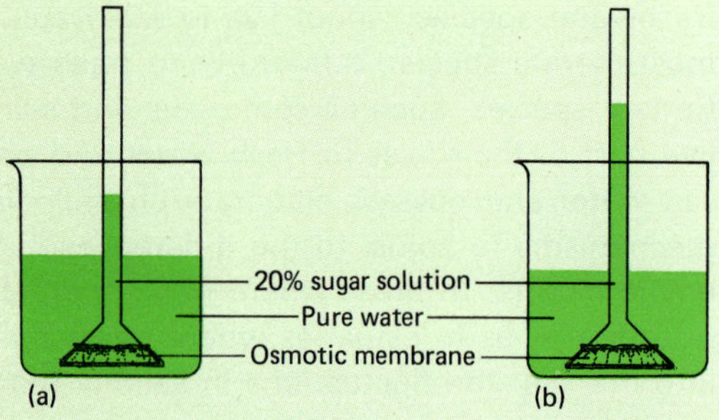

OSMOSIS

An osmotic membrane, such as cellophane, stretched across the mouth of the funnel will permit water to enter the funnel. The water moves from the beaker, where the concentration is high (100%), to the funnel, where the concentration is lower (80%). The movement will stop when an *osmotic equilibrium* is reached. In this instance an osmotic equilibrium may be reached before the concentrations of water on each side of the membrane are equal, as the weight of the sugar solution in the funnel prevents the further movement of water across the membrane.

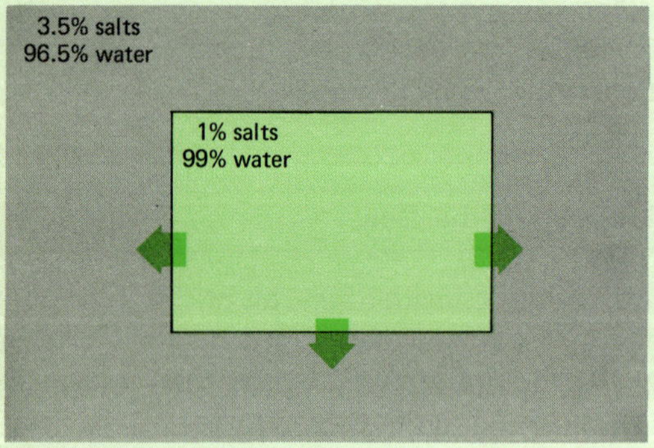

(a) Water moves out of cell

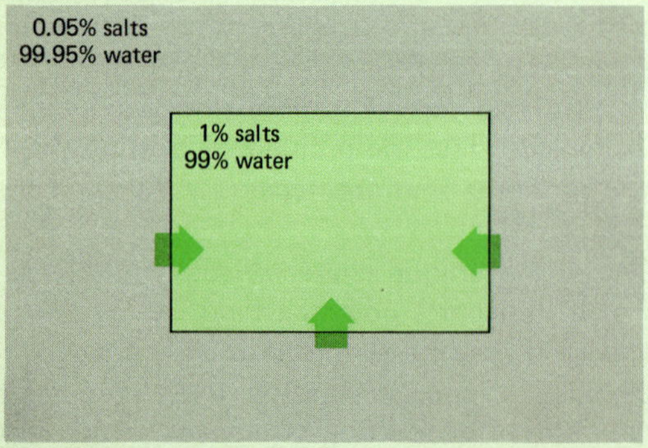

(b) Water moves into cell

This figure indicates the movement of water into or out of a hypothetical cell. The concentration of salts in (a) is approximately the concentration found in the marine environment. In (b) the salt concentration is approximately that which occurs in fresh water. A cell with a concentration of 1% of salts and 99% water would be affected very differently in sea water than in fresh water.

BOX 3.1

DIFFUSION

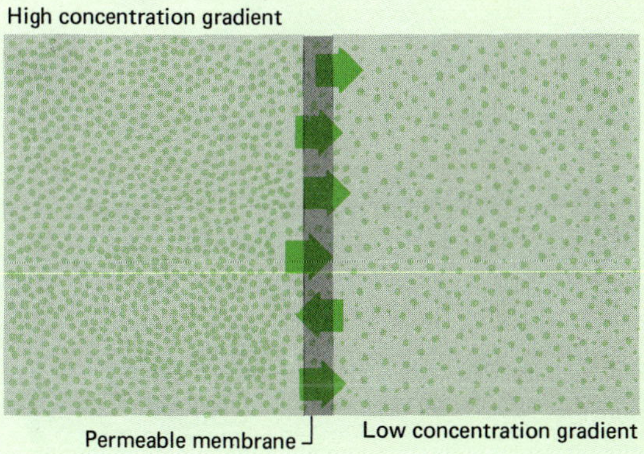

If some material can permeate a membrane, then it will have a tendency to move across the membrane until it is in equal concentrations on each side. As in osmosis, the movement is down a *concentration gradient*, that is, from higher to lower concentrations. As shown, the material actually moves through the membrane in each direction. Because there is more material on one side of the membrane, however, there is more material that can move through the membrane to the other side, and thus the net movement is to the region where concentration is low. When the concentrations on each side of the membrane are equal, the material continues to permeate it, but approximately equal amounts pass through the membrane in each direction and there is no net change in concentrations.

GILLS

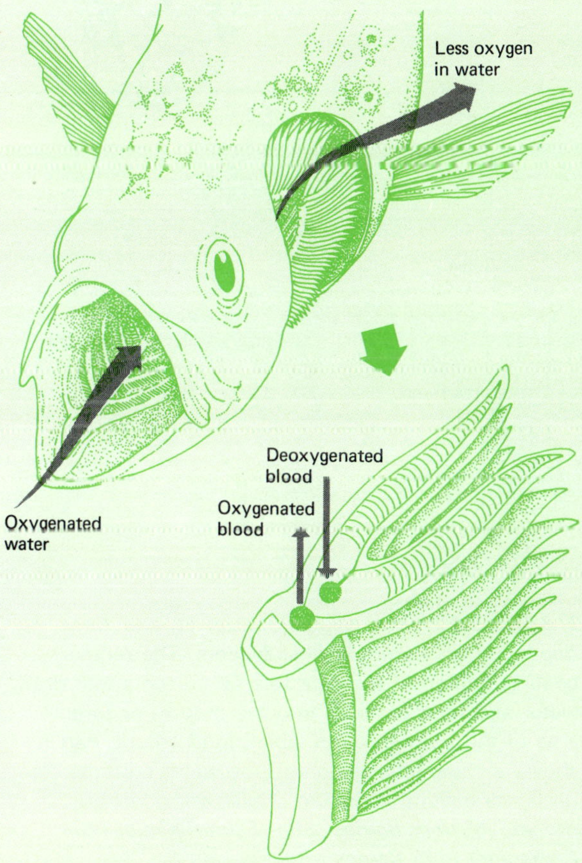

Although we cannot utilize it (and thus must carry it with us if we dive), oxygen is present in sufficient quantities in most bodies of water to support a variety of aquatic animals. Most small invertebrates exchange oxygen and carbon dioxide by diffusion through the surfaces of their bodies. Larger multicellular invertebrates, and nearly all vertebrates, however, possess gills, or structures that are similar to gills, for the exchange of gases. Although they differ in many respects from species to species, basically all gills operate in the same way. By presenting a very large surface area over which very large amounts of water can be passed, sufficient oxygen to meet respiratory needs diffuses out of the water through the gill membrane, and into the bloodstream of the animal. *(Redrawn from* Life on Earth, *by E. O. Wilson* et al. *Copyright © 1973 by Sinauer Associates, Inc.)*

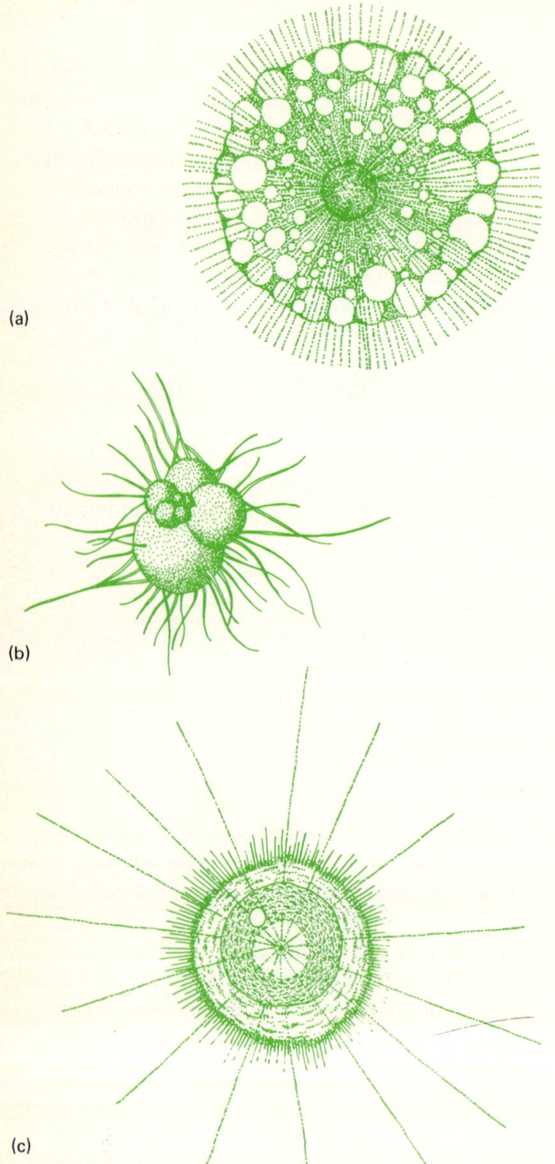

FIGURE 3.11 The radiating pseudopodia of (a) Radiolaria, (b) Foraminifera, and (c) Heliozoa are covered with a granular adhesive substance which allows the organisms to capture or trap other organisms that come into contact with the pseudopodia. *(Redrawn: (a), (c) from Barnes after Hall; (b) from Barnes after* The Invertebrates: Protozoa Through Ctenophora, *Vol. I, by Libbie Hyman. Copyright © 1940 by McGraw-Hill, Inc. Used with permission of McGraw-Hill Book Co.)*

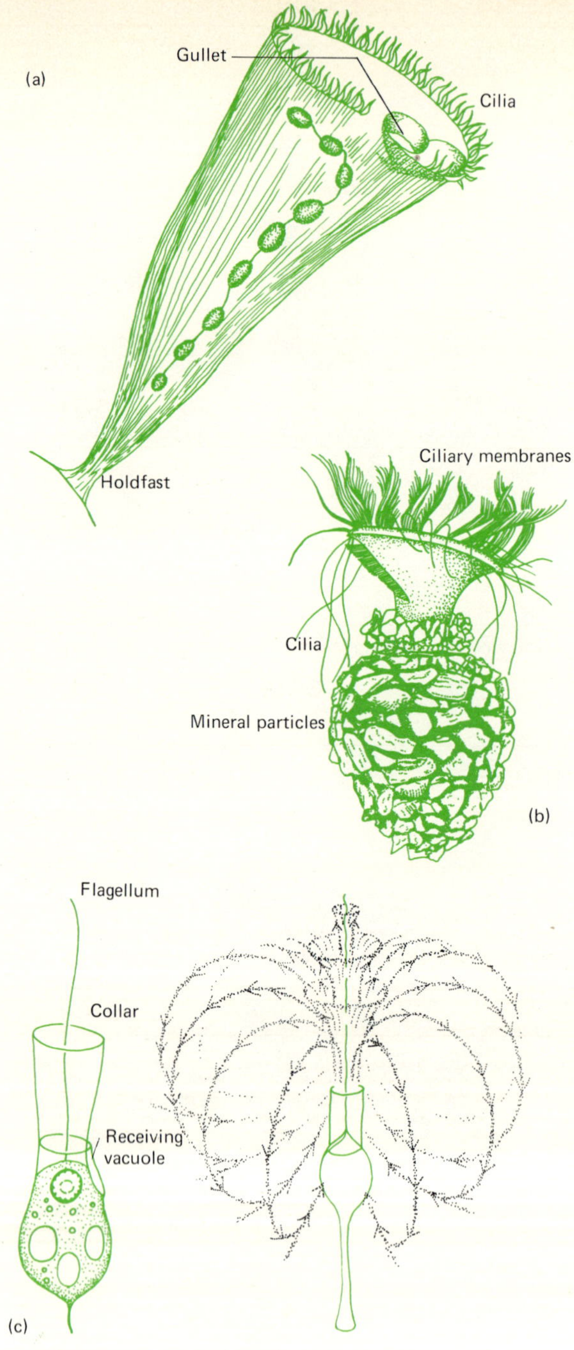

FIGURE 3.12 The ciliates *Stentor* (a) and *Tintinnopsis* (b) are common filter feeders. The regular beating motion of the cilia sets up currents that draw water and food particles into the mouth, or to an area of the body through which food enters. Part (c) shows a choanoflagellate and the water currents set up by its beating flagellum. *(Redrawn: (a) after Barnes; (b) from Barnes, after Fauré-Fremiet from Corliss; (c) from Barnes after Dogiel.)*

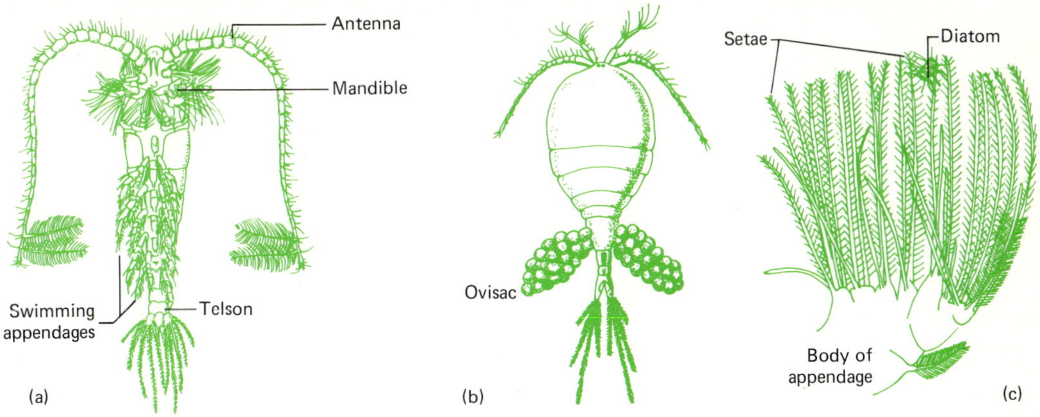

FIGURE 3.13 The copepods *Calanus* (a) and *Cyclops* (b), the latter with its large ovisac containing a number of eggs being hatched. When the eggs mature, anywhere from one half to five days after mating, they will be replaced by a new batch of eggs from the same mating as the last. Also shown in (c) is a detailed drawing of the filtering bristles of *Calanus* with a diatom, a single-celled plant, drawn in to indicate the sizes of particles a copepod can filter from water. The diatom at the top is 25 microns long. *(Redrawn: (a) after Meglitsch; (b) from Barnes; (c) from Barnes after Marshall and Orr.)*

finely divided material, by filtering it directly from the water or by extracting it from bottom sediments. Thus, the majority of marine animals are *filter feeders* or *sediment feeders.* Only the specialized predators high on the food chain can actually pursue prey in the way that land animals do.

Some adaptations to filter feeding are shown on pages 54–57. The simplest adaptations to filter feeding are seen among many protozoa, although they are more accurately described as random search than true filter feeding. The Radiolaria are adapted to "sticking" to food particles and ingesting them, and Foraminifera and Heliozoa have similar adaptations. (See fig. 3.11.) Ciliates, such as *Stentor* and *Tintinnopsis* (see fig. 3.12), are more truly filter feeders, passing water currents by their bodies and capturing detritus, phytoplankton, and other protozoa. A multitude of other protozoa are similarly adapted to filter feeding, and of course the protozoa themselves constitute particles of food for other organisms.

Filter feeding is by no means restricted to simple, single-celled organisms like protozoa. At the other extreme of complexity are the Crustacea, shelled animals such as lobsters, shrimp, and crabs. Ecologically, the most important crustaceans are small, shrimplike organisms found in all of the oceans of the world and called *copepods*. Nearly all copepods possess complex oarlike appendages that are used for swimming and also for passing water through comblike bristles on the appendages near the mouth. (See fig. 3.13.) These appendages catch the phytoplankton on which the copepods feed. A single copepod about one millimeter long may filter several liters of water in a single day. Other Crustacea are also filter feeders, many of them being superficially similar to copepods.

Intermediate between simple protozoa and more complex crustaceans are a group of animals called *coelenterates*. Jellyfish are a prime example of this group, although they are not strictly filter feeders. Most coelenterates possess stinging cells on their tentacles. These are used to capture prey that passively drift into contact with the coelenterate. (See fig. 3.14.)

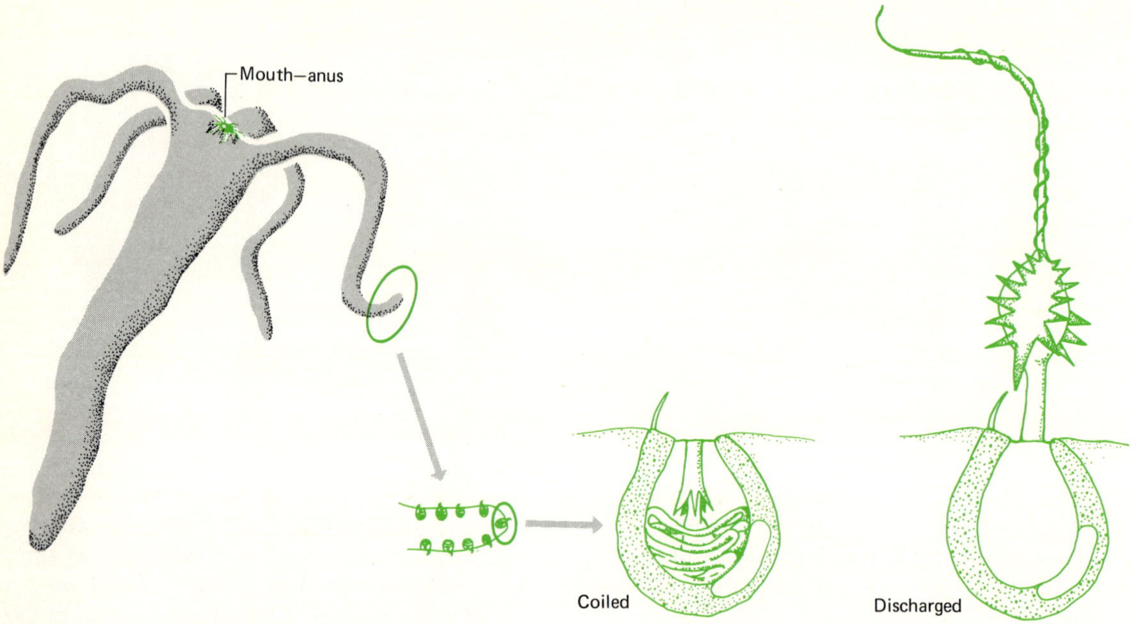

FIGURE 3.14 The tentacles of coelenterates such as the jellyfish or *Hydra* (shown) possess stinging cells which are capable of injecting toxin into prey and thus stunning or immobilizing it. The prey may then be taken into the gastrovascular cavity. *(Redrawn: coiled cell, from Barnes after* The Invertebrates: Protozoa through Ctenophora, *Vol. I, by Libbie Hyman. Copyright © 1940 by McGraw-Hill, Inc. Used with permission of McGraw-Hill Book Co.)*

FIGURE 3.15 The baleen plates in the mouth of the Atlantic Right Whale (a), (b) filter zooplankton from the water. The inner edges of the plates (c) are composed of the thick filaments that strain the plankton from the water. *((a) and (b) Courtesy of the American Museum of Natural History; (c) courtesy of Thomas Eisner, Cornell University.)*

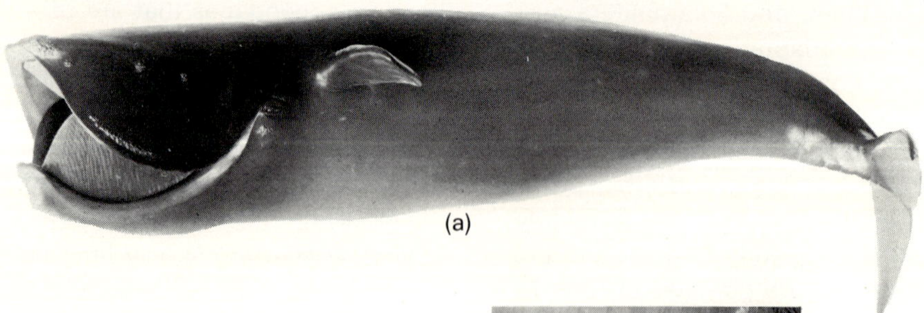

(a)

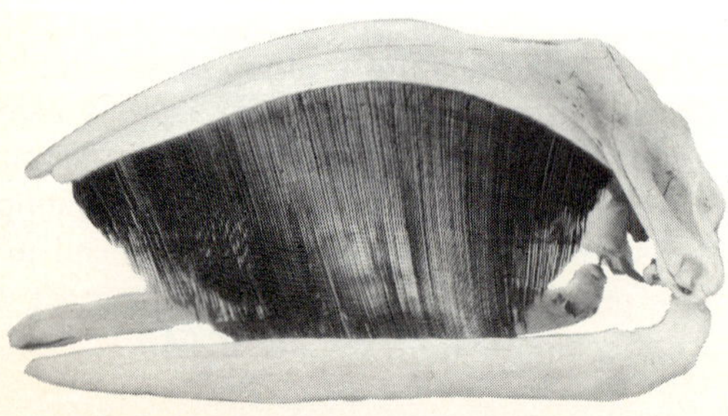

(b)

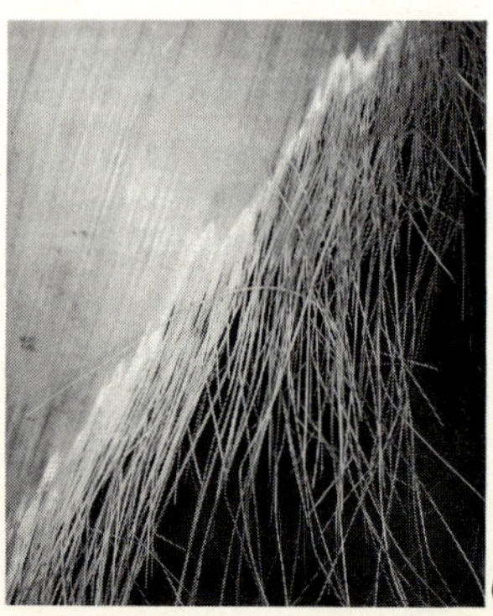

(c)

There are also vertebrate filter feeders. The gills of herring and anchovies are equipped with ''combs'' that remove plankton from the water that is passed over the gills. Even many whales are filter feeders. The baleen, or whale-bone, whales possess massive plates in their mouths that serve to filter crustaceans from the water. (See fig. 3.15.) In fact, one of the simplest major food chains in the world is based on filter feeding. In Antarctic waters, phytoplankton are eaten by filter-feeding shrimp called *krill*, and the krill in turn are eaten by antarctic baleen whales. The only creature that regularly hunts and eats the whales is man.

Many benthic animals are also filter feeders. Barnacles are crustaceans that are related to copepods, and after hatching from eggs, the larvae of barnacles very strongly resemble the larvae of copepods. The adult form of the barnacle is very different from the larval form, however, and leads a very different life. The adult barnacle has been described, very aptly, as lying on its back and kicking food into its mouth. When ready to assume the adult form, barnacle larvae cement their backs to a solid substrate such as a rock or the hull of a ship at anchor and secrete a hard shell; the former swimming appendages become adapted to comb detritus and small food particles from the water.

Perhaps the best examples of benthic filter feeders are sponges. (See fig. 3.16.) Although they are very simple organisms, sponges possess some specialized cells with flagella that are surrounded by collars. Each of the so-called ''collar cells'' contains a single flagellum that, when moved, sets up water currents that draw water through the body of the sponge. Particles of food contained in the water are trapped by other cells in the body and ingested in a way similar to that employed by many protozoa.

We could continue to present examples of other filter feeders, and some of the adaptations to this means of obtaining food and nutrients would appear quite bizarre indeed. We do not

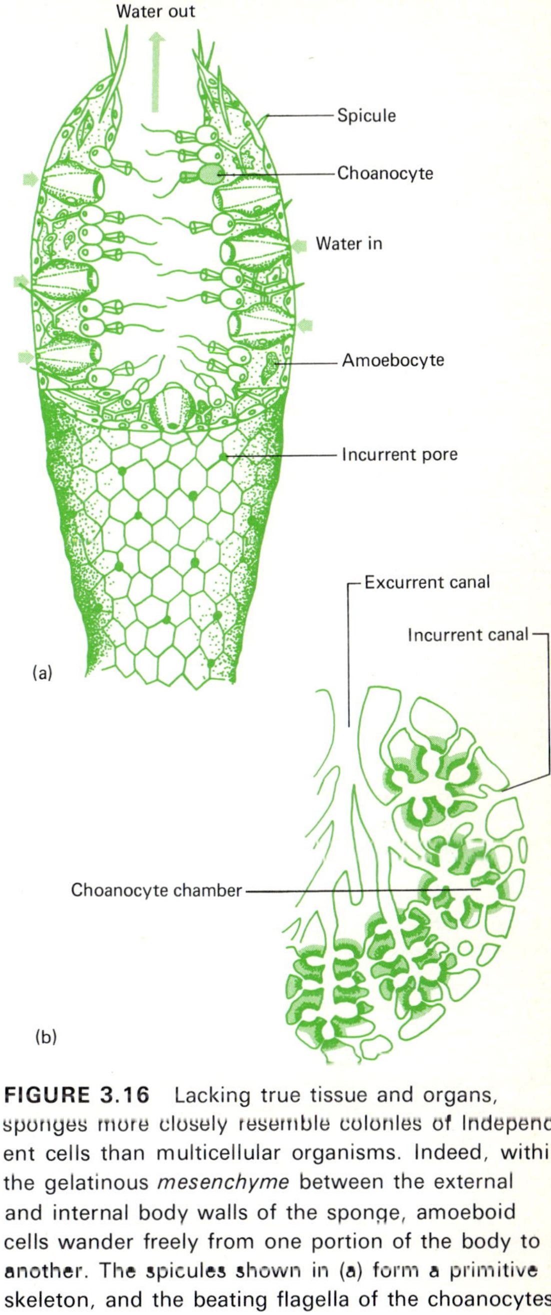

FIGURE 3.16 Lacking true tissue and organs, sponges more closely resemble colonies of independent cells than multicellular organisms. Indeed, within the gelatinous *mesenchyme* between the external and internal body walls of the sponge, amoeboid cells wander freely from one portion of the body to another. The spicules shown in (a) form a primitive skeleton, and the beating flagella of the choanocytes draw water and food into the body of the sponge. In the more commonly found body form of the phylum, such as the sponge shown in cross section in (b), the shaded areas represent the locations of the choanocytes. *(Redrawn: (a) from Barnes, modified from Buchsbaum; (b) from Barnes after* The Invertebrates: Protozoa through Ctenophora, *Vol. I, by Libbie Hyman. Copyright © 1940 by McGraw-Hill, Inc. Used with permission of McGraw-Hill Book Co.)*

wish to imply, however, that all marine animals are filter feeders. While this means of obtaining food is widespread, other means have certain advantages for some organisms. Any spade of mud from a tidal flat, for example, will yield a variety of organisms that pass mud through their guts and extract and digest the food contained therein. There are also animals that could best be described as grazers. Many snails, for example, possess a filelike organ called a *radula* near their mouths that is used to scrape algae from rocks, while sea urchins can move slowly over rocks scraping algae from the surface.

Then, too, many marine organisms are active predators. Most species of large squid, shrimp, and fish fall into this category, and some of their adaptations are bizarre. The "gulper eel," for example, lives at great depths, where meals are relatively infrequent. This "eel" survives because it has an extensible stomach that allows it to eat an organism its own size and then digest the meal over a long period of time. Obtaining food in total or nearly total darkness poses some obvious problems. Vision, for example, has little utility. In those regions where light is not completely filtered out, it is directed downward, and prey can best be seen as a silhouette when viewed from below. Thus, most predatory squid, shrimp, and fish at these depths have upward-directed eyes. Many potential prey species, however, have luminescent organs on their under surfaces that may serve to break up their silhouettes.

We have touched on only a few of the adaptations that have arisen to enable marine animals to fulfill one of the necessities of life—the need to obtain food. However superficial our treatment may have been, it should be clear that the adaptations are diverse and very different from those we can see in the terrestrial environment, and it should be equally clear that it is the environment itself that has dictated the adaptations.

On land, most animals must move about in order to obtain food. In the sea, some predators must search for their food, some benthic grazers move from rock to rock and some pelagic filter feeders require locomotion to maintain their position—that is, to keep from sinking. (See box 3.2.) But for most marine animals, the medium transports the food and nutrients necessary for survival. Filter feeders, therefore, can be much simpler animals, requiring few secondary adaptations for locomotion. In its larval stage, a barnacle has a complex nervous system, but when the larva settles and constructs a shell, the nervous system degenerates and becomes very simple. Neither the jellyfish nor the sea anemone has a central nervous system, but rather they have a diffuse net of nerves. Other features frequently found in the body plans of animals that search actively for food are lacking in many marine animals. The differentiation of the body into a head and tail, or *bilateral symmetry*, is frequently absent. Instead, many aquatic animals have *radial symmetry*. (See fig. 3.17.) This difference in body plan is correlated with the kind of life led by an organism. Animals that are *sessile*, or that remain seated in one place, are generally radially symmetrical, because changes in the environment, such as the arrival of food, are equally likely to come from any direction. Thus, many marine organisms are more analogous to terrestrial plants than to terrestrial animals. (See fig. 3.18.) Animals that must move about in search of food generally encounter changes in the environment in the direction in which they move. Thus, they have evolved a head in which are located mechanisms that serve to inform them of environmental changes. We shall see later in this chapter the profound implications that this has had in understanding the ancestry of organisms.

Locomotion in Marine Animals

One of the basic laws of physics states that every action has an equal and opposite reaction. This principle is the basis for most of the locomotion of aquatic animals. All pelagic organisms move in one direction by pushing water in the opposite direction.

(a)

Squid give an example of classical jet propulsion.

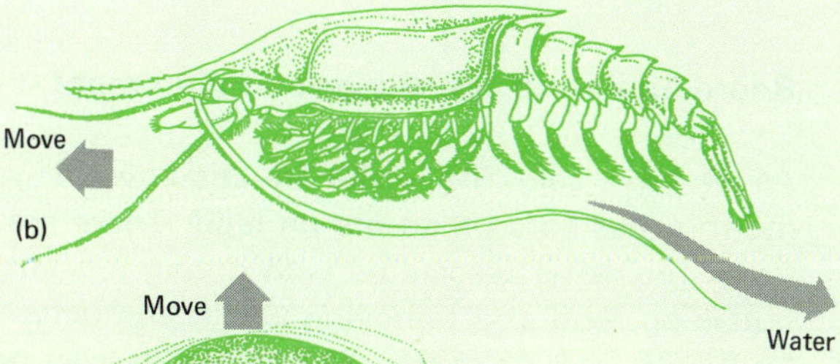

(b)

Shrimp possess a number of oarlike appendages used for propulsion. *(Redrawn: (a), (b) from Barnes.)*

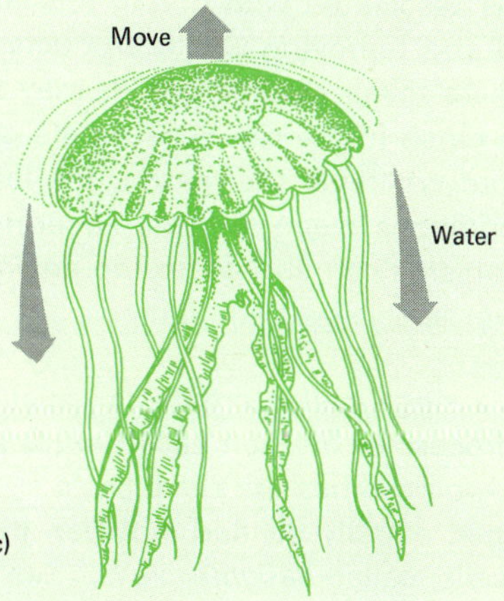

(c)

Jellyfish resemble an umbrella and by opening and closing can move in some direction. *(Redrawn: (c) from Barnes after* The Invertebrates: Protozoa through Ctenophora, *Vol. I, by Libbie Hyman. Copyright © 1940 by McGraw-Hill, Inc. Used with permission of McGraw-Hill Book Co.)*

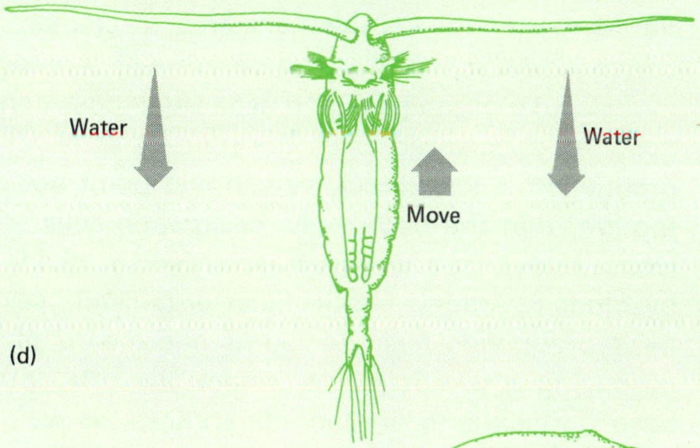

(d)

Copepods use their antennae as oars. *(Redrawn: (d) from Barnes after Cannon.)*

(e)

Fish move their bodies to push water backwards and thus propel themselves forward.

BOX 3.2

FIGURE 3.17 Radial symmetry in a sand dollar, *Echinarachnius parma. (Courtesy of Herman H. Giethoorn from National Audubon Society.)*

FIGURE 3.18 Radial symmetry in an anthozoan: a sea anemone, *Cerianthus sp. (Courtesy of Arthur W. Ambler from National Audubon Society.)*

Many of the same properties of the aquatic environment that play a large role in determining the adaptations of organisms to the problem of obtaining food also play an important role in determining the reproductive adaptations of organisms. The conditions of an aquatic environment are strikingly different from terrestrial conditions and have produced some adaptations that appear to the human eye to be curious indeed.

Reproduction in the Aquatic Environment

As we have said, life in the aquatic environment is less harsh than life on land. There are fewer problems associated with drying out, and obtaining food and removing waste products frequently require fewer adaptations. It may be that life is less harsh in the sea because life originated there, but whatever the reason, the less demanding conditions mean that the offspring of aquatic organisms can survive in the environment with fewer specialized adaptations.

To reproduce, a single-celled organism usually undergoes a process of division called *mitosis*, whereby the chromosomes are doubled in number and then equally divided between the two resulting cells, called *daughter cells.* (See fig. 3.19.) Thus the daughter cells have precisely the same genetic complement and, of course, chromosome number as the parent cell. Among multicellular organisms, and especially those composed of specialized cells, the process of reproduction is less direct. We should point out, though, that the body cells, or *somatic* cells (*soma* body), in a multicellular organism divide by mitosis. Among multicellular organisms, sexual reproduction has evolved as one means of producing new individuals; in fact, the sole "purpose" of sex is to produce differences, as we shall see in Chapter 6. Basically, sexual reproduction involves the production of a male *gamete*, the *sperm*, and a female gamete, the *ovum* or egg, which fuse to form the *zygote*, which then develops into a new individual. Each gamete contains only half as many chro-

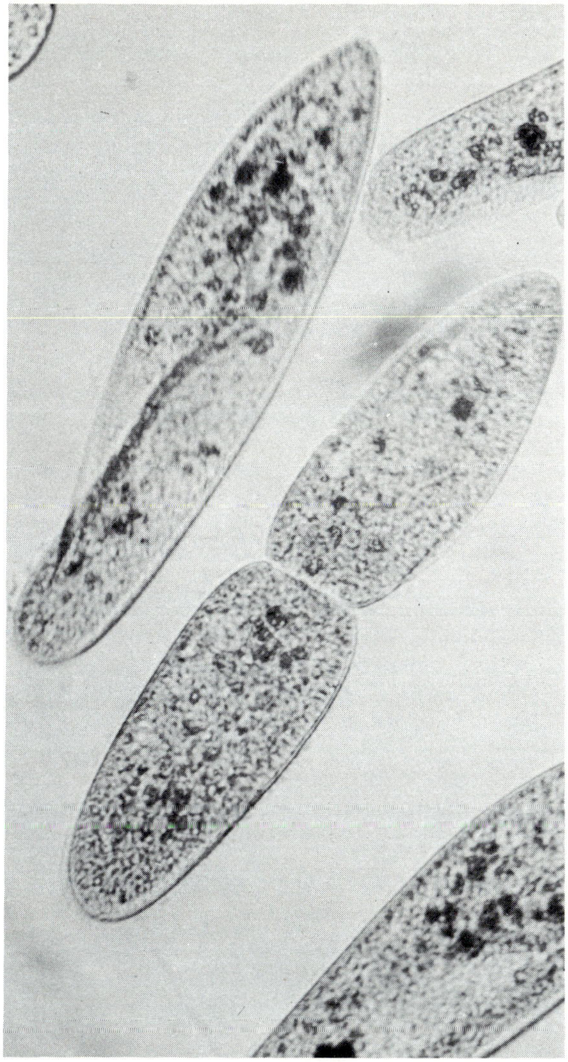

FIGURE 3.19 Photomicrograph of paramecia, single-celled organisms. The parent cell in the center is dividing to form two daughter cells, each with the same genetic complement as the parent cell. A mature *Paramecium* is shown at the upper left. *(Courtesy of Hugh Spencer from National Audubon Society.)*

mosomes as the somatic cells of the parent, and each gamete is produced by a division process called *meiosis*. We shall examine both mitosis and meiosis in greater detail in Chapter 6. It suffices to say here that mitosis results in no genetic change and meiosis always results in a change in the chromosome number. We shall restrict our discussion in this section to the problems associated with sexual reproduction in the aquatic environment.

Among many marine organisms, the fusion of male and female gametes, or *fertilization*, takes place outside the bodies of the parent organisms, although there are some species in which fertilization takes place within the body of one of the parents. In either case, however, some of the fertilized eggs or newly produced larvae must survive to produce adults, and then they must be dispersed to prevent overcrowding.

Consider the oyster, for example. During the spring, male oysters release large quantities of sperm into the water, and this triggers the release by females of large quantities of ova. Fertilization takes place in the water itself, and the fertilized eggs, which are now single cells, immediately begin to divide. Within a day or so, small ciliated larvae are produced that are capable of feeding. A small proportion of these larvae will settle and produce an oyster bed. However, the eggs and larvae are subjected to very heavy predation by fish that feed on them. Thus, in order that the species might survive, the parent oysters must produce tremendous numbers of eggs. On the average, only one larva out of one million ever settles and becomes an adult. The process of external fertilization is common among many marine organisms, including clams, many segmented worms, some crustacea, many fish, and echinoderms such as starfish and sea urchins. (See box 3.3.)

The Larvae of Marine Organisms

Each species of marine organism produces a distinctly different kind of larval offspring. Each of the larvae strongly reflects the niche of the offspring and the kind of life it will lead until it enters the adult phase of its life cycle.

The larval stages of the life of an organism are always the most vulnerable, and thus usually many larvae must be produced to withstand the great selective pressures that the population encounters. The quantity of larvae is often also a good indication of the health of an adult population. The study of marine larvae is therefore important and may be a good monitor of pollution or overexploitation. If the larval population of some species declines it may indicate that the parent stock is in trouble, that toxic substances are present, or that the adult population is being overfished. The problem, of course, is identifying the species of larvae. Many larvae were first described as adult members of plankton populations. The identification process requires careful laboratory culturing of eggs, often a very difficult task.

LARVAL FORM

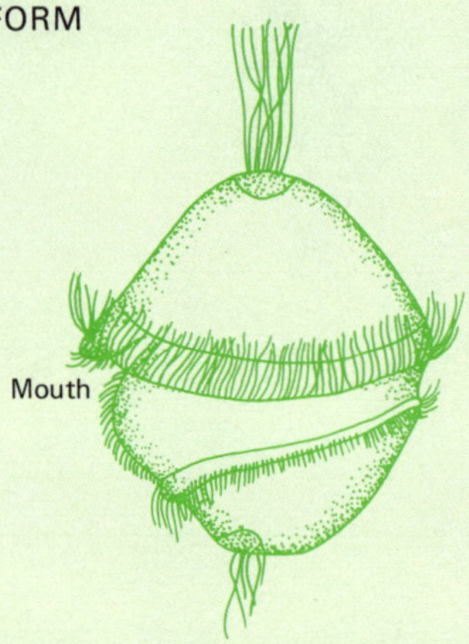

(a) Trochophore. *(Redrawn from Barnes.)*

ADULT FORM

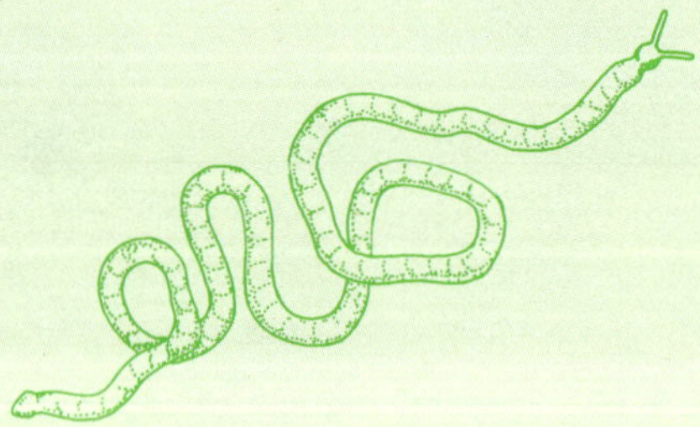

(a) *Polygordius,* an annelid. *(Redrawn from Barnes.)*

BOX 3.3

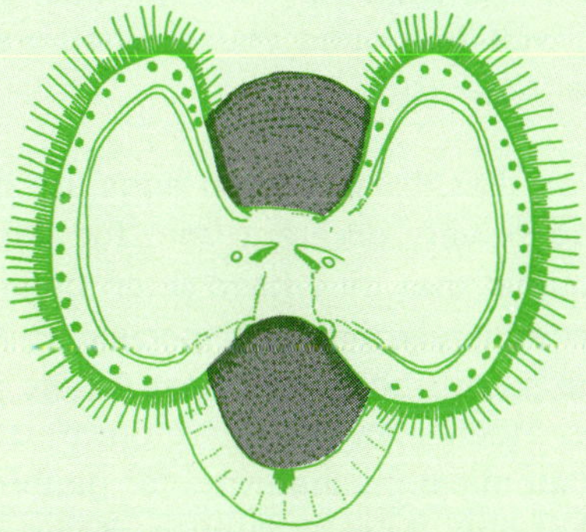

(b) Veliger. *(Redrawn from Barnes.)*

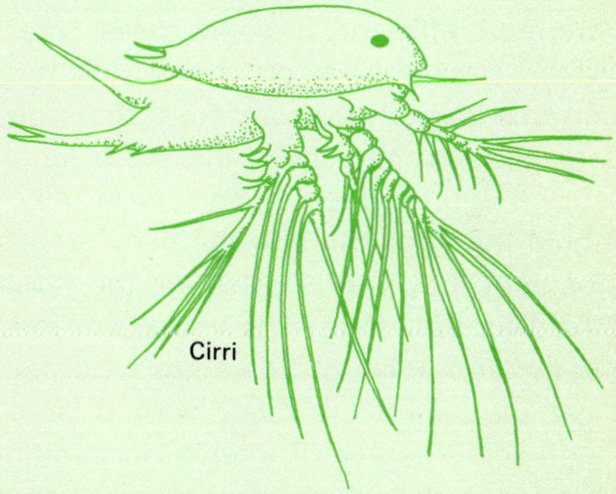

(c) Nauplius. *(Redrawn from Costlow in Etkin and Gilbert.)*

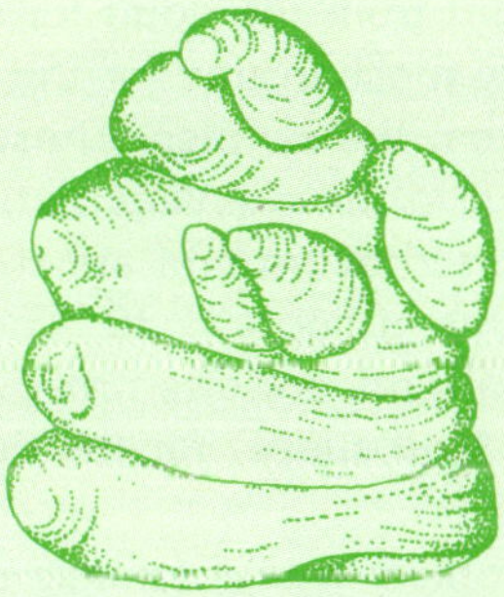

(b) *Crepidula,* a mollusc. *(Redrawn from Barnes after Coe.)*

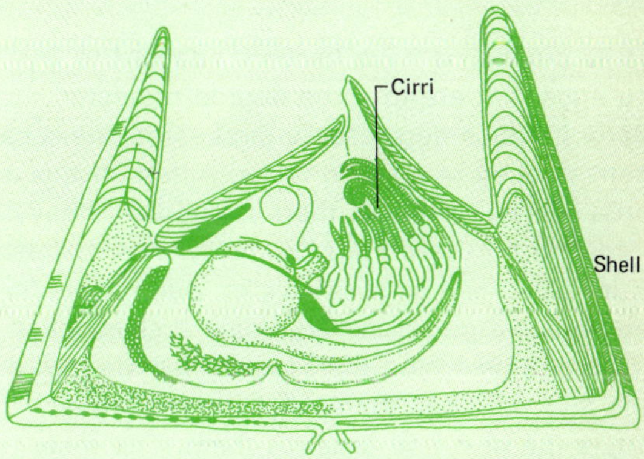

(c) *Balanus,* a crustacean. *(Redrawn from Barnes.)*

Although external fertilization is possible in the aquatic environment due to the nature of water, the process is "expensive," and too much is left to chance. Because of the vulnerability of eggs or larvae to predators, a great deal of energy is expended by the parent organisms in producing gametes that never develop into adults. Some organisms have evolved processes that are more efficient. In some cases, internal fertilization has evolved, often involving the transfer of a packet of sperm from the male to the female, who then stores the sperm and fertilizes each egg produced. The female may even hold the eggs through the early stages of development, but in few cases are the fertilized eggs held for very long. The young are released very early, and although there are no unfertilized gametes, there are many wasted larvae. Although some degree of internal fertilization has evolved in the aquatic environment among some species of fish, barnacles, squid, snails, and some segmented worms, this method of reproduction remains less common than external fertilization.

Some organisms have become adapted to laying eggs that are to some degree protected. Many produce egg masses, and some snails cement clusters of eggs to rocks. Although this affords some protection, there is still very heavy predation on these eggs. Some species of freshwater fish, such as sticklebacks, build nests for reproductive purposes. The male stickleback builds the nest, and then lures a female into it. She lays her eggs there, and the male then enters the nest and fertilizes them. The male then remains close to the nest, aggressively defending it against potential predators on the eggs.

The next evolutionary step toward protecting the young may seem obvious to a human being, but few marine animals have taken the step. The larvae may be retained through development within the body of the female, or in some cases the male, and then be released. Some snails, especially those subjected to periodic drying, have evolved in this way, as have some sharks, and some freshwater species such as the guppy have evolved to internal development of the young. But the parent's body is not any better than an egg case unless it can afford some protection to its young. Most cannot, and thus *viviparous* animals (those that bear living young) usually live in extreme environments or are capable of vigorous defense, as is the shark. We will examine the implications of bearing live, well-developed young when we treat terrestrial organisms.

The larvae that develop from eggs are highly diverse and so very different from the adult forms of the animal that they have often been identified as completely different species. This ought not to be surprising, since these larvae must survive on their own and therefore must have those anatomical mechanisms for gaining food that will enable them to survive. Being so different, not only anatomically but physiologically as well, they are subjected to different selective pressures from those operating on adults and they therefore may evolve, in a sense, separately from the adult population. However, although the larvae produced may be highly diverse and very different from the adult forms, there are two primary mechanisms for the development of larvae from the zygotes. These mechanisms are reflected in the cleavage patterns of the dividing fertilized ovum and are shown in their simpler forms in figure 3.20. These simplest cleavage patterns are often modified by the need for a food supply for the developing larva.

Another problem facing aquatic species is that of finding a mate. Of course, this is not a major problem for certain benthic or sessile species, such as oysters or clams, that live in large beds made up of males and females. Some pelagic species of fish live in schools that are also made up of males and females, and thus they too experience little difficulty in finding mates. In such instances, we should point out, there is little difference in body shape, size, or coloration between males and females. Among some species, however, schooling is rarely, if ever, observed, and thus other mechanisms must exist

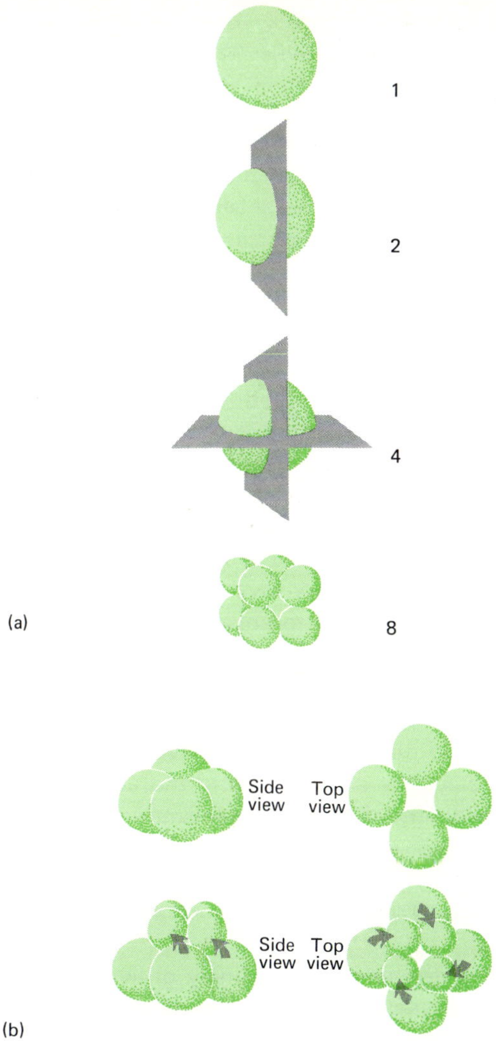

FIGURE 3.20 Cleavage patterns. As a fertilized ovum divides to form the zygote, the patterns by which the cells arise from other cells differ. *Radial cleavage* occurs when the fertilized ovum divides along lines at right angles to each other as shown in (a). *Spiral cleavage* takes place when the dividing cells overlap as shown in (b), where the arrows indicate which cells are derived from other cells. Such cleavage patterns seem to have some evolutionary significance. Spiral cleavage is characteristic of molluscs, annelid worms, and arthropods, important phyla that we shall discuss in detail later. Radial cleavage is characteristic of several phyla including echinoderms (starfish, for example) and chordates, the phylum of which man is a member. We shall return to the subject of cleavage and the subsequent development of the zygote in Chapter 6.

to bring the sexes together for mating. Among solitary, deep-ocean species, luminescence may serve as a means of identifying males and females of the same species. Among shallow-water species, differences in body size and color or pattern may serve to differentiate between males and females. Certain behavioral mechanisms also operate to bring the sexes together. The Bermuda fire worm, a segmented worm, undergoes an elaborate ritual at mating time. About half an hour after dusk, when the moon is in a certain phase, male fire worms swim to the surface of the sea and emit bursts of light from luminescent organs. Females are attracted to these bursts of light and swim toward them, also flashing their luminescent organs. The two sexes congregate and swim about each other releasing clouds of luminescent sperm and eggs.

Some adaptations have been evolved that make the search for a mate unnecessary. Barnacles, for example, are hermaphroditic; that is, each individual possesses male and female reproductive organs. Barnacles do not usually fertilize themselves, however. But the possession of both male and female organs insures that a potential mate is nearby. In some cases, some truly bizarre adaptations have evolved. Slipper shells live all their lives stacked in one place, and when young, they are always males. They enter a transitory period in their lives when the male reproductive organs degenerate, and then, depending on certain conditions, they either become females or resume being males. If an individual at this sexless stage is attached to a female, it will become a male again. If it is removed or isolated from females, it may become a female. If, however, there are large numbers of females present while the sexless individuals are in the transitory stage, many of the individuals will become males, although some will become females. Once an individual assumes the female sex, there is no turning back; no female can ever become a male under any known conditions. Among the other means of assuring that males and females will not have to search for a mate are adaptations such as that shown

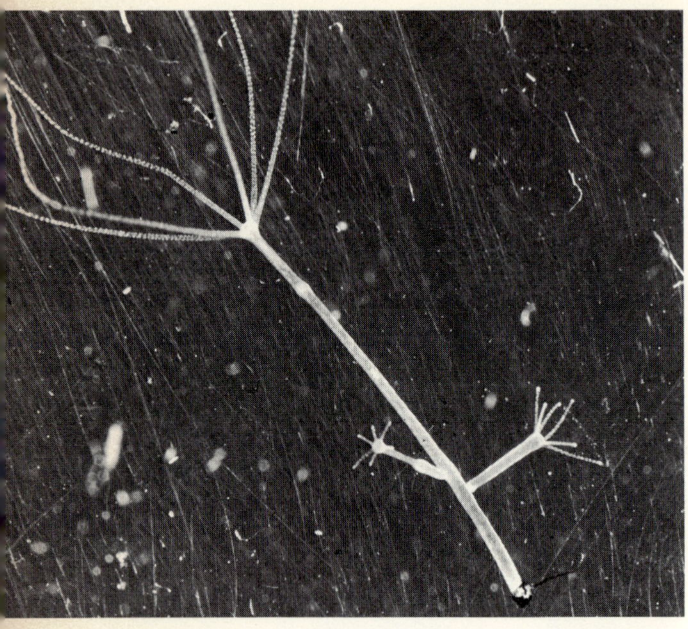

FIGURE 3.21 A coelenterate (*Hydra*) reproducing by budding. *(Courtesy of Lester V. Bergman & Associates, Inc.)*

by the marine worm *Bonella viridis.* The larvae of this species are sexless, and when they settle they become females with a body that is several centimeters long and a proboscis that is as much as a meter long. If a larva settles on the proboscis of a female, it becomes a male within about four days. It then leaves the proboscis and makes its way into the sexual organs of the female where, with other males that may already be there, it fertilizes the female. Because of its rapid development, the male never reaches a length of much more than one millimeter. Such males can be considered sexual parasites and are actually not uncommon among marine animals.

For reasons that we shall examine later in more detail, sexual reproduction is a desirable means of perpetuating a species; thus most, if not all, species have this mechanism, although sexual reproduction probably evolved only once. Of course, sexual reproduction is not the only way to reproduce, and many organisms simply reproduce by dividing into two individuals. Some simple multicellular organisms such as sponges and some coelenterates can reproduce by budding. In this case, a cell or cells on the external body wall of the organism begins to develop into a complete individual that ultimately separates from the parent organism. (See fig. 3.21.) In some cases, organisms can develop from unfertilized eggs through a process called *parthenogenesis.*

We have examined only two main adaptive mechanisms—feeding and reproduction—for several reasons. First, we can not consider all the adaptations that enable organisms to survive in the aquatic environment, and therefore it seems best to concentrate on the more important ones. Second, we wished to give the reader an appreciation of the very diverse adaptations that may arise in response to environmental requirements. Third, the adaptations that we have examined provide us with a better background for understanding the problems faced by terrestrial animals and plants. Last, they have helped us set the stage for a discussion of the evolution and phylogeny of marine organisms.

The Family Tree of Marine Animals

In view of the immense diversity of living things, it is necessary to categorize animals and plants into some sort of groups in order to examine them efficiently. The classification scheme in use today is a much modified version of one developed in the Eighteenth century by the Swedish botanist Carl von Linné (more commonly known by the Latinized form of his name, Carolus Linnaeus). The bases for the scheme of classification now in use are the known or strongly suspected evolutionary relationships between species. Using this system of classification, biologists group together such diverse organisms as squid, snails, and clams in the phylum Mollusca, or lobsters, copepods, and barnacles in the phylum Arthropoda. (See box 3.4.) Nevertheless, the relationships that exist within a phylum can be remote. In spite of some superficial resemblances between a

Classification

Since there are so many species of animals and plants, the common names of which can be misleading, biologists classify and name organisms on the basis of a system developed originally in the eighteenth century. The system recognizes very broad relationships as well as more refined ones and makes it possible for biologists from any part of the world to immediately and unmistakably identify the animal or plant under discussion. As examples of the system consider four very different animals, a human, a trout, a barnacle, and a lobster. There is only one human species, and so theoretically there ought to be no misnaming of man. Nevertheless, the classification scheme also identifies man's broader relationships with the living world. There are, however, many species of trout, and research or observations on one species may not hold true for all of them. This last statement also holds true for the other two examples.

Beginning with the broader relationships, we can indicate how these animals are superficially differentiated from each other. Obviously, all categories are not used in classifying all organisms.

	Human	**Trout**	**Barnacle**	**Lobster**
KINGDOM	Animalia	Animalia	Animalia	Animalia
PHYLUM	Chordata	Chordata	Arthropoda	Arthropoda
SUBPHYLUM	Vertebrata	Vertebrata	—	—
CLASS	Mammalia	Actinopterygii	Crustacea	Crustacea
SUBCLASS	Eutheria	—	Cirripedia	Malacostraca
ORDER	Primate	Clupeoidea	Thoracica	Decapoda
SUBORDER	Anthropoidea	Salmonoidea	Balanomorpha	Reptantia
SECTION	—	—	—	Macrura
SUPERFAMILY	Hominidea	—	—	Nephropsidea
FAMILY	Hominidae	Salmonidae	Balanidae	Homaridae
GENUS	*Homo*	*Salmo*	*Balanus*	*Homarus*
SPECIES	*sapiens*	*gairdneri*	*balanoides*	*americanus*

(The generic and specific names are always italicized, with the generic name capitalized.)

BOX 3.4

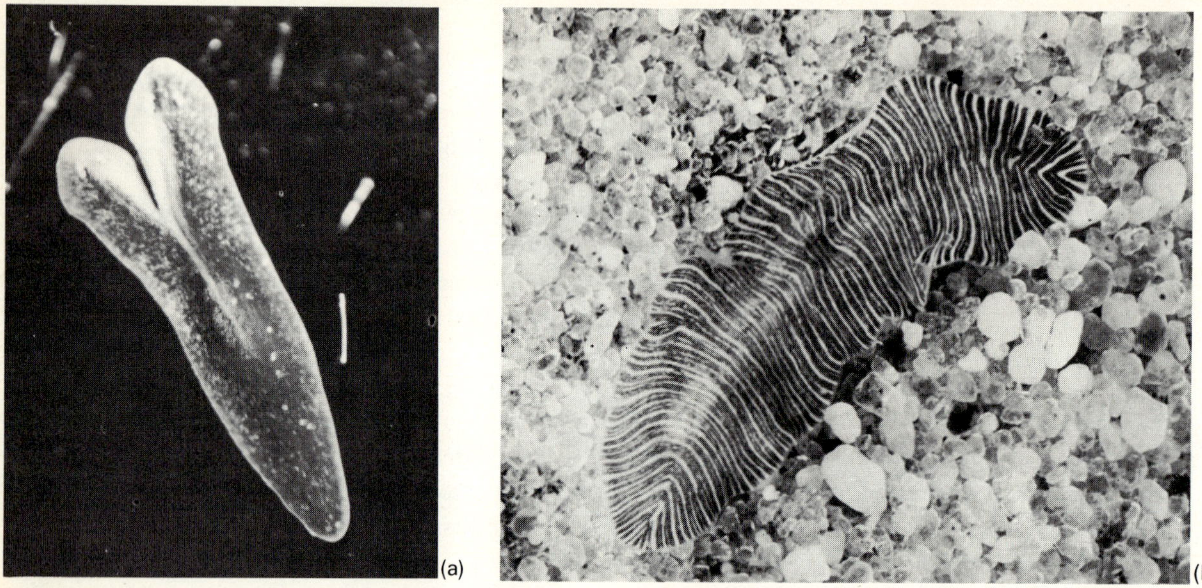

FIGURE 3.22 Flatworms. (a) *Dugesia* regenerating two anterior ends after being split longitudinally. (b) Many flatworms, especially marine forms such as this one *(Stylochus sp.),* are quite ornate. *((a) Courtesy of Carolina Biological Supply Company; (b) courtesy of William Hopkins Amos.)*

shrimp and a grasshopper, for example, the shrimp is considered to be more closely related to the barnacle than to the grasshopper. Many other equally unlikely relationships are often taken for granted by biologists. We cannot go into great detail about how these evolutionary relationships have been derived—many scholarly monographs have been written about phylogenetic relationships—but at least we can examine some of the reasons for them.

Clearly, each phylum and class contains members that have diverged to fit an immense variety of niches. Still, we shall see that the members of a phylum or class share certain features or that they share some features with a common ancestor. We have briefly indicated some of the different ways of ''making a living'' in the sea, and as we proceed through the phylogenetic relationships that follow, keep in mind that the characteristics of the differing groups of animals all represent different but successful solutions to the problems of surviving in the aquatic environment.

Perhaps the first question to ask is the most difficult to answer: How did multicellular animals arise? A number of theories attempt to answer this question. One theory maintains that some species of colonial protozoa developed specialization among some members of the colony. This specialization then led to an ability to reproduce as an organism. Sponges are often cited as examples of simple multicellular organisms that might have arisen from the colonial organization of protozoa.

A similar, but somewhat less widely held theory, is that organisms such as sponges arose as a result of symbiotic associations between different species of protozoa. The problem with this theory is that it does not satisfactorily explain how sponges can reproduce by means of a single cell. At any rate, the descent from some colonial ancestor is thought to have led to cellular specialization, and then to groupings of cells into specialized tissue, and ultimately to the evolution of organs from a specialization of tissue within the organism.

Another popular theory, quite different from the ones already mentioned, is that multicellular organisms arose from multinucleate protozoa. Protozoa exist today that have several nuclei and even exhibit a remarkable internal organization in spite of being unicellular. These protozoa are similar to some flatworms, and if flatworms are descended from some ancestral stock which also gave rise to a line in which the nuclei came to be separated by cell membranes, then one can imagine that the next step was the evolution of tissue, and the next, the development of organs. However, one could also consider multinucleate protozoa to be degenerate flatworms rather than ancestral flatworms.

Perhaps the most attractive theory is that multicellular organisms arose several different times, and perhaps some arose only later to become extinct. Thus, sponges may be derived from colonial protozoa, coelenterates from other colonial protozoa, and flatworms from multinucleate protozoa. According to this theory, sponges, flatworms, and coelenterates are only very distantly related.

Whichever theory is the correct one, and at present there is no way of knowing, it is generally accepted that all higher organisms are derived from the same ancestral stock—namely, flatworms. Coelenterates and sponges, highly successful and ancient groups that they are, are considered to be evolutionary "dead ends." We therefore will not discuss these latter phyla in the text.

The phylum Platyhelminthes, or flatworms, live in an extensive variety of environments. Some are planktonic, some can live in moist soil, and a large number of flatworms are parasitic, including the medically important tapeworms and flukes. The majority of flatworms, however, are free-living benthic forms, some burrowing in sediments, and some grazing on algae and detritus on the surface of sediments and rocks. Most species are small—a fraction of a centimeter long—although some reach lengths of 20 to 25 centimeters (8 to 10 inches). The flatworm most commonly seen in college biology laboratories is *Dugesia*, a freshwater genus more frequently called *Planaria*. (See fig. 3.22.)

Except for the parasitic forms, flatworms have a gut, although it usually has only one opening that serves both as mouth and anus. They do not possess a lined body cavity and are thus considered *acoelomate*. (See fig. 3.23.) They are also the simplest organisms with bilateral symmetry, or fore and aft orientation. It is speculated that bilateral symmetry resulted from locomotion and the pressure to evolve a concen-

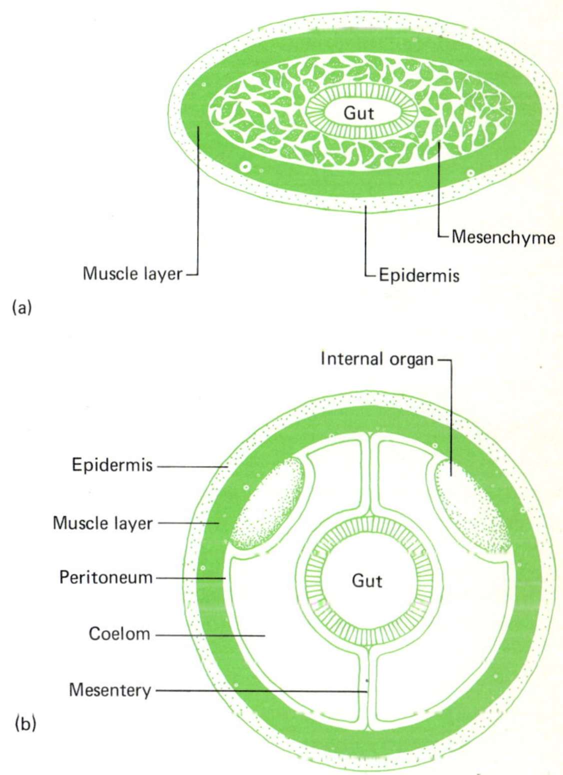

FIGURE 3.23 *Coelom* is derived from a Greek word meaning *hollow* and is generally used to refer to a lined, internal body cavity. The lining of the cavity, the *peritoneum*, separates the organs and the muscle-lined body wall from the digestive system. In an acoelomate animal, such as a flatworm, the gut is separated from the body wall by mesenchyme, a loose network of cells as shown in (a). *(Redrawn from Barnes.)*

tration of nervous and sensory organs in the forward or anterior end of the organism, or that part of the body which comes into contact with the environment first. Thus flatworms, although they lack the lined body cavity that characterizes higher forms, are similar to more advanced animals.

Some biologists have held that all the higher animals are descended from a so-called "planuloid organism" (see fig. 3.24) that was similar to the larva of some present day coelenterates. Bilateral symmetry then evolved in this larvalike organism, and it became the ancestor of the flatworm. The flatworms provided a body plan and set of organ systems that could be organized in a great number of ways. Nearly all the basic elements were present, except for a body cavity

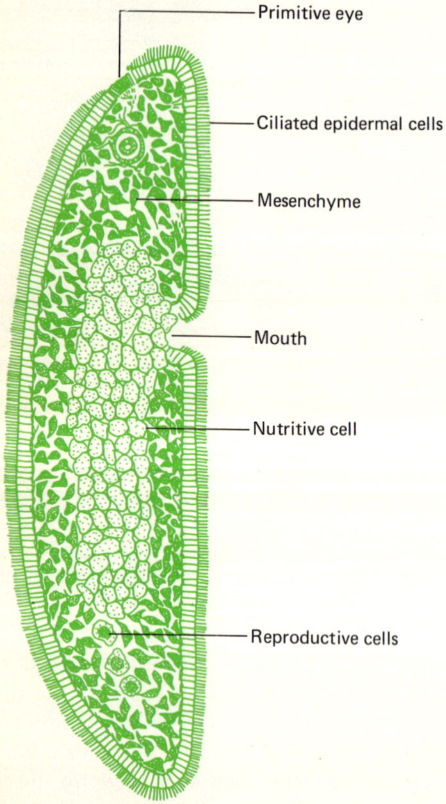

FIGURE 3.24 A planuloid organism. *(Redrawn from Barnes.)*

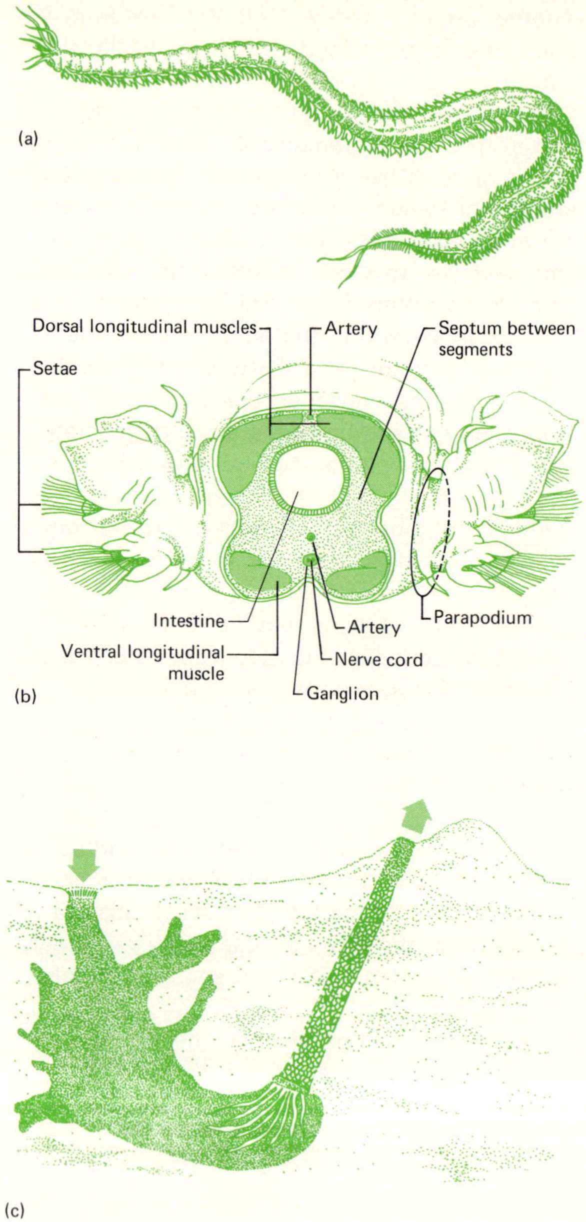

FIGURE 3.25 Many polychaetes, such as *Nereis*, (a) and (b), are errant, that is, free living; others are planktonic, and frequently transparent; and still others, such as *Pectinaria*, (c), live most of their lives in sand-grain or calcareous tubes entirely beneath the surface of the ocean bottom, obtaining their food from the water. *((b) Redrawn from Barnes after Snodgrass; (c) redrawn from Barnes.)*

and an exit from the gut. A variety of minor phyla have apparently derived from this line and are shown in box 3.5. Although some of these phyla, such as the nematodes, are ecologically important, all are evolutionary dead ends, and we shall concentrate on the five major phyla of animals: the annelids (segmented worms); the molluscs (snails, clams, and related species); the arthropods (crustaceans and insects); the echinoderms (sea urchins, starfish, and related species); and the chordates (the phylum of which man is a member). These five phyla comprise most of the important animals of the biosphere. (See box 3.6.)

There are two main evolutionary lines in these five phyla. The annelids appear to have given rise to the arthropods and molluscs, and some now extinct ancestor is thought to have given rise to the echinoderms and chordates.

Annelids are segmented worms. Each body segment contains a massive coelomic cavity between the gut and the body wall, a bundle of nerve cells called a *ganglion*, an excretory organ, musculature, and spines on the lateral outer surface. (See figs. 3.25 and 3.26.) It is on the basis of the spines that two of the major classes of the phylum are defined: the *oligochaetes* are annelids with a few small spines, while the *polychaetes* are annelids with larger, more numerous spines, usually on the flaplike extensions of the body wall called *parapodia*. The third major class in the phylum, the Hirudinea, are leeches, and are much less common. (See fig. 3.27.) Although the common earthworm, an oligochaete, is frequently used as an example of an annelid, it is somewhat atypical. The marine polychaetes are more representative of the phylum; they are a diverse class including many rapid-moving predatory or sediment-feeding species and a number of sessile species. On the whole, the annelids are a successful phylum, having to some degree radiated to almost every environment on earth.

FIGURE 3.26 Among the most beautiful of the marine invertebrates are the polychaetes called fan worms. These animals construct tubes, of sand and mucus or other materials, and then live in the tubes with their funnel-shaped fans protruding. The beating cilia that cover the fans draw in food which is captured and carried to the mouth of the worm. *(Courtesy of Douglas P. Wilson.)*

FIGURE 3.27 A leech, *Hellobdella fusca.*

Minor Phyla

Nemerteans (phylum Nemertea) resemble flatworms to some degree and commonly are found in marine muds and sands. All nemerteans are carnivorous or scavengers and possess a proboscis that can be rapidly "shot out" to capture prey. They possess both a mouth and an anus, an important advance over flatworms.

Rotifers (phylum Rotifera) are common freshwater filter feeders. The name of the phylum is derived from the appearance of the crown of cilia found in most species. When beating to draw in food, the cilia appear to be a rotating wheel. Most species of rotifers are planktonic, but there are also some colonial and some stationary species. One particularly unusual thing about some species of rotifers is that all members of the species have the same number of cells in their bodies.

The nematode worms (phylum Nematoda) are small, very common inhabitants of nearly every environment on earth from the soil of the forest floor to the ooze of the ocean bottom to the snow of alpine glaciers. The phylum includes species that are detritus feeders, scavengers, predators, and plant and animal parasites. Some important human parasites, such as roundworms and pinworms, are nematodes, and the human diseases trichinosis and elephantiasis are caused by nematodes. The phylum constitutes a classic example of a very successful, important, but inconspicuous group of animals.

The chaetognaths (phylum Chaetognatha), or arrowworms, as they are called, are a very poorly known but important group of marine predators. Chaetognaths rarely exceed three centimeters (a bit more than 1 inch) in length. There are only 50 species in the phylum, all but one genus of which are planktonic. Next to fish and marine arthropods, the chaetognaths are probably the most important predators in the ocean environment.

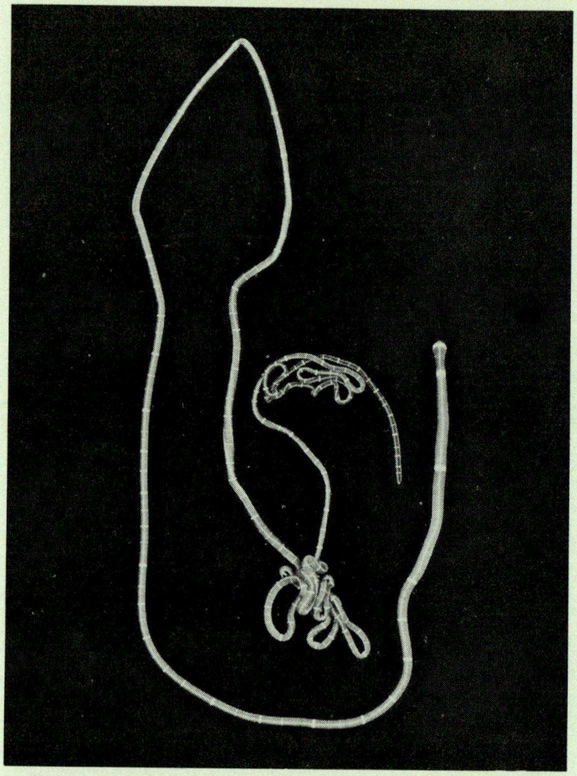

Nemertean
(Courtesy of Douglas P. Wilson.)

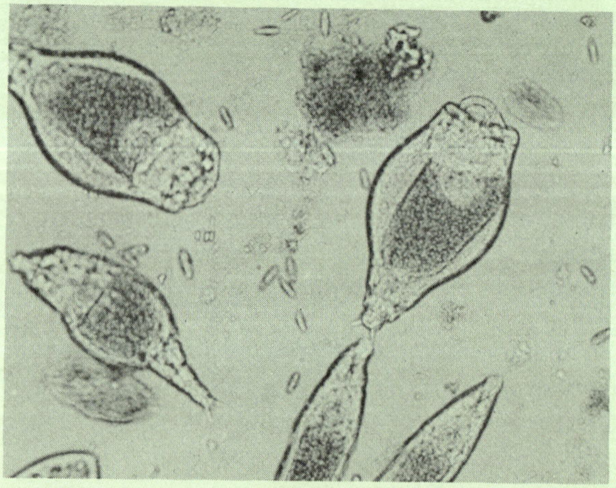

Rotifers
(Courtesy of Hugh Spencer from National Audubon Society.)

BOX 3.5

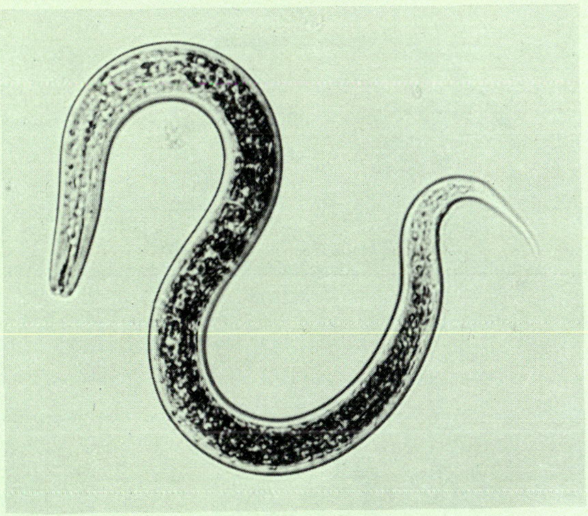

Nematode
(Courtesy of Grant Heilman.)

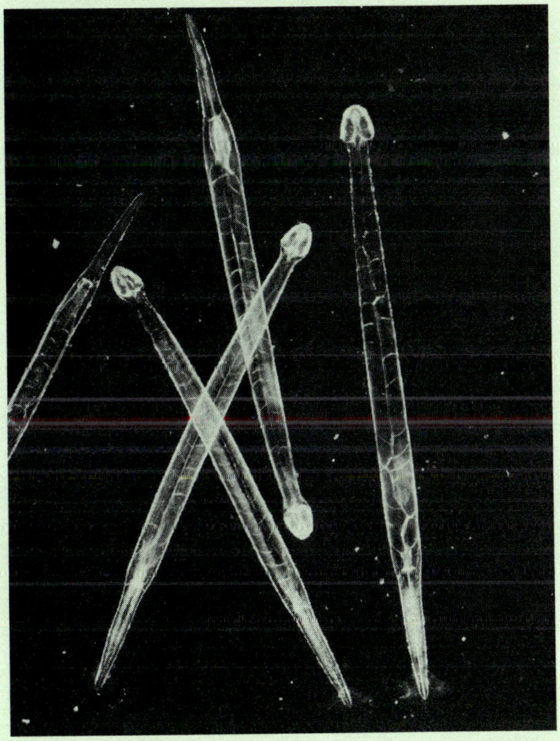

Chaetognaths
(Courtesy of Douglas P. Wilson.)

Animal Phylogeny

The proposed relationships shown in this diagram are not by any means the only proposal that can be defended. We will probably never be able to say with certainty that we can clearly trace the origin of many of the important phyla. The "branching" ancestors are unknown, for example, and in some instances we have had to hypothesize the existence of organisms for which we have little or no direct evidence. One thing may help strengthen some of our ideas about the ancestry of various animals, however. On rare occasions living individuals are found that belong to groups thought to have been long extinct. Such primitive individuals can often indicate relationships that fossils simply cannot reveal.

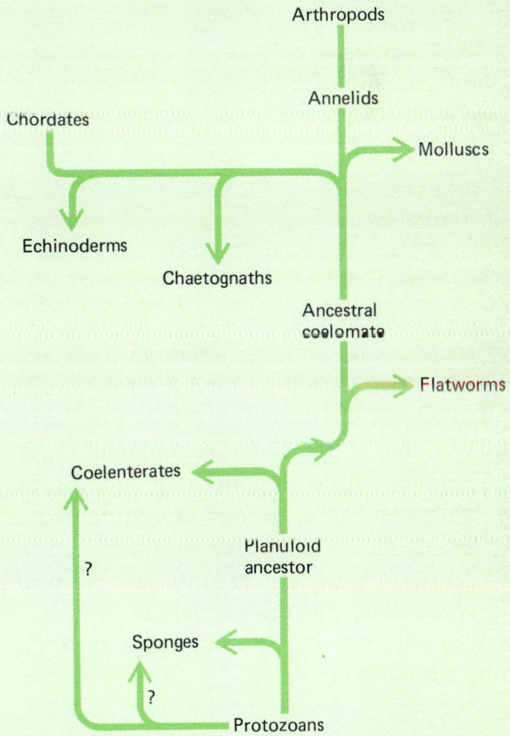

BOX 3.6

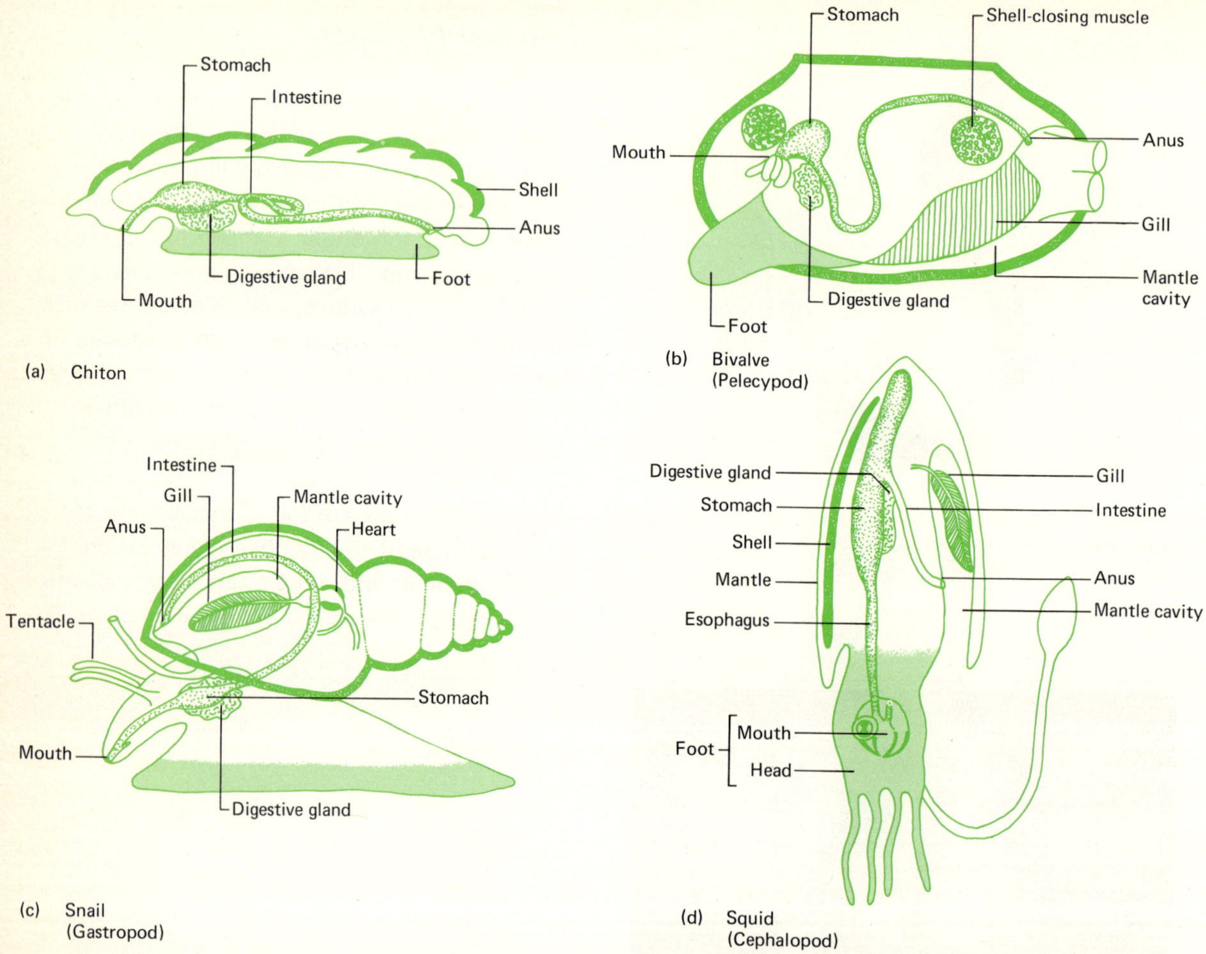

FIGURE 3.28 Chitons (a) are believed to be the common living molluscs most closely resembling the ancestral forms of the phylum. One of the main differences between chitons and the forerunners of the phylum is that the ancestor is thought to have had a single oval shell, not the eight shell plates that now characterize the chitons. The mantle (see (b), (c), and (d)) is characteristic of the molluscs and is the layer of tissue from which the shell is secreted. In some instances almost the entire body of the mollusc is within the mantle cavity. Another characteristic of the molluscs is the foot, which is used in most cases for precisely what its name suggests, locomotion. In one case, that of the squid, the foot is now the location of the head. *(Redrawn from* Invertebrate Zoology, *Second Edition, by Paul A. Meglitsch. Copyright © 1967, 1972 by Oxford University Press, Inc. Reprinted by permission.)*

The phylum Mollusca appears to be derived from an ancestral annelid, although from an examination of a clam or snail the relationship hardly appears obvious. The recent discovery of a mollusc that is somewhat exceptional in having a segmented nerve cord, segmented excretory organs and gills, as well as segmented muscles, confirmed a relationship between molluscs and annelids that had long been suspected. Both phyla have *trochophore* larvae, (see box 3.3) and this had strongly suggested their common ancestry.

The most primitive class of common molluscs, the *Amphineura,* are *chitons,* which still show an obvious bilateral symmetry and the typical calcareous (containing calcium carbonate) shell. (See fig. 3.28.) The class *Gastropoda* includes snails. If one imagines a chiton twisted around upon itself so that its anus is over its head and then further twisted so that the body mass fits a spiral shell on its back, then it is possible to see how the snail might have evolved. Such a body is obviously awkward and necessitated the evolution of a number of subsequent changes, but it gives the advantage of good protection for this proverbially slow-moving animal. (See fig. 3.29.)

Many snails graze on algae that are anchored to rocks and use a rasplike organ, the radula, to scrape off the algae. Many, however, have radiated to a number of other niches. Some are predatory and have a radula that can be used as a drill to get inside the shells of other snails or clams. Some are poisonous and are capable of ejecting a venomous barb. Some other snails have lost their shells and have become planktonic.

Members of class *Pelecypoda*, or *bivalves* such as clams and oysters, are much more conservative than the gastropods. Most are sedentary filter feeders, as we have already described, and thus many of their adaptations such as the nervous system have degenerated. Nonetheless, a certain amount of diversity has developed among the members of this class. Shipworms are bivalves that bore into the hulls of wooden ships; clams burrow into the sea bottom by using an organ called the *foot*; and oysters settle into a spot and remain in one place all their lives.

The most specialized class of molluscs is the *Cephalopoda*, which includes the octopus and squid. (See fig. 3.30.) This class, superficially, appears to bear no relationship to the other classes of the phylum; instead it shows a great amount of convergence with fish. Cephalopods have complex eyes, well-developed gills, and a complex circulatory system. Since they lead an active life, they have a complex nervous system

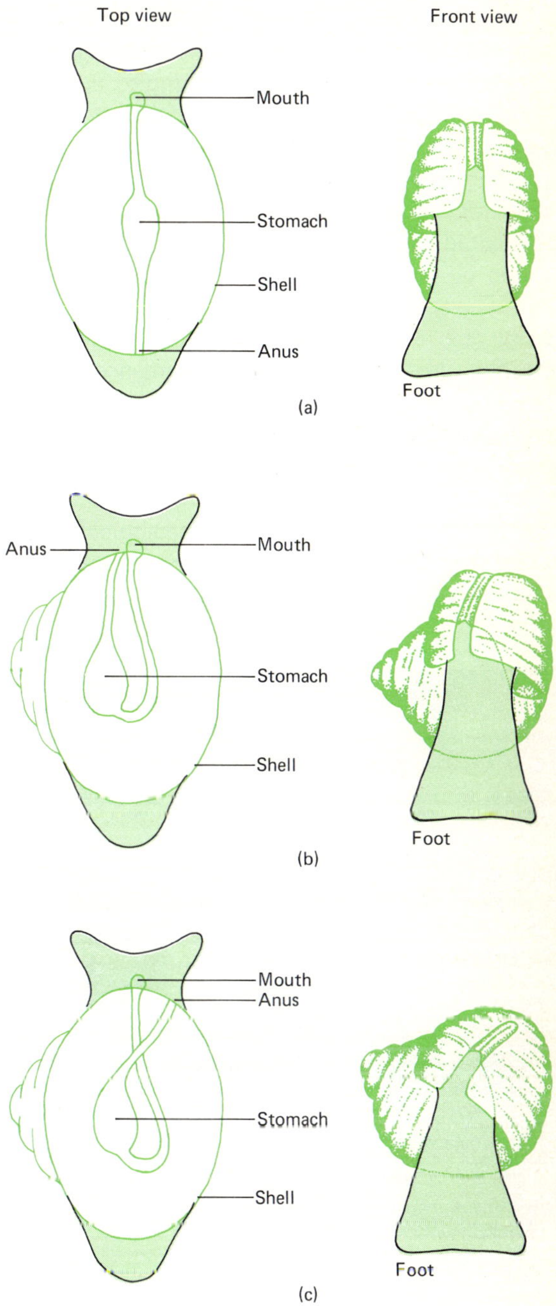

FIGURE 3.29 The presumed evolution of the snail shell. *(Modified from Barnes.)*

and exhibit some complex behavior patterns. An octopus, for example, can learn to distinguish between shapes set at different angles (see fig. 3.31) and when placed in a tank with a collection of materials, will gather the materials up and construct a nest. Bilateral symmetry in the cephalopods is secondary to other adaptations. The squid shows how much change can be effected by evolution.

By far the most diverse phylum of animals is the phylum *Arthropoda*. The two main classes of arthropods are the *crustaceans* and the *insects*, which evolved separately and which include many species. We will present a few illustrative examples. The only organisms rivaling the arthropods in radiation and ecological significance are the vertebrates, and the vertebrates place second.

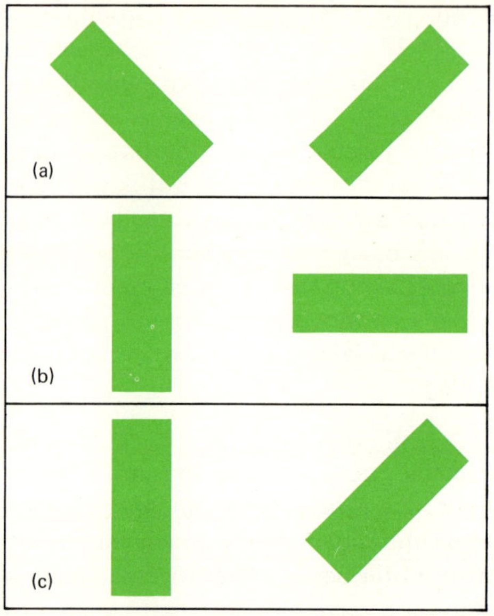

FIGURE 3.31 After training, octopuses were presented with shapes of a correct and incorrect orientation and made to choose between them. The percentage of errors made was taken as an indication of their ability to perceive differences in the orientation. In choosing between the shapes shown in (a), they erred 50% of the times, suggesting that they were incapable of perceiving the difference between them. The octopuses made the fewest errors with the shapes shown in (b), only 29%, and made errors in 30% of the tries with figures shown in (c). Their inability to learn the differences between these shapes seems to lie in the obliquity of the angles at which the shapes are oriented. The presentation of other shapes to the octopus confirms this and also suggests that the number of corners, length, and area play little role.

FIGURE 3.30 An octopus photographed from a research submarine at nearly 1000 m depth (3000 ft). This unusually large specimen is about 2.5 m long (nearly 8 ft). *(Courtesy of Ron Church, Photography Unlimited.)*

The derivation of the arthropods from the annelids is deduced from a variety of evidence. For example, the nerve cord of arthropods is generally ventral and segmented as in annelids. The appendages of arthropods are usually segmented and are thought by some biologists to be homologous to the parapodia seen in the polychaetes. The similarity to annelids is best seen in a primitive arthropod, *Peripatus*, which bears a very strong resemblance to annelids, although some biologists place *Peripatus* in its own phylum. (See fig. 3.32.)

FIGURE 3.32 *Peripatus sp.* Although in many ways it resembles an annelid, this is really closely related to the very primitive arthropods. *(Courtesy of Runk / Schoenberger from Grant Heilman.)*

We shall discuss insects in the next chapter, since they are terrestrial animals. The most important arthropods in the aquatic environment are the crustaceans. They constitute the major part of the pelagic biomass, and they are represented on all parts of the ocean bottom. We have already discussed the copepods, which are ecologically the most important crustaceans. Other typical crustaceans are shown in figure 3.33.

The echinoderms appear to be related not to the ancestral annelids, but to some other more ancient segmented organism. The larvae of echinoderms show a relation to vertebrates in the cleavage pattern of the egg, but the adults all have re-evolved secondary radial symmetry. The echinoderms provide a rich fossil record, dominated largely by crinoids. (See fig. 3.34.) Although they are fairly common, echinoderms are not as important to the ecology of the ocean as the other phyla we have described, and they are evolutionary dead ends.

The phylum Chordata, of which man is a part, is at present the most important phylum on the earth after the arthropods. We know frustratingly little about the origin of the chordates. We do know, however, that their ancestor was segmented and coelomate, possessed bilateral symmetry and a dorsal, segmented nerve cord, and may have been related to an ancestor of the echinoderms.

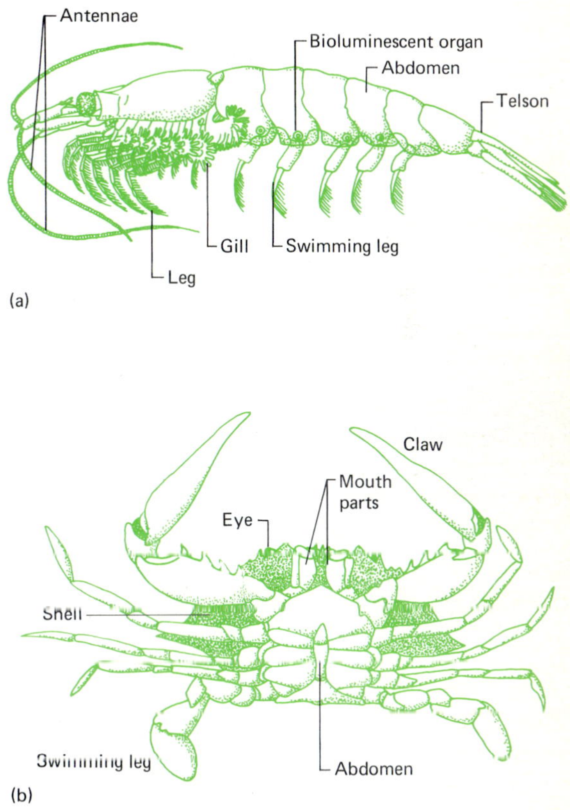

FIGURE 3.33 The external skeleton or *exoskeleton* of crustaceans is characteristic of all arthropods, and has evolved into an immense variety of structures. Shown above are two crustaceans, a shrimplike euphasiid (a) and a crab (b). Note the degree of specialization that has evolved in the exoskeletal structures of these animals. *(Redrawn: (a) from Meglitsch; (b) from Barnes.)*

(a)

(d)

(b)

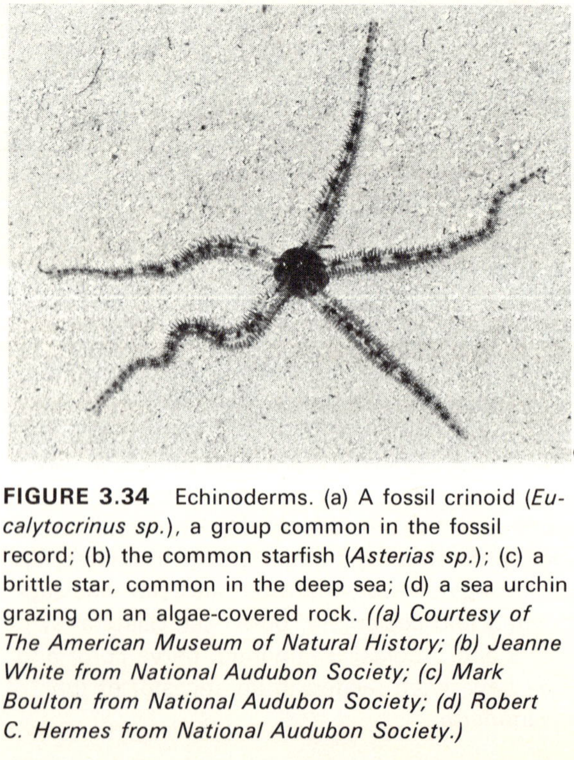

(c)

FIGURE 3.34 Echinoderms. (a) A fossil crinoid (*Eucalytocrinus sp.*), a group common in the fossil record; (b) the common starfish (*Asterias sp.*); (c) a brittle star, common in the deep sea; (d) a sea urchin grazing on an algae-covered rock. *((a) Courtesy of The American Museum of Natural History; (b) Jeanne White from National Audubon Society; (c) Mark Boulton from National Audubon Society; (d) Robert C. Hermes from National Audubon Society.)*

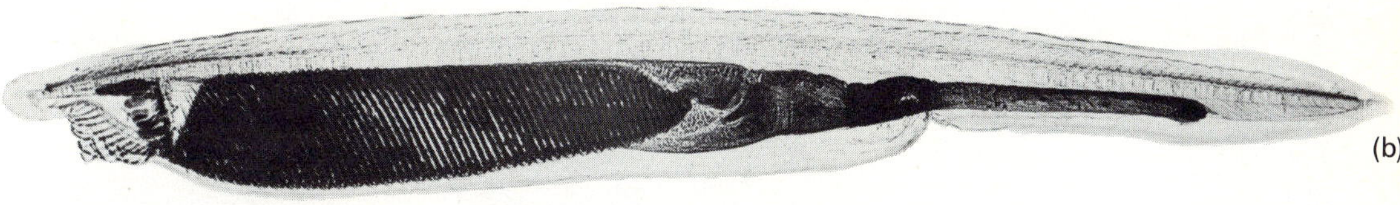

FIGURE 3.35 Although they hardly appear to be chordates, the animals shown above possess a notochord, the characteristic of the phylum. The sea squirt in (a) has a notochord in its larval stage, but loses it as it settles into the sedentary adult stage. *Amphioxus*, prepared and stained to show internal structure, is shown in (b). This chordate retains its notochord throughout its life. *((a) Courtesy of Douglas P. Wilson; (b) courtesy of Ward's Natural Science Establishment, Inc., Rochester, New York.)*

The phylum is named for a feature that is the main characteristic of the group. The *notochord* is a gelatinous supporting rod that lies below the dorsal nerve cord in some very primitive members of the phylum but exists only in the embryos of higher forms. In the higher forms, it is replaced by the vertebral column. Some primitive members of the phylum are shown in figure 3.35. Of course, they bear little resemblance to the ancestors of the fish, amphibia, and reptiles, but their larvae and the tadpolelike *Oikopleura* (see fig. 3.36) may be similar to the larvae of the ancestral form.

We shall postpone any discussion of the four terrestrial classes of the chordates (amphibians, birds, reptiles, and mammals) until the next chapter. The class Pisces, however, deserves mention here. Everyone is familiar with the basic body form of the fish. Its main adaptation is the muscular body that enables it to move easily through water. On the whole, there are few significant variations on this body plan. The segmented muscles, as shown in figure 3.37, pull against each other and are well adapted to the means of locomotion most suited to the aquatic environment. A certain amount of specialization has, of course, occurred, as in the sea horse. The sea lamprey, which is also something of an exception, is a fish that has become a parasite. However, if we look at a goldfish, a tuna, or a trout, we can see the same basic way of making a living. A tuna or shark may be a predator, a cod or eel a scavenger, and a herring or anchovy a filter feeder, but they all rely

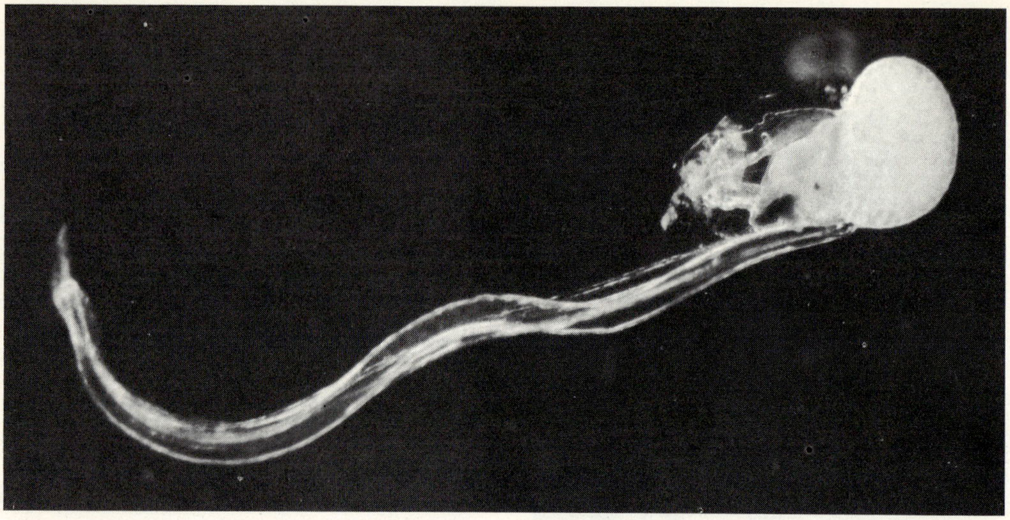

FIGURE 3.36 *Oikopleura sp.* An adult planktonic chordate that resembles the larvae of many other chordate species. *(Courtesy of Douglas P. Wilson.)*

on their ability to propel themselves through the water as their major adaptation. No fish is truly benthic, although many (described as *demersal*) live by swimming near the bottom and eating benthic materials.

As with the crustaceans, it is difficult to know where to start and stop in discussing the fish. One key point that must be made is that rapid locomotion requires a highly developed nervous system and good skeletal and muscular systems. In both the arthropods and chordates, these mechanisms have been of major importance in allowing the transition to land.

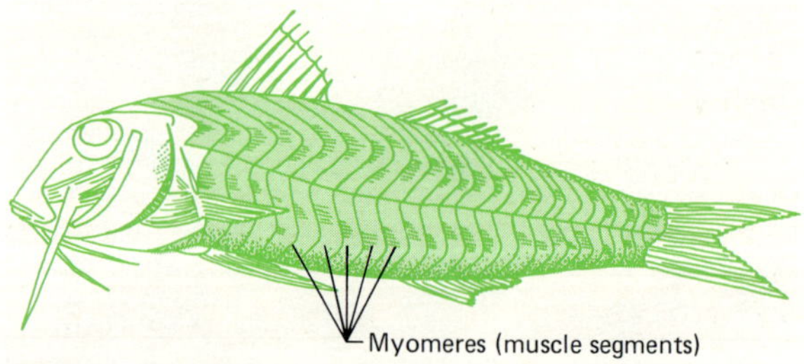

FIGURE 3.37 Segmentation in the muscles of a fish. *(Redrawn from Ihle.)*

Thus far we have restricted our discussion to the evolution of aquatic animals. Obviously, part of the purpose of this chapter is to provide an admittedly sketchy background and appreciation of the adaptations that are required for survival in the aquatic environment. In part, we also wish to indicate briefly the development of the world biota. We must not forget the plants, however, for they are as basic to the food chains of the sea as they are to the food chains of the land.

FIGURE 3.38 Flagellated plants. *(Redrawn from* The Algae: A Review, *by G. W. Prescott, Houghton Mifflin Company, Boston, 1968.)*

What plants must do to survive in an aquatic environment is simpler than that which animals must do, and thus there has been less adaptive radiation. But marine plants are extremely important and thus of interest. By far the most ecologically important plants in the sea are phytoplankton, or free floating, single-celled plants. Some of the larger attached plants are the forerunners of terrestrial plants, but they are present in a very limited area of the sea, always being found only in the shallow waters near coastlines. The phytoplankton not only form the basis of all life in the sea, but they also produce a large proportion of the world's oxygen supply and are thus nearly equal to all land plants in importance.

There are a number of different kinds of single-celled plants in the ocean, some of which fall into the diffuse area between plants and animals. (See fig. 3.38.) Two kinds of phytoplankton, *diatoms* and *dinoflagellates*, are of primary importance, with the diatoms being considered the more important of the two. Any sample of water taken from near the surface of any of the oceans of the world or from most bodies of fresh water will contain diatoms.

Diatoms are among the smallest living things; 250 of the largest placed end to end just measure an inch, and the smallest may be 1/100 of that size. Thus, to achieve any appreciable biomass requires immense numbers of individuals. Yet, in spite of their size, diatoms are intricately structured. Their basic form is determined by a skeletal structure called a *frustule*. The frustule is made of a silicon dioxide (SiO_2), as is glass, and is intricately sculptured. (See fig. 3.39.) The plan of this skeleton consists of two overlapping plates called *valves*, each covering half of the cell. When the cell reproduces, which it does by simple fission, each new cell gets one half of the frustule and forms a new inner one. (See fig. 3.40.) Since the cell obviously cannot diminish in size for an indefinite time, it soon rounds up and forms an *auxospore* which divides many times. The auxospore may form spores for sexual reproduction, or it may reproduce asexually; in either case, new large cells are produced.

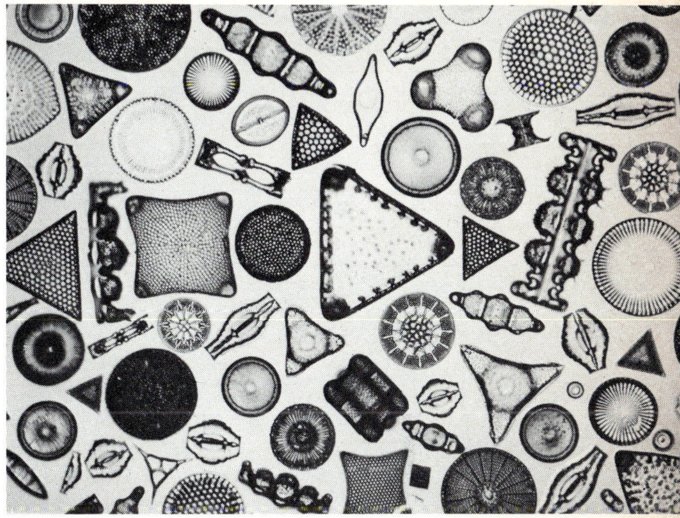

FIGURE 3.39 A variety of diatoms, prepared so as to show the structures of the frustules. The largest specimens are about one- to two-tenths of a millimeter in size. *(Courtesy of Carl Struwe from Monkmeyer Press.)*

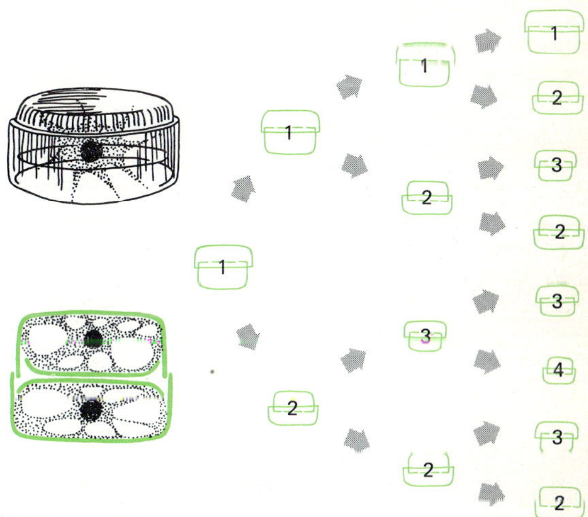

FIGURE 3.40 Reproduction in diatoms. *(From H. U. Sverdrup, W. Johnson, and R. H. Fleming,* The Oceans, Their Physics, Chemistry, and General Biology, copyright © *1942, renewed 1970. Reprinted by permission of Prentice-Hall, Inc., Englewood Cliffs, N.J.)*

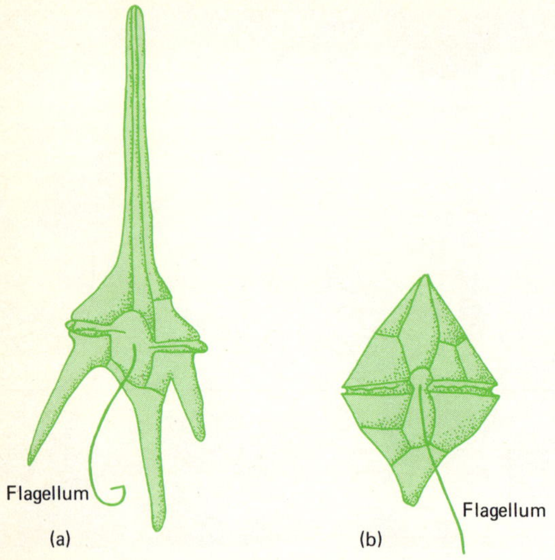

FIGURE 3.41 Dinoflagellates. (a) *Ceratium sp.;* (b) *Peridinium sp. (Redrawn from Prescott.)*

(a)

(b)

(c)

FIGURE 3.42 "Seaweeds" have a wide variety of forms depending on their adaptations. Corraline red algae (a) have calcareous skeletons and are small encrusting forms. Kelps (b) live below the tidal zone. They may be very large with complex structures and with flotation bladders. Kelps are responsible for almost all of the plant production in some coastal areas. *Fucus* (rockweed) (c) is the typical intertidal "seaweed" of the North Atlantic. It is adapted to withstanding strong wave movement. *(Courtesy of (a) Eileen Tanson from Tom Stack & Associates, (b) Ron Church from Tom Stack & Associates, (c) Douglas P. Wilson.)*

One of the major problems encountered by phytoplankton is being able to remain in the lighted zones of water where photosynthesis can take place. This is particularly problematical for an organism such as the diatom with its heavy frustule. The shape of the frustule itself may help the diatom remain near the surface, but more importantly, the diatoms store energy in the form of oils which, being lighter than water, provide the organisms with the necessary buoyancy and therefore control the depth at which the diatom floats.

The second most important group of phytoplankton are the dinoflagellates. These plants also possess an external skeleton, but unlike diatoms, the skeleton is made of plates of cellulose. More importantly, the dinoflagellates possess flagella that enable them to move through the water. (See fig. 3.41.) One of these flagella is directed backwards, thus propelling the cell, and the other is wrapped around the middle of the cell, causing it to rotate, which for hydrodynamic reasons assists in locomotion. The cells therefore can maintain their location in the most favorably illuminated regions of a body of water. Although asexual reproduction is predominant in the dinoflagellates, some sexual reproduction is suspected. Not all dinoflagellates are photosynthetic, since a majority of species in some genera do not contain chlorophyll. Many of the dinoflagellates can produce flashes of light. In the surface waters of the ocean, especially in tropical coastal waters, most of the luminescence is produced by dinoflagellates. Exactly what advantage this confers is speculated upon, but no one really knows. Some of the nonphotosynthetic species have pockets in their armor shells which contain symbiotic algal flagellates; having lost their own ability to photosynthesize, they harbor cells that can.

To most observers, diatoms and dinoflagellates seem less like "real" plants than do seaweeds, or macroscopic algae. There are several major groups of seaweeds, and they have been studied extensively. Some are shown in figure 3.42. Seaweeds are differentiated mainly on the basis of (1) the pigments they contain and (2) their methods of reproduction and life cycles.

The seaweeds, being continually bathed in a nutrient medium, have no need for many of the structures that characterize terrestrial plants. Structures such as leaves and roots for gathering energy and nutrients are unnecessary, as is an internal system for the transport of nutrients and energy. Thus, the body of the seaweed, the *thallus,* can be relatively amorphous. Seaweeds are also not posed with the problems of dispersing their young or providing very much protection for the young, and thus their reproductive organs can be relatively simple. Most seaweeds simply liberate motile cell gametes that fuse to form the adult plant. (See fig. 3.43.)

The cells of marine plants, with the exception, of course, of blue-green algae, are typical of the cells found in higher plants. The outermost boundary of the plant cell consists of a permeable cellulose cell wall similar in chemical composition and structure to the cellulose plates of dinoflagellates. Within the cell wall is the plant cell itself, a structure bounded by a cell membrane similar to that of animal cells and containing a nucleus, mitochondria as in animal cells, and chloroplasts that enable the cell to carry out photosynthesis.

The green algae (see fig. 3.44) exhibit the greatest diversity of form and are considered to be the stock from which higher plants evolved. Besides their evolutionary importance, of course, there are other reasons why algae are important. First, they provide a habitat for other aquatic organisms. A number of marine animals nest in beds of algae, and a number of animals that live in the intertidal zone are protected from drying by algae. Second, algae have a number of important economic uses. Ice cream is stabilized by algin, a product of red algae, which is also used in the commercial production of cheese. There are several other uses made of products of algae, particularly in the food processing industry. It has even been suggested

FIGURE 3.43 One of the most prevalent features of plant reproduction is the phenomenon known as *alternation of generations.* In the sense that it is used here, the term *generations* does not refer to the period between the births of the parents and the births of their offspring. *Generation* refers to a phase in the life cycle of a plant during which reproduction may be accomplished by sexual or by asexual means. The marine alga, *Ulva,* provides an example of the process. In many species the sexual, or *gametophyte,* and asexual, or *sporophyte,* stages are indistinguishable, but in many other cases the two adult forms of the plant may be so different that they have been classified as separate species. In some cases one generation is inconspicuous and the other dominant. As explained later, the chromosome number changes in this process.

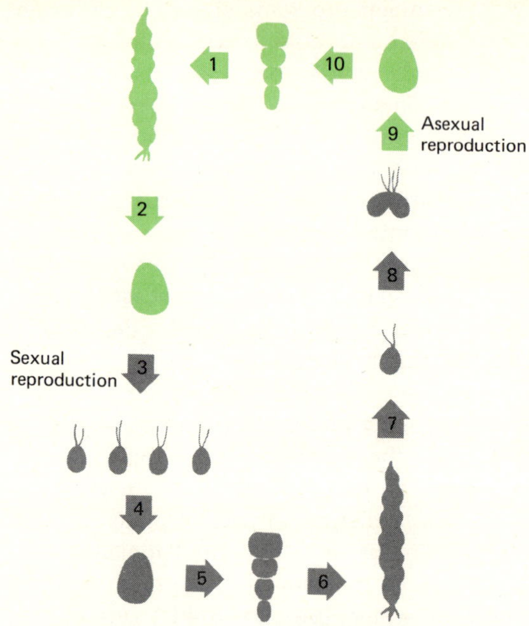

1) Sporophyte (asexual) stage with full chromosome complement
2) Single cell from the sporophyte stage undergoes meiosis
3) Spores formed with half the chromosomes as the adult sporophyte
4) Flagella lost
5) Mitosis
6) Multicellular gametophyte stage; each cell with half the chromosomes as in sporophyte stage
7) Gamete with two flagella liberated from the adult gametophyte
8) Fusion of two gametes
9) Zygote formed with full chromosome complement
10) Mitosis

FIGURE 3.44 Forms of green algae. (a) *Batophora sp.* (b) *Bryopsis sp.* (c) *Ulva sp.* (Redrawn from Prescott.)

that algae be cultured in closed ecosystems (such as space capsules and submarines) as a source of both food and breathable oxygen and as a means of removing excess carbon dioxide from the air.

That the oceans are the mother of life has become something of a cliché, but from the perspective of this chapter it should be clear how true the cliché is. We have devoted a considerable amount of space to aquatic and marine animals and plants for several reasons. First, it gives us some idea of the many ways in which evolution leads to diversity in the plant and animal kingdoms. Second, it provides us with a background against which we can contrast the problems faced by terrestrial organisms. Last, we should now be in a position to appreciate the incredible diversity of life that is involved in an ecosystem. To complete the portrait of the biosphere, we must now turn to the terrestrial environment.

Selected Readings

Barnes, Robert D. *Invertebrate Zoology.* Philadelphia: W. B. Saunders Company, 1968. A thorough and extensively illustrated text.

*Buchsbaum, Ralph. *Animals Without Backbones.* Chicago: University of Chicago Press, 1972. A classic, first published in 1948. Buchsbaum writes with great verve about a subject that is too often ignored by the general reader.

Bernal, J. B. *The Origin of Life.* New York: Universe Books, 1967. A good treatment of the alternative explanations for the specific processes that may have resulted in the first living cell.

Hardy, Alister. *The Open Sea.* Boston: Houghton Mifflin Company, 1971. A one-volume edition of the original two volumes. An excellent treatment of the physical conditions and natural history of marine life.

McConnaughey, Bayard. *Introduction to Marine Biology.* St. Louis: The C. V. Mosby Company, 1970. A sound text that concentrates on the animals.

Margulis, Lynn. *Origin of Eucaryotic Cells.* New Haven: Yale University Press, 1970. Dr. Margulis is one of the chief proponents of the symbiotic theory of the origin of higher cells; her book reviews in detail the evidence for the theory and suggests lines of research that may be fruitful in proving or disproving it.

Nicol, Joseph A. C. *The Biology of Marine Animals.* New York: John Wiley & Sons, Inc., 1967. A complete survey of adaptations to the marine environment, emphasizing physiology.

*Oparin, A. I. *The Origin of Life.* New York: Dover Publications, Inc., 1953. A theory of the origin of life based on biochemical principles and a consideration of competing theories.

Telfer, William, and Kennedy, Donald. *The Biology of Organisms.* New York: John Wiley & Sons, Inc., 1965. A brief and very useful survey of many of the topics covered in this chapter, and of several that will be covered in Chapters 4 and 5.

Barghoorn, Elso S. "The Oldest Fossils." *Sci. Am.,* Vol. 224 (May, 1971), pp. 30–42. Offprint number 895.

Echlin, Patrick. "The Blue-Green Algae." *Sci. Am.,* Vol. 214 (June, 1966), pp. 74–81. Offprint number 1044.

Isaacs, John D. "The Nature of Oceanic Life." *Sci. Am.,* Vol. 221 (September, 1969), pp. 146–160. Offprint number 884.

MacInnis, Joseph B. "Living Under the Sea." *Sci. Am.,* Vol. 214 (March, 1966), pp. 24–33. Offprint number 1036.

Margulis, Lynn. "The Symbiotic Theory of the Origin of Life." *Sci. Am.,* Vol. 225 (August, 1971), pp. 48–57. Offprint number 1230.

Satir, Peter. "Cilia." *Sci. Am.,* Vol. 204 (February, 1961), pp. 108–116. Offprint number 79.

*Paperback edition available.

adaptations to environments—land

chapter four The adaptations that are necessary to live on land developed in very few phyla, since many of the factors that regulate life on land require adaptations that are very different from those required in the sea. The transition from the sea to land may only be equalled in complexity and evolutionary change by the origin of life itself. The environment of terrestrial organisms is so very different from that of aquatic organisms that in order to understand the significance of the physiological and ecological adaptations of terrestrial animals and plants we must first look at the differences that exist between the two environments.

As we have seen, water provides support for aquatic organisms; it transports most of their nutrients, food, and often the reproductive products required for the survival of individuals and species, and drying is not usually a very serious problem in aquatic environments. Also, only benthic organisms are limited to life on a surface, and most aquatic systems are relatively uniform. Environmental changes that occur are usually gradual, temperature fluctuations are generally slight from place to place and time to time, and most of the ocean is uniformly dark.

These wings, antennae, eyes, sucking mouth parts, and legs all represent specializations of insects required for existence in the terrestrial environment. *(Courtesy of John Urban.)*

In contrast, consider the terrestrial environment. There is no buoyancy to support organisms; little or no food, nutrients, or reproductive products are transported through the atmosphere. (There is a clear exception to this statement concerning reproductive products, which we will consider when we examine the life of plants.) The water content of the atmosphere is very low and makes drying a major problem for terrestrial organisms. With the exception of birds and some insects, all terrestrial organisms are limited to life on the surface of the earth, and even birds and flying insects depend on the land for most of their food and nutrients, as well as for their reproductive activities. Moreover, the terrestrial environment is highly varied and discontinuous. Temperature fluctuations are often radical from place to place and from time to time in the same place. The amounts of rainfall, sunlight, and humidity also frequently change from day to day, season to season, and place to place. The soil, in contrast to water, does not consist of uniformly distributed materials; rather, it contains more or less clumped aggregations of resources that differ from one location to another. We would thus expect, and indeed find, that the organisms populating terrestrial environments are very different from those populating aquatic environments.

We mentioned earlier that only two kinds of animals made really successful transitions to land: the arthropods and vertebrates. In considering the arthropods, chiefly insects, we will examine only the adaptations that have evolved, not their evolutionary history. When considering the vertebrates, however, we shall not only examine the adaptations but will also see a progressive evolutionary change. We should point out here that although we shall consider animals first, it is obvious that plants were the first form of life to make the transition to land. Plants provided the terrestrial food resources that transitional animals depended upon for their survival.

The Insects

The exoskeleton of arthropods provided a built-in solution to part of the problem of support in terrestrial situations. (See fig. 4.1.) Even the marine crabs have partially invaded the land. Support alone, however, is insufficient, since on land there is no means of having food transported to an organism. An animal on land must be motile in order to find food and water.

Insects have adapted to the locomotory requirements of finding food in two ways. First, they can walk. Although this may seem to be a relatively simple adaptation, and although it is seen in other arthropods such as lobsters and crabs where it evolved independently, it is by no means a trivial solution to the problem. The solution was in part built into the exoskeleton and the basic body plan of arthropods, but walking requires a complex musculature and nervous integration and coordination of complex movements of three pairs of legs. Imagine trying to design a machine that could walk on six legs, much less run, crawl, change direction, and jump! It would be easier to design a swimming machine similar to a fish or a shrimp. Nevertheless, the arthropod exoskeleton provided an excellent mechanism for locomotion, and walking evolved independently in spiders, crabs, centipedes, millipedes, and other arthropods.

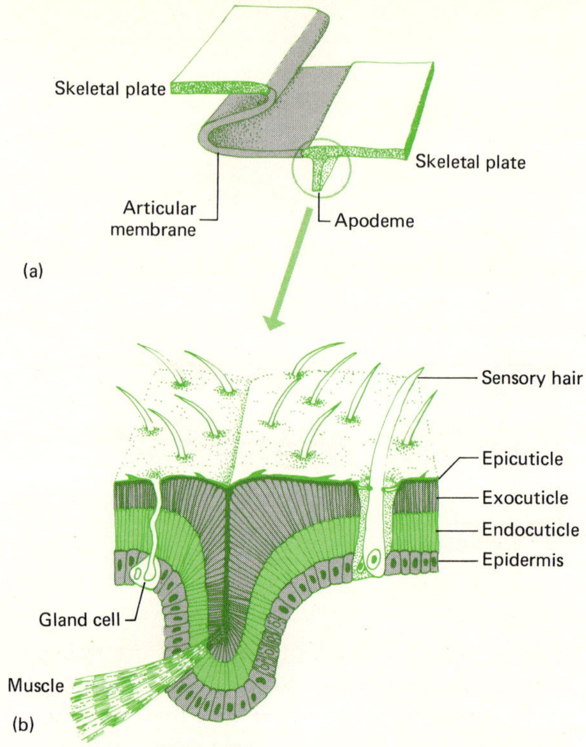

FIGURE 4.1 The exoskeleton of an arthropod is made up of plates which are joined by articular membranes that permit movement. Extending into the body of the insect are portions of the cuticle called apodemes, to which muscles are attached. Although some muscles are attached directly to the cuticle, some may attach to the skin (epidermis) that directly underlies the cuticle. *(Redrawn from* Invertebrate Zoology, *Second Edition, by Paul A. Meglitsch. Copyright © 1967, 1972 by Oxford University Press, Inc. Reprinted by permission.)*

The second solution, flight, was also made possible by the possession of an exoskeleton. The ability to fly greatly expands the range of an animal and allows it to exploit more potential food resources than an animal limited to locomoting overland, although this may not have been the reason that flight evolved. It is thought that wings may have evolved as shore-dwelling terrestrial arthropods were preyed upon by animals that came more recently from the sea. Ancient insects may have climbed trees to escape these predators, but to go from one tree to another would require a return to the ground and thus re-exposure to predators. A leap from one tree to another was one means of avoidance, and some selective advantage would accrue in those insects that could reach another tree without falling. Early wings may have evolved from extensions of the exoskeleton on the part of the body to which the legs were attached and may have been used simply to glide, as some squirrels today glide from tree to tree. The next step would have been the evolution of existing muscles to allow actual controlled flight. Among the insects alive today that are considered to be primitive fliers, dragonflies for example, the musculature associated with flight is quite different from that of the more recently evolved flying insects. (See fig. 4.2.) The wings of a primitive flier such as the dragonfly are attached to the body more closely together than in more advanced insect species, and the flight muscles are directly attached to the bases of the wings. The muscles are *synchronous*—that is, they contract once for each nerve impulse received—and thus the dragonfly beats its wings, at a relatively slow rate (about 30 beats per second). The flight muscles of more recently evolved insects, such as bees or flies, are *asynchronous*; that is, the muscles have a characteristic contraction rhythm that is not directly related to the number of nerve impulses received.

Whatever the means used by insects to reach food, the foods themselves are highly varied. Some insects subsist very well on paper, some on opium, and yet others on paintbrushes. There are insect herbivores that eat leaves or bark, or drink nectar from plants. There are also insect predators that feed on other insects or animals, as well as insect parasites. The adaptations of insects to these diverse sources of energy and materials are reflected in the shapes of their mouthparts. Again, as in locomotion, the means by which the problem of obtaining food has been solved have depended upon an elaboration of the exoskeleton of the arthropod. The exoskeleton thus not only provides support but also has inherent in it much evolutionary potential.

A major problem faced by terrestrial organisms is the availability of water. Many insects, of course, take in a large proportion of their food in liquid form. Butterflies that feed on nectar and mosquitoes that feed on human blood are examples. Others take in moisture with their food. But the precise control of the composition of body fluids is a major problem in living systems, as we shall see in detail in the next chapter, and thus the problem is to conserve water. The exoskeleton of the insect functions to prevent drying. The relative impermeability of the exoskeleton inhibits the exchange of carbon dioxide and oxygen, while serving to retain

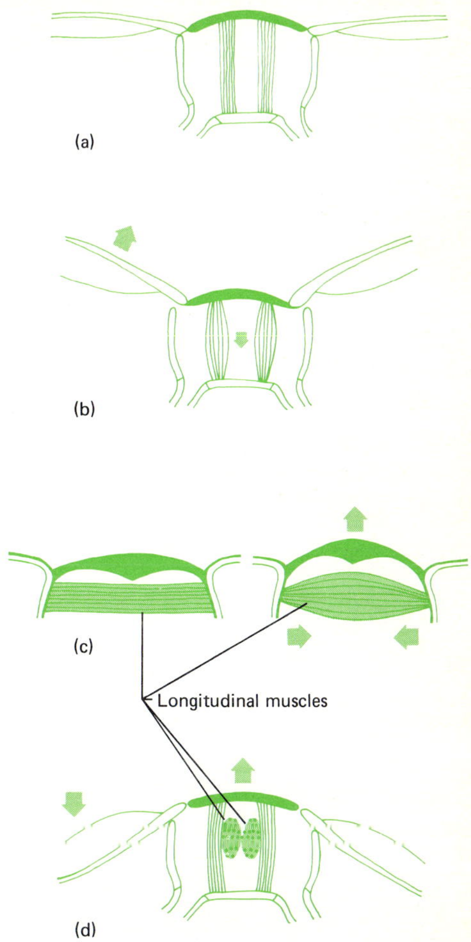

FIGURE 4.2 The muscles used in flying, among most species of insects, do not attach to the wings at all, but to the cuticle in the segment of the body where the wings attach. Two sets of muscles are brought into play to produce one complete wing beat. One set is attached to the dorsal and ventral plates of the body. When these muscles contract, the wings are raised as shown in (a) and (b). When the dorsoventral muscles are relaxed, however, the wings do not complete a full downstroke. Longitudinal muscles contract, raising the plate to which the wings are attached (c) and (d) and produce the downstroke. The wingbeat of some insects is as high as 1000 beats per second, and the ordinary mosquito has a wingbeat of about 600 beats per second. *(Redrawn from* Invertebrate Zoology, *Second Edition, by Paul A. Meglitsch. Copyright © 1967, 1972 by Oxford University Press, Inc. Reprinted by permission.)*

water. Crustaceans, as we have seen, exchange gases and water through the very large surfaces of their gills, but such exchange would produce too great a water loss in the terrestrial environment. Among the insects, the exchange of gases is accomplished by a system of waxy tubes that ramify throughout the body, delivering oxygen to the cells and carrying carbon dioxide away from them. These tubes, called *tracheae* (see fig. 4.3), are in contact with the external environment—the source of oxygen—through openings on the outside of the insect body called *spiracles,* which can be opened and closed by valves. The oxygen that enters the body reaches each cell and enters the cell by diffusion. The simplicity of the tracheal system has certain disadvantages. Gas exchange in insects is slow compared to that in vertebrates, and it serves to limit the body sizes that insects can achieve. Nevertheless, the tracheal system suffices to conserve water and to adapt the insect to life on land.

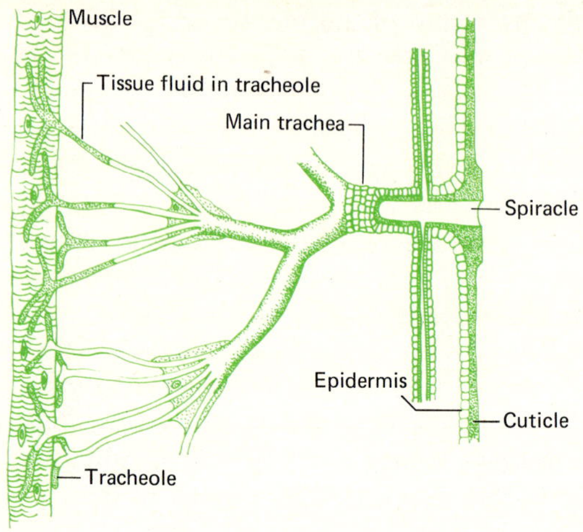

FIGURE 4.3 Insects, not having lungs, transport gaseous oxygen directly to the tissues. Air enters through the spiracle, passing into the trachea and tracheoles, where oxygen dissolves in fluid and eventually passes to the muscles. *(Redrawn from* Invertebrate Zoology, *Second Edition, by Paul A. Meglitsch. Copyright © 1967, 1972 by Oxford University Press, Inc. Reprinted by permission.)*

Another means by which water is conserved is found in the excretory system of the insect. Another set of tubules, called *Malpighian tubules,* branch off the digestive tracts of many insects. (See fig. 4.4.) These tubules are closed sacs that are bathed in the blood of the insect. Fluid moves across the closed end of the tubules from the blood. Near the end closest to the main digestive tract, waste products are removed from the blood while the water reenters the bloodstream. The concentrated wastes are still largely fluid, however, in the form of urine. The urine passes into the portion of the intestine called the *hindgut*, and there, as well as in the rectum, very large amounts of water are reabsorbed. The feces and urine produced by insects consequently are very dry.

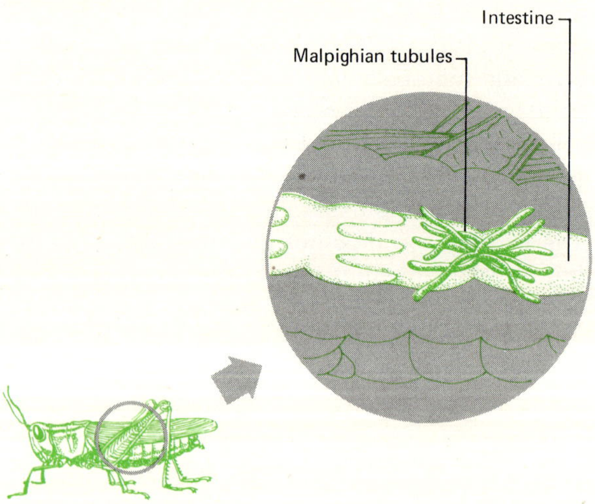

FIGURE 4.4 Malpighian tubules in the grasshopper.

Another of the problems associated with life on land is that the gametes, sperm and ova, cannot be expelled into the environment for fertilization. Almost invariably, wholly terrestrial organisms are adapted to internal fertilization, and the variety of means that have evolved to accomplish this among insects and other

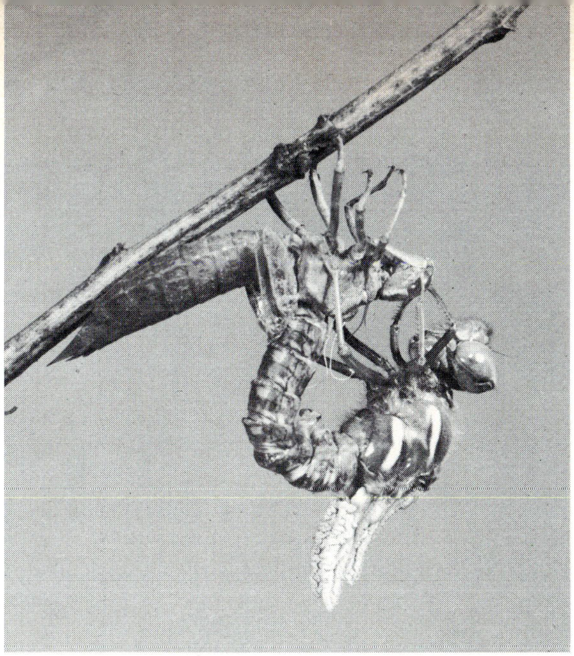

FIGURE 4.5 A dragonfly crawling from its molted exoskeleton. *(Courtesy of Robert C. Hermes from National Audubon Society.)*

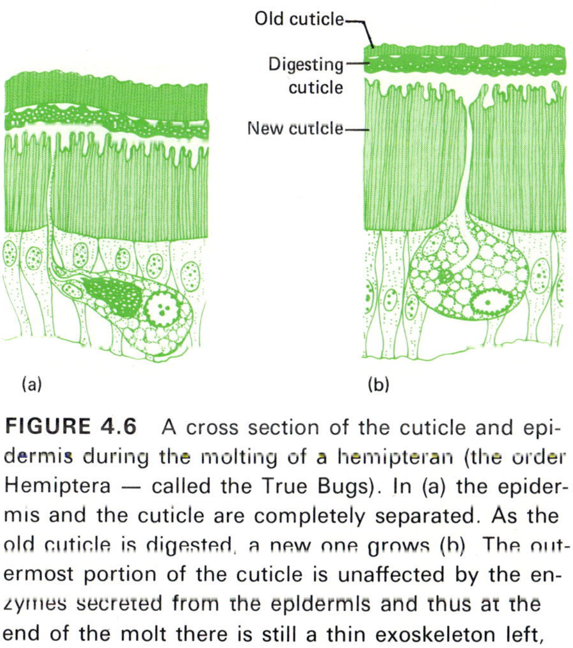

FIGURE 4.6 A cross section of the cuticle and epidermis during the molting of a hemipteran (the order Hemiptera — called the True Bugs). In (a) the epidermis and the cuticle are completely separated. As the old cuticle is digested, a new one grows (b). The outermost portion of the cuticle is unaffected by the enzymes secreted from the epidermis and thus at the end of the molt there is still a thin exoskeleton left, and the insect must break through and crawl out. Some species of insects, after the molt is complete, eat the recently shed skeleton and thus "recycle" the nutrients it contains. *(Redrawn from* Invertebrate Zoology, *Second Edition, by Paul A. Meglitsch. Copyright © 1967, 1972 by Oxford University Press, Inc. Reprinted by permission.)*

arthropods are staggering and often bizarre. In some cases, the male deposits sperm outside the body of the female, not necessarily even in proximity to any female. These sperm are deposited in packets called *spermatophores,* which, in some species, are placed on the top of a stalk like small plants. When a female of the same species encounters these spermatophores, she may take them into her *cloaca*, a cavity through which pass the products of the digestive, excretory, and reproductive systems. In one species of springtails the male at least comes into contact with the female. In this species, the male surrounds the female with stalked spermatophores, so that no matter which way she moves, she will take up the sperm. Among certain species of centipedes (which are not insects), the male deposits a stalked spermatophore that the female, when she comes upon it, bites off and stores in a special pouch in her "cheek". When she then lays eggs, she takes each one separately into her mouth, smears it with the stored sperm, and thereby fertilizes it.

The possession of an exoskeleton, while advantageous in several ways, is not without some drawbacks. The key problem is that of allowing room for growth. The relative rigidity of the exoskeleton has been overcome by the process of *molting*—that is, the periodic shedding of the exoskeleton. (See fig. 4.5.) Confined as they are in a rigid casing, insects must remain the same size during periods between molts. Among some insects, relatively little of the intermolt time is spent preparing for the next molt, but some other insects spend most of the intermolt period preparing for the next burst of growth. In either case, when the growth period commences, the insect does undergo growth within the hard exoskeleton and is "squeezed" by its own skeleton. The innermost layer of the exoskeleton is dissolved by secretions of the "skin" glands, while the rigid outermost layer remains unaffected. (See fig. 4.6.) The insect begins to grow a new exoskeleton while still inside the old one, and then splits the old skeleton open and emerges, soft and wrinkled from the pressures of having grown within the confines of the old skeleton. Soon after emergence, the new skeleton expands and hardens. Although some insects continue to molt throughout their lives, a winged insect must undergo one final molt to a winged form, since wings cannot be molted and grown anew. Thus, in the period preceding its final form, a winged insect is wingless.

This brings us to an important adaptation of insects: complex life cycles. After hatching from the egg, all insects are wingless and in one of three general forms. They may be small-scale replicas of the adult form of the insect, in which case the young merely grow by molting and never undergo substantial morphological change from hatching to adulthood and death. This life cycle is characteristic of most wingless insects. In contrast, other insects emerge as nymphs, in which case they approximately resemble the adult form that they will assume, although they lack wings and may have different body markings. As they grow, they pass through several nymphal stages until they reach the final molt and assume the adult winged form. (See fig. 4.7.) The third case is strikingly different from the first two: the insect emerges from the egg as a wormlike larva and grows by several moltings, usually without ever significantly changing its form. At the end of the larval state, the insect forms a *pupa* and enters an apparently dormant state during which vast biochemical changes take place as the tissue structure of the larva is totally reorganized. At the end of pupation, a winged adult insect emerges from the pupa in a form completely unlike any of the preceding stages. (See fig. 4.8.)

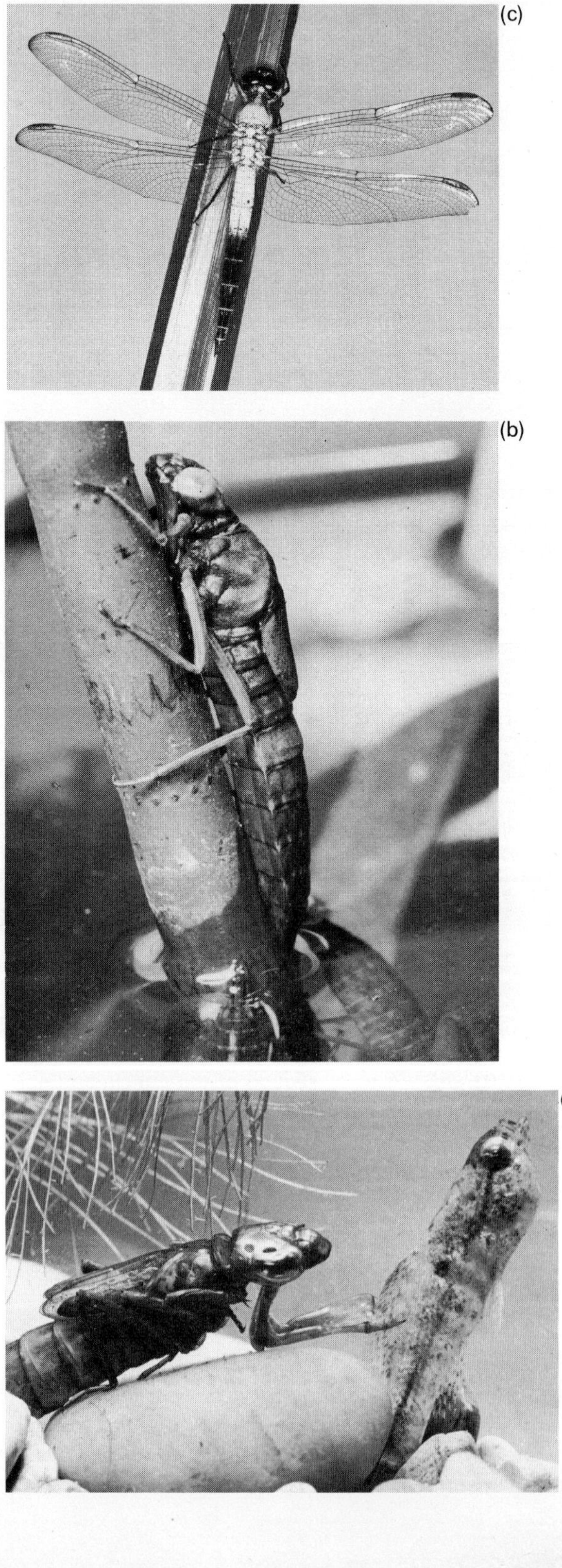

FIGURE 4.7 Incomplete metamorphosis. A dragonfly nymph is a very capable predator, as are the aquatic nymphs of many insects. It emerges from the water and molts to the adult, sexually reproducing stage, which is an active predator on smaller insects. *((a) Courtesy of Dr. Ross, E. Hutchins, State College, Mississippi; (b) Toni Angermayer, and (c) Jen and Des Bartlett, both courtesy of Photo Researchers, Inc.)*

(a)

FIGURE 4.8 Complete metamorphosis. The Cecropia moth starts life as an egg on wild cherry leaves (a). On hatching, the active, herbivorous larva, or caterpillar (b) undergoes several molts, finally spinning a cocoon of silk and forming a pupa inside (c). After many physiological changes, the adult (d) emerges. *(Courtesy of Muriel V. Williams, Harvard University.)*

(b)

(c)

(d)

The larvae of insects which undergo such complete changes contain within them the cells that will direct the development of an adult, but those cells do not grow during the larval stage. So-called *juvenile hormones* secreted from glands in the heads of the larvae inhibit the adult cells. At a particular time in the life of an insect, the juvenile hormones cease being secreted, and the cells that will produce the adult are released from inhibition. At that time, the pupa forms, the larval cells die, thus providing nourishment for the growing adult cells, and the adult body begins to take shape. When the reorganization of tissue is complete, the adult emerges from the pupa and takes up a life that is totally different from that of the larva. It may feed on different foods and be preyed upon by different predators. Compared to the larva, the adult can travel great distances and thus disperse the species and prevent overcrowding. Most importantly, the adult is capable of reproducing—a capacity that larvae lack. Some adult insects, in fact, are incapable of any activities except flying and reproducing. They possess no digestive systems, thus can not feed, and are, in effect, reproducing ''machines''.

The differences that exist between adult and larval insects are of importance not only for biochemical reasons but also because they show adaptations to different environments. The larvae of most species spend their lives in moist environments, whereas the adults spend theirs in drier ones. The consumption of different foods in the larval and adult stages reduces the competition between adults and larvae for food resources, and thus the young are indirectly protected by the parents.

One interesting example of even more highly evolved parental ''care'' is the behavior of some species of mud-dauber wasps. The female wasp builds a mud nest consisting of a number of separate compartments. She then captures and paralyzes spiders, places one in each compartment, lays an egg by each spider, and then seals the chamber. After the eggs hatch, the larvae feed on the spiders, pupate, and hatch into adults. A number of other parasitic wasps inject their eggs into caterpillars, the larvae of moths and butterflies, where the wasp larvae consume the caterpillars from the inside.

Obtaining food and reproducing have produced among the insects a variety of fascinating behavioral adaptations, ranging from the relatively simple behavior of mud-dauber wasps to the very complex societies of ants and some wasps and bees. The most thoroughly studied of the social insects is the honeybee (*Apis mellifera*). (See fig. 4.9.) In many ways the societies of honeybees, which may number as many as 80,000 individuals, behave as single organisms, because groups of individuals perform some tasks that are performed by organs in other organisms. Moreover, because only the queen reproduces and all of the members of the colony are therefore her offspring, the colony as a whole is selected for or against. In this way, too, the hive is like a single organism. The rest of the bees, or workers, perform a variety of duties, which among others include building new waxen cells for the deposition of food and eggs, feeding larvae, maintaining nest temperature, feeding the queen, removing dead bees and larvae from the nest, and cleaning cells to prepare them for reuse. Although there is a considerable division of labor in the colony, there are not subgroups of worker bees that perform only one function, such as feeding the queen. Rather, a worker performs different activities at different times during the six or so weeks that it lives. (See fig. 4.9.) The times during which the different activities are performed are correlated with changes in certain glands. The coordination of all these activities is produced in part by the exchange among the bees of a variety of chemical substances. For example, the queen produces a chemical called ''queen substance,'' * which inhibits the workers from building cells that will house new queens. As long as the queen is producing this substance in sufficient quantities for the workers

*The queen substance used by honeybees has been identified as 9-ketodecenoic acid.

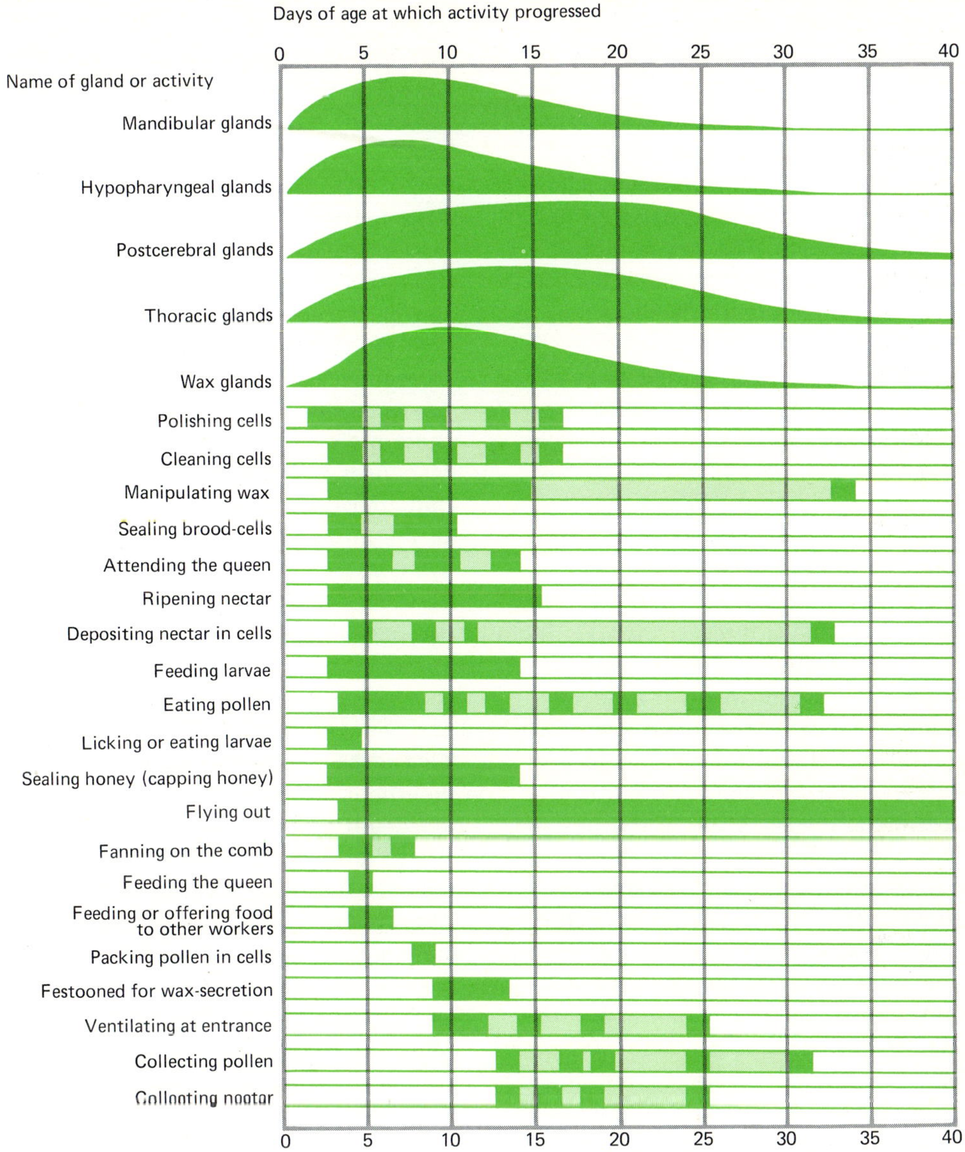

FIGURE 4.9 This diagram shows some of the correlations between the size of certain glands in the honeybee and the activity performed at a particular age. The shading indicates that the activity is not completely abandoned but is engaged in only intermittently. The mandibular and hypopharyngeal glands produce materials that constitute most of the food fed to larvae in a nest, and thus the bees spend most of the first few days of their lives feeding other bees, either workers, larvae, or the queen. After two weeks of life, however, the bee spends most of its time out of the nest, collecting pollen or nectar. *(Modified from G. E. King from E. O. Wilson,* The Insect Societies, *Belknap Press of Harvard University Press, 1971.)*

—that is, an average of 0.1 microgram per worker per day—no queen cells will be produced. In the spring, the old queen reduces her production of this substance, and the workers begin building larger cells to house incipient queens. These cells are filled with "royal jelly," a substance that has thus far proven to be extremely difficult to analyze. The larvae that hatch from the eggs deposited by the queen in the queen cells, having fed on royal jelly, develop into new queens. Before the new queens emerge, the old queen leaves the hive with a swarm of workers to find a new site. Before doing so, however, she lays unfertilized eggs in another set of cells, the drone cells, which are specially built by the workers. These eggs develop into adults without being fertilized. The drones are genetic males—in fact, the only males in a colony. When a queen emerges from her royal cell, she searches through the nest for other newly emergent queens and exchanges certain auditory signals with them. If a new queen encounters another new queen just as she is emerging from a cell, they will usually fight until one of them is killed. A new queen is then almost literally pushed out of the nest by the worker bees to begin a nuptial flight. The drones have already left the nest and swarm nearby. As the virgin queen approaches the males, she releases small quantities of a substance that is similar to queen substance and which the males can detect from about 10 meters. Mating takes place quickly and, for the male, fatally. In the words of one student of social insects, "The male literally explodes its internal genitalia into the genital chambers of the queen and quickly dies." * The new queen may make as many as three flights each day for three or four days, mating on each flight, until she has copulated with twelve or more males. She then flies back to the nest where she may make way for another virgin to make her nuptial flight or, as is usually the case, destroy the other virgins and take over the nest. Under favorable conditions, the new queen may lay up to 1000 eggs each day, all fertilized by sperm collected on her nuptial flights. Within a year, the colony size may double and a new division occur.

*E. O. Wilson, *The Insect Societies,* Harvard University Press, Cambridge, Mass., 1971.

An important point to be made about these behavior patterns is that they are *not learned.* The mud-dauber wasp, for example, after depositing the paralyzed spider and an egg with it, will seal the compartment, even if an observer were to remove the spider. To a large degree, the behavior of insects is controlled by hormones, but there are also many behavior patterns that are responses to information about the environment gathered by the nervous system. The responses to the information are built into the nervous systems and the hormonal excretions of insects, however, and are genetically programed. Thus the behavior is evolved, not learned.

One example of this may be seen from an experiment performed with two strains of bees. Within the hives from which these strains came, they behaved quite differently with regard to the death of larvae in the cells of the hive. One strain, termed "hygienic," opened the caps of the cells and removed the dead larvae. An "unhygienic" strain simply left alone cells containing dead larvae. Experimental interbreeding between the two strains showed that the hygienic behavior was dominant. (We shall see what this means in more detail in Chapter 6.) Further interbreeding produced four kinds of behavior among the offspring. One group of bees was hygienic and another unhygienic, but two new kinds of behavior arose. A third group of bees would open cells containing dead larvae but would not remove the corpses. The fourth group of bees would not open cells, but if presented with cells that were already open, they would remove the corpses. For reasons that we shall see in Chapter 6, this indicated that these behaviors were controlled by two genes. For now, the important point is that the bees did not learn anything but behaved only in ways determined by their genetic constitutions.

Fixed behavior patterns of this sort are as much an adaptation as the morphology of mouthparts or wings and stand in strong contrast to the behavior patterns of many vertebrates, especially mammals, where behavior frequently contains a large learning component. The complex behavior of social insects indicates, though, how extreme the adaptations that evolve may be.

Ecologically, the insects are by far the most important animals on land. They have solved the problems of the terrestrial environment in a variety of ways, ranging from morphological adaptations and reproductive mechanisms to very complex, evolved behavior patterns. We shall examine some of the physiological mechanisms involved in their adaptations in the next chapter, but here we should compare the adaptations of the insects with those of another successful terrestrial group of animals, the vertebrates.

The Vertebrates

The ways in which the vertebrates have met the rigors of terrestrial life differ sharply from those of the insects. Nonetheless, the two groups have converged in some respects. There are species of both insects and vertebrates that fly, just as there are examples from both phyla that walk or jump. The terrestrial environment has dictated, as we shall see, that some form of protection be provided for the eggs, and in many species of insects and vertebrates the adaptation that has arisen is the eggshell. Although there are fewer species of vertebrates than insects and frequently vertebrate populations are smaller, vertebrates have some advantages over insects in the terrestrial environment. One such advantage is that vertebrates can attain larger body sizes than insects. The exoskeleton of the insect, being essentially a rigid cylinder, grows relatively weaker as it grows larger and thus prevents insects from achieving very great sizes. One of the advantages of large body size is that it allows an animal to cover correspondingly larger areas of land in search of food. There are, as we shall point out, certain ecological constraints on the size to which an animal may grow, but the insects have the added constraint of an external skeleton. With few exceptions, such as the monarch butterfly, insects do not migrate over very great distances.

Another important adaptation of some vertebrates is *thermoregulation,* the physiological ability to maintain a high, constant body temperature. Since the terrestrial environment is characterized by significant fluctuations in temperature over time spans as short as twenty-four hours, thermoregulation confers a distinct advantage on vertebrates.

Many vertebrates, and particularly mammals, have adapted to caring for their young in a more direct way and over a usually greater proportion of their life span than even the most highly social insects. Care for the young is a particularly important characteristic, since it means that learned behavior may be transmitted from parent to offspring and thus sets the stage for the development of culture.

The first vertebrate to make the transition to a terrestrial environment was probably a fish with the capacity to breathe air, conserve body moisture, and propel itself across the ground with its fins. Precisely how and when this occurred is not known. We do know that some fish possessed muscular fins and lived in temporary pools, and they may have developed an outpocketing of the throat that allowed them to take in air from the atmosphere when the oxygen concentration in the water grew too low. Such fish might have evolved the ability to pull themselves from pond to pond with their fins—an inefficient means of locomotion, since it would entail dragging their bodies over the ground. If they did go from pond to pond, we can assume that those that were able to reach the other ponds probably possessed an ability to resist the drying effect of the atmosphere,

FIGURE 4.10 A mudskipper, a fish that uses its fins to locomote on land. *(Courtesy of Arthur W. Ambler from National Audubon Society.)*

FIGURE 4.11 The African lungfish, shown here in water, can survive long periods of drying by burrowing in the mud and breathing air using structures similar to lungs. *(Courtesy of Russ Kinne from Photo Researchers, Inc.)*

possibly by the possession of a relatively impermeable skin. There are, in fact, contemporary fish which possess each of these adaptations, but there is no single species which possesses them all. (See figs. 4.10 and 4.11.) These contemporary species are not related to the ancestral amphibian, but they do indicate that the potential for life on land existed in fish.

The selective advantage that was conferred on such an amphibious fish must have been great. Life in the sea had already reached a high level of diversity, but the land offered vast, unexploited opportunities for radiation. Of course, a number of problems had to be met: support and locomotion, drying, excretion, and reproduction. The potential for support and locomotion existed but was not efficient. The legs, or fins, located on the sides of the body had to pull the animal along the ground. (See figs. 4.12 and 4.13.) One amphibian, the frog, solved this problem by jumping, but the frog is a highly specialized, and therefore atypical, amphibian. The contemporary amphibian that is closest to the ancestral model is the salamander, and even this comparison is not very good. The limbs of early amphibians evolved from muscular fins and had five "fingers." They also had tails, which were dragged about but may have afforded some support.

FIGURE 4.12 Although we think of them as very different, fish and amphibia are constrained by similar skeletons and musculature, and the basic movement for locomotion is similar. *(From* The Vertebrate Body, A Shorter Version of the Second Edition, *by Alfred S. Romer. Copyright © 1956 by W. B. Saunders Company, Philadelphia.)*

The early amphibians, like their present-day descendants, probably had a permeable skin, and thus although they could exchange carbon dioxide and oxygen through the skin they also suffered water loss. They also lost water in the dilution of feces and urine. They were undoubtedly dependent upon ponds for reproductive activity and still fertilized the young outside the body of the parent. There was probably some specialization of the larvae, as there is today among amphibians. The larvae of frogs, for example, are strictly herbivorous, while adult frogs are carnivores. Actually, it is surprising that some amphibians escaped extinction. The "typical" amphibian no longer exists, and frogs and salamanders are highly specialized forms.

Obviously, a selective advantage would be conferred on those species that achieved greater independence from water and better locomotion. The reptiles, which are the descendents of amphibians, became very successful, although most are now extinct. The evolution of reptilian locomotion, which was a great improvement over amphibian locomotion, was accomplished when the body was raised from the ground. (See fig. 4.14.) The limbs could be used for carrying, as opposed to pulling. A key to the improved locomotion was in the curvature of the vertebral column, which evolved into an arch from which the body hung. The limbs developed muscles that permitted walking, and the shoulder and pelvic girdles evolved in a way that allowed the legs to carry the body. The ribs, while offering protection for the internal organs, also came to play an important role in the intake and expulsion of air from the lungs. A host of changes took place in the muscles, the skin, and the skeleton that allowed reptiles to exploit the terrestrial environment more efficiently than their predecessors.

The drying problem was overcome slowly as the lungs evolved to allow better gas exchange with less water loss. Once these functions could be performed internally, the skin could thicken. Wastes became more concentrated, further cutting down on water loss. Some reptiles need not drink water at all but subsist on the water that is a part of their food.

The freedom from the aquatic environment achieved by reptiles included freedom from returning to ponds for reproduction. But, of course, that posed the problem of protecting the eggs from drying. Although internal fertilization evolved among terrestrial organisms, most species continued to lay eggs in which the young developed. In the drying environment of land, such eggs must be deposited in ponds or other bodies of water or be protected from drying in some other way. The evolution of an egg covering, or shell, within which could be deposited a personal ''pond'' for each developing embryo, allowed increased freedom from the aquatic environment. The embryos in such eggs are supplied with a source of nutrients and energy—the yolk, a liquid medium in which to develop, and often a supply of oxygen as well. These materials are all contained within a system of membranes, each of which is specialized for its separate function. The fertilized ova is attached to a *yolk sac* in the genital tract of the female; the liquid medium, or *amnionic fluid,* is contained within the *amniotic membrane* and surrounds the embryo; and another membranous sac, the *allantois,* receives the waste products of the embryo. The whole set of membranes is enclosed in another membrane, the *chorion.* (See fig. 4.15.) Thus, within the genital system of the female, an entire environment is created for the development of the fertilized ova, from zygote to individual offspring.

FIGURE 4.13 A salamander, an amphibian, moves by dragging its belly across the ground. *(Courtesy of Alvin E. Staffan from National Audubon Society.)*

Given their reproductive emancipation from water, the reptiles possessed the potential to radiate into every environment on earth and into a multitude of niches. Some early reptiles, such as ichthyosaurs, readapted to aquatic environments, but others, such as pterodactyls, be-

FIGURE 4.14 A crocodile, a reptile, can lift itself from the ground and truly walk. *(Courtesy of Leonard Lee Rue III from National Audubon Society.)*

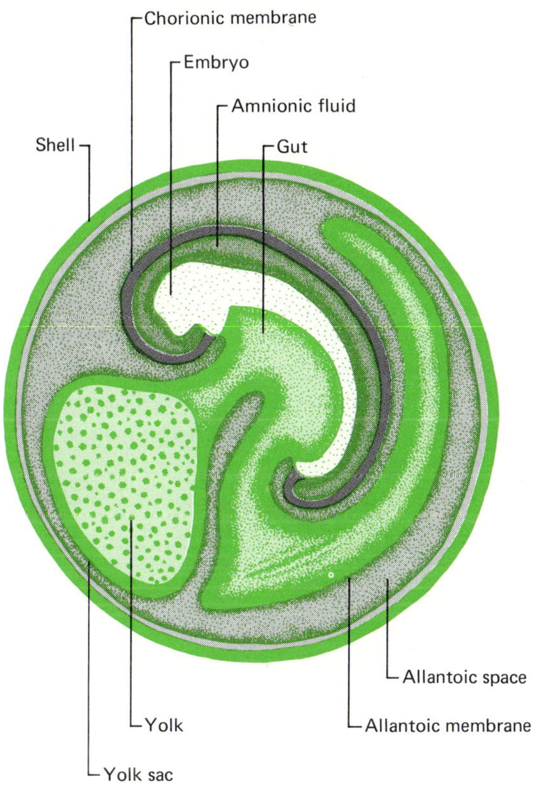

FIGURE 4.15 An embryo developing within an egg. *(Redrawn from Romer.)*

came flying predators. There were also a variety of vegetarians and insectivores. These early reptiles were usually very large (see fig. 4.16), but more importantly, they were the first truly successful terrestrial vertebrates.

Some problems remained, however. Eggs that are rich in nutrients and energy are a resource for predators. Some of the early reptiles may have protected their eggs to some degree, but for the most part, they probably did so to no greater extent than do modern reptiles, and that is very little. Most modern reptiles simply deposit their eggs and leave them. One example of the protection of the eggs among reptiles is particularly striking. The females of some marine species of turtles return to land to lay their eggs each year. These females compete vigorously for the nesting locations that seem to offer the best protection for their eggs, and if two females choose the same spot, they will fight. After the female has laid her eggs, she remains at the nest for a time; if another female attempts to use the nest, the first will try to drive her away. While such fights have been in progress, predatory birds have been observed to enter the nest, break open the shells, and begin feeding on the eggs. The fights last only a short time, and if the first female succeeds, she returns to the nest where the eggs are being fed upon by

FIGURE 4.16 The reptiles evolved adaptations for life in nearly every environment. *(Courtesy of the American Museum of Natural History.)*

the birds. She does not attempt to drive the birds off. She simply leaves the nest and the eggs to the birds and returns to the sea to reproduce again a year or so later.

Another important problem of the terrestrial environment for which a solution did not evolve among the reptiles is fluctuating temperature. Since the chemical reactions of the body that provide the energy needed for growth and maintenance (*metabolism*) proceed most efficiently within specific temperature ranges, an organism that is able to regulate its internal temperature can remain active through a wider range of environmental temperatures than an animal lacking this ability. The internal temperature of the latter rises and falls as the environmental temperature changes. When the body temperature falls, the metabolism of the animal slows down, thus forcing it into inactivity. Very high temperatures create different problems that force the animal to seek cooler locations, such as a shady spot. Thus, some reptiles have evolved behavioral mechanisms that allow thermoregulation. Nevertheless, reptiles remained *poikilothermic* (*poikilos* varied) as opposed to *homeothermic* (*homeo* same). Since they did not evolve physiological mechanisms for thermoregulation, they have not been able to maintain activity in the cooler times of the day, such as the early morning or the night, and thus could not utilize the resources of the environment as fully as animals that were capable of physiological thermoregulation.

One line of reptiles evolved in a way that solved at least one of these problems. This line gave rise to birds. The reproductive adaptations of birds are not very different from those of reptiles; both lay eggs. But birds have evolved the ability to control internal body temperatures at high, constant levels. Naturally, such thermoregulation involves some costs. It requires much more energy, since producing the heat requires energy, and therefore homeotherms require a high energy input. A reptile can subsist on a single meal for a very long time, whereas homeothermic animals must eat frequently to keep up their high metabolic rate and therefore require greater quantities of oxygen and must dispose of greater quantities of carbon dioxide. The loss of heat must also be minimized. Homeothermy, we might note, is not especially advantageous in an aquatic environment, since the environmental temperature is relatively stable; it is obviously very difficult to achieve, because water conducts heat very well. On land the situation is very different, and the evolution of homeothermy required a whole host of adaptations. Feathers provide some protection against heat loss, and they are thought to have evolved from modified scales on the flight surfaces of the ancestors of birds. Feathers are indeed amazing organs, as figure 4.17 shows. Flight confers numerous advantages, as we have seen for insects, although flight in birds probably evolved as a means of increasing the efficiency of predation, rather than as a means of escaping predators. Early birds had teeth characteristic of predators, but these teeth have now been lost. Once flight had evolved, it allowed the birds to range over greater areas in search of food, and if they were poikilothermic they could migrate and thus overcome the limitation of a changing climate.

The adaptations of flight and feathers meant that food gathering was not a major problem for birds and that heat loss was to some degree minimized. Thus, the high energy requirements of homeothermy were attainable. The other adaptations that were required soon followed. The blood of birds is particularly efficient in transporting oxygen; their lungs are relatively large for their small bodies, even extending into contiguous bones. This reduces the weight of the bones, thus permitting more efficient flight. Correspondingly, the bones evolved maximum strength for their weight.

Thus, birds were in an advantageous position to radiate into a large variety of environments and niches. Modern birds are herbivorous, insectivorous, and carnivorous, and there are also flightless birds, such as the ostrich and emu, as well as truly aquatic flightless birds, such as

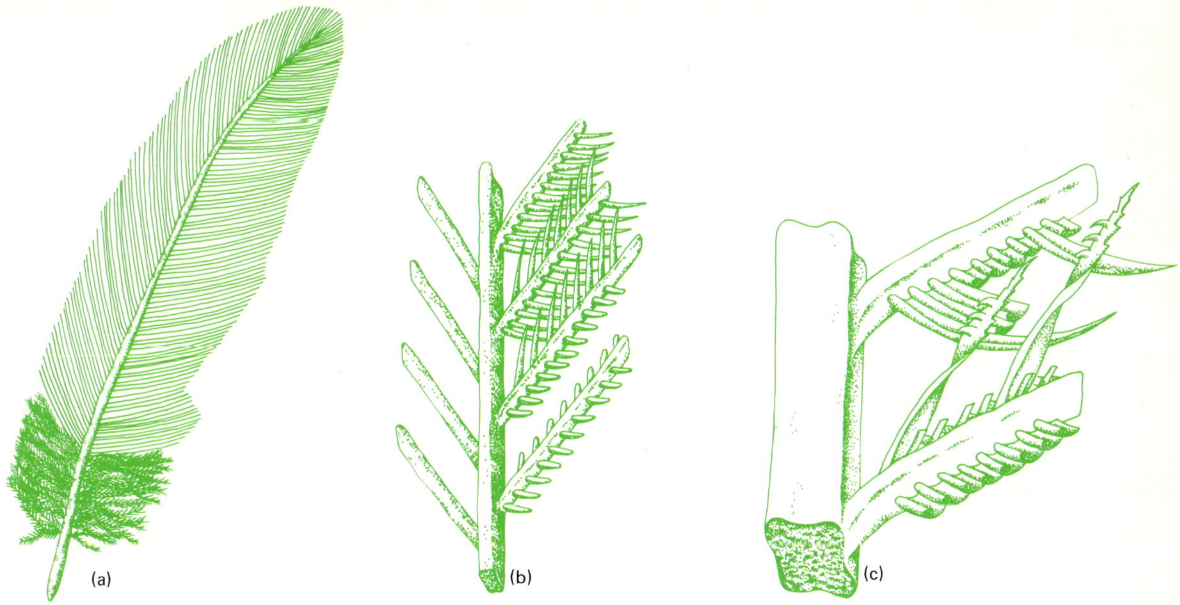

FIGURE 4.17 A feather is composed of a shaft with attached soft, flexible barbs (a). The barbs have smaller barbules attached (b), and these in turn are held together by hooklike projections (c). *(Redrawn: (a) and (b) from* Biology *by Karl Von Frisch. (Harper & Row, 1964), p. 60; (c) from Von Frisch after W. Jacobs.)*

penguins. But birds have become extremely specialized for flight, and they have not evolved a solution to the problem of the vulnerability of their eggs.

Another line of reptiles gave rise to mammals, a class of thermoregulating animals that, with two or three notable exceptions, do not lay eggs. There is no way of knowing exactly when the mammals arose from their reptilian ancestors, and in any event, any such estimation would be artificial. They arose, as did the reptiles from the amphibians, and the birds from reptiles, through the long, slow process of evolution. The original mammals were probably creatures about the size of a mouse or rat that subsisted on plants or insects. Some of the early mammals escaped competition with their contemporaries by taking to life in the trees, and this brought about the evolution of grasping structures at the ends of the animals' limbs. Treetops are a precarious environment, requiring a great deal of coordination of the muscles used in locomotion and an ability to see things in three dimensions so as to perceive the distance from one branch to another. These early mammals began to evolve complex brains to integrate the information of their environment. This set the stage for the evolution of man.

The very earliest mammals, however, were probably all ground dwellers and probably still laid eggs. Nevertheless, the crucial evolutionary adaption of thermoregulation had certain reproductive advantages. As pointed out above, one of the requirements of homeothermy is the minimization of heat loss. It is not known whether early mammals had hair or fur, but all contemporary mammals do, and it serves as an insulation against both heat loss and overheating. In wet climates, though, fur alone is a poor insulator, since water may seep under the fur, reach the skin, evaporate, and thus cool the animal. Hair oil, however, prevents the penetration of water. Thus, the evolution of skin glands that produce oil is favored as a means of attaining thermoregulation. It is thought that at some time during the evolution of mammals, some of the oil-producing skin glands might have become specialized for the production of milk—that is, might have become *mammary glands.* The mammary glands of spiny anteaters (echidnas) and duck-billed platypuses, which are considered to be premature mammals and are called *monotremes*, are very rudimentary. They consist of little more than specialized pockets in the skin from which milk seeps to be lapped up by the young. Both of these species, it is important to note, bear strong resemblances to reptiles in the structure of the bones of their shoulders and pelvises, and *both lay eggs.* Even the reproductive structures of female echidnas and platypuses are very similar to those of reptiles. The eggs are produced in the ovary and are passed into a tubular uterus and thence into the cloaca.

Monotremes share a basic characteristic of all mammals; they care for their young, even during the period when the young are still developing in their eggs. This limits the amount of time the female may devote to gathering food, and less primitive mammals have avoided the egg altogether, thus freeing the female to gather food. In these groups of mammals, the embryo develops to some degree within the body of the female.

The first group that evolved the internal development of the young were the marsupials. Female marsupials do not possess a cloaca, the wastes produced in the digestive system being excreted through an anus. However, female marsupials do possess a urogenital canal through which pass the sperm and living young, as well as the liquid wastes of the excretory system. Sperm deposited in the urogenital canal of a female marsupial pass into a lateral vagina that leads to the uterus. Although there is thus a connection between the lateral vagina and the urogenital canal, the first offspring born to the female in some species of marsupials does not travel through the lateral vagina. Rather the offspring breaks through the walls separating the vaginal sinus and the urogenital canal and thus is expelled from the body. Thus the lateral vagina is used only to deliver sperm to the uterus and is really only a duct for that purpose. In some species of marsupials, the connection that is established between the uterus and the urogenital canal after the first offspring are born heals over, although in some other species it remains open. (See box 4.1.)

Once the marsupial offspring leave the urogenital canal, they make their way into a pouch that has evolved on the bodies of female marsupials. Within the pouch are well-developed nipples to which the young attach themselves firmly—so firmly, in fact, that removing the offspring can sometimes damage the nipples. Clearly the evolution of mammary glands in marsupials is more advanced than in monotremes. Thus freed of guarding her eggs, the female marsupial may engage in food gathering. (See fig. 4.18.)

Marsupials are also more efficient at thermoregulation than are monotremes. In addition to the hair oil glands, they possess a limited ability to perspire, which serves to cool the body. Being more efficient, marsupials radiated into a large number of environments. They were later outcompeted by another form of mammal, but in Australia marsupials were isolated some 60 million years ago when Australia and other land masses were probably severed. Marsupials in Australia have radiated into niches corresponding to those filled elsewhere by mice, rats, dogs, wolves, and bears, and there are marsupial herbivores, carnivores, and omnivores. Elsewhere in the world, the only marsupial that did not become extinct after the next step in mammalian evolution occurred was the opossum.

The evolution of efficient mammals brought to an end a period of evolutionary history that had been dominated for millions of years by reptiles. Precisely why the great reptiles disappeared is still unknown; climate may have been an important factor, but the competition from mammals was almost certainly a factor, and an important one.

The Uterus in Mammals

The female genital organs of marsupials are intermediate between those of reptiles and placental mammals. (Recall that the reproductive organs of the monotremes are basically reptilian.) Among the major differences between the marsupials and the reptiles is that in the former a vagina has evolved for the delivery of sperm. In the marsupial reproductive organs shown, those of an opossum, sperm travel through the lateral vaginae, into the uterus, and fertilize the ovum in the oviduct. The developing embryo passes into the uterus where development continues until the time of birth is near. In some marsupials, after the embryo passes into the vaginal sinus, the sinus expands, extending downwards, and the embryo must break through the walls of the vaginal and urogenital sinuses before birth takes place. Among some other marsupials, the vaginal sinus has evolved into a tubular organ with an opening in the urogenital sinus, thus avoiding such ruptures.

Among placental mammals, the vaginal sinus has disappeared, and the vagina serves for the delivery of sperm and as a birth canal. The most primitive uterus in the placental mammals is the duplex uterus, found in many species of rodents and bats. Most other placental mammals possess bicornuate or bipartite uteruses. Among the higher primates, the fusion of the two uteruses is complete, as seen in the diagram of the simplex uterus. Note also that the higher primates lack a urogenital sinus. In higher primates the urethra, through which liquid wastes are passed, makes no connection with the vagina.

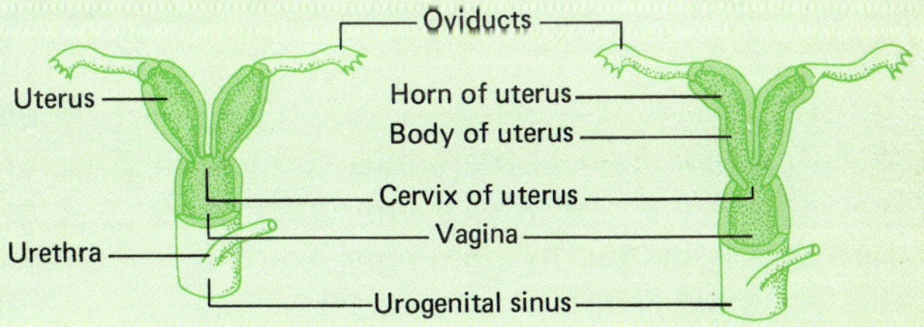

(b) Duplex uterus (c) Bipartite uterus

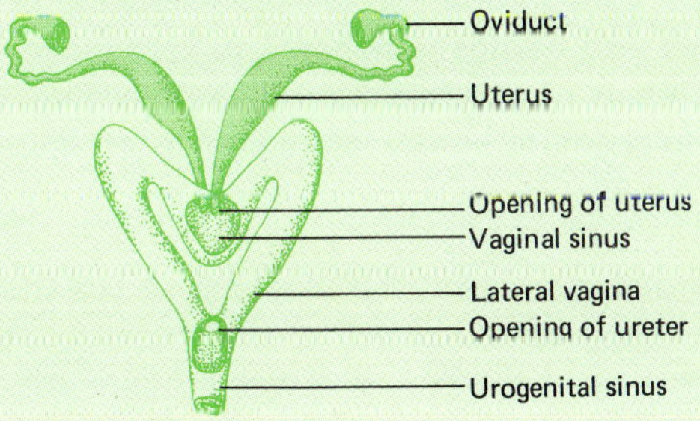

(a) Uterus of primitive mammal, a marsupial

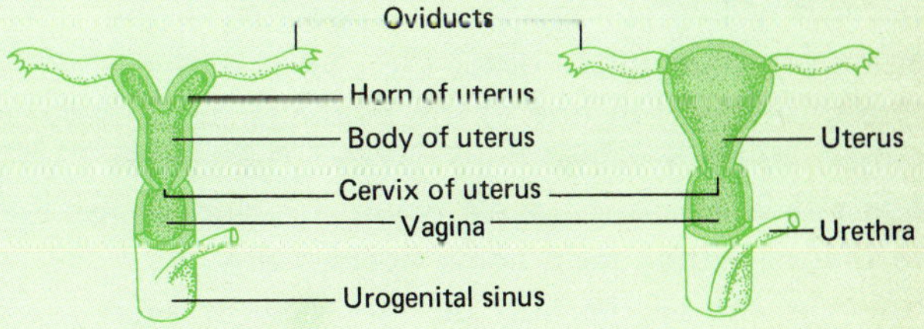

(d) Bicornuate uterus (e) Simplex uterus

(Redrawn from Romer.)

BOX 4.1

FIGURE 4.18 Some marsupials maintain their young to a very late stage of development. This has obvious advantages for protection of the young and mobility of the mother. *(Courtesy of Lisl Dennis from Black Star.)*

The latest development of mammals came with the evolution of the *placenta,* a reproductive structure that is unique to mammals. Female placentates, moreover, do not possess a urogenital canal. Instead, wastes are excreted from apertures different from the vagina, which is itself specialized for the reception of sperm and the birth of live young. Sperm deposited in the vagina of a placentate enter the uterus and then an oviduct. In one of the oviducts, an ovum may be fertilized by a sperm. The zygote begins dividing as it travels down the oviduct to the uterus, although in some species the growth of the zygote may be very slow or even cease temporarily. In any event, the zygote is attached to a yolk sac, and the membranes that characterize an egg are added as the dividing zygote makes its way from the oviduct to the uterus. Once in the uterus, the embryo becomes attached to the uterine wall, and the placenta begins to develop. A combination of tissues contributed by the mother and the embryo, the placenta serves as an exchange organ for nutrients and wastes. Soon blood vessels from the embryo are sent out and connect with the blood supply of the uterus, forming an umbilical cord through which the exchange of nutrients and wastes is henceforth carried on. Some details of the structure of the placenta and of placentate embryos are shown in figure 4.19.

In placental mammals, the placenta serves the same purpose as the pouch of marsupials, but the evolution of the placenta was a major step forward, as it allowed for the birth of relatively well-developed offspring. The descendants of the first placentate radiated into nearly every environment on the earth. However, in Australia marsupials became geographically isolated from competition with placentates. The only placentates that reached Australia until the eighteenth

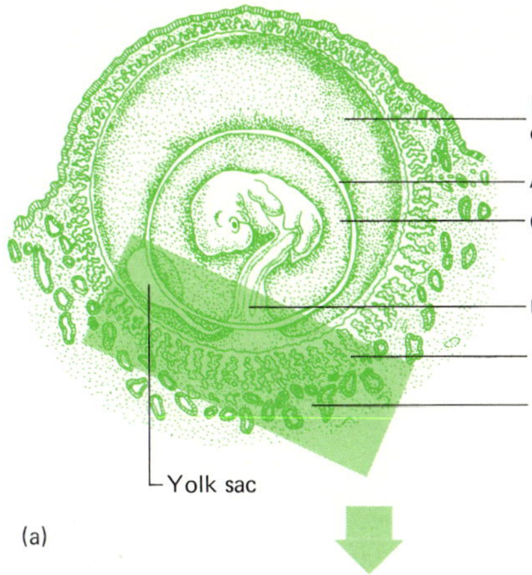

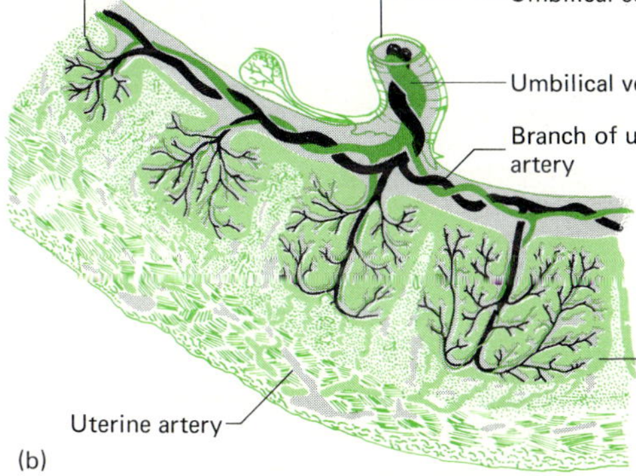

FIGURE 4.19 Shown in (a) is the embryo of a primate within the uterus. In (b), a section of the placenta is shown. The diagram in (b) shows the placenta in several stages of development, with the earlier stages to the left and progressing to the later stages on the right. Note that the maternal bloodstream makes no direct connection with the embryonic bloodstream. The exchange of materials between the mother and the embryo takes place largely by diffusion through the chorionic villi. *(Redrawn: (a) from Romer; (b) from B. M. Patten,* Human Embryology, 3d ed., *copyright © 1968 by McGraw-Hill, Inc., New York. Used with permission of McGraw-Hill Book Company.)*

century A.D. were primitive tribes of man and a small rodent. Placentates also dominated South America until 3 million years ago when the land bridge at the Isthmus of Panama was reestablished. After some 60 million years of separation, the placentates of Central and North America migrated into South America and outcompeted the marsupials there, driving most of them to extinction. (See box 4.2.)

Although thermoregulation was of great evolutionary importance in the success of mammals, another adaptation, parental care of the young, was of equal significance. This adaptation has had three important effects. First, it reduced infant mortality. We have seen that many marine invertebrates must expend enormous numbers of sperm and ova in order to produce a single viable individual. Although the fertilized ova of such animals develop rapidly, the production of gametes and particularly ova in such large numbers is relatively inefficient. Even terrestrial animals such as insects frequently ''waste'' large amounts of energy in producing the numbers of offspring necessary to insure the survival of the species. The care of the young that is characteristic of mammals enables them to utilize the energy available to them more efficiently and to insure that a large proportion of the offspring produced will survive to reproductive age and thus carry on the species.

The second important effect of caring for the young is that it allows learned behavior to be passed from the parent to the offspring. The behavior of the offspring is therefore not wholly determined by hormonal or genetic influences. There are, to be sure, components in the behavior of any mammal that are influenced by genetic or hormonal factors, but there are important components of learned behavior which are associated with brain structure, a topic that we shall examine in the next chapter.

In a sense, the ability to transfer knowledge from generation to generation is analogous to the inheritance of acquired characteristics. It

Placental Orders of Mammals

The class Mammalia is highly diverse, consisting of eighteen orders and some 4500 species that have adapted to life in both aquatic and terrestrial environments. Indeed, fully one-fifth of the species of mammals, members of the order Chiroptera (bats), are flying animals and to some extent compete with birds. By far the largest order is Rodentia, which accounts for nearly half of all mammalian species. Rodents are also probably the most numerous mammals in the world. Shown in this box is a selected group of species representing the major orders of mammals.

Chiroptera (about 900 species) consists of the bats, most of which are nocturnal and live in tropical or subtropical environments. Although there are some species, such as the one shown here, which subsist on fruits and seeds, the majority are insectivores. *(Courtesy of Arthur W. Ambler from National Audubon Society.)*

Insectivora (about 400 species) is probably the order most like the ancestors of the class, which includes among others moles, hedgehogs, and shrews. The tree shrew, shown here, is an interesting animal that has proved a taxonomic puzzle, since it is sometimes included as a member of the primate order and considered as a very, very distant forerunner of man. *(Courtesy of Arthur W. Ambler from National Audubon Society.)*

BOX 4.2

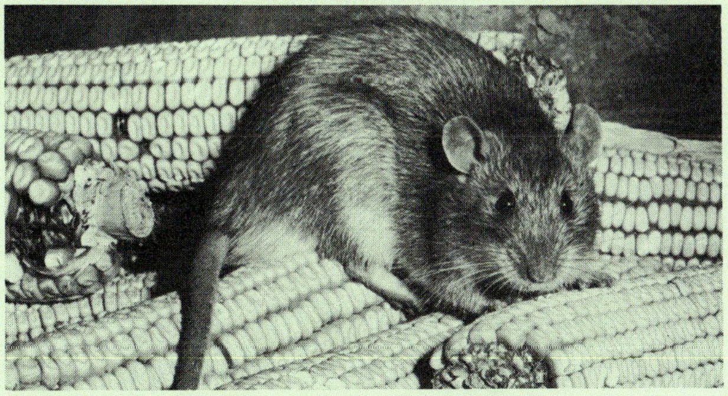

Rodentia (about 2100 species) includes rats, mice, squirrels, porcupines, beavers, and numerous other less well-known examples. Rodents live in very nearly every environment on the earth, from hot to cold and humid to dry. *(Courtesy of Leonard Lee Rue III from National Audubon Society.)*

Carnivora (about 880 species) includes some very familiar species such as wolves, coyotes, raccoons, cats, and seals. As the name of the order indicates, most species are carnivores, although some are scavengers and will eat almost anything. *(Courtesy of Joseph Van Wormer from National Audubon Society.)*

Artiodactyla (ca. 200 species) includes cattle, bison, pigs, deer, antelopes and giraffes. Most of man's domesticated food animals are members of this order of even-toed ungulates, (hoofed mammals) as are some of his favorite game animals. *(Courtesy of Joseph Van Wormer from National Audubon Society.)*

has produced in a relatively brief time (in terms of evolution) the sort of explosive* evolution of a mammal having the greatest capacity for learning—man. The manifold ecological and evolutionary consequences of this ability in man, both for himself and for other species, will be the subject of a large proportion of the later chapters of this book.

The third important effect of caring for the young is a consequence of the second effect and is the evolution of mammalian social behavior. Societies have evolved among insects, but the differences that exist between insect and mammalian societies are crucial and are more striking than the similarities. The key difference is that insect societies have no cognizance of kinship relationships that exist between individuals. Although all of the insects in a nest of honeybees, for example, are directly related to each other, no personal recognition is extended between the individuals. On the average, an adult honeybee lives for only six weeks, and in a population of as many as 80,000 individual workers, the recognition of individuals may simply be impossible. Moreover, the population in a honeybee society undergoes rapid change at all times as large numbers of adults die and are replaced by newly hatched adults. Overall, the conditions that exist in a colony of honeybees make such recognition inefficient. The queen is recognized to some extent, but the basis of recognition is a set of chemicals that she exudes, and it has been shown experimentally that such chemicals can be extracted and absorbed into a dummy of inert material, which may then be "recognized" by the workers in the hive as though it were a queen.

In contrast, mammalian social groups are based on individual recognition of members of the group and often on a knowledge of the relationships that exist between them. A large variety of signals that permit such recognition have evolved, including visual, auditory, and chemical scents. Status in the social group has become important among many mammals, and such signals also serve to define social statuses. The relationships between parents and offspring are specific and frequently serve to bind the social group together. Parent-offspring interactions are also of relatively long duration and allow a process of socialization to take place during which offspring learn, frequently through an activity called play, their status in the group and their status among other individuals of the same age. Although in some mammalian social groups there are "specialists" at defense or foraging, such roles are not usually well defined, and the individuals are never morphologically differentiated from the rest of the group. Finally, all adult members of the mammalian social group can reproduce.

The social behavior of mammals is a subject of considerable complexity and great diversity. Most mammals are small, nocturnal, nonsocial animals. Only rarely are large social groups found among small mammals, the clearest exceptions being species such as prairie dogs that live where food resources, such as grass, are abundant and densely distributed, and mammals that are highly mobile although small, such as bats. One of the reasons for the nonsociability of small mammals is strictly ecological. A small animal of any kind has limited mobility; it cannot travel any distance in search of food that would require the expenditure of more energy than there might be in the food it would obtain. If the animal were twice as large, however, the distance it could travel would be twice as great, and the energy expended would also be twice as great, but the area over which it could range would be four times as large. In opposition to the pressure for larger size thus exerted by increased foraging area is the pressure of the greater probability of encountering predators in the enlarged area. It has been speculated that one of the reasons that larger mammals are more social is that predators may be distracted by larger numbers of individuals and not be able to single out any individual as prey. In

*In the words of one student of evolution, George Gaylord Simpson, evolutionary explosions are conceivable if you can imagine an explosion that takes millions of years to occur and makes no noise.

FIGURE 4.20 A baboon troop in Nairobi Park, Kenya, with the typical grouping of males, females, and young. The dominant males are in the center, females and young nearby, and the subordinate males serve as a vanguard. *(Courtesy of Irven DeVore, Harvard University.)*

fact, many predators that subsist on individuals that are members of social species rarely attack the group; rather, they attack the young that may wander from the group or the old or sick that may be expelled from, or fall behind, a moving group.

A mammalian social group may consist of a single female and her offspring, this being the smallest possible unit that would permit the species to survive, with males and females coming together only for mating and then separating. Another level of complexity may be seen in some species in which several females and their offspring gather into large groups, with the mother and offspring being the basic unit and the males leading separate lives, sometimes in unisexual herds. During the breeding season, the males associate themselves with a female group and defend their positions against other males. Much less common among mammals is a slightly more complex social grouping in which the male is a part of a family unit, and both the male and female provide for the young. Some of the most complex mammalian societies may be found among the order of mammals to which man belongs, primates, in which permanent male-female groups form. These consist of a number of adult males and adult females living in troops with their offspring, and in which the care of the young is a concern of many of the members of the group. In some such groups, the males may act in concert to defend the females and the infants. These groups may number more than 100 individuals, and social conflicts may arise that require the control of behavior among the individuals that make up the group.

As an example of the latter sort of mammalian society, consider a troop of baboons. The baboons of East-Central Africa have been carefully studied. They live in the open savanna, an area characterized by grassy plains and interspersed clumps of trees and patrolled by a number of predators. In this environment, the baboon has evolved into a fiercely aggressive social primate that travels in large troops. (See fig. 4.20.) At the figurative, and often the literal center of a troop of baboons are found the dominant adult males which have achieved their positions through sheer strength and fighting ability, either singly or in particular combinations of individual males. Near them are the females and infants, and still further from the center, but near the dominant males, are the younger juveniles and then the juveniles. At the periphery of the troop are the lower-ranking adult males. This arrangement has not evolved by chance. The dominant males in the center of the troop, by virtue of their position and their ability to defend that position against other males, usually do most of the copulating with females, especially at times when such copulation is most likely to result in fertilization. The males at the periphery have either finished reproducing or have not yet had an opportunity to reproduce. Should a predator attack suddenly, these peripheral individuals are the most likely to be killed. If they are too slow to react, they, and therefore also their genes, will be eliminated from the population. For the most part, these peripheral males serve as guards for the troop as a whole. When a predator is spotted, the peripheral males issue warning barks that alert the troop. The dominant males may then join the subordinates to defend against attack, as the females, infants, and juveniles retreat to the nearest trees. Such groups of male baboons have been observed to drive off potential predators merely by a display of aggression. (See fig. 4.21.) When the troop is not moving, the dominant males occupy the most desirable feeding and sleeping sites, and the subordinate males are left with the less desirable ones.

FIGURE 4.21 It is easy to understand the effectiveness of a baboon's threat display. *(Courtesy of Irven DeVore, Harvard University.)*

Although both baboons and humans are primates and thus share many characteristics, the living primates that are most closely related to man are chimpanzees, which behave quite differently from baboons. We will not examine the behavior of chimps except to note two things about them: *first,* chimpanzees live not in the open savanna, but rather in open forests, an environment containing fewer large predators and more potential places to which the chimpanzee may retreat when threatened; *second,* as a consequence of living in such an environment, chimpanzees, in spite of being highly gregarious, do not live in troops that are rigidly hierarchical. Even troops of baboons that live in the forest environment are much less rigidly structured than savanna-living troops. We shall consider more fully the implications of these behavioral differences when we examine the evolution of man.

Thus far we have discussed the evolution and adaptations of some important groups of animals that have made the transition to terrestrial environments, and we have seen some of the ways in which they have adapted to living on land. Clearly, the transition to land could not have been made by animals had not plants first colonized the terrestrial environment and thereby provided a source of food. Many of the problems that confronted transitional animals are the same as those encountered by the plants that preceded them, and we must now see how plants solved those problems.

Plants on Land

As we saw in Chapter 3, most marine plants are relatively simple. They are supported by the buoyancy of water; water transports nutrients and reproductive products to them and transports waste products away, and the marine plant needs few adaptations to resist drying. Each of these conditions is lacking on land, and thus transitional plants faced several of the same problems that had to be faced by transitional animals.

Rather than begin by tracing the evolution of terrestrial plants from the most primitive to the most advanced, let us instead consider the structure of an advanced plant, a maple tree. Such an examination will reveal a variety of adaptations to the terrestrial environment that may be seen in one form or another in many plants and will enable us then to review more economically some primitive plants and consider how the transition to land may have occurred.

In the terrestrial environment, water and nutrients are present in the soil, but light is present only above the soil; thus plants are required to live both above and below the soil surface. In a maple tree, as in most other plants, the roots provide an immense absorptive surface for the intake of water and nutrients, and the leaves provide a large surface for the absorption of light and gases for photosynthesis. Connecting these two systems is a trunk that provides support. Although roots to some extent also serve to support a tree, their primary function is to serve as entry points of water and nutrients. Consider, however, that the cells in the roots of a tree require the energy that is made available by photosynthesis, and the leaves require the water and nutrients taken in at the roots. The exchange of materials in the tree has been accomplished by the evolution of an internal transport system in the trunk called the *vascular system.* In many plants, the transport system that has evolved consists of two sets of channels in the trunk or stem, one of which, the *xylem,* is specialized for the transport of water and nutrients and the other, the *phloem,* for the transport of energy-rich compounds synthesized at the sites of photosynthesis. Each system is essentially a one-way track for the distribution of materials. (See fig. 4.22.)

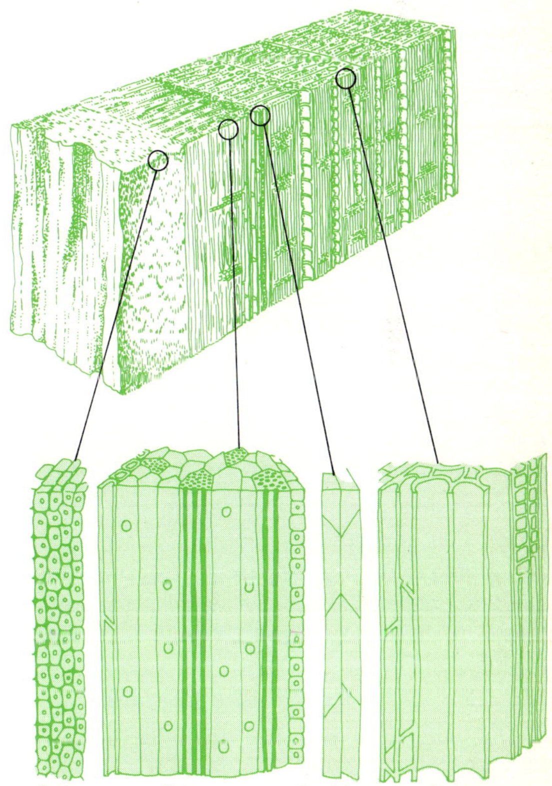

FIGURE 4.22 A tree trunk is protected on the outside by bark. Inside of that, phloem cells conduct water and photosynthetic products of the leaves to the roots. The xylem, separated from the phloem by a cambium layer, conducts nutrients and water absorbed by the roots up to the leaves. *(Redrawn from BSCS, 3d ed.)*

Most of the reproductive adaptations of aquatic plants are relatively simple, but among terrestrial plants a great variety of reproductive adaptations exist. A maple tree, of course, reproduces sexually. Male gametes fertilize female gametes to produce seeds, which are embryos. The pollen of many plants, from which the male gamete is derived, may be transported in two ways: by the wind, in which case the plant has relatively inconspicuous flowers or no flowers at all; or by insects and birds, in which case the plant usually has flowers that are readily apparent and may also produce substances that attract specific insects or birds. In some cases, however, a plant may possess conspicuous flowers and still not depend on another organism for pollination. A pea plant, for example, is self-pollinating, and the pea blossom is constructed in such a way that cross pollination is unlikely.

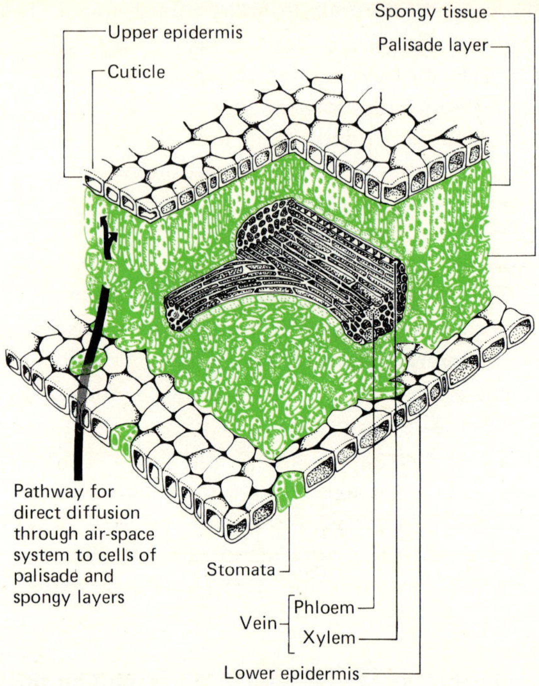

FIGURE 4.23 The structure of a leaf. The leaf is protected by a layer of cells, the cuticle. Photosynthesis takes place in the spongy and palisade layers, which receive nutrients through the xylem and discharge photosynthetic products through the phloem. *(Redrawn from BSCS, 3d ed.)*

While the roots function to provide water for the plant, the leaves serve as organs in which the photosynthetic production of starch takes place and provides energy for the plant. To do this, however, the plant must take up carbon dioxide from the atmosphere and release oxygen and water vapor. The general structure of the leaf is shown in figure 4.23. On the surface of the leaf are numerous small openings called *stomata,* (sing. *stoma*) which serve to admit carbon dioxide and emit oxygen and water vapor. Each stoma is surrounded by specialized *guard cells* that, depending on the water pressure in the leaf, cause the stomata to open and close. When water pressure is high, the stomata remain open; when the pressure falls, the stomata close. The mechanism for opening and closing stomata is shown in figure 4.24. Thus, indirectly, the amount of water in the plant is an important factor in determining how much photosynthesis may take place. Stomata also play a role in controlling the rate at which the plant loses water. The movement of water vapor out of the plant is called *transpiration,* and most of the transpiration of the plant takes place through the stomata. Transpiration can be affected by a number of factors, including heat and wind and, under some conditions, the plant transpires more water than is taken up by the roots. When this occurs, the lack of water in the leaf reduces the water pressure in the guard cells, and the stomata close. As long as there is sufficient soil moisture, the water deficit in the leaves may be made up; the stomata will reopen, and photosynthesis will resume. However, if there is a serious lack of soil moisture the stomata will close, but since water can still escape through the cuticle, the leaf will wilt. (See fig. 4.25.) If dry conditions persist for a significant length of time, the plant will die—not only because the closed stomata restrict the amount of carbon dioxide available in the leaf but also because photosynthesis itself requires water.

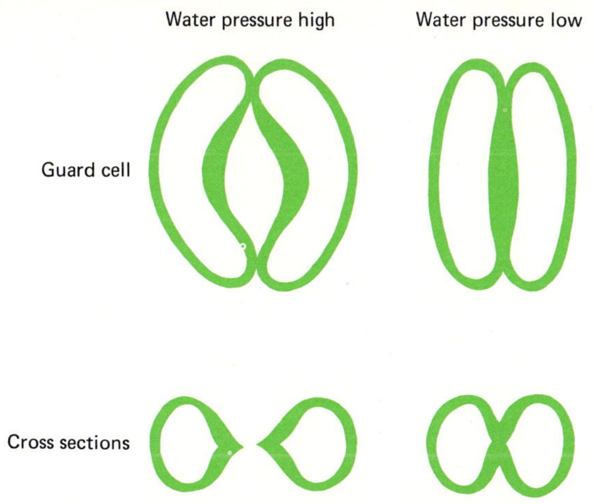

FIGURE 4.24 Stomata open and close by a mechanism that offers us a glimpse of how the form and structure of specialized cells contribute to the function of an organ. (The leaf, of course, is an organ of photosynthesis.) One wall of each of the cells surrounding the stoma is thicker than the other walls and thus is affected differently by water pressure in the cells. When water pressure is high, the thinner walls of such "guard cells" bulge outward more readily than the thick walls, pulling the stoma open. When the pressure falls, the thicker walls have a tendency to "straighten out," thus closing the stoma.

FIGURE 4.25 With a medium less concentrated than the cell fluid, cells absorb water and become turgid. With water loss, cells first wilt, and then contents pull away from the cell wall as the cells plasmolyze. *(Redrawn from* The Living Plant, *Second Edition, by Peter Martin Ray. Copyright © 1963, 1972 by Holt, Rinehart and Winston, Inc. Reprinted by permission of Holt, Rinehart and Winston, Inc.).*

The adaptations of a complex plant such as a maple tree are shown in box 4.3. The same general plan is found in most plants that have a vascular system and produce seeds. Clearly, such a complex organism could not have sprung into existence in its present form, and we must now turn briefly to the evolution of plants.

The earliest plants to make the transition to land are now largely extinct, but certain groups of contemporary plants are believed to be representative of primitive terrestrial plants and to have descended from them. These plants are the liverworts and mosses, *bryophytes,* which are relatively simple when compared with the plants we have been discussing. (See fig. 4.26.) They possess few specialized tissues for support and none for the transport of water and nutrients. Individual cells of a bryophyte can absorb water and nutrients from the air or, by diffusion, from other nearby cells in the plant. For the most part, bryophytes are restricted to living in moist environments such as bogs, and because they lack specialized supportive tissue, they are usually very small, low lying plants. Some bryophytes are capable of absorbing very large amounts of water and of effectively creating a very watery environment in which to live. Bryophytes do possess primitively specialized tissue for the transport of the products of photosynthesis, but when such specialization exists, it

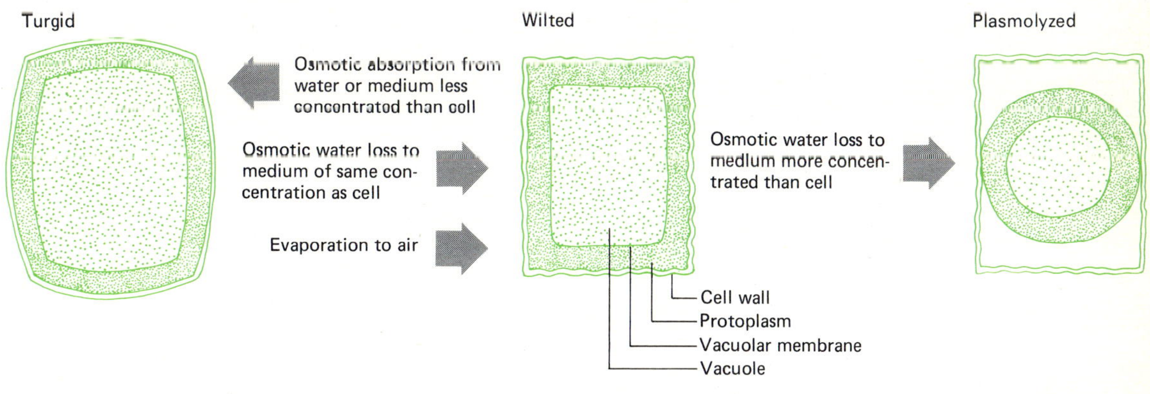

is very crude and cannot be described as a true vascular system.

Reproduction in the liverworts and mosses involves the alternation of generations that we described in the preceding chapter. Unlike the marine plant described there, however, the gametophyte generation is dominant, and the sporophyte generation is always attached to the gametophyte. Moreover, the bryophytes are members of a class of plants called *Embryophyta,* meaning that the embryo forms within the body of the parent plant. Among embryophytes, multicellular sexual organs have evolved—the *antheridium* as the male organ, and the *archegonium* as the female organ. (See fig. 4.27.) Both of these organs are frequently found on the body of the same plant, although in some cases there are distinct male and female plants. The life cycle of a bryophyte is shown in figure 4.28. In the plant body of the bryophyte shown (*Ricciocarpus*), small furrows exist wherein are found the reproductive organs. The sperm that are released from the antheridium are transported to the archegonium by water that collects in these furrows. When a sperm reaches an egg and fertilizes it, the resulting embryo undergoes a number of mitotic divisions. A globular mass of identical cells arises, but soon the outermost cells form a capsule that encloses the whole

FIGURE 4.26 *Marchantia*, a liverwort, showing the vegetative body or thallus and the star-shaped spore producing organs. *(Courtesy of Carolina Biological Supply Company.)*

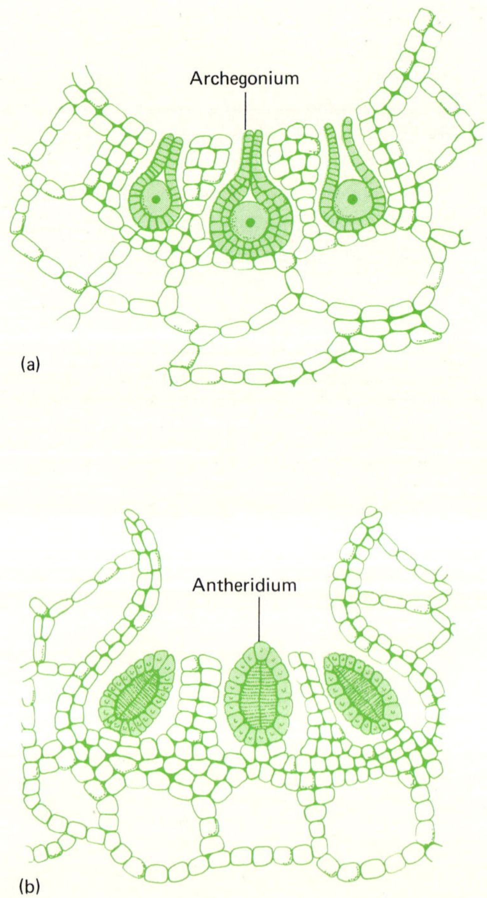

FIGURE 4.27 Antheridium and archegonium in an embryophyte.

Transport in Plants

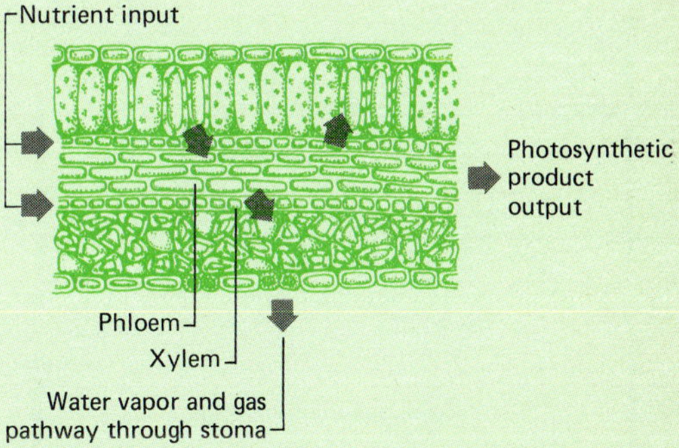

The leaf makes possible an exchange of water vapor, oxygen, and carbon dioxide with the atmosphere.

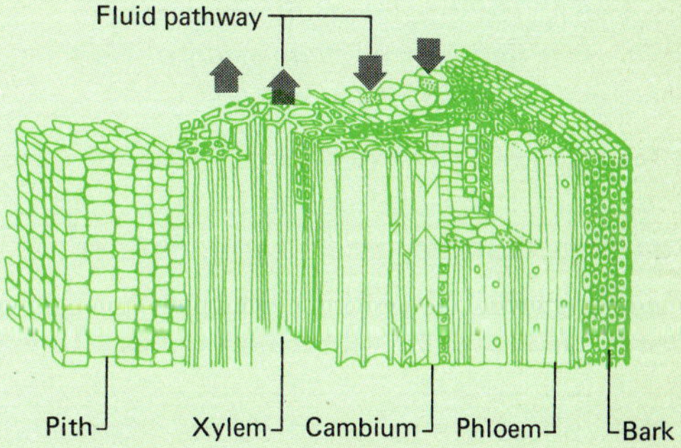

The xylem and phloem transport materials from the soil and materials from the leaves back and forth between the roots and the tree crown.

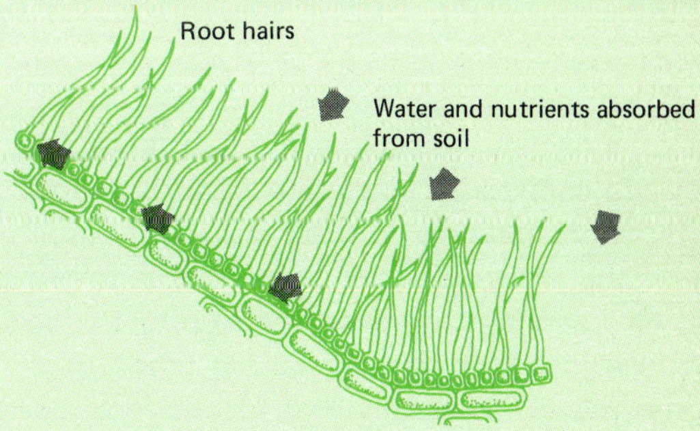

Roots extend into the soil to absorb water by osmosis across the root cell membranes. Root hairs vastly expand the amount of surface of the roots. One four-month-old rye plant, for example, was found to have roots that measured a total of 387 miles in length and presented a surface of 6875 sq. ft.

BOX 4.3

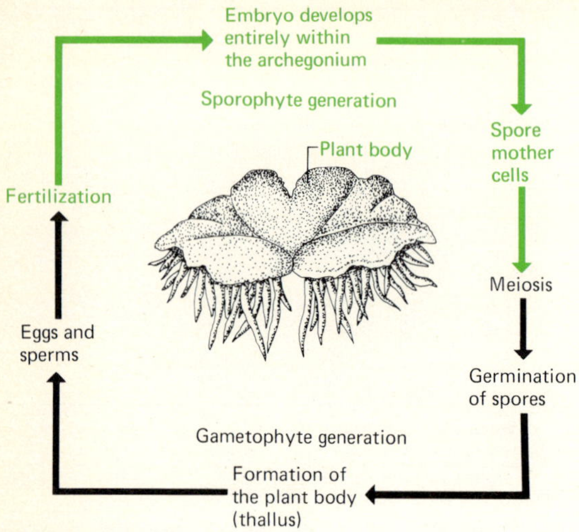

FIGURE 4.28 The life cycle of *Ricciocarpus. (From* Botany, *5th ed., by Carl L. Wilson, Walter E. Loomis, and Taylor A. Steeves. Copyright 1952, © 1957, 1962, 1967, 1971 by Holt, Rinehart and Winston, Inc. Redrawn by permission of Holt, Rinehart and Winston, Inc.)*

embryo within the archegonium. The cells of the embryo then become ''spore-mother cells,'' or cells from which the spores are derived. Each spore-mother cell divides twice to produce four spores, the capsule breaks down, and the spores are released as the plant body decays. In the next growing season, the spores germinate, and the cycle begins anew.

Although they are common in many parts of the world, bryophytes are limited to very moist environments, or to growth during a moist season of year. Many species of bryophytes, such as Sphagnum moss (often called peat moss), are of considerable ecological importance, as they are capable of absorbing great amounts of water, thus creating covered bodies of water in which the decay of other organisms is promoted and also serving to prevent erosion of the soil. Because they lack the specialized tissue of higher plants, especially that for the internal transport of water, the sizes which they may achieve are very limited. Of equal importance is the fact that the bryophytes depend upon water for the dissemination of the products of reproduction—either sperm or spores. Although they have achieved some independence from the aquatic environment, the bryophytes are *not* entirely terrestrial.

Vascular plants, having tissue specialized for the transport of water and for the transport of the products of photosynthesis, have been much more successful. Most vascular plants are less dependent on the presence of water for the dissemination of their reproductive products and thus have been able to radiate into a greater proportion of the terrestrial environment than the bryophytes.

There are four surviving groups of vascular plants; the *Psilopsida*; the *Lycopsida* (club mosses); the *Sphenopsida* (horse tails); and the *Pteropsida*. We will not consider the first three here, for although they do represent varying degrees of successful adaptation to the terrestrial environment, they are nevertheless evolutionary dead ends. They do possess vascular systems and rudimentary roots but, on the whole, have remained relatively primitive.

The other group, Pteropsida, includes most of the plants with which we are familiar and nearly all the plants that are of great importance to man, so it is necessary to examine it in greater detail. The simplest of these plants are the ferns. Ferns possess fully developed vascular systems and thus can grow larger than the club mosses, horse tails, and Psilopsida. In the tropics, ferns may reach as much as 60 feet (20 meters), although the ferns more commonly found in North America are much smaller and rarely exceed 3 or 4 feet (about one meter) in height. The life cycle of a fern includes alternating generations, and spores are most frequently grown on the underside of the leaf of the fern. When a spore germinates, a small, heart-shaped plant grows, bearing male and female sex organs and producing the gametes from which the familiar *fronds* of ferns grow. (See fig. 4.29.)

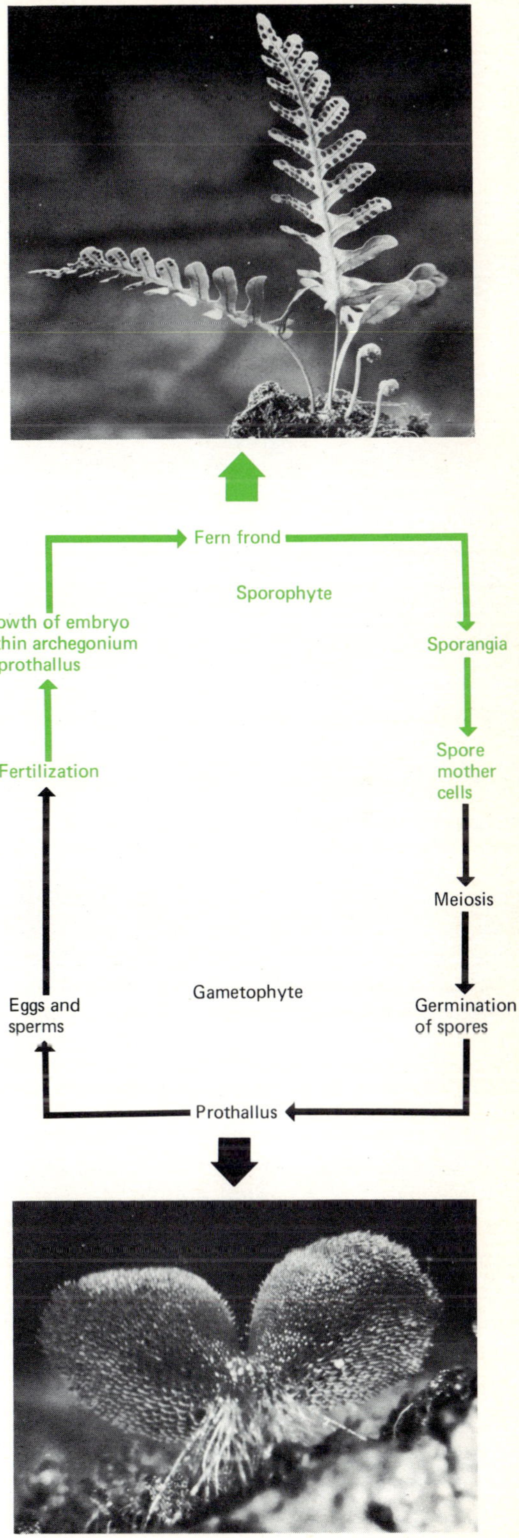

FIGURE 4.29 The life cycle of a fern. *(Courtesy of Hugh Spencer.)*

Far more advanced and far more important both ecologically and economically are the seed-bearing gymnosperms and angiosperms. The term *gymnosperm* means "naked seed" and refers to the exposed position of the seed in the female reproductive organ. The most modern and successful gymnosperms to have evolved are the conifers: pines, spruces, firs, cedars, hemlocks, and sequoias. The reproductive organs of the conifers are the cones which characterize the order and from which the order is named. In most species of conifers, each tree bears both pollen cones and seed cones. The smaller male cones expel their pollen to be distributed by the wind to the seed cones in a way analogous to the shedding of sperm in water. Of the billions of pollen grains scattered, only a few ever settle on a seed cone to fertilize an ovum. Each scale of a female pine cone bears two ovules. The ovules of pines contain a small amount of pollination fluid suspended near the end of the ovule that serves to retain pollen that reaches the seed cones. After pollination, the liquid withdraws into the ovule, carrying the pollen grains to a small pore, the *micropyle,* at the end of the ovule. In the micropyle, the pollen germinates, producing pollen tubes that grow through the tissue of the ovule. The sperm then move through the pollen tubes to fertilize the ovum. Usually, more than one pollen grain reaches the micropyle, but after the first ovum is fertilized, any other sperm or ovum in the ovule degenerates as the embryo, or seed, develops. Among species of pine, this seed then develops a seed coat and a wing, is released from the cone, and drifts to earth on the wind, at a distance several feet from the parent tree. The whole process of reproduction, from the production of pollen and ovule to the falling of the seed, takes two growing seasons, since the ovum is not fertilized until almost a year after pollination. (See figs. 4.30 and 4.31.)

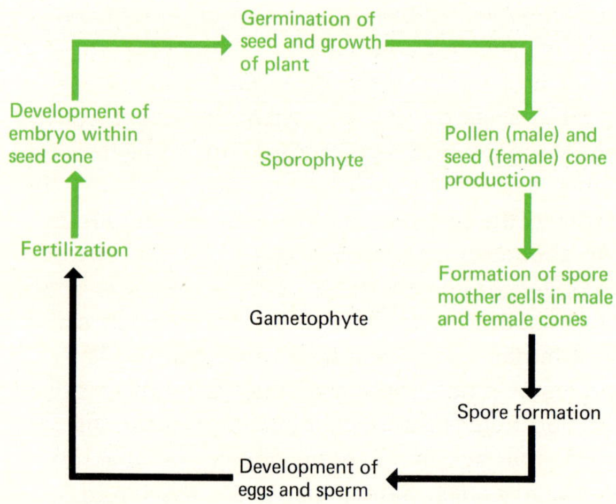

FIGURE 4.30 The life cycle of a white pine. *(Photo courtesy of Hugh Spencer.)*

Another feature of the conifers is the needlelike leaf that characterizes most of the families in this order. As shown in figure 4.32 the pine needle consists of a central bundle of cells in which the vascular system is found. Although it is not exactly homologous to the xylem and phloem system that ramify into the leaves of the maple, the system in the pine needle serves essentially the same functions with different groups of cells specialized for the transport of different materials. Surrounding the central vein of the needle are the cells in which photosynthesis takes place. The outer covering of cells is hard but possesses stomata. The pine needle is very resistant to freezing, as are the leaves of most conifers, and thus these plants are able to carry on photosynthesis throughout the year in cold climates. The needles are shed regularly, but not all at once as in maples, and conifers are thus termed "evergreens."

Although gymnosperms are widespread and of great economic importance to man, the *angiosperms* are perhaps even more widely distributed and of greater importance, since very nearly all human nutritional needs are met by them.

FIGURE 4.31 (a) The seed-bearing female cones of white pine (*Pinus strobus*), blown to the ground in a storm. Seeds from these cones will allow regrowth of the damaged forest. (b) The small pollen-bearing male cones of the same species. *((a) Courtesy of United States Department of Agriculture; (b) courtesy of Grant Heilman.)*

Like gymnosperms, angiosperms are seed-bearing plants, but they possess a unique reproductive adaptation, the flower. It must be said here that all angiosperms possess flowers or reproductive structures that are based on the same general plan as a flower, although the blossom that we usually designate as a flower may be very inconspicuous or even totally unrecognizable to all but the trained observer. Another important feature of the angiosperms is the fruit, which is an adaptation to the problem of dispersing seeds and is based upon the structure of the flower.

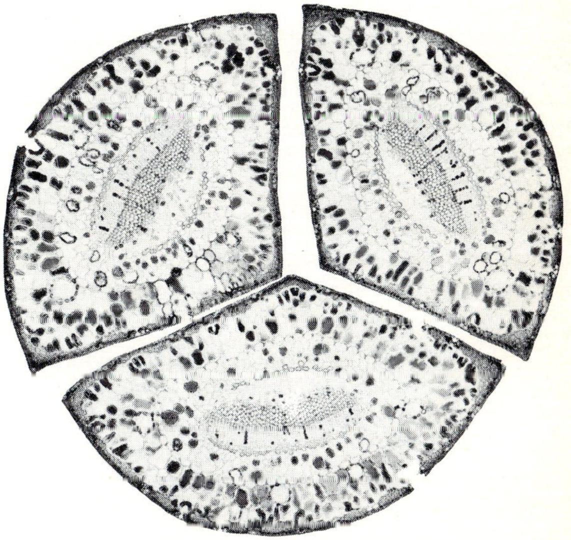

FIGURE 4.32 A cross section of a bundle of three pine needles. The central vascular tissue transports materials in and out of the needle. The whole structure is protected by a hard layer of outer cells. *(Courtesy of Carolina Biological Supply Company.)*

Let us look first at the generalized structure of a flower. (See fig. 4.33.) At the base of the flower are the *sepals,* which are believed to have evolved from leaves and which cover the bud as it is developing in the plant. The petals are often the most conspicuous parts of a flower. They are frequently colored and sur-

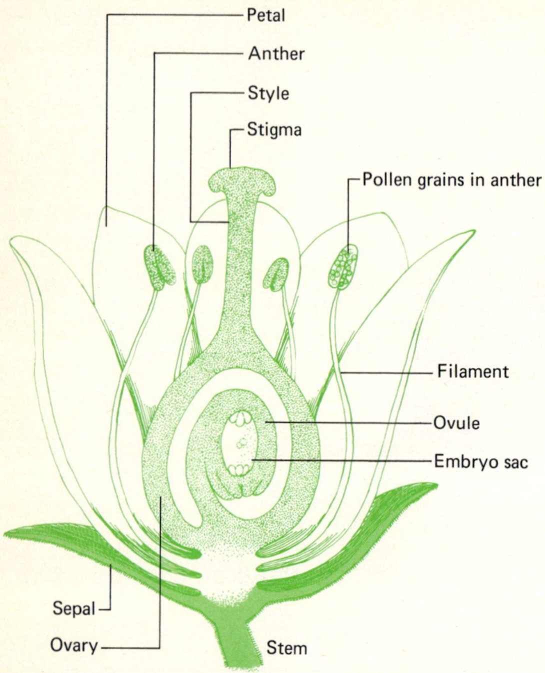

FIGURE 4.33 Generalized structure of a flower. *(Redrawn with permission of Macmillan Publishing Company, Inc., from* Development in Flowering Plants *by John G. Torrey.)*

round the actual reproductive structure of the angiosperm. Within the petals are the male sexual organs, the *stamen,* consisting of a stalk—the filament—on which is located the *anther*, the pollen-producing organ of the plant. The female organ is called the *carpel* and consists of the *stigma*, where pollen settles, the *style*, through which pollen must travel, and the *ovary*, wherein are contained the *ovules*. The carpel has been likened to a leaf that has been folded to enclose the ovary, and indeed, in plants with a single carpel, a cross section view supports this suggestion. The number of sepals, petals, stamen, and carpels in a flower varies from species to species, but they are usually present in specific ratios, as shown in box 4.4.

An important difference between gymnosperms and angiosperms lies in the means by which pollination occurs. In many species of angiosperms, the pollen is carried from plant to plant not by means of the wind, but by various insects, birds, or other animals. In many angiosperms, the color of the petal attracts specific insects to feed on the pollen, which is a rich source of protein. As they are feeding, however, some pollen is deposited on the insect body. When the insect alights on another plant, the pollen is rubbed off on the stigma, and the plant is pollinated. Some angiosperms produce a sugar solution called *nectar,* the smell of which attracts specific insects that pick up a load of pollen when they come to drink the nectar. The pollen is then deposited on another plant as the insect feeds there. Some plants have evolved forms that strongly resemble female insects and that attract male insects, which in attempting to copulate with the plant, receive a deposit of pollen to carry to another plant.

The pollen of angiosperms, like that of gymnosperms, are spores. One of the chief differences between the reproductive adaptations of these two classes, however, is that when pollen is deposited on a stigma of an angiosperm, the whole process of seed production is very rapid. In some species gametes are produced from the gametophyte and fuse to begin seed production in a matter of hours, while in others this occurs within a matter of days. Overall, however, reproduction is much more rapid in the angiosperms than in the gymnosperms, and this in part accounts for their success. The life cycle of some angiosperms is shown in box 4.5.

In many angiosperms, after fertilization has taken place, the ovary develops into a fruit containing the seeds. The fruit may be fleshy and moist, as in oranges, cucumbers, or pears, or it may be dry, as in the pea, corn, or peanut. The evolution of fruit-bearing angiosperms provided both protection for the seeds and a means of dispersing the seeds. Some fruits ripen and burst, thus spreading their seeds, but some edible fruits can be even more widely dispersed. Animals may be attracted by a fruit and eat it. The seeds pass unharmed through the digestive system of the animal and are deposited in its feces, frequently at great distances from the parent plant. Some angiosperms produce burrlike fruits, which may adhere to the coat of a passing animal and thus be carried many miles from the parent plant. Another means of dispersal may be seen in the coconut. The fruit produced by a coconut palm consists of a hard, leathery shell within which is enclosed a single large seed. The coconut sold at a supermarket is not the fruit, but the seed. The large outer shell is filled with a fibrous tissue that permits the fruit to float. After the fruit falls from the tree, it may be washed out to sea and carried by ocean currents to a location a considerable distance from the parent plant. As the seed germinates, it grows out through a pore in the hard inner-shell and through the fibrous covering until it reaches the air; it then puts down roots, sends out leaves, and becomes established as a young coconut palm. (See fig. 4.34.)

The seeds of all angiosperms possess *cotyledons,* or seedling leaves, which serve either for photosynthesis or for food storage. It is on the basis of the cotyledon that the angiosperms are divided into two classes: *monocotyledons,* or plants with one cotyledon, and *dicotyledons,* or plants with two cotyledons. The differences between monocots and dicots, as they are called, are manifold, as shown in box 4.4. The dicots are by far the larger group, containing over 200,000 species. The monocots contain over 50,000 species, although some, like the grasses, are very successful. It is believed that the monocots are the more advanced class and include such important families of plants as the lilies, the grasses, the palms, and the orchids. One order of the monocots, the grasses (Graminae), includes approximately 7000 species and supplies the basic food resource for the human population. Wheat, corn, barley, oats, rye, and rice are all species of grasses. Many of the animals that man raises for food rely primarily on species of grasses for their growth.

It would be impossible in a book of this size to survey the entire plant or animal kingdom and to deal with all of the adaptations that have arisen as provisions for survival. We must look more closely at the ways in which the adaptations that have allowed species to survive have been expressed in the physiological mechanisms of animals and plants. We can then turn to the subject of genetics in order to see how these mechanisms are transmitted from generation to generation. With a basic understanding of the mechanisms of physiology and genetics, we will then be in a position to understand evolution in detail and to see how species, populations, and communities interact with, and are adapted to, the world.

FIGURE 4.34 A germinating coconut. *(Courtesy of G. R. Roberts.)*

Monocots and Dicots

Generally, dicot flowers possess floral organs in fours and fives, or in multiples of four and five. In some cases there may be a large number of stamens. Among the monocots, the floral organs are present in threes or multiples of three. *(Photos courtesy of Grant Heilman.)*

The leaves of dicots are usually more sharply lobed than are the leaves of monocots, and have veins radiating from a central vein. Monocots, for the most part, possess leaves with smoother edges and parallel veins. *(Photos courtesy of Grant Heilman.)*

BOX 4.4

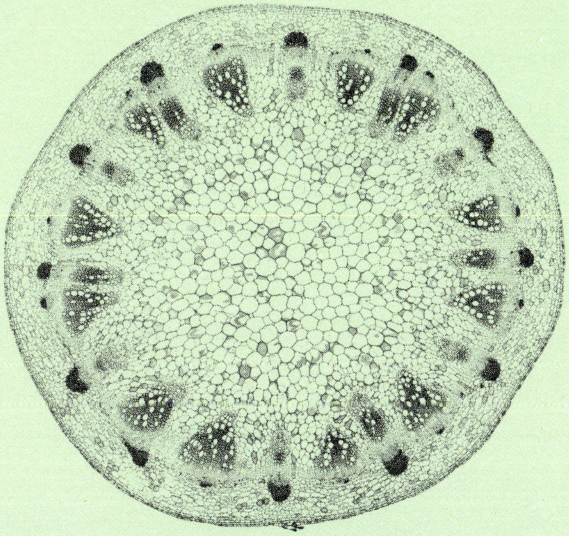

The stem of a dicot has the vascular system organized in a ring around its outer perimeter. In the monocots, vascular bundles are distributed throughout the stem. *(Photos courtesy of Carolina Biological Supply Company.)*

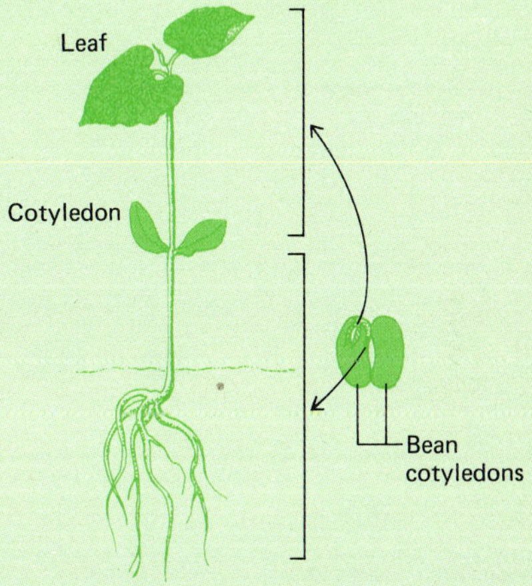

Seeds of dicots, have two cotyledons, whereas those of monocots have only one. *(Redrawn from BSCS, 3d ed.)*

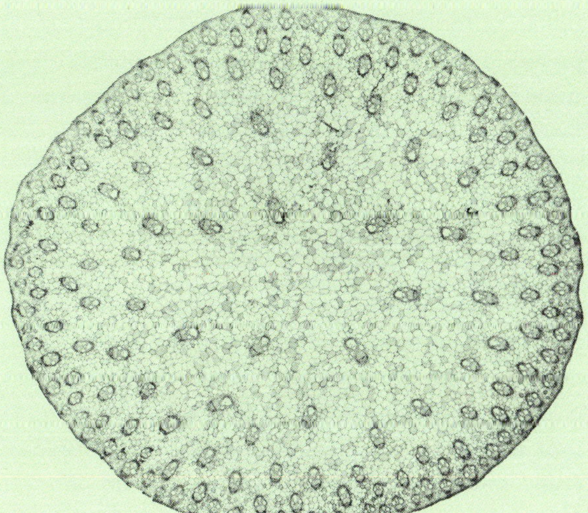

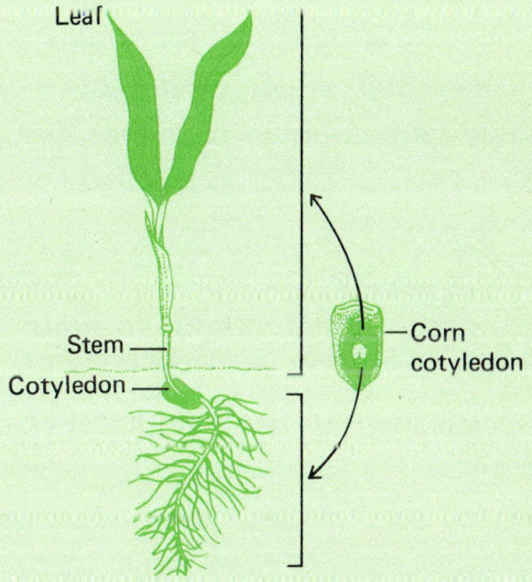

Although some differences exist between the roots of monocots and dicots, only the differences between the flowers, leaves, stems, and seeds are shown.

Life Cycle of an Angiosperm

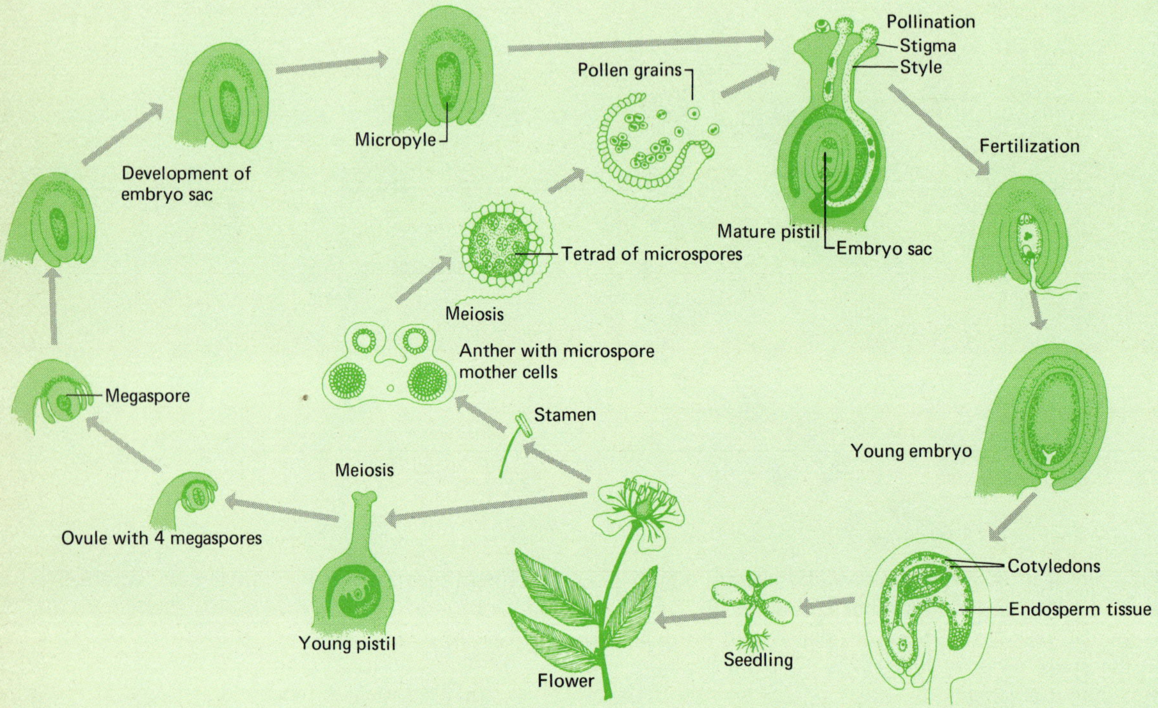

In the developing flower of an angiosperm an ovule arises within which certain cells give rise to a female gametophyte or *megaspore.* This megaspore divides twice to give rise to four cells, three of which deteriorate as the embryo sac develops within the ovule. The remaining megaspore then undergoes three more divisions giving rise to eight nuclei, not all of which necessarily are contained within cell walls; they may exist as free nuclei within the embryo sac. Two of these free nuclei become so-called *polar nuclei;* three remain at one end of the sac and apparently serve no function. The other three remain at the end of the embryo sac where the micropyle is located, and one of these three becomes an egg. The other two nuclei at the micropyle end of the embryo sac appear to be functionless. Meanwhile, within the anther of the flower, although not necessarily the same flower or the same plant, pollen is formed from *microspores.* When the pollen is deposited on the stigma, a pollen tube grows through the style, and sperms are formed in the pollen, two of which penetrate the micropyle and enter the ovule. One of the sperms fertilizes the egg, while the other migrates to the polar nuclei, which by now may have fused. If the polar nuclei have not fused by the time fertilization occurs, they fuse with the sperm. In any case, the fertilized polar nuclei begin dividing to form what is called the *endosperm tissue,* which becomes an important source of food for the developing embryo. This ''double fertilization'' appears to be necessary for the complete development of a seed. As the embryo develops, the *cotyledon,* or seedling leaf, develops, which later will serve as an organ of photosynthesis in the developing seed, or as a supplement of the food supply formed by the endosperm tissue.

BOX 4.5

Selected Readings

*Delevoryas, Theodore. *Plant Diversification.* New York: Holt, Rinehart and Winston, Inc., 1966. The emphasis in this book is on selected topics in the evolution of plants.

*Eibl-Eibesfeldt, Irenäus. *Ethology: The Biology of Behavior.* New York: Holt, Rinehart and Winston, Inc., 1970. A superb book on animal behavior by one of the foremost ethologists in the world.

Evans, Howard Ensign. *Life on a Little Known Planet.* New York: E. P. Dutton & Co., Inc., 1968. A very personal account of the insects, by an entomologist who truly enjoys his work and can convey that enjoyment to the general reader.

Gordon, M. S., Bartholomew, G. A., Grinnell, A. D., Jorgenson, C. B., and White, F. N. *Animal Functions: Principles and Adaptations.* New York: The Macmillan Company, 1972. A solid text emphasizing vertebrate organ systems and the relationships of physiology to the external environment.

*Ray, Peter M. *The Living Plant.* New York: Holt, Rinehart and Winston, Inc., 1972. A brief, good survey of plant biology stressing the physiological activities of plants.

*Romer, Alfred S. *The Procession of Life.* New York: Doubleday & Co., Inc., 1972. A book for the general reader. Romer covers both invertebrate and vertebrate evolution from a relatively nontechnical point of view.

Romer, Alfred S. *The Vertebrate Body.* Philadelphia: W. B. Saunders Company, 1970. In this standard text, Romer treats vertebrate evolution and comparative anatomy more technically, but in a clear and easily read manner.

*Schmidt-Nielsen, Knut. *How Animals Work.* New York: Cambridge University Press, 1972. Although dealing with the animal kingdom as a whole, this imaginatively written work stresses vertebrates.

Wilson, Edward O. *The Insect Societies.* Cambridge, Mass: Harvard University Press, 1972. This comprehensive treatment of the natural history, physiology, behavior, population dynamics, and evolution of social insects has been called a masterpiece. Of particular interest is the final chapter, "The Prospect for a Unified Sociobiology."

Bogert, Charles M. "How Reptiles Regulate Their Body Temperature." *Sci. Am.*, Vol. 200 (April, 1959), pp. 105–108. Offprint number 1119.

Johansen, Kjell. "Air Breathing Fishes." *Sci. Am.*, Vol. 219 (October, 1968), pp. 102–111. Offprint number 1125.

Taylor, T. G. "How an Eggshell Is Made." *Sci. Am.*, Vol. 222 (March, 1970), pp. 88–95. Offprint number 1171.

Wigglesworth, V. B. "Metamorphosis, Polymorphism, Differentiation." *Sci. Am.*, Vol. 200 (February, 1959), pp 100–102. Offprint number 63.

*Paperback edition available.

(Diagram adapted from Botany, Principles and Problems *by E. W. Sinnott and K. S. Wilson, Copyright © 1963 by McGraw-Hill, Inc., New York. Used with permission of McGraw-Hill Book Company. Drawing in lower right showing the developing embryo redrawn with permission of Macmillan Publishing Company, Inc., from* Development in Flowering Plants *by John G. Torrey.)*

The sensitive binocular vision of an owl, combined with its other sense organs, central nervous system, and locomotor systems, make it a very effective predator. The physiological processes required for energy-gathering and transport of materials throughout the body are basic to an organism's ability to survive within its environment. *(Courtesy of Eric Hosking from National Audubon Society.)*

the physiological mechanisms of survival

chapter five Clearly, many of the adaptations that have enabled plants and animals to survive are the evolved responses to the conditions of the environment. As conditions vary, for example between aquatic and terrestrial environments, correspondingly different adaptations have evolved, even in some closely related species. However, many of the conditions of different environments are remarkably similar and have therefore produced similar adaptations even in totally unrelated species. Regardless of whether environmental conditions are alike or not, certain very basic problems exist for all species. For example, all species must obtain food, convert it to available energy, and reproduce. We have seen some of the means whereby these problems have been solved by some species of plants and animals, but for the most part, we have examined only the external manifestations of the solutions. To comprehend the complexities of evolution more fully, it is helpful to understand something about the internal workings of organisms—that is, their internal structures and how those structures operate to permit the organisms to survive.

Within the scope and purpose of this book, it would be impossible to carefully examine *all* of the anatomical and physiological mechanisms that contribute to survival. Thus, we have chosen to consider but a few, although perhaps the most basic and important ones. The first that we shall consider involves the systems concerned with the acquisition of, and response to, information about the environment. We shall then discuss some of the systems involved in the transport of materials within the bodies of animals.

Information and Behavior

Simple organisms, such as protozoa, usually live in environments where the supply of food is continuous and abundant. Like sedentary filter-feeders, such organisms do not need complex bodily systems to aid in the search for food. As the dimensions of the problem grow more complex, as when food is available only intermittently or in a dispersed fashion, organisms exhibit more complex adaptations. And, as the activity associated with the search for food increases, so too does the likely number of contacts with predators. Thus, the location and nature of food and the absence or presence of predators are pieces of information that can be crucial to survival.

The Use of Environmental Information

Possessing information is not, in itself, sufficient to insure survival; it must be acted upon. Therefore, the organism must have some means of responding to information. The ability to gather and respond to environmental information has been provided for by the evolution of nerve and muscle systems.

Environments are saturated with information in the form of radio waves, x rays, and other electromagnetic phenomena; with sounds, temperature information, and changes in air and water pressure; and with changes in the chemical composition of air and water. No animal could possibly take in all of the information available, nor would it need to. It is necessary, however, that an organism extract the information required for survival, which is usually only a small portion of all that is available. The evolution of adaptations that would serve to increase the amount of information gathered would increase the probability of survival only if the adaptations were accompanied by the evolution of other adaptations that permitted appropriate responses to the increased information.

It should be understood that the environmental information crucial to survival is not restricted to that which concerns the external environment. The state of the internal environment of an organism is equally important. While most information about the state of the external environment is gathered and responded to by nerve and muscle systems, much of the information about the internal environment of an animal is handled by the endocrine system. Although nervous systems and endocrine systems operate together to coordinate and control environmental information and the responses thereto, we shall discuss only nerve and muscle systems.

Reception, Transmission, and Response

Basic to an understanding of the evolution and operation of nervous systems is an understanding of *irritability*, a universal characteristic of protoplasm, which is the living substance of all cells. Defined as a *capacity to respond to a stimulus*, irritability is basically a chemical phenomenon, and may be broken down into three steps.

The first step is the *reception of a stimulus*. Any change in an environment can be a stimulus, but not all environmental changes are stimuli to any one organism. The chemical nature of the protoplasm determines whether environmental changes are stimuli, and thus two organisms in the same environment may both be subjected to the same change in conditions, while only one of them may respond.

The second step is the *transmission of a signal*. If the protoplasm is stimulated at any point—that is, if a change is effected in the protoplasm by the reception of a stimulus—the change will usually be detected by the surrounding protoplasm, which will then also change. The repetition of this process may ultimately involve all of the protoplasm in the cell. This simple mechanism for transmitting a signal is thought to be the basis upon which complex nervous systems were organized.

The third step is the actual *response by an effector*. Since the reception and transmission of stimuli are carried out in simple organisms by the same material that makes the response, this step is usually easier to see in complex organisms in which the effectors are usually specialized cells. The action produced as a response to a stimulus differs from one organism to another, and even the same organism may respond differently to the same stimulus at different times. A predator, for example, will respond to the sight of its prey quite differently when it is hungry than when it has just finished a meal. Thus, in large measure, the response depends upon the *state* of the organism rather than the *nature* of the stimulus. (One may argue, of course, that the sight of prey is not a stimulus to a satiated predator, but although there is no *apparent* response, the fact that the prey is *seen* is a biological response and can be described in terms of chemical changes, as we shall see.)

The Evolution of Nervous Systems

Although some ciliates, such as *Paramecium,* possess a bodily network to coordinate the movement of cilia (see fig. 5.1) and some other protozoa possess some sensory organelles specialized for the reception of stimuli, we must turn to organisms of a more complex nature to see how nervous systems may have evolved and how they function. Before doing so, however, we should briefly outline the kinds of cells that have evolved to perform the three steps discussed above.

Cells specialized to sense environmental changes are called *receptors*. In a complex or-

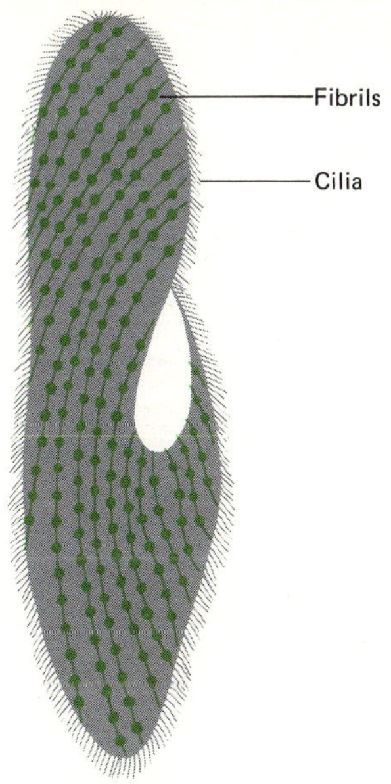

FIGURE 5.1 The movement of cilia in *Paramecium* is controlled by a network of fibrils. *(Redrawn from BSCS, 3d ed.)*

ganism there may be a number of different kinds of receptors. Those which receive light stimuli are called *photoreceptors*. *Mechanoreceptors* receive stimuli that take the form of movement, such as touch or sound. Information about the properties of molecules in the environment is sensed by *chemoreceptors*. Taste and smell are examples of information sensed by chemoreceptors. Information about the internal environment is supplied by both chemoreceptors and mechanoreceptors, and in animals with muscles, it is in part supplied by a form of mechanoreceptors called *stretch receptors,* which are adapted to indicating the positions of muscles at any given time.

The transmission of the information gathered by receptors is carried out by *neurons*. Although all neurons are thought to function in essentially the same way, they exhibit a high degree of morphological diversity. (See fig. 5.2.) Each consists basically of a cell body containing the nucleus and a variable number of fiberlike projections that extend from the cell body. The longer projections, or *processes* as they are usually called, are the *axons*. The shorter and usually more numerous processes are *dendrites*. With some notable exceptions, the stimuli received by neurons are delivered to the dendrites and are then transmitted to the next neuron (or an effector) along the axon. The point of contact between two neurons is called the *synapse*. (See fig. 5.3.) Neurons do not make direct contact with each other; rather, they terminate at the synapse, at a very short distance from each other, and the space across which the stimulus is transmitted is called the *synaptic cleft*.

For the most part, the only *effectors* that we shall consider are muscle cells, although in some of the simpler organisms there are no muscle cells. In such organisms, certain cells are capable of contracting, and these *contractile cells* function analogously to the muscle cells of more complex organisms.

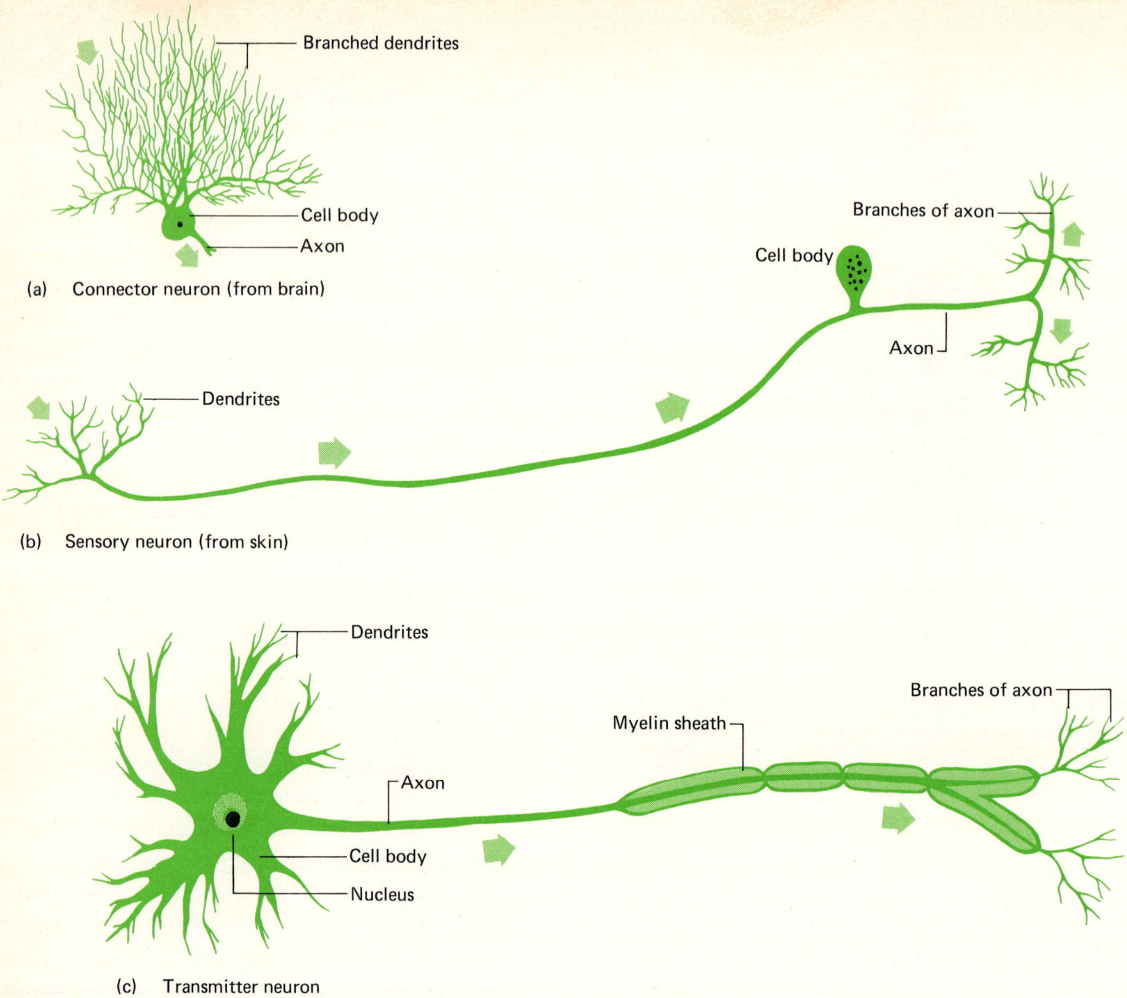

FIGURE 5.2 Neurons may be very diverse, with specific structures as described in the text. The myelin sheath of very long neurons serves to make the conduction of a nerve impulse much more rapid. *(Redrawn from BSCS, 3d ed.)*

FIGURE 5.3 A nerve impulse is conducted across a synapse between the axon branches and dendrites. *(Redrawn from BSCS, 3d ed.)*

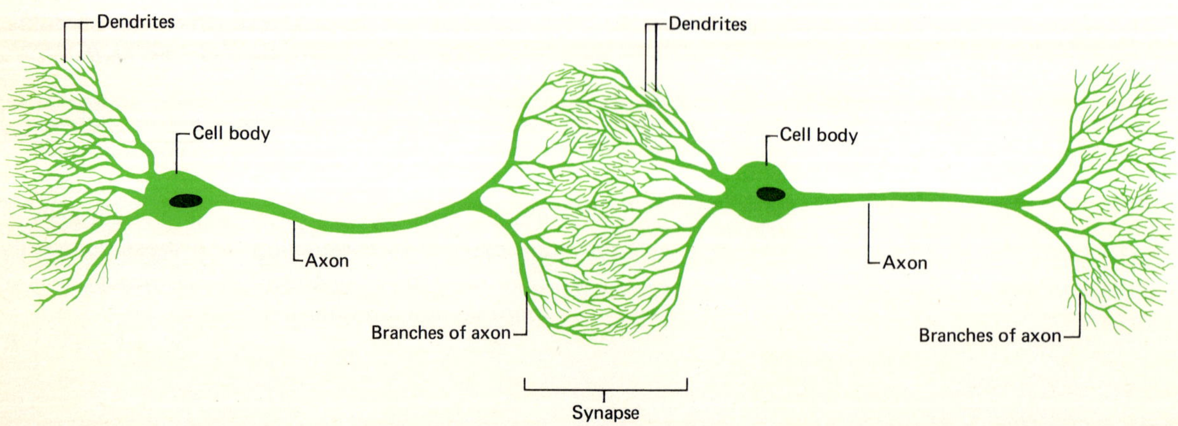

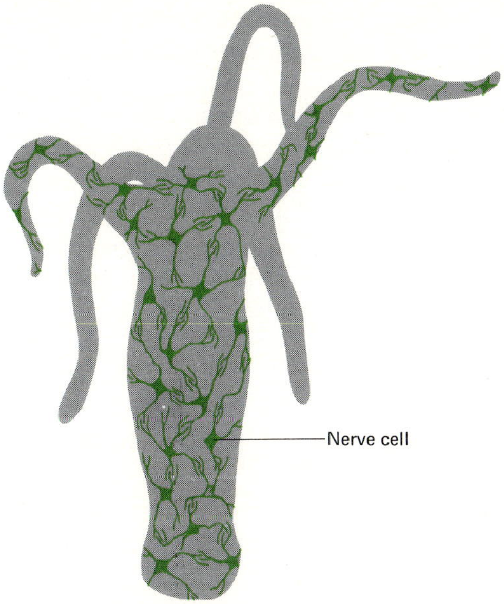

FIGURE 5.4 Coelenterates, such as *Hydra,* have a very diffuse network of nerves. *(Redrawn from BSCS, 3d ed.)*

Invertebrate Nervous Systems

Proceeding from very simple to more complex organisms, the first which can truly be said to possess a nervous system are the coelenterates. A freshwater coelenterate, *Hydra*, provides an excellent example. *Hydra* can best be described as having a *nerve net.* (See fig. 5.4.) Distributed throughout the outer layer of cells in *Hydra* are numerous receptor cells capable of sensing changes in pressure and chemicals in their environments. This information may be fed to a nearby contractile cell or may be transmitted to a neuron in the nerve net to be relayed to other contractile cells. The neurons in the nerve net of the *Hydra* appear to differ from those in many higher organisms in that they seem to transmit information in both directions. The neurons of most other organisms are unidirectional; that is, they transmit a stimulus only in one direction.

A gentle touch delivered to the body of *Hydra* will produce a small local contraction of the body in the area touched. If a stronger stimulus is delivered, a larger portion of the body will contract, and if the touch becomes a rough poke, the whole body contracts. Thus, *Hydra* is capable of producing a graded response which reflects the sizes of the stimuli that are delivered. The sensory cells and neurons of coelenterates are supplemented by specialized effector cells, *nematocysts,* in the tentacles. These *independent* effectors can both receive and respond to stimuli independently of the nerve net. When stimulated, nematocysts discharge threadlike processes that may be barbed or sticky and can coil around food or prey or attach to objects as the coelenterate moves from place to place. Often, as in the Portuguese man-of-war, these effectors can inject toxins capable of killing prey.

The very simple system that has evolved in coelenterates allows for a variety of responses to stimuli, and although we may consider the behavioral repertoire simple, it is nevertheless sufficient for the survival of these organisms.

A more elaborate nervous system is found in the flatworm. The nerve cells are generally elongated and gathered in bundles that lie roughly parallel along each side of the flatworm and join at the head and tail. Branching off from these nerve bundles are single neurons that bridge the body of the worm just under the surface of the skin. (See fig. 5.5.) At the anterior of the flatworm *Dugesia,* commonly called planaria, there is a concentration of photoreceptors, which gives it the appearance of having two eyes. These "eyes" are no more than light sensitive spots that can detect the presence of light in much the same way that a human can if the eyelids are closed. The flatworm can also detect the presence of food through chemoreceptors located in its anterior end. Although capable of detecting several kinds of information, the flatworm has a limited repertoire of responses. To integrate information of several sorts and to coordinate more complex responses requires a somewhat more highly organized aggregation of neurons than the flatworm possesses. Such an aggregation of neurons is called a *brain.*

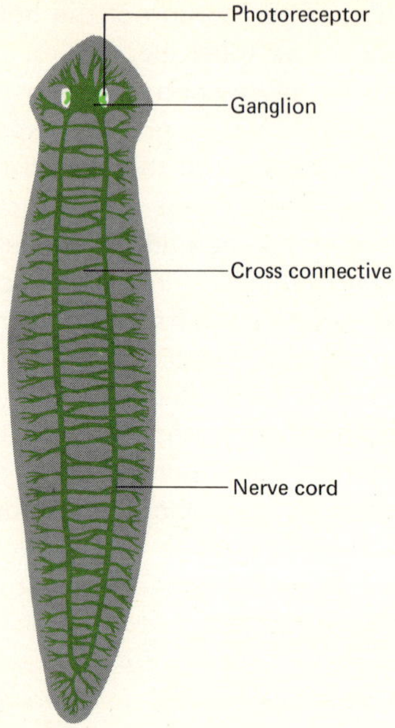

FIGURE 5.5 The organization of nerves in a planarian. *(Redrawn from BSCS, 3d ed.)*

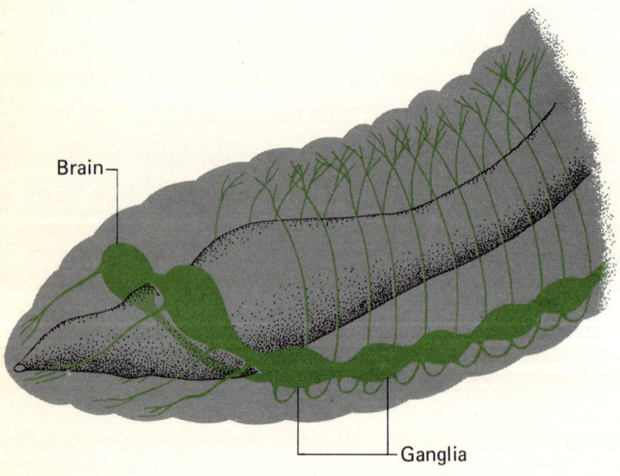

FIGURE 5.6 The nervous system of the earthworm.

Said this way, it sounds as though brains are nothing more than larger bundles of nerves. In one sense this is true. Primitive brains begin to appear in the annelid worms, such as the common earthworm. The function of the brain is to receive information fed into it by the nervous system and then to direct responses to the information, and since a flatworm can do this, we must justify saying that a flatworm lacks a brain. The general structure of the nervous system of an annelid is shown in the figure of the earthworm (fig. 5.6). It consists of a ventral nerve cord that runs the length of the body. As in the flatworm, there is an aggregation of neurons at the head that is greater than the aggregation along the cord. Two important differences exist, though, between flatworms and annelids. In the flatworm, the nerve cord is relatively uniform in thickness, and at its termination, it is only slightly thicker than it is along the cord. In the segmented body of the earthworm, the nerve cord thickens in each segment into aggregations of neurons called *ganglia* (the anterior of a flatworm contains a ganglion also), and at the anterior of the annelid, two ganglia join to form an aggregation of neurons that is significantly larger than any of the other ganglia. This thickening may be considered a brain in that most of the annelid's activity is centrally coordinated at that point.

Thus, we see an evolutionary trend toward greater centralization of the nervous system. The ganglia of the annelid receive information about the environment from mechanoreceptors in each body segment and can initiate a response in that segment. But if each ganglion were equal, the responses to stimuli would be chaotic as each segmental ganglion "ordered" a response to the limited amount of information that was available to it. The function of the brain is to receive information about the state of the entire relevant environment, coordinate the information, and initiate a response for the whole organism. The segmented mechanoreceptors of the earthworm are augmented by chemoreceptors and photoreceptors in the rest of the body, and although each segmental ganglion has some limited autonomy, most of the information supplied by the receptors is integrated in the brain.

The mechanoreceptors of the earthworm are sensitive to touch and contain small hairs that receive stimuli and may be able to detect vibrations in the ground. The chemoreceptors are especially sensitive to the presence of water and are predominantly located in the anterior segments. When an earthworm comes out of its burrow, these chemoreceptors are among the first to receive stimuli. If the ground around the burrow is dry at one spot, the worm will seek out a moister area, and if it does not encounter any sufficiently moist ground, it will withdraw. If these anterior chemoreceptors are removed or anesthetized, the worm will crawl over completely dry ground. The photoreceptors are largely concentrated in the anterior and posterior segments and are almost indistinguishable from skin cells. Since most of the earthworm's life is spent below ground, there is not very strong selective pressure for light detection. Since the anterior segments are the first out of the burrow, there is adaptive significance to the concentration of photoreceptors there. Since it is also useful to know when the entire body is below ground, the concentration of photoreceptors in the posterior segments is similarly significant.

Generally speaking, receptors have aggregated at the anteriors of active animals, for that is the part of the body that makes initial contact with the environment as the animal moves forward. Important ganglia and brains also tend to have evolved in regions of the body close to the sources of information, and thus they too are usually located in the anterior ends of most animals.

There are some interesting exceptions to the last rule. In some insects, for example, large numbers of ganglia are concentrated in the thorax, the portion of the body where the wings and legs attach. The brains of these insects, while sharing many important coordinating functions with the thoracic ganglia, have remained small. The large thoracic mass of ganglia in these insects is probably due to the important role that they play in controlling the motor activity (movement) of the legs and wings. It may also be due partially to the location of many sense organs on the legs. For example, many flies ''smell'' food with their feet, and some insects receive sound through mechanoreceptors located on the sides of the thorax.

The central nervous systems of flatworms and annelids show a trend toward greater centralization that provides a more complex pattern of responses to environmental changes than observed in the coelenterates. An exhaustive survey of the nervous systems of organisms would be inappropriate here, but it is important to understand that the evolution of increasingly complex systems permitted species to adapt to new environments and to exploit the energy resources that those environments contained. Consider the difference between an earthworm and a sponge, for example. The sponge is able to respond only in a very limited way to a change in its environment. Should the sponge's environment cease delivering food, the sponge would die. The earthworm is relatively freer than the sponge from this sort of environmental dependence and moves about in search of new energy sources. In order to do this, however, it must acquire, transmit, and respond to environmental information. Neither form is really free from environmental constraint; *no* organism ever really is. But the earthworm has evolved to the point where it can utilize a broader range of the environment than can a sponge.

Vertebrate Nervous Systems

A roughly similar comparison can be made between earthworms and simple vertebrates, and as you would expect, there are some very important differences between the central nervous systems of vertebrates and invertebrates. The vertebrate, for example, always possesses a single, hollow nerve cord. Although the earthworm also has a single cord, many invertebrates have double nerve cords, and none have hollow ones. The ganglia of vertebrates are not usually as well developed as those of invertebrates. Most importantly, although a number of functions are performed by the spinal nerve cord of the vertebrates, the degree of centralization that has evolved among vertebrates is far greater than that of almost all invertebrates. The vertebrate brain is the central controlling organ for all bodily processes.

With one critical exception, most of what we know about the function of the nervous system has been gleaned from a study of vertebrates. Partly because of this, but also partly because it is of inherent interest to us, we shall examine the vertebrate nervous system in more detail than we have the nervous systems of invertebrates. Therefore, we shall not consider the progression of vertebrate nervous systems from simple to more complex ones. Rather we shall deal with the actual method (to the extent that we understand it) of gathering and processing environmental information.

Reception of Stimuli

Many vertebrates (and some higher invertebrates) rely heavily on visual information about their environment, and thus the organization of photoreceptors has evolved to a high degree. This is especially true of humans, and approximately three quarters of all the receptors in the human body are located in the eyes. (See box 5.1.) The vertebrate eye has often been compared to a camera, and while the eye is far more complex than a camera, certain basic principles are found in both. To focus the eye, some vertebrates, such as fishes, move the lens backward and forward, just as a camera is focussed by moving the lens nearer or farther from the film. Light enters a camera through a variable aperture called the *diaphragm*, and in the eye light passes through a variable aperture called the *iris diaphragm*. But there are more dissimilarities between a camera and an eye than there are similarities. Many vertebrates, for example, do not move the lense of the eye at all; rather, they change the thickness of the lens itself through the contraction of muscles attached to the lens. Since it is not an inflexible solid, but an organ the shape and optical properties of which can be altered, the lens of an eye is quite different from the lens of a camera. The iris is a specialized form of muscle that by contraction and relaxation permits more or less light to enter the eye. But one of the most important portions of the eye, for our purposes, is not the lens or iris, but the region where the image admitted to the eye is received, the *retina*.

The Structure and Function of the Retinal Field

Although there is a great deal more to be learned about the retina, it offers us an instructive example of the reception of a stimulus. One of the first things that should be pointed out about the vertebrate retina is that it is "inside out": that is, the cells where an image is received are overlaid with other nerve cells that conduct the stimuli out of the eye and to the brain.

The retinal field is made up of two kinds of photoreceptors that are called, because of their shapes, *rods* and *cones*. Rods are sensitive to low intensity light, and some nocturnal vertebrates have retinas that contain only rods, thus allowing them to see clearly in dimly lighted places. Cones, in contrast, are sensitive to light of higher intensity, provide for resolution of the image received, and sense color.

The human eye contains approximately 100 million rods and 7 million cones. The density of cones is greatest in a small area of the retina called the *fovea*, a slightly depressed point on the retina on a direct line with the center of the lens. Thus, the sharpest image is received at the fovea. The density of cones decreases greatly a short distance from the fovea, and the rest of the retina contains cones that are relatively evenly dispersed. At the exact center of the fovea there are no rods at all, but at a slight distance from the fovea the density of rods increases rapidly and then decreases slowly toward the peripheral regions of the eye. Thus, the ''best'' information about an object is received when its image is in the center of the visual field. Objects whose images fall farther from the center are still perceived, but in order to perceive them most clearly, the eye must be turned until the image is received at the fovea. This is accomplished in man by the simple arrangement of muscles attached to the eye itself, but in some vertebrates it is necessary to move the entire head, and in some others, the entire body.

The stimuli received by rods and cones are transmitted to neurons called *bipolar neurons*, and the bipolar neurons transmit the stimuli to *ganglionic neurons*, which lead out of the eye to the optic nerve. It appears that there are several different kinds of bipolar and ganglionic neurons in the eye, the exact structures and functions of which have yet to be determined. It is known that in the peripheral region of the retina, many rods transmit stimuli to a single ganglionic neuron in a ''party line'' arrangement. At the center of the retina, however, there are as many ganglionic neurons as there are cones, and this has led some to state that the foveal cones have ''private lines'' to the optic nerve. Recent evidence indicates that this is not so, that the action of each ganglionic neuron in this region is influenced by a number of cones, and that each cone influences a number of ganglionic neurons. To further complicate the problems of understanding the retina, it is strongly suspected that rods and cones are capable of interacting with each other.

Vision as a Chemical Process

Although a great deal of work has been done to determine exactly what happens when light strikes rods or cones, there are still many areas of uncertainty about the process. For example, it is believed that each rod or cone becomes adapted during the life of the organism to receiving stimuli in a highly specific pattern and that calcium is somehow implicated in the process, but it is not yet at all clear how these cells can achieve this high degree of specialization.

Electron microscopic and biochemical investigations have indicated that rods and cones contain large amounts of proteins and chemical pigments believed to be embedded in an internal membrane of the receptors. This membrane is highly folded and gives the appearance of being arranged in stacks within the rods and cones. (See box 5.1.) One pigment called *retinal* has been isolated from rods and is known to be bonded at each end to certain proteins. The protein-pigment complex is called *rhodopsin*. When struck by light, the chemical bonds of the pigment are believed to change, causing the molecule to ''unfold'' and break loose from the protein at one end. (See fig. 5.7.) At body temperature, this is an unstable state, and the molecules of pigment and protein undergo further changes in their structure that cause the pigment to break free of the protein entirely and then to return to its original condition, fully bound to the protein. If, however, the pigment is struck again by light before it has broken free of the protein, it is driven back to its original, fully bound condition before it can go on to break free. This allows the rod to prepare for the reception of another stimulus more quickly than if the pigment had to go through the whole process of breaking free and then returning to the fully bound position. In all, it takes about one second for the molecule of pigment to break free of the protein and about 5 minutes for it to return to the original state. This is part of the reason why it takes several minutes for a person's eyes to become accustomed to the dark after leaving a brightly lighted place, and why a brightly lighted place seems brighter when a person enters from a dark place.

The changes that take place in rhodopsin are stimuli, of course, and must be transmitted to the brain. At the base of the stacked membrane in a photoreceptor such as a rod, is a cilium that projects into the body of the receptor. It is

The Eye

Within the retina of an invertebrate like the octopus, images fall directly on the photoreceptors and are conducted to the brain by nerve fibers that lie under the receptors. In the diagram shown, such an eye is represented by (a), with the colored line indicating the receptors. In (b) the "inside out" eye of a vertebrate is shown. Images reaching the retina of a vertebrate must pass through the nerves that will conduct impulses out of the eye. As a consequence of this structure, vertebrates have a blind spot in each eye at the point where the nerves must pass through the retinal field. You can test yourself for this blind spot by placing one hand over your right eye and fixing your left eye on the letter "(b)". Start with the page about a foot from your eyes and move your head forward slowly. When the image of "(a)" falls on the blind spot it will disappear from your peripheral view. Test the other eye by covering it with your hand and viewing the other letter.

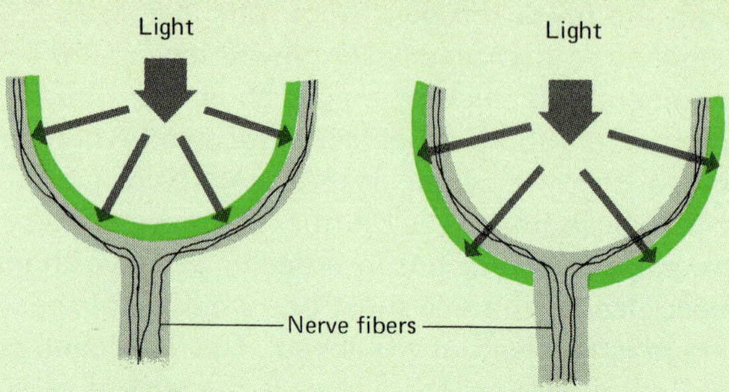

(a)
Octopus eye

(b)
Vertebrate eye

The density of the cones (black line) in the eyes is greatest near the center, giving the best resolution and the best sensitivity to color when an object is viewed straight on. The rods, on the other hand, have their greatest density near the periphery of the eye, providing for good peripheral vision in dim light. For example, it is often easiest to see a dim star when you are not looking directly at it, but see it through "the corner of your eye." Obviously, there are neither rods nor cones where the optic nerve enters the eye; hence the blind spot.

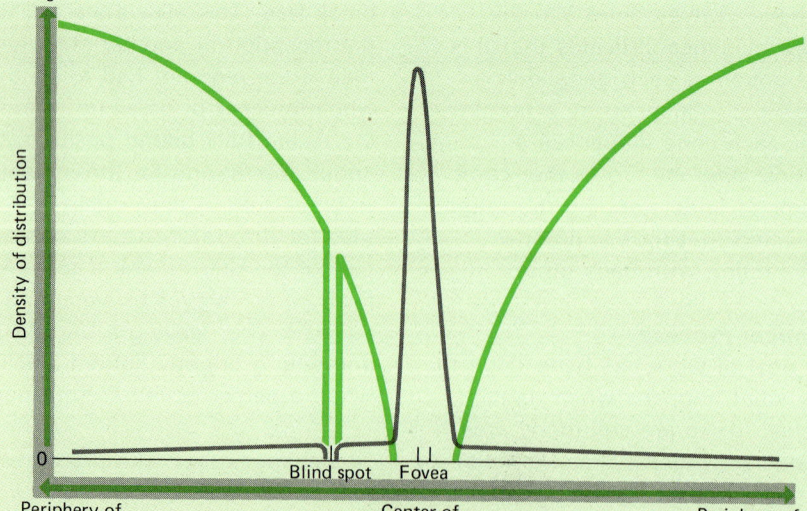

(c)

(Redrawn from Cornsweet after Pirenne.)

BOX 5.1

The structure of the vertebrate retinal field is shown in (d), which is, of course, much oversimplified.

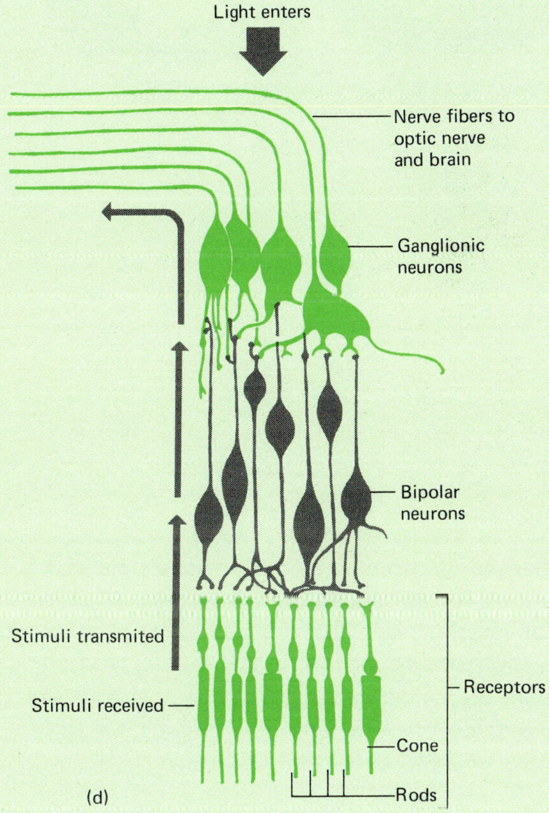

(d)

(Redrawn from Cornsweet after Wald.)

An electron photomicrograph of the cone of a frog retina showing the stacked membranes, or lamellae, and the cilium.

(e)

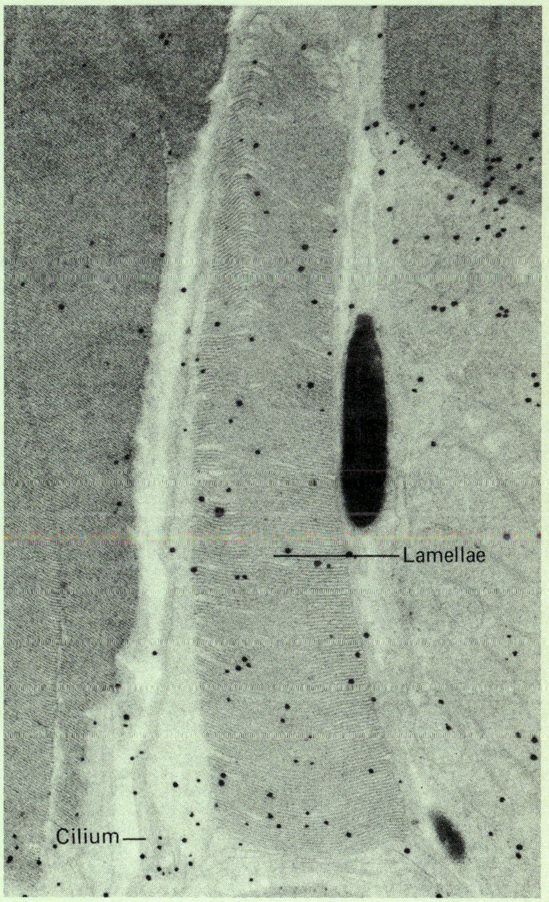

(Courtesy of Richard W. Young, University of California, Los Angeles.)

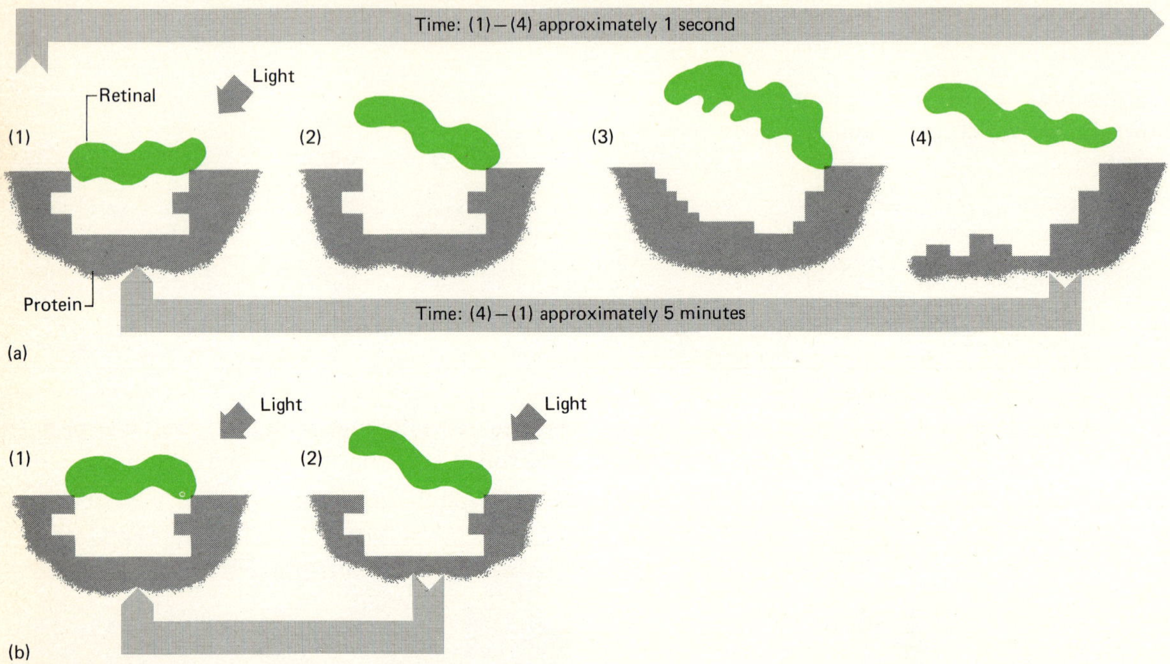

FIGURE 5.7 The reactions of rhodopsin with light. Retinal separates from its attached protein when struck by light (a): (1), (2), and at body temperature it undergoes further changes (a): (3), (4). If not exposed to light again, it returns to its original state (a): (4) to (1). If struck by light when in condition (a): (2) it returns immediately to its original state (b) (see text).

believed that this cilium somehow transmits the stimulus from the light-sensitive portion of the photoreceptor to the body of the cell for further transmission to the central nervous system. But in order to understand the transmission of a stimulus, it is helpful to consider briefly the electrical phenomena that occur in living systems.

Bioelectricity

The nuclei of all atoms carry a positive electrical charge (+), and the electrons that orbit the nuclei carry a negative charge (–). Under some conditions, atoms and molecules may gain or lose electrons and thus acquire a net negative or net positive charge. Such atoms or molecules are called *ions*. Because opposite charges attract each other, an ion that is negatively charged is attracted to the positive pole of a battery, or the *anode*, and is termed an *anion*. Conversely, positively charged ions are attracted to the *cathode*, or the negative pole, and are therefore called *cations.* To repeat, *anions are negatively charged ions, and cations are positively charged ions.*

Electricity tends to flow from areas of negative charge, where there is an excess of electrons, to areas of positive charge, where electrons are scarce. Thus, electrical charges tend to achieve neutrality. The flow of electrons is called a *current*, and it takes place along a *conductor*. A piece of copper wire, for example, conducts electrons very efficiently from areas of negative to areas of positive charge. The difference in charges between two points on a conductor is a force that is referred to as the *electrical potential* of the conductor or, if "electrical" is understood, simply as the *potential*.*

* Electrical potentials are measured, most commonly, in *volts*.

Investigations have shown that cell membranes often act as conductors, and research into the electrical properties of cells, including neurons, has also revealed that the interior of the cell is negatively charged relative to the fluid medium in which the cell is bathed. In other words, both the interior of the cell and the intercellular fluid are negatively charged, but the interior is even more negatively charged. Thus, a potential exists that we may refer to as the *membrane potential*. While the cell is at rest, this potential is called the *resting membrane potential* or simply the *resting potential*. The resting potential is brought about partially by the concentration within the cell of a number of negatively charged organic ions that are too large to diffuse out of the cell. These organic anions (A^-) alone, however, do not account for the relative negativity of the cell interior and the resting potential. Three other important ions have been found to be involved: cations of potassium (K^+); cations of sodium (Na^+); and an anion, chloride (Cl^-). The concentrations of these ions on either side of the resting cell membrane are very unequal, as shown in table 5.1.

TABLE 5.1 Concentrations of K^+, Cl^-, Na^+ *(in moles per milliliter)*

	Frog Muscle		*Squid Axon*	
	Internal	External	Internal	External
Potassium (K^+)	140.0	2.5	400	10
Chloride (Cl^-)	3 – 4	120.0	40 – 100	540
Sodium (Na^+)	9.2	120.0	50	460
Resting membrane potential	—90 mV		—60mV	

These differences and the existence of the organic anions produce resting membrane potentials so small that they are measured in thousandths of volts, or millivolts (mV). The resting membrane potential is derived by arbitrarily setting the charge in the intercellular fluid at zero, and thus the interior of frog muscle cells is 90mV *more* negative than the intercellular fluid and the resting membrane potential is —90mV. The resting membrane potential of squid axon is —60mV.

(Adapted from Bernard Katz, *Nerve, Muscle, and Synapse,* New York: McGraw-Hill Book Company, 1966, p. 43.)

Because the large organic anions are confined within the cell, and since K^+, CL^-, and Na^+ can readily diffuse across the cell membrane, the movement of one of these smaller ions into or out of the cell would seem to be favored to achieve electrical balance. The flow of K^+ out of the cell would be favored by diffusion, but because this would further increase the interior negative charge of the cell, it is resisted. The flow of Cl^- into the cell is favored by diffusion but is resisted by the electrical imbalance that exists. The flow of Na^+ is favored by *both* chemical and electrical forces. The problem facing investigators, then, is to explain how Na^+ is kept in electrical and chemical imbalance across the cell membrane. (See fig. 5.8.)

Active Transport and the Sodium Pump

Although sodium can permeate the cell membrane by diffusion, analyses of living cells revealed that sodium appeared not to flow through the membrane in very significant quantities. It was learned, however, that if the metabolism of the cell was stopped, the concentration of sodium on the interior of the membrane underwent a rapid increase. This and other evidence led investigators to postulate the existence of a *sodium pump* that *actively transported* this ion out of the cell. The active transport of chemicals across membranes is widely known in living systems and is instrumental in maintaining necessary imbalances of materials in cells. The key difference between active transport mechanisms and passive diffusion is that the former *require that energy by expended by the cell,* whereas materials that move by passive diffusion require no expenditure of energy. When the metabolism of the cell is stopped, the cell stops working and expending energy, active transport mechanisms cease functioning, and materials that were formerly kept out of balance on either side of the membrane achieve an equilibrium by diffusing through the membrane. In many cases, the restoration of a balance leads to death, and thus active transport mechanisms can be of critical importance to survival.

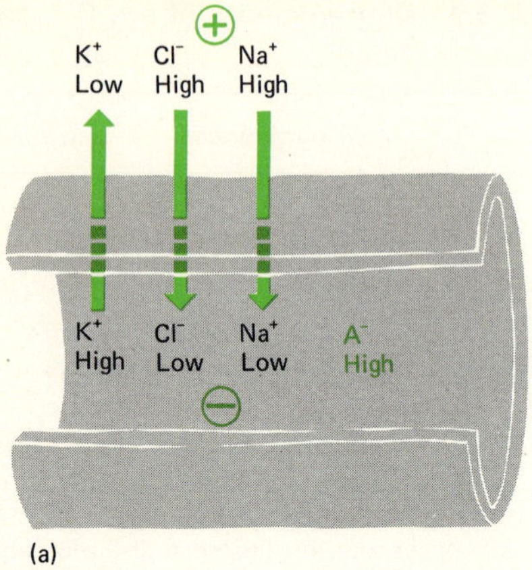

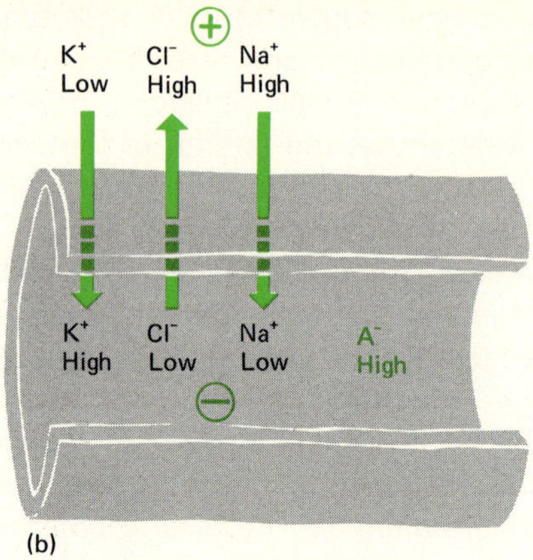

FIGURE 5.8 The diffusion of potassium (K), chloride (Cl), and sodium (Na) ions is favored in the directions shown in (a). The movement favored by electrical forces is shown in (b). These tendencies for movement "cancel each other out" for K and Cl, but a "pump mechanism" is required to keep the internal concentration of Na at a low level. The internal negative charges is maintained by anions (A^-).

The maintenance of a sodium imbalance on either side of the cell membrane is an example of active transport, and the sodium pump is the mechanism that is responsible for maintaining the imbalance. The sodium pump, of course, is not thought to be a structural entity in the cell, such as the organelles that we mentioned in Chapter 3 (mitochondria, for example). Rather, it is thought that within the cell, proteins react with sodium ions that have diffused through the membrane, and then the sodium-protein complex reacts with the chemicals of the cell membrane and is extruded through the membrane and into the intercellular fluid. When the complex reaches the intercellular fluid, further chemical reactions take place, and the sodium dissociates from the protein, which then reacts again with the membrane and re-enters the cell. Thus, the same protein molecule may carry a large number of sodium ions out of the cell.

Thus the resting potential is maintained by the action of a sodium pump, and it should be clear that "resting" is something of a misnomer, since chemical work is being done to keep the cell in a state of rest. We should also point out that resting potentials may be observed in many cells, not just neurons. A photosynthetic unicellular organism, *Nitella*, has been shown to have a resting potential, and changes in the concentrations of substances in the environment of *Nitella* bring about changes in the resting potential of the organism, thus signalling it to respond in some way.

With this understanding of the resting membrane potential we can return to the problem of how stimuli are delivered, conducted, and transmitted from neuron to neuron in the nervous system.

Conduction of the Nerve Impulse

Let us assume for the time being that a stimulus has been delivered from a receptor to a neuron. Recall that we said that the signal, or *nerve impulse* as we shall now call it, is received at a dendrite and transmitted from the termination of the axon. The problem then is to understand how the impulse is conducted from one end of the nerve to the other. Obviously, conduction must be associated with the resting potential, and the cell must be required to do work. The work that the neuron does is called *firing*, and it should be understood that in order for a cell to fire, a stimulus must be large enough, or enough stimuli must be presented within a short enough space of time, to bring about a firing. Thus, the neuron has a *firing threshold* below which small stimuli, although they may effect small changes in the neuron, will not cause it to fire. Once the firing threshold is reached, a large change takes place in the resting potential in the area where the stimulus is presented. The permeability characteristics of the membrane change in that location, and large amounts of sodium flow into the nerve cell. This has the effect of decreasing the relative negativity of the cell interior, and may even occasionally cause the interior of the neuron to become relatively positive and the exterior relatively negative. At the same time, potassium begins to flow out of the cell, although at a lower rate than the incoming sodium. By some as yet unknown mechanism, the local permeability of the membrane to sodium suddenly changes, and no more sodium enters the cell, although the membrane's permeability to potassium continues for a short time. During this period, called the *absolute refractory period,* no stimulus, however strong, can cause the cell to fire again in this location. While the inflow of sodium ions is blocked, the sodium pump continues to operate and removes the sodium excess that has built up in the neuron. The outflow of potassium ions is reversed, and the combined inflow of potassium and outflow of sodium restores the resting potential of the membrane. During the time that the resting potential is being restored, the membrane enters a *relative refractory period,* during which the nerve may fire again, but only if a larger stimulus is presented to it than before.

Because the interior and the exterior of the neuron carry different relative charges, the membrane is said to be *polarized*. When a stimulus is delivered to the membrane, causing the inflow of sodium to begin and the electrical charges to begin to change on either side of the membrane, the membrane is said to be *depolarizing* in the location where the stimulus was received. The effect of a local depolarization of the membrane is the induction of depolarization in adjacent locations. (See figs. 5.9 and 5.10.)

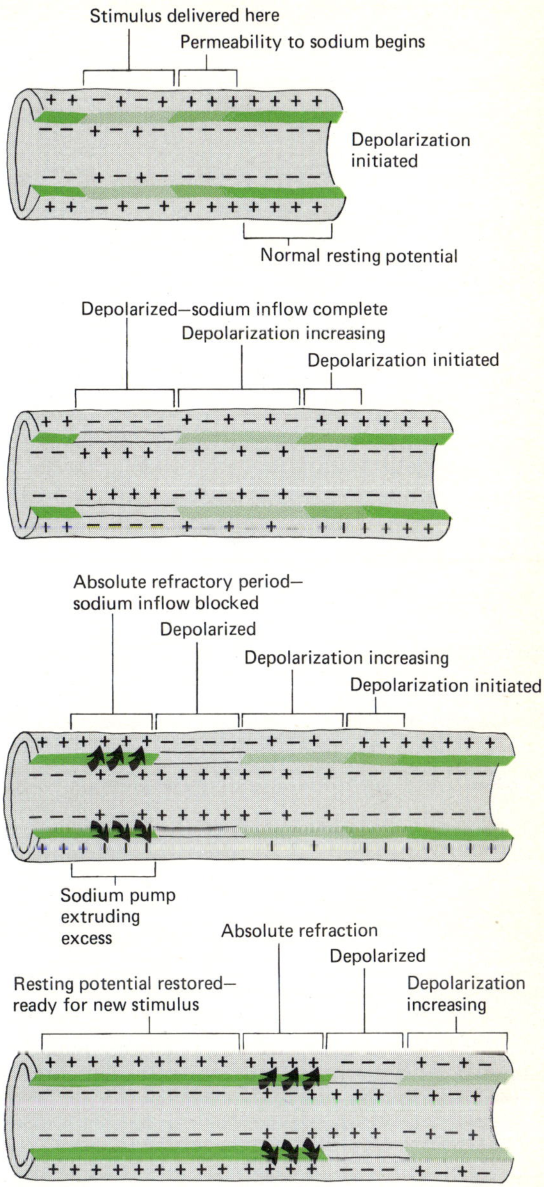

FIGURES 5.9–5.12 The changes in polarity, movement of sodium, and refractory period associated with the initiation and conduction of a nerve impulse (see text).

Theoretically, then, a nerve impulse can travel in two directions from the location where a stimulus is presented. However, as we pointed out earlier, with some exceptions the nerve impulse usually travels in only one direction. It is at this point that the absolute refractory period is important in the conduction of the impulse. As we said, shortly after firing the membrane becomes momentarily impermeable to sodium. Thus, a depolarization of the membrane cannot occur again at a location where a stimulus has just been presented, because sodium inflow is blocked at this point, or at least cannot occur until the resting potential is restored or the relative refractory period begins. (See fig. 5.11.) However, in the location adjacent to the one where a stimulus has been presented, a depolarization is beginning and sodium inflow can occur, and thus the impulse is conducted along the neuron as a wave of depolarization, followed by sodium inflow, followed by a local refraction of the membrane that determines the direction in which the impulse is conducted. (See fig. 5.12.)

Under laboratory conditions, stimuli are often presented to the axon and can be measured as they travel in both directions along the membrane. Within the body of a vertebrate, however, the stimulus is usually delivered to a dendrite and thus can travel only in one direction.

It would seem that the wave of depolarization that reaches the end of the axon might induce a depolarization in the membrane of the next nerve cell and in that way transmit the impulse from neuron to neuron. That, however, is *not* the case, and we must turn our attention now to the way in which the impulse crosses the synaptic cleft.

Transmission of the Nerve Impulse

A series of careful experiments on the biochemical and electrical nature of the transmission of the nerve impulse indicate that when an impulse reaches the synapse, a *transmitter substance* is released from the end of the neuron. It appears that such transmitter substances serve to increase the permeability of the membrane in the next neuron, and the increased permeability initiates a depolarization wave that is then conducted along the neuron. Thus, the transmission of a stimulus is accomplished primarily by specific chemical substances.

Electron microscopic studies of the endings of neurons have revealed the presence of roundish sacs, or *vesicles*, near the synapse, which are believed to be the sites from which transmitter substances are released into the synaptic cleft. (See fig. 5.13.) Interestingly, one substance long known to man was instrumental in learning about the transmission of the nerve impulse. *Curare* is a deadly poison applied by some South American Indians to their arrows. When

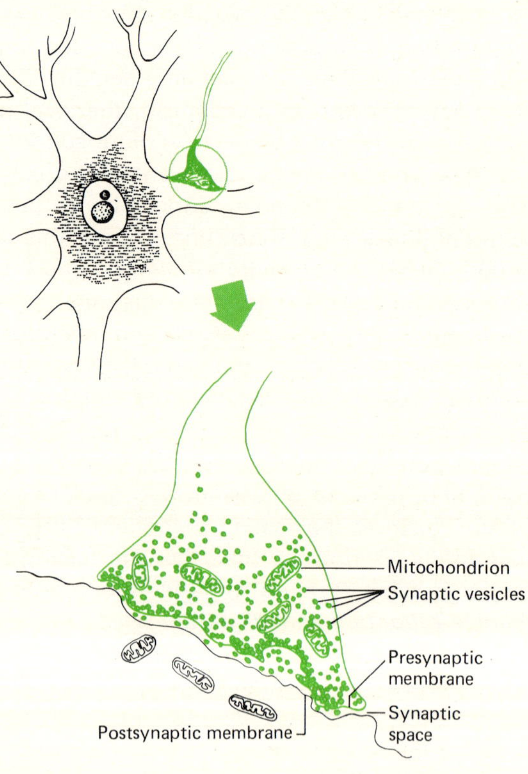

FIGURE 5.13 When a nerve impulse travels across a synapse, the material in the synaptic vesicles is released into the synaptic space, stimulating the postsynaptic membrane. This initiates the conduction of a nerve impulse in the second cell. *(Adapted from the drawing by Anthony Ravielli by permission of Charles Scribner's Sons from* The Shape of Intelligence *by H. Chandler Elliott, illustrated by Anthony Ravielli, Copyright © 1969 by H. Chandler Elliott. Illustration copyright © 1969 by Anthony Ravielli.)*

introduced into the bloodstream, it has the effect of preventing muscle cells from reaching a firing threshold. By treating muscles with curare, investigators have been able to observe the operation of the neurons that synapse with muscles without having the firing of the muscle cell interfere with the measurements being made.

The special regions of muscle where nerve and muscle synapse are called *endplates*. These regions have been shown to be extremely sensitive to the application of transmitter substances. One such known substance is *acetylcholine* (ACh). It has been shown that nerve cells constantly release ACh in small amounts, probably from the vesicles mentioned above. It has also been found that another substance, an enzyme called *acetylcholinesterase* (AChE), is found in the synaptic cleft. The function of AChE is to break down ACh. When the nerve impulse reaches the synapse, though, far more ACh is released into the cleft than AChE can break down. Thus, the permeability of the membrane at the dendrite of an adjoining neuron, or the endplate of a muscle cell, is changed, causing that cell to reach its firing threshold. The function of AChE is the very practical one of insuring that any ACh that is in the synaptic cleft after an impulse is transmitted is broken down so that the nerve will not fire twice when only one stimulus is presented to it. AChE also prevents the small amount of ACh that is constantly being released from the neuron from accumulating and causing a nerve impulse to begin when no stimulus has been presented.

Thus, the reception, conduction, and transmission of stimuli can be seen as a series of chemical or electrochemical events that take place within the body of an organism. We shall not dwell here on the nature of the important chemicals involved (enzymes, proteins, and pigments), but we shall see the important roles that they play in the lives of organisms when we examine the process of genetics in Chapter 6 and the processes of energy conversion and use in plants and animals in Chapter 9.

The problem that remains for us to discuss here is how nerve impulses that are received from the many sense receptors in the body of an animal are coordinated and then acted upon by effectors. As we have pointed out, complex organisms receive greater amounts of information from their environment and have larger repertoires of responses. In general, such repertoires have required the evolution of highly centralized nervous systems in which the increased amounts of information are transmitted, integrated, and responded to appropriately. The integrative function is performed by the brain, and thus we must now consider the organization and structure of the vertebrate brain.

The Brain—Central Controlling Organ of the Vertebrate Body

Early in the evolution of the vertebrates, the hollow nerve cord underwent some specialization as three areas evolved, represented by three fluid-filled enlargements of the nerve cord called the *forebrain, midbrain,* and *hindbrain*. As evolution proceeded, the hindbrain underwent more specialization as one part, the *medulla*, evolved as a coordinating center for the control of certain so-called "unconscious" functions such as breathing, digestion, and circulation. The other part of the hindbrain became the *cerebellum* and came to control such things as balance and muscular coordination. The midbrain became the center for the reception of visual information. Like the hindbrain, the forebrain underwent a specialization into two parts. The first was the *cerebrum*, which became specialized for the reception of olfactory information (smell), and from which the cerebral hemispheres of higher animals have evolved. The other part of the forebrain became the *thalamus* and *hypothalamus*. The thalamus is a relay center for the transmission of information that is received from sensory stimuli and is then passed on to the cerebral hemispheres, as well as for the transmission of information from the cerebrum to the spinal cord. The hypothalamus lies below the thalamus and is believed to be the central control organ for emotions and for activities or states such as sex, hunger, thirst, and pain. The subsequent evolution of the vertebrate brain was based essentially on this plan, but the most significant changes that have taken place have been in the cerebrum. (See fig. 5.14.)

In fish and amphibians, the cerebrum has remained largely associated with olfactory functions. Among reptiles and then mammals, an important departure took place as a region *on the surface* of the brain evolved not to receive information or to relay signals, but to receive and coordinate information from other regions of the brain. This *associative* function became increasingly important as the brains of vertebrates evolved. Studies made to map the location of various functions of the vertebrate brain have indicated that, in humans, proportionally more of the surface of the brain is devoted to association than to sensory reception or the control of motor activities. A great deal of investigation of the human brain has been undertaken by neurosurgeons and neurophysiologists, and it has suggested that of the portions of the cerebrum that are devoted to motor activities, a significantly large proportion is concerned with the movement of the muscles of the face and hands. Similar studies are presently underway to map the associative areas of the human brain, but the problems facing investigators are immense. The human brain contains at least 30 billion neurons, and the cell bodies and dendrites of each neuron synapse with many other neurons. In some cases, a neuron in the human brain may receive as many as 1000 synapses. In addition to the neurons in the brain, there are also certain cells called *glial cells* that were once believed to provide support and nutrients to the neurons. Now there is some evidence that glial cells may also participate in the electrical and biochemical activities of the brain. The number of potential interactions in the human brain between the neurons alone is stupendous, and in the opinion of some biologists, higher nervous systems may be beyond the capacity of man to comprehend; thus, the human brain may be incapable of understanding itself. In spite of that possibility, research continues, as it should, for although a complete understanding of the human brain may be beyond our capacity, many investigators are convinced that there is a long way to go and a great deal more to be learned before we reach the point at which further study will no longer contribute to further understanding. (See box 5.2.)

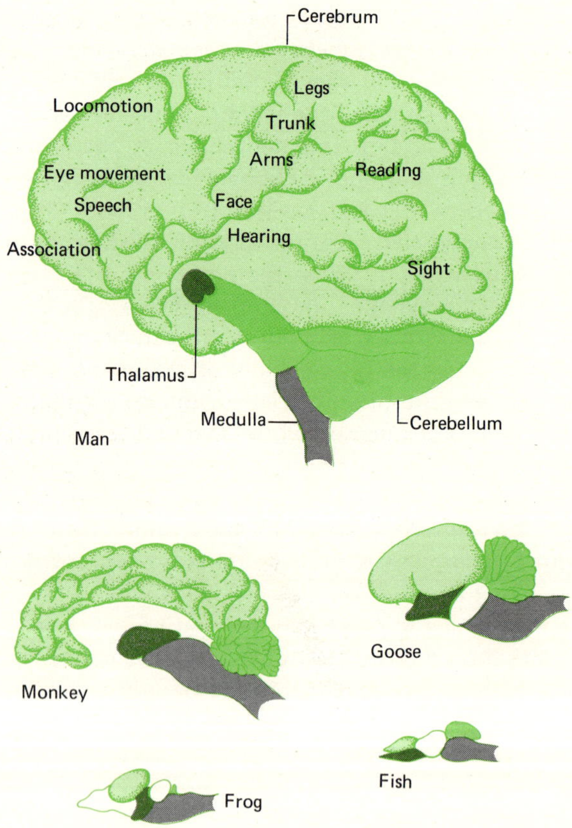

FIGURE 5.14 The parts of the brain in primitive and advanced vertebrates, and the areas of specialization in the cerebrum of a human. *(Redrawn from BSCS, 3d ed.)*

Understanding the Brain

We know less than we would like to about the functioning of any human organ, but perhaps the most mystifying of all is the human brain. We are reasonably certain that ours is the most complex brain to have evolved, and in the last twenty years alone we have made amazing strides toward understanding what has been poetically described as the ''net that secretes mathematics.'' We have nevertheless barely begun.

There are several approaches to studying brain function. Mapping studies, such as we mentioned, constitute one avenue of research; another is the study of brain biochemistry. Later in this book we shall discuss the structure and function of chemical compounds known as enzymes, a class of proteins that serve to drive chemical reactions to completion. We shall also discuss lipids, organic compounds of which fats and oils are made. Every cell, of course, contains both enzymes and lipids; the cells of the nervous system are particularly rich in lipids and a number of brain disorders have been traced to failures of lipid metabolism. Many cases of mental retardation, for example, are the result of a buildup of lipids in the central nervous system, called lipid storage diseases. Such diseases have in some instances been shown to be caused by hereditary failures to produce enzymes which would break down accumulations of lipids, while other lipid storage diseases result from a failure to inhibit the synthesis of lipids. One such disease, which we shall consider in more detail later, is Tay-Sachs disease, a hereditary condition which produces mental retardation, blindness, and paralysis, leading to death before the age of four.

As we shall see in the next chapter, much of our understanding of genetics has been gleaned from the study of mutations—in a sense, from errors in the transmission of biological information. Just so, much of our understanding of the brain is being obtained from studies of brain disorders, errors in the functioning of the brain. It is a sad fact, from which we can only take comfort in realizing that such understanding is incidental to attempts to alleviate human suffering.

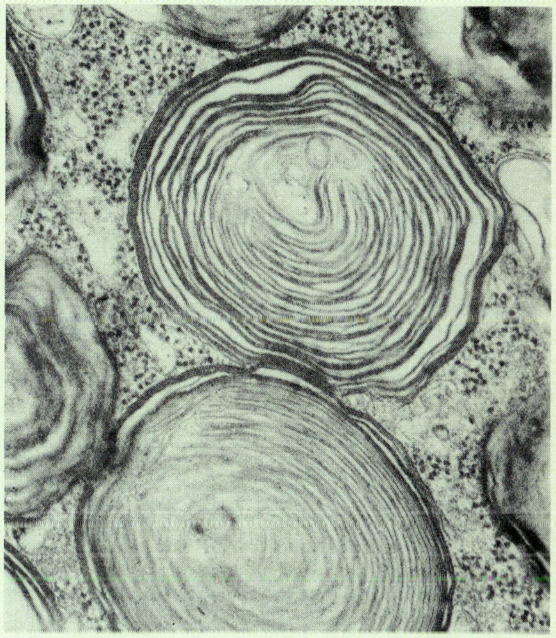

These bodies of laminated membranes, composed of lipids, accumulate in nerve cells of individuals with Tay-Sachs disease. *(Courtesy of John S. O'Brien, University of California, San Diego. Reprinted from* Advances in Human Genetics, *Vol. 3, 1972, p. 45, courtesy of Plenum Publishing Corporation.)*

BOX 5.2

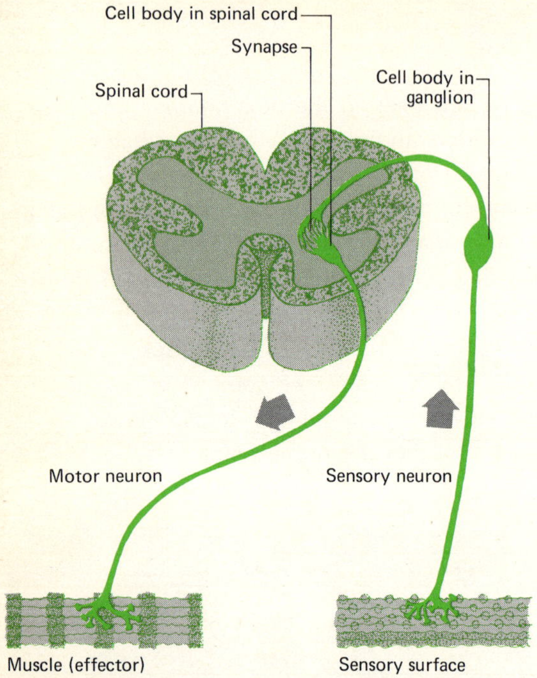

FIGURE 5.15 In the simplest reflex arc, a nerve impulse travels from a sensor to the spinal cord, where it synapses with a motor neuron and is carried to an effector. At the same time a signal is carried along the spinal cord to the brain. *(Redrawn from BSCS, 3d ed.)*

After stimuli are received in the brain and integrated with the information stored and being received there, they are acted upon. We must now complete our consideration of the progress of stimuli through the nervous system.

The System of Effectors

The responses to stimuli are often manifested in some sort of movement on the part of an organism. In a simple organism such as the hydra, the motor response to a stimulus is carried out by contractile cells. In higher organisms, the highly specialized cells that mediate movement are muscle cells. The examination of a relatively familiar experience can serve as an example of a motor response to a simple stimulus. Consider what happens when a physician taps the flexed knee of a patient to test the "knee-jerk" reflex. The tap and the jerk of the lower leg constitute a *reflex arc*, the simplest forms of which are made up of a receptor innervated by a neuron that synapses with another neuron innervating an effector. (See fig. 5.15.) The knee-jerk reflex arc is slightly more complex. The stimulus is delivered to a tendon that is innervated by a stretch receptor. The nerve impulse that is created by the changed position of the tendon travels to the spinal cord on the axon of the stretch receptor, which synapses with two neurons: one is a *motor neuron*, or a neuron that conducts a stimulus to an effector that induces movement; the other is an *association neuron,* which delivers the stimulus to the brain for integration. The motor neuron conducts the nerve impulse to the motor endplate of a muscle in the leg which responds by contracting and causing the lower leg to jerk. The association neuron keeps the subject aware of what is taking place. Although the knee-jerk reflex is a response to an intentionally delivered stimulus, it does indicate the role of muscles as effectors and so serves to introduce us to the problem of how muscles operate.

We may categorize muscle into two classes based on the form of control that is exercised over each. The first category is *involuntary muscle*, and the operation of these muscles cannot

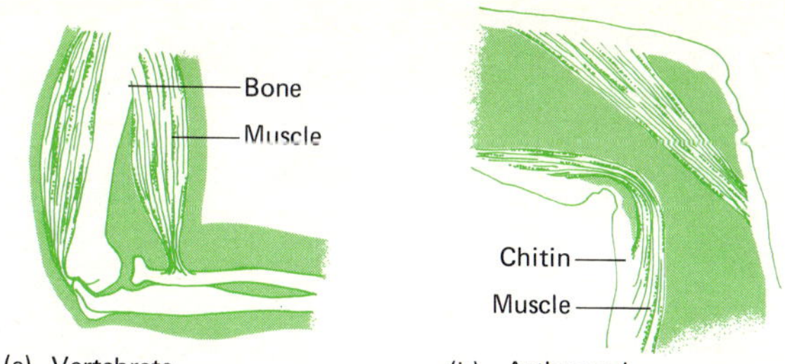

FIGURE 5.16 In a vertebrate, muscles attach to an internal skeleton, and in an insect to an exoskeleton. Examine what muscles control the movement of your elbow joint (a), and compare that with what must happen for the equivalent movements in an insect (b). *(Redrawn from BSCS, 3d ed.)*

usually be consciously controlled by the organism. Involuntary muscle is found in the internal organs of the body; for example, lining the wall of the stomach or blood vessels or making up the major portion of the tissue of the heart. Although involuntary muscles may be innervated by neurons, they may also be under the control of chemicals secreted from glands or may even be capable of functioning for a time in the absence of any connection with the body of an organism. Heart muscle tissue from chicks, for example, can be seen to continue to ''beat'' when removed from the chick and kept in a solution to which certain nutrients have been added. The other class of muscle is *voluntary muscle* (sometimes also called *skeletal muscle*) and is primarily under the conscious control of the organism. As its alternative name indicates, this is the muscle tissue that is primarily concerned with the movements of the skeleton (See fig. 5.16.)

The cells which compose muscles are traditionally categorized into three groups; *smooth muscle* cells, *cardiac muscle* cells, and *striated muscle* cells. (See fig. 5.17.) Most of the involuntary muscle in the body is made up of smooth muscle cells organized into sheets of muscular tissue. Striated muscle cells, which are named for their appearance under a light microscope, are organized into elongated bundles connected to the skeleton. The operation of smooth muscle is under the control and coordination of the medulla, whereas striated muscle is controlled from other parts of the brain. One more difference is that smooth muscle cells contain only one nucleus, while striated muscle cells may contain several nuclei. Cardiac muscle was thought for a long time to be significantly different from either smooth muscle or striated muscle, but recent evidence indicates that it shares some characteristics with both. Like smooth muscle, cardiac muscle cells are organized into sheets rather than bundles, but like striated muscle, cardiac muscle cells contain multiple nuclei. Although cardiac muscle is striated, it is under the chemical and neural control of the medulla.

Not all animals possess all types of muscle cells. Insects, for example, often contain only striated muscle cells, even in their internal organs. Some invertebrates possess only smooth muscle, and some possess smooth and striated muscle. Generally speaking, striated muscle is adapted to rapid contraction but cannot remain contracted for very long. Smooth muscle contracts slowly but may remain contracted for very long times.

Clearly, striated muscle cells are an adaptation to the external environment, not only because their rapid action allows rapid responses but also, obviously, because they are primarily involved with the movement of the body. The slower contraction time of smooth muscle is well suited to the internal environment, and the combination of characteristics exhibited by cardiac muscle is also clearly adaptive. Its organization into sheets allows it to form the heart, and its resemblances to both smooth and striated muscle allow for a variable rate of contraction; thus, it may be thought of as capable of responding to both the internal and external environment. Because most of our understanding of the characteristics of muscle has been the product of the study of striated muscle cells, we shall restrict our subsequent discussion to this kind of muscle.

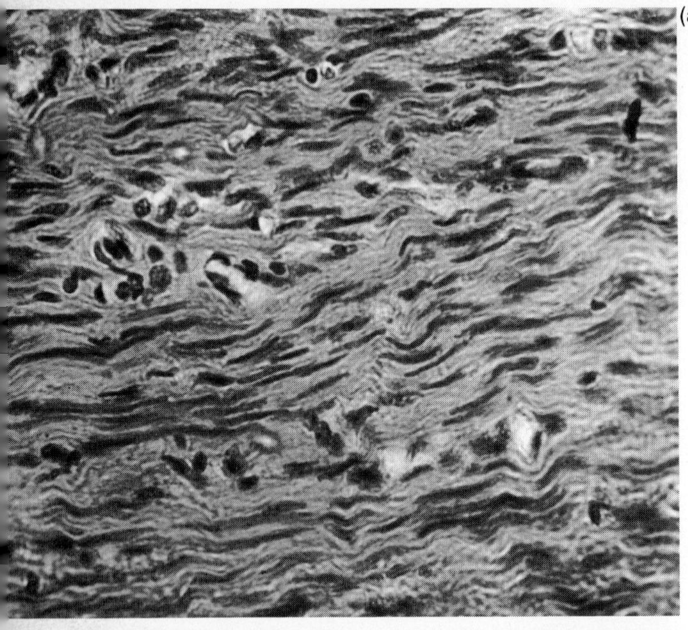
(a)

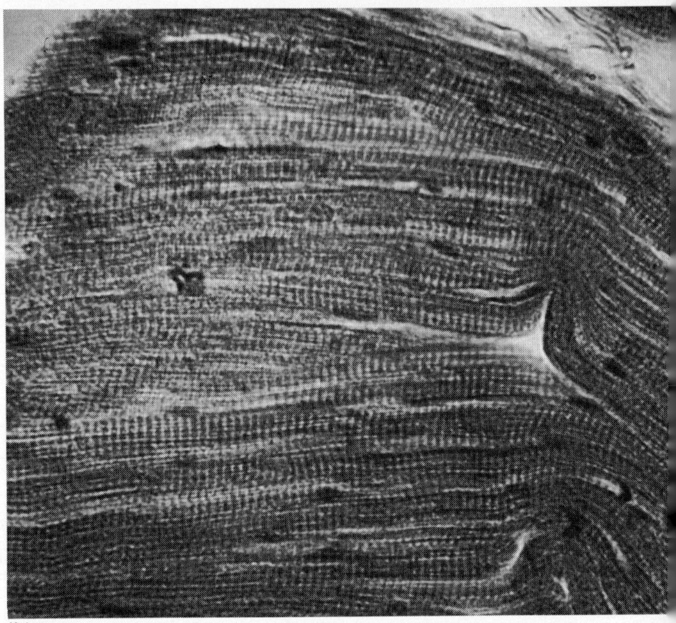
(b)

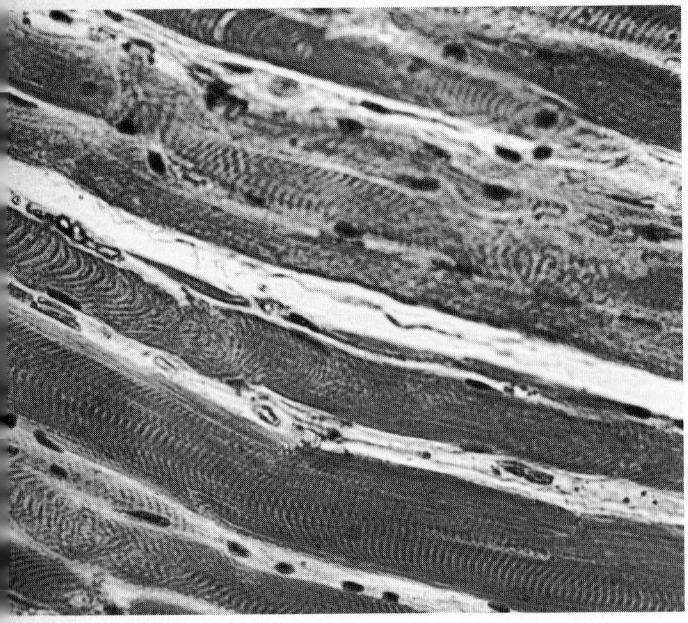
(c)

FIGURE 5.17 (a) Smooth, (b) cardiac, and (c) striated muscle. *(Courtesy of Runk/Schoenberger from Grant Heilman.)*

Striated muscle cells are organized into elongated bundles surrounded by a sheath of connective tissue which attaches at each end of the large bundle to ropelike connective tissues called *tendons.* The tendons, in turn, attach to the skeleton. The movement of the body is produced by the contraction of the bundle of muscle cells, hereafter referred to simply as a muscle, which pulls on the tendon. Since the tendon can only pull the skeleton, two muscles must be involved for each movement.

Each muscle is innervated by many motor neurons, each of which synapses with a number of fibers. This complex of motor neurons and fibers is called a *motor unit.* As a single nerve impulse is delivered on the axon of a motor neuron, all of the fibers that synapse with that neuron contract. As the number of motor neurons delivering impulses increases, greater numbers of fibers contract. When the muscle is fully contracted, nearly all of the motor neurons that innervate the muscle are firing. Thus, the contraction of the muscle can be graded, allowing a variety of responses to stimuli. Consider the difference, for example, between lifting a pencil from the floor and placing it on a desk, as opposed to lifting a dictionary from the floor and raising it to the same height. Since the pencil offers less resistance—that is, is lighter—relatively few motor units must be involved. Raising the dictionary involves the participation of a larger number of motor units. If the same number of motor units were involved in both cases, either the dictionary would have to be raised much more slowly, if it could be raised at all, or the pencil would be raised very rapidly. From what we have discussed so far, however, it is not perfectly clear why this should be so. Why, for example, could it not be that all the fibers in a bundle would contract, but at a slower or faster rate depending upon the amount of resistance offered?

The answer is that the contraction of a fiber is an *all-or-none response*; either a fiber contracts and contracts fully, or it does not contract at all. When it does contract, it does so at the same characteristic rate every time. Thus, in order to achieve a graded response, more or fewer motor units must be in operation at any time. We must now examine how a fiber contracts in order to complete our understanding of the progress of a stimulus through the nervous system.

Contraction of a Muscle Fiber

Electron microscopic studies have revealed the presence within the fiber of repeating units of *fibrils,* which are made up of *thick* and *thin filaments.* (See figs. 5.18 and 5.19.) When a fiber is relaxed, these filaments are arranged in a distinctive pattern shown in figure 5.20(a). After contraction, the pattern changes greatly, *giving the appearance that the filaments have slid past each other.* (See fig. 5.20(b).) Biochemical studies have shown that the fibrils contain large amounts of long, chainlike protein molecules. About half of the protein in a fibril is *myosin,* about a quarter is *actin,* and the remainder is made up of a number of other proteins, the most abundant of which is *tropomyosin.* Myosin and actin are believed to be the proteins of which, respectively, the thick and thin filaments

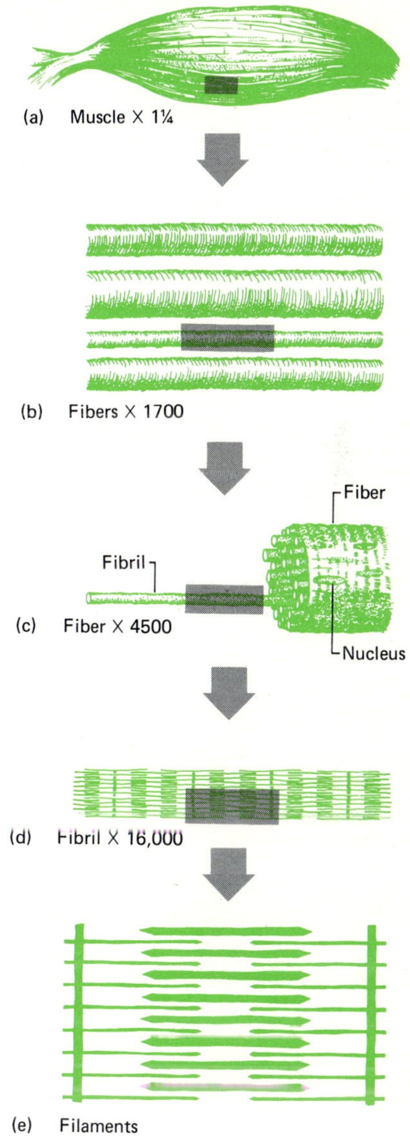

FIGURE 5.18 A whole muscle (a) is comprised of bundles of fibers (b) which in turn are made up of fibrils (c). The fibrils themselves are composed of filaments (d), (e) and these in turn are made up of actin and myosin, as shown in Figure 5.19. *(Redrawn from BSCS, 3d ed.)*

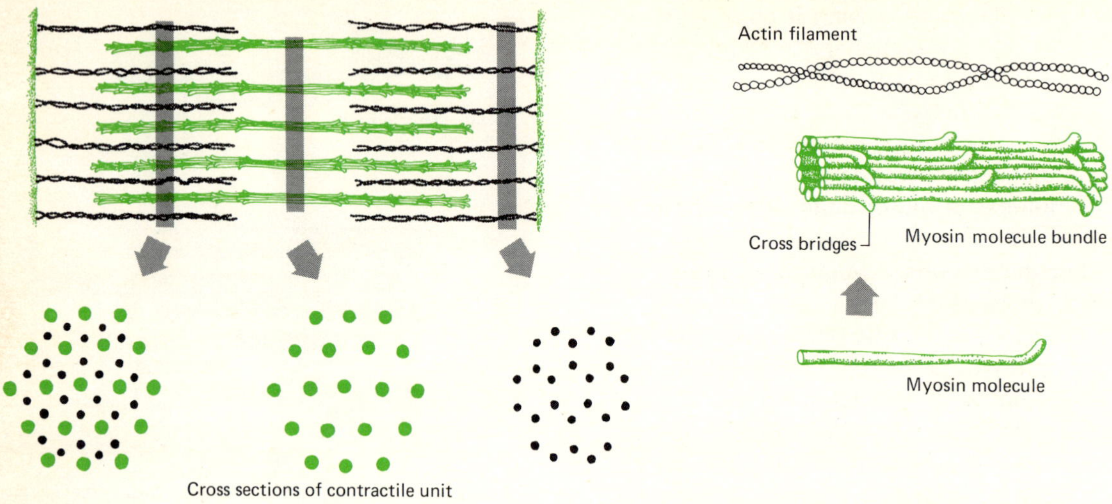

FIGURE 5.19 Fibrils of muscle fibers are made up of repeating units structured of actin and myosin. As shown in the diagram, the actin and myosin molecule bundles interdigitate in a very regular, geometric pattern.

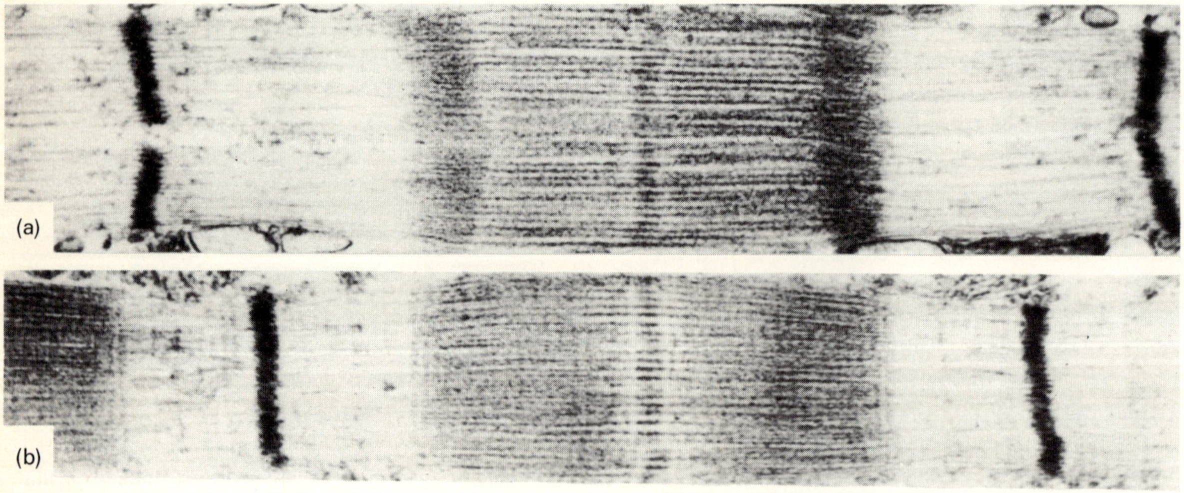

FIGURE 5.20 Contracted muscle and relaxed muscle. As muscle in the relaxed state (a) contracts, the filaments of actin and myosin slide past one another giving the appearance of the electron micrograph in (b). *(Courtesy of Hugh E. Huxley, Medical Research Council, Cambridge, England.)*

are made, and these two proteins are thought to be the most active in the contraction of the fiber. Actin and myosin have been isolated from fibrils, and when placed in a solution, they form a complex called *actomyosin.* When energy is then added to the solution in the form of a chemical substance we shall hear more about later (ATP), the molecules of actomyosin can be seen to contract. Thus, it appears that all that is needed for contraction is actomyosin and chemical energy.

The behavior of actomyosin in the fibril is described by what is called the *sliding filament theory of contraction.* In the electron micrograph in figure 5.21 small projections of the myosin molecule can be seen to form "cross bridges" connecting the myosin to actin to form actomyosin. It is believed that as energy is supplied to the fiber, these cross bridges react chemically in such a way that the thin filaments of actin are drawn along the surface of the myosin molecule in a manner shown in figure 5.22.

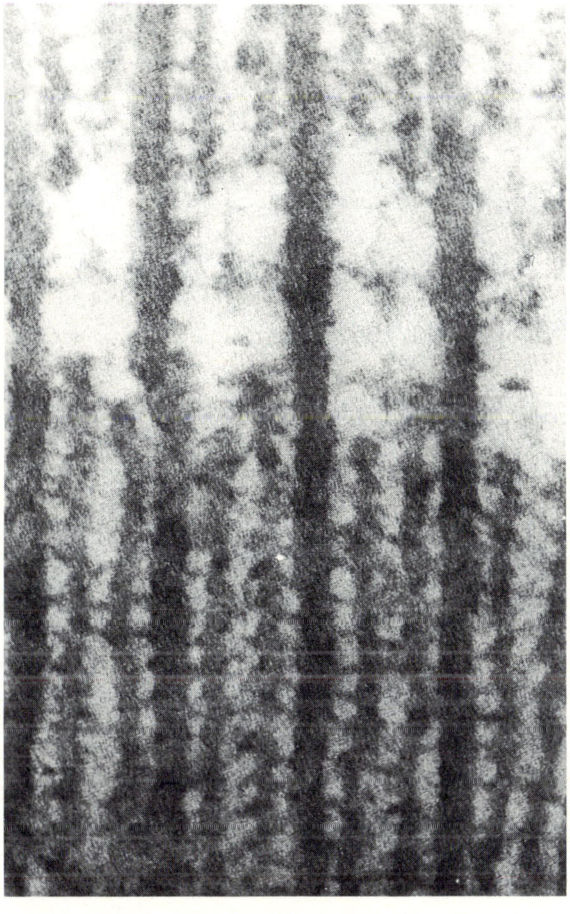

FIGURE 5.21 The thick filaments of myosin are connected to thin filaments of actin by crossbridges. As shown in Figure 5.22, these cross bridges are thought to result in the contraction of the muscle. *(Courtesy of Hugh E. Huxley, Medical Research Council, Cambridge, England.)*

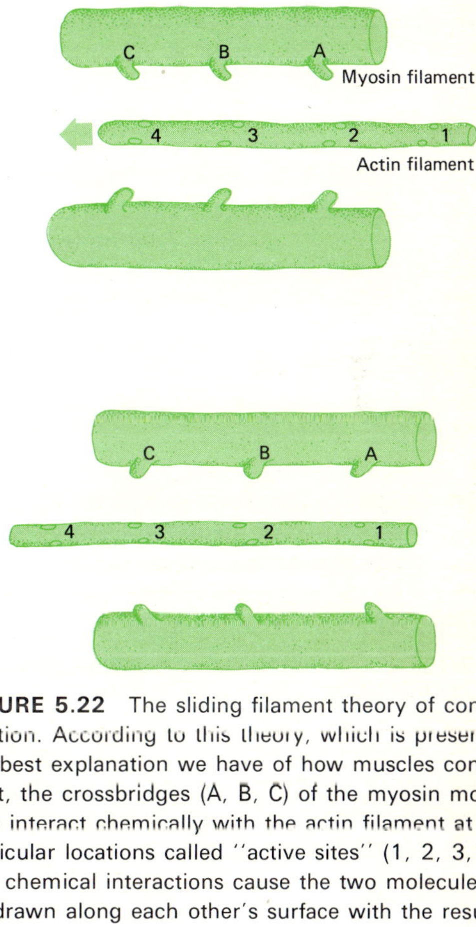

FIGURE 5.22 The sliding filament theory of contraction. According to this theory, which is presently the best explanation we have of how muscles contract, the crossbridges (A, B, C) of the myosin molecule interact chemically with the actin filament at particular locations called "active sites" (1, 2, 3, 4). The chemical interactions cause the two molecules to be drawn along each other's surface with the result that the contractile unit contracts.

What appears to happen when a nerve impulse reaches a muscle endplate is that ions of calcium (Ca^{2+}, indicating the lack of two electrons) are released into the fiber from areas of the fiber membrane. These calcium ions undergo chemical reactions with the chemicals in the fiber, and these reactions release energy. This released energy then drives the chemical reactions of actin and myosin to produce the contraction of the fibril. Because fibrils are so small, the contraction of a single fibril could not produce the movement of a fiber, let alone an entire muscle; but the process is repeated simultaneously thousands of times in many fibers to bring about a contraction of the entire muscle.

Clearly then, the progress of a stimulus through the nervous system of an animal requires energy at nearly every step. Therefore, in the next section of this chapter, we must take up the problems faced by organisms in the acquisition of energy and the transport of materials associated with delivering energy to cells.

The Transport of Materials

To survive, all animals must take food into their bodies to be converted there into energy for growth, tissue maintenance and repair, reproduction, and utilization in the nervous system. The food taken in is nothing other than chemical compounds organized into a particular form. The problem is that the form in which the food is taken in is not the form in which it can always be used. For example, one of man's most important foodstuffs is rice, an energy rich *carbohydrate* made up of many molecules of *sugar.* In order for a human being to extract the energy in rice, however, the carbohydrate must be broken down, or *digested,* into molecules of sugar that can enter the cell. Once in the cell, the sugar can be further broken down and converted into other energy-rich-compounds for use in the metabolic processes of the cell. Proteins, fats, and oils also contain energy, and these chemicals must similarly be digested into their constituent parts before they can be used in cellular activities. Thus, food must undergo a series of chemical changes before the organism can use its energy or incorporate it into the body.

Single-celled organisms live in environments in which food is abundant and usually continuously available to be taken into the cell. Thus, they are faced only with the problems of effecting the necessary chemical changes to convert food into energy. Multicellular organisms, however, are also faced with the problem of transporting the food to the cells of the body, some of which may be at considerable distances from the point at which food is taken in. We have already seen that plants have a similar problem in the terrestrial environment. Thus, as in terrestrial plants, systems to serve the internal transport of materials have evolved in many multicellular animals. These *circulatory systems* vary considerably from animal to animal, and although they serve an analogous function in the lives of all multicellular animals, they also differ significantly from the systems that have evolved in terrestrial plants.

We should also point out here that the transport of materials is as basic to the functioning of ecosystems as it is to the functioning of individual organisms. We shall see just how basic in Chapter 10.

The conversion of food to energy is called *respiration* and, in most organisms, requires the intake of oxygen (O_2) and produces water (H_2O) and carbon dioxide (CO_2). As we have seen, some animals are adapted to retaining large volumes of the water that is produced in respiration; many, of course, are not so efficient. Carbon dioxide, on the other hand, must be disposed of in some way. Many multicellular organisms have circulatory systems that serve to both transport food and oxygen to the cells of their bodies and carry away excess water and carbon dioxide. As we have seen, insects possess two separate systems to perform these functions.

Digestion in Simple Animals

Since the intake of food is the beginning of the process, let us look first at how some simple organisms distribute food through their cells. Unicellular organisms, of course, need only to bring food into the cellular body. Some, amoebae for example, simply surround food particles with their bodies. Some protozoa, on the other hand, possess an oral groove lined with cilia that beat with a regular motion drawing food particles to an area of the body where the food enters and is surrounded by a vacuole. The food is broken down into its molecular constituents by chemicals (enzymes) secreted into the vacuole, and the molecules of food diffuse out of the vacuole and into the cytoplasm. A phenomenon called *cytoplasmic streaming* distributes the digested molecules of food throughout the cytoplasm and also carries the vacuole to another area of the body where any undigested wastes are expelled. CO_2 simply diffuses out of the cell into the environment.

Some aquatic multicellular organisms, such as sponges, take food into the body through incurrent pores, where it is taken up by the collar cells in a way analogous to that in protozoa.

Coelenterates exhibit the beginnings of true multicellularity. The coelenterate body consists of two layers of cells organized around a *gastrovascular cavity* (*gastro* meaning "stomach"; *vascular* meaning "circulatory vessel"). Food that is brought into the gastrovascular cavity is in part digested there by enzymes secreted by some of the internal cells. (See fig. 5.45.) Such *extracellular* digestion is a major evolutionary departure and, in one form or another, is characteristic of most higher multicellular organisms. In the coelenterates, however, extracellular digestion is incomplete. Only the proteins in the food are digested in the gastrovascular cavity. Food particles are then taken up by the inner lining of cells and digested in food vacuoles. Waste products are eliminated through the single aperture, which acts both as mouth and anus. Since most coelenterates are relatively small, the products of digestion are distributed to the contractile and receptor cells in the outer lining of the body by diffusion. It should be noted that the means of digestion employed by coelenterates allow them to take into their bodies food particles that are, relative to body size, much larger than those available to protozoa and sponges. Thus, they may exploit more of the resources in their environments.

Some other adaptations to the transport of materials in more complex organisms are shown in box 5.3.

Digestion in Higher Multicellular Animals

In order to understand more fully how food is processed for conversion to energy, we shall turn our attention to higher organisms, and primarily to man. Like many invertebrates, man possesses the means to grind his food—namely, teeth. As you would expect, human teeth are adapted to the food resources upon which man depends. Carnivores generally have sharp teeth adapted to tearing and shredding flesh. Herbivores have teeth adapted to grinding. Omnivores, such as man, have teeth of both sorts. Some carnivores, such as snakes, do not use their teeth for the breakdown of food at all but use them to impale food and transfer it to other parts of the system for digestion. Thus, the teeth of a snake are generally sharply pointed and curve backwards. Some animals, such as the hagfish, have teeth on their tongues and use the combination of organs to manipulate food and to break it down at the same time. Man, of course, uses his tongue to manipulate his food, to communicate by talking, and as a sense organ.

In higher animals, the enzymatic breakdown of food frequently begins in the mouth. In humans, starches are subjected to the action of *amylase,* an enzyme that breaks starches down into sugars. As the food is chewed the amylase in the saliva breaks down more and more of the starches, and other substances lubricate the food to facilitate its easy passage to the next stage of the system. The food is passed into the storage organ, the stomach, through a tubular

Digestive System of the Flatworm and the Earthworm

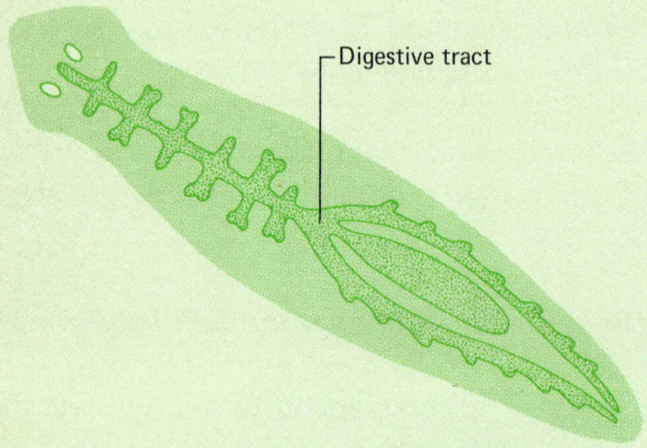

Digestive tract of flatworm. The folding of the tissue lining the gastrovascular cavities of flatworms greatly increases the surface area of the cavities. This allows for the presentation of more food to the cells of the body and thus the extraction of more energy from the food supply.

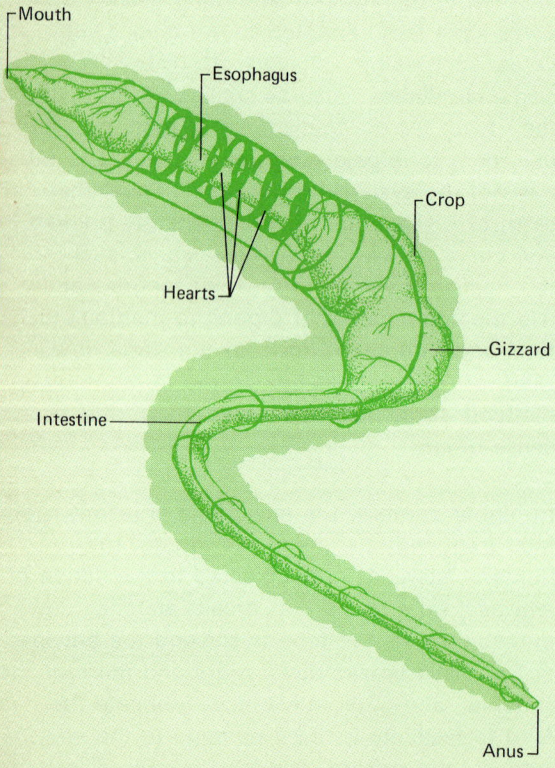

The earthworm possesses a crop, in which food may be stored, and a gizzard, in which it is ground mechanically by muscular contractions. The masticated food is extracellulary digested in the intestine, and the products of digestion are reabsorbed at the intestinal walls. Water is also reabsorbed in the intestine.

Although the earthworm possesses a circulatory system through which digestive products are transported, it exchanges oxygen and carbon dioxide by diffusion through its outer cuticle.

The crop allows the earthworm to practice discontinuous feeding. Thus, it is able to take in large amounts of food to be released and utilized as needed. In an environment where food is dispersed, discontinuous feeding permits the organism to survive between meals and to devote time and energy to such activities as searching for a mate or rearing the young.

BOX 5.3

connector, the esophagus. Once food has passed through the esophagus, the tube is closed by a ring of muscle that prevents the food from re-entering the esophagus during the next stage in the process. Lined with three layers of muscle, each oriented in a different direction, the stomach begins to undergo a series of strong contractions when food is introduced into it. While the stomach is grinding the food, the enzymatic breakdown of the food is continued. The enzymes in saliva serve only to break down starches, and that is continued in the stomach. But now some proteins are also broken down. In man, the digestive enzyme *pepsin* begins the process of protein breakdown in the stomach.

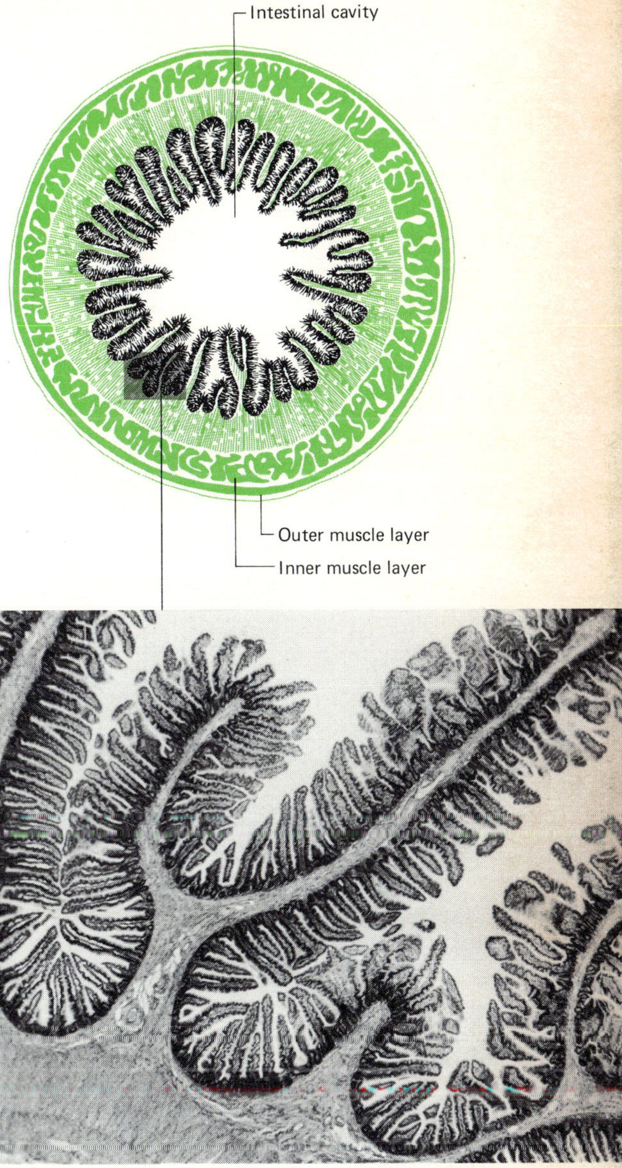

FIGURE 5.23 Cross section of the small intestine. Just as the root hairs of the plant provide a tremendous absorptive surface within a relatively small space, the villi, microvilli, and highly folded interior of the small intestine provide a very large surface area for the digestion of food. *(Redrawn from BSCS, 3d ed.; photo, courtesy of Carl Struwe, Monkmeyer Press.)*

The greatest digestive activity takes place in the small intestine, which is the next part of the digestive tract. The mixture that leaves the stomach is still a largely undigested mass, but when it reaches the small intestine, a number of enzymes are secreted by the intestinal wall and by other organs. The digestion of starches is completed here, and most of the fats, oils, and proteins are also broken down in the small intestine. The breakdown of proteins and fats is carried out by a large number of enzymes. Each enzyme may break down only a portion of a protein or fat, and thus the number of enzymes that the organism must synthesize in its body can be very large.

The Structure of the Small Intestine

Like the gastrovascular cavity of the flatworm, the interior of the small intestine is not a smooth-walled space; a human intestine, contains many minute fingerlike projections called *villi.* (See fig. 5.23.) Each villus is made up of cells that themselves have an immense number of closely packed cylindrical outgrowths called *microvilli.* (See fig. 5.24.) The combination of villi and microvilli provides an extremely large surface area over which food must pass. The length of the intestine also has adaptive significance with respect to the food resource that the organism depends upon. The intestines of strict herbivores, for example, are very long relative to body size. The length is correlated with the difficulty of digesting plant material. The plant cell wall is made up primarily of cellulose, which most animals cannot digest. The greater length of the small intestine in a herbivore is an adaptation that permits the food to come into contact with a far greater surface area and to be acted upon by enzymes for a longer period of time. Conversely, strict carnivores have relatively short intestines, since they encounter no similar difficulty in digesting the food they procure. A particularly interesting example of the differing lengths of the intestine in herbivores and carnivores can be seen in the frog. Frog larvae are herbivores and possess highly coiled small intestines that are relatively longer than the intestine of the carnivorous adult frog. (This is also an excellent example of protection of the young by adaptation to a different niche from that occupied by the adult.) Man, as an omnivore, has an intestine of an intermediate length, relative to body size, of about twenty-three feet.

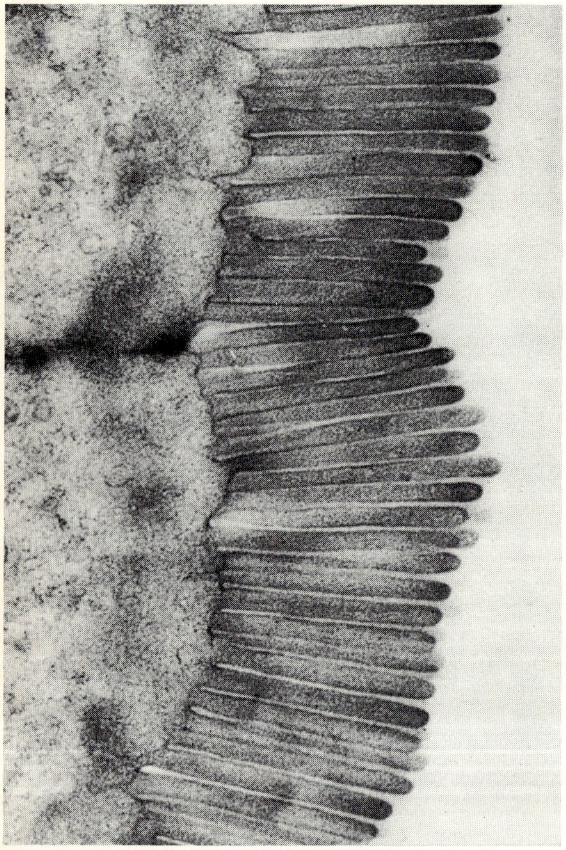

FIGURE 5.24 A large surface area in the intestine allows for high rates of uptake of the products of digestion. This surface area is provided in part by the villi, and is increased by the cylindrical outgrowths from the villi, called microvilli, that are shown here. *(Courtesy of Don W. Fawcett, Harvard University Medical School.)*

The presence of adaptive structures for the conversion of particles of food to molecules in the small intestine is complemented in many organisms by the presence of intestinal bacteria or protozoa. In fact, there are bacteria in the intestines of all animals. These microorganisms often serve to digest cellulose and thus make more of the food available to the host organisms. Cows and other ruminants, for example, support large populations of bacteria in their digestive tracts, and these bacterial populations break down cellulose, material that the cow is incapable of digesting. A number of insects that feed on wood also harbor bacteria or protozoa that digest cellulose.

The enzymatic breakdown of starches, proteins, and fats is carried out in the small intestine in a soupy mixture made up largely of water. Some of the water passes through the intestinal wall, but most of it continues through the system into the large intestine. Entering the large intestine, the food undergoes little further digestion. At this stage, most of the water is reabsorbed into the body. If it were not, severe water loss would occur in the form of diarrhea. Occasionally this happens when for some reason the large intestine becomes irritated and moves food through too rapidly to allow sufficient water reabsorption to take place. Should the food move too slowly, too much water would be absorbed resulting in constipation.

The Movement of Molecules from the Small Intestine

Within the intestinal villi are beds of *capillaries*, the smallest blood vessels in the body. (See (b), box 5.4.) Such beds also occur in other parts of the body, especially in the lungs, and it is through these capillaries that the products of digestion and respiration are exchanged. Although some nutrient materials make their way into the system from the stomach and large intestine, by far the greatest amounts diffuse into the bloodstream in the capillary beds of the small intestine.

Although diffusion is a slow process, the capillaries ramify into virtually every part of the body, and thus no cell is without some contact with a capillary. In the human body, for example, there are approximately 53,000 miles of capillaries. The capillary wall is highly permeable to the products of digestion and to oxygen, water, and carbon dioxide. The blood itself contains some very large entities, such as red and white blood cells and large proteins, none of which can diffuse through the capillary. The products of digestion and some water diffuse out of the capillaries and into the intercellular fluid bathing the cells and are taken up by the cells as their physiological states demand. At the same time, the waste products of respiration diffuse out of the cell and enter the capillaries to be carried away.

The capillary system feeds into larger blood vessels, the *venules* and *arterioles,* which in turn feed into still larger vessels. Ultimately flowing into the largest vessels in the body, the veins and arteries, the blood is delivered to the heart. In man, the blood that reaches the heart from the intestines contains a rich supply of nutrients from the extracellular digestion of food in the small intestine, but it also contains a large amount of carbon dioxide, the product of the respiration of cells in the intestines and in the blood vessels themselves. Lacking the oxygen required for the further respiration of the nutrients it contains, the blood enters the left side of the heart and is pumped from there into the lungs. In the capillary bed of the lungs, the blood gives up its accumulated carbon dioxide and picks up oxygen. It also supplies nutrient molecules to the cells of the lungs. The now oxygen-rich blood is drawn into the right side of the heart where it is pumped with its food supply to the rest of the body. (See box 5.4.)

The circulatory system of a vertebrate is a closed one; that is, the blood being pumped throughout the body remains within the arteries, veins, and capillaries, and only the nutrients and gases necessary for respiration or resulting from respiration ever enter or leave the system. In some invertebrates, however, open circulatory systems can be found. In some insects, for example, there is usually only one well-defined blood vessel that runs the length of the body. (See fig. 5.25.) This vessel serves as a heart and is pierced by several small openings called *ostia*. When the vessel contracts, blood is forced out and forward into a cavity where it bathes the insect's organs directly, and then it flows backward to supply the more posterior organs. When the vessel relaxes, the blood is drawn in again at the rear. As a system for transporting nutrients, open circulatory systems tend to be slower than closed ones, and it might seem that such very active animals as insects would be at a disadvantage without a rapid means of transporting nutrients to their cells. However, the movement of the blood can

FIGURE 5.25 While the vertebrate circulatory system is completely closed, that is, all blood flows through vessels, that of arthropods is distributed throughout the body, where it flows freely between the tissues.

The Flow of Nutrients and Oxygen

The digestion of food in the small intestine produces molecules of sugar, proteins, and fats which diffuse through the intestine wall and into the capillaries in the intestinal villi. (a) The products of digestion are carried to cells, where they diffuse out of the capillaries and into the cells. At the same time, the waste materials of intracellular processes diffuse out of the cell and into the capillaries. In (b) are shown the major capillary beds of the body and the flow of nutrients and gases in the blood. Note that as the blood leaves the capillary bed of the intestines it lacks the oxygen needed for the respiration of the nutrients that it contains. After passing through the liver, where certain wastes and impurities are removed, the blood enters the heart, is pumped to the lungs, and there exchanges carbon dioxide for oxygen. Part (c) shows the flow of blood through the heart. If we imagine the entire cycle as beginning in the intestine, then we can follow the flow of nutrients and gases through the whole body.

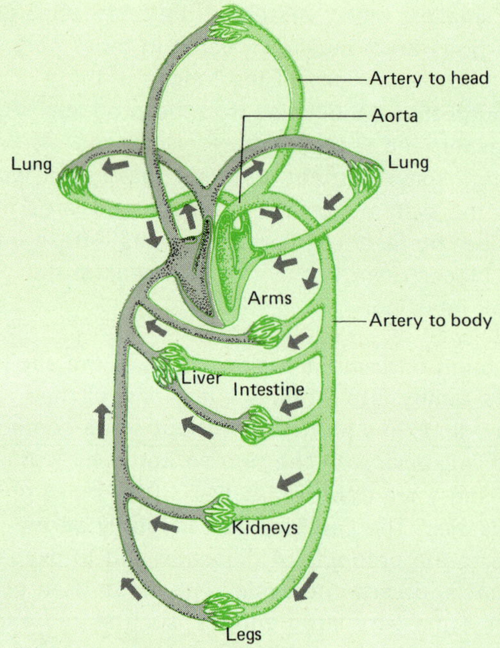

(b)

(Redrawn from BSCS, 3d ed.)

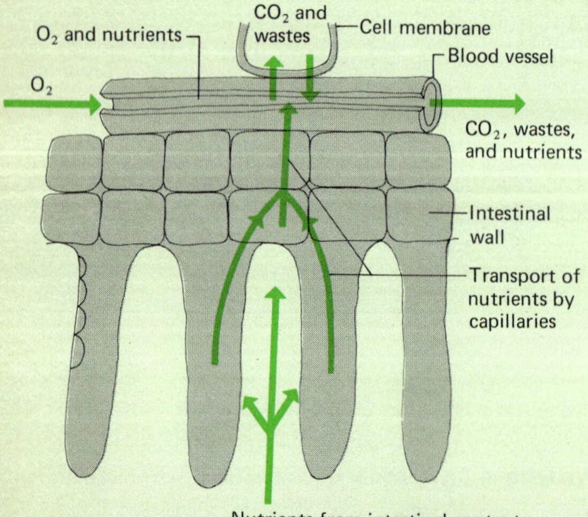

(a)

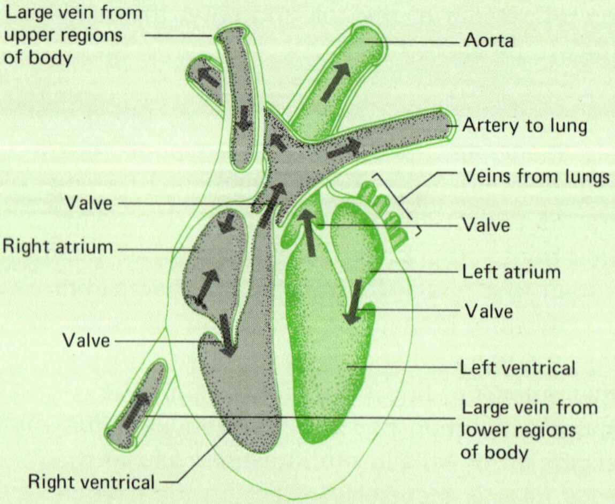

(c)

BOX 5.4

be accelerated as activity increases, since the muscles and organs begin moving rapidly, bringing about a mixing action in the blood-filled cavity. Thus, activity itself causes the blood to move more rapidly, and when the insect is flying or running the needs for nutrient delivery can be efficiently met.

Since oxygen and carbon dioxide play such important roles in the use of energy by the cell, we should examine more carefully how these gases are exchanged in the body.

The Exchange of Gases

Each human red blood cell contains approximately 280 million molecules of *hemoglobin*, a complex molecule made up of four protein chains (two different pairs) and four *heme* groups, which are organic structures containing iron. (See fig. 5.26.) Because of its structural relationship with the protein chain, each iron atom in the heme group is capable of loosely combining with a molecule of oxygen. Thus, a single molecule of hemoglobin may combine with four oxygen molecules, becoming what is called *oxyhemoglobin*. The combination of oxygen and hemoglobin changes the shape of the larger molecule slightly, and makes it appear redder; thus, arterial blood is redder than venous blood.

Whether oxygen and hemoglobin combine depends, in part, on the partial pressure of oxygen in the medium in which hemoglobin is transported. (The partial pressure of a particular gas is arrived at by determining the total pressure of all gases in a container multiplied by the percentage of the volume that is occupied by the gas under consideration.) When the partial pressure of oxygen is high, oxyhemoglobin forms readily; when it is low, oxygen dissociates from hemoglobin. Thus, when hemoglobin enters the capillaries of the lungs, where oxygen is a major component of the gases drawn in from the atmosphere and thus has a high partial pressure, oxyhemoglobin forms. In the capillary beds in tissue, the partial pressure of oxygen is low, and the oxyhemoglobin disso-

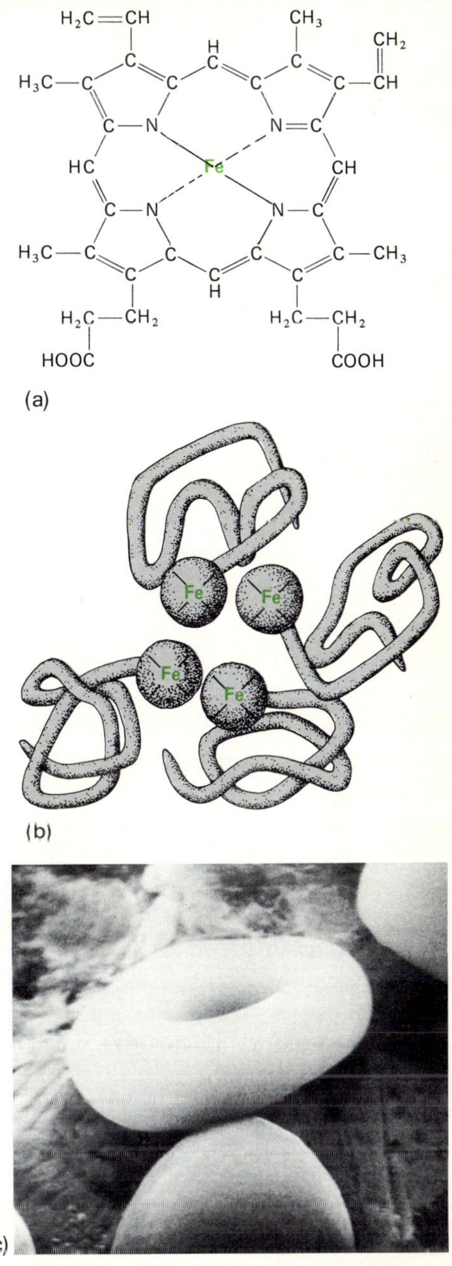

FIGURE 5.26 (a) The heme group. (b) Each molecule of hemoglobin consists of protein chains attached to a heme center, which consists of four heme groups. The actual molecule is much more highly folded than indicated here. (c) A scanning electron photomicrograph of a red blood cell, which contains the hemoglobin. *((c) Courtesy of Patricia Farnsworth, College of Medicine and Dentistry of New Jersey.)*

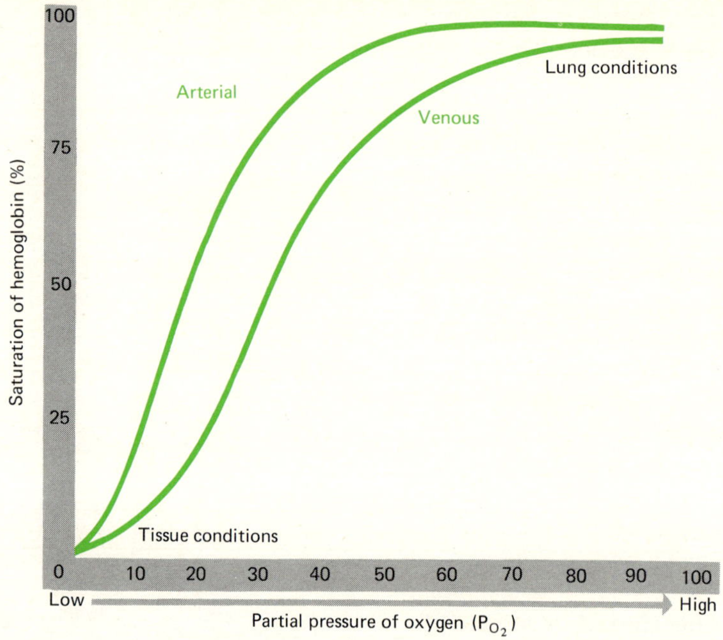

FIGURE 5.27 Dissociation curves of hemoglobin. In the high CO_2 conditions of venous blood, a particular partial pressure of oxygen results in less saturation of oxyhemoglobin formation than it does in arterial conditions.

ciates. Thus, it is possible to draw dissociation curves to indicate the percentage of oxyhemoglobin dissociation in the lungs and tissue. Such curves are shown in figure 5.27.

Note in the dissociation curves that oxyhemoglobin dissociates readily in the tissue, thus releasing oxygen for the process of cellular respiration. Note also that the carbon dioxide content of the blood affects the dissociation of oxyhemoglobin. In the venous blood the high carbon dioxide concentration promotes dissociation, while at the same time, the low carbon dioxide concentration of the arterial blood promotes oxyhemoglobin formation. Thus, both oxygen partial pressure and carbon dioxide concentration in the blood facilitate the delivery of oxygen to respiring cells.

The carbon dioxide released by respiring cells also enters the blood, where some of it dissolves in the fluid. But carbon dioxide is not very soluble, and it reacts with water in the blood to form *carbonic acid* (the same acid that is in club soda):

$$H_2O + CO_2 \rightarrow H_2CO_3$$

Unless the blood is very acid, however, carbonic acid is not a great deal more soluble than carbon dioxide. Under the normal conditions of the blood—that is, neither too acidic nor too alkaline—another reaction takes place to form *bicarbonate* ions that are very soluble:

$$H_2CO_3 \rightarrow HCO_3^- + H^+$$

The hydrogen ions (H^+), however, raise the acidity* of the blood, and since an increase in blood acidity can prove harmful or even fatal, some other reaction must occur to prevent death. Hemoglobin has an affinity for hydrogen

*The acidity or alkalinity of solutions is measured in units of pH (Chapter 11). A neutral solution, one that is neither acid nor alkaline, is said to have a pH of 7. If a solution has a pH higher than 7, it is alkaline; a solution whose pH is lower than 7 is defined to be acid.

ions, combining with them even more readily than with oxygen. Thus, at the sites where oxyhemoglobin dissociates most readily, free hemoglobin is available to prevent the acid level of the blood from rising. The carbon dioxide is carried to the lungs in the form of bicarbonate, and the hydrogen ions are carried to the lungs in association with the hemoglobin of the venous blood. In the lungs, the hydrogen ions are released from the hemoglobin and co-react with the bicarbonate to form carbonic acid, which then breaks down into carbon dioxide and water by the reversal of the reactions shown above. The carbon dioxide then diffuses out from the body.

Homeostasis

Obviously, the processes that take place in the body of any organism are precisely controlled events or series of events. As we have seen, these events produce a state of what can be called equilibrium. But that term is misleading, since it implies that no change takes place. In fact, we have seen that the body tends in some ways to reduce equilibrium. The resting membrane potential, for example, is a precisely controlled imbalance. The term that more accurately describes the processes that take place within the body of a living organism is *homeostasis*, or the *maintenance of a steady state* (*homeo* same + *stasis* state or condition).

The internal control of an organism's physiology is brought about by homeostatic mechanisms. As an illustration, consider the mechanism involved in the entry of a simple sugar, *glucose*, into a cell. After a meal, the level of glucose in the blood rises. The increased blood glucose concentration stimulates the production in the pancreas of a hormone, *insulin*. This hormone has many effects in the human system, but its most important one is to stimulate the entry of glucose into cells, particularly muscle cells. When the glucose concentration rises, insulin production is stimulated, and the increased insulin concentration causes the glucose concentration in the blood to fall. The lower glucose concentration inhibits insulin production in the pancreas and the insulin concentration in the blood falls.

This is an example of a feature common to most homeostatic mechanisms—*negative feedback*. A feedback mechanism is one in which the output of the system is "fed back" to the system as input and thus controls the system. In negative feedback mechanisms, the output that returns as input inhibits the functioning of the system. See figure 5.28 for a diagrammatic representation of the insulin–blood-glucose system as a negative feedback mechanism.

FIGURE 5.28 The removal of blood glucose causes reduction of insulin secretion, causing more glucose to be retained in the blood, and an intermediate amount of insulin to be secreted. This results in a steady state of insulin and glucose concentration, and illustrates the way negative feedback can maintain homeostatic steady state conditions.

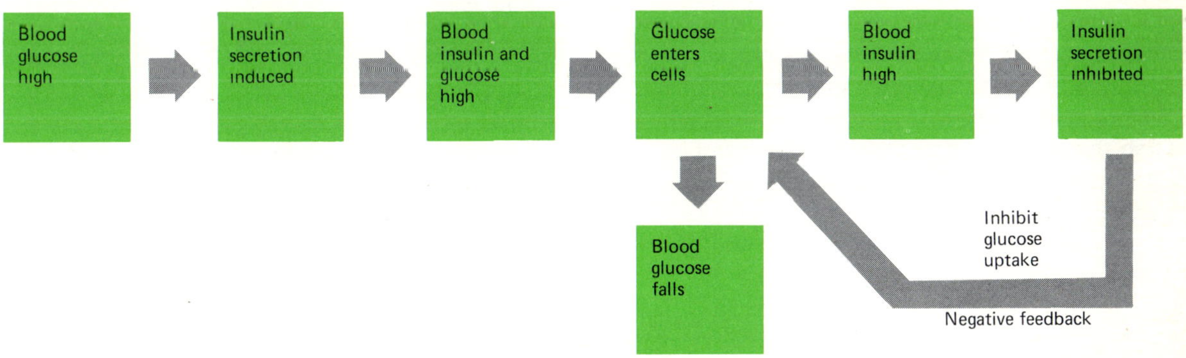

In conclusion, we can see that there are two environments that an organism is adapted to by natural selection: an internal one and an external one. As we shall later see, homeostatic mechanisms not only control the internal environment of an organism but may also be observed to exercise control over biological communities in a number of ways, including the regulation of species populations or the flow of chemicals in the environment.

We ought not to close this chapter without saying that we have only scratched the surface of the many physiological mechanisms that provide for the survival of organisms. The control of rates of breathing and the heart beat, the hormonal control of many of the internal activities of organisms, the excretion of liquid and solid wastes, the processes of nutrient transport in plants—all of these and many more have gone untreated, not because they are unimportant, but because our purpose has not been to exhaustively treat all the evolved adaptations of animals and plants. Rather, it was our intention to treat only a very few systems which we consider to be especially important in allowing species to survive and evolve and to show how these mechanisms are the evolved products of natural selection.

We have not discussed at all how these mechanisms are inherited or transferred from generation to generation, or how changes occur in the genetic composition of species to produce these adaptations. How, for example, can the complexities of the nervous system be transmitted from generation to generation with such precision? What is there about the transfer of genetic information from cell to cell that allows each cell of the lining of the small intestine to be essentially identical to every other cell there and yet so thoroughly different from a muscle fiber of a photoreceptor? And how are the structures of proteins determined, and what might the structures of proteins have to do with the process of natural selection? We shall now turn our attention to the subject of genetics in order to partially answer some such questions.

Selected Readings

*Gregory, R. L. *Eye and Brain.* New York: McGraw-Hill Book Company, 1966. A brief, extensively illustrated book that treats the biological bases and psychological effects of vision.

*Jennings, J. B. *Feeding, Digestion, and Assimilation in Animals.* Elmsford, N.Y.: Pergamon Press, Inc., 1966. A concise overview of the evolution, natural history, and laboratory work done concerning this subject.

*Katz, Bernhard. *Nerve, Muscle, and Synapse.* New York: McGraw-Hill Book Company, 1966. This is a brief but authoritative treatise on neurobiology. It is recommended to the student who wishes a rigorous text.

Prosser, C. Ladd, and Brown, Frank A. *Comparative Animal Physiology.* Philadelphia: W. B. Saunders Company, 1961. A complete and comprehensive treatment of physiological mechanisms.

Smith, C. U. M. *The Brain: Toward an Understanding.* New York: G. P. Putnam's Sons, 1970. This book assumes no knowledge of cell biology, and thus contains complete discussions of the cell, cell membranes, and cellular physiology as preparatory material for the discussion of neurons and sense organs, and their role in perception and nervous integration.

*Smith, Homer W. *From Fish to Philosopher.* Garden City, L.I.: Doubleday Publishing Company, 1959. An entertaining yet sound account of the evolution of higher animals with emphasis on osmotic regulation and the kidney.

*Wilentz, Joan S. *The Senses of Man.* New York: Thomas Y. Crowell Company, 1968. Written for the layman, this book provides a complete and accurate picture of human perception.

Mayerson, H. S. "The Lymphatic System." *Sci. Am.*, Vol. 208 (June, 1962), pp. 80–90. Offprint number 158.

* Paperback edition available.

Montagna, William. "The Skin," *Sci. Am.*, Vol. 212 (February, 1965), pp. 56–66. Offprint number 1003.

Perutz, M. F. "The Hemoglobin Molecule." *Sci. Am.*, Vol. 211 (November, 1964), pp. 64–76. Offprint number 196.

Rasmussen, Howard, and Pechet, Maurice M. "Calcitonin." *Sci. Am.*, Vol. 223 (October, 1970), pp. 42–50. Offprint number 1200.

Smith, Homer W. "The Kidney." *Sci. Am.*, Vol. 188 (January, 1953), pp. 40–48. Offprint number 37.

Wood, J. Edwin. "The Venous System." *Sci. Am.*, Vol. 218 (January, 1968), pp. 86–88. Offprint number 1093.

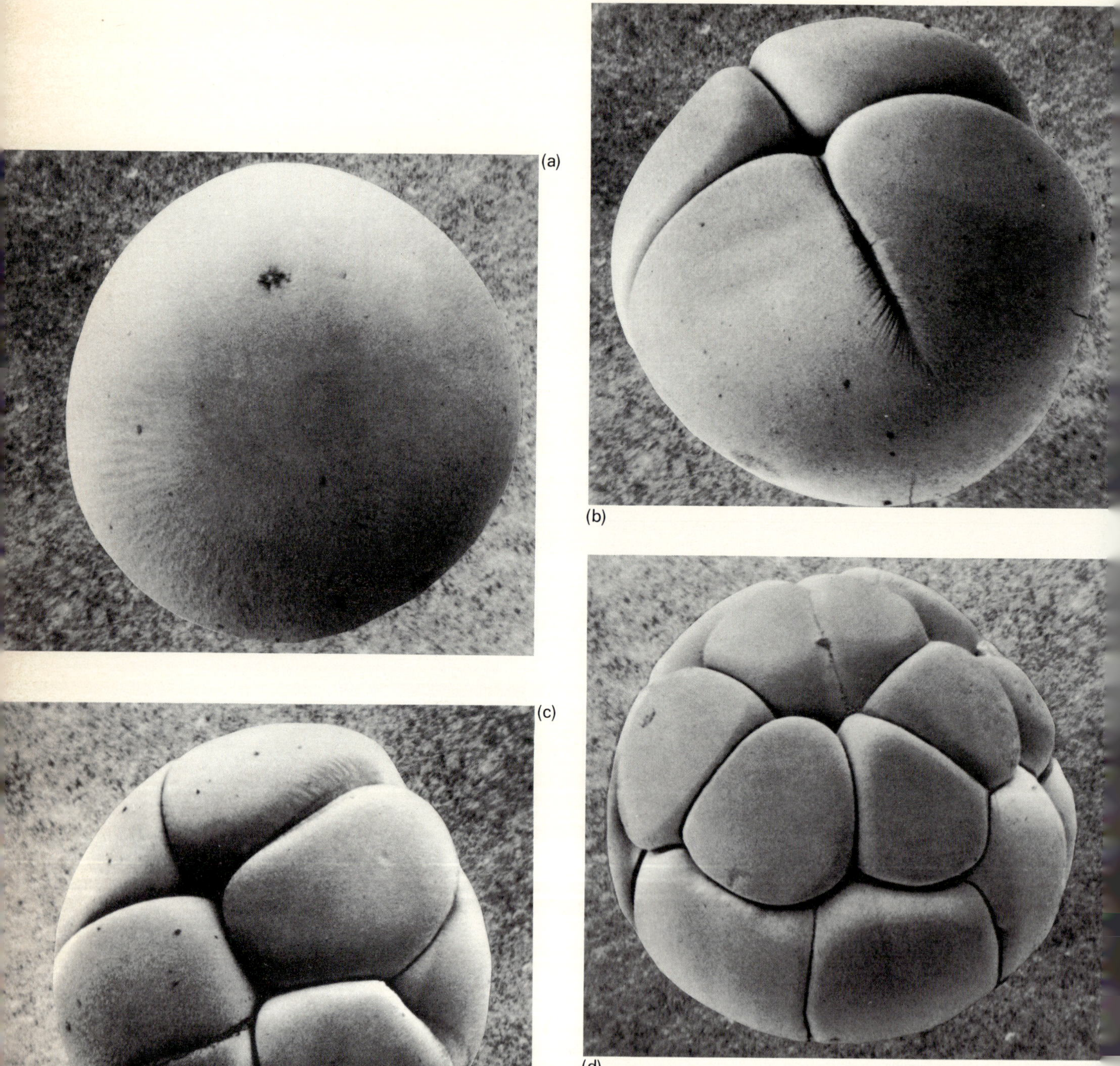

The start of an individual: The fertilized egg of a frog dividing to form the early stage of a blastula. Scanning electron photomicrographs. *(Courtesy of L. M. Beidler, Florida State University.)*

genetics and inheritance

chapter six A full understanding of the process of evolution demands an understanding of the structure and actions of genes. Since living organisms are very complex, ordered aggregations of chemicals, the question of what a gene is and how it operates is a chemical one. The chemical that we now know to be the gene is *deoxyribonucleic acid,* or *DNA.* Located in the chromosome, this self-replicating macromolecule contains all of the information necessary to direct the synthesis of every chemical and thus of every cell, tissue, and organ in the body of an organism. Furthermore, DNA may ultimately determine when those events shall take place; thus, it precisely directs how the individual will develop. In some (probably most) cases, DNA even determines exactly how the organism will behave in response to its environment. Obviously, DNA is important in the process of evolution, for it is DNA and genes that allow adaptations to be passed from generation to generation.

Changes may take place in the structure of a DNA molecule, and as the molecule replicates itself, the changes are passed on to other molecules of DNA. These changes may be so small that they cannot be detected by any known microscopic means or so large that they can be readily observed; both large and small changes are generally classified as *mutations.* Ultimately, mutations may be translated into differences in the structures of the chemicals that make up the body of an organism. Some mutations can occur in any of the dividing cells of the body of a multicellular organism (*somatic mutations*), but of most interest to us here are mutations in the DNA of the cells that contribute to the production of a new organism (*germinal mutations*). Germinal mutations, unlike somatic mutations, are reproduced in every cell of the new organism and result in an offspring that differs from its parents. This offspring will be affected differently by its environment, and if it survives and reproduces, it will pass on the mutation to its offspring. All the genetic material in a species population defines what that species is, and as the individuals are selected and their genes eliminated or propagated in the population, the characteristics of the population change and the species evolves.

If only one individual gave rise to another, then a species would be very difficult to define. Mutations would accumulate, and the individuals would gradually bear less and less relation to each other. Sexual reproduction provides species with a mechanism whereby the genes of one parent are combined with the genes of the other and the offspring share the characteristics of both. Sexual reproduction serves to "mix" the genetic composition of a population, prevents most individuals from departing very far from the species' norm, and thus conserves the characteristics of the population.

If there were little or no differences in the genetic constitutions of the members of a population, then there would be little variation for selection to work upon. If all members of the species were identical and the environment changed radically, the species might be completely wiped out. The variable genetic constitution of a species serves to prevent its extinction. Thus, mutations provide a source for variations in a species, and sexual reproduction serves to recombine the changes produced by mutations and prevent individuals from accumulating the changes. Of course, should the mutation be detrimental to the individual carrying it, that individual and therefore the mutation would be eliminated from the population by selection.

Inheritance by Cells

We have said that it is necessary to understand modern, biochemical genetics in order to understand the evolution of species and of ecosystems. Genetic information is transferred from individual to individual, but only as it is transferred from cell to cell. Spermatozoa and ova receive genes from the cells of the parents, and when the fertilized ovum divides to produce the millions of cells in a new individual, those genes are inherited by each newly formed cell. Thus, inheritance by cells is basic to inheritance by individuals.

The Chemicals of Inheritance

By the middle of the nineteenth century, biologists had concluded that all living things were composed of cells and that all cells arose from preexisting cells. But if this theory was valid, then two questions about inheritance might be asked: Does the genetic material exist in each cell? And, if so, where does it reside? Several fantastic theories were advanced that hypothesized the existence of the genetic material outside the cell, but they need not detain us here. Logic seemed to dictate that the genetic material resided in the cell, and the more fruitful lines of investigation focused on seeking an answer to the second question.

Even as early as Darwin's time, some biologists guessed that the nucleus of the cell contained the material of inheritance. Two lines of investigation resulted: the first was a study of the chemistry of the cell, particularly of the nucleus; the second was a study of the structure of the cell. But the early investigations of the chemistry of the nucleus were flawed. The material that was finally isolated was a mixture of protein and nucleic acids, and the determination of which, if either, was the material of inheritance had to await more refined isolating techniques. However, before the nineteenth century ended, the basic chemical constituents of the nucleic acids, *deoxyribonucleic acid* and *ribonucleic acid,* had been identified and described.

Deoxyribonucleic acid had been found to contain the following molecules: two *purines* (double-ring compounds), *adenine* and *guanine*—

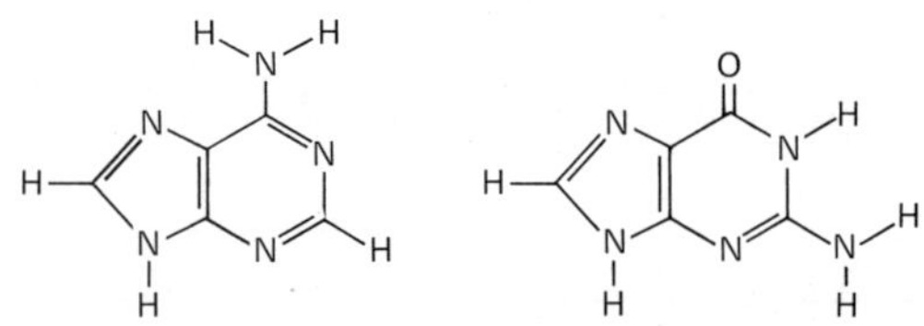

Adenine Guanine

and two *pyrimidines* (single-ring compounds), *thymine* and *cytosine*—

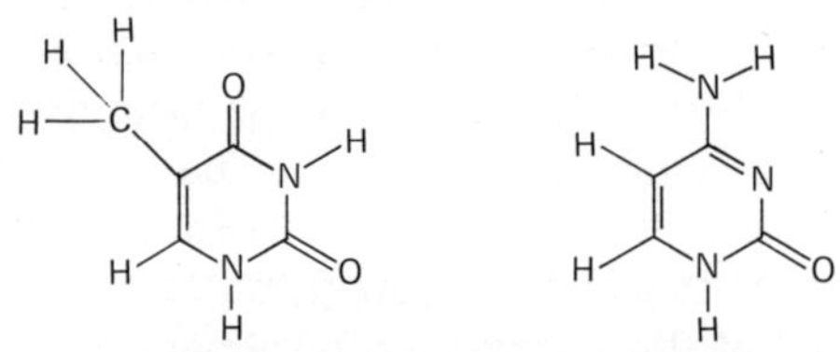

Thymine Cytosine

DNA was also found to contain a sugar molecule, *deoxyribose,* and a *phosphate* molecule.* Ribonucleic acid (RNA) was found to contain the purines, adenine and guanine, and one of the pyrimidines, cytosine; but it contained a different pyrimidine, *uracil,* in place of thymine, and a different sugar, *ribose,* in place of deoxyribose.

While some nineteenth century biologists were seeking answers to the question of heredity in

*As we shall see in chapter 9, biological molecules that are classified as acids are characterized by a carboxyl group (COOH). Nucleic acids are designated as such by virtue of the phosphate, which is also known as phosphoric acid.

the chemistry of cells, others were studying cell structure. It was found that certain portions of the cell reacted readily with certain chemicals, and that these reactions produced visible color changes, or *stains,* in various parts of the cells. Staining permitted biologists to view portions of the cell that had previously been transparent under the microscope, and by the late 1870's, the process of cell division by *mitosis* was described for the first time. It was seen that during the mitotic division of cells, the material in the nucleus underwent some striking changes. At the beginning of mitosis, the contents of the nucleus appeared to be a relatively uniform substance that took a stain quite well and was thus called *chromatin,* or ''colored substance.'' As mitosis proceeded, the chromatin seemed to undergo a contraction and finally formed a number of visible, organized bodies that came to be called *chromosomes,* or ''colored bodies.'' (See fig. 6.1 and 6.2.) Rather than go into the details of the whole process in the text, we refer you to box 6.1. The important point is that *mitosis is the mechanism whereby the genetic material of a cell is divided equally between the parent and the daughter cells.*

Before the 1880's ended, a very few biologists had come to believe that the material found in the nucleus, chromatin, was probably one or both of the nucleic acids. By the turn of the century, the foundations had been laid for a chemical understanding of the nature of heredity. But consider the foundation: It was hypothesized that the chromosomes might be the carriers of hereditary information and it was hypothesized that the chromosomes might be

FIGURE 6.1 The chromosomes of a human male and a human female. The human chromosome number is 23, indicating that each cell in the body of a human being contains 23 *pairs* of chromosomes. In a male and a female 22 pairs of chromosomes are the same, and are called *autosomes.* One pair differs. In the male there are two dissimilar sex chromosomes, *X* and *Y*. In the female there are two *X* chromosomes (♂ *Courtesy of Carolina Biological Supply Company;* ♀ *courtesy of D. S. Borgaonkar.)*

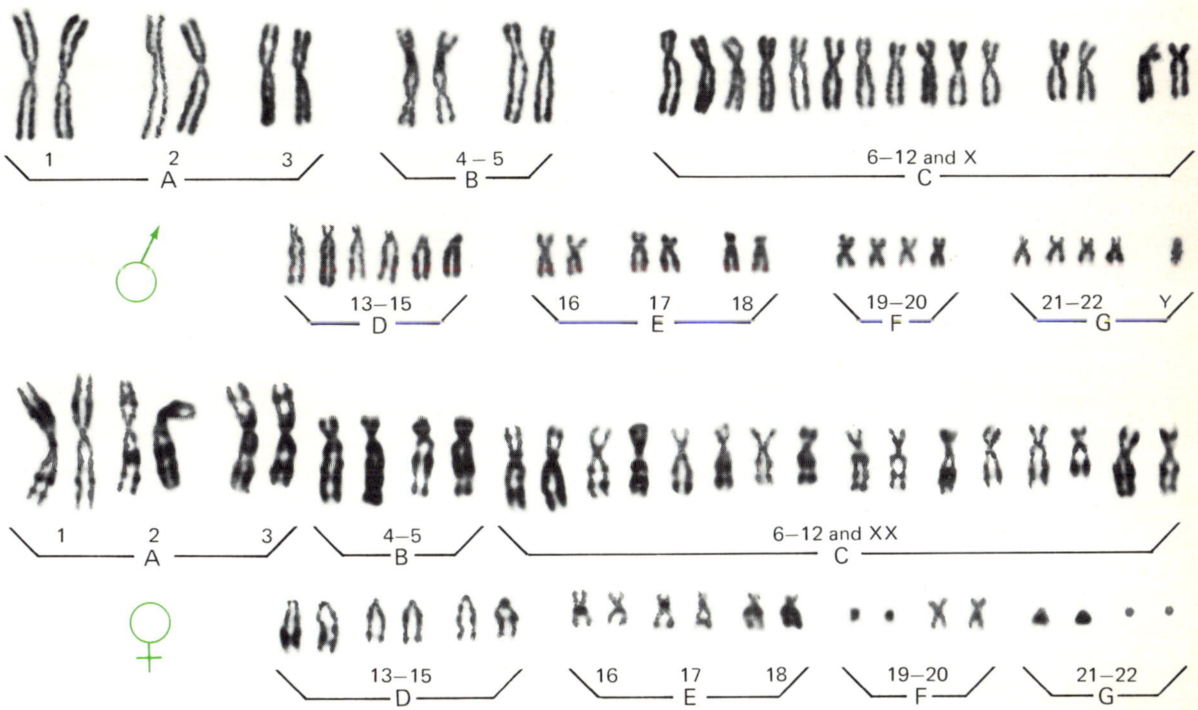

(a) (b) (c)

FIGURE 6.2 The chromosome: (a) When the cell is in the resting state the chromosome assumes an extended form. Viewed under a light microscope the extended chromosomes are called the *chromatin.* (b) In the early stages of division chemical changes in the nucleus induce the chromosome to begin contracting, forming the *chromatids.* (c) At the beginning of mitosis, the chromosomes are visible as compact bodies. After the cell divides and the new nuclear membrane begins forming, the chromosomes reverse this process. While passing back to chromatin, the chromatids replicate.

made of nucleic acid. A great deal of work remained to be done to validate these hypotheses, and we should reemphasize that very few biologists thought that these ideas were worth pursuing.

The Chromosome Theory of Inheritance

At the turn of the century, the experiments of Gregor Mendel (see box 6.2), an Augustinian monk, were discovered almost simultaneously by three botanists: DeVries, Correns, and Tschermak, seeking corroborative evidence for research they had been doing. The results of Mendel's research had lain unnoticed for nearly thirty-five years in an issue of an obscure journal. The experiments indicated that (1) the characteristics of parents were inherited by offspring *as discrete units*, and (2) each parent apparently contributed only half its genetic material to the offspring. It also appeared from Mendel's experiments that although the characteristics of one parent might seem to disappear from the phenotype of its offspring, the genes (Mendel called them ''factors'') were not lost, but only hidden from view. Under certain conditions, the characteristics would reappear in the phenotypes of some of the descendants of the offspring.

The discovery of Mendel's evidence, when coupled with some evidence provided by nineteenth century cell biologists, led directly to the formulation of the *chromosome theory of inheritance.* The fundamental points of the theory can easily be summarized. First, it was known that the only cells that constituted a physical connection between generations were the gametes. Second, sperm cells were known to consist almost entirely of a nucleus. What little other material there is in a sperm is lost as it matures. Therefore, it was believed that the nucleus probably contained the hereditary material. Third, within the nucleus, chromosomes were known to behave in precise and predictable ways during cell division. Fourth, Mendel's findings and chromosome behavior seemed to match. (We shall elaborate on this point later in this chapter.) Therefore, it was hypothesized that chromosomes were probably the carriers of hereditary information.

Mitosis

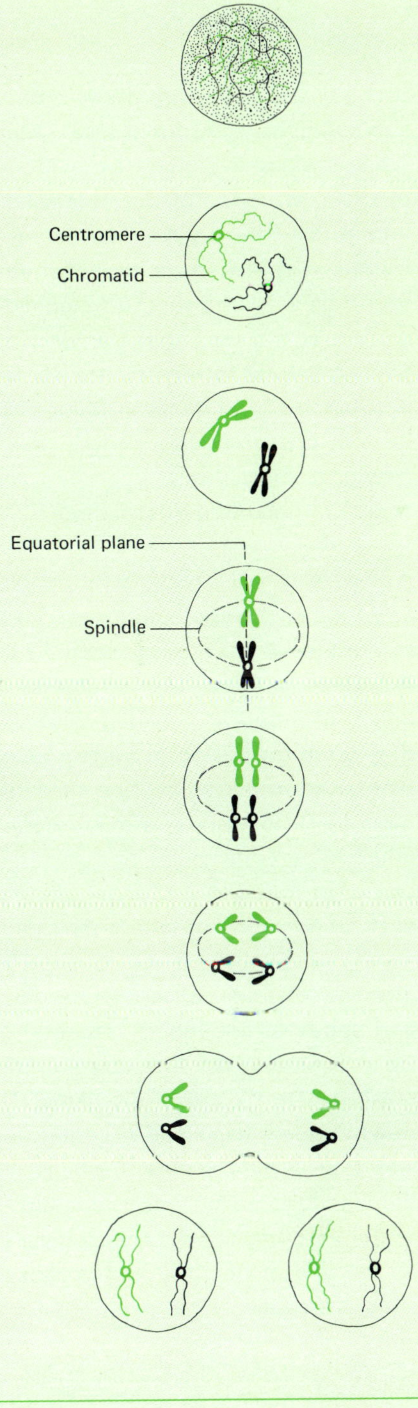

(a) Prior to division the chromosomes are in the fully extended form and the cell nucleus appears to contain relatively evenly distributed material, the chromatin. For illustrative purposes, only the nucleus is shown.

(b) As mitosis gets underway, the chromatin can be seen to give rise to chromosomes, each with two chromatids attached at the *centromere.* The centromere plays a key role in mitosis, controlling the distribution of chromosomes to the daughter cells. Should the centromere fail to divide, one cell will be deprived of a chromosome and the genes it contains.

(c) Soon after the chromatids become visible the chromosomes become compact bodies.

(d) The spindle appears, and the chromosomes line up on the equatorial plane.

(e) The centromeres divide.

(f) The chromosomes move to opposite poles of the nucleus. On occasion, a chromosome may fail to move to the pole of the spindle and one of the daughter cells receives an extra chromosome.

(g) A new nuclear membrane is formed around each of the developing daughter cells.

(h) When division is complete, each of the new cells enters a phase during which the chromatids replicate connected by the centromere. During replication the chromosomes appear as chromatin.

BOX 6.1

Gregor Mendel

The contribution that Gregor Mendel made to the study of genetics is recognized now as one of the cornerstones of modern biology. Unfortunately, Mendel himself was never to appreciate it as such.

After several years of assiduously recording the results of his breeding experiments, he presented his conclusions at a meeting of the Natural History Society of the small town in Eastern Europe where he was working as a part-time teacher of high school physics and mathematics. (The town was then known as Brünn and was in Austria. It is now Brno, Czechoslovakia.) On the night Mendel read his paper, several scientists, including some biologists, were in the audience. After listening to him document the nature and results of his research, none of those present asked a question or made a comment. Many years later, after Mendel's contribution was recognized, the one or two persons who had been in the audience and were still alive could not recall that anyone had been at all impressed, excited, or even interested by what Mendel had said.

Mendel himself forwarded a copy of his report to a prominent botanist, Carl Nageli, and tried to convince him of its validity, but Nageli dismissed Mendel's work as unsound. Subsequently, Mendel turned his attention to other matters and died in 1884, nineteen years after publishing his report in the Journal of the Natural History Society of Brünn. Before his death, convinced of the importance of his work, he said, "My time will come."

It came in 1900 when three botanists almost simultaneously "rediscovered" Mendel's paper while seeking corroborative evidence for work of their own. One of the botanists, ironically, had studied for his doctoral degree with Nageli. It is also ironic that Mendel's work had found its way into the Encyclopaedia Britannica current at the time that Darwin was searching unsuccessfully for exactly such information to buttress his arguments in favor of natural selection. Encyclopedias, in this case unfortunately, are rarely ever used by scientists as a source of scientific information.

It is hard to say exactly why people paid no attention to Mendel's findings. Two possible reasons have been suggested. At that time, biologists were seeking an answer to the mechanism of heredity in biological terms. Thus, they were interested in finding some sort of biological entity, something that might perhaps be visible under a microscope and to which they could point and say, "*That* is the carrier of heredity." Mendel offered them no such thing; nothing which they could, so to speak, grasp. In fact, he couched his entire argument in statistical terms. He described ratios and spoke of undefined "factors" of heredity. He was among the first to apply mathematical reasoning to biological problems. Though it is now a widely used and accepted method, it was then a radical idea. Perhaps, then, Mendel was ignored because he could offer no describable thing to which biologists could refer as the unit of heredity and because he took such a radically different approach to the problem.

BOX 6.2

As with so many "discoveries" in science, the chromosome theory was by no means immediately accepted. One problem with the theory was that there were so few chromosomes in the cells of an individual and so many characteristics that must be accounted for. It would seem to follow from the theory that each chromosome must contain the information for many characteristics, and if so, then many characteristics must be inherited together by an individual. The great American geneticist Thomas Hunt Morgan raised this objection to the chromosome theory eight years after it was first proposed. Five years later, however, Morgan's own research convinced him that many characteristics were indeed to be found on the same chromosome, and, as we shall see, he abandoned his earlier position.

DNA: The Chemical of the Gene

Although the study of genetics was ready to yield a chemical explanation of the gene, research into the nucleic acids as the carriers of hereditary information came nearly to a standstill as biochemists, cell biologists, and geneticists turned their attention to the other chemical constituents of the nucleus, proteins. A chemical analysis of the nucleic acids had indicated that they were too uniform in their chemical composition to account for the complexity of the gene. The analysis indicated that a *base* (adenine, guanine, cytosine, and thymine or uracil) bonded with a sugar and a phosphate to form what was called a *nucleotide.*

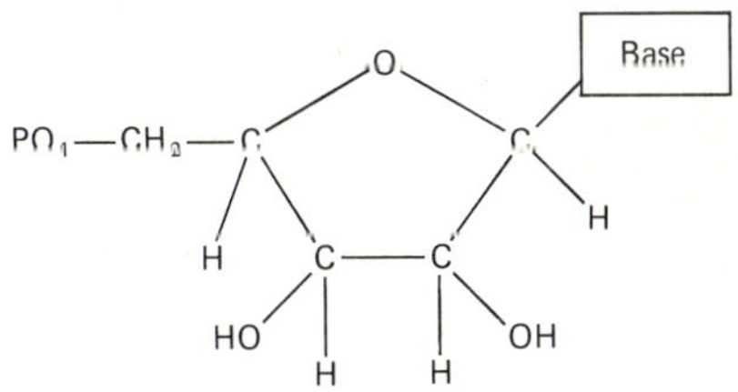

But the bases were *apparently* always present in approximately equal quantities, and thus it was incorrectly concluded that the nucleic acids were made up of repeating units of four nucleotides. It was difficult to imagine that such a uniform macromolecule could be at the cornerstone of life's immense diversity. Analyses of proteins had indicated, though, that these chemicals differed from each other considerably, and thus proteins seemed sufficiently diverse to account for the variety of living forms.

Some twenty years after nucleic acids had been dismissed as the carriers of hereditary information, a bacteriologist, while investigating the disease pneumonia, discovered that live nonvirulent strains of the bacteria *pneumococcus,* when injected into mice with dead virulent strains of *pneumococcus,* could somehow be transformed into living virulent forms. After such injections, the mice would die of pneumonia. When autopsies were performed, the mice were found to contain no signs of the nonvirulent forms of bacteria but were filled with the living virulent strain. (See fig. 6.3.) The investigator was unable to explain how this transformation took place. Subsequent investigators showed that it was not necessary to inject the bacteria into an experimental animal at all. The transformation could be brought about in a culture by adding a solution made from heat-killed, dissolved virulent bacteria to live nonvirulent bacteria. The chemical (if that was what it was) that could bring about the transformation was capable of withstanding both the heat and solvents used to dissolve the virulent bacteria.

In 1944, the "transforming factor" was identified as DNA. All that was necessary to transform *pneumococcus* from nonvirulent to the virulent strain was to isolate DNA from the virulent strain and add it to a culture of nonvirulent bacteria. Such experiments began to direct attention once again toward nucleic acid, and especially DNA, as the carrier of genetic information.

Other evidence began to accumulate. One very important clue was the discovery that although the chemical composition of DNA was some-

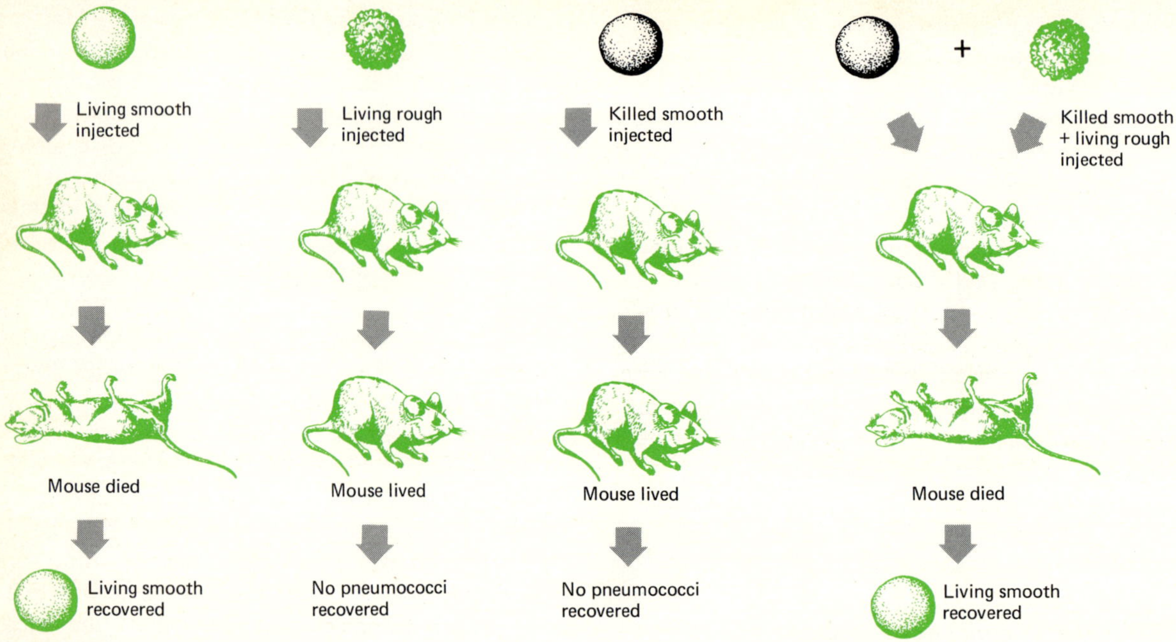

FIGURE 6.3 Transformation. The smooth strain of pneumococcus is virulent and causes death in the experimental animal. The rough strain is not virulent. When both are injected together, even though the smooth strain is dead, the animal dies, and no rough strain is recovered. How the nonvirulent rough strain could be transformed into virulent, smooth pneumococci was an inexplicable problem to the experimenter who discovered this phenomenon. *(Redrawn from* Genetics *by A. M. Winchester, Houghton Mifflin Company, Boston, 1972.)*

what uniform, there were significant differences in the relative quantities of the bases in a molecule of DNA. For example, the chemical composition of DNA from, say, a calf's thymus gland was significantly different from the chemical composition of DNA taken from another species. Thus, DNA appeared to be a sufficiently variable molecule to account for the diversity of living forms. Moreover, the relative quantities of the bases found in the DNA from any particular species even suggested certain things about the nature of the structure of DNA. (See table 6.1.)

The analytical techniques available to investigators in the mid-forties were considerably more accurate than those in use thirty-five years earlier. Obviously, from table 6.1 we can see that the relative quantities of adenine and guanine and of thymine and cytosine are complementary; that is, they are present in the molecule in approximately equal amounts. This suggested that adenine and thymine were always bonded to each other and that cytosine and guanine were always bonded to each other as well. Nevertheless, at the time this discovery was made, proteins were still contenders for the role of the gene, and it was still necessary to show that DNA was passed from cell to cell and determined the characteristics of the offspring.

TABLE 6.1 Base Composition of DNA Chemical Components *(in percentages)*

Source	Adenine	Guanine	Cytosine	Thymine	Ratio of A/T	Ratio of G/C
Man	30.4	19.6	19.9	30.1	1.01	1.02
Calf	29.1	21.2	21.2	28.5	1.01	1.00
Hen	28.8	20.5	21.5	29.2	0.99	0.95
Salmon	29.7	20.8	20.4	29.1	1.02	1.02
Crab	47.3	2.7	2.7	47.3	1.00	1.00

Further evidence of the role of DNA came from a group of biologists working with certain viruses. These viruses were first discovered during World War I by a French physician who called them *bacteriophages,* which simply means ''bacteria eaters.'' (See fig. 6.4.) Although they were too small to be seen by ordinary microscopic means, the effect of phages on bacteria could be readily detected. It was learned that if bacteria infected by phages were left alone, the bacteria would sooner or later burst open, (or *lyse*), and release a number of new phages. It also became clear that mutants occasionally arose in the populations of phages produced in infected bacteria and that these mutants were sometimes incapable of attacking bacteria. More importantly, if two different phages infected the same bacterium, some of the phages produced at the end of the infection cycle were found to share the characteristics of the two original phages; in other words, they appeared to exchange genetic material. It was determined by chemical analysis that phages consisted only of protein and nucleic acid (RNA or DNA), and thus the field of contenders for the role of the gene was without question narrowed to these two chemicals.

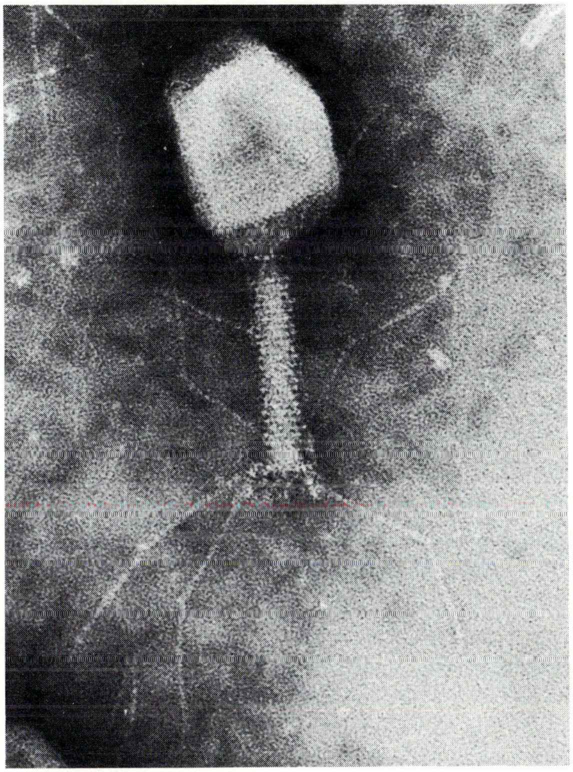

FIGURE 6.4 An electron photomicrograph of the T2 bacteriophage, an organism which has been of basic importance for understanding modern genetics. *(Courtesy of the Virus Laboratory, University of California, Berkeley.)*

One important difference between protein and DNA made it possible to settle the question in the minds of most biologists. DNA was known to contain no sulfur, and protein to contain no phosphorus. One team of investigators grew a common bacterium, *Escherichia coli*, on a medium containing radioactive forms of these two elements ($^{35}SO_4$ and $^{32}PO_4$) and then infected these bacteria with phages. At the end of the infection cycle, all the newly produced phages were thus labeled with radioactive sulfur and phosphorus taken up from the bacteria. They next infected a fresh batch of bacteria with the labeled phages. There was some reason to believe that a phage was a sort of ''hypodermic needle'' (see fig. 6.5) filled with DNA and that when it infected a bacterium it somehow injected its DNA into the bacterium and left all the protein

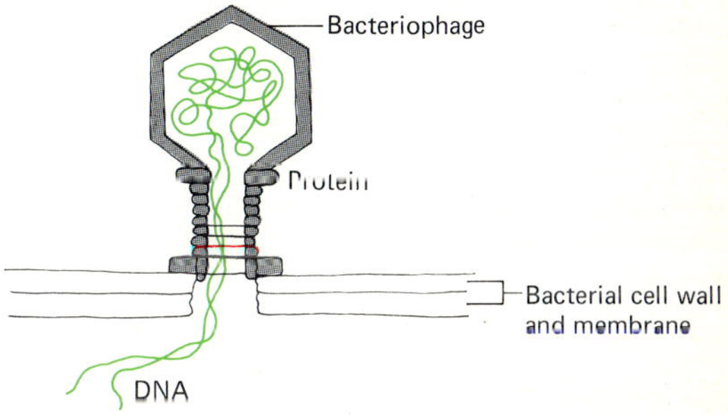

FIGURE 6.5 The ''hypodermic'' phage. Well before it had been determined that the phage was made up of a protein coat within which the phage DNA resided, the model of such a structure had been postulated. The phage, and indeed all viruses, present a problem in defining living things. They are incapable of reproducing themselves without using the metabolic structures of the cells they infect, yet are capable of reproduction in association with a living cell.

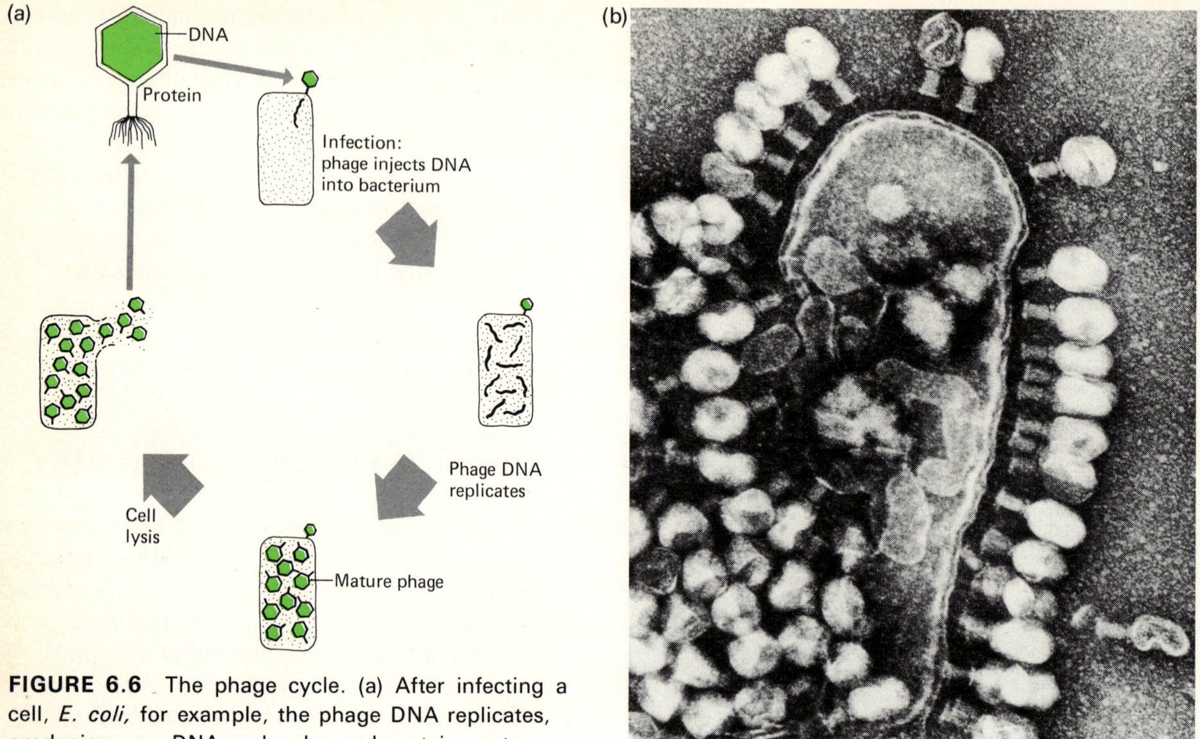

FIGURE 6.6 The phage cycle. (a) After infecting a cell, *E. coli,* for example, the phage DNA replicates, producing new DNA molecules and protein coats using the metabolic apparatus of the host cell. Within 30 minutes after the injection of DNA, 50 to 100 new viruses may be present in the bacterium. (b) The electron photomicrograph shows the attachment of the bacteriophage particles to the cell wall. *((a) Redrawn from* Genetics *by A. M. Winchester, Houghton Mifflin Company, Boston, 1972; (b) courtesy of Lee D. Simon, The Institute for Cancer Research, Fox Chase, Pennsylvania.)*

outside. If so, then soon after the infection cycle began, all the radioactive sulfur could be separated from the infected bacteria, and after the cycle was completed, all the radioactive phosphorus would emerge from the bacteria in new phages. After the necessary separations were performed, it was indeed found that the bacteria contained the radioactive phosphorus, and the radioactive sulfur remained outside the bacteria. Thus, apparently the protein served as a vehicle for the transport of DNA from cell to cell, and once a cell was injected with DNA, the protein served no further use. (See fig. 6.6.) This evidence, coupled with the evidence of DNA as a transforming factor, clearly indicated to most observers that DNA was the chemical of which genes are made.

The One Gene–One Enzyme Hypothesis

To understand the operation of the gene, we must examine some experiments performed in the 1940s by two geneticists, George Beadle and Edward Tatum. The usual approach to the study of genetics, until Beadle and Tatum performed their experiments, had been to choose a genetic trait and attempt to trace its chemistry. These two investigators chose to turn that approach around—that is, begin with known chemical reactions and examine the genetic effects. It was known that certain enzymes acted as *catalysts* in living systems; that is, the enzymes *participated in a chemical reaction to accelerate the rate at which the reaction took place but were themselves in no way affected by the reactions.* The roles of some enzymes in some organisms had been worked out before Beadle and Tatum began their studies, but there were still many that acted in completely unknown ways. The hypothesis that Beadle and Tatum began with was that the genes somehow directed the production of enzymes. The enzyme would then interact with the chemicals present in the organism to assemble other chemicals, each larger chemical molecule being constructed of smaller molecules. Each step in the assembly process was theorized to be carried out with the aid of an enzyme. Any alterations in the genes of an organism, then, could be detected by alterations in the series of chemical reactions that were known to occur during the life of the organism.

The experimental organism Beadle and Tatum worked with was a simple bread mold, *Neurospora crassa.* This mold can synthesize all the amino acids necessary to build proteins and thus can grow on a medium in which a minimum of nourishment is present. Moreover, if some amino acid is present in the medium, the mold will forego synthesizing it and will use the amino acid in the medium. Another characteristic of this mold makes it particularly well suited for genetic research: It can reproduce asexually, and thus any mutation can be readily detected, since it cannot be attributed to the genes of another parent; the mold can also reproduce sexually, the importance of which will be apparent later.

To test the hypothesis that each gene controls the production of a single enzyme, Beadle and Tatum produced mutants by irradiating cultures of *Neurospora* with X rays. They then cultured spores (see fig. 6.7) from the X-rayed molds on media containing all the amino acids, sugars, and vitamins that the mold was known to require for survival. This step allowed any potential mutants to grow normally by utilizing the materials from their environment; the presumed mutants were thus allowed to express themselves phenotypically. Next, the experimenters removed samples from each of the cultures and recultured them on media containing some simple sugars and inorganic salts but no vitamins or amino acids. On such a *minimal medium*, any mutant unable to synthesize a compound necessary to survival would not grow, and thus Beadle and Tatum could identify potential mutants produced by the X rays. Once

FIGURE 6.7 Mutant culture, *detection of mutation in Neurospora.* The failure of X-rayed *Neurospora* to grow on a minimal medium indicates that a mutation has arisen, but to determine the nature of the mutation requires further testing. Samples from the suspected mutant cultures must be transferred to a number of media each of which lacks only one amino acid or vitamin. For illustrative purposes, only three of more than twenty-five cultures are shown. In this case, the failure of the suspected mutant to grow on a medium lacking arginine indicates that the mold is unable to synthesize this amino acid. Since the synthesis of an amino acid or vitamin requires several steps, the next stage in the process is to determine which step is not being performed by the mold. *(Redrawn from* Genetics *by A. M. Winchester, Houghton Mifflin Company, Boston, 1972.)*

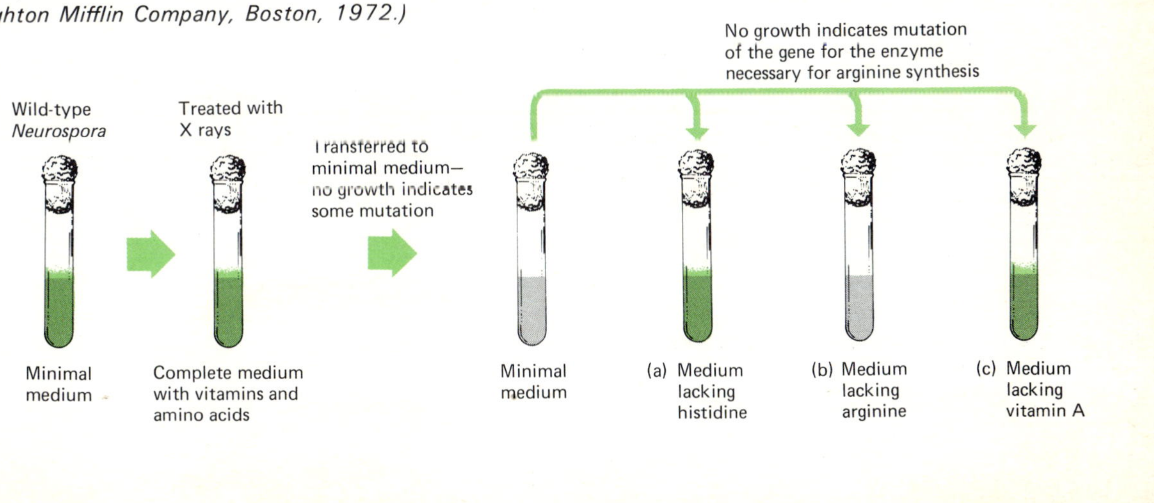

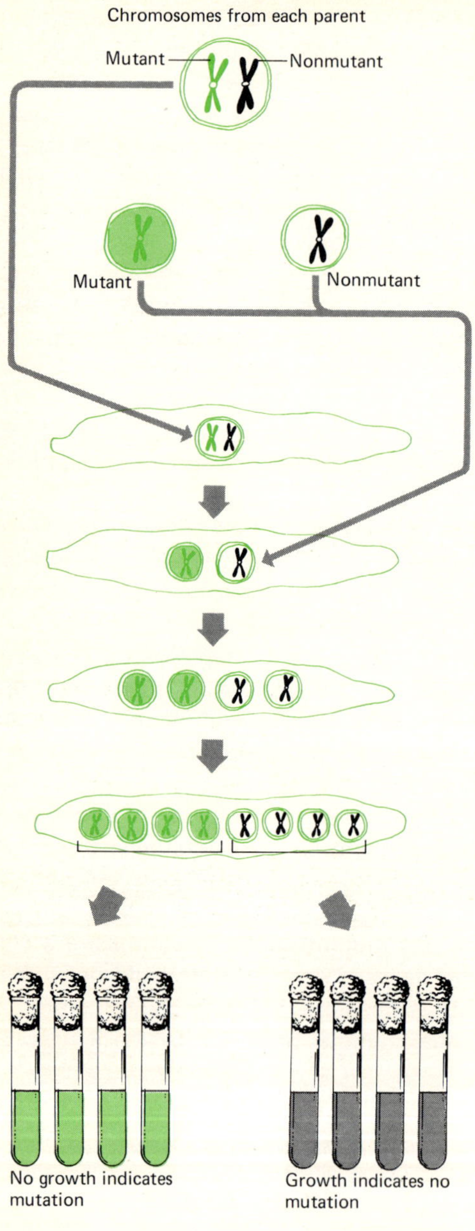

FIGURE 6.8 After allowing suspected mutants to crossbreed with nonmutant *Neurospora* and produce fruiting bodies, the thesis that one gene controls the mutation can be tested. Within the ascus the zygote contains chromosomes from each parent, one mutant and one nonmutant. The first division of the zygote is such that each daughter cell gets only one chromosome, either carrying the mutant gene or carrying the wild type gene. The next two divisions are mitotic, and the cells that arise are identical to the parent cells. The result of these divisions is that half the cells in the ascus carry the mutant gene and the other half the wild type one. The removal of these spores from the mature fruiting body, and the culture of the spores on media lacking the amino acid or vitamin that the mutant is incapable of synthesizing will show that half the spores are mutant and half carry the wild type gene.

the mutants were identified, the question remained of which amino acid or vitamin a mutant was incapable of synthesizing. To solve this problem, Beadle and Tatum once again took samples from the original culture, now suspected to be a mutant, and began a new culture on a minimal medium to which had been added one vitamin or one amino acid. Thus, if a mutant was able to grow on a medium containing the amino acid arginine and on no other medium, it was identified as incapable of synthesizing arginine.

Once Beadle and Tatum had identified the nature of a mutant mold, they were still faced with the problem of knowing whether more than one gene was involved. To determine this, Beadle and Tatum were able to take advantage of the sexual phase of the life cycle of the mold. When *Neurospora* cross breed, the nuclei of the gametes fuse to form a zygote with twice as many chromosomes as in the asexual stage. (See fig. 3.43, p. 84, to see the life cycle of *Ulva* and alternation of generations.) The cell with the fused gametic nuclei then undergoes three divisions to produce eight spores. In the first two divisions, the number of chromosomes in the daughter cells is reduced to half as many as were in the zygote. The third division is mitotic, with the result that each spore has only half the number of chromosomes as the cell from which it is descended. This process takes place within a minute sac called the *ascus*, which is so narrow that as the daughter cells arise, they cannot slip past each other. Placed under a microscope, the ascus can be split open and the spores removed in a precise order determined by the process of division, and then the spores can be cultured on a medium with the culture vessels in the same order as the spores in the ascus. (See fig. 6.8.) Thus, an experimenter can count the actual number of differing offspring and does not have to rely on statistical techniques to determine the percentage of offspring carrying a particular genotype.

After producing and analyzing many mutants, Beadle and Tatum let them interbreed with normal forms of *Neurospora.* They reasoned that if only one gene was involved in producing a mutation, exactly half the spores in an ascus would carry the mutation, while the other half would be normal. After removing the spores from the asci, they did in fact find that only half of them were mutants and that each gene was responsible for a single mutation. Using similar methods of genetic analysis, they were able to determine not only the nature of the mutation but also exactly which steps in the synthesis of some proteins and vitamins were missing as a result of the mutation. (See fig. 6.9.)

Although the Beadle-Tatum experiments were said to validate the ''one gene–one enzyme'' hypothesis, subsequent research has indicated that single genes appear to direct the cellular manufacture not only of enzymes but also of other proteins, and thus the hypothesis has come to be called the ''one gene–one polypeptide'' hypothesis. (For reasons which we shall discuss later, proteins are also called polypeptides.)

The Structure of the Gene

Molecular biologists had come to believe that the structure of a molecule strongly reflected its function, and they thus turned their efforts to determining the structure of DNA. It was not long after the experiments showing that phages inject their DNA into *E. coli* that the structure of DNA was at last determined.

In 1953, an American biologist, James D. Watson, and an English biophysicist, Francis H. C. Crick, announced that they had solved the riddle of the structure of DNA. By the time that Watson and Crick began their work, it was already known that DNA was a large, threadlike molecule. Other studies had indicated that it was helically structured. Given the information available about the molecule, Watson and Crick set about the task of constructing a physical model of DNA by using copper and wire models of deoxyribose, phosphate, and the bases. However, they were unable to assemble the

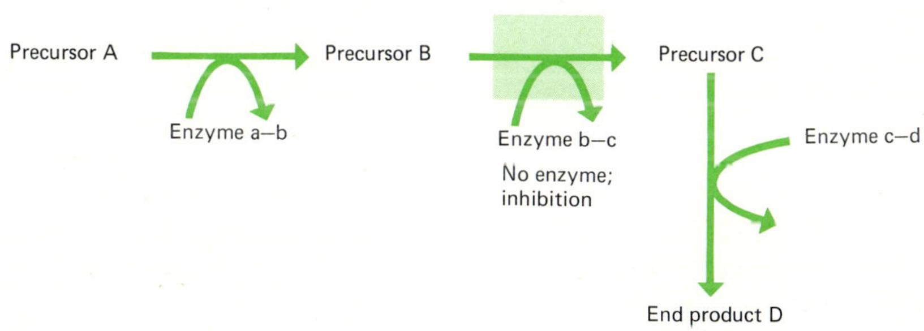

FIGURE 6.9 Within the body of any animal or plant, series of chemical reactions take place in a specific order as chemical molecules are assembled and reassembled for use. Each of the steps in such metabolic pathways is mediated by an enzyme. Thus, according to the Beadle-Tatum hypothesis, a mutation in a gene that controls the production of some enzyme would show up as a failure at some particular point in the metabolic pathway. All of the known precursor molecules might be produced, but the single reaction affected by the inoperative enzyme would inactivate all subsequent reactions. In the simple hypothetical pathway shown, a mutation in the gene for enzyme b–c would inhibit the syntheses of both *precursor C* and the *end product* of the pathway.

parts into the appropriate helical structure until they learned that the bases could be structured in two different ways. Both structures were known to exist in similar chemicals, but Watson and Crick were working with only one of the possible alternatives. After learning of the other possibility, Watson, unable to wait for metal models to be made, cut out cardboard models to test the new information. He soon found a way of placing the different structures together that fit all the other available data. Although Watson and Crick performed no major experiments as such, their analysis and their insights into the available biochemical data brought about one of the most important discoveries of twentieth century biology.

FIGURE 6.10 The DNA molecule consists of two sugar and phosphate strands wound in a helix, and held together by paired molecules of adenine and thymine and of cytosine and guanine. Adenine and thymine always pair together as do cytosine and guanine.

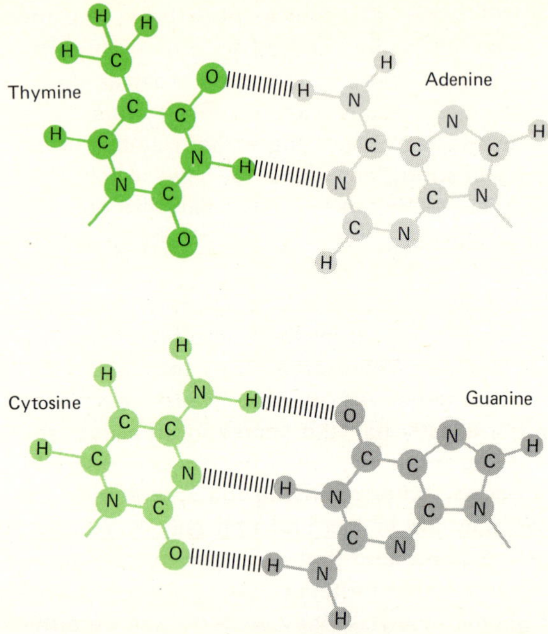

FIGURE 6.11 The bonding of thymine to adenine and guanine to cytosine in DNA, called base pairing, is brought about by a large number of weak attractions for H atoms.

Watson and Crick proposed that a DNA molecule consists of two sugar-phosphate strands wound about each other in the form of a double helix. (See fig. 6.10.) The strands are connected by the bases adenine, guanine, cytosine, and thymine. The sizes of the bases indicated that a purine is always bonded to a pyrimidine, and this information conformed to the earlier work on the composition of DNA (see table 6.1). The chemical structures of the bases allow only one possible combination of bonding between a purine and a pyrimidine, and thus adenine always bonds to thymine, and guanine to cytosine. (See fig. 6.11.) The bases are held in position by weak chemical attractions between hydrogen atoms. Although these ''hydrogen bonds'' are weak, their large number makes the molecule very stable. Such an arrangement of bases, called ''base pairing,'' suggests immediately how the DNA molecule carries the information necessary for the development of a whole organism. If the two strands are separated in such a way that each

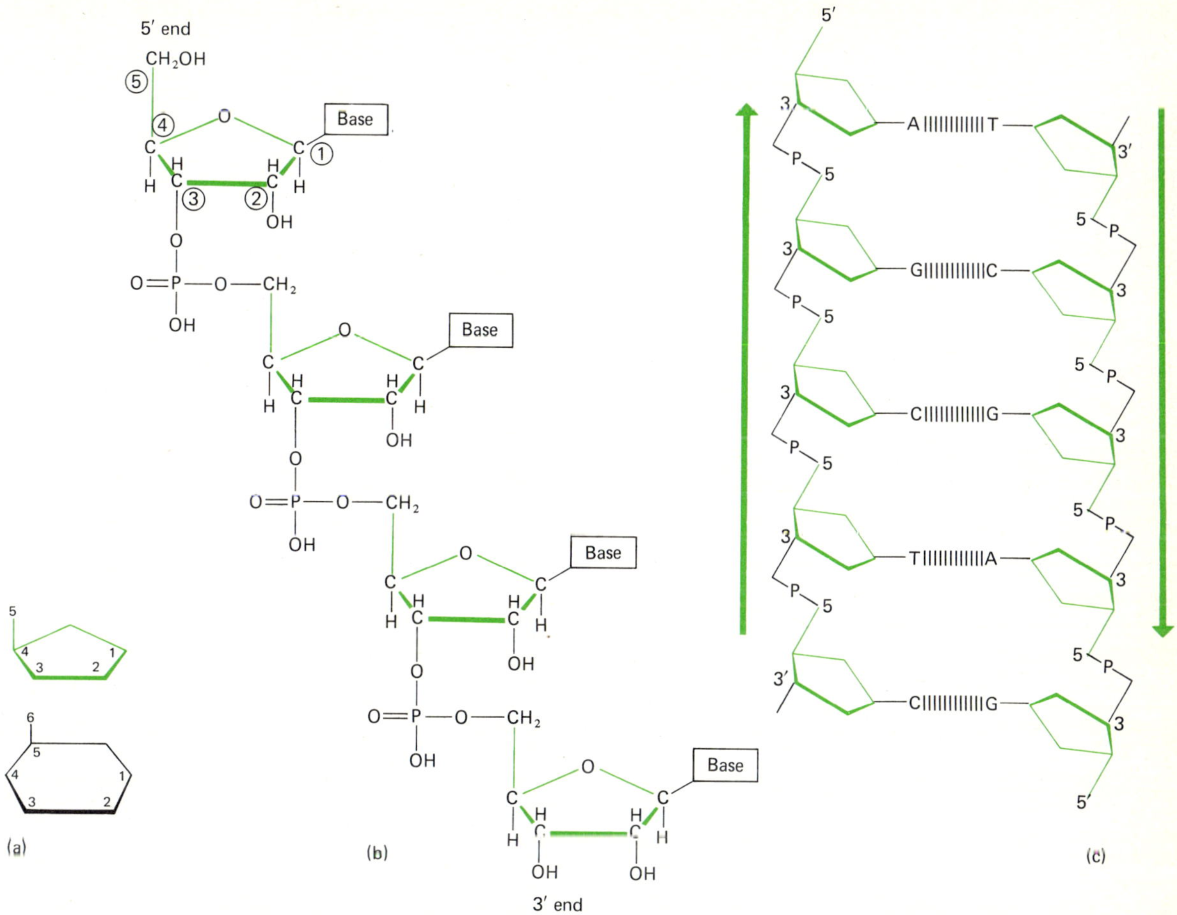

FIGURE 6.12 As shown in (a), when sugar molecules such as ribose and deoxyribose form rings, the carbon atoms are assigned numbers to identify their locations. Using this numbering system, the sugar-phosphate chain of DNA and RNA can be seen to have a 5′ and a 3′ end. In DNA, the structure of the bases and the bond locations of the bases to the chain make the complementary chains run in opposite directions. The actual bond locations are shown in (b), and the orientation of the chains in (c).

strand has a series of bases bonded to it, each single strand will "read" differently. (See fig. 6.12.) But if a new strand is synthesized from free nucleotides in the cell, the newly synthesized strand can only be assembled in one way, and that way is determined by the order of the bases on the "old" strand. For example, if the order of the bases on one strand is TTAGCCT, then AATCGGA is the only order in which new bases can be added. Therefore, the molecule of DNA serves as a *template* for the assembly of another molecule exactly like itself. Moreover, the structure of the sugar-phosphate chain allows the bases to be arranged *in any order* along the chain. This suggests that the genetic information in a molecule of DNA may be carried in a "code" that is spelled out in the order of the bases.

The Genetic Code and Protein Synthesis

Taking into account Beadle and Tatum's "one gene–one polypeptide" hypothesis, the next important question could be framed. How does the sequence of bases in a molecule of DNA direct the synthesis of the enzymes and proteins that constitute the cell? The evidence available suggested that DNA was not, in fact, directly implicated in such syntheses but that the other nucleic acid, RNA, was. In cells where a great deal of protein synthesis occurs, such as those of the liver and pancreas, large amounts of RNA were found, while in cells where the synthesis of proteins proceeded at a relatively lower rate, the amounts of RNA detected were significantly lower.

Moreover, several kinds of RNA have been detected. They differ not in chemical composition, but in structure. The first is called *ribosomal RNA* (rRNA) and is found as a structural component of the *ribosome,* the cellular organelle where protein synthesis is known to take place. As of now, however, the role of rRNA remains unknown. The second kind of RNA is called *transfer RNA*, or tRNA. Structurally, tRNA is a small molecule, averaging only eighty nucleotides. It is now known that tRNA functions in the cell to bind free amino acids and carry them to the ribosome, where they are assembled into proteins. Third is *messenger RNA,* or *mRNA.* Messenger RNA, like the others, is synthesized from nucleotides in the cell on a template of DNA. (See fig. 6.13.) All three are synthesized and transcribed from DNA by the enzyme RNA polymerase.

It is thought that the molecule of DNA in the nucleus somehow separates into two strands and that RNA is synthesized from free nucleotides on one of the strands of DNA. After synthesis, mRNA moves to the ribosomes with its coded instructions for the assembly of a protein from the amino acids carried to the ribosomes by tRNA. (See fig. 6.14.)

This very broad outline of protein synthesis was apparent by the early 1960s, and it was then possible to turn to another question of critical importance: What is the significance of the sequence of the bases in DNA? Biologists had begun to call the sequence the "genetic code," and it was clear that an important step would be taken when the code was finally broken.

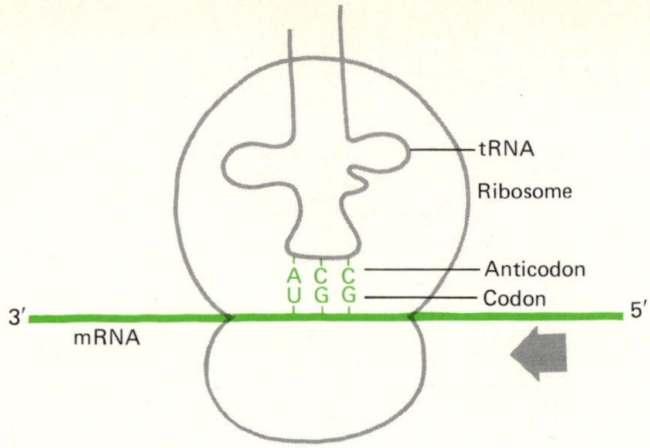

FIGURE 6.13 The anticodon of transfer RNA (tRNA) attaches to the codon on a strand of messenger RNA (mRNA) inside the ribosome. Since the tRNA is attached to a specific amino acid, when a series of tRNA's "line up" along the mRNA the order of amino acids in a strand of protein is determined.

Two developments that preceded the breaking of the code should be mentioned. First, it was found that many of the reactions and syntheses that occur in living cells (*in vivo*) could be made to take place under precisely controlled conditions in the laboratory in test tubes or culture dishes (*in vitro*). This was an important development, since within the interior of a cell, so many chemical reactions occur simultaneously that it is often extremely difficult to isolate the one that an experimenter wishes to observe. Second, it was found that for protein synthesis to proceed it was not necessary that the RNA used be from the same organism as the other materials needed for synthesis. Thus, for example, RNA from mouse liver cells could direct protein synthesis *in vitro* even if the ribosomes and other materials were derived from bacteria.

Marshall Nirenberg, a biologist at the National Institutes of Health, performed a crucial experiment leading to our understanding of the genetic code. Since the introduction of "foreign" RNA could direct protein synthesis *in vitro,* Nirenberg reasoned that it might be possible for synthetic RNA—that is, RNA that had been synthesized in the laboratory—to also direct protein

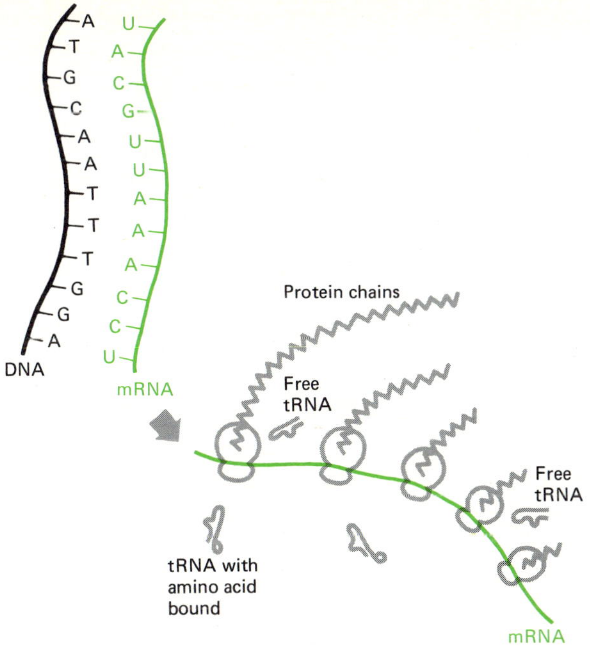

FIGURE 6.14 Transcription and translation. Molecules of RNA, in this case mRNA, are assembled, or *transcribed,* along a DNA template from free nucleotides in the cell. The mRNA then travels out of the nucleus of the cell, where it interacts with ribosomes to *translate* the message coded upon it. Translation is carried out as molecules of tRNA, with amino acids bound to them, react with the ribosome and the mRNA molecules, sequentially attaching amino acids to the growing protein molecule.

synthesis *in vitro.* Early forms of synthetic RNA were difficult to use, however, for there was no way of knowing the order of their bases and, therefore, the "message" coded on the molecule. The problem was solved when an RNA was synthesized that contained only one base, uracil. (This RNA was called *synthetic poly-U.*) Nirenberg prepared twenty test tubes, each containing all the elements needed for protein synthesis. Of the twenty amino acids in each test tube, only one was radioactively labeled. He then introduced the synthetic poly-U RNA into each tube. He was able to detect nothing in nineteen of the test tubes, but in one of them a polypeptide chain consisting entirely of one amino acid was detected. He had asked the question, "Uracil?" The answer was, "Phenylalanine."

But, of course, the question remained of how many bases were involved in the message. Did, for example, each uracil add one phenylalanine to the chain, or did it take two, three, or more uracils to produce phenylalanine? Those who had given thought to the nature of the genetic code had concluded that it probably employed three bases. Such a "triplet" code seemed the most economical one. If the code were a doublet one, there would not be enough messages available to code all twenty amino acids. If there were four bases and the code employed two, then the maximum messages it could carry would be sixteen, (4^2). A triplet code could produce sixty-four messages (4^3), well in excess of the necessary twenty.

Within a year after developing his technique, Nirenberg had worked out the codes for about half of the amino acids. Using the techniques of organic chemistry, he and the inventor of synthetic poly-U worked out a way to synthesize mRNA molecules that were only three bases long and in which the order of the bases was known. Using these triplets, he was able to work out the entire genetic code. Using somewhat different techniques, another investigator arrived at the same code that Nirenberg had worked out, thus confirming the validity of Nirenberg's studies.

The process by which DNA directs the life of the cell can be summarized as follows. First, the two helical strands of DNA are thought somehow to separate. Along one strand of the DNA molecule, a molecule of mRNA is assembled from free nucleotides in the cell. The nucleotides are arranged on the mRNA molecule in an order determined by the order of the bases in the DNA molecule. Within the genetic code are several sequences of bases that at one time seemed to code no particular amino acids and thus were referred to as "nonsense" sequences.

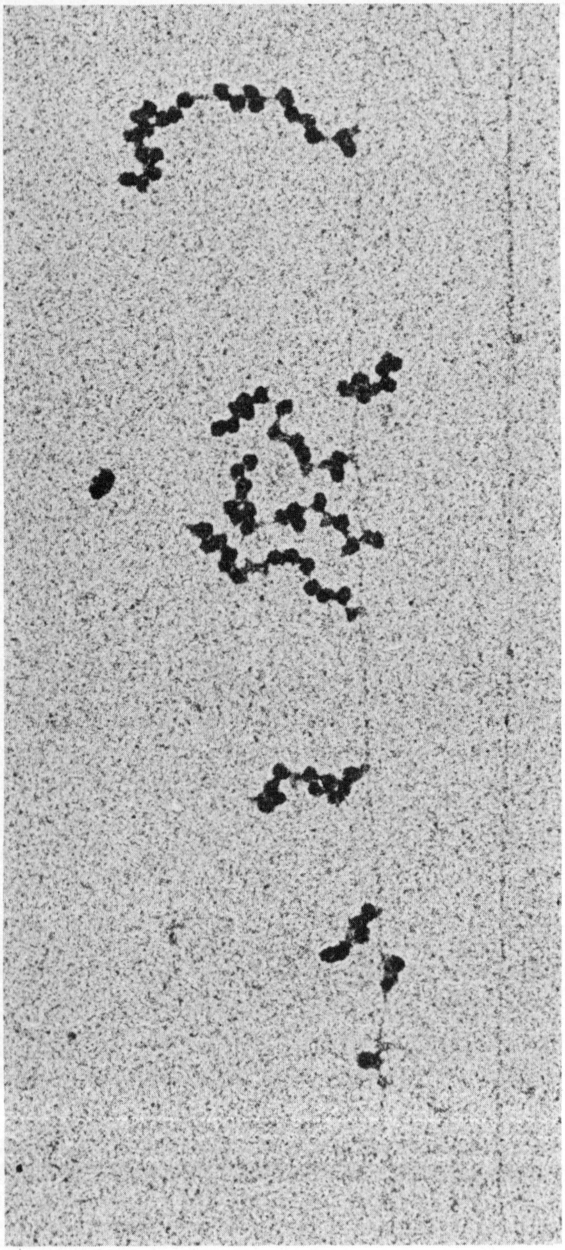

FIGURE 6.15 Ribosomes attach together along mRNA to form strings of "beads." These in turn are attached to the DNA of the chromosome of the bacterium *E. coli. (Courtesy of Oscar L. Miller, Jr., University of Virginia. Reprinted from* Science, *Vol. 169, 25 Mar. 1970, p. 394. Copyright © 1970 by the American Association for the Advancement of Science.)*

It is now believed that some of these sequences are not nonsense at all, but that they specifically identify the beginnings and ends of the mRNA molecule. The subsequent triplets of bases in the mRNA molecule are called *codons,* and each specifies the attachment of an amino acid to the growing protein. After synthesis on the DNA template, mRNA moves out of the nucleus of the cell and interacts with ribosomes in the cytoplasm. Molecules of tRNA, which are also synthesized on the DNA template, bind free amino acids in the cytoplasm. At another location of the tRNA molecule is the *anticodon,* a sequence of three bases, usually located in the conventional representation of the molecule in the loop that is opposite the site where the amino acid is bound. When the mRNA attaches to a ribosome, it passes through a "groove." On the surface of the ribosome, molecules of tRNA with amino acids bound to them react with the molecule of mRNA, the anticodon "reading" the codon and attaching its amino acid to the growing chain that forms the protein. Electron microscopic evidence has shown that ribosomes are often found in the cell in lines that appear to resemble beads on a string. (See fig. 6.15.) It has been learned that the "string" is mRNA, indicating that the same mRNA molecule is involved in the simultaneous synthesis of many protein molecules.

This model of protein synthesis has come to be called the "central dogma" of molecular biology. It is often shown as a diagram that indicates the key steps in the process. (See fig. 6.16.)

Although the broad outline of protein synethesis has been worked out, there is still a great deal to be learned, and there is a considerable amount of perplexity concerning the structure and mode of operation of the ribosome. Although the ribosome clearly plays a crucial role in protein synthesis, the structure of the proteins found in the ribosomes and of ribosomal RNA are largely unknown, and it does not appear

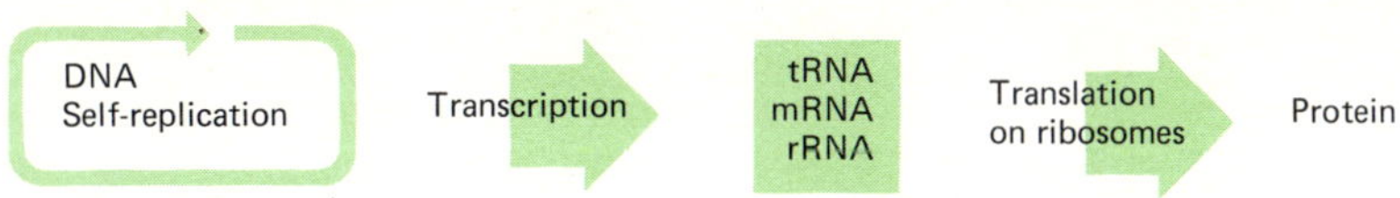

FIGURE 6.16 "Central dogma" of molecular biology.

that they will be worked out for several more years. Although many of the problems of protein synthesis center around an understanding of the ribosome, there is also the problem of how the DNA molecule separates, if it does. Moreover, if DNA separates, the information that would be transcribed on the mRNA would be different if the mRNA were synthesized on each strand. For example, the DNA nucleotide sequence AGC is transcribed on RNA as UCG and codes for cysteine, whereas the complementary sequence on DNA (which would be on the other strand of the same molecule) is TCG, which transcribes AGC and codes for serine. Whether only one strand of DNA is transcribed or both strands are transcribed and only one produces an RNA molecule that is inoperative has not yet been clearly determined, although it appears almost certain that only one DNA strand is transcribed. Although in the last twenty years our understanding of the nature of genetics has deepened considerably, much more remains to be learned.

Sexual Reproduction and Genetics

By considering the genetics of cells first, we have left out an important part of the story. We have seen that it is the gene, or the material of which the gene is made, that directs the synthesis of the proteins that make up the body of an organism, but how is the genetic material of two parents put together to form the genetic material of the offspring? To understand this, we must return to the ideas of Gregor Mendel. An understanding of this process will provide us with the groundwork for a fuller appreciation of evolution, adaptation, and the development of the living world.

Mendelian Genetics—Inheritance from Two Parents

The basic ideas that Mendel developed grew out of some seven or eight years of meticulously breeding pea plants. The fact that he chose the pea plant is considered by some to have been a happy accident, although some insist that Mendel was an astute observer who realized that several characteristics of the pea plant were well suited to genetic research. One of the key characteristics is that the blossom of the pea plant is structured in such a way as to prevent cross-pollination, thus allowing an experimenter to rigorously control the breeding of plants. The blossoms are also usually red or white, the offspring of red-blossomed plants almost invariably possessing red blossoms, the offspring of white-blossomed plants, white blossoms. Six other characteristics of the plant also "breed true."

Dominants and Recessives

In the first of his experiments, Mendel cross-pollinated white-blossomed and red-blossomed plants. In the first generation, now designated as F_1 (for *first filial*), all the offspring had red blossoms. In crosses between plants with yellow and green seeds, all the seeds were yellow in the F_1 generation. In each cross, one of the characteristics disappeared from the phenotypes of individuals in the F_1 generation. Mendel designated those characteristics that remained visible as *dominant* and those that disappeared as *recessive.* Rather than cross-pollinate the plants again, Mendel let the F_1 individuals self-pollinate. In the F_2 generation, the white blossoms reappeared, as did all the other recessive traits. Moreover, in each case in the F_2 generation, the dominant trait appeared in a direct ratio of 3:1 to the recessive trait. (See table 6.3.)

Table 6.3 Mendel's Experimental Results

Trait	*Dominant / Recessive*		*Appearance in F_2 Dominant / Recessive*	*Ratio*
Seed Shape	Round	Wrinkled	5474:1850	2.96:1
Seed Color	Yellow	Green	6022:2001	3.01:1
Seed Coat Color	Grey	White	705:204	3.15:1
Pod Form	Inflated	Constricted	882:299	2.95:1
Pod Color	Green	Yellow	428:152	2.82:1
Flower Position	Axial	Terminal	651:207	3.14:1
Height of Plant	Tall	Short	787:277	2.84:1

Each of the traits listed breeds true, which is one of the reasons that Mendel's choice of the pea plant as an experimental subject was such a wise one. Since his notes on the characteristics of the plant consist of nearly 400 pages, it is not likely that the choice was a "happy accident."

Mendel reasoned that the ratio could be accounted for *if genes came in pairs.* It is now obvious that this is a result of the behavior of chromosomes.

Meiosis

In order to understand Mendel's results better, we should examine more carefully how gametes are produced. Gametes arise by a division process called *meiosis.* The actual steps of meiosis are presented in box 6.3. Meiosis differs from mitosis, which we have already examined, in several important respects. In a meiotically dividing cell, the homologous chromosomes arrange in pairs and line up in a plane as in mitosis, but the *centromeres do not divide.* When the daughter cells are formed, each receives a whole, double-stranded chromosome, but since the centromeres have not divided, each receives only half the chromosomes present in the original cell. A resting phase occurs, during which there is *no replication of the chromosomes in the two newly arisen cells.* Another division then takes place during which the centromeres divide, and each of the four newly arisen cells receives a single-stranded chromosome. In other words, the second division is essentially a mitotic division. The important feature of the whole process is that *each of the four cells that results contains only half the number of chromosomes as were present in the original cell.*

In the human body, each cell contains twenty-three pairs of chromosomes, or a chromosome number of forty-six. The sperm and egg cells, however, contain only twenty-three chromosomes. If we designate the number of pairs of chromosomes as n, then we may designate the species' chromosome number as $2n$. Meiosis reduces the chromosome number to $1n$. A cell with the full complement of chromosomes is called a *diploid cell* ($2n$), and one with half the chromosomes is called a *haploid cell (n).* Thus, meiosis may be described as producing haploid gametes from diploid cells. (We should recall here our earlier discussion of plant life cycles and remind the reader that haploid cells may divide mitotically to produce an individual organism. Many microorganisms also spend most of their lives in the haploid state, the diploid state occurring only briefly after the fusion of gametes that then rapidly undergo meiosis.)

The Principle of Segregation

Although Mendel could not explain the action of genes in terms of meiosis, he was able to witness the effects of meiotic division, and he

Meiosis—The Formation of Germ Cells

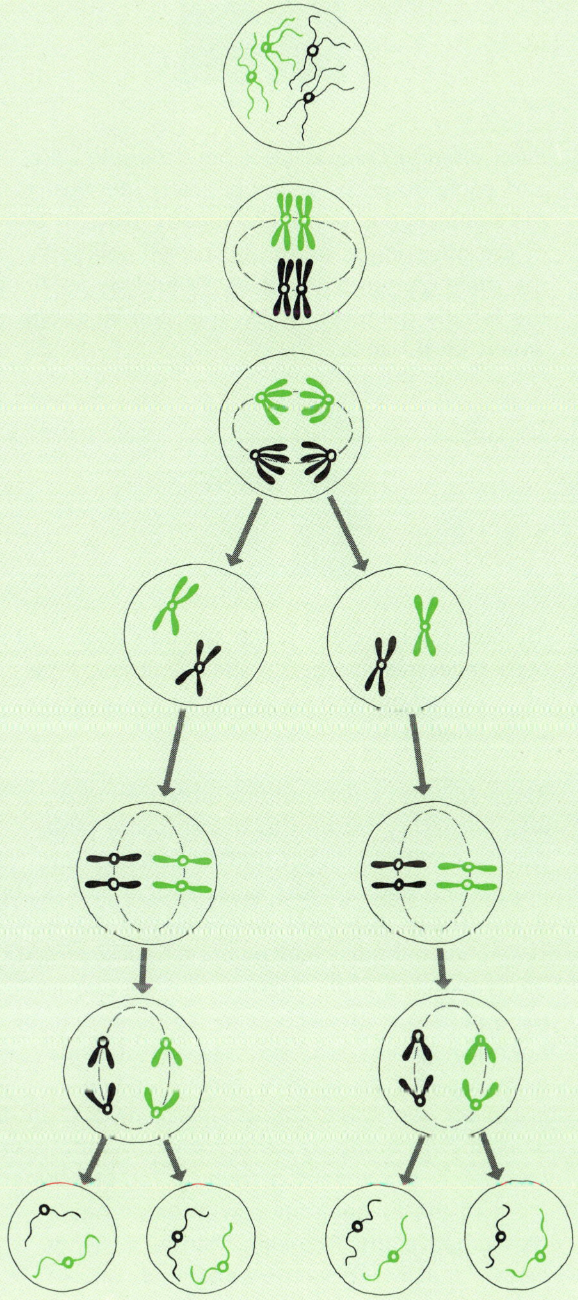

(a) Early in meiosis the chromatin organizes into chromosomes with visible chromatids.

(b) The pairs of homologous chromosomes line up on the equatorial plane, but unlike mitosis, the *centromeres do not divide* at this time.

(c) The homologous chromosomes are drawn along the spindle to the opposite poles of the dividing nucleus with each future daughter cell receiving one homologue.

(d) The new nuclear membrane is formed, but now each daughter cell contains only half the chromosomes present in the original cell; they are *haploid.*

(e) The daughter cells enter a resting phase and before the chromosomes shorten into compact bodies again, *the centromeres divide.*

(f) After the centromeres have divided another division takes place which is essentially a repetition of the events of mitosis. The result of this division is the production of four haploid cells containing only one of each homologous chromosome. When fertilization takes place the full set of chromosomes is restored. In the human female, only one of the four haploid cells will become an ovum. The other three fail to develop fully and become so-called polar bodies which are sometimes seen attached to an ovum.

BOX 6.3

called those effects the *principle of segregation.* He thus indicated that *at some time before reproduction, the pairs of genes controlling differing phenotypic characteristics were separated from each other.*

At this point, we should introduce some new terminology that will allow us to discuss segregation more economically. The alternative forms of a particular gene are called *alleles.* A gene, then, may be understood to possess two or more alleles. (Within an individual organism, any gene for any trait has only two alleles; when we discuss populations, we shall see that more than two alleles of a gene may exist in a population.) When an organism has two identical alleles for some trait, it is said to be *homozygous* for that trait (*homo* same + *zygon* yoke), and when it has two different alleles for some trait, it is said to be *heterozygous* (*hetero* other). A dominant allele is usually symbolized by an upper-case letter, and a recessive allele for the same trait is symbolized by a lower-case letter. In the pea plant, for example, the allele for red-blossom color is symbolized as *W*, and the white allelle as *w.* Using this shorthand, we can designate the genotype of a homozygous pea plant with a red blossom as *WW.* and the genotype of a homozygous plant with a white blossom as *ww.* Since white blossoms are recessive, it is easy to designate their genotypes as *ww.* However, it is difficult to determine the genotypes of plants with red blossoms, since they may be either *WW* or *Ww.*

That was precisely the problem presented to Mendel in his first series of experiments. Since the original plants all bred true, they were homozygous for blossom color or some other characteristic. But what were the genotypes of the red-blossomed plants in the F_1 generation? Understanding meiosis, we can see what Mendel could not. The genes of the red-blossomed parent (*WW*) separate (or segregate) into two *W* gametes. The white-blossomed parents segregate and produce two *w* gametes. There are four ways in which these gametes can combine, all of which are identical:

	W	*W*
w	*Ww*	*Ww*
w	*Ww*	*Ww*

Each offspring would be a heterozygote, *Ww,* and each would be phenotypically identical to the homozygous dominant parent. Now, if one of the offspring is permitted to self-pollinate, the male gametes would be *W* and *w,* as would the female gametes. Thus, four combinations would be possible:

	W	*w*
W	*WW*	*Ww*
w	*Ww*	*ww*

(By convention, dominance symbols always precede recessive symbols.) The result is a ratio of three red-blossomed plants to every white-blossomed one.

Since Mendel knew nothing about meiosis, it was necessary for him to devise some other way to determine what had happened. The method he devised was quite simple and is still in use today. After all, no amount of microscopic examination can reveal whether a gamete is carrying a dominant or a recessive allele for a particular characteristic. To determine the unknown genotype of an organism, Mendel performed a *test cross,* or *a cross between the organism with the unknown genotype and an organism with a known genotype.* In Mendel's experiments, this meant crossing a red-blossomed plant of the F_1 generation with any white-blossomed plant. Actually, Mendel was not certain that he knew the genetic constitution of the white-blossomed plant; all that he could reason was that *if* the genes were discrete units and could separate before reproduction, and *if* there were recessive and dominant forms, then a phenotype that showed a recessive form must contain two recessive alleles. Thus, the test crosses that he performed were made not only to determine the genotypes of red-blossomed plants but also to help prove his hypothesis that genes segregated. If an hypothesis is to be valid, the experimenter must be able to make accurate predictions based upon it. Mendel predicted that if genes segregated, the results of a test cross between red plants of the F_1 generation and plants with white blossoms would produce red-blossomed and white-blossomed plants in approximately equal proportions. According to the hypothesis, each of the plants in the F_1 generation carried one red and one white allele. Each gamete, then, would carry either a *W* or a *w*. The white-blossomed plant could produce only *w* gametes. Thus, the four combinations that could occur would be

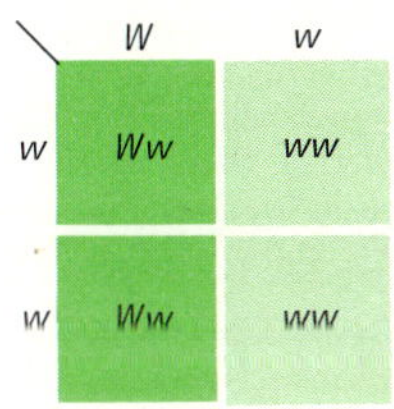

and the offspring would be genotypically *Ww, Ww, ww,* and *ww*. Mendel made the necessary crosses and found that his hypothesis was confirmed. Test crosses with other characteristics of the pea plant further confirmed it. Although he would never know it, the chromosome theory of inheritance fit well with the segregation and recombination of genes that Mendel had described in his experiments.

Independent Assortment

Among the characteristics that Mendel observed in the pea plant were certain traits that consistently appeared together. For example, the seed of a plant seemed always to be either round and yellow or wrinkled and green. After a cross, all the F_1 offspring had round, yellow seeds, and thus green color and wrinkled shape were designated as recessives. The genotypes of the parental plants, then, were *RRYY* and *rryy*. After self-pollination in the F_1 offspring, however, a new ratio appeared in the F_2. Of the 556 seeds produced, 315 were round and yellow, and 32 were green and wrinkled; but in the remaining seeds, a new combination of parental genotypes appeared. There were 101 wrinkled, yellow seeds and 108 round, green ones. By rearranging these numbers slightly, we can find the 3:1 ratio predicted by the principle of segregation. The total of the round seeds was 423 (315 + 108), and the total of the wrinkled seeds was 133 (101 + 32); similarly, there were 416 yellow seeds and 140 green seeds. The ratio of those offspring showing the parental genotypes was 9:1, and including the new combinations, we can fill out the ratio as 9:3:3:1 (423:108:101:32). Thus, as long as the plants were allowed to self-pollinate, characteristics that appeared consistently together in the phenotypes would continue to appear together. After interbreeding was allowed to occur and then was followed by self-pollination, the F_2 offspring indicated that two genes were involved in producing these characteristics and that these genes could be expressed in the phenotype *independently* of each other. Because of this independence of characteristics, Mendel described the phenomenon as the principle of *independent assortment.* (See fig. 6.17.)

The Determination of Sex

Near the end of the nineteenth century, one investigator noticed that in certain insects the gametes did not always receive an exactly equal number of chromosomes during meiosis. He was puzzled by this apparent inconsistency but was not able to determine its significance. He labeled the extra chromosome as the *X* chromosome. Soon after his "discovery," he left the field of genetics entirely.

Later, it was shown that sex determination was dependent upon a particular pair of chromosomes. It was noted that in the body cells of male organisms of some species, one pair of chromosomes differed from one pair in a female of the same species. In the exceptional pair, the female appeared to have two identical chromo-

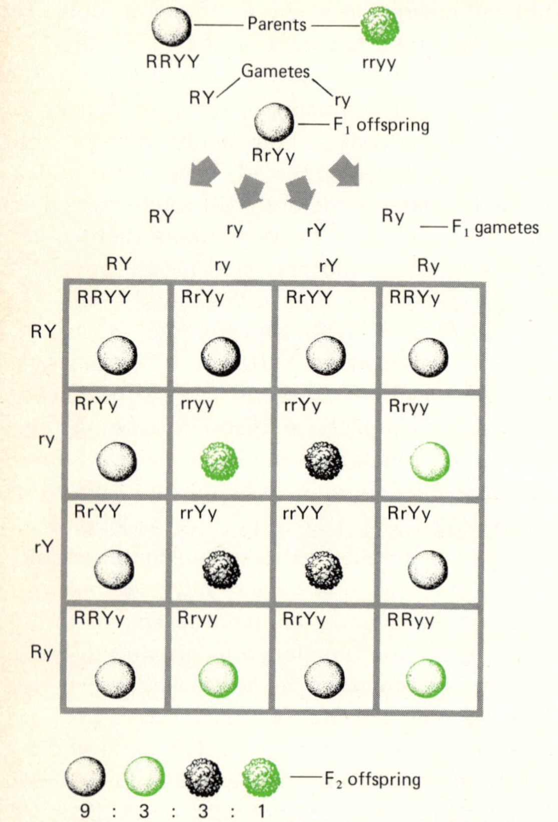

FIGURE 6.17 Independent assortment. The production of gametes in the parental generation, when the parents are homozygous (one dominant and the other recessive), can only result in haploid sperm or ova. Consequently, all the F_1 offspring from a mating of such parents would be heterozygous, and all would show the dominant traits. The heterozygous offspring, however, can produce homozygous and heterozygous gametes. When self-pollination occurs, as if heterozygotes crossbreed, the recombination of characteristics produces new phenotypes in the F_2 generation. The *Punnett Square* shows the possible combinations of gametes and thus the frequency of the offspring.

somes, and the male had one chromosome like those of the female and one which was slightly different. Experiments definitively linked these differing chromosome pairs to the determination of sex, and the odd chromosome of the male came to be called the *Y* chromosome. (A reexamination of the earlier investigator's work revealed that the *X* chromosome that he had labeled belonged to the female and that the males of the species he was examining contained no information for sex determination.) Thus, a female possesses two *X* chromosomes and can produce only *X* ova. The male possesses *X* and *Y* chromosomes (or only a single *X* chromosome) and can produce sperm with *X* or *Y* chromosomes (or no information). If an *X* sperm fertilizes an ovum, then, the offspring will be a female (*XX*), whereas if a *Y* sperm fertilizes an ovum, the offspring will be a male (*XY*). (If a sperm carrying no sex information fertilizes an ovum, the offspring will be male.)

Thus, we can see how the number of males conceived is statistically equal to the number of females:

♂ \ ♀	*X*	*X*
X	*XX*♀	*XX*♀
Y	*XY*♂	*XY*♂

Sex Linkage

Shortly after the sex determination experiments were performed, one of the great geneticists, Thomas Hunt Morgan, founded a laboratory at Columbia University. Using the tiny fruit fly, *Drosophila,* (see fig. 6.18) as an experimental organism, Morgan had been breeding flies for nearly a year, when one appeared that had white eyes. Since *Drosophila* normally have red eyes, this "one in 2 million" greatly aroused Morgan's curiosity. The white-eyed fly was a male, and Morgan immediately bred the male

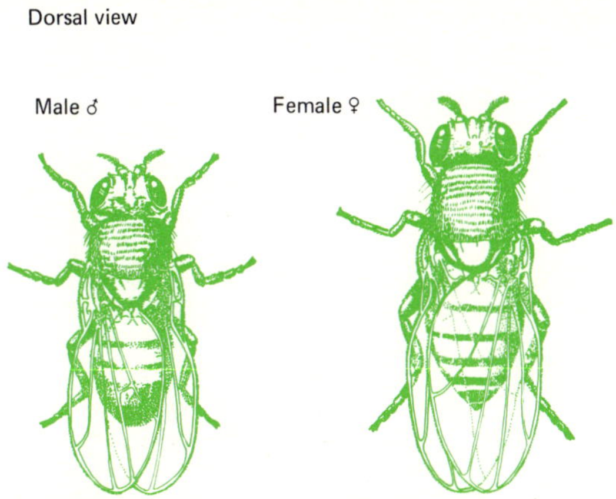

FIGURE 6.18 *Drosophila melanogaster.*

with a normal (or *wild type*) red-eyed female. In the F_1 generation, all the offspring were red-eyed, and therefore the allele for white eyes had to be recessive. In the F_2 generation, the white eyes reappeared but *only in male flies.* Certain experiments that Morgan was familiar with had suggested that certain characteristics in some organisms did not assort independently but seemed always to be linked. Morgan began to suspect that he had a similar situation on his hands but that the characteristics were *sex-linked.* To test the idea that sex and eye color were linked, Morgan then crossbred the white-eyed males of the F_2 generation with red-eyed females. If the characteristics were sex-linked, then *all* the males in the F_3 generation should be white-eyed. But the mating produced

129 red-eyed females
132 red-eyed males
86 white-eyed males
88 white-eyed females

The ratio of males to females is nearly 1:1, but the ratio we should expect from independent assortment does not fit our expectation at all, suggesting that the characteristics are linked. (The expected 3:1 ratio of red to white eyes does not fit either, but *suppression* of the recessive characteristic can explain this. That is, some fertilized ova may have been suppressed and not have developed into flies.) But how could the white-eyed females be explained? Morgan guessed that it was because the chromosome (or whatever it was that determined heredity*) carried *no* information for eye color.

Thus, the genotype of the original white-eyed male was X^wY^0, with the superscript indicating linkage and the 0 indicating "no information." The original female was X^WX^W, and all F_1 offspring would have red eyes; but all the F_1 females would be heterozygotes, X^WX^w, as would all the F_2 males, X^WY^0. Thus, we can see why all the F_2 white-eyed flies were male:

♀ \ ♂	X^W	Y^0
X^W	X^WX^W	X^WY^0
X^w	X^WX^w	X^wY^0

Given that the original white-eyed male was mated to one of his daughters, we can see how the white-eyed females arose.

♀ \ ♂	X^w	Y^0
X^W	X^WX^w	X^WY^0
X^w	X^wX^w	X^wY^0

Morgan's assumption that the characteristics were *sex-linked*—that is, that *the genes that controlled them were on the sex chromosome*—was correct, but linkage was not quite as he had first envisioned it. Within a year after Morgan published his paper describing his experiments, a number of other sex-linked characteristics were discovered in *Drosophila*. Moreover, this and other evidence changed

*Recall that Morgan did not at first subscribe to the chromosome theory of inheritance on the grounds that it would mean that each chromosome must carry numerous characteristics. These experiments were the beginning of his conversion to the theory.

Morgan's mind about the chromosome theory of inheritance.

Crossing Over

As progress was made in the understanding of chromosome behavior, it became clear that certain genes always appeared to travel together. These groups of genes became known as *linkage groups.* As Morgan continued his study of *Drosophila* genetics, it became very evident that there were four linkage groups in the fruit flies, and since there were also four pairs of chromosomes, it seemed obvious that the linkage groups were located on the chromosome. There were, however, some problems with this idea, since it had been observed that a gene occasionally appeared to change linkage groups. The microscopic examination of chromosomes seemed to offer an explanation for linkage-group changes, though, since visible breaks in chromosomes had been observed. Moreover, during meiosis, the chromosomes did not always lie side by side but could sometimes be seen twisted about each other. From the microscopic evidence and from the changes that were seen to take place between linkage groups, it was reasoned that the chromosomes were capable of breaking and exchanging parts. (See fig. 6.19.)

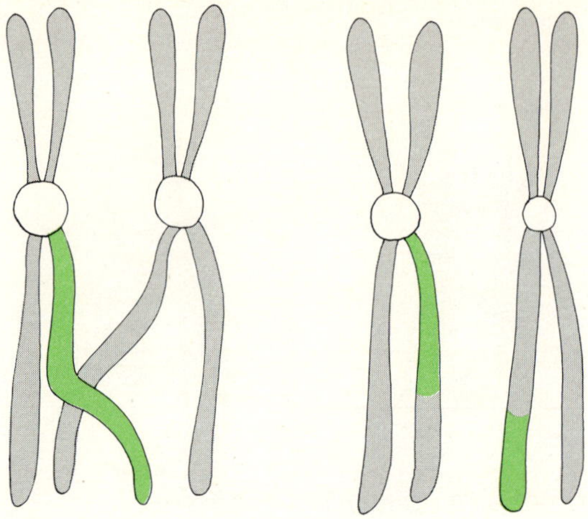

FIGURE 6.19 Crossovers. During meiosis crossovers occur frequently, thus providing greater potential for variation of the offspring.

If such *crossing over* occurs, how is it expressed in the phenotype of an organism? Consider the events that occurred in Morgan's laboratory as these ideas were being formulated.

Wild type *Drosophila* have grey bodies and long wings. When two mutants with black bodies and short wings arose, they were crossed with wild type individuals and produced a F_1 generation that was phenotypically of the wild type. Thus, black bodies and short wings are recessive. If F_1 offspring were interbred, one of two things should happen: the characteristics could assort independently if they were controlled by genes on separate linkage groups, or they could produce a ratio of 3:1 (grey-long : black-short) if they were linked.

When the necessary interbreeding had taken place, some individuals arose that could not be explained by Mendelian principles. Although the ratios were close to those expected if the genes were linked, there were a few flies with black bodies and long wings and a few others with grey bodies and short wings. This indicated that either the genes were *not* on the same chromosome or, if they were, that they had somehow changed linkage groups. Morgan suspected that the latter was the case and used a series of letters to explain such exchanges. If one chromosome contained the letters *ABCDEFGHI* and the other *abcdefghi*, and if crossing-over occurred in the middle of the chromosome, the result would be chromosomes with the series *ABCdefgHI* and *abcDEFGhi.*

Such crossing over is quite difficult to see in the chromosomes themselves, and as yet, no one has been able to show how the crossing over that does occur takes place with such a precisely regulated exchange of material. What, for example, keeps the chromosomes from becoming scrambled in this way?

ABCdefghHI / abcDEFGi

Although crossing over has only rarely been observed, enormous amounts of data have accumulated in the years subsequent to Morgan's work that verify the occurrence of crossing over. In fact, it appears that such events occur at nearly every meiotic division of a cell. We shall not dwell here on the evolutionary significance of linkage groups and crossing over except to point out that as a part of the apparatus of sexually reproducing species, along with the combination of parental genotypes, the potential for variation within a species is greatly increased by this phenomenon.

Chromosome Maps

The existence of crossing over suggested that each gene on a chromosome could be found at a specific location (later to be known as its *locus*) within the chromosome. If so, the problem was to determine the order in which the genes were arranged on the chromosome, or devise *chromosome maps.* It had been noted that changes in linkage groups usually occurred with a relatively predictable frequency. For example, the test crosses of *Drosophila* for body color and wing size always produced 41.5% each of grey-long and black-short flies and 8.5% each of grey-short and black-long flies. Such relatively stable frequencies suggested that the percentages of crossing over probably could be correlated with the distances between the genes on the chromosome. Since chromosomes are so very small, it would be impractical to try to use any such normally accepted measurement as millimeters; therefore, an arbitrary unit of distance was set. The unit chosen was based on the *percentage of crossing over per hundred fertilized eggs.* Thus, if two linked genes were found to cross over 10% of the time (crossing over occurs 10 times per hundred eggs), such genes were said to be ten units apart.

Consider, for example, two linked traits, A and B. If test crosses indicate that crossing over occurs between these traits in 12% of the cases, then we would say that they are twelve units apart. Now consider another trait in the same linkage group, which we will call C. If test crosses show that A and C cross over in 22% of the cases, then A and C are 22 units apart. But what is the relative position of B? If additional crosses indicate that B and C cross over in 10% of the cases, then B falls between A and C.

A————B
12 units

A——————C
22 units

B———C
10 units

A————B———C
12 units 10 units

If, however, the tests indicate that B and C are 34 units apart, then B is 12 units to the left of A on the map.

A————B
12

A——————C
22

B————A——————C
12 22

Although this is the basic mechanism underlying the construction of chromosome maps, the subject is considerably more complex. Double crossing over can occur, or crossing over can occur and then be partially reversed. (See fig. 6.20.) Any number of factors can complicate the construction of such maps. (See fig. 6.21.)

The frequency with which crossing over occurs involving any two genes on a chromosome is partially correlated with the distances of the genes from each other. Genes that are close together are more likely to segregate together when crossing over occurs. However, genes that are very far apart on the chromosome exhibit somewhat less predictable crossing-over frequencies than might be expected due to (1) the many loci between them where breaks may occur and (2) the frequencies of multiple crossing-over that can occur between widely separated genes. As important as the distance in

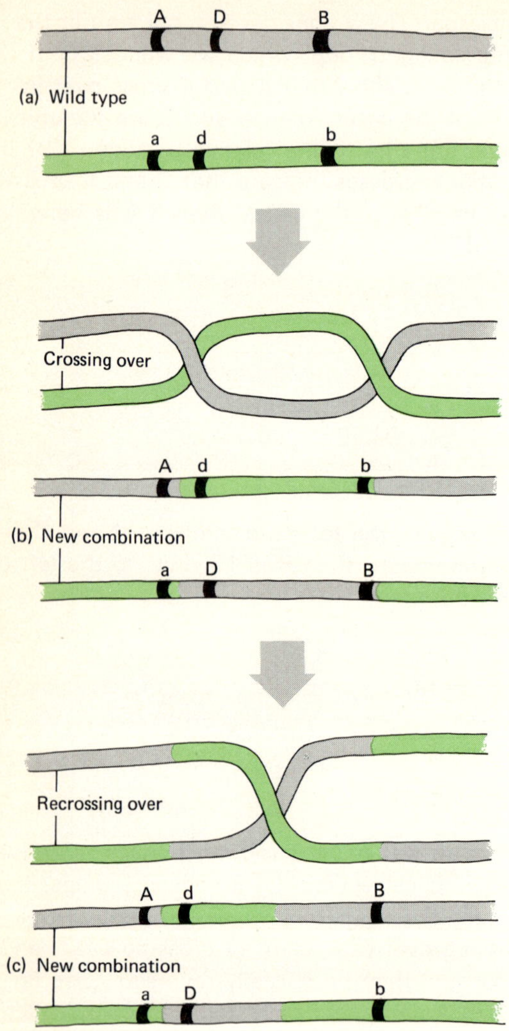

FIGURE 6.20 Double crossover and recrossover. The meiotic division of cells is marked by frequent crossovers of the sort shown in Figure 6.19 and even more complicated events such as shown here. Double crossovers and recrossovers can complicate the construction of chromosome maps, but they serve as a potentially important source of genetic variation above and beyond independent assortment.

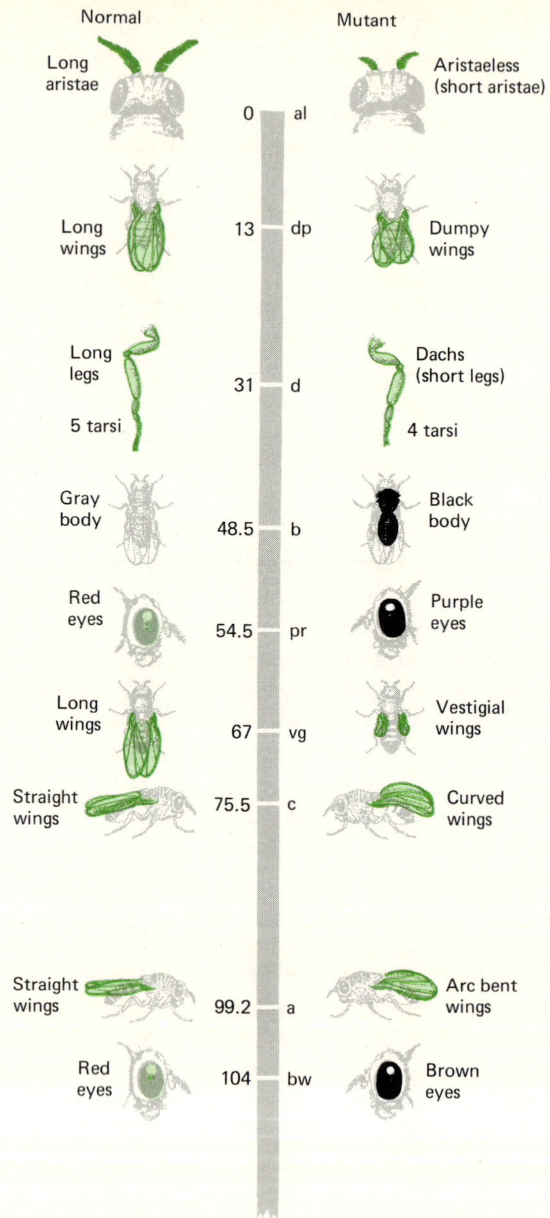

FIGURE 6.21 Studies of crossing over have enabled the traits of *Drosophila* to be mapped on its chromosomes. A typical map is shown. *(From James D. Watson,* Molecular Biology of the Gene, *Second Edition, copyright © 1970 by J. D. Watson; W. A. Benjamin, Inc., Menlo Park, California.)*

affecting the percentage of crossing over is the nature of the structure of the chromosome itself. Within the chromosome, there are usually areas where breaks occur relatively often; thus, even if two genes are very close together, if each is on a different side of such a location, they will be seen to cross over frequently. Therefore, these genes will appear to be located at a greater distance from each than, in fact, they are.

Why Is There Sex?

We have seen that the sexual process is basic to the determination of the genotype of an organism and that sexual reproduction produces offspring that differ genetically from their parents. Geneticists have used the process to learn about the gene, its location on the chromosome, and how the gene determines the characterictics of the organism. But we must also ask why sex is so important to the survival of a species; that is, why has sex evolved?

Asexual reproduction is, of course, more efficient. It takes only half as many individuals to produce an equal number of offspring. In spite of this, we may assume that since so many species reproduce sexually, there must be some selective advantage to doing so. Asexual reproduction has at least one drawback: the genes available to the offspring are restricted to the genes available to its single parent. Excluding the possibility of mutation, the offspring are identical to the parent. Since mutation is known to occur at a very low rate, the variability in an asexually reproducing population would be very low, allowing selection to have profound effects. In a sexually reproducing population, the offspring share the genetic features of two parents, and the recombination of parental genotypes produces offspring that differ from both parents. Thus, the variability in the sexually reproducing population is increased.

Moreover, if mutations occur in an asexually reproducing population, the mutation is passed on to the offspring intact. The offspring could evolve in a different direction from the parents or other members of its own generation. As mutations accumulated, the population could develop into a number of exceedingly different kinds of organisms. In a sexually reproducing population, the effects of mutations are lessened by the contribution of genes by two parents. Thus, the population may evolve, but it is prevented from departing very far from that which is normal for the species.

The Expression of the Genes—From Zygote to Adult

We have seen that the sequence of nucleotide bases in the molecule of DNA determines the structures of the basic chemicals of life: proteins, and especially enzymes. We have also examined how genes are located on chromosomes and how the chromosomes are passed from parent to offspring, and we have seen that sexual reproduction is an important process for evolution. But thus far, we have assumed that it is the genes that determine the characteristics of the adult organism; that is, we have assumed that genes determine the characteristic phenotype of a species. We must therefore examine the process of development and the role that the genes are thought to play in the process.

How an organism develops from a single cell has been observed repeatedly, and an as yet unproven model of gene action has been proposed to explain development. Before we examine the model, however, we should briefly review some basic developmental processes. We should note, too, that many of the processes that will be described in some detail are not repeated exactly in every multicellular organism. There are many similarities, of course, but from species to species, there are even more differences.

Fertilization and Early Development

In a sexually reproducing organism, the penetration of the ovum by a sperm is the beginning of development. Although the nuclei of the sperm and the ovum must usually fuse for the zygote to receive a full complement of chromosomes, penetration alone can often trigger development. In some cases, fertilization is not even necessary for the development of a complete multicellular organism. Frog eggs can be induced to develop into larvae simply by pricking them with a fine glass needle dipped in frog's blood, by subjecting the eggs to mild shocks, by changing the salt content of the surrounding water, or occasionally, merely by shaking them. A few eggs so treated will grow into viable tadpoles. Similar methods have produced adult rabbits from unfertilized eggs.

Normally, however, gametic nuclei must fuse and endow the zygote with the full chromosome complement. In animals, it appears that once the zygotic nucleus is formed, the cytoplasm of the cell plays a very important role in development. The cell undergoes a rapid series of mitiotic divisions, although the ball of cells that is developing remains the same size, and the cytoplasmic material of the zygote is progressively apportioned to more and more individual cells. This series of *cleavages* continues as the cells organize into a hollow sphere called a *blastula,* which has a fluid-filled interior known as a *blastocoel.* Soon the sphere begins a stage called *gastrulation* with infolding of some of the cells. (See fig. 6.22.)

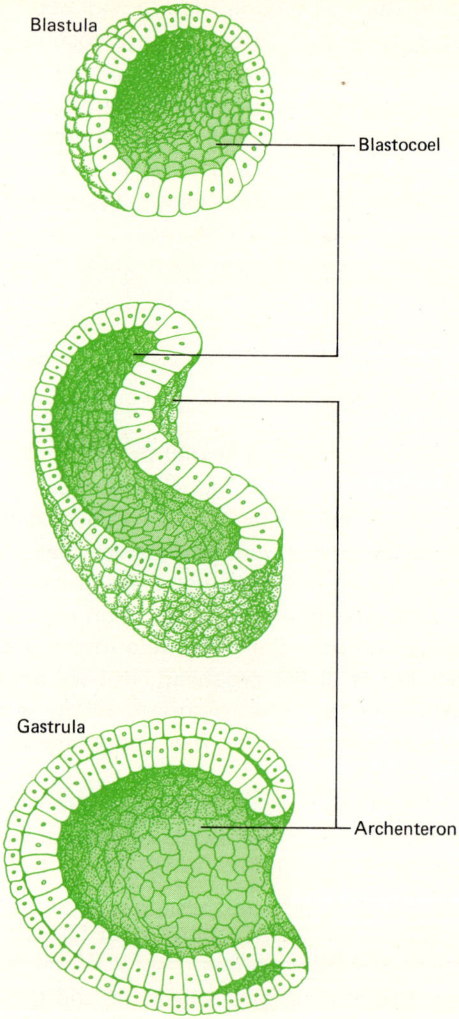

FIGURE 6.22 The formation of a gastrula from a blastula. The archenteron will become the gut of the adult. *(Redrawn from Curtis, 1968, after Huettner.)*

Cleavage and the formation of a blastula and gastrula can be seen in the embryonic development of many species. However, most similarities often end here, and in dealing with different species, embryologists often find themselves dealing with very differently expressed phenomena. One of the controlling factors in the course of development is the amount of yolk that the egg contains. In those species whose eggs contain very little yolk, such as the sea urchin, the process is as shown in figure 6.22. Since development is very rapid, there is enough yolk for the embryo to survive. A frog egg contains an intermediate amount of yolk, and a chick hatches from an egg with a very large amount of yolk. Some of the differences that result from the yolk content of the egg are shown in box 6.4.

The biological function of the stages of cleavage, blastulation, and gastrulation is to bring masses of cells into the proper position for them to differentiate into the highly specialized cells

Gastrulation in Amphibian and Chick Eggs

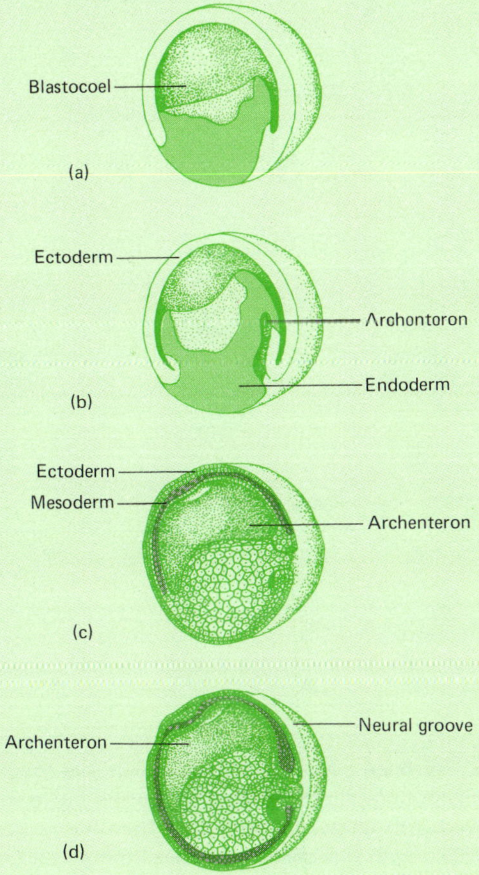

Figures (a)–(d) show the process of gastrulation in an amphibian egg. The blastocoel that remains at the beginning of gastrulation in the amphibian egg is relatively smaller than that of the egg of an echinoderm since some of the cells have been dividing more rapidly than others. The formation of the archenteron begins, and as it proceeds, three definable structures arise, the ectoderm, the mesoderm, and the endoderm. The archenteron will ultimately become the gut and digestive system made up of cells derived from the endoderm. The ectoderm is the outer layer of cells and will give rise to the skin, central nervous system, and sense organs such as the eyes and ears. The cells of the mesoderm give rise to the skeleton and connective tissue. The neural groove will sink into the interior of the embryo and form the notochord and ultimately the spinal chord and brain.

The development of an embryo from an egg with a relatively large amount of yolk, as in a chicken egg, (e)–(g), is strikingly different from the development of an amphibian egg. Cleavage takes place only within a small part of the avian (bird) egg, in the *cytoplasmic disc.* Ultimately the endoderm separates from the yolk and the cells of the upper layer give rise to the ectoderm and mesoderm as the cells fold inward along a line running through the middle of the embryo (h). This fold produces a visible line on the surface of the cytoplasmic disc, which is called the primitive streak.

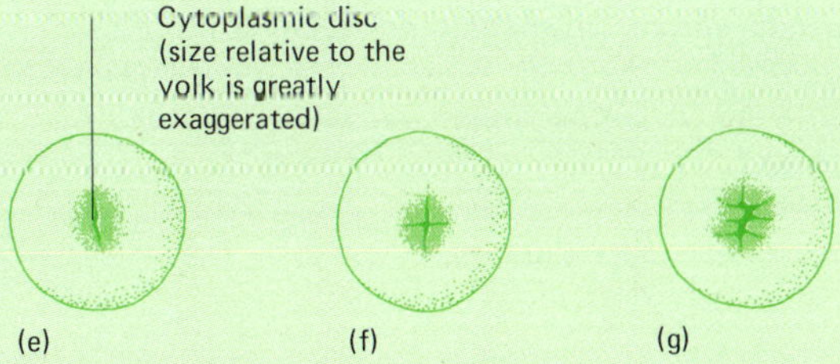

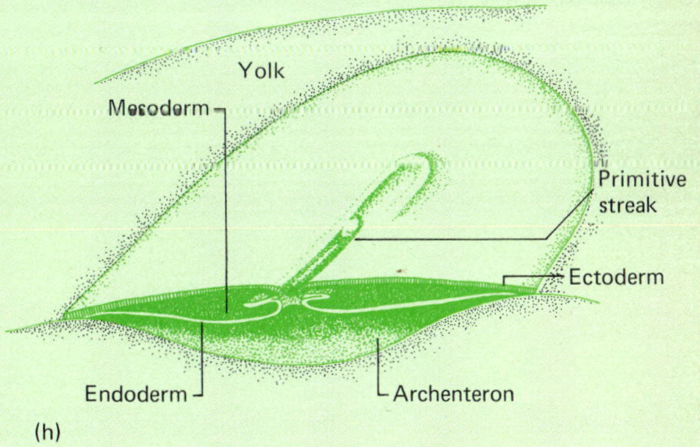

(Redrawn: (a), (b) from Saunders after Nelsen; (c), (d), (h) from Curtis, 1968, after Huettner; (e), (f), (g) from Saunders.)

BOX 6.4

of the adult organism. This differentiation is the crux of the problem in understanding development. How do cells differentiate? Since they are all dividing by mitosis, each is genetically identical to all of the others. Each carries a gene for every characteristic that will be found in the adult organism and is theoretically capable of becoming a neuron, a skin cell, or even an entire organism. But at some time in development, this broad ranging capability is lost. When and how does this happen?

The Nucleus in Control of Development

In a simple and ingenious experiment performed by the embryologist Hans Spemann early in the twentieth century, it was strongly suggested that the nucleus directed the development of the embryo. Spemann tied a human hair around the middle of an egg of a newt and

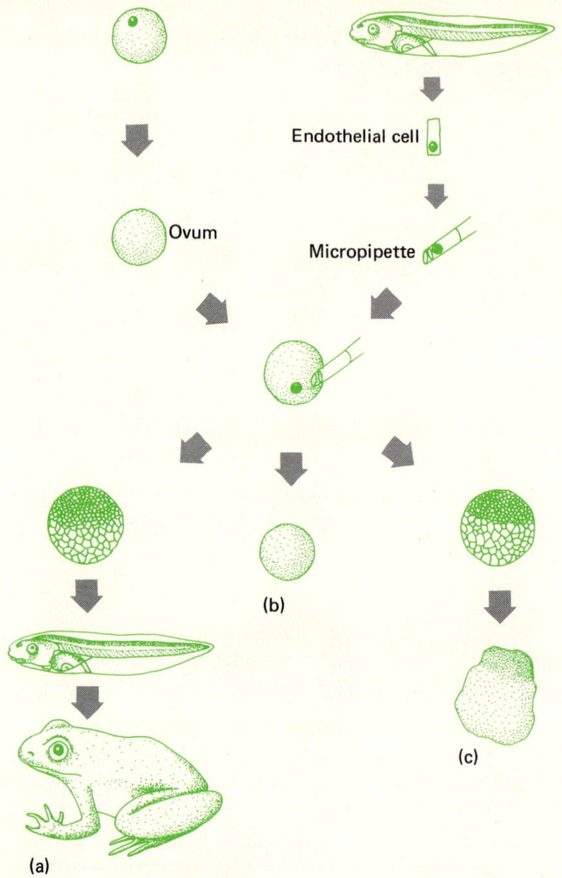

FIGURE 6.24 Nuclear transplant. In this experiment an unfertilized frog ovum has the nucleus removed with ultraviolet radiation. Then the nucleus is replaced with one removed intact from an endothelial cell from the intestine of a frog embryo. In a few instances the cell will develop into a mature frog as in (a). Frequently, no divisions at all occur, (b), and in a few instances, (c), an abnormal embryo is formed. *(From "Transplanted Nuclei and Cell Differentiation" by J. B. Gurdon. Copyright © December 1968 by Scientific American, Inc. All rights reserved.)*

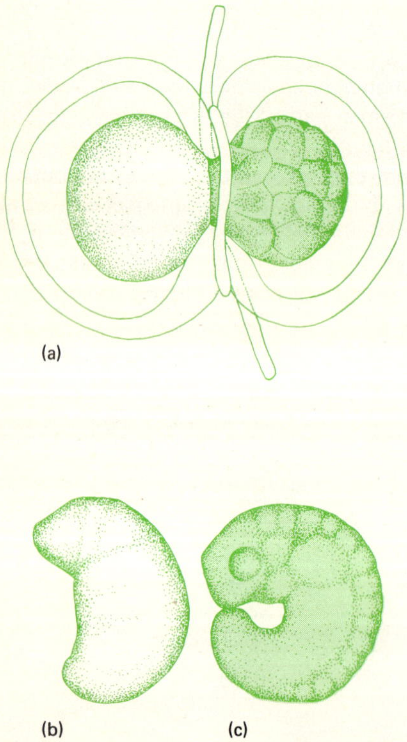

FIGURE 6.23 Spemann's experiment with a newt's egg. *(Redrawn from Balinsky after Spemann.)*

tightened the hair until the egg was almost severed, carefully restricting the nucleus to one side. (See fig. 6.23.) In the nucleated side of the egg, cell division proceeded normally, but in the side without a nucleus, no division took place. As the number of cells increased in the nucleated side, the nuclei there grew progressively smaller. Then Spemann allowed one, and only one, nucleus to slip through the noose. He then tightened the hair and severed the two halves. In many instances, the recently nucleated eggs went on to divide and develop into an embryo.

Subsequently, in the early 1950s, the techniques of nuclear transplant were developed sufficiently to permit embryologists to see what would happen if a nucleus were removed from a well-developed blastula, or even an embryo, and transplanted to an egg. By subjecting an egg to ultraviolet radiation, they were able to remove its nucleus and replace it with one taken from a frog embryo that contained about 16,000 cells. (See fig. 6.24.) Presumably, the cells were already well along in the differentiation process. Still, a number of their transplanted nuclei were able to direct the development of the eggs into embryos. Usually, the more well-developed—that is, older—the embryo, the less likely it is that one of its transplanted nuclei can induce development. Precisely why this is so is as yet unknown. Whether it is because the nucleus is more easily damaged in the transplantation process or because it has permanently lost some genetic capability is a major question that remains to be answered.

We have mentioned that the cytoplasm of the egg plays an important role in the development of the embryo. What is the evidence for this? Spemann noticed that the newt's egg exhibited what he called a "grey crescent" located slightly below the equator of the egg. (See fig. 6.25.) He also noticed that if the egg were divided in such a way that each portion contained material from the crescent, both portions would develop into complete embryos. On the other hand, if one half of the egg were to receive no part of the crescent, that half would continue

FIGURE 6.25 Grey crescent. The fertilized ova of some vertebrates show a pigmented crescent-shaped region that is apparently of considerable developmental importance. If the material in the crescent is not apportioned to one-half of the egg at cleavage, the portion lacking the material will not develop into an embryo.

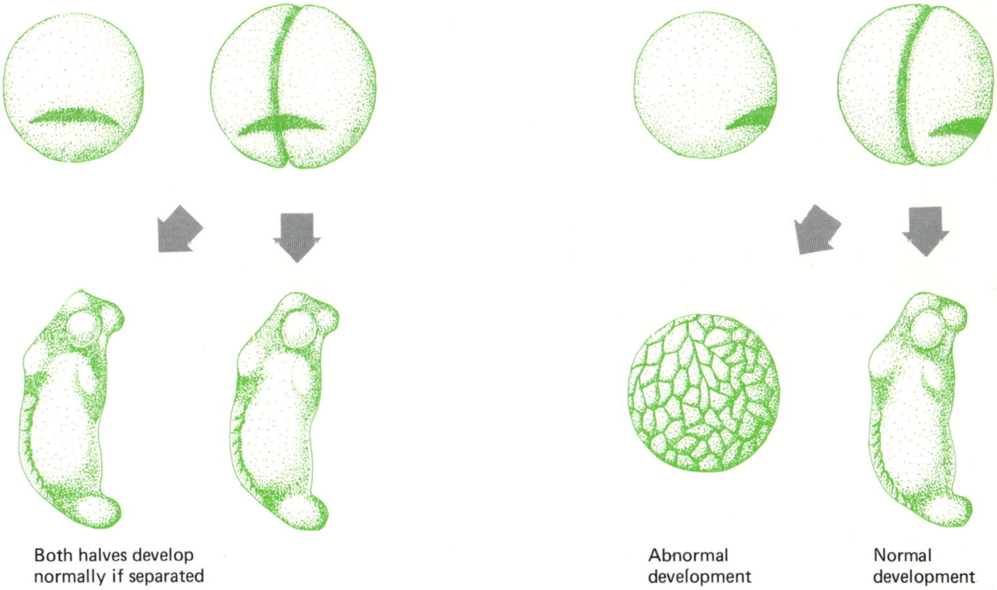

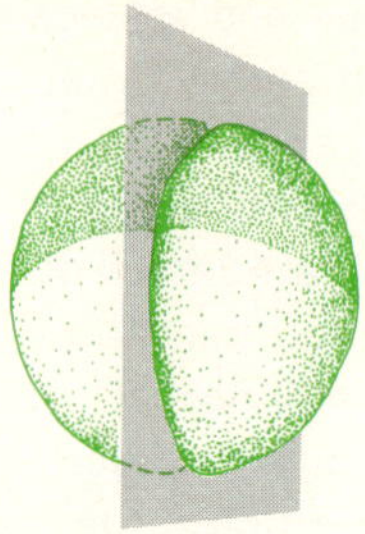

(a)

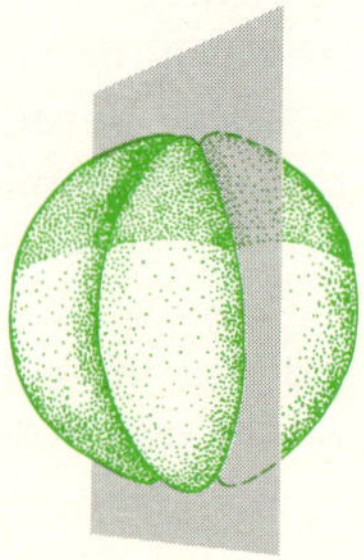

(b)

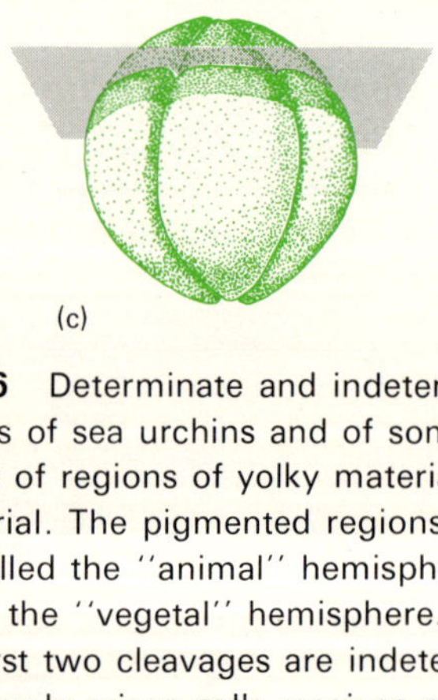

(c)

FIGURE 6.26 Determinate and indeterminate cleavage. The eggs of sea urchins and of some vertebrates consist of regions of yolky material and of pigmented material. The pigmented regions are sometimes called the "animal" hemisphere, and the yolky regions the "vegetal" hemisphere. In the sea urchin, the first two cleavages are indeterminate and each of the newly arisen cells receives material from pigmented and yolky regions. The planes of cleavage, in such instances, are as shown in (a) and (b). The third cleavage plane would be as shown in (c). Once the third cleavage takes place, the internal contents of the cells are different, and these internal cytoplasmic differences seem to play a large role in the expression of the genes.

to divide but would not differentiate into an embryo. He concluded, and further studies have confirmed, that cytoplasmic, not nuclear, differences determined the direction of development.

Exactly how important the cytoplasm is was shown in a series of experiments on sea urchin eggs. After the first cleavage had taken place, the cells were separated by shaking them in a vial of sea water. Each went on to develop into a normal embryo. If separated after the second cleavage, all four grew normally. After the third cleavage, however, when eight cells were present, separation by the same method produced abnormal embryos. We can understand this if we look at cleavage as being of two types: *determinate* and *indeterminate.* When the egg undergoes indeterminate cleavage, each cell receives a portion of cytoplasmic material identical to that which each of the other cells receives. When the cleavage is determinate, the cytoplasmic material of one cell differs from that of another, and the cells develop differently. In the zygotes of sea urchins, the first two cleavages are indeterminate and the third is determinate. (See fig. 6.26.) Thus, we can visualize the egg not as a uniformly constituted cell, but rather as one in which several "ingredients" are distributed in unequal shares in various locations. Experiments have also shown that if the cytoplasmic material is relocated within the zygote, or if cells are relocated within a developing embryo, an organ may develop in an inappropriate location. Careful investigation of the eggs of frogs and other organisms show that the cells from one portion of the egg contain high concentrations of pigments, while those from another portion contain little or no pigmentation. Other distinct differences are suspected but have yet to be conclusively demonstrated.

How the Genes Control Development

It appears, then, that the environment conditions the expression of the genes. In the zygote, the immediate environment is the cytoplasm. At fertilization, the cytoplasm consists of materials contributed largely by the parents; as development proceeds, gene products are added in the form of newly synthesized proteins. It is believed that as the cells develop and differentiate, they produce chemical substances that serve to inactivate some genes or to promote the activity of others. A great deal of what is known about the control by genes of developing cells has been extrapolated from studies of *Escherichia coli,* the common human intestinal bacterium. Still open to question is whether the data that have been derived are valid in multicellular organisms. Let us examine briefly how the genes governing the metabolism of certain sugars are controlled and operate in *E. coli.*

Perhaps because these bacteria are often found growing in environments where glucose, a simple sugar, is abundant, they can grow in cultures containing glucose as the only source of carbon and energy. However, if *E. coli* are cultured on media containing a more complex sugar, lactose, they are found to contain a large quantity of an enzyme, β-(beta) galactosidase, which serves to split lactose, a *di*saccharide, into glucose and galactose, two *mono*saccharides. When grown on media containing either glucose or galactose, the bacteria produce only basal levels (about 3 molecules per cell) of the splitting enzyme, but when lactose is added to the medium they can be induced to synthesize about a thousand times as much (about 3000 molecules per cell). Thus lactose is an *inducer* for the synthesis of β-galactosidase. Lactose also induces the synthesis of two other enzymes, namely galactoside permease which transports the lactose across the membrane of the cell into its interior, and galactoside transacetylase, the function of which is unknown. A study of mutations in the genes of *E. coli* indicated that the production of all three enzymes could be located in certain genes. Some mutant forms were incapable of producing any of the enzymes, and were thus unable to grow on media in which lactose was the sole source of carbon and energy. Another class of mutants was unable to "turn off" the three genes, even though lactose was absent from the growth medium. Still other strains of *E. coli* have been found that produce very large amounts of the enzyme, up to 15% of the total protein content of the cells. Such "super producing" mutants, by the way, grow poorly, apparently because they synthesize β-galactosidase at the expense of other proteins.

Such studies have led to the formulation of the *operon theory* of gene regulation. The operon is composed of different regions of nucleotide bases: (a) the *structural* genes, (b) the *regulator* genes, and (c) the *operator.**

The structural genes contain the sequence of nucleotide bases that code for the synthesis of mRNA by mRNA polymerase, and thus for the synthesis of specific proteins. An operon may contain only one structural gene, but it may also contain two, and in some cases it may contain several adjacent genes which function coordinately in the biosynthetic or metabolic pathways of the cell.

Regulator genes consist of a sequence of bases for the transcription of mRNA molecules that produce *repressors.* Repressors are proteins which bind to the operator and stop the transcription of mRNA at the structural genes, thus inhibiting the synthesis of the proteins coded for by such genes. Should a regulator gene mutate, the structural genes quite likely will continue to synthesize proteins that are not needed by the cell.

*The operon is also defined by some geneticists as a coordinate unit of transcription from which one mRNA molecule is transcribed for one, two, or several adjacent genes that act together in some metabolic pathway.

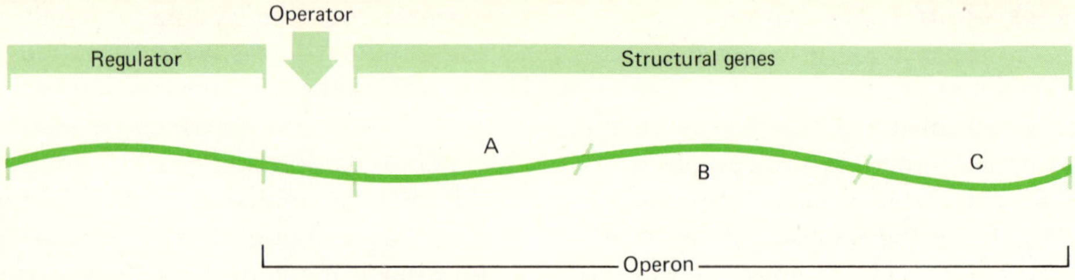

FIGURE 6.27 The operon theory of the gene. Although the regulator is shown here as a nucleotide region adjacent to the operator, it is not known that they are adjacent. Some evidence indicates that they may be separated by only a few nucleotides, but in some instances there is reason to believe that they may be located some distance from the operators and the structural genes. In *E. coli* the regulator gene is slightly over 1000 nucleotides long, while the operator is only a few nucleotides in length. The *E. coli* structural gene for the production of beta galactosidase is over 3500 nucleotides in length. *(From James D. Watson,* Molecular Biology of the Gene, *Second Edition, copyright © 1970 by J. D. Watson; W. A. Benjamin, Inc., Menlo Park, California.)*

The operator is *not* a gene, since no protein is synthesized from this sequence of bases. It is a section of DNA at which the repressor binds. It is at the operator that the transcription of DNA by RNA polymerase is blocked, thus preventing the production of the proteins coded upon the structural genes. Should a mutation occur at the operator that prevented or impaired the binding of a repressor, then the structural genes would continue to transcribe molecules of mRNA. These steps are represented in figure 6.27.

The system controlling the metabolism of lactose in *E. coli* is the "lac operon," consisting of an operator (*o*), a regulatory gene (*i*), and three adjacent genes coding for the synthesis of β-galactosidase (*z*), galactoside permease (*y*), and galactoside transacetylase (*a*). Repressors may exist in an active or inactive state. In the lac operon system, the repressor that is synthesized from the regulatory gene is active, and thus binds at the operator, stopping further transcription. When lactose, an inducer, is present in the medium, it binds to the repressor and inactivates it, thus allowing transcription to proceed and the enzymes of the structural genes to be produced for the metabolism of lactose. It appears that glucose serves generally to inhibit *E. coli* from producing β-galactosidase and the other enzymes in the lac operon. In the presence of glucose and lactose, for example, *E. coli* first utilizes only the glucose, and the enzymes necessary for the transport and further metabolism of lactose are not produced. (See fig. 6.28.)

The lac operon system is a form of gene regulation called *negative induction,* and has been characterized as a form of double negative. The repressor has a negative effect on the structural genes, while the inducer negates the effect of the repressor. The two negatives make a positive, and the structural genes are turned on. Exactly how the inducer inactivates the repressor is still unknown, but it is thought that the shape of the repressor is altered when an inducer binds with it, and that such conformational changes prevent the repressors from binding at the operator.

Negative induction is not the only known, or hypothesized, control system in the expression of genes. Although we shall not discuss them in detail, at least three other forms of control have been proposed or shown to exist. In *positive induction* systems, an inducer binds with a repressor, and the inducer-repressor complex then binds with a nucleotide region to initiate the synthesis of enzymes at the structural genes. In *positive repression* systems, the repressor inactivates the inducer, so that the operon cannot function and is thus repressed. The control of

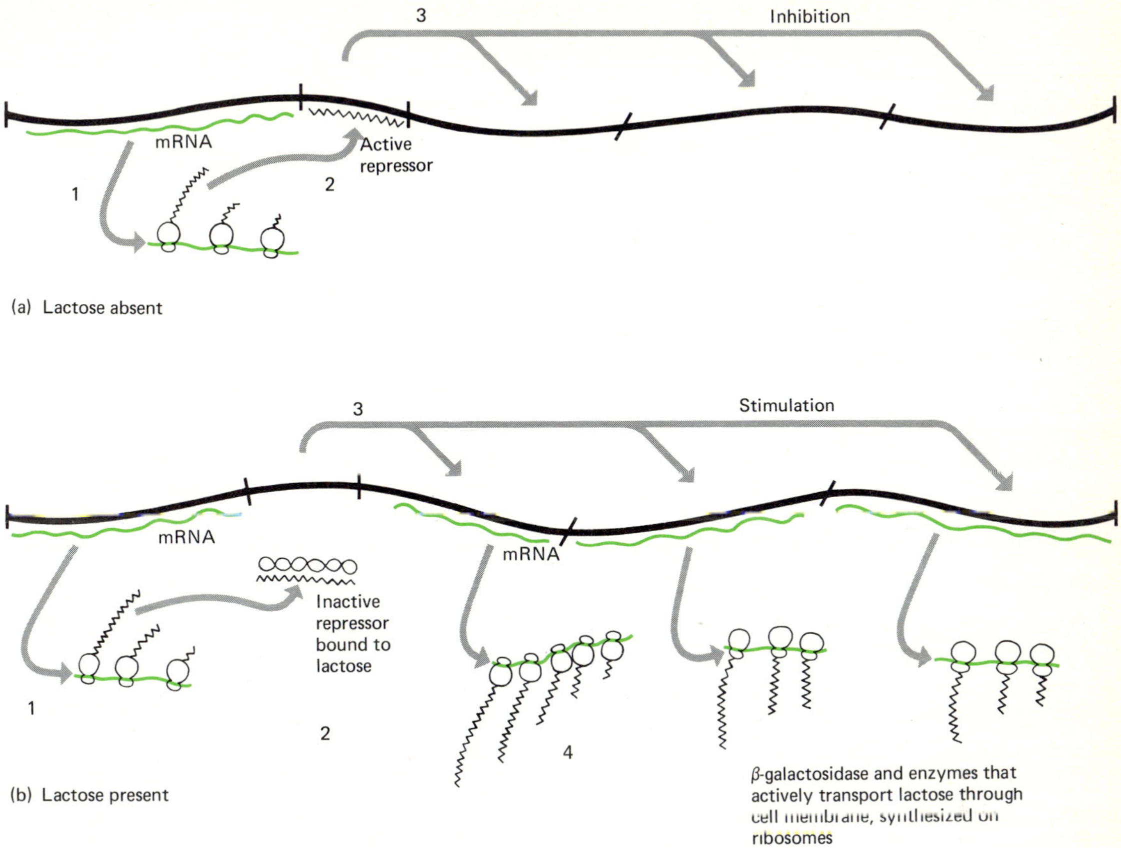

FIGURE 6.28 The repression and activation of genes. *(From James D. Watson,* Molecular Biology of the Gene, *Second Edition, copyright © 1970 by J. D. Watson; W. A. Benjamin, Inc., Menlo Park, California.)*

most of the pathways by which molecules are synthesized is via *negative repression;* the operon remains functional *unless* it is switched off. Repressors produced by the regulator gene are inactive and require the presence of *core*-pressors to activate them, thus enabling them to switch off the operon.

It is hypothesized that a similar set of circumstances exists in the developing cells of a multicellular organism, although as yet very few enzymes have been isolated that can be shown to serve in switching genes on and off. The cytoplasmic contents of the cells in developing embryos differ minutely, and it is thought that these cytoplasmic differences may play some role in the repression and expression of the genes. As the number of cells grows, different cells or groups of cells are known to produce different enzymes and proteins that may also serve to repress or express specific genes. It is possible, too, that enzymes and proteins produced by some cells are transmitted to other cells and repress or activate genes there. Thus, if the nucleus of a developing cell is transplanted to another developing cell, the cell with the new nucleus may not develop into an embryo, since some genes may have been repressed and no proteins exist in the cytoplasm of the new cell to inactivate the repressor.

We should repeat that the model of gene action that has been developed is the result of studies of single-celled organisms and may not be applicable to cells in multicellular organisms. The problems associated with validating the model for multicellular organisms are large, and it has even been estimated that no one will ever be able to work out all of the details of the differentiation of cells in a higher plant or animal. Moreover, since development will probably be attacked as a problem in molecular genetics, there is the additional obstacle that the genetics of the frog, salamander and the sea urchin, organisms that embryologists have worked with for many years, are very poorly known.

However, the implications of such models of gene action are immense. In part, they indicate that the environment of the nucleus exerts some control over the expression of the genes. Thus, the development of the phenotype upon which selection will act is conditioned, to some extent, by the environment from the moment of fertilization. Although the genes determine the potentialities of an organism, the environment is the factor that conditions the final outcome of gene action.

The expression of the genes, whether regulated or not, determines the constitution of the individual and thus the constitution of the population and the niche. Ultimately, then, the expression of the genes determines the adjustments of the communities in which the populations are found. Thus, all of the foregoing material is critical to an understanding of evolution and ecology, although the fine details of the interactions of genes and communities are a very long way from being worked out. Biologists will continue to explore life on a variety of fronts, as they should, perhaps to the point where we may someday be able to integrate our understanding of life from the molecular level to the level of the global ecosystem. We must turn now, however, from the individual organism in the laboratory to the population in the ecosystem.

Selected Readings

Balinsky, B. *An Introduction to Embryology.* Philadelphia: W. B. Saunders, Company, 1970. A clear and lucidly illustrated general text.

*Levine, Robert Paul. *Genetics.* New York: Holt, Rinehart and Winston, Inc., 1968. A concise and rigorous text that covers classical and chemical genetics.

*Monod, Jaques. *Chance and Necessity.* New York: Random House, Inc., 1972. Not the easiest of books to read, but well worth the effort. Using the evidence of molecular biology and genetics, Monod considers the possibility that the universe itself is without a cause, and the meaning that such an idea may have for man.

Moore, Ruth. *The Coil of Life.* New York: Alfred A. Knopf, Inc., 1961. A highly readable history of genetics.

Saunders, John W. Jr. *Patterns and Principles of Animal Development.* New York: The Macmillan Company, 1970. A new and excellently illustrated treatment of embryology for the introductory student.

Stern, Curt. *Principles of Human Genetics.* San Francisco: W. H. Freeman and Company, 1960. An excellent text for the introductory student.

*Watson, James D. *The Molecular Biology of the Gene.* New York: The Benjamin Company, Inc., 1970. A book that is almost required reading for anyone interested in genetics and molecular biology. Watson covers a great deal of the known chemistry of the gene, and suggests what some future experimentation is likely to find. Another of Watson's books, *The Double Helix,* is a personal account of the search for the structure of DNA.

*Schrödinger, Erwin. *What Is Life?* New York: Cambridge University Press, 1944. This book, less than 100 pages long, is often given credit for starting the molecular "revolution" in biology. In it, a physicist speculates on the possibility of deriving physical laws from a study of genetics.

Britten, Roy J., and Kohne, David E. ''Repeated segments of DNA.'' *Sci. Am.,* Vol. 222 (April, 1970), pp. 24–31. Offprint number 1173.

Crick, Francis H. D. ''The Genetic Code: III.'' *Sci. Am.,* Vol. 207 (October, 1962), pp. 67–72. Offprint number 1052.

Gurdon, J. B. ''Transplanted Nuclei and Cell Differentiation.'' *Sci. Am.,* Vol. 219 (December, 1968), pp. 24–35. Offprint number 1128.

Kornberg, Arthur. ''The Synthesis of DNA.'' *Sci. Am.,* Vol. 219 (October, 1968), pp. 64–70. Offprint number 1124.

*Paperback edition available.

Marine iguanas of the Galapagos Islands, although evolved from a strictly terrestrial species, are the only marine lizards in the world. They are active swimmers and feed exclusively on submerged seaweeds. This illustrates the evolutionary change and adaptations possible when a population is isolated in a ''new'' environment. *(Courtesy of Ira Kirschenbaum.)*

population genetics and evolution

chapter seven By now it should be clear that our previous discussion of evolution was a greatly simplified one. The mechanisms of genetics are very complex and are still not fully understood, and the translation of a genotype into its phenotype during development involves some of the most complex chemistry known. The process of evolution involves not only those mechanisms but also selection in the natural habitat. But without evolution, there would be no adaptation of species, no natural communities, no biotic habitat, no biosphere, and no man. If man is to live with, preserve, and use the biosphere for his cultural and economic benefit, he must understand evolution, the process that produced it as a balanced system. An understanding of evolution allows him to appreciate more fully the delicate nature of that balance and the ease with which it may be destroyed.

As the mechanism for transmitting adaptations, genetics is of great importance to the survival of an individual. To understand evolution, however, we must go from the individual, or the laboratory population, to populations living in nature. We must consider the genotypes and phenotypes (and hence adaptations) of whole populations and whole species. We must also consider the following: how the genes of a population or species are selected; how the variations that selection acts upon are produced; how selection acts on a population; how races, subspecies, and species arise, and how this produces different niches; how the niches produce an ecological balance; and, finally, how man affects this balance. Some of these topics will be examined later in connection with the biology of populations and communities, but several can best be looked at here.

The Genetics of Populations

The number of individuals in a species may range from a few thousand to billions. The death or survival of a single individual and its genotype is of negligible importance to the evolution of a species as a whole. Many persons find that idea disagreeable, since the life of an individual man is generally considered valuable or even sacred. But only man possesses the conscience of culture; he alone knows he is mortal; only man agonizes over the survival of an individual (when it suits him). The words *cruel* and *tragic*, however, have no biological significance. The process of evolution is a matter of the births and deaths of large numbers in large populations. It is only when the statistics of populations are considered that survival and selection are important for evolution.

The Gene Pool

An individual possesses one genotype. The species can also be thought of as having a genotype, but in this case we call it the species' *gene pool.* The gene pool is defined as *the totality of genes that exists in a population.* Serving as a temporary caretaker of the genes received from his parents, the individual represents a very small proportion of the genes in the gene pool. In all species but man, when the individual has finished reproducing, whether once or twenty times, its genotype no longer affects the evolution of the gene pool. In most species, that rarely happens, because death usually occurs before an individual has lived out its entire reproductive life. Man used to be included in this generalization, but as his average life expectancy has increased over time, he has become an exception.

Sexual reproduction, by independent assortment and recombination of parental genotypes, produces over many generations a mixing of the genes in the gene pool and determines the particular arrangement and number of genes that produce the phenotype of a species. A *deme*, or a *particular and usually local population of a species,* has a phenotype and a genotype that we may think of in statistical terms. It is the statistical survival of the genes in the gene pool, as we shall see, that determines the niche of the species. We must therefore study how the genes in a deme vary from generation to generation.

A Note on Probability and Statistics

Since evolution can be understood in statistical terms, it would be useful to review some basic principles of probability and statistics. The number of times in our lives that we make statements of probability or take some action based on probabilities is (probably) so high as to defy counting. We decide to cross a particular street at a particular time based on the probability that we will reach the other side safely; we make that decision on the basis of known locations of moving automobiles, estimates of their probable speed, and the probability that we will not fall or move so slowly that we will be struck by one of them. In short, we assess as much as possible of the information available concerning the outcome of an action or event. If our assessment of the information is faulty, if we fail to consider an important piece of relevant information, or if the conditions change sufficiently before the action is completed, then our prediction, to some degree, will be invalid.

In everyday language, we usually express probabilities in vague terms, referring to "a good chance" that something will happen. (Perhaps the classic example of a vague probability statement is the apocryphal weather forecaster's prediction that "there will be rain tomorrow, now and then, here and there.") For the purposes of understanding genetics and evolution, we must be more precise.

To begin simply, consider the problem of predicting whether a pregnant woman will have a boy or girl. Since there is only one possible event, the birth of a child, and two equally probable alternative forms that the event may take, we say that the probability that she will have a girl is one over two, or 1/2. Of course, the probability that she will have a boy is also 1/2. Now consider the probability of rolling either a 5 or a 6 on a single throw of a die. The probability of throwing a 5 is 1/6 (one outcome of six possible events), and the probability of rolling a 6 is also 1/6; thus, the probability of rolling a 5 or a 6 is 2/6, or 1/3. In terms of probability theory, the foregoing examples illustrate the *rule of the sum,* which states that *the probability that mutually exclusive events will occur is equal to the sum of their separate probabilities.*

Another rule of probability that we will see frequently in the following pages is the *rule of the product*, which states that *the probability that each of two or more independent events will occur is equal to the product of their separate probabilities*. What is the probability that a pair of parents will have two boys? Assuming that the events are independent—that is, that twins will not be born to the mother—the probability of two boys being born to a pair of parents is 1/2 times 1/2 (the product of the separate probabilities), or 1/4.

Now look at a slightly more complex question. What are the probabilities that a pair of parents will have two boys, a boy and a girl, or two girls? We have already computed the probability of the mother having two boys as 1/4. The probability that she will have two girls is the same: 1/4. But there are *two* ways in which a pair of parents may have a boy and a girl. They may have a boy first and then a girl, or a girl first and then a boy. Therefore, the probability of having a boy and a girl is $2 \times (1/2 \times 1/2)$, or $2 \times 1/4$, or $1/2$. Thus, the probabilities are

two boys	1/4
a boy and a girl	1/2
two girls	1/4

Notice that these probabilities add up to one.

A general formula exists that provides a convenient shortcut to the sometimes tedious computation of probabilities:

$$(a + b)^n$$

In this formula, a is the probability of one event, b is the probability of the other event, and n is the number of independent events under consideration. Using the terms of our previous example this statement becomes

(probability of boy + probability of girl)^2.

Such an equation follows regular rules when it is stated in an expanded form, and for the most part, the only expansion that we shall use in this book will be that in which n is equal to 2. When expanded, $(a + b)^2 = a^2 + 2ab + b^2$. Using the probabilities of two parents giving birth to two children, we can see how the equation expresses the probabilities that we derived earlier:

$$\begin{aligned}(\tfrac{1}{2} + \tfrac{1}{2})^2 &= (\tfrac{1}{2})^2 + 2[(\tfrac{1}{2})(\tfrac{1}{2})] + (\tfrac{1}{2})^2 \\ &= \tfrac{1}{4} + \tfrac{1}{2} + \tfrac{1}{4}\end{aligned}$$

Now suppose that we are faced with unequal probabilities. For example, what are the probabilities that two offspring born of heterozygous parents will show the dominant characteristics? To begin with, we know that there are four possible recombinations of gametes that can result from a mating between two heterozygotes: *AA, Aa, aA,* and *aa*. In other words, the probability that any single offspring will show the dominant is 3/4 and that it will show the recessive is 1/4. Substituting these probabilities into the general equation and expanding it, we can find not only the probability of two offspring showing the dominant but also the probability of two offspring showing the recessive:

$$\begin{aligned}(\tfrac{3}{4} + \tfrac{1}{4})^2 &= (\tfrac{3}{4})^2 + 2[(\tfrac{3}{4})(\tfrac{1}{4})] + (\tfrac{1}{4})^2 \\ &= 9/16 + 6/16 + 1/16\end{aligned}$$

The answer to the question "What are the probabilities that two offspring will show the dominant?" is $9/16 + 6/16 = 15/16$.

If we consider only single cases, like the one just presented, we should not be surprised to find that occasions will arise in which two heterozygotes produce one offspring showing the dominant and one showing the recessive. However, if we view a very large number of cases, we will find that the results will conform to the probabilities that we have just derived. When the probabilities of large numbers of cases are taken into consideration, we say that we are dealing with *statistics*. When we toss a coin once, we say that the probability of its coming up heads is 1/2 and of its coming up tails is also 1/2. Were we to toss a coin one thousand times, we should then expect to get heads 500 times and tails 500 times. If you were to actually try such an experiment, you might find, however, that heads would come up 497 times, while tails appeared 503 times. Statistically, you would say that the frequency of heads was .497 and of tails .503. Such deviations from the expected frequencies of .500 are wholly acceptable to statisticians. The expected outcome of such a trial is based on an infinite, or unlimited, number of tosses, and we accept the deviation as being caused in part by the finite (limited) nature of the trial experiment. Were we to make several trials of one thousand tosses, we might find that each thousand tosses gave us slightly differing results. We consider the trials to be subject to *errors of sampling*—that is, *subject to chance deviations from the expected values.*

That brings us to an important question: How large a deviation from the expected values can we accept as due to sampling errors alone? Any answer, of course, is arbitrary. If we were to toss a coin 1000 times and get 800 heads, we would have good reason to suspect that there was something about the coin that was out of the ordinary or that the tossing itself was being somehow controlled. On the other hand, *if in repeated trials a deviation can be shown to occur more than once in twenty times,* then it is usually accepted by statisticians as being due to errors of sampling. In our coin tossing experiment, the deviation from the expected frequency of 500 heads or tails was three. If we repeated the experiment twenty times and found that in fifteen out of the twenty trials, or 0.75 of the times, the deviation was as much as three, then we would accept it and say that the trials fit our expectations. If the deviation occurred only once (0.05), then we would say that it was a "significant" deviation and that perhaps something more than chance was operating. Should the deviation occur only once in one hundred trials (0.01), then it would be highly significant, and we would consider it very likely that something other than chance was operating. *The more frequently the deviation occurs, the more likely it is that it is due to errors of sampling.*

The Hardy-Weinberg Law

Since the gene pool of a population varies over time, we must examine what determines the relative proportions of alleles in the population. Intuitively, we can understand that both mutation and selection operate to bring about changes in the gene pool, but before we can discuss the effects of mutation and selection, we must know something about the proportions of alleles in the gene pool as they are expressed at some particular time. We must find some starting point from which we can measure the changes brought about by mutation and selection.

As a starting point from which we may go on to consider a real population, we may use a *model population*, that is, a population that does not exist in nature but serves as a comparison for a natural population. Such models are usually carefully defined and operate only under very specific conditions. The value of a model is that it allows us to pinpoint our inquiries about the nature of the real world. If the model is valid and there is a discrepancy between the model and the real population, then the conditions of the model may be examined to determine what is happening in the real world.

The Hardy-Weinberg Law is a model to describe what will happen to the relative proportions of alleles in a population from generation to generation. In a population described by the Hardy-Weinberg Law, several conditions must be met. First, *no mutations* may take place. Second, *no selection* takes place. Third, *no individuals may migrate* into or out of the population, thus adding or subtracting alleles. Fourth, *mating must be random,* which means that there is an equal probability that *any individual will mate with any other individual*, providing only that they are of opposite sexes. Summarized then, the conditions of the Hardy-Weinberg Law are, no mutation, no selection, no migration, and random mating.

The Hardy-Weinberg Law predicts that *under these conditions, no changes will take place in the relative proportions of alleles in a population from one generation to the next.* Let us consider what this means and how it can come about.

Consider a hypothetical population in which at one locus there are only two alleles, *A* and *a*, which may combine to form *AA, Aa,* and *aa.* At meiosis, these allele pairs will produce only *A* and *a* gametes. Homozygotes in the population can produce only one kind of gamete, but half the gametes of heterozygotes will carry *A* alleles and the other half will carry *a*. At mating, then, the number of alleles that will be available for recombination will depend on the numbers of homozygotes and heterozygotes:

The number of *A* gametes = twice the number of *AA* individuals + the number of *Aa* individuals.

The number of *a* gametes = twice the number of *aa* individuals + the number of *Aa* individuals.

To learn the *proportions* of the alleles we would need to know the size of the population. We could then determine the proportion of *A* alleles by dividing their number by the total number of all alleles. But since we have no population size to begin with, let us say that the proportion of *A* is equal to *p,* and that the proportion of *a* is equal to *q*. We call these proportions the *gene frequencies* of *A* and *a*.

Now, what happens to these gene frequencies when mating takes place? If mating is random, any single gamete has an equal chance of fusing with any other gamete of the opposite sex. If 8/10 of the gametes are *A*, then $p = 0.8$ and *q* must equal all the rest of the gametes, or $1 - 0.8 = 0.2$. Thus $p + q = 1$. Then, if random mating takes place, the probabilities of the fusion of any two gametes can be calculated:

Probability of Fusion

A with *A* = $0.8 \times 0.8 = 0.64$
a with *a* = $0.2 \times 0.2 = 0.04$
A with *a* = $2(0.8 \times 0.2) = 0.32$

If the gene frequency of *A* is *p*, and that of *a* is *q*, then the probability of random mating producing

$$AA = p \times p = p^2$$
$$Aa = (p \times q) + (q \times p) = 2pq$$
$$aa = q \times q = q^2.$$

These calculations may also be presented in a device similar to a Punnett Square, such as used in the last chapter. However, it must be understood that this device does not represent the probable outcome of a single mating between two parents. Rather, it expresses the statistical probabilities of the results of random mating in a very large population.

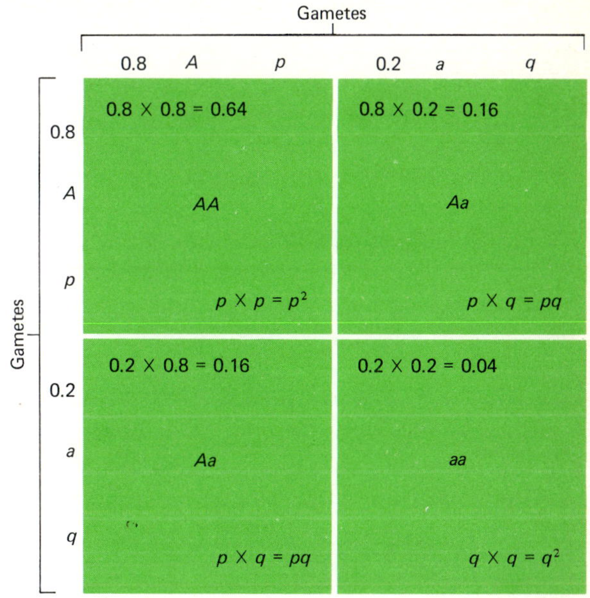

Thus,

$$p^2 + 2pq + q^2 = 1$$
$$0.64 + (0.16 + 0.16) + 0.04 = 1$$
$$AA + 2Aa + aa = 1.$$

Now, if the population is large enough and mates randomly, the proportions of the offspring in the next generation should conform to the probabilities we have calculated. In our example, 0.64 of the offspring would carry the genotype *AA*, 0.32 would carry the genotype *Aa*, and 0.04 would carry the genotype *aa*, and these add up to 1. (Said differently, the proportion of *AA* is p^2, the proportion of *Aa* is $2pq$, and the proportion of *aa* is q^2. Since these must add up to 1, $p^2 + 2pq + q^2 = 1$.)

But what is the gene frequency in this generation? The frequency of *A* is $p^2 + \frac{1}{2}(2pq)$ which is equal to $p^2 + pq$. Since $p + q = 1$, it follows that $q = 1 - p$. Therefore, the frequency of *A* is $p^2 + p(1 - p) = p^2 + p - p^2 = p$, and the frequency of *A* does not change from the previous generation. If *p* does not change and $p + q = 1$, then *q* could not change either.

In other words, according to the Hardy-Weinberg Law, if selection and mutation do not take place, if mating is random, and if migration into or out of the population does not take place, then the gene frequencies in a population will not change from one generation to the next.

To clarify this abstraction, we should see how it might work in describing some populations. First, consider a hypothetical population consisting of 1000 individuals. Assume that the population contains 600 *AA*, 200 *Aa*, and 200 *aa* individuals. The most probable number of *A* gametes will be 1200 + ½(400) = 1400; the most probable number of *a* gametes will be 400 + ½(400) = 600.* We may determine the frequencies of *A* and *a*, by dividing the total number of alleles into the numbers of *A* and *a* alleles. Thus, $p = 1400/2000 = 0.7$ and $q = 600/2000 = 0.3$. Then $p^2 = 0.49$, $2pq = 0.42$, and $q^2 = 0.09$. Now, suppose that in the next generation the population size is the same as in the last, 1000 individuals. What is the gene frequency of *A*?

$$\frac{490 + \frac{1}{2}(420)}{1000} = \frac{490 + 210}{1000} = 0.7 = p$$

and of *a*?

$$\frac{90 + \frac{1}{2}(420)}{1000} = \frac{90 + 210}{1000} = 0.3 = q.$$

Whereas in the parental generation there were 600 *AA* individuals, there are only 490 in this generation, and though similar changes have taken place in the numbers of *Aa* and *aa* individuals, the *frequencies* of *A* and *a* have remained unchanged. If you wish to try further calculations for subsequent generations, you will see that the frequencies of *A* and *a* will remain unchanged and that the distributions of genotypes and phenotypes will remain the same. You will also see that if you were to arbitrarily change the frequencies in one generation, the new frequencies remain unchanged in subsequent generations.

* We are discussing the gametes that actually participate in the formation of a zygote, of course. Were we to consider the gametes actually produced, further, more complicated calculations might be required. The average human male, for example, releases more than 200 million sperm in each ejaculation, only one of which is required to fertilize an ovum.

The consideration of a specific human population will enable us to further show how the Hardy-Weinberg Law may be applied. Within the blood of higher vertebrates are compounds called *antigens* and *antibodies.* An antigen may be any large molecule that can induce the synthesis of an antibody. The function of an antibody is to remove an antigen from the blood. Since antigens are proteins that have been synthesized in the body of a particular animal, the structures of antibodies must be very precise and must be able to interact only with "foreign" antigens introduced into the blood. When human blood is injected into rabbits, it induces the synthesis of certain antibodies in the rabbit. Since the rabbit antibodies are designated as M and N, the antigens in the human blood are also designated as M and N. These human antigens are controlled by alleles which, for convenience, we shall designate as M and N also.

Certain tests will reveal the presence in the blood of these antigens, and it has been found that human blood may contain only M or only N or both. A very small population of 208 Syrian Bedouins was found to contain these antigens in the following distribution:

M	119
MN	76
N	13

Assuming that those individuals with M or N are homozygotes and that those with both M and N antigens in their blood are heterozygotes, we may designate the genotypes as MM, MN, and NN. Now we can determine the frequencies of the alleles for M and N. The frequency of M is

$$\frac{2(119) + 76}{416} = \frac{314}{416} = 0.76 = p$$

and the frequency of N is

$$\frac{2(13)+76}{416}=\frac{102}{416}=0.24=q.$$

Substituting those frequencies into the Hardy-Weinberg equation we find that

$$\begin{aligned} p^2 &= 0.76 \times 0.76 = 0.5776 \\ 2pq &= 2(0.76 \times 0.24) = 0.3648 \\ q^2 &= 0.24 \times 0.24 = 0.0576. \end{aligned}$$

We may now determine the expected distribution of genotypes in the Bedouin population by multiplying p^2, $2pq$, and q^2 by the population number:

$$\begin{aligned} 0.5776 \times 208 &= 120.14 \\ 0.3648 \times 208 &= 75.87 \\ 0.0576 \times 208 &= 11.98. \end{aligned}$$

These figures approximate a ratio of 120 MM : 76 MN : 12 NN, which is very close to our expected values. Had the population been sufficiently large, we would probably have come even closer.

Once an allele is established in a population, it will remain in the population at the same frequency under the stipulated conditions—that mating is random, that no new genes carrying new alleles for that character are introduced into the population, and that selection and mutation do not occur. These conditions are rarely, perhaps *never*, found in nature. Of what value then, is the Hardy-Weinberg Law? Essentially, it is a starting point from which to determine whether anything is happening in a population. If analysis indicates that the frequency of a gene is changing from generation to generation, then it can be assumed that evolution is taking place within that population's gene pool.

Obviously, if the Hardy-Weinberg Law operates under the very specific conditions that we have mentioned, then any changes in the gene pool of a population must be brought about by the suspension of one or more of those conditions. In short, in a population that "obeys" the Hardy-Weinberg Law, evolution never occurs. Thus, to understand evolution, we must examine the effects on a population of mutation, selection, and nonrandom mating. For our purposes, we can consider two of the conditions to be the same—namely, nonrandom mating and selection. If mating is not random, those individuals that do not mate are selected against just as effectively as if they had never lived to mate. We will, then, examine next the effects of mutation and selection on the genetic equilibrium of a population, for the evolution consists of the changes brought about in a population's gene pool.

The Effects of Mutations

Hugo DeVries, one of the codiscoverers of Mendel's paper, introduced the term *mutation* into the language of biology to describe the *reproducible changes that take place in the genes of organisms* and that can give rise to newly different offspring. Ironically, most of the examples that DeVries used to support his thesis were not mutations at all, but rather new combinations of chromosomes. DeVries carried the idea of mutations as the basis of evolution too far, however, and insisted that a new species could arise by the introduction of a single mutation in an organism. His theory, called *saltationism* (L. *saltare* dance) to indicate that evolution was brought about by great forward leaps, has been disproved, with one exception, and we now understand that mutations introduce into populations very small changes that may accumulate over long periods of time. Because mutations are the source of variations that selection acts upon, they are of considerable importance in the study of evolution.

Ultimately, all mutations are changes in DNA. However, mutations may involve greater or lesser amounts of DNA and may therefore be classified as (1) mutations involving a *single base*, or *point mutations*; (2) mutations involving *parts of chromosomes*; (3) mutations involving *whole chromosomes;* and (4) mutations

TABLE 7.1 Mutation Rates

Organism	*Trait*	*Mutations per 100,000 (10^5) Gametes or Cells per Generation*
E. coli	Streptomycin resistance	0.00004
	Phage (TI) resistance	0.003
Diplococcus pneumoniae	Penicillin resistance	0.01
Zea mays (corn)	Shrunken seed	0.12
Homo sapiens	Hemophilia	2.5–3.2
	Muscular dystrophy	4.3–10
	Tay-Sachs disease	1.1

involving *whole sets of chromosomes*. (See box 7.1).

A mutation in the genetic constitution of an individual is a *reproducible* event. In a population, mutations are *recurrent* events as well; that is, they occur at periodic intervals which may be measured in terms of the numbers of gametes produced by the population (see table 7.1). In some instances, mutations are also *reversible*. Thus, if A mutates to a, then a may *back mutate* to A. Because of the ability of mutations to be reversed, the situation may arise in which mutation and back mutation may balance each other in a population. Let us consider how this may occur.

If q is the frequency of a in a generation, then $(1 - q)$ is the frequency of A. The change in the frequencies of a and A in the next generation will depend on how frequently A mutates to a, thus adding alleles, minus the rate at which a back mutates to A and subtracts alleles. If we make the rate at which A mutates to a equal to u, and the rate of back mutation equal to v, then the change in A will be equal to the rate at which mutation occurs in q, or vq, minus the rate of back mutation of a to A, or $u(1 - q)$.

That is, the change in A is

$$vq - u(1 - q).$$

It may sometimes be the case that mutation and back mutation balance each other out after a time, and the population achieves a state of equilibrium. Although such an equilibrium is unlikely to occur, we can express this idea mathematically in the following equation, using u and v as before to represent the rates of mutation of A to a and a to A, respectively:

$$vq = u(1 - q).$$

To clarify this, we may substitute some mutation rates of unequal nature. Say that A mutates to a 100 times as fast as a mutates to A, or $u = 100v$. These unequal rates will bring about an equilibrium in the frequencies of A and a in a population when the frequency of a is equal to 0.990099. How can that be? We can restate the above equation as

$$\begin{aligned} vq &= u(1 - q) \\ vq + uq &= u \\ q(v + u) &= u \\ q &= \frac{u}{u + v} \end{aligned}$$

and substitute $u = 100v$ as follows:

$$\begin{aligned} q &= \frac{100v}{100v + v} = \frac{100v}{101v} = \frac{100}{101} \\ &= 0.990099 \end{aligned}$$

Therefore, when the frequency of a is 0.990099 the population will be in equilibrium under mutational pressure. This is so because as A mutates to a at a constant rate, there will be fewer alleles of A to mutate to a, while at the same time each succeeding generation will have a larger proportion of a's to mutate to A's. This is not to say that the reversibility of mutations means that changes in gene frequencies due to mutation cannot be observed in a real population. We stress again that evolution usually proceeds at a slow pace, and the possibility that changing gene frequencies can be brought about by mutation is real.

In all of the foregoing, we have presented the gene as having a single allelic alternative. Unfortunately, the picture can be somewhat more complicated when we consider that although an individual has only two alleles for any characteristic, there may be more than two alternatives for that characteristic when a whole population is considered. For example, while blue eyes and brown eyes predominate in the human species, there are other alleles that produce green eyes. The rates of mutation that the geneticist must sometimes use, then, must each be made up of several component rates.

The impact of mutation on the genetic equilibrium of a population is quite small and as we have just seen, can ultimately bring about another state of equilibrium. We should never confuse the equilibrium described by the Hardy-Weinberg Law with mutational equilibrium. The conditions necessary for the genetic equilibrium described by the Hardy-Weinberg Law probably *never* occur in the natural world. Selection and mutation always operate to some degree, and mating is rarely truly random. Under Hardy-Weinberg conditions, equilibrium is maintained because no changes of any sort are taking place. Under mutational equilibrium, the changes themselves produce the equilibrium.

We have said that mutations can bring about reproducible and recurrent changes in the gene pool of a population, but we have shown that the reversibility of such changes may ultimately restore genetic equilibrium in the population. Any further changes in a gene pool that is in equilibrium under mutational pressure must be from another source, then, and that other source is selection.

The Effects of Selection

The effects that selection may have on the frequencies of alleles in a population can be dramatic, as we saw in our discussion of the peppered moth population in the Manchester area between 1850 and 1900. For the most part, however, the changes that selection produces are gradual and take place over a long span of time.

For illustrative purposes, we will consider an extreme case—a population in which mating is random and mutation and back mutation have brought about an equilibrium in the frequencies of the alleles A and a. At the time we begin our observations of this hypothetical population, we find that $A = 0.7$ and $a = 0.3$, and we note that phenotypically there are only two expressions of the gene—that determined by the dominant (in the homozygous or heterozygous condition) and that determined by the recessive. Since the frequencies of A and a are 0.7 and 0.3 ($p = 0.7$ and $q = 0.3$), $q^2 = 0.09$, and the Hardy-Weinberg equation tells us that within this population the recessive will be found in a homozygous condition in 9 individuals out of every 100. We also know that $p^2 = 0.49$ and $2pq = 0.42$ and therefore that 91 (42 + 49) individuals out of every 100 will have the phenotype determined by the dominant allele.

Now imagine that the environment changes radically and that as a consequence individuals with the *aa* genotype do not survive and mate; that is, *selection* against the recessive phenotype is *complete.* The population will then consist of only two interbreeding genotypes, *AA* and *Aa*. These genotypes will then constitute the whole *breeding* population, and we can determine the proportion of each genotype by adding the two original proportions of breeders together and dividing that sum into the original proportion of each breeding genotype alone. Thus, $AA = 0.49/(0.42 + 0.49) = 0.54$ and $Aa = 0.42/(0.49 + 0.42) = 0.46$. The frequency of gametes would be

$$A = 0.54 + \tfrac{1}{2}(0.46) = 0.77 = p \text{ and}$$
$$a = 0.00 + \tfrac{1}{2}(0.46) = 0.23 = q.$$

We can derive the proportions of the genotypes of the offspring of the breeding population in the next generation by substituting the new fre-

Mutations

Point Mutations

For illustrative purposes we have included both strands of the DNA molecule in this sequence of a hypothetical gene.

DNA	ATG	CGT	AGC	TGG	ATC	GAT	TAA
DNA	TAC	GCA	TCG	ACC	TAG	CTA	ATT
mRNA	AUG	CGU	AGC	UGG	AUC	GAU	UAA
Amino acids	met	arg	ser	cys	ileu	asp	terminate

A *base change* in one nucleotide changes the amino acid at one location of a protein. Such a mutation is called a *missense mutation.*

DNA	TAC	GTA	TCG	ACC	TAG	CTA	ATT
mRNA	AUG	CAU	AGC	UGG	AUC	GAU	UAA
Amino acids	met	hist	ser	cys	ileu	asp	terminate

The *addition* of a single base may change the sequence of a number of amino acids in the protein. This is an example of a *frame shift mutation.* The deletion of a single base can have a similar effect. Because so few amino acids in an enzyme are involved in the active site of the enzyme, frame shift mutations usually produce inactive enzymes.

DNA	TAC	GCA	TCG	TAC	CTA	GCT	AAT
mRNA	AUG	CGU	AGC	AUG	GAU	CGA	UUT
Amino acids	met	arg	ser	met	asp	arg	leu

Translocations

In some cases, because more than one codon can translate for a single amino acid, some missense mutations can occur, but remain undetectable.

Translocations, inversions, and deletions can all occur during meiosis. Of the two possible recombinations shown in the figures for translocations, (b) may give rise to serious consequences, since the segment lacking a centromere may easily be lost during cell division and the segment with two centromeres may not divide properly and thus not be replicated.

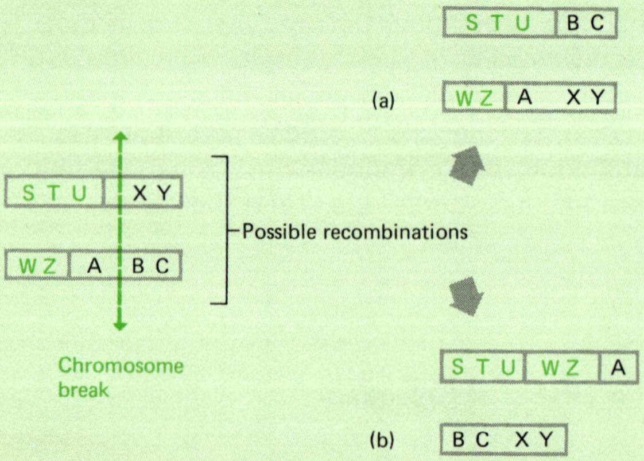

Deletions

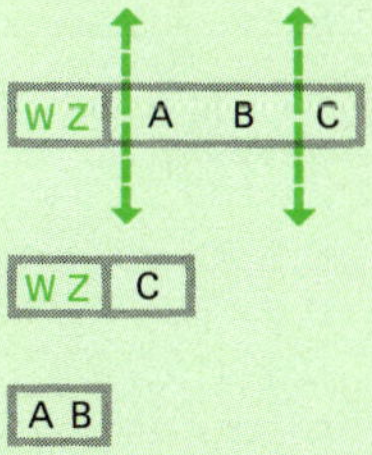

The deletion shown may also be deleterious or lethal for the same reason that combination (b) above may be; the chromosome fragment lacking a centromere may be lost during division. One human genetic condition is believed to be caused by a deletion in chromosome 5 and causes serious mental retardation. Because of the sounds that infants with this condition make when they are crying, the condition is called *cri du chat,* the "cry of the cat."

Inversions

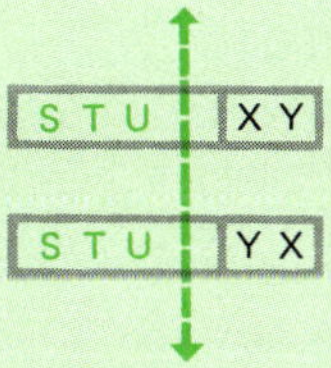

Inversions are often not lethal.

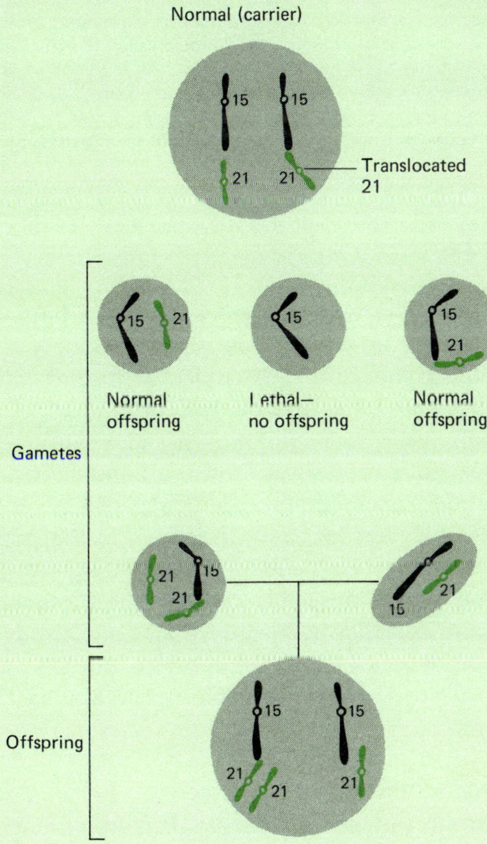

A translocation involving the human chromosomes 15 and 21 gives rise to a condition called *trisomy 21,* in which the offspring of a female carrier of the condition may receive three of chromosome 21. This condition is called Down's Syndrome after the physician who described it, but is more popularly known as "mongolism." Among the symptoms of the condition are severe mental retardation, an eyefold that gives the condition its name, and a susceptibility to leukemia.

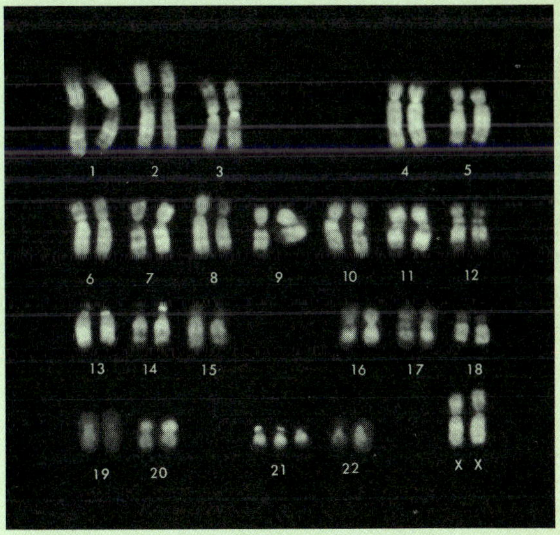

(Courtesy of Dr. Irene A. Uchida, McMaster University, Ontario.)

quencies into the Hardy-Weinberg equation:

Proportion of AA =

$$p^2 = 0.77 \times 0.77 = 0.5929$$

Proportion of Aa =

$$2pq = 2(0.77 \times 0.23) = 0.3542$$

Proportion of aa =

$$q^2 = 0.23 \times 0.23 = 0.0529$$

Thus, selection has changed the proportions of the genotypes and the gene frequencies, and this has resulted in a change in the proportions of the genotypes (and phenotypes) in the next generation. Recurring selection would continue to lower the value of q, but the a alleles would, to some extent, remain preserved in the heterozygote. (See table 7.2.)

TABLE 7.2 Elimination of a Recessive Allele Under Complete Selection

Frequency	*Generation*
0.500	0
0.333	1
0.250	2
0.200	3
0.167	4
0.100	8
0.083	10
0.045	20
0.031	30
0.024	40
0.020	50
0.010	100
0.005	200
0.001	1000

A recessive gene, under complete selection (s = 1) may remain in a population for many generations, as this table shows. Notice that although the gene is eliminated rapidly at first, the rate at which it is eliminated also declines rapidly. At lower coefficients of selection, the elimination of a completely recessive gene proceeds even more slowly. For example, if s = 0.01, it has been calculated that to reduce the frequency from 0.25 to 0.0001 will take 1,000,411 generations.

(From Theodosius Dobzhansky, *Genetics of the Evolutionary Process,* New York: Columbia University Press, 1970, p. 103.)

Relative Fitness

What we have done is to assign to each genotype a value for *relative fitness*; that is, we have said that the ability to survive of individuals carrying specific genotypes is

$$\begin{array}{rl} p^2 & 1 \\ 2pq & 1 \\ q^2 & 1 - s \end{array}$$

where s is called the coefficient of selection. In the preceding example, since selection against aa was complete, s was equal to 1. Thus, $q^2 \times (1 - 1) = 0$, and none survived to reproduce. Under the same coefficient of selection (s = 1) in the next generation of our example population, a^2 will equal 0.038. Beginning as we did with a frequency of a = 0.3, and thus equal to 9 individuals out of 100, we have seen that selection reduces the incidence of that phenotype in the population in the next generation to 6 individuals (0.059) out of 100, and then to 4 (0.038) out of 100. Even in the face of severe selection, then, the recessive allele is by no means immediately eliminated from the population (see table 7.2.). The highly deleterious allele is "protected" in the heterozygote.

In general, selection is not very often so severe, but at the same time, the recessive allele in the heterozygote is not completely without some kind of effect, often a deleterious one. However, the selective pressure against the heterozygote may be so low that it becomes very difficult to measure. The subject is further complicated when you consider that even though severe selection may be eliminating the homozygous recessives from the population, mutation may continue to produce recessives at a rate sufficient for them to remain in the population at very low frequencies for a long time. It is theoretically possible, too, that the mutation rate and selection will achieve an equilibrium and produce a population in which the frequency of the recessive allele, while low, is constant.

We have thus far treated only the effects of selection on the recessive allele. It is important to understand that selection can work against the dominant allele as well and that when this occurs, the frequency of the dominant will decline in the population much more rapidly than will a recessive under the same coefficient of selection. One reason for this is that the dominant allele will be expressed in the phenotypes of a larger proportion of the population. If we were to use the same gene frequencies and the same coefficient of selection used above, we would see that all of the dominants would disappear from the population in one generation, for the only breeding genotype would be *aa*. Thus, the only source of *A* would be from the mutation of *a* to *A*. One of the reasons for the moth population in Manchester undergoing such a radical redistribution of gene frequencies in so short a time was that selection shifted from the phenotype of the homozygous recessive to the phenotype determined by the dominant. An important point to be made at this time about selection and selective pressure is that they are *not always constant.* As the Manchester moth population so dramatically shows, selection may shift from one characteristic to another, or as some aspect of the environment is altered, the alteration may serve to raise or lower the value of the coefficient of selection. All of the lichens on all of the trees in the Manchester area did not die at exactly the same time, but as more and more did die, and as the trees became blacker, the coefficient of selection on the white moths rose higher over several generations. Conversely, the coefficient of selection for the black moths declined over the same period of time. Thus, *as the environmental conditions change over several generations, the relative fitness of any allele correspondingly changes.*

Evolution consists of the changing frequencies of genes in populations. Though we have said that such changes can reach an equilibrium, we must point out that this is not a likely event. First, it is rare to find a population in which mating is truly random. Nor can the introduction of new alleles be completely prevented in most instances. Even if mutation achieves an equilibrium, no population is completely free of new alleles introduced by the immigration into the population of individuals from other populations. At the same time, alleles can be lost from a population as individuals carrying them emigrate from the population. Most importantly, selection will continue to operate, and as the environment changes, the selective pressure on a population will change. In spite of all of the changes, the whole system is in relative balance. The equilibrium in the species populations of the biosphere is the product of a vast series of genetic changes taking place at the present and that have taken place in the past.

Genes in Populations—Sickle Cell Anemia

Perhaps the best understood expression of homozygous dominants, recessives, and heterozygotes in populations in the real world is in the disease called *sickle cell anemia.* It illustrates both how a gene affects the biochemistry of an organism and how its expression in a population is important to evolution.

Sickle cell anemia takes its name from the peculiar sicklelike shape assumed by red blood cells of persons who have the disease. (See fig. 7.1.) It is caused by a recessive gene in which the amino acid sequence of the hemoglobin molecule is altered by a single amino acid substitution. As we have seen, hemoglobin contains four polypeptide chains. Two are called *alpha chains* (α), each of which is 141 amino acids long. The other two are *beta chains* (β) and are 146 amino acids in length. In sickle cell anemia, in one location on each beta chain, valine is substituted for glutamic acid. Such a substitution can result from a point mutation in one nucleotide of the DNA molecules, altering the mRNA codon for valine from UGU to AGU. Thus, of a total of 574 amino acids, one nucleotide change can result in a different amino acid in two locations on the whole hemoglobin molecule. In the homozygous condition, this minute alteration usually results in death before adulthood.

As we have seen, hemoglobin binds oxygen, which supplies the respiratory needs of the cells. As exercise, or even movement, increases,

Tay-Sachs—A Genetic Disease

The human implications of topics such as the Hardy-Weinberg law are not always readily apparent. Consider, though, that certain human diseases result from the occurrence within an individual of two identical alleles at the same locus. Using techniques that include, or are extensions of, the Hardy-Weinberg law, the incidence of such genetic conditions can often be calculated. Once such calculations are made, we may be able to predict, or at least make better estimates of, the impact of the disease under varying conditions; plan for health care facilities more efficiently; and more realistically evaluate the allocation of resources toward the relief or prevention of the disease. As a case in point, let us consider the genetic condition known as Tay-Sachs disease (TSD).

At birth, infants who are homozygous for an autosomal recessive allele that produces TSD appear to be normal, healthy, attractive babies with clear, translucent skin and long, curly eyelashes. Rarely, however, do they live much beyond their third birthdays; death usually intervenes by about the fortieth month of life, often from pneumonia. Because such infants fail to produce an enzyme that breaks down a certain lipid (lipids are compounds of which fats and oils are made), it accumulates in the neurons of the brain and central nervous system. The characteristic symptoms of TSD appear as early as the sixth month. Beginning with an impairment of motor function, the disease progresses rapidly through blindness, general paralysis, and mental deterioration. By the sixteenth month, often sooner, hospitalization is usually required. The cost of treatment ranges from \$10,000 to \$50,000 per year, with the average closer to the higher figure than the lower one.

The allele which causes the disease occurs in all human populations at a very low frequency, but in one population its frequency is substantially higher. Populations of Jews of eastern European ancestry, known as Ashkenazi Jews, show an incidence of the allele that is ten times higher than other Jewish or non-Jewish populations. Among Ashkenazi Jewish populations, the estimated ratio of heterozygotes to homozygous dominants is 1 in 30: In other populations about 1 in 300 persons are heterozygous for the allele. We can translate such heterozygote frequencies into gene frequencies and show that the disease is likely to occur 100 times more frequently in Ashkenazi Jewish populations than in others. If one of every thirty Ashkenazi Jews is a heterozygote for this allele, then any sample we took from such a population would probably contain 0.03333 heterozygotes for each homozygote. Making the sample size equal to 1, we find the gene frequencies, p and q, as follows:

$$p = \frac{2(0.9667) + 0.03333}{2} = 0.98335 \text{ and}$$

$$q = 1 - p = 0.01665$$

In non-Ashkenazi populations we proceed the same way, expecting that testing would reveal 0.00333 heterozygotes for each homozygote, and therefore that

$$p = \frac{2(0.9967) + 0.00333}{2} = 0.99837 \text{ and}$$

$$q = 1 - p = 0.00163$$

BOX 7.2

Since the disease occurs in homozygous recessives, q^2 is equal to the probable incidence of the disease under perfect Hardy-Weinberg conditions. Thus in Ashkenazi Jewish populations, the incidence of TSD will be 0.0002722, and in non-Ashkenazi populations will be 0.0000026. From these values alone we cannot predict the numbers of cases of TSD that will actually occur in either population, however. Recall that the conditions of the Hardy-Weinberg law are very stringent and that the law is a useful starting point in the analysis of a population, but it describes no real population that we know of. A human population is subject to mutation, back mutation, and selection, and mating is not always random. In the case of TSD, selection against the homozygous recessive is complete ($s = 1$), which affects mutation rates, since the number of recessive alleles is diminished each time a TSD infant dies, and thus the only source for back mutation is in the heterozygotes. Moreover, the intermarriage of Ashkenazi Jews with non-Ashkenazi Jews, or with non-Jews, tends to alter minutely the allelic frequencies in each population. Taking into account all the factors that impinge upon the incidence of TSD, it is calculated that approximately fifty TSD infants will be born in 1973, of whom between forty and forty-five will be Ashkenazi Jews.

When such diseases are primarily confined to an identifiable population, as is sickle cell anemia in certain African populations, or as cystic fibrosis is in Anglo-Saxon populations, genetic screening programs may be conducted and may be very effective in relieving certain psychological and social burdens that the conditions entail. Screening programs for TSD can be particularly effective, since the identification of heterozygotic individuals is highly accurate and the genotype of a fetus can be determined from analysis of cell cultures derived from amniotic fluid or by analysis of the amniotic fluid itself. The high level of confidence in the results of such tests permits sound genetic counseling before reproduction, and the determination of fetal genotypes can be made while therapeutic abortion can be prescribed without endangering the safety of the mother. The cost of setting up a screening program for identifying TSD heterozygotes is about $65,000. One such program, continuing now in the Washington-Baltimore area, has been made self-sufficient through the use of a $5 testing charge, to be paid voluntarily. Since hospitalization for TSD may average $35,000 per year, the prevention of two cases can pay for the initial costs in the first year. Screening programs for other genetic conditions are also being conducted on an ongoing basis, notably in the case of sickle cell anemia. Given the difficulties of positively identifying fetal genotypes in many instances, other genetic screening programs may not be said to be as "cost-effective" as some TSD programs. Properly conducted, however, such programs can produce social benefits that simply cannot be measured in dollars and cents alone.

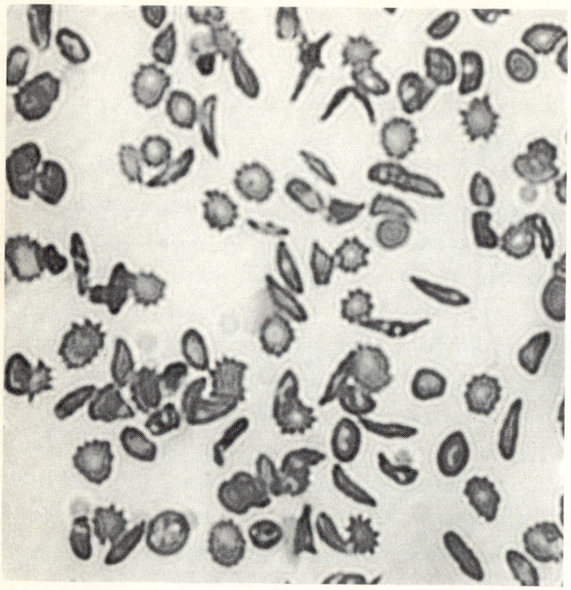

(a)

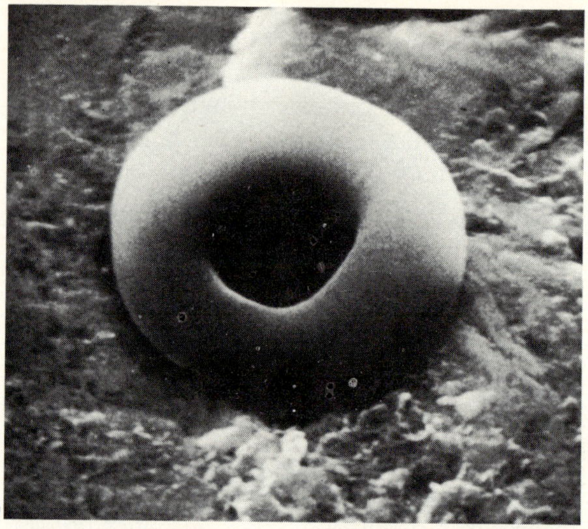

(b)

(c)

FIGURE 7.1 Blood cells from an individual with sickle cell anemia. (a) Light photomicrograph of a "smear" showing distorted sickle cells and normal cells. (b) A scanning electron photomicrograph of a normal red blood cell. (c) A similar view of a sickled cell. *((a) Courtesy of Runk/Schoenberger from Grant Heilman; (b) and (c) courtesy of Patricia N. Farnsworth, New Jersey College of Medicine.)*

the cells require greater amounts of oxygen than in their "resting" state. The defective hemoglobin molecules produced by the sickle allele, however, cause red blood cells to assume their sickle shape when the partial pressure of oxygen falls. These sickled cells do not move through the capillary system as easily as normal cells; they create painful blockages in the capillaries and reduce the amount of oxygen available to cells by preventing the passage of normally shaped red blood cells.

Generally, anemia is defined as a condition brought about by an individual's inability to produce a sufficient number of red blood cells to carry sufficient oxygen to the cells of the body. Thus, sickle cell anemia differs as to the reason for the failure of oxygen to reach cells, but it has the same general effect.

We can symbolize the alleles for hemoglobin as *S* and *s*. (See table 7.3.) A person with the genotype *SS* produces normal hemoglobin and is not affected by sickle cell anemia. Genotypes with *ss* alleles produce no normal hemoglobin and are afflicted with severe sickle cell anemia, a lethal condition. The heterozygotes, *Ss*, produce both kinds of hemoglobin; they are not completely free of the disease, but they are not as severely affected by it either. The heterozygote has a phenotype called *sickle cell trait;* its

TABLE 7.3 Sickle Cell Genotype, Phenotype, and Incidence of Disease

Genotype	*Phenotype*	*Malaria*	*Sickle Cell Anemia*
SS	Normal	Susceptible	Not affected
Ss	Sickle Cell Trait	Not susceptible	Very slightly affected
ss	Sickle Cell Anemia	—	Lethal—death at an early age

cells will be sickled under a low partial pressure of oxygen, but the sickling is not as extensive, severe, or frequent as that of the *ss* genotype. Individuals with the *ss* genotype usually die before adulthood, except in parts of Africa where the disease is prevalent, and health care facilities are poorly staffed and equipped. There, death occurs before adolescence. (See fig. 7.2.)

It would seem, then, that the *SS* genotype is the best adapted one, but under some conditions, this is not so. In the areas of Africa where the *s* allele is widespread, the incidence of Falciparum malaria is too. This disease is caused by a mosquito-borne protozoan (*Plasmodium falciparum*) that parasitizes man by growing in red blood cells after a mosquito bite. The protozoan is best adapted to normal red blood cells (*SS*) and is not adapted as well to the cells of a person with the sickle trait (*Ss*). Thus, individuals with normal hemoglobin provide an optimal environment for the protozoan, while those individuals with the sickling trait provide a poorer one. The consequences should be clear. In areas where the malaria is prevalent, those individuals carrying the *SS* genotype are susceptible to malaria and are thus selected against. Those individuals with the *ss* genotype die of sickle cell anemia before they can reproduce. Thus, both homozygotes (*SS* and *ss*) are subjected to selective pressure, though of differing degrees. The heterozygotes (*Ss*) have the sickle trait and are slightly affected by the disease, but they are not as susceptible to malaria as the homozygous dominants. Thus, a lethal gene, in this case *s*, can be and is maintained at a high frequency under certain conditions.

There are several important points to be made about this example. First, it shows that *we cannot always view dominant and recessive alleles as absolutely dominant or recessive*. In this case, the heterozygote is not phenotypically identical to the homozygous dominant, but falls somewhere between the extremes of both ho-

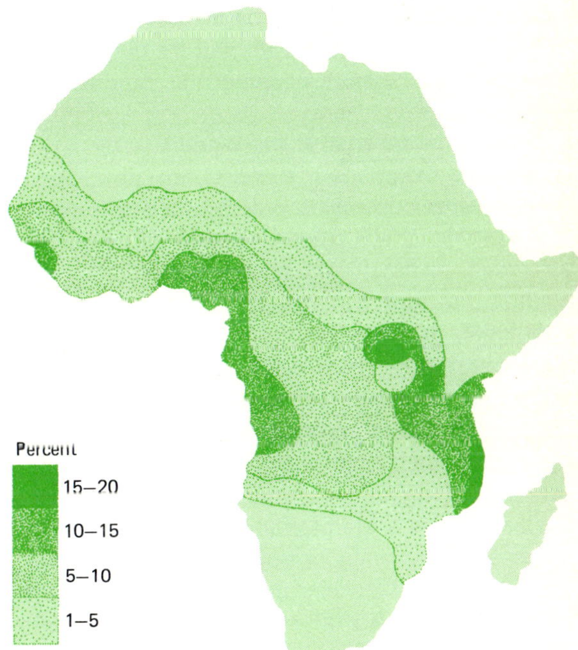

FIGURE 7.2 Incidence of sickling gene in Africa. *(Redrawn from BSCS, 3d ed.)*

mozygotes. That is often the case, since one allele will code the synthesis of one protein, the other allele will code a slightly different protein, and the body will manufacture and utilize both. Next, a human population living in a part of the world where Falciparum malaria persists will suffer a high rate of death among infants and preadolescent children if the gene for the sickle trait is present. In the absence of the gene, however, the same population would be susceptible to epidemics of malaria, and none of the individuals in the population would be free from the possible effects of malaria. Malaria can be fatal, and when it is not, it is usually chronic and interferes with reproduction.The presence of the sickle cell trait in such a population insures that some individuals, the heterozygotes, will reproduce normally and thus provide, to some extent, for the survival of the population as a whole. Third, and of great importance, *the variation that produces the fittest individuals is preserved in the heterozygote.* Under certain environmental conditions, should the variation not exist, the population would face severe selection. Just as the heterozygotic individuals of the moth population preserve the allele that determines that a peppered moth will be black, so does the heterozygote preserve the variation that provides fitter individuals among a human population in an area of the world where Falciparum malaria is prevalent. Any population of organisms possessing only one allele for a given characteristic risks extinction if the environment exerts strong selective pressure against that trait. Thus, the importance to the population of the preservation of variation in the heterozygotes can readily be seen. Sickle cell anemia presents us with a human example of heterozygote superiority. Some argue that it is the only such example in human populations, although evidence for others is accumulating. Thalassemia is a Mediterranean form of anemia which seems to parallel sickle cell anemia, and recently the exploration of the genetics of schizophrenia has raised the possibility that heterozygotes for schizophrenia may enjoy some sort of selective advantage. Tay-Sachs disease may provide another case, as there is tentative evidence that suggests that Tay-Sachs heterozygotes may be resistant to tuberculosis.

Saying that the heterozygote is the means by which variations are preserved in a population is a generalization that is subject to some exceptions. Occasionally, the heterozygote may be selected against more severely than either homozygote. There are also instances in which the heterozygote and recessive homozygote are wiped out almost immediately. We will examine some of these exceptions in later chapters, but they are worth mentioning here. The exceptions show that our present knowledge is inadequate to the task of deriving fully accurate generalizations about the nature of evolution. As our knowledge increases, it will undoubtedly be necessary for us to revise our generalizations or abandon them entirely to create new ones.

The Expression of Variation—Phenotypes

It has been said that it is the phenotype that is the target of selection but that it is the genotype that is selected. In the preceding section of this chapter we discussed the frequencies of the alleles of a single gene in the population and saw how the selection of the genotype affects the gene pool. However, *the phenotype is not simply the sum of single genes determining single characteristics*.

Pleiotropic Genes

A single gene may code a single protein, but since that protein may activate or participate in a number of metabolic processes, the gene will affect each of the processes. A gene that affects several characteristics is said to be *pleiotropic* (Gr. *pleios* many). For example, in the flour moth (*Ephestia kuehniella*), the gene for eye color is known to control ten other characteristics. The basic effect of a mutation in this gene is to render the moth incapable of synthesizing a chemical precursor to tryptophan. It appears that every biochemical pathway in which tryptophan plays some role is affected by such mutations. Not only is the eye color affected, but also such widely varying traits as mating behavior, copulatory success, the rate of overall development, oxygen consumption, and the rate of preadult mortality. In short, a single gene affects several visibly expressed and nonexpressed phenotypic characteristics. Thus, the survival of a particular phenotype depends not only on its outward appearance but also on its physiological makeup. When we add to this the fact that the phenotype is the result of many genes working together, we see that a very large number of phenotypic variations are possible in a population. On the other hand, a single phenotypic characteristic may be the result of the action of more than one gene. As an example, consider body size. The rate of bone growth is determined by one set of genes, the time at which bone growth ceases by another set, and the ability to assimilate nutrients that contribute to bone growth controlled by still another set. Thus, all genes involved in metabolism affect the size and rate of growth.

We have already said that the environment is an important factor in determining the phenotype. If the individual cannot get the nutrients necessary for bone growth, the genetically determined rate at which he can assimilate such nutrients makes no difference at all; he will not grow. If he can utilize such nutrients very efficiently but can obtain them only in small quantities, he will probably grow slowly anyway. Almost any measurable variation is subject to this generalization, whether it is the length of a cat's tail or the size of a pine tree. Even the less obvious variations such as metabolic rate, vitamin requirements, or permeability of the cell membrane are the result of the interactions of several alleles and the interaction of the individual with his environment.

Phenotypic Variations and Unity of the Genotype

The phenotypic differences in a population are usually continuous; that is, throughout the population there are many intermediate stages between the extremes. These phenotypic differences are usually distributed throughout the population in such proportions that a plot of the numbers of individuals against the range of variations under consideration would derive a normal distribution curve. Within the population, there is usually an ''optimal'' phenotype that is the best adaptation to the particular environment in which the population lives. Deviations from the optimum are less well adapted, and the greatest deviations are subject to the strongest selection. Thus, the extremes of the population represented by the ''tails'' of the curve, are selected against. (See fig. 7.3.) Should selection against the extremes decrease or stop entirely, those individuals will increase as a proportion of the population and the curve will ''flatten out.'' (See fig. 7.4.) Such flattened curves often describe populations of organisms

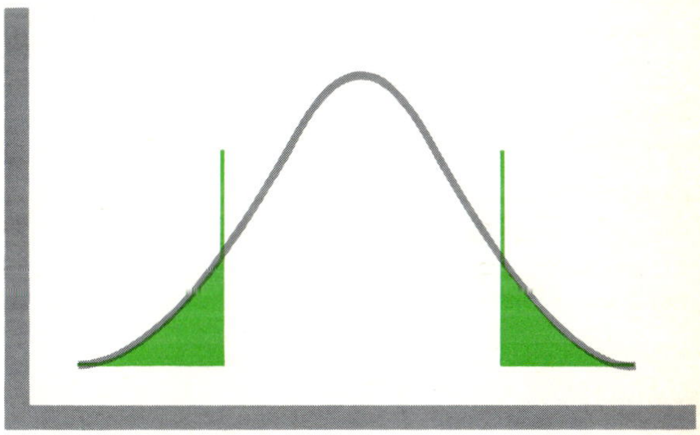

FIGURE 7.3 Selection against the extremes.

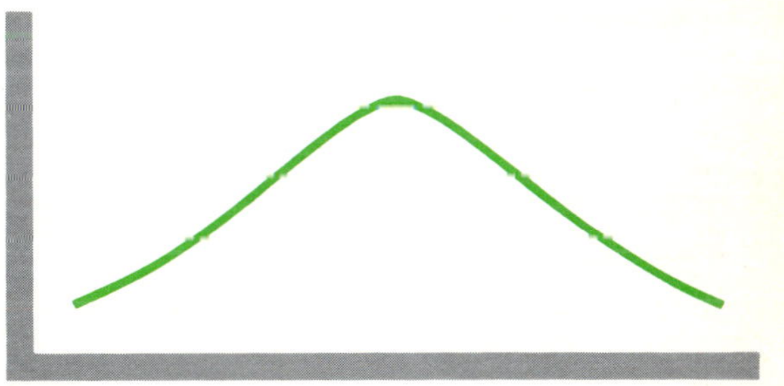

FIGURE 7.4 When selection against the extremes ceases, the curve tends to ''flatten out.''

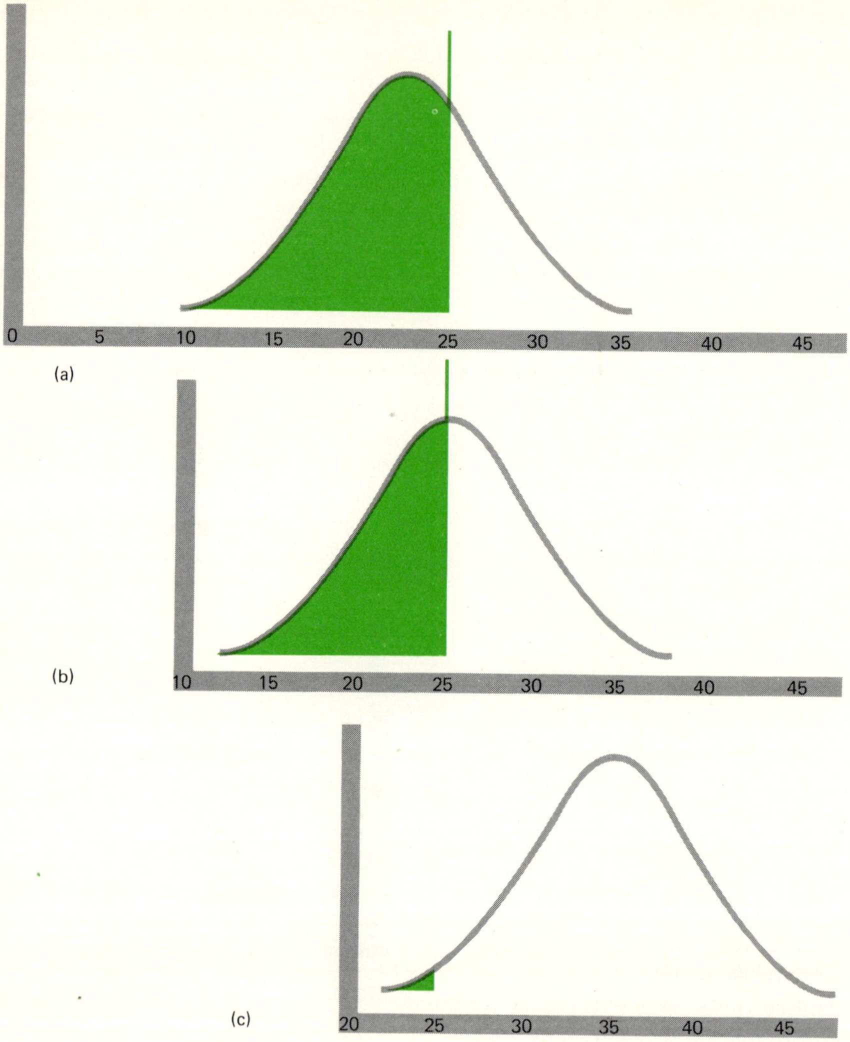

FIGURE 7.5 Directional selection occurs when a major proportion of the population is disadvantaged, individuals in that portion of the population are selected against, and thus eventually the curve stabilizes with only the "tail" of the population being selected against.

that have been introduced into a new habitat where selection against the formerly best adapted proportion of the population is sometimes greater than that against either of the extremes or where selection against the extremes is less strong than it had been in the old habitat. In either case, such flattened curves are only temporary. Sooner or later, selection will resume, and some individuals in the population will be subjected to it at a greater rate than others. When that happens, the curve may shift in one direction or another, with the mean approaching the new, better adapted characteristic. (See fig. 7.5.) As the population adjusts, the gene frequencies will readjust, and directional selection will bring the population into a more stable condition as stabilizing selection is reestablished.

As an example of the amount of variability that a population may contain, of how a single characteristic may be affected by several genes, and of how intense selection may result in several correlated effects, we will examine an experiment performed by two English geneticists, Mather and Harrison. Mather and Harrison chose an apparently insignificant characteristic of *Drosophila*—the number of bristles on one segment of the flies' abdomens—as a characteristic upon which they would base selection. They began with a population of flies which showed a mean bristle number of 36. They then separated all flies with more than 36 bristles from those with less than 36. Those with more than 36 bristles became the "high" selection line, and the others became the "low" line. None of the flies from the high line were permitted to mate with flies from the low line, and vice versa. In each succeeding generation, the flies were selected for an increase or decrease in bristle number. In the high selection line, only the flies with the greatest number of bristles were permitted to reproduce; in the low line, only those with the lowest number of bristles were allowed to reproduce. After twenty generations, the bristle number in the high line reached a mean of 56, and after thirty generations, the low line had achieved a mean of 25. (See fig. 7.6.) Thus, within the original population, there was a considerable amount of genetic variation that only became apparent under intense selection.

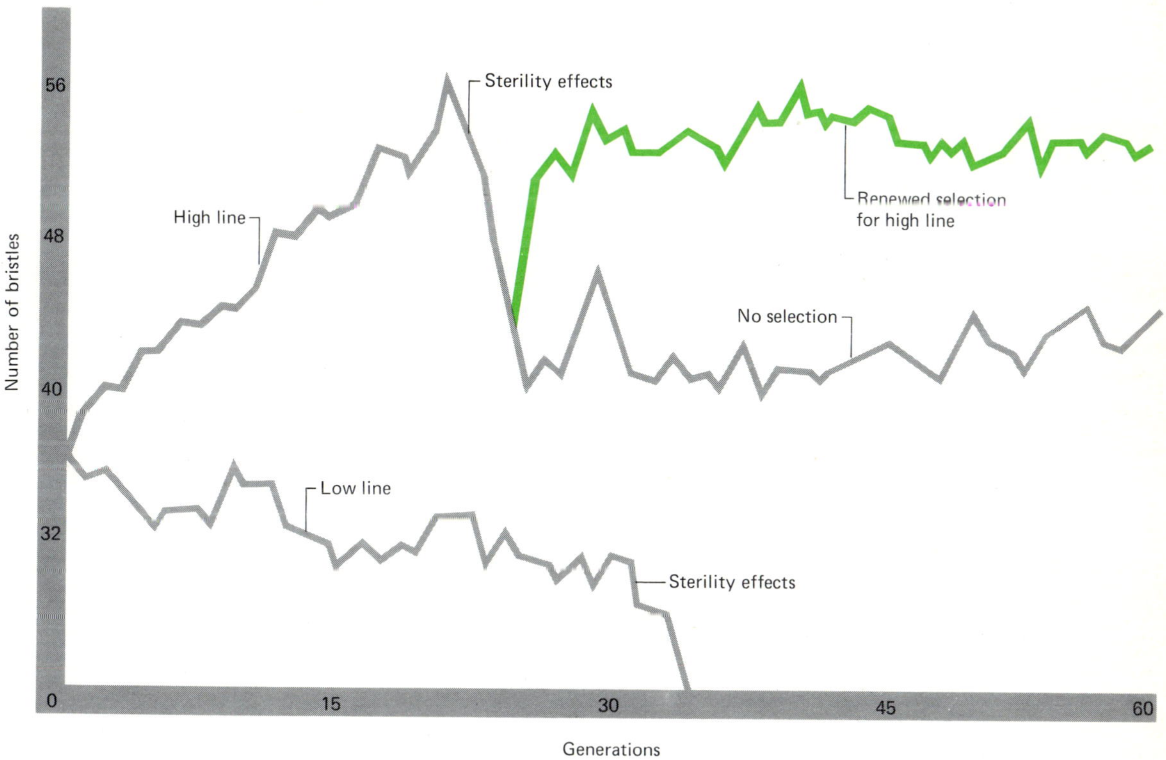

FIGURE 7.6 Graph of Mather-Harrison experiment. *(Redrawn from* Populations, Species, and Evolution *by Ernst Mayr. Copyright © 1970 by The Belknap Press of Harvard University Press.)*

However, after twenty generations of interbreeding in the high line, and thirty in the low line, sterility began to affect reproduction. Mather and Harrison then turned their attention solely to the flies in the high line. They discontinued selection with these flies, and after five or six generations of interbreeding the line showed a constant mean of 40 bristles, just slightly above the mean of the original population. The experimenters then took a sample from this population and resumed selection for high bristle number. The mean in this sample soon rose to 56 again, but this time there were no correlated effects on reproduction. After almost thirty more generations, Mather and Harrison ceased their observations, with the high line still showing a mean bristle number of 56.

Further studies of the flies by other investigators have shown that *on each chromosome there is at least one gene that affects bristle number, and on some chromosomes there are several such genes*. Thus, selection for bristle number affects the organisms' entire genotype, and it is not surprising that fertility should also be affected by selection for such an apparently insignificant trait. After selection was resumed on the high line, the genotype apparently had adjusted to the effects of selection and permitted the flies to reproduce.

Because the genotype is an integrated whole, the phenotype is often a compromise. Selection is rarely constant over time or over the geographic range of a species, and as the selective pressure varies from season to season or from place to place, it works on different genes at differing rates. Consider, for example, the evolution of the horse. When its feeding habits changed, longer teeth became an advantage. Such a change, however, produced selection for larger jaws and stronger jaw muscles and, therefore, a larger, heavier skull supported by larger neck muscles. Attendant on these changes were probable changes in the intestinal tract of the ancient horse as well. In other words, a rapid evolution in tooth length alone was not sufficient for the new feeding habits and would have been opposed by selection until all the other changes had evolved. Indeed, the varying selective pressures that equipped the horse to survive by changing its feeding habits permitted the teeth to grow at a rate estimated to be no greater than one millimeter each million years.

We must understand, then, that the importance of a single gene in the gene pool of a population depends on its contribution to the total fitness of the genotype and, thus, the phenotype; that is, how the single gene "fits" into the genotype and how it interacts with the rest of the genes will determine survival. When we also consider the pleiotropic action of some genes, we see that survival involves the effects of genes on the whole phenotype of an organism, not merely on the visible manifestations of gene action.

Polymorphism

If we consider certain characteristics within some populations, we find that the variations in the population do not form any intermediate stages. Rather, the variations in one portion of the population are distinctly different from those shown in another portion. We call this phenomenon *polymorphism*, and a polymorphic population may exhibit *discontinuous genetic variations in body form, coloration, or even physiology*. Since it is relatively easier to see discontinuities in body form or coloration than it is to detect physiological differences, most cases of polymorphism are described in terms of these more obvious characteristics.

Occasionally, polymorphic populations have been described as separate species. For example, among Australian hawks, there are both white and grey individuals, and in San Diego County, California, king snakes (see fig. 7.7) are both striped and ringed. In both cases, naturalists thought that they were observing separate species. However, investigators found nests of hawks containing both grey and white offspring. The offspring of ringed female king snakes were found to be both ringed and striped, and

FIGURE 7.7 These two California king snakes are of a single species (*Lampropeltis getulus*), although they appear very different because the species is polymorphic. *(San Diego Zoo photo.)*

the offspring of striped females, both striped and ringed. The human species is polymorphic for the blood groups A, B, and O, and some white human populations are polymorphic for hair color.

One of the most widely cited examples of polymorphism concerns the common British land snail *Cepaea nemoralis.* In this species, an individual may or may not have a series of bands on its shell, and whether or not they are banded, the snails may have a shell that is pink, brown, or yellow. (See fig. 7.8.) Since the polymorphic variations are genetically determined, can the variations in *C. nemoralis* be explained in terms of mutations? Insofar as that is the origin of the variation, yes. But in polymorphic populations, the proportions of the population represented by the variants is far too large to be maintained by mutation alone.

FIGURE 7.8 Banded and unbanded forms of *Cepaea nemoralis. (Courtesy of P. M. Sheppard, University of Liverpool, and Worth Publishers, Inc.)*

Polymorphism is of interest, then, because there must be some combination of selective forces acting upon the genes to maintain the discontinuity. We have already seen such a combination in our examination of sickle cell anemia, which is an example of what is called *balanced polymorphism*—that is, a form of polymorphism in which the heterozygote has a selective advantage over either of the homozygotes. In the case of the snails, polymorphism is maintained by the selective advantage of the variants under different climatic or geographic conditions. In this instance, it was found that individuals living on different backgrounds were protected from predation by their color and marking. The most common predator on *C. nemoralis* is the song thrush. This bird, hunting visually, captures a snail, carries it to a large stone where it cracks open the shell and then eats the flesh. The stones, called *anvils*, provided investigators with a convenient tool for measuring predation rates on the different kinds of snails. (See fig. 7.9.) Examining the shells found at the anvils at particular times of the year, they computed the relative proportions of different colors and of banded and unbanded shells found. They then computed the relative proportions of the different colors in the local population of snails. They found that in the spring, the proportion of yellow shells found at the anvils was significantly higher than the proportion of yellow individuals in the local population. In the summer and early fall, they found that the proportions of pink and brown shells at the anvils were significantly higher than the proportions of these individuals in the population. They concluded that in the spring, the selective pressure on yellow snails was higher than that on pink and brown snails, whereas in the summer and early fall, the selective pressure shifted and was higher on pink and brown snails than on yellow ones. Using the same technique, they compared the pressure on banded and unbanded snails and found that the banded snails were subjected to greater pressure in the areas where the background was relatively uniform than in areas where the background was mottled.

FIGURE 7.9 Anvil stone littered with cracked *Cepaea nemoralis* shells. *(Courtesy of Mark Williamson, University of York, and Worth Publishers, Inc.)*

Though we would expect the banded and unbanded snails ultimately to become confined to areas where they were under less intense selective pressure, migration often causes a mixing of phenotypes. There is also some evidence that the genes that control banding and color also affect the snails' physiology. Yellow snails, for example, have been found to be more tolerant of temperature extremes than pink or brown snails. Thus, several factors—color, background, time of year, and physiological constitution—all play a role in maintaining the discontinuity of phenotypes in this population. How long can such a situation last? The fossil record in England indicates that *C. nemoralis* has been polymorphic for many thousands of years.

Were we to attempt to draw distribution curves for *C. nemoralis* shell color or banding, or for bristle number in the population of *Drosophila* that Mather and Harrison selected, what might they look like? Since selection was for extremes in the original populations, it was the middle of the curve that was selected against. The result of such selection against the mean produces curves that look like this:

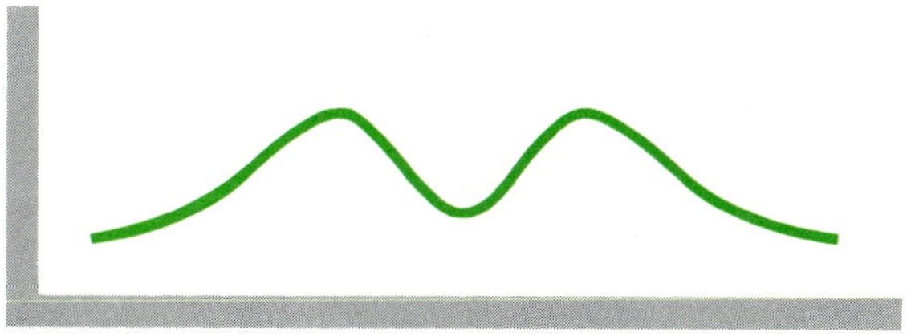

The curve represents a polymorphic population, and we could construct similar curves for other polymorphic populations *if we choose certain morphological characteristics*. As long as the population continues to interbreed, the curve will be maintained. However, it has been suggested that speciation can begin in this way if the population is in a state of transient polymorphism.

Relation of the Phenotype to the Genotype

We are now much closer to an understanding of the relationship between the phenotype and the genotype. In order to understand the basic principles of heredity, we began by studying single genes as they affected single characteristics. Obviously, that discussion was greatly simplified. The target of selection, the phenotype, is the result of many genes interacting together, each gene, in most cases, affecting more than one characteristic. Thus, the genotype of an organism cannot be realistically viewed as a series of beads on a string. During development, the interaction of the genes are most striking, and the differentiation of cells produces the distinctive external features of the organism but no less importantly, its overall physiological makeup as well. The polypeptides produced by the genes can cause a wide array of physiological effects, promoting or inhibiting growth, reproduction, tolerance of temperature extremes, and so on. Each of the genetically produced effects in the phenotype, whether visible or not, affects the probability of the organism's survival in the environment or environments in which it is found. We should recall Morgan's early comments about the chromosome theory of inheritance. He doubted the validity of the theory on the grounds that many characteristics would have to be determined by one. But now we have seen that *not only can genes be carried together on the chromosomes but characteristics can also be carried together on the gene*.

Our discussion of genes in populations has been largely limited to the effects of single genes, but it must now be clear how much of an oversimplification that is, too. The contribution to fitness that any single gene makes in a population is a function of its interaction with all the genes in all the genotypes of the population. How any gene is expressed in the phenotype of any individual in a population will be minutely or greatly different from the way in which it is expressed in any other individual. Thus, the selective value of any gene in a population is an average of the selective values of all its expressions in the population. We have only touched here on the mathematical nature of population genetics. Many sophisticated mathematical techniques have been developed to describe theoretical evolutionary phenomena, and indeed such theorizing has proved a rich and challenging source of biological knowledge. But clearly, there are many factors that greatly complicate the analysis of the effects of genes in populations, and, as a result, the application of such mathematical tools in field studies has been relatively infrequent.

Thus far, we have seen how genetic changes relate to the evolution and adaptation of single species in single environments. We have not yet explained how the many diverse species that we examined in Chapters 3 and 4 might have arisen, nor have we examined the effects on genes of environments that differ geographically. Clearly, if we are to understand the relationships between evolution and ecology, we must understand these other phenomena.

The Galapagos finch has provided what is probably the best example of the process of formation of species. *(Courtesy of Ira Kirschenbaum.)*

the evolution of species and ecosystems

chapter eight Individuals of a species that live in differing locations will be subjected to differing selective pressures that are characteristic of the environments and communities in which they are found. They will thus have a tendency to diverge in their niches, but since their gene pools are continuous, the divergence will not be very wide. Thus, species adjust to different environments. We must look now at how geographically variable evolution operates and examine how the different species that make up biological communities arose.

Geographic Variation

If we were to choose some morphological or physiological characteristic and examine its variation over the entire range of a species, we could probably derive a normal distribution curve. We would be ignoring, however, differences in the environments and, thus, differences in the selective pressures under which the demes of the species were living. For example, a frog that lives in Vermont is subjected to a very different set of environmental circumstances than a frog of the same species that lives in North Carolina. The differences in the environments and selective pressures operating on the demes of species produce differences in the gene frequencies over the range, but an exchange of genes between demes takes place as long as the populations continue to interbreed. This exchange of genes produces what is called a *gene flow.*

Gene Flow

As a hypothetical example of geographic variation and gene flow, imagine two mountains separated by a valley. Mountain A is 1500 feet higher than mountain B and has a greater average rainfall at the summit. Two populations of a single species of rodents live on the mountains, and each is adapted to the respective amounts of rainfall by the production of skin oil which coats the fur and protects the rodents from the rain. Rodents on mountain A produce more such oil than those on mountain B, and the summits constitute optimal environments for the populations. As the populations at the summits increase, some animals migrate down the mountains into drier environments to which they are less well adapted. At the zone where rodents from mountains A and B meet, they interbreed and exchange genetic material. A *gene flow* takes place. (See fig. 8.1.) In the valley, we would find that the distribution of variations in hair oil production of the offspring of interbreeding rodents from the two summits would differ somewhat from that of the animals at the summits. (See fig. 8.2.) The gene flow occurring between the two populations produces intermediate forms.

Should the valley between the two mountains be an environment that is intolerable to the animals from either summit, or if for some reason the valley cannot be crossed, then we say that the populations are *allopatric;* that is, they have *differing ranges.* Allopatric populations do not interbreed and produce a gene flow, because some external barrier exists that prevents interbreeding. In contrast, if the two populations

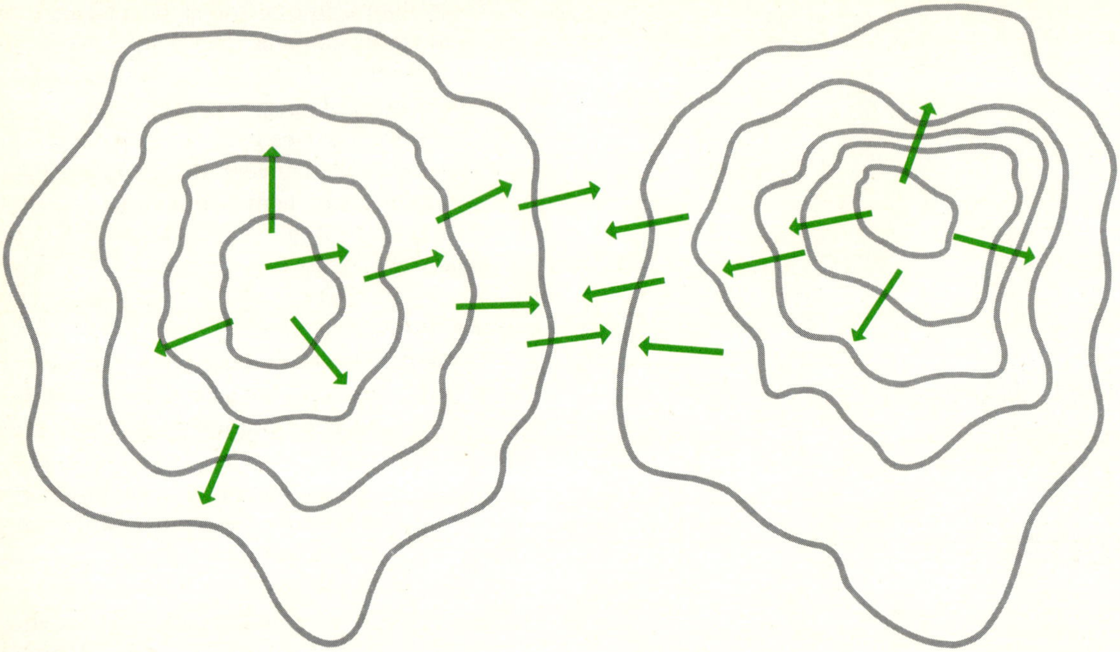

FIGURE 8.1 Gene flow.

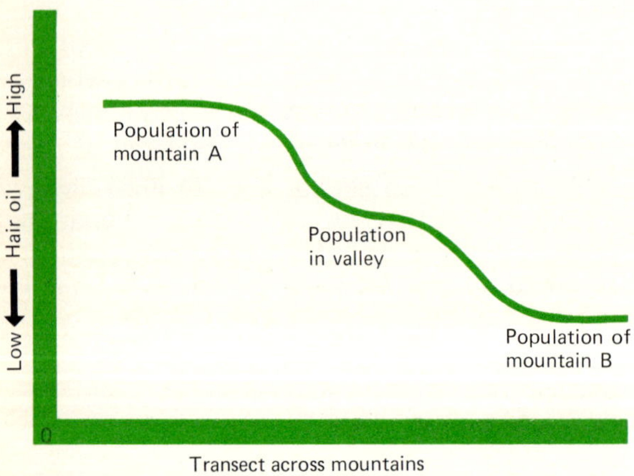

FIGURE 8.2 A distribution curve for a characteristic involved in gene flow.

meet in the valley, then they are said to be *sympatric*; they have *overlapping ranges.* (See fig. 8.3.)

Our example, of course, is an oversimplification, for selection works on all variations, not merely one, and not just in one geographical area. Just as the animals at the summits would be subjected to different selective pressures from those affecting the animals in the valley, the animals living on the mountainsides would be subjected to different ones, too. Within the limits of a species' range there are usually several intermediate conditions. Thus, between populations living at the outermost geographical limits of the range, there are a number of intermediate populations, each with its own gene pool and characteristic gene frequencies. Since most environmental change, and biological change as well, is continuous along a gradient,* the selective pressures change continuously and produce continuous change in the phenotypes and gene pools of the species.

* A gradient is defined as the change in some variable per unit of distance.

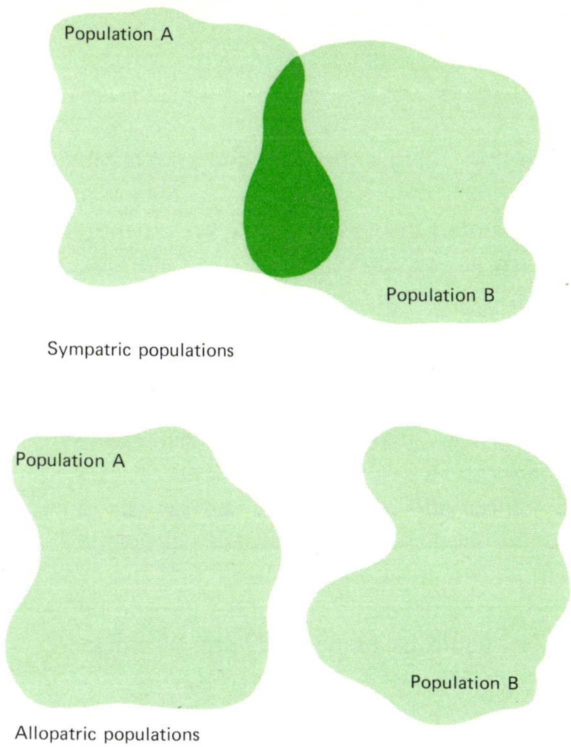

FIGURE 8.3 Allopatry and sympatry.

Such continuous genetic changes along an environmental gradient are called *ecoclines,* or if they refer to morphological changes only, *clines.*

We call the variations found along ecoclines *ecotypes,* of which there are basically three varieties: *climatic, edaphic,* and *biotic.* Climatic ecotypes are selected by factors associated with the climate in which the population is found. (See box 8.1.) Edaphic (Gr. *edaphos* bottom) ecotypes are selected by nutrient availability in the soil, by the availability of soil moisture and other soil characteristics. Biotic ecotypes are selected by responses to biological features of the environment, such as the relative presence or absence of predators, competitors, or food organisms. In the case of some plants, a biotic ecotype can be selected by the presence of a browsing or grazing animal. A plant that grows upright in an area where grazing pressure is high will be selected against, while variants that lie more closely to the ground will be selected for. As in any case, for reasons that we have already mentioned or we will soon see, it is not easy to pinpoint a single adaptation or a single selective factor and say that it alone makes the difference between survival and death.

Subspecies and Biological Races

Ecotypes and ecoclines are not the only kinds of geographic variations. There are less gradual examples of geographic variation—namely, *subspecies,* or *races.* In this case, the variation is not as gradual as in the case of clines. There may be a clear clinal variation over a part of the species range for some distance, but then it becomes more abrupt. After the abrupt step, it might resume a clinal variation, but at a differing rate. (See fig. 8.4.) Some biologists call the populations on either side of the abrupt step *subspecies* or *biological races.* These terms might also be used to describe some more clearly separated populations, such as the trout in two different streams. A more mundane example of subspecies, or races, might be the house fly populations of Boston and San Francisco. The populations that are usually designated as subspecies or races *do not share a common gene pool with other populations of the species but are thought to be capable of interbreeding with the rest of the populations in the species if the opportunity presents itself.* The problem with such designations is that the distinctions are often based on a very few characteristics. Consideration of other characteristics will sometimes give a very different picture of the populations. If somehow brought together, subspecies would interbreed, and the distinguishing characteristic would probably disappear. Subspecies may be contrasted with polymorphic populations in which interbreeding is regular and the different characteristics are maintained.

Human races should be mentioned here. Races share the characteristics of the species, and though individuals from different geographical areas may be very different, there can be no absolute distinction, because intermediates can

Ecotypes

First proposed by a Swedish botanist, the concept of the *ecotype* was put on sound experimental footing by three American botanists. The Americans gathered representative specimens of the plant *Achillea* from locations in California starting on the coast at Bodega and extending in a line approximately two hundred miles inland. Such a transection includes several distinctly different environments ranging from coastal lowlands to high mountain meadows and the plainlike floor of the San Joaquin Valley. As expected, the plants gathered from very different environments exhibited distinct phenotypic variations. The question, of course, was how much of the variation was due to environmental and how much to genetic factors. After carefully selecting plants that were truly representative of the environments from which they were taken, the botanists transplanted the specimens to an experimental garden at Stanford University in Palo Alto, California. Grown under uniform environmental conditions the plants retained certain phenotypic resemblances to plants grown in the locations from which they were taken. Particularly striking were the similarities in height between plants grown in the garden (as shown in the illustration) and plants at the sample locations. Further studies revealed certain physiological differences between the plants, and the experimenters concluded that distinct genetic populations could be found in different locations through the species' ranges.

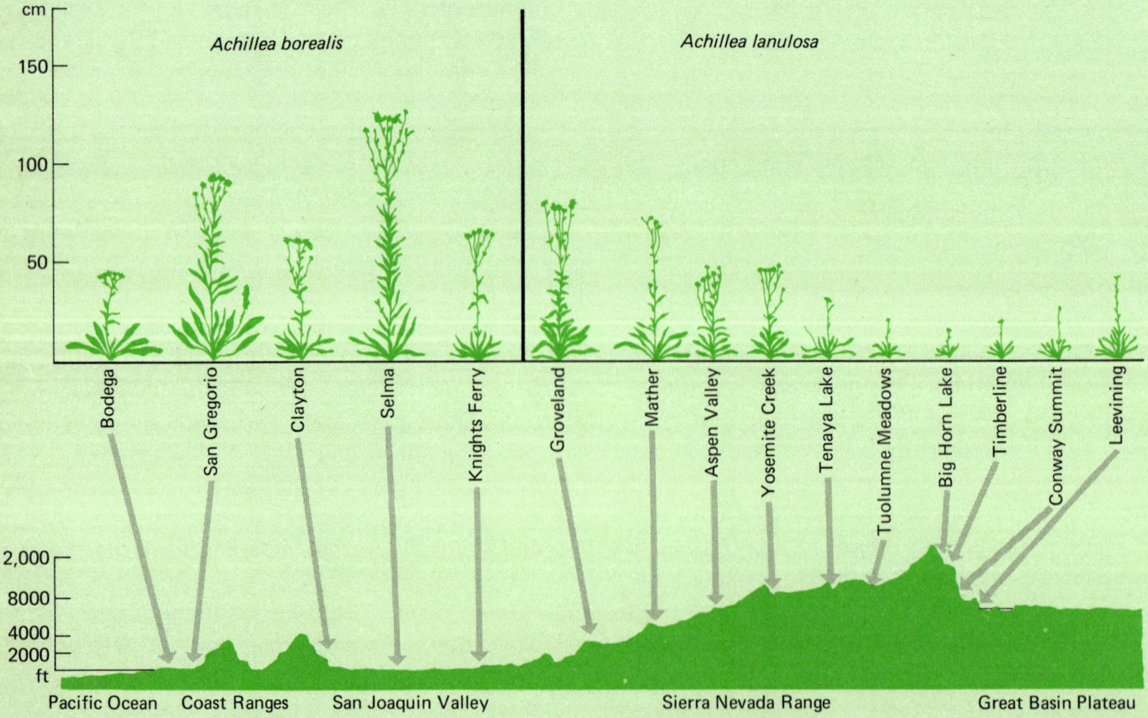

(Redrawn from Clausen et al.)

BOX 8.1

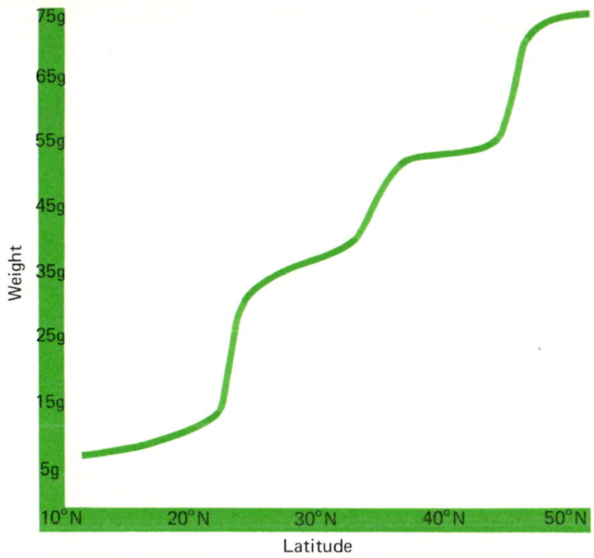

FIGURE 8.4 By choosing some characteristic, in this instance body weight at maturity of some hypothetical animal, and measuring the characteristic over the range of the species, it may be found that pronounced variations exist in the incidence of the characteristic in populations from different parts of the range. This does not conclusively demonstrate the existence of biological races, since such phenotypic differences may be attributable to purely environmental factors such as the length of the growing season. If genetic factors can be strongly correlated with the differences, then one may tentatively conclude that the populations do constitute races.

always be formed by interbreeding populations. Biologically, no race is ''better'' or ''worse'' than another; they are different because they evolved, and are evolving, in different areas. However, races do differ; to deny that is to deny that selective pressures differ between areas and that isolation prevents mixing of gene pools. Whether or not these differences may be reflected in behavior is a subject of great controversy and one that we will examine later.

Genetic Drift

The variations that we see in a species reflect variations in the frequencies of certain alleles in the gene pools of the species populations. Usually, changes in frequency result from selection, but under certain conditions, chance can play a role in altering gene frequencies. When a gene is *eliminated from a population or increases in frequency by random events, rather than by selection,* we say that *genetic drift* has occurred.

The conditions under which genetic drift can occur are rather stringent. First, the population must be small. For example, imagine a human population of 1000 persons in which there is only one allele for some particular trait. If, for some reason, the individual carrying that allele does not mate and reproduce, then the allele will be lost to the population. There are several things that might interfere with the spread of the allele from this one heterozygote. For example, the person may be selected against for some characteristic not highly associated with that characteristic to which the unique allele makes a contribution, or the individual may mate with a sterile individual or may mate and produce only one offspring which would have only a 0.5 probability of receiving the allele. Second, the population must be isolated from gene flow with other populations. Thus a ''lost'' allele cannot be carried into the population from outside. The third condition is that the genes affected by genetic drift be of little selective value to the population.

At one time, genetic drift was assigned an important role in the process of evolution, but the conditions under which genetic drift can occur are so stringent that they almost never occur at the same time. The first two are not exceedingly rare. We have already suggested that Darwin's finches arose from a small isolated population of finches. It is the third condition that is critical. If a gene has high selective value and genetic drift is in play, then in extreme cases the population may become extinct. Usually, when genetic drift occurs, it produces trivial differentiation. It is extremely difficult to show that some gene has no selective value at all. For example, examinations of blood groups among Hutterites, a religious sect that established colonies in North America in the late nineteenth century after migrating from Switzerland by way of Russia, revealed that in some small colonies, the incidence of blood group B was zero, and that these colonies also varied widely from nearby non-Hutterite populations in the frequencies of A, M, and N blood groups as well. It was long assumed that this was the result of genetic drift, on the grounds that blood groups were of little selective value. However, it has been shown that the blood groups in the ABO system, at least, have adaptive significance. Thus, the differences in the frequencies of the alleles that code for blood type in the Hutterite populations can be attributed, to some degree, to natural selection, although it is probable that random events may also have contributed to their gene frequencies.

Factors Causing Selection

Until now, we have generally treated selection in a simple way, as a single physical or biological feature of the environment. But selection is not so simple. When we consider that the target of selection, the phenotype, is the complex product of the interaction of many genes, each of which is subject to differing selective pressures, then the selection of an individual, or a number of individuals, must result from several aspects of the environment. We usually somewhat artificially break down the factors of the environment into two categories: *biotic* and *abiotic*. Biotic factors are simply the other organisms in the environment of the species under consideration. Abiotic factors are those elements of the physical world in which the organism lives, including rainfall, the availability of sunlight, minerals in the soil, temperature ranges, and so on.

Rarely if ever do either biotic or abiotic selective pressures act alone. On a forest floor, for example, the amount of sunlight available to plants is determined by the latitude of the forest, the cloud cover, the time of day, and the canopy formed by the crowns of trees. To study selection, investigators often hold all but one factor constant and concentrate their attention on the effects of that one factor. Although this often produces a fair picture of the effects of that factor, it is misleading insofar as it fails to consider those other factors that may, at the same time, be affecting the organism under study. In the example used earlier of the common British land snail, we saw that the selective value of banding and color changed as the seasons changed, but we also noted that predation was only one of the things that may affect the snail. The ability of the yellow snails to withstand heat and cold may also be a factor in their survival.

Biotic Selection

Later, when we examine specific communities of organisms, we will look in detail at some of the abiotic factors affecting selection. Although we will also look in more detail at some of the biotic ones then, we shall examine some of the main ones here. All biotic selection is the result of the species' biotic habitat. Defining the *habitat* simply as *where, and the conditions under which, an organism lives,* the biotic habitat refers to the other organisms that live there. The biotic habitat, then, would include, among other things, food, predators, competitors, parasites, and cover. The availability of biological resources and the presence or absence of predators or competitors will determine where an organism may survive and, thus, are considered selective forces. Should one population of organisms, such as trees, be removed from an area, then all of the other populations that depend on the trees as a resource will be selected against.

The two major biotic interactions that result in selection are those between the eaters and the eaten and those between organisms in competition with each other for limited resources. In the first category are predator-prey, herbivore-plant, and host-parasite relations. The first of these is often interpreted in behavioral terms. The evolution of hunting behavior allows a predator population to more efficiently capture its prey. But at the same time, the prey population may evolve more efficient escape behavior. The evolution of escape behavior may, in turn, lead to new adaptations in the predator population that increase its hunting efficiency. Some behavioral adaptations may not appear to be at all adaptive, but upon close examination, they may be revealed to be quite important to survival. For example, the black-headed gull always removes the bits of shell remaining in its nest after a chick has hatched. Usually, the shells are removed within an hour after hatching takes place. Though it might seem that this bit of behavior has no selective value, it was found that the shell removal behavior is highly adaptive. Careful observations revealed that predators on gull eggs used the broken shell fragments as signals to look for unbroken eggs nearby. The gull's removal of the shell bits, it was concluded, increased the chances that the predator would fail to identify the location of unbroken eggs. Thus, this apparent ''housecleaning'' reduces the selective pressure of the predator on the gull's young.

Herbivores exert selective pressure on plants, and the evolution of thorns and spines in some plants is thought to be a response to that selective pressure. But plants without thorns are plentiful, too, and this does not mean that they have not evolved adaptations to permit survival. Numerous instances may be found in the plant kingdom of species that are distasteful or even toxic to a number of herbivorous species. Cows, for example, will not eat buttercups unless there is nothing else available, because one of the first products of the chemical breakdown of buttercups is a strongly irritant and distasteful (to cows) oil. Some species of buttercups contain another chemical substance that can lead to fatal convulsions. Other plants, when eaten by some vertebrates, can produce heart attacks, and one plant that man prizes for its fragrant blossoms, the oleander, can be lethal to a man if he eats as little as a single leaf.

In predator-prey and herbivore-plant relations, the predator and the herbivore are selective forces, although we have tried to show that the prey and the plant can survive through adaptations that reduce selective pressure. One of the most spectacular and striking examples of how host-parasite relations produce adaptations and ecological balance is provided by the Australian rabbit population. A small population of rabbits (two dozen) was introduced into Australia in 1859, a gift from a gentleman in England to a friend. The rabbits were set loose, and Australia provided an ideal habitat for them. Within a few years, they had multiplied enormously, and in a drive to kill them in 1865, some 20,000 were bagged on the estate of the man who had let them free. They were equally abundant elsewhere, and nothing seemed to succeed in controlling the population. When the rabbits were shipped from England, they carried two common rabbit parasites with them, a louse and a mite, but they did not carry a third common parasite, a particular species of flea. Around 1920, a South American biologist suggested that a virus, myxoma, be introduced into the Australian rabbit population, but nothing was done about the suggestion until the early 1950s. The disease caused by this virus, *myxomatosis,* affects South American rabbits only mildly. When inoculated into the European rabbit, from which all the Australian rabbits were descended, it causes serious infections and death. In European rabbits, however, the disease is spread by the rabbit flea, which lands on the large sores (lesions) caused by the infection (see fig. 8.5) and takes up the virus while extracting blood from the rabbit. The flea then leaps to another rabbit to which the virus is transferred as the flea feeds. When introduced into a laboratory population of Australian rabbits, the disease did not spread readily to other rabbits, because there were no available fleas in Australia. However, it was soon found that mosquitoes worked just as well as fleas to spread the myxoma virus, and a small population of infected rabbits was released. At first, the results were exactly as predicted. The mosquito-borne viruses caused widespread epidemics of the disease and drastically reduced the rabbit populations. With the onset of cold weather, the epidemics ceased as the mosquitoes died, but the disease broke out again after a spring rain the following year.

FIGURE 8.5 A very sick rabbit, infected with the strain of myxoma virus originally introduced to Australia. It was infected in the laboratory ten days previously. *(Courtesy of Frank Fenner, The Australian National University.)*

Soon, however, rabbits with some resistance appeared, and it was noted that their offspring tended to be resistant too. But selection had worked in two ways. The most virulent strains of myxoma had killed the rabbits too quickly for the most lethal form of the disease to spread, since mosquitoes do not light on dead animals. Thus, the virulent form of viruses were killing themselves. The least virulent strains did not produce many sores, and thus the mosquitoes did not carry many of the least virulent strains from rabbit to rabbit. Moderately virulent strains, however, kept the rabbits alive long enough for the lesions to develop and for many mosquitoes to carry the disease. Therefore, the strain of virus that was most widely distributed was the moderately virulent one. Thus, a certain amount of genetic resistance arose in the rabbits as a response to the selective pressure of the virus, but strains of virus also arose that were adapted to the problems posed by the method of transmission.

The initial result of the infection of Australian rabbits with myxomatosis was one that is common when a parasite is newly introduced into a species; when the parasite has evolved a new and virulent strain; or when the host is not the principal host species of a parasite. Selection is usually severe, but ultimately, a balanced relationship evolves between host and parasite that allows both to survive. Humans, for example, support a harmless bacterial flora both on the skin and in the intestines; this bacterial flora was once, in all probability, composed of harmful species of microorganisms, but it has *coevolved* with mankind.

Competition is also an important selective factor. Given a limited resource, all organisms that depend for their survival on the resource will compete for it. For example, if mice are a food resource for foxes and owls, the two predators will compete for the mice. They will not often confront each other, but they will nonetheless be competing. Competition is by no means limited to biotic resources; it also includes abiotic resources such as light and space. Under a heavy forest canopy that prevents a great deal of light from reaching the forest floor, only those plants capable of utilizing a small amount of light most efficiently will survive. In fact, trees reach the heights that they do as a result of selection and evolution produced by competition for light. The competition for light produced the internal transport system described in Chapter 4. Some plants that did not undergo such specialization have evolved in ways that permit them to utilize less light. Plants may also compete for space, and spatial competition is sometimes related to the need for light. The more space available, the more leaf area the plants may expose for photosynthesis. Botanists have begun only within the last fifteen years to find instances of competition in the plant world in which plants of one species directly compete with those of another species by releasing toxic chemicals that inhibit the growth of other species. In Santa Barbara, California, for example, it was shown that one shrub, hard chaparral, gives off toxic compounds that are washed into the soil by rain water. These compounds prevent the seedlings of grasses from growing near the chaparral. After a fire that destroyed the parts of the chaparral above ground, the grasses were released from inhibition and grew profusely for a few years, until the chaparral was reestablished and subjected the grass to new inhibitions. (See fig. 8.6.) Generally speaking, the severest competition takes place between closely related species, since they often have very similar requirements for survival.

The ways in which these and other selective forces operate together are often very complex and difficult to record. At any one time, a population may be subjected to a combination of selective factors. As the gene frequencies in the population change in response to the selective pressures, however, the intensities of the pressures themselves change in response to the altered gene frequencies. Thus, whether a species population can survive in a particular environment depends on all of the forces of selection acting together and the response of the population to those forces. The black-headed gull mentioned earlier is selected not only on the basis of its shell-removal behavior but also by the availability of food, nesting sites, average mean temperature, and a host of other factors.

Evolution of Behavior

As we have seen, the survival of the individual matters most to the individual. To the species, it is the survival of large numbers of individuals sharing a particular set of characteristics that is important. As a rule, individual animals behave in ways that insure their own survival and often also the survival of their own young. Thus, their genes remain in the gene pool. But there are a number of instances in nature of individuals that exhibit "altruistic" behavior. Certain birds, for example, have evolved warning calls that inform other members of the population that predators are in the immediate vicinity. Such warning calls, however, may betray the location of the caller and, therefore, momentarily expose the caller to a greater selection pressure than the other birds in the population. Since much of the behavior of birds is genetically determined and stereotyped (that is, invariable), how could such behavioral genes persist in a population? It would seem that the callers would be eliminated from the population rather soon, leaving the population to evolve some other adaptation to keep the predator-produced pressure in balance. However, it has been shown that many of the

(a)

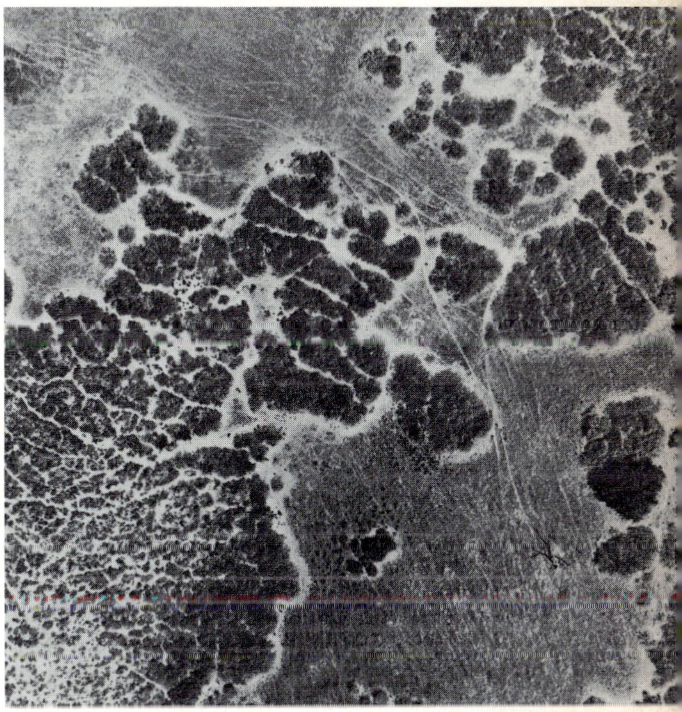

(b)

FIGURE 8.6 (a) Grasses cannot grow in a zone surrounding hard chaparral shrub *(Salvia leucophylla)*. (b) As seen in this aerial photograph, extensive areas of land covered by the shrub and grass show bare margins. *(Courtesy of C. H. Muller, University of California, Santa Barbara. (a) Reprinted from the* Bulletin of the Torrey Botanical Club, *Vol. 93, p. 334; (b) reprinted from* Science, *Vol. 143, 31 Jan. 1964, cover; copyright © 1964 by the American Association for the Advancement of Science.)*

individuals warned of the presence of the predator by callers are related to the caller. Since they are relatives, they share the genes of the caller, to some extent, and the altruism gives a selective advantage to carriers of the trait. How rapidly or extensively such genes will spread in a population depends on the difference between the risk to the caller and the increased probability of survival among the kin being warned.

Our understanding of "laws" or general principles in biology is not so well-developed as our understanding of the "laws" of physics or chemistry due to the immense complexity and range of adaptations and interactions between organisms and their physical and biological environments. The ecologist or applied biologist must be a good naturalist and, thus, able to see living events in the broadest possible context. We can describe some processes in great detail at a physical and chemical level, but often our descriptions of the processes are not universally valid for all organisms. Each species survives, but each survives by means that are unique to the species. And when a species does not survive, its extinction is also a unique case. In many ways, then, each case must be judged on its own merits.

We have seen that variation, selection, and adaptation occur in populations and that this produces differentiation within the populations and along clines. A major topic remains: Why are there not simply one plant, one herbivore, one carnivore, and one decomposer? How do species arise? Why are there so many species? Unless we answer these questions, we cannot understand the formation of ecosystems, niches, or ecological balance.

Speciation

Darwin, in spite of entitling his book *On the Origin of Species,* never really answered the question of how species originated. He described natural selection, which we now understand as the force that underlies speciation. The mechanism that was crucial to understanding the process of speciation was unknown to Darwin and most of his contemporaries. This critical factor, an understanding of genetics, was restricted to a handful of people who had not been able to see its importance as a tool to explain the Darwinian hypothesis, and most of those familiar with Mendel's accomplishments were largely unimpressed by them.

As we have seen, a great deal of genotypic and phenotypic variation may occur between ecotypes or along ecoclines, but the exchange of genetic material, gene flow, prevents the populations from separating into new species. Gene flow insures that while the species' niche, physiology, and morphology may vary, they remain distinct and definable in a manner characteristic of the species. For speciation to occur, evolution must proceed without gene flow; parts of the gene pool must be somehow *reproductively isolated* from each other. The origin of species is initiated when a population is prevented from interbreeding with other populations of the same species.

Reproductive Isolation

We must give some attention, then, to the ways in which a population can initially become reproductively isolated. The usual method, which some argue is the only method among animals, is geographic isolation. There are several forms of geographic isolation, the most obvious ones being islands and bodies of water, such as lakes, ponds, or rivers, that are not interconnected by a system of tributaries. A mountaintop may also isolate one population from another if the terrain around the mountain prevents migration. Isolating mechanisms may be biological as well. The broad grassland in the middle of the North American continent separates the forests of the east and west and prevents a gene flow from taking place between the trees of those forests.

Over geologic time, land masses are not stationary; they move, change their forms, and their climates fluctuate. The process is taking place at this moment, as anyone living in the earthquake zone in California can attest. Also, in other parts of the world major geological changes are taking place, with the most spectacular example being what geologists believe to be the birth of a new ocean in the Afar triangle, a region of the Red Sea area. Less spectacularly, the continent of North America is moving westward at the rate of one centimeter per year. These changes in the surface of the earth produce geographic features that affect the evolution of new species. Recent changes such as those that split Great Britain and Ireland from the continental land mass of Europe have brought about divergent species on those islands. With the retreat of the glaciers, lakes were formed, and new species of fish evolved in those lakes, although in some cases, it has been found that they were not really new species. Even though these fish were geographically isolated and even appeared to be different species, when man began to reintroduce them into other lakes, they were found to be capable of mating and reproducing with the populations already in the lakes.

The real test of whether speciation has taken place is not the geographical separation of a population, but whether it will interbreed with another population and produce surviving, reproducing offspring. If that is still possible, the populations are still members of the same species. When speciation takes place, the new species' adaptations very frequently include the evolution of a new niche. Without the genetic separation of speciation, populations have the same niche (allowing, of course, for local variations in the niches of differing demes) and thus the same ecological effects. The evolution of a new species, involving as it does the contribution of a new niche, has different ecological consequences.

How speciation occurs in the oceans presents a particular problem, since the oceans interconnect and of course within an ocean the water masses are continuous. Temperature differences in the oceans could produce ecoclines, but such differences do not seem abrupt enough to produce isolation. Nevertheless, the oceans contain a great diversity of species. A partial explanation for the evolution of so many species may be found in the action of ocean currents that may create a one-way gene flow, but the action of currents as geographic isolators is questionable. It may be that most species arose in bays or other coastal areas where they were isolated and unable to interbreed. It may also be that the deep basins and trenches that are found in the oceans are analogous to mountaintops. No one has yet satisfactorily explained how speciation takes place in oceans.

For one population simply to be isolated from another is not enough to cause speciation. The process demands time, as well. Complete reproductive isolation and speciation, then, are a matter of time and space. Often, it will appear that two species exist, when in fact, they are still capable of interbreeding and thus are not distinct species. We have already mentioned the example of fish in different lakes. Consider the recently discovered population of humans in the Philippines which was isolated from the rest of the human population for an estimated 400 years. Although to our knowledge no one has yet tested the extent of their reproductive isolation, there is no reason to believe that they would be incapable of interbreeding with other human populations. In terms of human generations, the Philippine population was isolated from the rest of the species for a mere twenty generations, which is hardly enough time on an evolutionary scale to produce speciation. Where geographic isolation and sufficient time have produced a "species" that appears quite different from other species in the genus, the taxonomist and the ecologist are faced with the problem of determining whether it is, in fact, a true species and thus has a different ecological role in its community.

In time, the subtle effects of selection will be felt by an isolated population, and if the population survives, time will often insure that the genetic divergence of characteristics will prevent effective interbreeding should it come into contact with the species from which it derived. Selection alters the morphology, physiology, biochemistry, and behavior of the population. Given enough time, these aspects of the population will diverge sufficiently to make it impossible for the two species to interbreed. This does not necessarily mean that they will not mate, or at least attempt to mate, but it does mean that if mating takes place, no viable offspring will result, or if the offspring are viable, they will be incapable of reproducing.

Polyploidy

In some instances, speciation is the product of a mechanism that is not at all related to geographic isolation and, in terms of time, is nearly instantaneous. This process is known as *polyploidy.* Although polyploidy is not unknown in the animal kingdom, it is *far rarer* there than in the plant kingdom. The product of polyploidy is an offspring so different genetically from its parents that it can interbreed neither with them nor with other individuals genetically like them. Since it can interbreed with other individuals like itself, it is considered a new species. Polyploidy can occur when *normal meiosis* fails to take place and the gamete remains in the diploid condition. If it then fuses with another diploid gamete that has arisen the same way, a *tetraploid* organism grows (tetra = four). In some cases, the diploid gamete may fuse with a normal haploid gamete producing *triploid* organisms. Several varieties of apple are triploids. Another way for polyploids to arise is for two organisms of differing, but usually related, species, having the same chromosome number, to crossbreed and produce a hybrid. If the hybrid is not sterile, as most hybrids are, it is incapable of interbreeding with either of the parent species. Polyploids can be induced by the application of chemicals, such as colchicine, and some plant breeders have produced some useful hybrids by inducing polyploidy. In one case, however, a cross between cabbages and radishes, both having nine chromosomes, was tried in the hope of producing a plant with the edible roots of radishes and the edible leaves of cabbages. Unfortunately, the offspring resulting from the cross had the roots of a cabbage and the leaves of a radish.

There is some evolutionary significance to polyploidy, for in the event of a large environmental change, a plant that can give rise to a polyploid has more "flexibility" than one that cannot. The potential for the rapid production of new forms endows a species with some selective advantage. Once the polyploid state is achieved, however, a certain amount of "inflexibility" sets in. Studies have shown that the impact of mutation on polyploids is somewhat less dramatic than in diploids, but in most cases polyploids grow more slowly.

Some biologists believe that speciation can occur by other means than those that we have thus far described. The mechanisms that have been proposed as other means of speciation are currently under debate, but they deserve to be mentioned briefly. The first is a process that is similar to polyploidy in that hybrid forms are produced by parents of different species and the offspring cannot interbreed with either species, but are capable of interbreeding among themselves. Other biologists contend that instances occur in which the "tails" of the normal distribution curve are favored, and the intermediate variations are selected against. In that case, the two extreme forms become separate species. In a small population, genetic drift may play a significant role in speciation, producing new species with some rapidity. None of the proposed ideas has yet to be proved conclusively, and some evidence brought forward to support them has sometimes been interpreted as supporting natural selection as the underlying cause of speciation.

We have not defined species very rigorously thus far, though our discussion of speciation makes clear the essentials of the currently held biological definition of the term. A species is *a reproductively isolated interbreeding aggregation of populations.* Thus, the species shares the gene pool of many populations that may be geographically isolated from each other, but *gene flow takes place between the populations,* even though some populations may never interbreed directly with some others. Reproductive isolation simply means that individuals of one species cannot interbreed successfully with individuals of another species. Since reproductive isolation is the crux of the definition of a species, we must examine the mechanisms whereby the isolation is maintained.

Reproductive Isolating Mechanisms

There are basically two kinds of mechanisms that have evolved to prevent interbreeding between species to insure that they remain productively isolated: prezygotic (or premating) mechanisms and postzygotic (or postmating) mechanisms. Prezygotic mechanisms prevent interspecific (between species) mating, and postzygotic mechanisms insure that if mating takes place and a zygote is formed, the offspring will not be viable or capable of reproducing.

We shall first examine the prezygotic mechanisms. The first is simple; mating does not take place because potential mates do not meet when they are ready for mating. Mating in most organisms is restricted to certain times of the year. If the members of one species occupy the same habitat as members of a closely related species, but each has a different breeding season, then when one species is prepared to mate the other will not be. A classic example of differing breeding seasons among closely related species is provided by several species of frogs found near Ithaca, New York. (See fig. 8.7.) To this kind of isolation, which we call *seasonal isolation,* we can add *habitat isolation.* Habitat isolation means simply that two species occupy different habitats and never come in contact with each other. Thus, interbreeding can never take place.

FIGURE 8.7 The mating timetables for several species of frogs and toads in upstate New York indicate that although some species mate at approximately the same time, their preferred breeding sites differ. The spatial and temporal differences that exist, of course, are augmented by certain behavioral differences, so that even should males and females of different species find themselves in the same location at the peak of their breeding season, differences in courtship calls and courtship behavior prevent mating. *(Redrawn from Wallace and Srb.)*

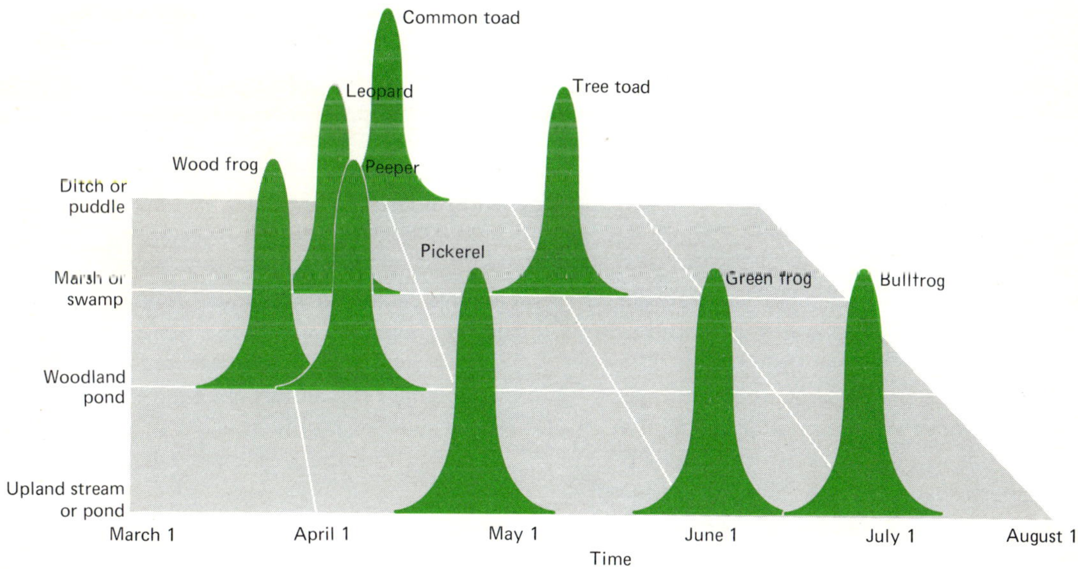

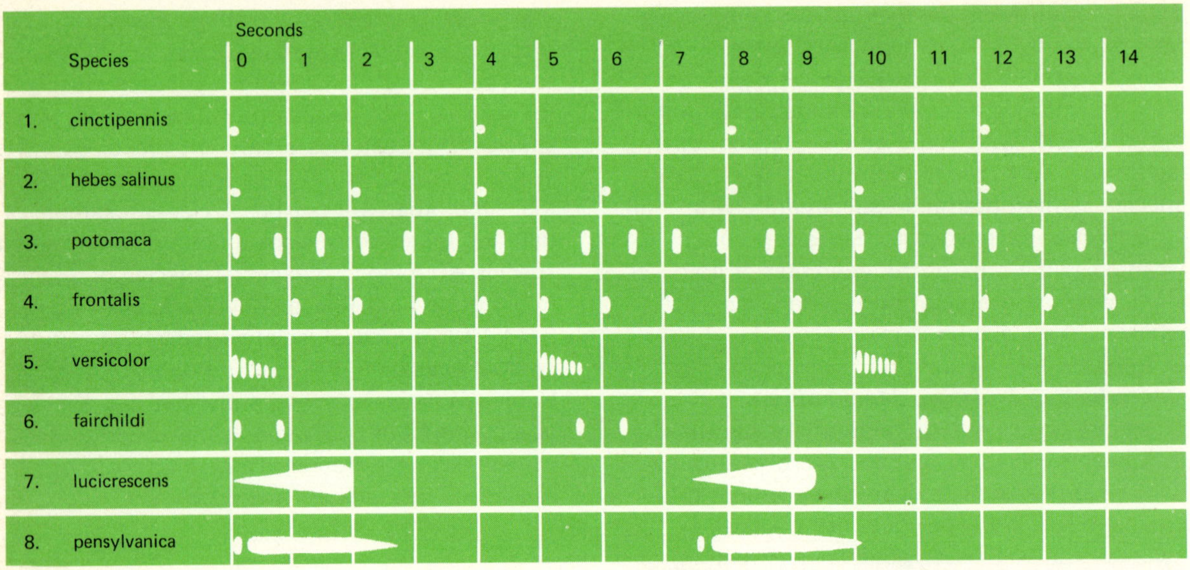

FIGURE 8.8 The flash patterns of North American fireflies. Each of the species of *Photuris* has a usually distinctive pattern of flashes, which differ in length, intensity, and color. The length and height of the marks in the graph above indicate the pattern and intensity of the flash. Only eight of the suspected sixteen species of *Photuris* are shown. Until the flash differences were identified, it was thought that only three or four species made up the genus. *(Redrawn from Mayr.)*

Among animals, one of the most important mechanisms preventing interspecific breeding is *behavioral isolation.* The males of almost every species exhibit courtship behavior that stimulates females of their own species. Usually, only females of the same species are responsive to these behavioral patterns. Thus, reproductive isolation is accomplished on the basis of the production and reception of signals by the potential sex partners. When the signals, or stimuli, are incomplete or inappropriate, the female will not respond, and mating will not take place. Since this is so large a class of isolating mechanisms, it is useful to break it down even further and discuss *visual, auditory,* and *chemical* stimuli.

Visual stimuli may take the form of courtship displays, as may be observed in certain species of birds or fish. In the peacock, for example, the fanning of the male's brightly colored tail feathers sexually stimulates the female. Other birds, such as the prairie grouse, spread their wings and assume a posture suggesting that they are about to take flight. In the stickleback, the male performs a zig-zag "dance," leading the female to his nest where she lays her eggs and he fertilizes them. One of the most widely cited examples of visual stimuli is exhibited by the firefly. Each species of firefly produces a light that is distinctive in terms of color, timing, and pattern. (See fig. 8.8.)

Auditory stimuli take the form of songs, calls, or other acoustical signals. The songs of birds are highly specific and often attract only females of the singers' species. Often, the song serves as a prelude to visual signals. After attracting the female with the song, the male begins a courtship display which ultimately is consummated in copulation. Certain crickets are known to have evolved highly specific signals that stimulate only members of their own species.

Chemical stimuli are particularly important among mammals and some insects. These signals, called *pheromones,* are usually olfactory, but they may occasionally involve actual contact between individuals. Perhaps the best studied sex pheromones are those secreted by the female gypsy moth. When ready for reproduction, the female releases a quantity of the pheromone that can be detected by the male gypsy moth as far as a mile from the source. The male's antennae are adapted to detecting the pheromone in the atmosphere when as few as forty molecules per second are intercepted. The male is attracted by the pheromone and locates the female by flying in a direction dictated by the quantity of the molecules reaching his antennae. As he draws closer to the female, the number of molecules reaching the antennae increases tremendously, indicating that he is on the right course. The experimental study of sex pheromones is difficult, since it involves the necessity of eliminating other sensory mechanisms such as sound, sight, and touch to determine whether the olfactory signals are the only ones at work. Experiments to determine the specificity of such signals have often shown that the pheromones can be detected by individuals of other species and can attract males of other species. To insure the reproductive isolation of the species, other isolating mechanisms must then come into play.

In some marine organisms, such as the polychaete worm *Grubea clavata,* the female releases a small quantity of egg protein into the water which causes the male to release a small amount of sperm. The sperm liquid contains a chemical substance that induces egg-laying in the female. Within a minute after she lays her eggs, the male ejaculates a massive amount of sperm to fertilize them. Marine and aquatic organisms are faced with quite a problem in the production of new individuals. If the female expels her eggs, or the male his sperm, at the wrong time, the eggs or sperm can be washed away by currents before fertilization. To insure the success of the species, they must achieve a method of reproduction in which the egg and sperm are released at approximately the same time. *G. clavata* have done so through the evolution of chemical signals, whereas other species of organisms, such as some pelagic fish, have done so through the evolution of courtship behaviors that bring the male and female into synchronous spawn.

Mechanical isolation is a third kind of reproductive isolation in which two species cannot interbreed because of differences in the structures of their genitalia. This was at one time thought to be the most important basis of reproductive isolation, since the copulatory equipment of most species of insects that had been studied was so different. It is true that in some species, the differences are so extensive as to cause injury and sometimes death to individuals of different species if they attempt to interbreed. However, detailed comparisons of nearly 700 species of sphinx moth (Sphingidae) showed that 48 of them had identical genital structures, while in about 50% of the species, the geographic variation in color was found to be correlated with variation in genital structure. Such an obviously important bodily structure is subject to selection, and large variations will be strongly selected against. Small variations that do not interfere with reproduction will not be selected against as strongly, and in many groups of insects that have been studied, the morphology of the genitalia does not appear to have changed considerably during speciation.

In the event that interbreeding does take place between species in spite of the mechanisms described, certain postmating mechanisms come into play. The first of these is *gametic mortality.* After copulation, sperm may encounter in the genital tract of the female chemical substances that will kill them. In work on fruitflies, one experimenter found that the vaginal walls of female fruitflies inseminated with sperm of another fruitfly species swelled enormously, thus killing the sperm. In some cases, the sperm simply die because they cannot penetrate the egg of the other species.

Should insemination result in the formation of a zygote, its development might be irregular and may cease at any stage between fertilization and adulthood. This is called *zygotic mortality.*

Should the zygote develop, however, the hybrid offspring is often inviable, born weak or malformed, frequently dying before reaching reproductive age. In this case—*hybrid inviability*—the hybrid makes no contribution to the gene pools of either of the species that crossed. Another form of reproductive isolation takes place when the hybrid is vigorous but sterile. Finally, though some hybrids survive and even reproduce, the offspring of hybrids are very frequently inviable, sterile, or ill-adapted and thus subject to greater selection pressure than either of the parent species.

By now it should be clear that there are many complex, evolved mechanisms to prevent crossbreeding of closely related species. Why have these evolved? What selection has led to these mechanisms? First, an individual that breeds with another and produces infertile offspring is at a competitive disadvantage. It wastes effort on activity that does not propagate its genes. If it mates only once, its genes are lost just as surely as if it had died or never mated. Second, if the offspring is fertile, it is likely not to be as well or as specifically adapted as either of its parents, and thus interbreeding will lead to loss of adaptive characteristics. There is, thus, strong selective pressure for the evolution of mechanisms that prevent interbreeding of organisms with different adaptations or niches.

The Interactions of New Species

Obviously, the biosphere is not made up of isolated species. The significance of a species is its role in the community to which it contributes its niche. Presumably, as new species evolve, they interact with those species that already exist in a community, and the species and the community coevolve. What happens, though, if the original barrier that supplied the conditions for speciation breaks down? Often, the new species is reintroduced into an old community, which has changed, and this means that it interacts with what are effectively new species. Or, the old community may not have changed at all, but the adaptations that have evolved in the new species may no longer "fit" the old community; thus, the new species interacts in a new way with the community. Often, the new species must interact with its parent species.

To examine the interactions of new species, we must first review the ways in which the new species might be reintroduced into a community. Remember that in most cases, geographic isolation is thought to be necessary for speciation. There are a number of ways in which reintroduction can come about. First, the original barrier that produced the geographic separation can break down. For example, South America was at one time connected to North America, but the two continents were separated completely, over 60 million years ago. They remained separated for approximately 60 million years until the land bridge formed by the Isthmus of Panama arose some two to three million years ago. In the time that the continents were separated, species evolved that were able to interact when the land bridge was reestablished. In the time since the contact was reestablished, many North American species moved into South America, where they competed with South American species and ultimately brought about the extinction of many of the South American ones.

Climatic changes that serve to isolate populations that go on to become new species may also be reversed, permitting the newly evolved species to reestablish contact with each other. The history of the earth has been marked by such climatic "comings and goings." Climatic changes accompanied the ice ages, each of which ultimately reversed itself, leaving in its wake newly evolved species adapted to the climatic conditions prevalent during the glacial periods. When the glaciers retreated, they opened the areas they had covered to recolonization by the species that had moved south.

Once the newly evolved species is reintroduced, what happens? If reproductive isolating mechanisms do not break down—that is, if speciation is complete—the new species and its established counterpart may continue to diverge, or they may also converge. The first requirement for the reintroduced species is availability of resources in the habitat. If resources are available and not already being used, there is no problem. But unused resources are not common. Usually, the new species finds that the resources it needs are already being used by the original species, if it is still present, and perhaps by many other species as well. Thus, the two species compete. The more similar the needs of the two species, the more intense will be the competition. Competition can be a very powerful selective force, and there may be a number of possible outcomes. We will look at a few of the possibilities in some detail.

First, the new species may outcompete the old species and bring about its local extinction. What we usually see in nature are the results of competition, not the actual competitive activity. This is so because competition usually centers around the utilization of resources. However, certain observations of two species of barnacles, *Chthamalus stellatus* and *Balanus balanoides,* made by J. H. Connell, indicate that some such competition occurs. Connell chose to study the location of barnacles on rocks along a portion of the coast of Scotland. By placing a pane of glass over the rocks, he was able to record the location of the barnacles by species. Upon returning to the rocks at a later date and placing the marked glass over the same rocks, he was able to see and record the changes that had taken place. The *Chthamalus* barnacle fastens itself to rocks in the high part of the intertidal zone, the *Balanus* in the lower part of the zone. (See fig. 8.9.) In the larval stage of their development, both species resemble tiny shrimp, and before they can assume their adult form, they must cement themselves to a rock. The larval *Chthamalus* very often do so in the lower zone, and Connell recorded their presence there. Upon later investigation, however, he found that either the larvae had been crowded

FIGURE 8.9 Barnacle interactions observed by Connell. *(Redrawn from MacArthur and Connell after Connell, 1961.)*

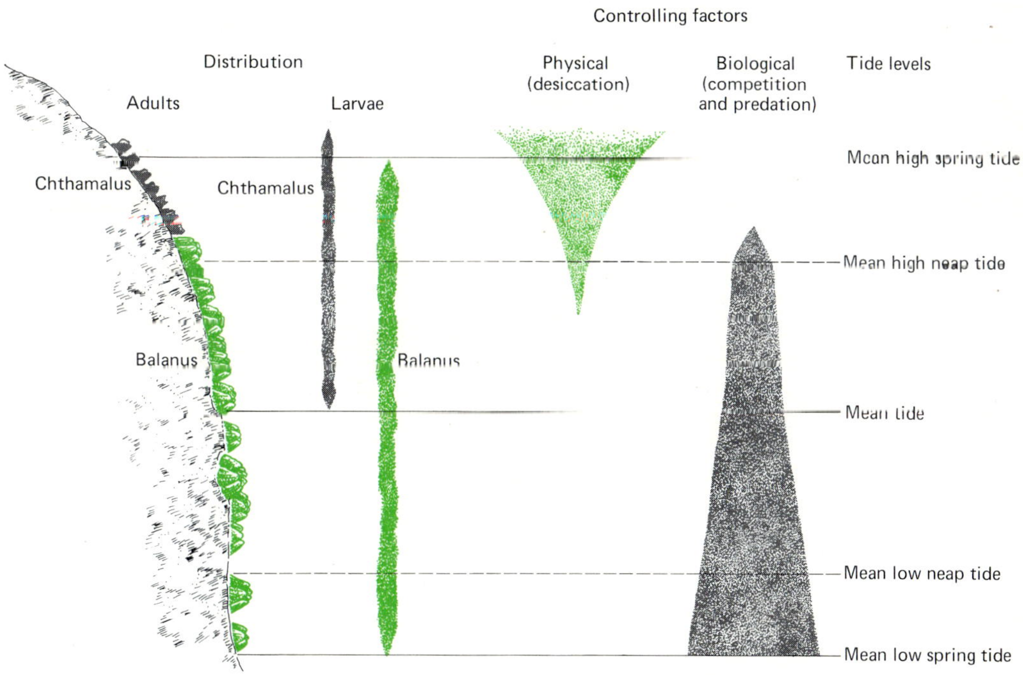

off the rocks or *Balanus* had grown over them. To determine whether it was actually competition that eliminated the *Chthamalus* or whether there were some other limiting factors at work as well, Connell isolated *Chthamalus* from contact with *Balanus* and found that the former grew in the lower zone with no difficulty. Thus, he proved that the reason *Chthamalus* were not found in the lower zone in the presence of *Balanus* was that they were ''outcompeted'' by *Balanus.* In this case, it appears that the competition is direct, but it may be seen as a competition for space, which is a limited resource. The faster-growing *Balanus* ''wins,'' because it occupies the space more rapidly in the lower zone. Connell also found that in the higher zone, *Chthamalus* outcompeted *Balanus,* because it did not need another resource, water, as much as *Balanus.* In the higher zone, *Balanus* was subjected to greater exposure to the air and thus to drying out.

Another result of the reintroduction of a species may be that the established species outcompetes the new one, which then suffers local extinction. This occurs frequently, simply because the old species is so well adapted to its environment that it makes the most efficient use of the resources available. The newer species is also often at a clear numerical disadvantage. If the numerical difference is great, the less populous species almost invariably loses.

It should be noted that biologists believe that severe competition takes place only in the initial stages of interspecific contact, and is often quickly replaced by a new balance of species in which competition is avoided. The fact that competition is frequently avoided in no way reduces its importance. In fact, the relative infrequency (if it is infrequent) of competition would seem to support the view that natural selection favors the evolution of mechanisms that reduce competitive severity.

One of the other possible effects of the reintroduction of species is *niche partitioning.* In this case, the old species might ''surrender'' a portion of its niche to the new species. Predators, for example, rarely depend upon a single spe-

FIGURE 8.10 Although superficially similar, the feeding habits of warblers living in pine trees differ sufficiently to prevent competition. *(Redrawn from Helena Curtis,* Invitation to Biology, *Worth Publishers, Inc., New York, 1972, p. 459.)*

cies as a source of food. Most predators have several target populations, and when faced with competition from other predators for one of the targets, they can shift their primary dependency to another one. Naturally, this can take place only when the niche of one of the species is sufficiently broad to permit the species to relinquish a part of it without suffering a decline in population that would threaten survival. Examples of *niche partitioning* are difficult to document, because like examples of competition, we often see only the results. In North America, there are five or six species each of tits (*Parus*) and of warblers (*Dendroica*) that can coexist in the same piece of woodland because of the specialization of their niches. It is assumed that this was made possible because the different species were able to share their niches.

The cases of *Parus* and *Dendroica*, however, may be ones in which no competition took place because of already well-defined niches. It is possible that a habitat may be heterogeneous enough to support two or more similar species if their niches are different. In this case, the reintroduction of a newly evolved species will not create a competitive situation. Robert MacArthur has carefully studied several species of warblers in North America, and he found that although they occupy the same habitat, their niches are sufficiently well defined to prevent any competition. The warblers have very similar food requirements, but each species feeds in a particular part of the forest canopy, and there is a minimum of overlap. In a given tree, one species of warbler may be feeding on the inside lower branches of the tree, while a warbler of another species is feeding on the outside upper branches. (See fig. 8.10.) In one of the more arresting cases of niche specialization, two species of parasitic wasps with identical requirements were observed to coexist for more than seventy generations in the same habitat without competing. Observations have led to the conclusion that one species is successful when the host organism is at a low population density, and the other is successful when the population is at a high density.

As the process of speciation occurs over and over again and new species are reintroduced into communities, the new species bring about specialization in the community and they also *increase the diversity of the community.* If the available habitat contains exploitable resources, and if the conditions of the physical environment are not too rigorous, a very diverse community will result. The new species may even provide a new habitat for other new species and further increase the diversity. Such highly diverse communities are, as we will see in Chapter 14, very well balanced, both physically and biologically. Just as cells are highly organized aggregations of specialized chemicals, and organisms highly organized aggregations of specialized cells, diverse communities are highly organized aggregations of specialized species populations. For reasons that we shall see frequently in later chapters, such organization is of *great* importance to man.

Selected Readings

The readings for this chapter and the last (chapter seven) are combined here, since this chapter is an extension of chapter seven, and because the topics treated in these two chapters are covered in their entirety by most of the books that are listed below.

*Brosseau, George E., Jr., ed. *Evolution,* Dubuque, Iowa: William C. Brown Company, Publishers, 1961. This anthology of original sources includes several items of note, such as extracts from Lamarck, Lyell, Wallace, and Malthus.

Ehrlich, Paul R., and Holm, Richard. *The Process of Evolution.* New York: McGraw-Hill Book Company, 1963. This is a general modern treatment of evolution for the introductory student.

Dobzhansky, Theodosius. *The Genetics of the Evolutionary Process.* New York: Columbia University Press, 1970. A rigorous and authoritative discussion of evolution by one of the world's foremost students of the subject.

Lerner, I. Michael. *Heredity, Evolution, and Society.* San Francisco: W. H. Freeman and Company, Publishers, 1968. A clearly written, fascinating discussion of human genetics and evolution.

MacArthur, Robert, and Connell, Joseph. *The Biology of Populations.* New York: John Wiley & Sons, Inc., 1966. This is a first-rate interesting text, with many clear examples.

Mayr, Ernst. *Animal Species and Evolution.* Cambridge, Mass.: Harvard University Press, 1963. A comprehensive and absorbing treatment of evolution written by a leading taxonomist. A shorter version, under the title *Populations, Species, and Evolution,* is available as a paperback.

Simpson, George Gaylord. *Horses.* New York: Oxford University Press, Inc., 1951. A classic work on the evolution of a single animal.

*Stebbins, G. Ledyard. *Processes of Organic Evolution.* Englewood, N.J.: Prentice-Hall, Inc., 1971. This brief, basic account deals particularly well with the subjects of speciation and plant evolution.

Wickler, Wolfgang. *The Sexual Code.* Garden City: Doubleday Publishing Company, 1972. Stressing the evolution of behavior, and especially sexual behavior, Wickler draws widely on examples of evolution in the invertebrates and vertebrates.

Wallace, Bruce. *Topics in Population Genetics.* New York: W. W. Norton & Company, Inc., 1968. This text was written as a springboard for the author's students, to enable them to go on to more advanced topics in evolution. A background in mathematics is virtually a prerequisite.

*Wilson, Edward O., and Bossert, William. *A Primer of Population Biology.* Stamford, Conn.: Sinauer Associates, 1971. This book contains what may be one of the best brief treatments of population genetics and evolution available. It is lucid and concise, although the student should have a background in mathematics to realize the full scope of the presentation.

Allison, Anthony. "Sickle Cells and Evolution." *Sci. Am.,* Vol. 195 (August, 1956), pp. 87–8. Offprint number 1065.

Cavalli-Sforza, Luigi Luca. " 'Genetic Drift' in an Italian Population." *Sci. Am.,* Vol. 221 (August, 1969), pp. 30–37. Offprint number 1154.

Wills, Christopher. "Genetic Load." *Sci. Am.,* Vol. 222 (March, 1970), pp. 98–107. Offprint number 1172.

*Paperback edition available.

part three
the functioning of organisms in ecosystems

All of the energy used in the biosphere is derived from the sun, shown here in eclipse. *(Courtesy of Ira Kirschenbaum.)*

energy—the fuel of organisms

chapter nine From the beginning, we have stressed the importance of energy in living systems. In some instances, we described the adaptations to energy-gathering among herbivores and carnivores or the processing of environmental information as an adaptation to locating food (energy) and avoiding predators (that is, becoming food). In our discussion of the operation of genes, one of the substances that we dealt with was glucose, described as a "source of energy" for *E. coli*. In our discussion of plants, we saw that the production of energy-rich compounds by photosynthesis was an extremely important activity.

It should be pointed out that none of the activities that we have thus far described can be accomplished without an input of energy. The transmission of a nerve impulse at the synapse, the contraction of a muscle, the transcription of a molecule of RNA on the DNA template, the movement of flagella by a dinoflagellate, and the production of a seed in a gymnosperm are all activities that require the expenditure of energy. It follows, then, that energy is required for the survival of organisms and, therefore, for the functioning of communities and ecosystems. In short, energy is required to drive all of life's processes. In order that we may grasp some measure of understanding of life, then, we must first understand something about energy.

Energy, Work, and Heat

Energy is the capacity to do work. Work is something that can be measured and is defined as *the product of a force exerted to move an object some distance*. Implicit in the definition of energy as the "capacity" to do work is an important idea: Although some object might have had energy applied to it, the energy might not necessarily produce immediately measurable work; that is, *energy can be stored.* A coiled spring, for example, is one to which energy has been applied and in which energy has been stored, but not yet released. *Energy that is stored in a system and that may be made available to do work* is called *potential energy.* As the coiled spring unwinds, its potential energy can be used to do work. Obviously, from our definition, we cannot say that potential energy performs any work. To differentiate between potential energy and the energy that actually performs work, we use a different term to describe the latter—that is, *kinetic energy.* Kinetic energy is the *energy of matter that is in motion.* As a coiled spring unwinds, its potential energy is converted to kinetic energy.

We shall not dwell at length on the properties of atoms. For our purposes, it is enough to know that each chemical element possesses a characteristic quantity of energy. When atoms come together to form molecules, they may be acted upon by certain forces that cause them to bond together. The position of the atoms in a molecule in relation to those forces determines the potential energy of the molecule. Because such chemical bonds define the potential energy content of the molecule, we say that energy can be stored "in the bonds." *When certain chemical bonds are broken, energy is released*. The quantity of potential energy in a molecule depends upon the atoms that it contains and the strength of their attractions to each other. Some molecules, such as H_2O and CO_2, possess relatively little potential energy, whereas the chemical bonds in a molecule of glucose, as we shall see, possess a considerably greater amount of potential energy.

It should be understood that *work is not energy, but a form of energy transfer*. There is another form of energy transfer that is of great importance in living sytems—namely, heat. *Heat is the motion of molecules*. Even the most apparently motionless object we might observe is in motion at the molecular level. The movement of molecules is an essential feature of the *kinetic molecular theory,* which states that in a body at constant temperature the average motion of molecules is constant and their direction is random, and that in an object the motion will increase as heat is added. The *temperature* of any object is a measurement of the motion of its molecules. If heat is added to the object, the speed at which its molecules move will increase; energy has been applied, and work is done. The work performed is measured by an increase in temperature that reflects the increased velocity of the molecules.

The First and Second Laws of Thermodynamics

The study of the relationships between energy, work, and heat is called *thermodynamics* and has been formalized into a series of laws, the first two of which are of particular importance in understanding the operation of biological systems.

The *first law of thermodynamics* states that *energy can never be created or destroyed, but only converted from one form to another*. Implicit in the first law is the idea that there is a finite amount of energy in the universe. It can be shown that, for all practical purposes, all of the nonnuclear energy that is available on this planet has come from the sun. The energy conversions that take place on the sun produce light (and other electromagnetic radiation) which is itself a form of energy. When light energy reaches the earth, some of it is converted into the potential energy of chemical bonds by photosynthesizing plants. By tracing the subsequent fate of the potential energy that is stopped (or *fixed*) by plants, we can see the importance to man of the careful management of his energy resources.

Note that we have said that *some* of the light energy from the sun is fixed by photosynthesis. Obviously, some of the light falls on locations where there are few or no plants. Although such is the case, another law of thermodynamics tells us that we cannot even expect all of the light that reaches a plant to be converted into the potential energy of chemical bonds. The *second law of thermodynamics* states that *more energy is required to do work than can be stored or used in the work*. This implies that some energy is ''lost'' as energy transfers take place; the first law tells us that energy cannot be lost. Energy is never lost in any energy conversion or transfer; however, some energy is *converted into a useless form*. Recall our earlier discussion of the resting potential of the nerve cell membrane. We saw there that the sodium pump functions to maintain an imbalance of materials across the cell membrane; that is, it maintains a disequilibrium that allows the organism to function. When a nerve cell is killed, sodium flows rapidly into the cell until an equilibrium is established on each side of the membrane. In thermodynamic terms, what has happened is that the energy available to the cell has been converted into a useless form. Lacking usable energy to maintain the action of the sodium pump, an equilibrium is established, and no more work is done. A key implication of the second law of thermodynamics is that as greater amounts of energy are converted into a useless form, the amount of disorder in a system has a tendency to increase, atoms or molecules tend to move randomly and at equal rates, and little or no work can be done. The system runs down. Living systems, as we shall see, consist of aggregations of highly ordered chemicals, organized and structured to perform the work that is necessary for survival. At the

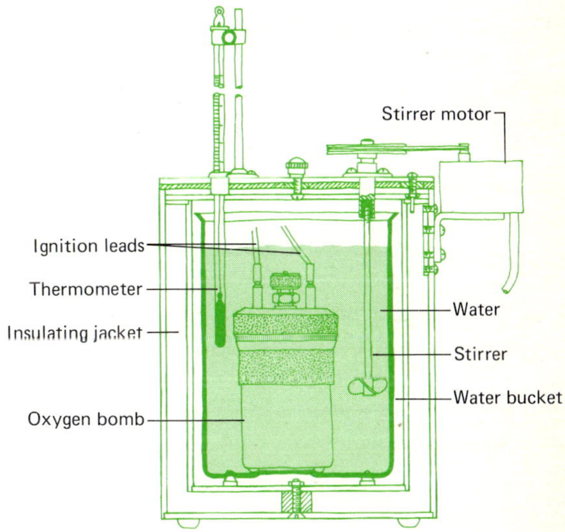

FIGURE 9.1 A bomb calorimeter is a device used to determine the energy produced by combustion of various materials. The heat liberated by the combustion is measured by the rise in temperature of the surrounding water. *(Redrawn, courtesy of Parr Instrument Company.)*

subcellular level, survival is a matter of carrying out a large number of energy-requiring and energy-yielding chemical reactions, but before we turn to the work of surviving, we shall consider the ways in which we can measure work in concrete terms.

Units of Measurement for Energy

Since energy, work, and heat are so profoundly related, the common means of measuring energy in a system is in terms of heat. (See fig. 9.1.) One measure of energy is in *gram calories* (abbreviated as *cal* or *gcal*). A gram calorie is defined as the *amount of energy required to raise the temperature of one gram of water one degree centigrade.** The "calories" that the general public is accustomed to dealing with, those used by commercial food processors and dieticians, are *kilogram calories (Cal* or *kcal),* the *amount of energy required to raise the temperature of one thousand grams of water one degree centigrade.* If a doughnut, for example, is said to contain 135 calories, it means that the complete release of energy in the doughnut by burning would raise the temperature of 135 kilograms of water one degree centigrade. (Presumably, food processors will continue to mean kcal when they say calorie, since the effect on sales of saying that a doughnut contained 135,000 calories might be too much to bear.) These units of measurement are useful in discussing biological energy production and transfer, whether we are discussing these events at the subcellular or ecosystem level.

*Sometimes the initial temperature of the water is defined. Thus, a cal raises the temperature of one gram of water from 14.5°C to 15.5°C. This is necessary because the heat capacity of water changes slightly as its temperature changes.

Biological Systems

To survive and reproduce, an organism must perform work, and among the many kinds of work that must be done are muscle contraction, digestion, and very importantly, the control of the thousands of chemical reactions that take place in the body of any animal or plant. The organism must perform to (1) regulate and insure the proper functioning of the body, (2) form the structural units of the body, and (3) store energy for future use. All of this obviously requires the organism to gather and process energy, and at the subcellular level, this is carried out by a number of highly organized and complexly structured chemicals. (See fig. 9.2.) At the risk of being redundant, we may restate the previous idea this way: Living systems require a constant input of energy, and require it because of the phenomena described by the second law of thermodynamics.

The many chemical reactions in the body of an organism may be grouped into two classes: those that require an input of energy, and those that release potential energy. It is important to understand that *the potential energy released by chemical reactions can be used to drive reactions requiring energy*. Indeed, the second law tells us that energy-requiring reactions will not take place unless the energy released in other reactions is greater. That portion of excess energy which is not used in the work of synthesizing molecules by energy-requiring reactions usually becomes heat. The input of energy into an organism, then, is used in reactions that either produce chemicals in which energy is stored or perform the work that the physiological state of the organism demands, while some energy, in accordance with the second law of thermodynamics, becomes heat.

ATP

Perhaps the most important molecule in the energy-processing systems of organisms is *adenosine triphosphate*, or *ATP*. Frequently called the universal energy currency of the cell, ATP is found in *every living thing*. The role of ATP is critical in the energy transformations that are necessary to sustain life. It is made up of three molecular components with which we are already familiar from our study of the gene: the purine base adenine, ribose, and phosphate. The bonding of adenine and ribose pro-

FIGURE 9.2 Although the burning of methane (CH_4) is not typical of energy release in cells, it illustrates some events in the process of oxidation of organic compounds. This gas is produced by the chemical degradation of organic matter such as plant material or feces. When burned it gives off a greenish light, and because it is a primary ingredient of marsh gas and often burns under the natural conditions existing in a swamp, it has sometimes given rise to false reports of flying saucers and UFO's.

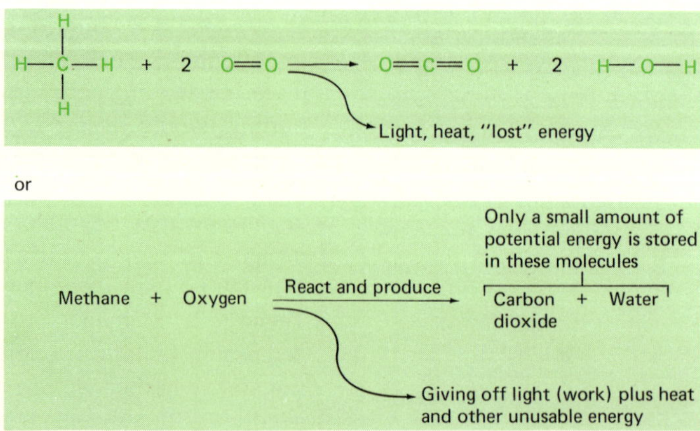

duces a compound called *adenosine*. (See fig. 9.3.) When three phosphates bond to the ribose side of adenosine, the ATP molecule is complete. When the bond between adenosine and the first phosphate is broken, the amount of energy released is not particularly large. However, the bonds between the central phosphate and the first, and between the central phosphate and the terminal one are so-called ''high energy'' bonds; when either of these bonds is broken, the amount of energy released is large. For that reason, they are denoted by the symbol ~ rather than the straight line that usually signifies a chemical bond. In the energy transfers of the cell, the bond between the central and terminal phosphates is usually the only one that is broken. When that occurs, ATP becomes *adenosine diphosphate,* or *ADP.* ATP is synthesized in the cell from ADP and inorganic phosphate (P_i) by a process called *phosphorylation*, and indeed, phosphorylation refers to the addition of phosphate to any molecule.

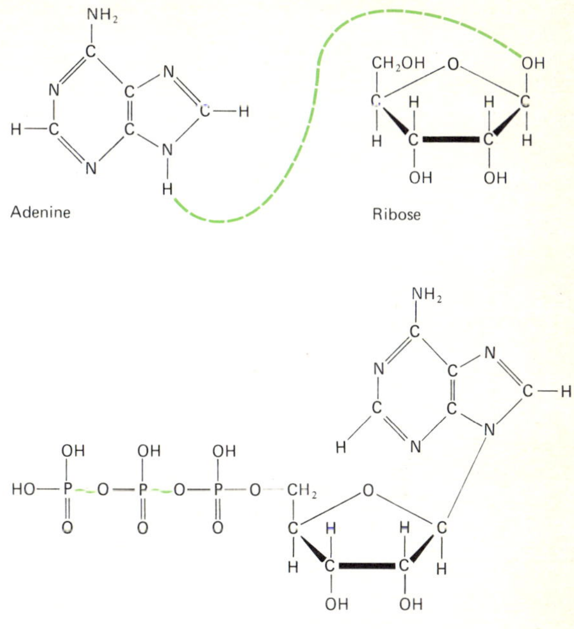

FIGURE 9.3 The joining of adenine and ribose forms adenosine, and the addition of three phosphates results in adenosine triphosphate or ATP. The two end phosphates are joined by ''high energy'' bonds (~), enabling the ATP to transport energy in the cell.

Energy Transfer and Food Chains

Although we have thus far concentrated on molecules and energy transfers in the cell, let us draw back for a moment and examine one of the consequences of the second law of thermodynamics for larger systems. (Although the same consequence is true for subcellular systems as well.)

Because more energy is required to do work than can be stored or used in the work, any transfer of energy is less than 100% efficient. This is just another way of saying that when energy transfers occur, some part of the energy initially present in the system becomes heat. In biological systems, all energy transfers start with the potential energy fixed in photosynthesis. Thus, the efficiency of photosynthesis is critical to determining the total amount of energy available to all living forms.

Actually, to calculate the efficiency of any living system is simple in principle. After first determining the amount of energy available to do work, the amount of energy used to do work and the amount of energy stored are determined. Then, a calculation of the efficiency of the process can be made by simple arithmetic:

$$\frac{\text{energy stored} + \text{energy used to do work}}{\text{total energy available to do work}} \times 100 = \text{efficiency (\%)}$$

For example, assume that 100 kcal of light energy reaches a plant in some specified time period. If, during that time, the plant can store and use 5 kcal, we calculate its efficiency as

$$100 \times \frac{5 \text{ kcal}}{100 \text{ kcal}} = 0.05 \times 100 = 5\%$$

The energy in living systems is transferred from one organism to another along a *food chain*, a term describing the order in which one organism eats another. This is something of an oversimplification, however, since it implies a straight line. In fact, the relationships between the eaters and the eaten are more accurately described by the term *food web*. A food chain is shown in the diagram below; the direction of the arrows indicates the direction in which energy travels:

$$A \rightarrow B \rightarrow C \rightarrow D \rightarrow E$$

A food web, on the other hand, is represented by quite a different diagram. (See fig. 10.14.) Each time one organism eats another, it both uses and stores some of the energy originally stored in the organism eaten. Obviously, all food webs must start with the potential energy fixed in photosynthesis.

Of the light that reaches a plant, some heats the surface of the leaf, some is reflected, and some passes directly through the plant. The amount that can actually be utilized by the plant varies, of course, depending on the plant itself and a variety of other factors such as water, nutrient availability, and temperature. The amount of light used in photosynthesis can range from a low of 1% to a high of 30%, but high figures such as 30% are *very* rarely achieved, usually only under very favorable laboratory conditions.

Assume, for example, that a particular food chain begins with a plant that can fix energy with 10% efficiency. If 1 million kcal of light energy are available to the plant, then only 100,000 kcal will be captured. Most of the energy captured will be used in metabolism, and only a small portion will be stored. Let us say that the amount the plant can store is also 10%. That means that only 10,000 kcal are available to other organisms in the community. If each successive organism to which the energy is made available is only as efficient as the plant, then clearly, there will be an immense decrease of available energy as it moves through food chains. At certain steps in the process, evolutionary adaptations have provided for increased efficiency, but it is thought that efficiency rarely exceeds 30%. Thus, an insect may be adapted to utilizing the molecules of a plant with a minimum amount of molecular restructuring and therefore a corresponding increase of efficiency. Similarly, when the insect is eaten by a bird, the bird may be able to utilize the molecules in the insect without conversion and thus not have to expend energy in restructuring molecules.

Although this discussion of food chains and webs is oversimplified, as we shall see in Chapter 10, it serves to emphasize the necessity of understanding the structure of biological molecules and the roles that such molecules play in the energy transformations of living systems.

The Chemicals of Organisms

Although we have frequently defined food as a source of energy, we have not said how energy is stored in foods or how it is released. It should be clear from our previous discussion that the energy value of foods depends on their molecular structure, and so we shall now examine the structures of the chemicals that constitute food, as well as the chemicals that are important in synthesizing and breaking down molecules of food.

It has been estimated that 99% of the matter in the universe can be accounted for by six chemical elements: hydrogen (H), nitrogen (N), oxygen (O), carbon (C), neon (Ne), and argon (Ar). Neon and argon are inert gases that, until recently, were not known to form compounds with any other elements. The other four elements are much more reactive, carbon being capable of forming at least half a million different compounds. These four elements, C, H, O, and N, constitute approximately 99% of the elements in living systems. The three important groups of biological molecules that we will discuss in this chapter, *carbohydrates, lipids,* and *proteins,* contain these four elements in abundance and are found in all living things. As we have seen, the nucleic acids also contain large amounts of them.

Carbohydrates

Carbohydrates are an important class of energy-storing compounds which contain carbon, as well as hydrogen and oxygen in the same proportion as in water (thus, *hydrate*). The chemical formula for a carbohydrate is $(CH_2O)_n$, with the subscript *n* signifying some number larger than two.

The simplest carbohydrates are sugars called *trioses* $(CH_2O)_3$. A single molecule of sugar is called a *monosaccharide* or, sometimes, a *simple sugar.* If a monosaccharide contains four carbons, it is called a *tetrose*; if it contains five carbons, a *pentose*; if six, a *hexose*. Biologically, trioses, riboses, pentoses, and hexoses are of greatest importance, and we shall restrict our discussion to these molecules. (See fig. 9.4.)

When two sugars bond together, they form a *disaccharide*, and as many more sugars are added, a *polysaccharide* is formed. Probably the

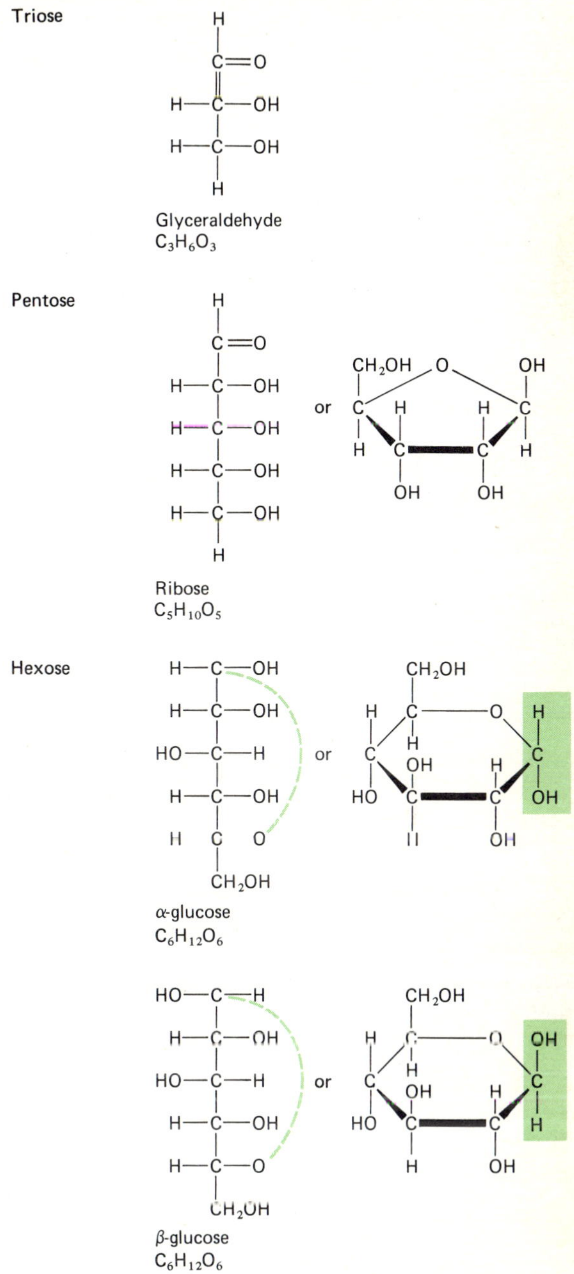

FIGURE 9.4 A triose, a pentose, and a hexose. The molecules may have linear or ring forms.

most familiar disaccharide is *sucrose*, or common table sugar, which is made up of one glucose molecule and one fructose molecule. The sugar found in grapes and some other fruits is *maltose*, a disaccharide consisting of two glucose units. Disaccharides are formed when two monosaccharides undergo a *condensation reaction.* In such reactions, a molecule of water is removed from the two sugars, and they form a new bond where the water molecule was removed. (See fig. 9.5.)

The monosaccharide of greatest interest to us is *glucose*. Glucose occurs in two forms, alpha glucose and beta glucose. Condensation reactions between units of the first form produce a polysaccharide we call *starch,* as shown in figure 9.6. Because of the slightly different structure of the second form of glucose, a series of condensation reactions will produce another polysaccharide, *cellulose.* (See fig. 9.7.)

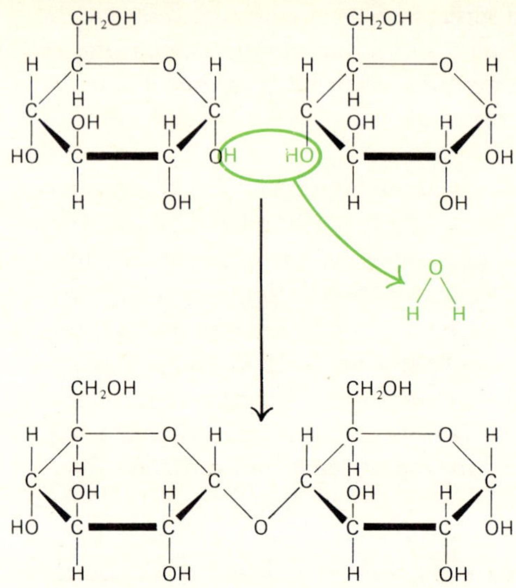

FIGURE 9.5 Condensation reaction.

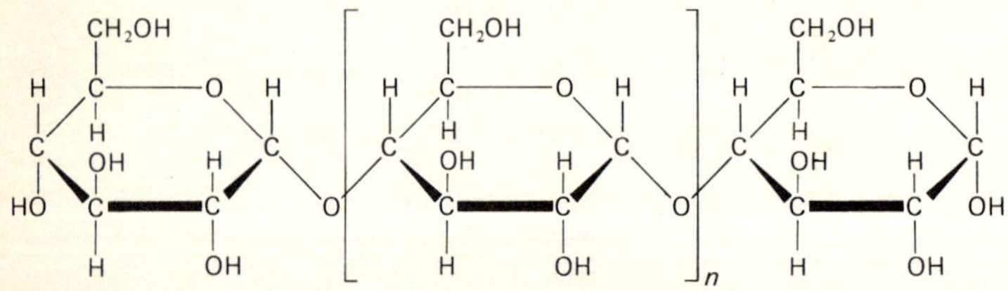

FIGURE 9.6 The chemical structure of starch. Starch, a large polysaccharide, consists of repeated units of α-glucose bonded together by condensation reactions in the manner shown in figure 9.5. The bracketed segment indicates some number, *n*, of linked monosaccharides. Contrast the structure of starch with that of cellulose in figure 9.7.

FIGURE 9.7 The chemical structure of cellulose. Unlike starch, cellulose is constructed of repeating units of β-glucose. Note the differences in positions of the hydrogen and oxygen bonds in comparison with starch.

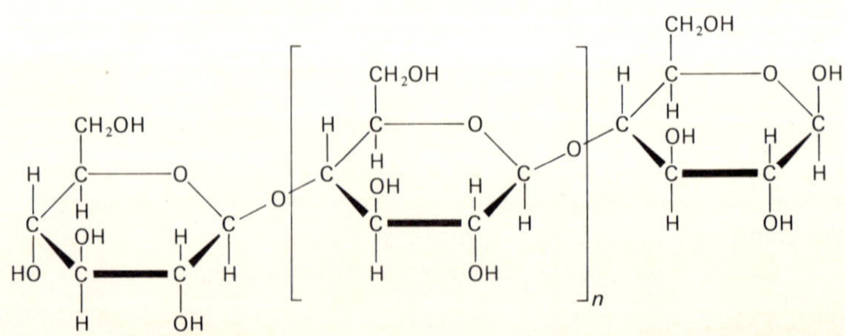

Both of these polysaccharides are synthesized by plants, and both require approximately the same amount of energy for synthesis. The importance of cellulose, as we have seen, is its structural contribution to the plant cell wall. Very long chains of cellulose are synthesized, and these chains are organized into bundles, rather as a rope is constructed of smaller fibers. The long bundles of cellulose are then deposited on the exterior of the cell to provide its characteristic rigidity. Once formed, the bonds of a molecule of cellulose are very stable and are insoluble in water. The resistance of cellulose bonds to breakdown means that potential energy is more easily released from starch molecules than from cellulose, since starch is more soluble. Some bacteria and protozoa have evolved means of releasing the energy stored in cellulose, but most animals utilize the more soluble starch as a source of energy.

In animals, a polysaccharide called *glycogen* plays approximately the same role as starch does in plants and is sometimes called "animal starch." Like starch, glycogen is a carbohydrate consisting entirely of glucose molecules bonded together by condensation reactions. Both glycogen and starch are large, branched molecules, with the branches being the result of condensation reactions that take place between different carbons of the glucose units, as shown in figure 9.8. One of the key differences between glycogen and starch is that the latter is less highly branched than the former.

FIGURE 9.8 Formation of branched chains of starch and glycogen. When a reaction takes place between the hydroxyl of the first carbon in a molecule of glucose and the six carbon in another, a branch point is formed as shown. This branch is characteristic of amylopectin, a form of starch. Similar branches form glycogen, and the key difference between starch and glycogen is the number of branch points.

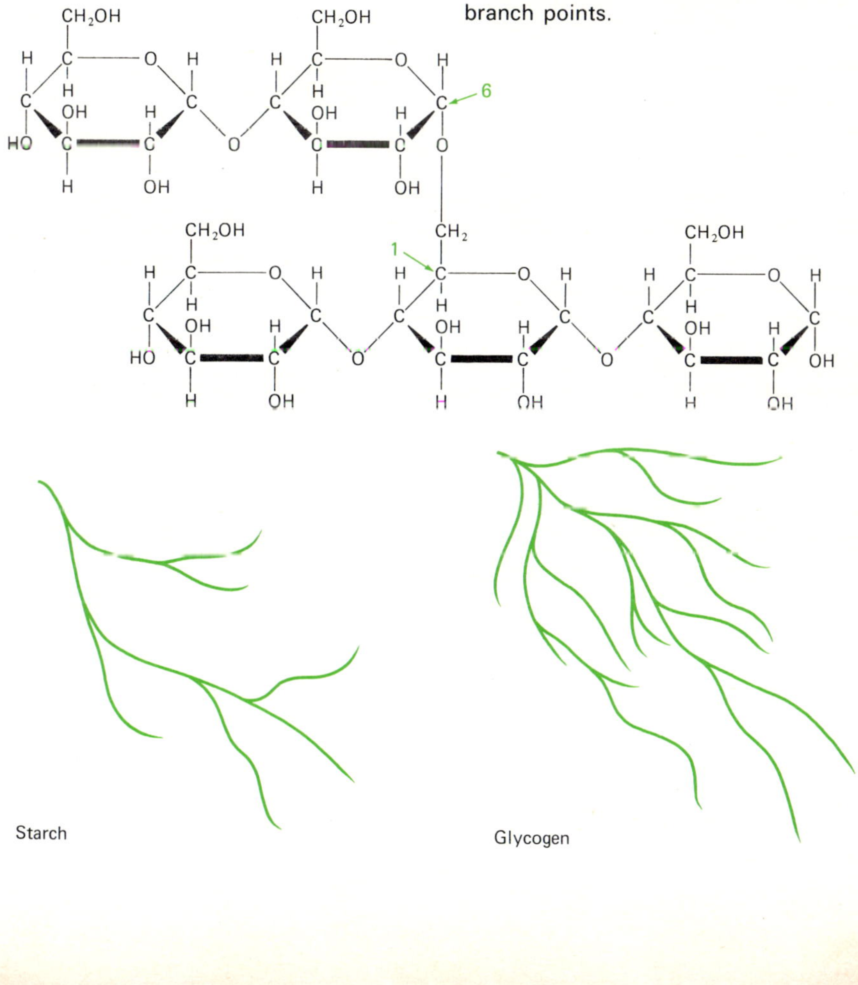

Once energy is stored in the bonds of a polysaccharide, such as starch, it can be retrieved by the addition of a molecule of water at the location where the condensation reaction took place. This reaction is called *hydrolysis*. We may write the reactions for the synthesis of starch and for its hydrolysis as

$$2C_6H_{12}O_6 \rightarrow C_{12}H_{22}O_{11} + H_2O$$
$$C_{12}H_{22}O_{11} + H_2O \rightarrow 2C_6H_{12}O_6$$

In a laboratory, a biochemist can synthesize several carbohydrates, but when he does so, they are found to be about equally divided between "right-handed" and "left-handed" forms, called respectively *D* and *L*. (See figs. 9.9 and 9.10.) In nature, almost all biologically active molecules are found in the *D* form. Whether this has been true since living systems arose on this planet, or whether molecules were once found in both forms and the *D* configuration somehow "won" the evolutionary struggle, is unknown. It is known, however, that although the contents of two molecules might be identical, the biological activity of the *D* form differs from that of the *L* form. As far as we know, plants and animals in nature can utilize only molecules in the *D* configuration. If molecules of both configurations were synthesized in living systems, two alternative situations could exist. First, plants and animals would require two metabolic systems, one for *D* molecules and one for *L*. Second, plants that synthesized only *D* molecules could be eaten only by animals with metabolic systems adapted for utilizing *D* molecules, and *vice versa* for *L* molecules. It would be interesting to speculate on what either situation might be like, but all in all, the situation that has evolved is a most economical one.

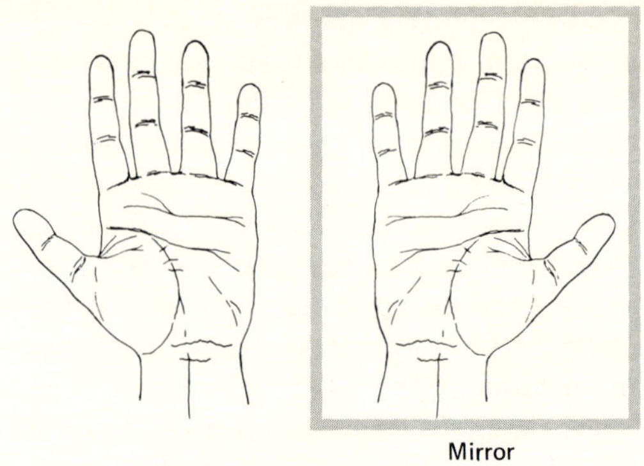

FIGURE 9.9 Structure and mirror images. Your right hand is a mirror image of your left hand, but they are structurally different. There is no way that you can move one hand so that it will be identical in configuration to the other. Similarly, although two molecules may contain the same elements bonded to one another, they may be structurally different because they are mirror images of one another, as in figure 9.10.

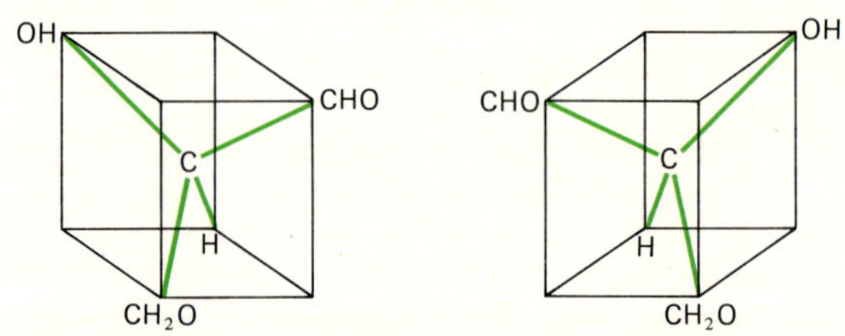

FIGURE 9.10 Although these two molecules of glyceraldehyde ($C_3H_6O_3$) have the same chemical elements, they are mirror images of one another. By convention, one form is considered to be right-handed *(D)* and the other left-handed *(L)*.

Lipids

Lipids are another important class of energy-storing molecules found in living systems. In addition to differences in their molecular structures, lipids differ from carbohydrates in that they may also contain nitrogen (N) and sometimes phosphorus (P), as well as large amounts of C, H, and O. Moreover, the proportion of oxygen in relation to hydrogen is much lower in lipids than it is in carbohydrates. Lipids also differ from most carbohydrates in that they are insoluble in water, but soluble in oils. Basically, there are two components in most lipids: *glycerol* and three *fatty-acid chains*. A fatty acid is a long chain of carbon and hydrogen with a carboxyl (COOH) group attached to one end. (See fig. 9.11.) A condensation reaction takes place between the hydroxyl (OH) group of the glycerol and the hydroxyl group at the carboxyl end of the fatty acid and thus forms the components into a lipid. (See fig. 9.12.)

There are an immense number of fatty acids, containing from four to more than twenty-four carbon atoms. When the fatty acid contains the maximum number of hydrogen atoms that can be bonded to the carbons in the chain (four bonds for each carbon), it is said to be *saturated* and, at room temperature, is usually a solid. Suet, for example, is a saturated lipid. When the carbon atoms within the fatty-acid chain form double bonds with each other and therefore do not bond with hydrogen, the fatty acid is said to be *unsaturated* and, at room temperature, is usually an oily liquid, such as corn oil.

Lipids are an essential part of the diet of any organism and may be taken in directly or synthesized by the cell from available nutrient molecules. Like carbohydrates, lipids are synthesized by condensation reactions and broken down by hydrolysis.

Phospholipids, which are important constituents of the cell membrane, are similar to the lipids

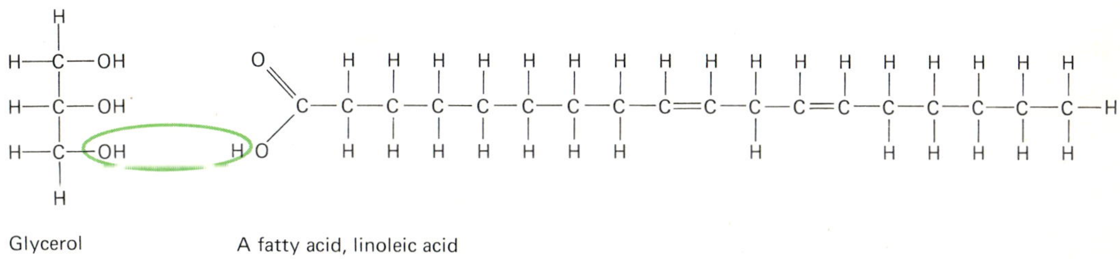

FIGURE 9.11 Glycerol and a fatty acid, showing the condensation site.

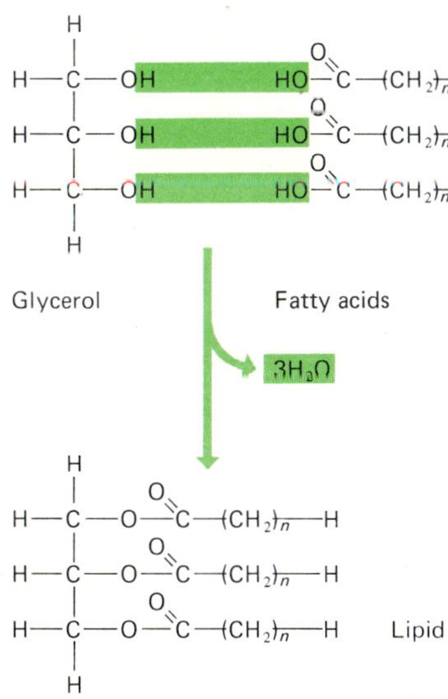

FIGURE 9.12 The synthesis of lipids, with three fatty acids joined to a glycerol molecule (CH_2). The subscript *n* refers to a number of connected CH_2 groups joined end-to-end as in figure 9.11.

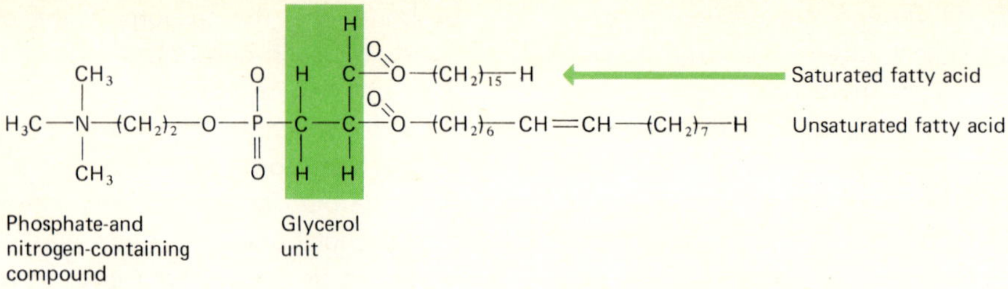

FIGURE 9.13 Phospholipid. One important property of the phospholipid shown is that the phosphate-containing end of the molecule is hydrophilic, that is, ''water loving,'' while the fatty acid end of the molecule is ''hydrophobic,'' or ''water fearing.'' Within the cell membrane, phospholipids thus orient themselves in a particular direction forming a bimolecular layer of lipids sandwiched between layers of proteins. The arrangement is shown in figure 9.14.

FIGURE 9.14 Structure of the cell membrane. Although it is yet to be conclusively proved that the cell membrane is constructed as it is shown here, this model continues to be supported by direct and indirect experimental and physical evidence. The concentration of water inside and outside the cell causes the phospholipids to arrange themselves with the hydrophilic heads in the direction of the water.

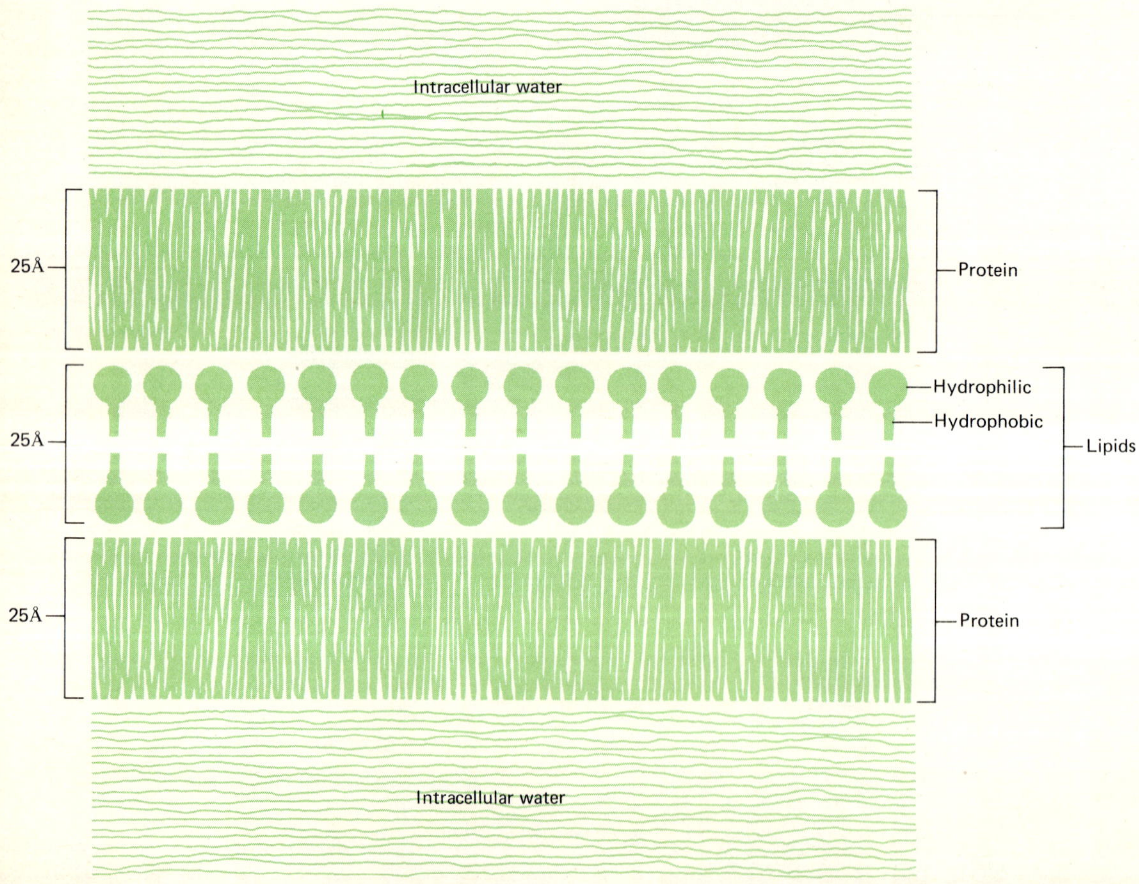

already described, except that a group containing phosphorus and nitrogen replaces one of the fatty acids bonded to the glycerol. (See figs. 9.13 and 9.14.)

Steroids are classed as lipids because their solubility characteristics are similar. However, unlike other lipids, they are not derived from glycerol and three fatty acids. Rather, they are characterized by four interlocking rings of carbon, with varying side groups attached. Biologically, the steroids include important vitamins and hormones and are frequently important structural elements of the cell membrane. The steroid shown in figure 9.15, *cholesterol,* forms the basis for many other steroids.

Proteins

Although proteins may be a direct source of energy for organisms, their importance in the energy transfers that take place in the cell lies not as much in their role as a source of energy as in the storage and release of energy from other molecules. Thus, although we shall have occasion to mention other functions of proteins as we progress toward an understanding of their structure, our primary concern shall be with those proteins that are involved in energy transfer.

Proteins differ from carbohydrates in several important ways. First, proteins are made up not of repeating units of the same molecules, but of repeating units of twenty different amino acids. Second, proteins assume several complex and markedly different shapes depending upon their function. Third, and most important, proteins are usually species-specific. All plants contain essentially identical starches, just as all animals contain essentially identical glycogen. Starch and glycogen perform the same function in different organisms. A protein that plays the same role in the lives of two different species, however, is usually structured differently in each species. The structural differences may be minute, but they are often substantial. Even within the same species, variations in protein structure can be found from one individual to another. As we have seen, no two individuals possess precisely the same genotypes (possibly excepting identical twins), and since the genes direct the synthesis of proteins, the order of amino acids in the proteins synthesized by any individual may be minutely different from that in another individual. We should recall in this context our

FIGURE 9.15 Cholesterol. Cholesterol is one of the most common steroids found in nature. Indeed, it forms the basis for most other known steroids, such as the sex hormones estrogen and progesterone, as well as a number of important drugs such as cortisone and the group of drugs called corticosteroids.

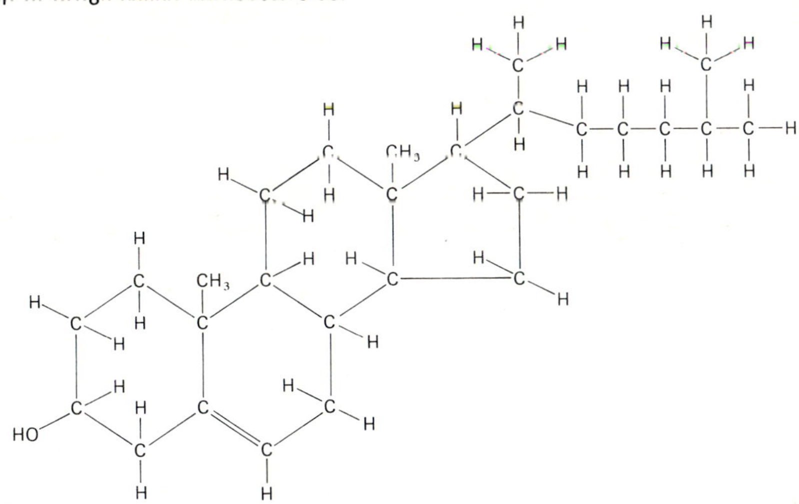

discussion of sickle cell anemia in order to point out that even within a single individual, we may find the same protein in more than one form. In short, the phenotypic variations found in a population reflect the different structures of the proteins produced by the individuals in the population.

We have mentioned amino acids frequently, and since they are the building blocks of proteins, we should begin with their structure. An acid, as we have pointed out, is characterized by a carboxyl group. Amino acids take their name from another functional group, the *amine group* (NH_3). These two groups bond with a carbon atom, and the remaining two bonding locations of the carbon atom are taken up by a single hydrogen atom and what is called a *radical* Ⓡ, a side group that determines which amino acid the basic structure will be. (See fig. 9.16.)

The structures of the twenty amino acids found in nature are given in box 9.1. It is not necessary to memorize these structures; they are presented here simply to give you an idea of how similar they are and how they proceed from simple to more complex structures.

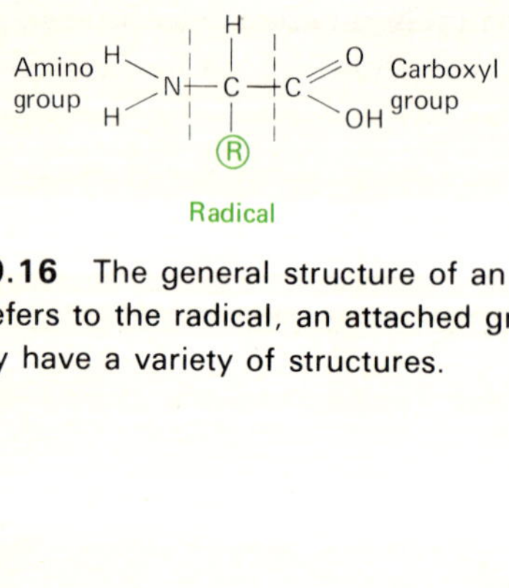

FIGURE 9.16 The general structure of an amino acid. Ⓡ refers to the radical, an attached group which may have a variety of structures.

The size of a protein is frequently expressed in terms of its *molecular weight* (MW), which is simply the weight of all the atoms in the protein. The molecular weight of proteins ranges from around 5000 to as much as 500,000 or more for a large protein such as myosin (MW 493,000). Insulin, a small protein (MW 5700), contains 51 amino acids, while some large proteins may contain as many as 10,000. Moreover, the number of different proteins is immense. In the intestinal bacterium *E. coli*, between 2000 and 3000 may be found. It is little wonder that so many geneticists believed that the proteins, and not the nucleic acids, carried the key to an understanding of the gene.

Like carbohydrates and lipids, proteins are formed by condensation reactions and are broken down by hydrolysis. The bond that is formed between the amino acids in a protein is called the *peptide bond* (hence the term *polypeptide*). Peptide bonds are always formed between the carboxyl and amine groups, and thus a polypeptide is characterized by a free amine (called the N-terminal) and a free carboxyl (called the C-terminal) at opposite ends. (See fig. 9.17.)

Considering the size of a protein, the determination of its structure is no easy task. The chemical formulas for a number of proteins were established in the nineteenth century. Hemoglobin, for example, was listed in 1866 as $C_{600}H_{960}N_{154}Fe_1S_3O_{179}$, but the determination of the arrangement of nearly 2000 atoms was far beyond the capability of nineteenth century chemical techniques. In fact, the complete structure of a protein was not worked out until the 1950s. Consider the problem: If a *small* protein contains fifty amino acids and there are twenty different amino acids to work with, the number of possible combinations is staggering. The analyst who tackles the structure of a protein must (1) determine how many amino acids are in the protein; (2) determine which amino acids are present (not all proteins contain all amino acids); (3) determine how many of each amino acid are present; (4) determine the cor-

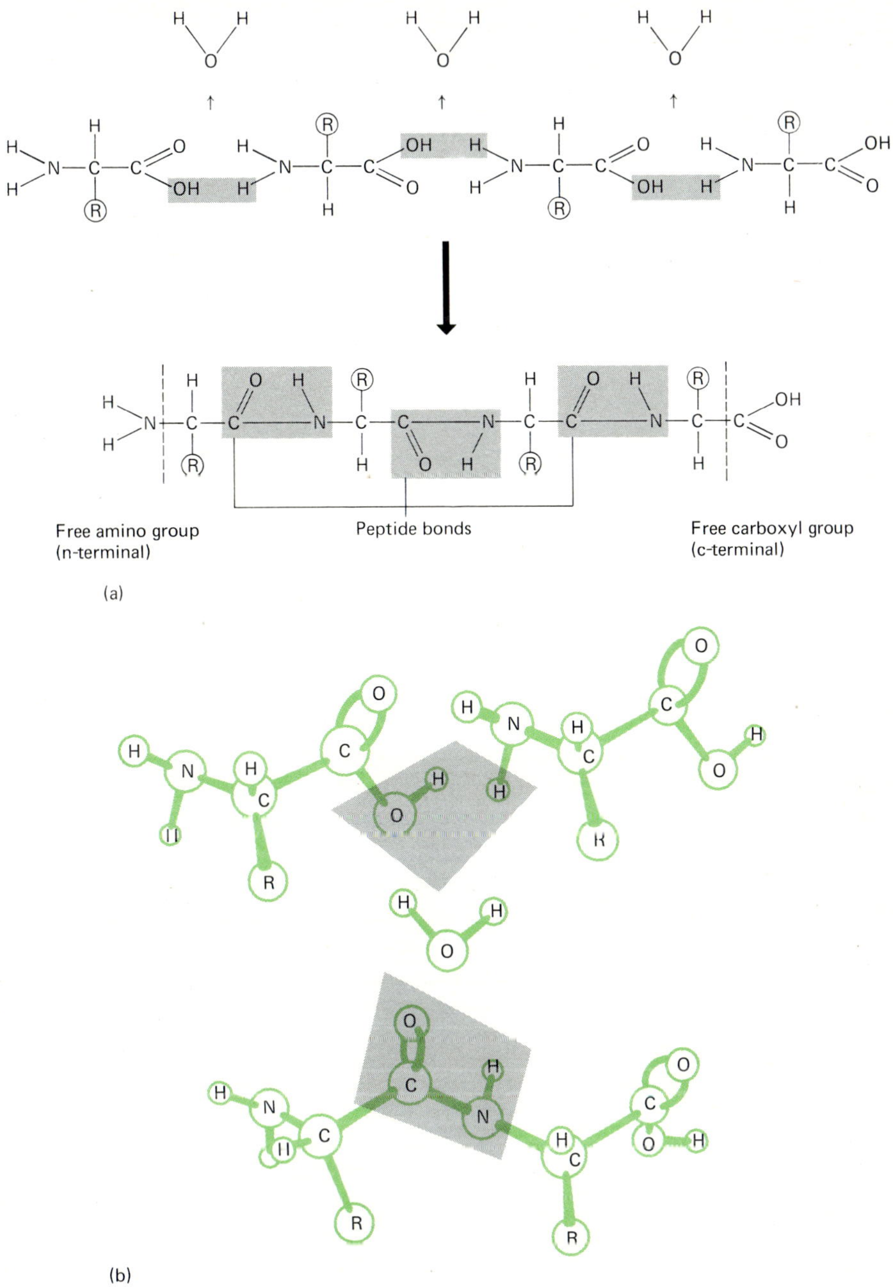

FIGURE 9.17 The peptide bond formed by condensation reactions between amino and carboxyl groups of amino acids. The formation of the peptide bond, like the formation of all chemical bonds, affects the structure of the molecules. Usually, the graphic representation of the peptide bond and the resultant polypeptide is done in a nonprojective way, as in (a). However, the three-dimensional representation (b) shows how the reactions change the conformation of the molecules.

The Structure of Amino Acids

In this box are shown each radical (—Ⓡ) with the amino acids formed. Thus, when the radical is hydrogen (—H) the amino acid is glycine.

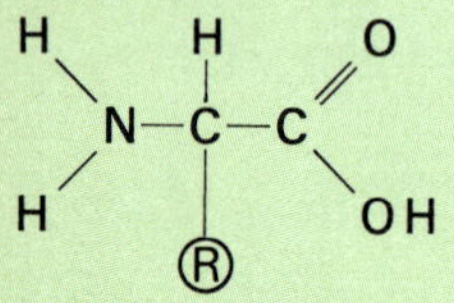

RADICAL	NAME
—H	Glycine
$-CH_3$	Alanine
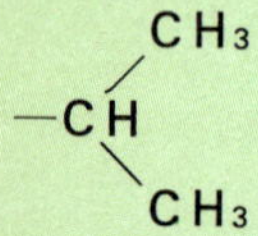	Valine
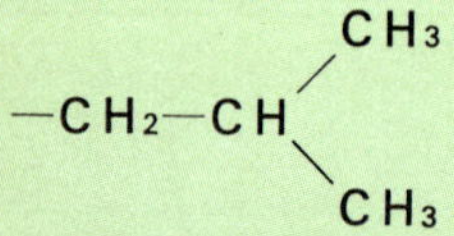	Leucine
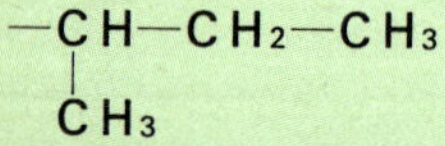	Isoleucine
$-CH_2-OH$	Serine
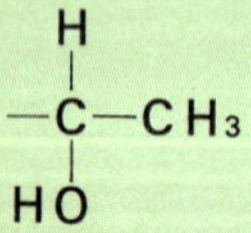	Threonine
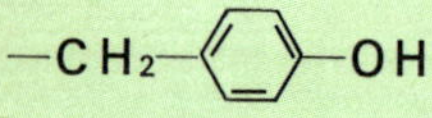	Tyrosine
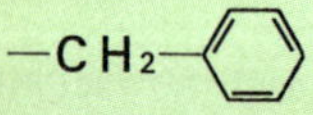	Phenylalanine
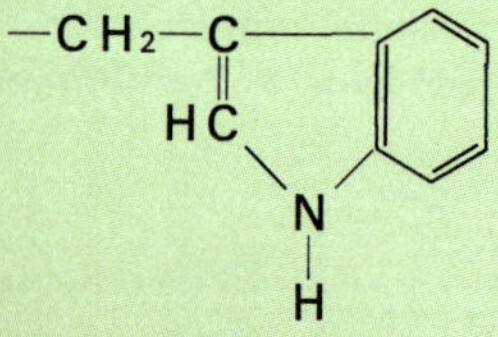	Tryptophan

RADICAL	NAME
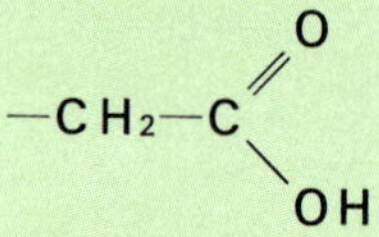	Aspartic acid
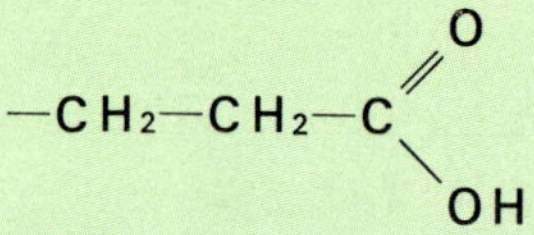	Glutamic acid
$-CH_2-CH_2-CH_2-CH_2-NH_2$	Lysine
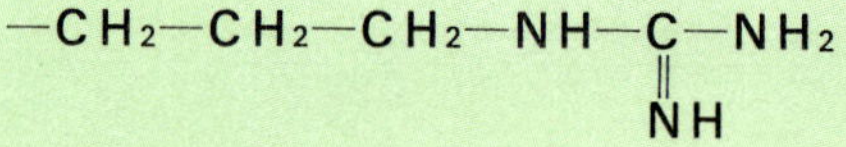	Arginine
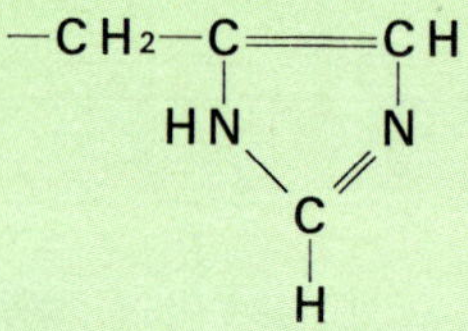	Histidine
$-CH_2-SH$	Cysteine
$-CH_2-CH_2-S-CH_3$	Methionine
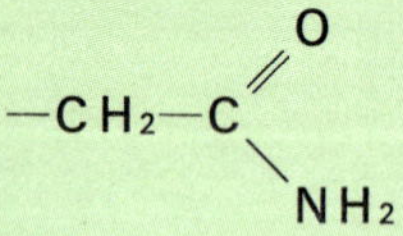	Asparagine (AspNH₂)
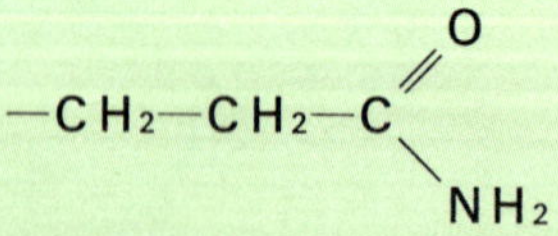	Glutamine (GluNH₂)
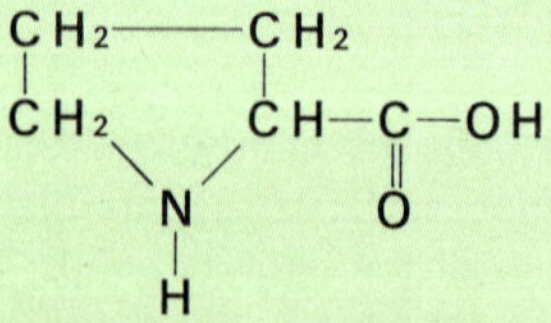	Proline

BOX 9.1

rect sequence of amino acids; and (5) establish how that sequence determines the natural shape of the protein.

The first determination of the structure of a protein was accomplished by a British biochemist, Frederick Sanger, for the protein insulin. Prior to its structural analysis, the chemical formula had been worked out as $C_{254}H_{377}N_{65}O_{75}S_6$. It had also been learned that insulin contained fifty-one amino acids. As proteins go, it offered a relatively accessible, small problem. But the analysis had never been done before, and it took Sanger ten years of painstakingly careful analysis and the assistance of several talented biochemists to solve the structure of a small protein. First, he subjected the molecule to intense hydrolysis, thus separating each amino acid. Of the twenty possible amino acids, only seventeen were present. Next, a more gentle hydrolysis yielded fragments of two, three, four, or more amino acids. Analysis of the fragments revealed the amino acid content, but not the sequence. Remember, in a polypeptide chain, there is an N-terminal and a C-terminal. Sanger developed a means of identifying the N-terminals of the fragments with a chemical "marker." This marker bonded to the amino group and, when the polypeptide chain was crystallized, identified the N-terminal by color. Then, by hydrolysis, the amino acid containing the marker could be removed from the fragment and identified separately. If, for example, a fragment contained four amino acids, A, B, C, and D, and the marker bonds with D, then D is the first amino acid in the fragment. After removing D, the reapplication of the marker will identify the next amino acid in the fragment. By sequentially adding the marker and removing it from fragments of the polypeptide, the entire sequence of the fragment can be established. Let us say that it is DBCA for the fragment we have hypothesized. Similar analyses of other fragments will produce similarly identified fragments. The remaining problem is to determine how the fragments fit together. By studying the *overlaps* in the sequences of the fragments identified, the sequence of the polypeptide can be worked out. For example, if the sequences of six fragments are found to be DBCA, STX, XYDB, AEF, BCAE, and TXY, although apparently a jumble of sequences, they can be put into an order that will reveal the sequence of a portion of the molecule that is ten amino acids long.

				D	B	C	A		
		X	Y	D	B				
	T	X	Y		B	C	A	E	
S	T	X					A	E	F
S	T	X	Y	D	B	C	A	E	F

Early in his research, Sanger applied the marker to the whole molecule and found that it bound to not one, but two amino acids. This strongly suggested that the insulin molecule contained not one, but two polypeptide chains. If two chains were present, the problem was determining how they were linked together. This was not easily solved, but it finally became clear that the linkage depended upon the structure of one amino acid, cysteine. If cysteine loses the hydrogen atom bonded to sulfur, it can bond with other amino acids. If two cysteine molecules lose their hydrogen atoms, they can bond to each other and form what is called a *disulfide bridge*. (See fig. 9.18.) The chemical formula of

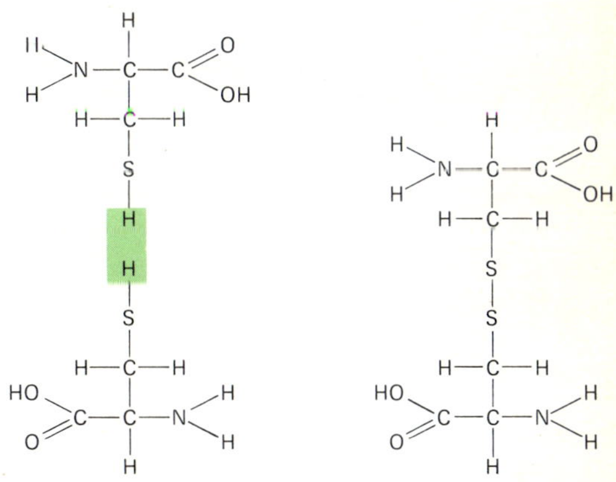

(a) (b)

FIGURE 9.18 Cysteine (a) and a disulfide bridge (b) joining two molecules of cysteine.

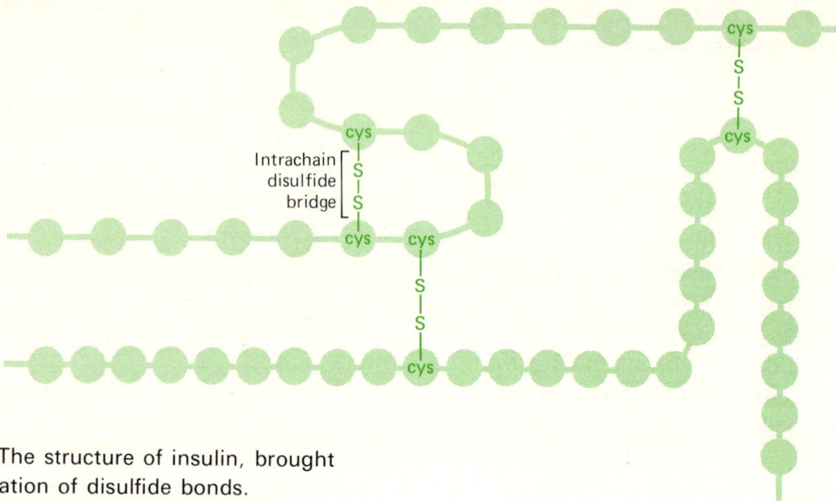

FIGURE 9.19 The structure of insulin, brought about by the location of disulfide bonds.

FIGURE 9.20 The helical (a) and pleated sheet (b) conformation of proteins. *(Reprinted from Linus Pauling,* The Nature of the Chemical Bond. *Copyright 1939 and 1940 by Cornell University. Third edition © 1960 by Cornell University. Used by permission of Cornell University Press.)*

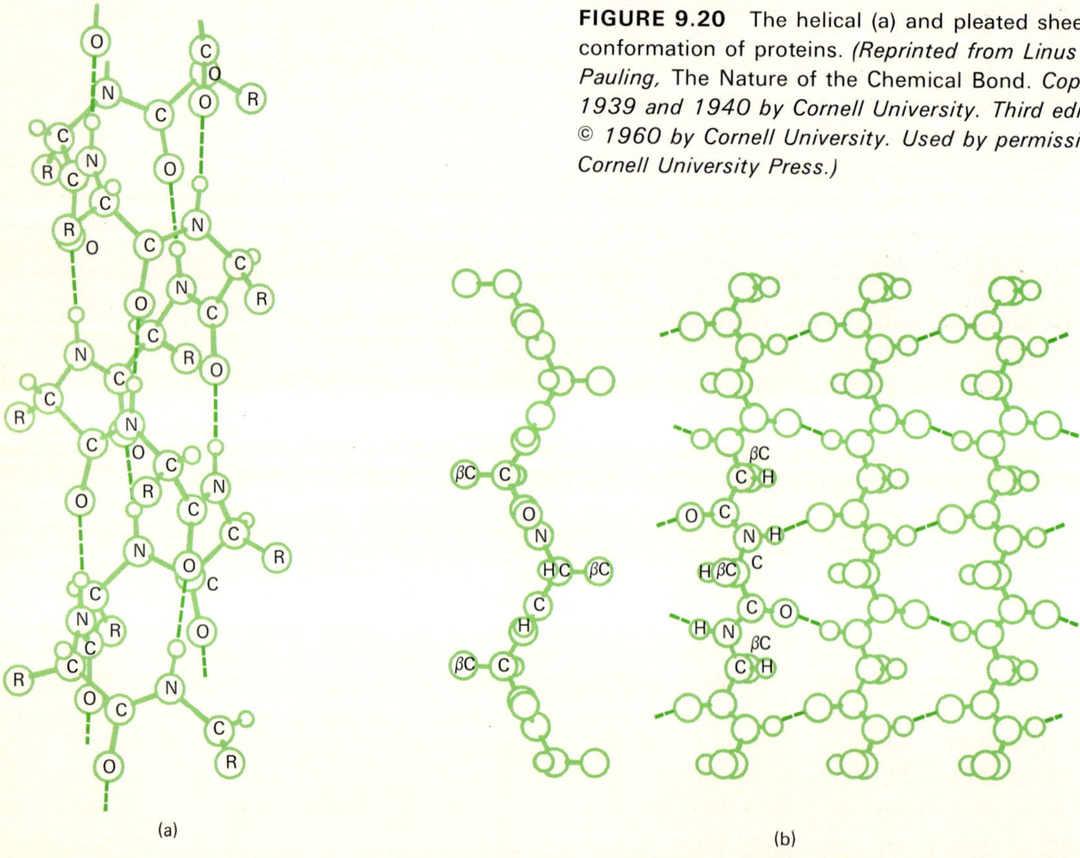

insulin suggests the existence of six cysteine molecules and three disulfide bridges. Such bridges could serve to hold the two polypeptide chains together. As Sanger learned, however, only two bridges connect the chains, one is an intrachain bridge that brings about a folding of one of the chains. (See fig. 9.19.) In 1958, four years after he completed his analysis of the structure of insulin, Sanger was awarded the Nobel Prize.

Classified in terms of structure, there are two major kinds of protein: *fibrous* and *globular*. When the oxygen atom of the carboxyl group and a hydrogen atom of the amine group exert a weak attraction for each other, the chains form a fibrous conformation, shown in figure 9.20 both as a helix and a ''pleated sheet.'' Alternatively, when weak attractions occur between the hydrogen atoms of the radicals, they bring about a complex folding of the molecule that gives it an approximately globular shape, as in figure 9.21. The shape of the protein is apparently very important in determining how it functions.

We said at the beginning of this section that the proteins of greatest interest to us are the ones that are implicated in storing and releasing energy from other molecules. The proteins that perform these functions are enzymes, to which we must now turn our attention.

FIGURE 9.21 Some proteins, such as lysozyme, shown here as a structural model, may have extremely complex configurations. Such proteins are referred to as globular proteins. *(Courtesy of J. A. Rupley, University of Arizona. Reprinted from* J. Biol. Chem., *Vol. 243, 1968, p. 1663.)*

Enzymes—Proteins That Control Energy in the Cell

In order to make a chemical reaction occur, energy, in some form, must be applied. One of the most common means of doing so is to add heat. However, because of the weak attractions of hydrogen atoms that hold proteins in their native shapes, the addition of heat in the body of an organism can be disastrous. A common protein, albumin, provides a familiar example. Albumin is the protein that makes up the largest share of egg white. When it is heated, the weak hydrogen bonds are broken and the albumin assumes a random shape, forming a coagulate that is insoluble in water. When the heat is removed, the albumin does not resume its native structure. This is an example of *denaturation*, a process that is usually irreversible. (See fig. 9.22.)

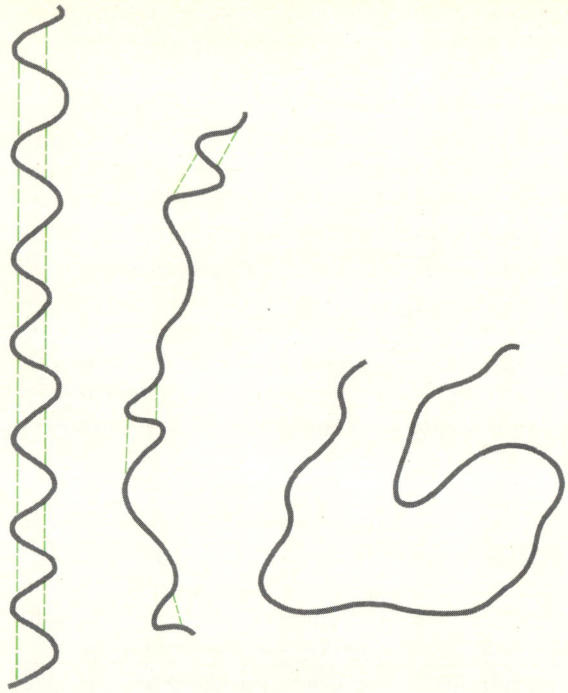

FIGURE 9.22 Denaturation. The weak hydrogen bonds that maintain the shape of a helical protein may break as heat is added, and the protein will gradually assume a random shape.

The energy that is required to start a chemical reaction is called *activation energy.* In the preceding example, heat was the activation energy that produced a denaturation of albumin. The function of enzymes is to lower the amount of activation energy required to drive chemical reactions and thus to enable an organism to carry out the many reactions necessary to maintain life within a temperature range that the body can tolerate. The amount of activation energy required to bring about a reaction depends, of course, upon the reactants, and it differs as the conditions under which the reaction is made to take place differ. Some of the factors that affect the quantity of activation energy necessary to drive a reaction are heat, pressure, and the concentration of the reactants or other products that will affect the reactants. It should be understood that activation energy is *not* a new form of energy being introduced here; it is a *quantity* of some form of energy.

The characteristic of many enzymes that is of overriding importance is that they are *catalysts,* or substances that react with other substances to bring about a chemical reaction, without undergoing any change themselves. Gaseous hydrogen and oxygen, for example, can react with the surface of a piece of platinum and then, while bound to the platinum, react with each other to produce water and electrical energy. The platinum is a catalyst and is in no way changed.

Or consider the reaction used in the Haber process to produce ammonia for industrial use:

$$N_2 + 3\,H_2 \xrightarrow[\text{iron oxide catalyst}]{\text{500°C, 15,000 lb pressure}} 2\,NH_3$$

To produce ammonia commercially requires very high temperatures, immense pressure, *and* a catalyst. But certain groups of bacteria do it regularly at *much* lower temperatures and pressures, through reactions that are mediated by enzymes!

The condensation reactions of carbohydrates, lipids, and other proteins are mediated by enzymes, and after the molecules of glucose are removed from starch, the potential energy that is stored in glucose is released in a series of reactions, all brought about by enzymes. Although the reactions that are necessary to sustain life might take place in cells under the conditions of concentration that exist there, in the absence of enzymes, it is thought that they would proceed so slowly that appreciable accumulations could not be possible without a great slowing down of the metabolism of an organism.

At the present, a detailed knowledge of the structure of some enzymes is available, and it appears that they work in a way similar to the platinum in our example; that is, the reactions take place on the *surface* of the enzyme. Unlike platinum, however, the surface of an enzyme is not homogeneous; it is not the same at any location. The reactions seem to take place only at a particular locus on the surface of an enzyme called its *active site.* The active site of the enzyme has a characteristic shape that complements the shapes of portions of the reactant molecules, or *substrates,* as a lock-and-key complement each other. (See fig. 9.23.) When the enzyme and substrate are brought together, they form an unstable complex that places a strain on the bonds of the substrate. The weakened bonds are susceptible to attack by free atoms or molecules in the cell. In hydrolysis, for example, the strained bonds of a disaccharide interact with water in the cell and restore the bonds that characterize the two monosaccharides.

Enzymes perform both kinds of reactions that we have discussed thus far—namely, condensation reactions and hydrolysis. Thus, they may be roughly classified as either *adding* or *splitting enzymes*, depending upon whether they add a fragment to a molecule or remove one. If the enzyme is active in hydrolysis, it is said to be *hydrolytic*; if it hydrolyzes the peptide bond, it is *proteolytic*. Since the pathways by which both large and small molecules are synthesized and degraded are often different, different en-

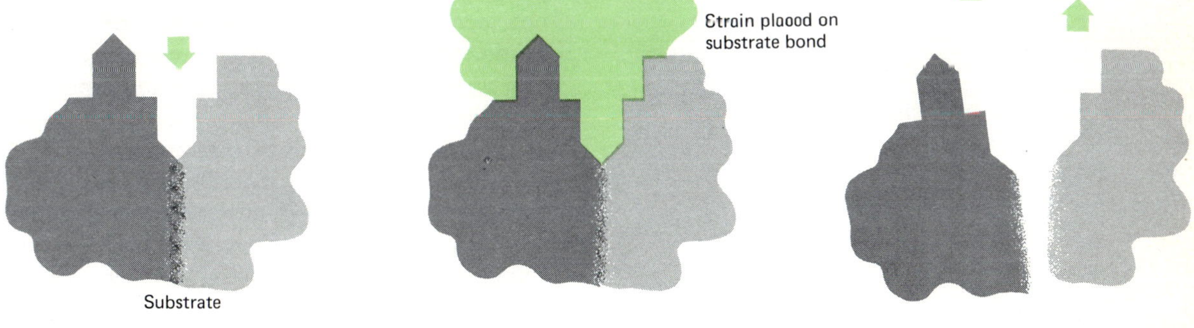

FIGURE 9.23 Enzyme-substrate interaction.

zymes may be necessary for the two processes. Occasionally, however, the same enzyme can perform both functions.

A simplified view of the structure of the enzyme *chymotrypsinogen* is shown in figure 9.24. Although it is an enzyme, it is catalytically inactive. It becomes active when another enzyme reacts with it and splits the chymotrypsinogen molecule. In its active state, chymotrypsinogen is called *chymotrypsin*, and it is thought that the folding of the molecule brings the amino acids *histidine* (57) and *serine* (195) into proximity on the surface of the molecule. Chymotrypsin is not a very specific enzyme, but it usually serves to hydrolyze the peptide bond on the carboxyl sides of *tyrosine* and *phenylalanine*.

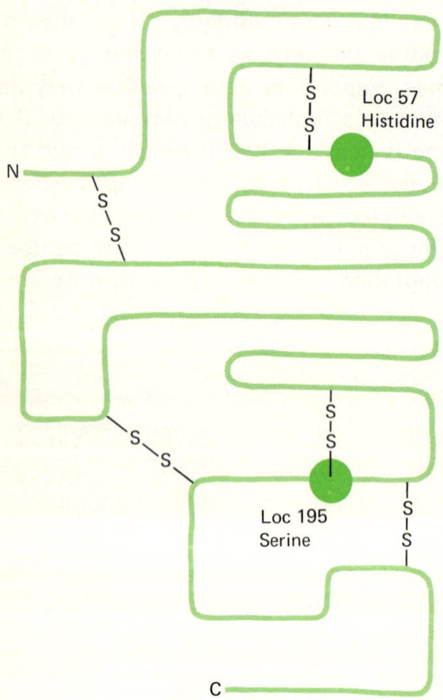

FIGURE 9.24 Chymotrypsinogen. *(From James D. Watson,* Molecular Biology of the Gene, *Second Edition, copyright © 1970 by J. D. Watson; W. A. Benjamin, Inc., Menlo Park, California.)*

An understanding of the basic structures and general functions (see box 9.2) of the chemicals we have thus far presented tells us, nevertheless, very little about the means by which organisms transfer and use energy. As we shall have reason to point out in our discussion of energy in ecosystems in the next chapter, photosynthesis is one of the key processes, if not *the* key process, in maintaining life on this planet. It is the source of almost all of the energy available to all species of plants and animals, and thus, we shall now turn our attention to details of the process of photosynthesis.

Energy Input to Organisms—Photosynthesis

As we mentioned in Chapter 3, photosynthesis is thought to have arisen relatively soon after life began on earth. Had it not, life probably could not have achieved its present highly diverse aspects. It is even speculated that life may have arisen on earth several times, only to disappear, and that it was not until photosynthesis evolved that life was able to sustain itself and continue to evolve. Though these speculations have been brought up earlier, we repeat them here to indicate the essential importance of photosynthesis to all of life.

Photosynthesis is the process by which organisms synthesize energy-rich compounds from light and from the relatively energy-poor molecules of carbon dioxide and water. Stated chemically, the reaction may be written

$$nCO_2 + nH_2O \longrightarrow (CH_2O)_n + nO_2$$

Thus, some number (n) of molecules of carbon dioxide react with water to produce carbohydrate and molecular oxygen. Still, the reaction shown is a great oversimplification. The plant is not a black box into which CO_2 and H_2O may be inserted and $(CH_2O)_n$ and O_2 retrieved. There are other requirements, and there is another product of the reaction. The requirements include light energy and a chemical, *chlorophyll*. In addition to the carbohydrate and oxygen, the

Enzymes and the Brain

We have dealt with the role of enzymes in energy conversions; it should be understood that they play other roles in living systems. At the present, for example, a great deal of exciting research is centered on the role of enzymes in psychiatric disorders such as mania and depression. We all experience periods of gloom and elation; in some persons such states may become so severe or prolonged that they impair the ability to carry on "normal" functions. That such disorders might be traced to biochemical malfunctions was once dismissed as unlikely. Recent research indicates the idea is not at all farfetched.

Three so-called brain amines, norepinephrine (NE), dopamine, and 5-hydroxytryptamine (5HT) have been isolated since 1950 and implicated in cases of depression. Tests of the cerebrospinal fluid of patients suffering from depression have shown lower than normal levels of breakdown products of both dopamine and 5HT, suggesting that the metabolism of these amines may offer clues to the treatment of such states. Researchers working on 5HT have reasoned that the low levels of 5HT metabolites could be explained as a deficiency in synthesis of 5HT caused by a lack of synthetic enzymes, or as a deficiency in the concentration of tryptophan, the basis of the 5HT molecule, or by a reduction in nerve activity, leading to reduced 5HT synthesis. Experiments seem to rule out the first two possibilities, leaving reduced nervous activity as an important mechanism to be explored.

Such reduced activity could be caused by a reduction in the sensitivity of receptors to transmitter substances. Cutting the nerve supply to tissue results in a great increase in the sensitivity of the tissue to transmitters; the tissue becomes supersensitive. Recent experiments utilizing supersensitized brain tissue from rats have shed a bit of light on the role of adenyl cyclase. This enzyme degrades ATP to cyclic AMP (adenosine monophosphate). It seems that within cell membranes, adenyl cyclase is acted upon by transmitters and then initiates reactions producing cAMP, which then brings about other reactions in the cell. Operations were performed on the brains of rats in which the nerve supply to certain cells on one side of the brain was cut, leaving the other side of the brain in a normal condition. The cells involved in this experiment are sensitive to norepinephrine as a transmitter that activates adenyl cyclase. Such tissue was then removed and treated with NE, with both supersensitive and normal tissue subjected to the same amount of the transmitter. The samples of tissue were then examined for cAMP concentrations and it was found that the cAMP level in supersensitive tissue was significantly higher than in the normal tissue. It was concluded that cAMP may be a reliable index of sensitivity to NE and possibly to the other brain amines. The meaning of the experiments becomes clear when compared with data from patients suffering from mania and depression. Urinalysis of manic patients reveals high levels of cAMP, while that of depressive patients indicates significantly low levels. It is presumed that such concentrations reflect higher receptor activity among manics and lower activity among depressives. Whether the changes in activity are the result of changes in the transmitter or defects in the receptors has yet to be worked out.

Generalizing from rats to humans is at best a risky business, but such research is leading to an understanding of the brain of a relatively complex animal, and could ultimately produce results with profound human implications.

BOX 9.2

reaction produces water. Including the necessary factors changes the reaction to

$$nCO_2 + nH_2O + \text{light} \xrightarrow[\text{chlorophyll}]{} (CH_2O)_n + nH_2O + nO_2$$

It would seem that there is an unnecessary step in the reaction, for water is both a reactant and a product of the reaction. Early investigators of photosynthesis thought that the carbon in CO_2 simply split off and bonded with the H_2O to form $(CH_2O)_n$ and O_2. It seemed a reasonable and economic possibility.

It was learned, however, that some photosynthetic bacteria produce sulfur instead of O_2 as a product of photosynthesis, after taking in hydrogen sulfide, H_2S, in place of H_2O. It seemed plausible to hypothesize that the O_2 produced in photosynthesis might come from H_2O, not CO_2. To test the hypothesis, investigators provided plants with normal H_2O and CO_2 that contained a heavy isotope of oxygen, O^{18}. (Normal oxygen is O^{16}.) The products of the reactions were then studied, and it was found that all of the heavy isotopes were associated with the H_2O (as H_2O^{18}) and with the $(CH_2O)_n$ (as $(CH_2O^{18})_n$). None of the O^{18} was found in the O_2. Therefore, H_2O did not merely flow through the reaction unchanged, but was somehow split and recombined with CO_2. There is an important difference between the energy-processing potential of photosynthesis in which such a flow of water occurred and photosynthesis in which water is taken in, split, and then recombined. Breaking the bonds of H_2O releases some energy to do work, energy that would not otherwise be available. Now look at the reaction in figure 9.25; it shows the "fate" of the elements involved.

$$nCO_2 + nH_2O \rightarrow (CH_2O)_n + nO_2$$

FIGURE 9.25 Movement of carbon and oxygen during photosynthesis.

From what we have presented thus far, it appears that photosynthesis is a single reaction. In reality, it is a complex series of reactions. We have shown only the initial input and final outcome of a number of chemical changes that occur in the reactants. Before considering these reactions, however, we should discuss the meaning of two terms, *oxidation* and *reduction*. Reduction is the addition of an electron to an atom or molecule, and oxidation is the removal of an electron. Since the transformation of energy into matter and matter into energy plays virtually no role in living systems, the electrons added to atoms and molecules in living systems must come from other atoms or molecules by the process of oxidation. In short, *for one molecule to be reduced, another must be oxidized, and vice versa*. In living systems, reduction and oxidation reactions usually involve hydrogen. Having only one electron, hydrogen "donates" it relatively easily to other atoms, especially oxygen. In fact, oxidation takes its name from oxygen, which has an affinity for electrons. In the forthcoming discussion of the reactions of photosynthesis, we shall generally mean the *addition of hydrogen when speaking of reduction* and the *removal of hydrogen when speaking of oxidation*. Thus, we can see that photosynthesis involves both the reduction of CO_2

$$CO_2 \xrightarrow{\text{reduced}} (CH_2O)_n$$

and the oxidation of H_2O

$$H_2O \xrightarrow{\text{oxidized}} O_2$$

In a plant, the CO_2 that is taken in is reduced by a series of complex reactions that also serve to oxidize H_2O. The remaining oxygen, all from CO_2, is then reduced to form new H_2O which is given off by the plant.

The reduction of CO_2 is not as simple as it sounds; it is accomplished by a series of chemical reactions in which a molecule of monosaccharide is produced and food, therefore, is synthesized in the plant. It should be kept in mind that the basic process is one whereby energy is transferred from molecule to molecule.

Photosynthesis and Light

Photosynthesis requires light as a source of activation energy. We may think of the light that arrives on the earth from the sun as being delivered in packets (or *quanta*) of light particles, called *photons*. These packets are not identical, since they differ in the amount of energy that they possess. Since light can also be thought of as traveling in continuous waves rather than discrete particles, the energy value of light is described in terms of wavelengths. The *shorter* the wavelength of a particle, the *greater* its energy; the *longer* its wavelength, the *lower* its energy. All of the visible light that reaches this planet falls within a very narrow range of wavelengths when compared to the total spectrum of energy radiated from the sun. (See box 9.2.)

When light strikes an object, some of the particles are absorbed, some pass through, and some are reflected. Depending upon the wavelength of the particles that are reflected, the object appears to be a particular color. If all light is reflected, the object appears white. If the object reflects only particles in the green portion of the visible spectrum, it appears green, indicating that all other particles are being absorbed or passing through.

Early in the research on photosynthesis (around the time of the American Revolution), it was learned that only the green portions of plants are involved in photosynthesis. It was later learned that the green color is produced by the pigment *chlorophyll*. Although there are several different kinds of chlorophyll, the most common is chlorophyll *a*. Less common, but essential to the process, is chlorophyll *b*. (See fig. 9.26.) Both forms are found in chloroplasts, and each makes a distinct contribution to the process. Since chlorophyll reflects green when it is illuminated with white light, it absorbs more or less all of the other wavelengths of light in the visible spectrum. Laboratory analysis has re-

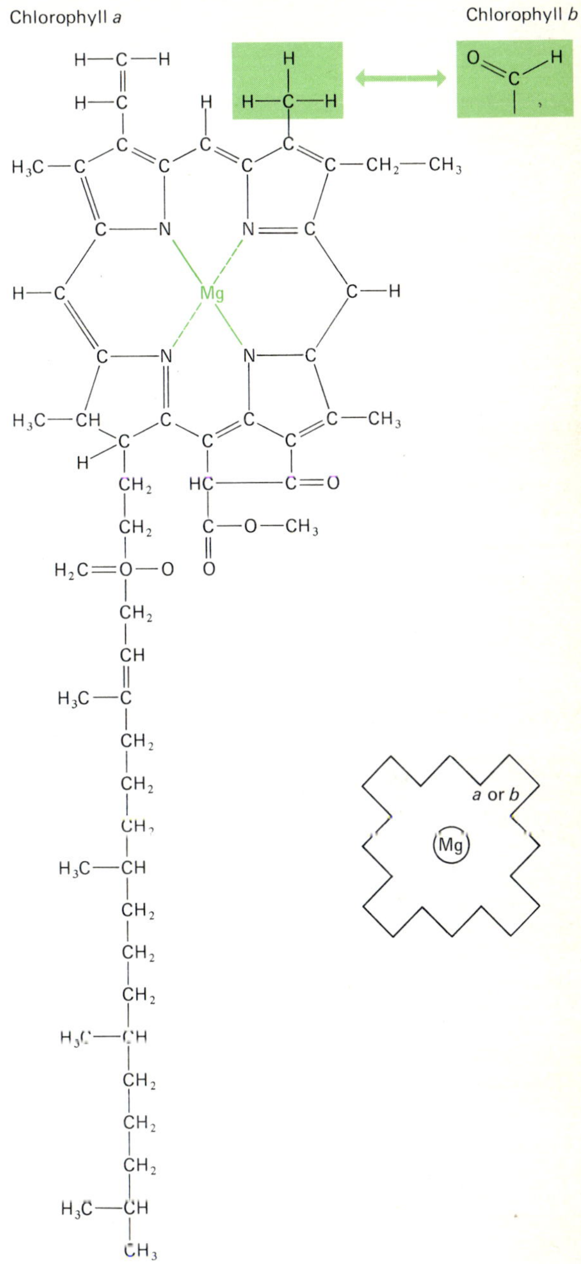

FIGURE 9.26 Structure of chlorophyll. The molecule shown in full is chlorophyll *a*. When the CH_3 in the shaded portion of the diagram is replaced with the CHO (to the right) then the molecule is chlorophyll *b*. For the sake of simplicity, the stylized figure will be used in other diagrams and will be identified as chlorophyll *a* or *b* in the upper right-hand corner.

The Electromagnetic Spectrum

We have used the term *radiation* to describe the adaptation of species to various environments; it is also used to describe the process by which energy travels in space. The electromagnetic spectrum is a schematic device by which energy is classified according to its wavelength. Figure (a) shows a diagrammatic definition of wavelength and a table of measurements. As we can see from figure (b), visible light constitutes but a small portion of the electromagnetic spectrum. Yet from within this narrow band of radiations comes very nearly all of the energy used by man, and virtually all of the biological energy on the planet. To describe the region of the electromagnetic spectrum from 3900 to 7600 angstroms as the ''visible'' spectrum is slightly anthropocentric, since other animals sense other portions of the spectrum. The honeybee, for example, is incapable of detecting radiations in the far red region, but can see in the ultraviolet region between 3000 and 4000 angstroms. Figure (c) shows a marsh marigold as seen by the human eye, *left*, and by the eye of an insect, *right*.

(a)

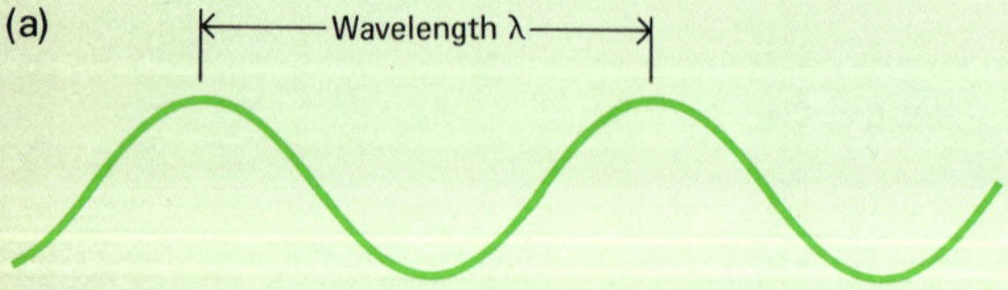

Wavelength is measured from crest-to-crest, and the shorter the wavelength the greater the energy. Most of the energy we shall be concerned with is in the region from 3900 Å to 7600 Å. The table below indicates the relative sizes of the units we are discussing.

39.37 inches =	1 meter (m)
1 meter =	1000 millimeters (mm)
1 millimeter =	1000 microns (μ)
1 micron =	1000 nanometers (nm)
1 nanometer =	10 angstroms (Å)

(b)

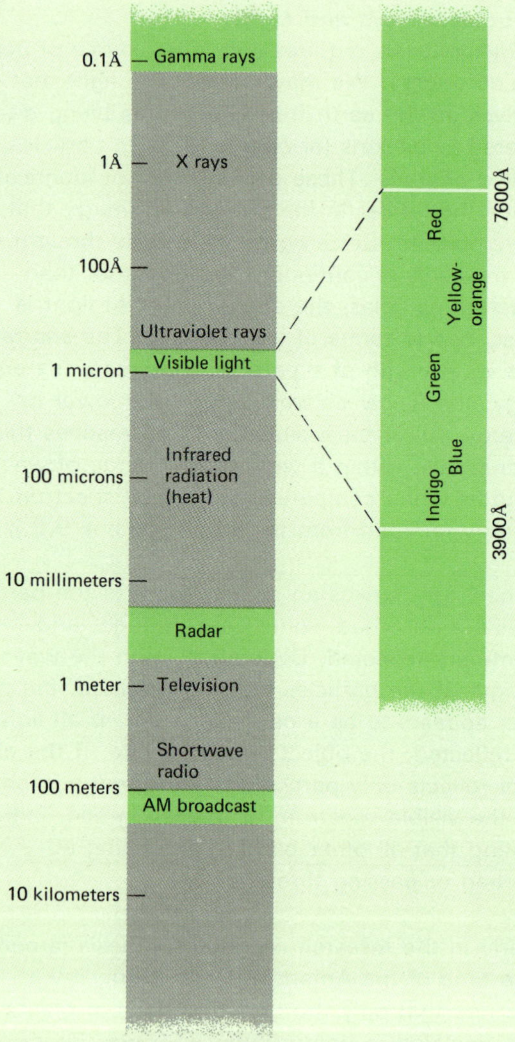

(c)

(Courtesy of Thomas Eisner, Cornell University.)

vealed that chlorophyll absorbs relatively more of the higher energy light in the spectrum. (See fig. 9.27.) Other pigments in the plant absorb more of the lower energy light and pass it on to chlorophyll. In that way, the plant is able to make use of a greater amount of energy than if it were to contain chlorophyll alone.

FIGURE 9.27 Only part of the light absorbed by a leaf is used in photosynthesis. The light absorbed at different wavelengths by chlorophyll *a* and by beta-carotene determines the effect of light on different wavelengths on the rate of photosynthesis. *(Redrawn from* The Living Plant, *Second Edition, by Peter Martin Ray. Copyright © 1963, 1972 by Holt, Rinehart and Winston, Inc. Reprinted by permission of Holt, Rinehart and Winston, Inc.)*

The precise event that occurs when a photon strikes a molecule of chlorophyll is not known. It is believed that the photon strikes an electron in chlorophyll and moves it further from the nucleus of the atom of which it is a part. By thus altering the position of the electron relative to the nucleus, the photon stores potential energy in the electron. Since a system in which potential energy is stored is less stable—that is, more readily undergoes change—than a system in which little potential energy is stored, the electron would soon fall back to its former position near the nucleus. Thus, the potential energy could be converted to kinetic energy and could be used to do work. However, in chlorophyll, if

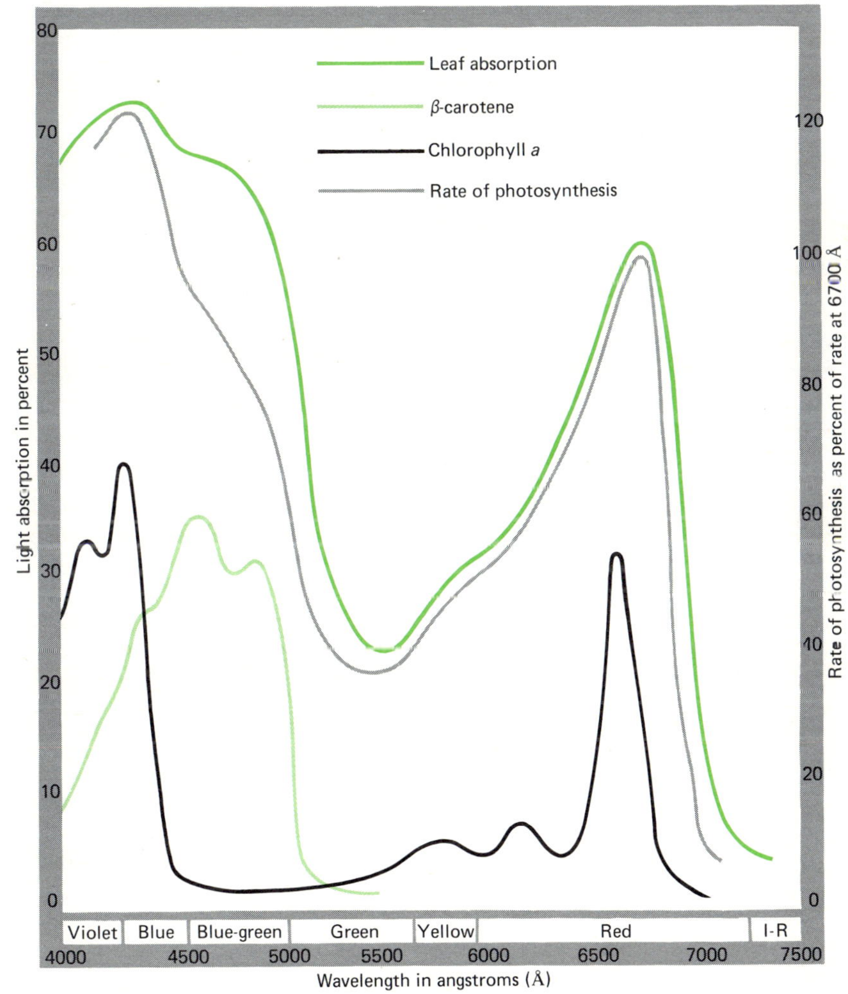

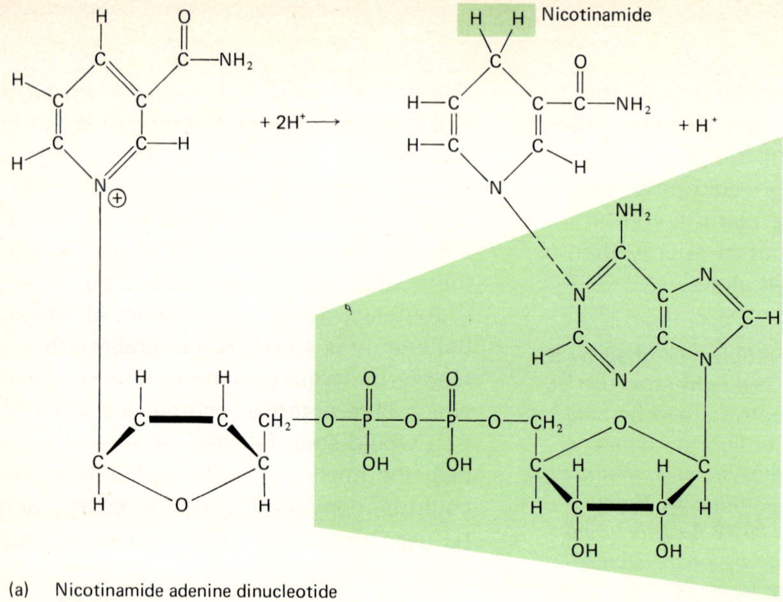

(a) Nicotinamide adenine dinucleotide (NAD)

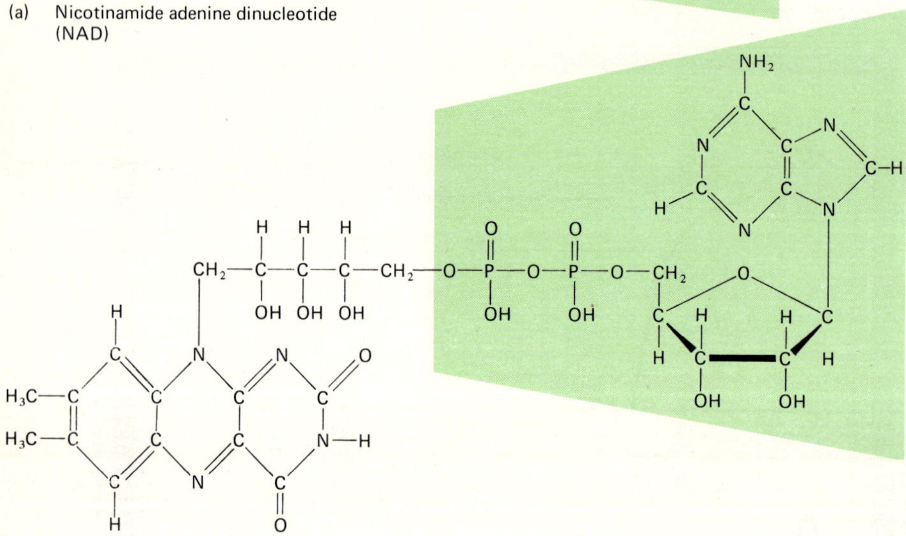

(b) Flavin adenine dinucleotide (FAD)

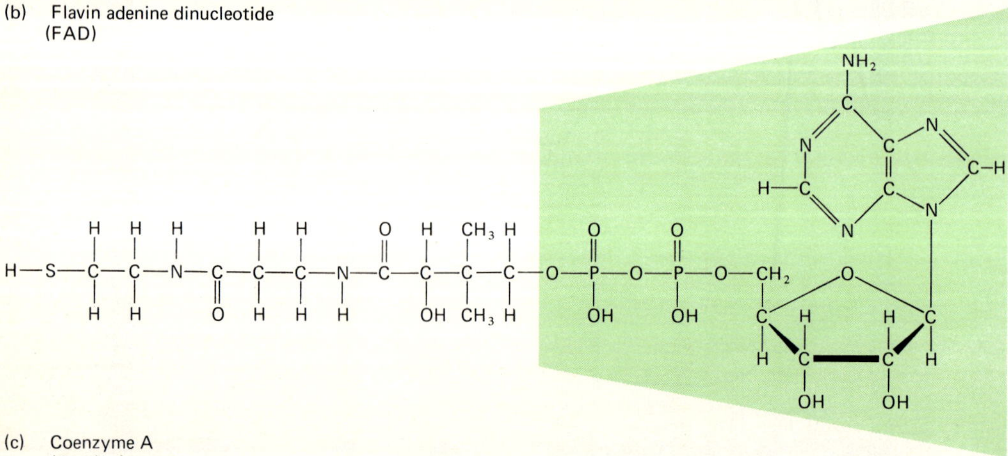

(c) Coenzyme A (CoA)

the electron were to fall back to its former position, the work that would be done would produce light. In the plant, then, something must come into play to keep the energy captured by chlorophyll from being lost as light.

Though it is by no means fully understood, what is thought to happen in the plant is that the electron is attracted to another molecule with a strong affinity for electrons. Once "captured," the electron slightly changes the structure of the molecule of which it is now a part, thus changing its energy value. It is then passed on through a series of molecules in an electron-transfer system made up of molecular *electron carriers*. Each transfer involves small changes in the structures of the carriers and, thus, changes in their energy content, and each transfer of the electron releases some energy that can be used to do work. Finally, the electron reaches its initial energy state and is returned to a molecule of chlorophyll of the sort from which it was removed.

The Chemistry of Photosynthesis

In order to view the process of photosynthesis in any detail, we shall have to first introduce several chemical compounds that play important roles in the energy conversions of photosynthesis. As we have already mentioned, enzymes play key roles in bringing about the many chemical reactions that characterize living systems. In many cases, however, enzymes cannot perform their roles without the aid of a class of chemicals called *coenzymes*. In the many oxidation and reduction reactions that are involved in photosynthesis, coenzymes serve to accept electrons from hydrogen and transfer them to intermediate compounds. Thus, the coenzyme nicotinamide adenine dinucleotide (NAD) may be written NAD_{red} or NAD_{ox} to indicate whether it is reduced (carrying an electron) or oxidized (lacking an electron for transfer). The structures of NAD and two other important coenzymes, FAD and CoA, are shown in figure 9.28.

FIGURE 9.28 NAD, FAD, and coenzyme A. Note the basic similarity of these molecules and their similarity to ATP. Also note that, as described in the text, the addition of two hydrogen ions to NAD produces a reduced form of NAD. The oxidized form of NAD is sometimes written as NAD_{ox} or NAD^+ and the reduced form may be written as NAD_{red} or $NADH + H^+$.

Another important molecule is *cytochrome*, of which there are several varieties. Portions of the cytochrome molecule are similar to chlorophyll. Like coenzymes, cytochromes can accept or donate electrons in oxidation-reduction reactions, with different cytochromes binding the electrons with different relative strengths. Cytochromes thus form important links in the electron-transfer system. Depending upon their source, cytochromes vary widely in size, ranging from MW 14,000 to MW 240,000.

Ferredoxin is believed to be the molecule to which the chlorophyll molecule donates its electron, and from which the electron is passed into the electron carrier system of cytochromes. Ferredoxin is an iron-containing protein with MW 5500 when isolated from photosynthetic bacteria. Ferredoxin isolated from spinach leaves is slightly more than twice the size of bacterial ferredoxin and contains fewer iron atoms.

We have already introduced the other important molecule, ATP. The synthesis of ATP from ADP and phosphate occurs in all living systems, but only in photosynthesis is the necessary energy for phosphorylation provided by light.

The Light Reactions—Cyclic and Noncyclic Phosphorylation

As we understand it at present, the first stage of photosynthesis is as follows: When a photon is absorbed by chlorophyll *a*, it raises the energy level of one electron which is then accepted by ferredoxin (Fd), thus reducing ferredoxin and oxidizing chlorophyll *a*. From reduced ferredoxin the electron is accepted by NAD^+, which then becomes $NADH + H^+$. The electron is then transferred to cytochrome *b*, and at this transfer, the electron loses some energy which is used to synthesize one molecule of ATP.

NAD+ then reacts with cytochrome *b* and transfers the electron to cytochrome *f*, releasing more energy and synthesizing another ATP molecule. Then, in a lower energy state, the electron is once more transported to a molecule in the carrier system, and from there it is returned to a molecule of chlorophyll *a*. (See fig. 9.29.) Because the electron makes a complete cycle beginning and ending with chlorophyll *a*, and phosphorylates two ADP molecules to ATP in the process, and because the process is begun by light, this reaction series is called *cyclic photophosphorylation*.

You will note that we have thus far not mentioned CO_2 and H_2O. Daniel Arnon, one of the leading investigators of photosynthesis, showed in 1954 that the chloroplast could synthesize ATP in the presence of light and in the absence of CO_2. Obviously, if the final products of photosynthesis are O_2, H_2O, and reduced CO_2, then something more must occur than we have heretofore presented.

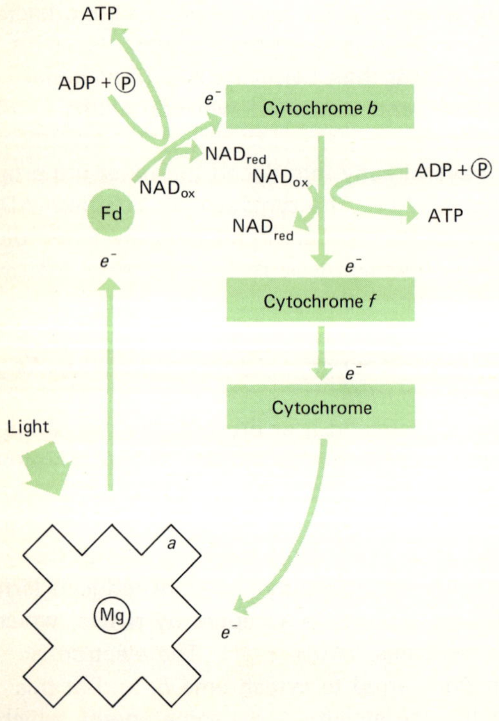

FIGURE 9.29 Cyclic photophosphorylation.

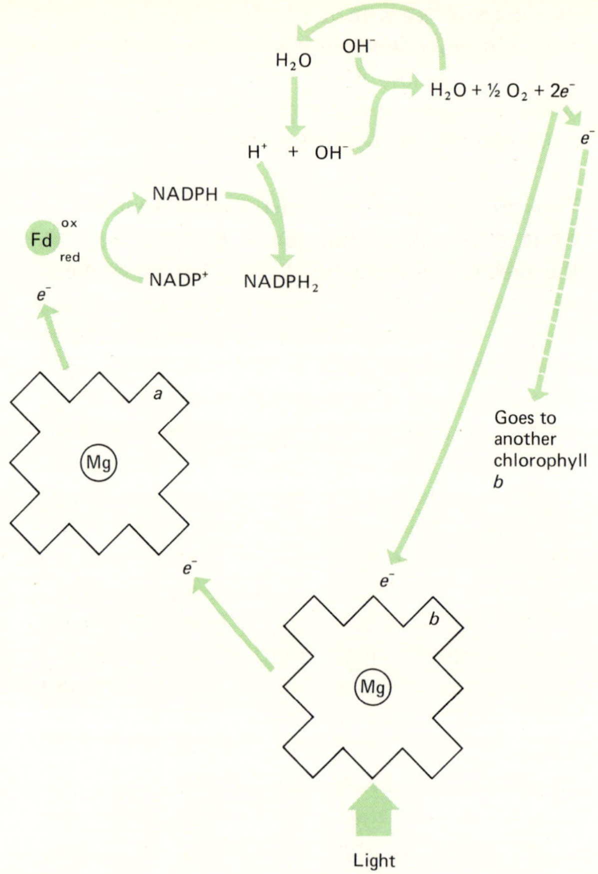

FIGURE 9.30 Noncyclic path of electron transfer in photosynthesis. The reaction of the hydroxyls is postulated, but seems the most likely course of events. It should be noted that no ATP is synthesized in this reaction series. Since ATP is a reduced form of ADP and P_i, though, the reduction of NADP in the noncyclic pathway provides reducing power for later reactions. Chlorophyll *b* is not the only pigment in the plant cell that is capable of transferring electrons to chlorophyll *a*. Other pigments, such as β-carotene, from which vitamin A is derived, also capture the light energy of the sun and transfer electrons to chlorophyll *a*. Because these "accessory" pigments absorb light at differing wavelengths from chlorophyll, the plant makes use of a greater proportion of the energy available to it than if chlorophyll were the only pigment it contained.

The next stage of photosynthesis is the one in which H_2O enters the process, and it is called *noncyclic photophosphorylation*. It is also at this stage that chlorophyll *b* becomes important. When struck by a photon, an electron in chlorophyll *b* is raised to a higher energy level and then reacts with chlorophyll *a* to store energy in an electron there. The electron of chlorophyll *a* is then accepted by ferredoxin, as in cyclic photophosphorylation, and is replaced by the electron from chlorophyll *b*. Chlorophyll *b* now lacks an electron and must accept one from some other molecule. It appears that this is accomplished by the splitting of water. It is postulated that the water splits into H^+ and OH^-. The hydroxyl ion then is thought to react with another available from an earlier reaction; $OH^- + OH^- \rightarrow H_2O + \frac{1}{2} O_2 + 2e^-$. The "extra" electrons (e^-) from this reaction are then accepted by molecules of chlorophyll *b*, which have donated electrons to chlorophyll *a*. We now have ferredoxin in a reduced state and a hydrogen ion from the splitting of water. Another coenzyme, $NADP^+$, which is similar to NAD^+, now enters the sequence. $NADP^+$ is capable of carrying two electrons in its fully reduced state. It reacts first with the reduced ferredoxin to become NADPH and then with the hydrogen ion from the split water to become $NADPH_2$. Thus the ferredoxin is returned to the oxidized state and the cycle can begin anew. The products of this reaction series are one-half of the molecular oxygen given off by the plant, water, and, very importantly, $NADPH_2$. Figure 9.30 shows a summary of these reactions, and in figure 9.31, both cyclic and noncyclic photophosphorylation are combined.

FIGURE 9.31 Summary of cyclic and noncyclic pathways.

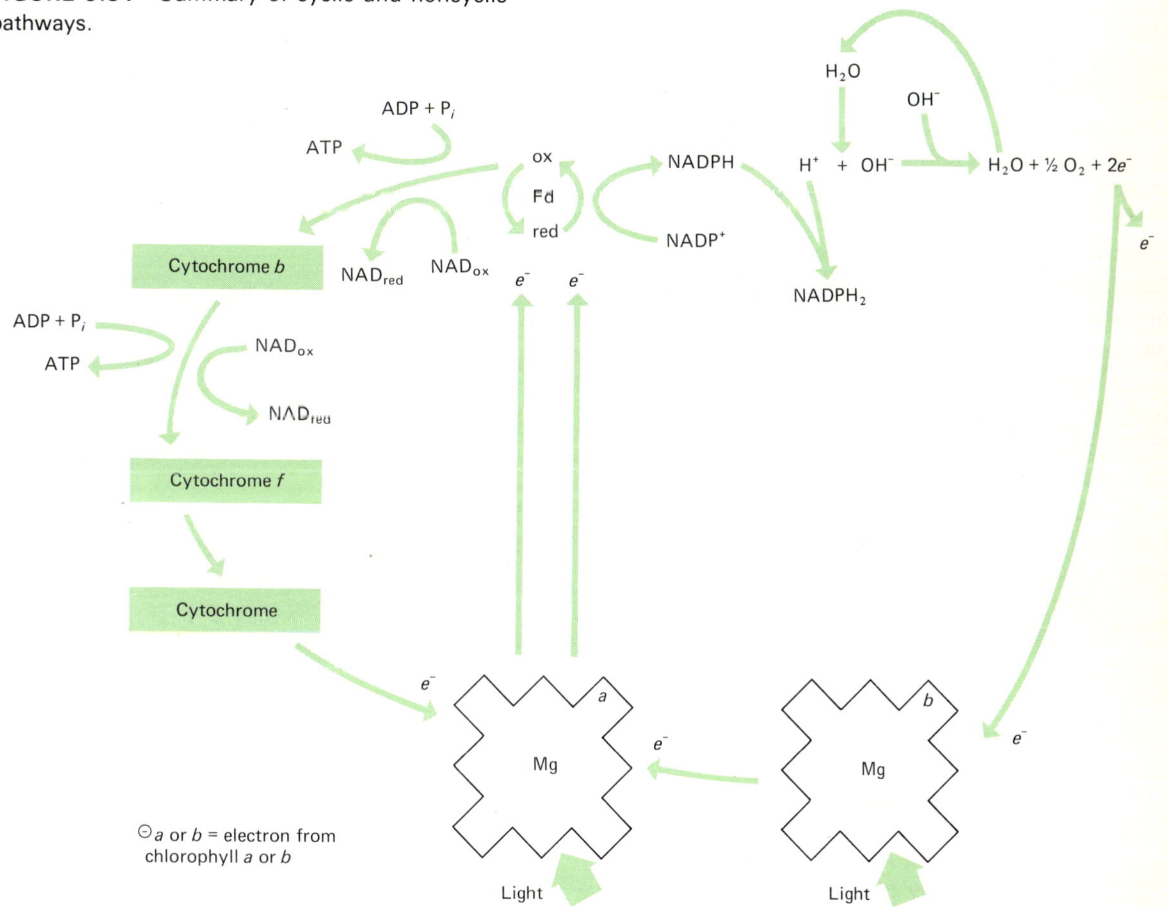

The Dark Reactions

We must now account for the reduction of CO_2 to $(CH_2O)_n$, and for the production of H_2O, the other waste product of the plant. Because cyclic and noncyclic photophosphorylation are initiated by light energy, they are called the *light reactions.* The reduction of CO_2 does *not* require light to proceed, and it is therefore called the *dark reaction*. Without the input of light energy, of course, some other form of activation energy must be supplied. It has been shown that the two molecules of ATP generated in photophosporylation provide the activation energy for the dark reactions and that $NADPH_2$ provides the hydrogen necessary for the reduction. The series of reactions by which these results are produced was worked out by Melvin Calvin of the University of California at Berkeley and is consequently called the Calvin Cycle.

Calvin worked out the cycle by providing *Scenedesmus*, a photosynthetic alga, with CO_2 containing a radioactive form of carbon, C^{14}. After permitting the algae to take up the $C^{14}O_2$, Calvin immersed them in alcohol, thus killing them and inactivating the enzymes that catalyze photosynthetic reactions. Then, by careful chemical analysis, the radioactivity was located, and the compounds in which it was found were identified. By killing the algae at shorter and shorter times after the $C^{14}O_2$ was introduced, Calvin was able to identify the first key compound of the cycle, namely 1–5 *ribulose diphosphate.* In summary, if the time after C^{14} introduction was short, the radioactivity was found in compounds formed early in the sequence; if the time interval was longer, the C^{14} was found all along the sequence.

The first reaction in the cycle is between $C^{14}O_2$ and ribulose 1–5 diphosphate, a five-carbon sugar with two phosphates bonded to it. The product of this reaction is an unstable hexose that breaks down almost immediately into two trioses called *3-phosphoglycerate*. Then, 3-phosphoglycerate reacts with ATP to produce a diphosphoglycerate and ADP. The diphosphoglycerate is reduced by $NADPH_2$ to form *phos-*

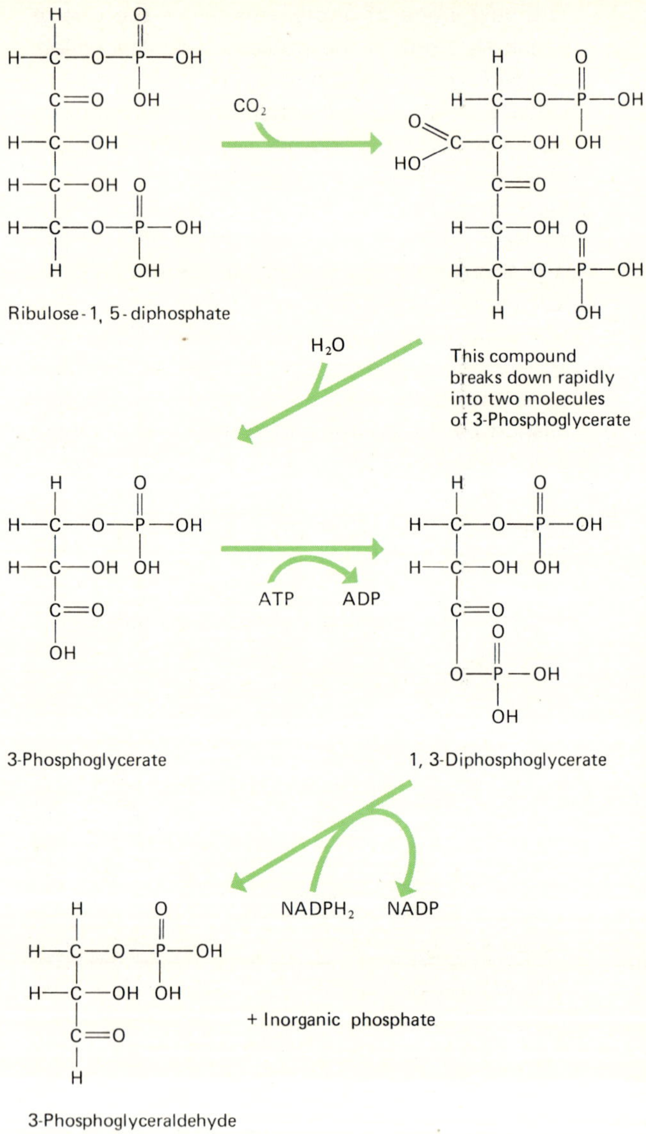

FIGURE 9.32 Initial reactions of the Calvin Cycle.

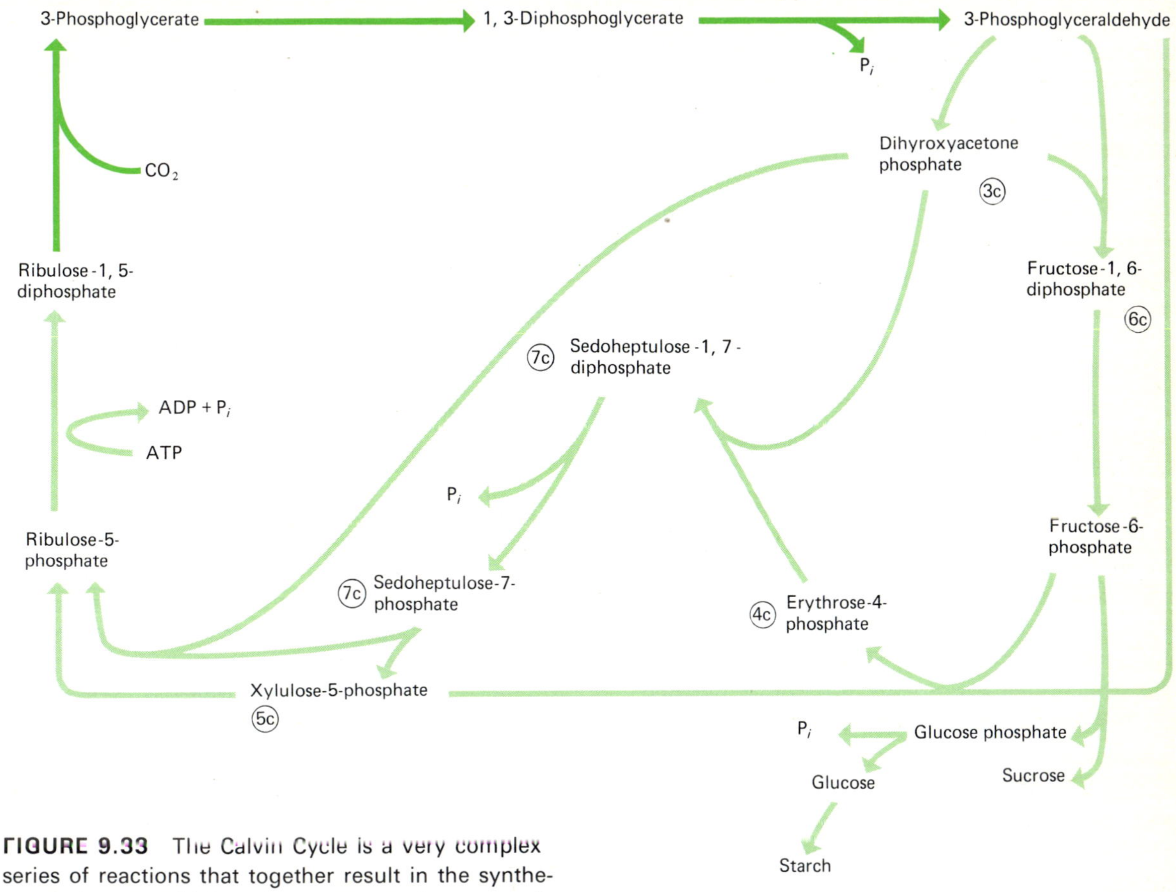

FIGURE 9.33 The Calvin Cycle is a very complex series of reactions that together result in the synthesis of glucose, which is the precursor of starch and all of the other energy-containing compounds in the plant. The initial reaction shown in figure 9.32 leads to a series of conversions of various sugars containing 3, 4, 5, 6, and 7 carbon atoms. The entire system forms a complex cycle.

phoglyceraldehyde. This compound is then used in the synthesis of new molecules of ribulose diphosphate, and most importantly, it is dephosphorylated to form glucose. (See fig. 9.32.) The Calvin Cycle is shown in figure 9.33. The cycle requires that six molecules of CO_2 be taken up with six of H_2O and that the reduction of CO_2 use up three molecules of ATP and two of $NADPH_2$. Phosphoglyceraldehyde is eventually used to synthesize glucose, in the very complex reaction series shown.

The Importance of Photosynthesis

Directly or indirectly, the intricate series of reactions that constitute photosynthesis is basic to all living systems on this planet. It is estimated that more than 98% of the energy that reaches the surface of the earth is not used in photosynthesis, and yet of all the energy presently used by man, less than 0.007% is derived from nonphotosynthetic sources (for example, the internal heat of the earth, called *geothermal* energy, or the energy generated by moving water). Coal, oil, and natural gas, the so-called *fossil fuels*, are all products of past photosynthesis. Until recently in human history, this "capital" went untouched as man powered his work by nonphotosynthetic energy sources, such as windmills and sailing ships, or by relatively recently stored photosynthetic products, such as wood, burned to heat homes, smelt iron, or fire steam engines. Some estimates of the extent to which we have dipped into this accumulated capital in recent times indicate that we are rapidly depleting it.

But even if we were not depleting our capital, there would be another reason for emphasizing the importance of understanding photosynthesis. What we have seen here, at the molecular level, is the continual cycling of materials from product to reactant and back to product. As we shall see again and again, such cycling is a basic operating principle in biological systems. Materials are used and reused to transfer the initial input of energy into appropriate chemical bonds. We often hear people speak of the "balance of nature." It is often unclear exactly what that balance consists of, although it is generally understood to refer to whole animals and plants and their physical surroundings. This is one expression of nature's balance, and it is an extremely important one, since the resources of this planet are limited. But the same phenomenon is expressed at every level in living systems, from molecules to the global ecosystem.

After energy from the sun is taken in and stored by plants, it must sooner or later be used to do work. Thus, we must turn to the problem of releasing energy from the compounds in which it is stored by plants.

Energy Use by Cells

While the primary result of photosynthesis is the reduction of CO_2, the reverse is true of *respiration*, the *process whereby cells make use of energy-storing molecules*. A carbohydrate is oxidized to form CO_2 and H_2O. As we would expect, this is not a single-step reaction, but rather a series of reactions. The energy must be released slowly and in a controlled manner if it is to be utilized efficiently by the cell. If it were not, organisms would have to obtain far larger amounts of energy and be able to tolerate higher temperatures to maintain their characteristic metabolic rates.

The first step in the process of cellular respiration is the breakdown of the carbohydrate into its simplest form. (Carbohydrates are by no means the only source of energy an organism may utilize. Lipids and proteins are also important, not merely as supplements to carbohydrates, but as necessary constituents of the diets of organisms. For our purposes, however, we shall concentrate on carbohydrate metabolism.)

Glycolysis

Once it is synthesized, glucose is a very stable compound, and energy must be invested to initiate its breakdown. Nearly all of the energy-releasing reactions of the body require some amount of activation energy to release the greater amounts of energy stored in molecules and allow their entry into the mainstream of metabolism. In the initial stages of photosynthesis, we saw that ATP was generated, later to be consumed in the synthesis of glucose. In respiration, ATP provides the activation energy required for the degradation of glucose. Like all biological reactions, this is mediated by an enzyme. The first reaction to take place is between glucose and ATP, mediated by the enzyme *hexokinase* to produce *glucose-6-phosphate* and ADP. The next reaction uses no ATP, but the glucose-6-phosphate is converted to *fructose-6-phosphate,* which then reacts with ATP and *phosphohexokinase* to phosphorylate the fructose compound to *fructose-1-6-diphosphate.* In both of the phosphorylations, the energy released is largely in the form of heat, but the fructose-1-6-diphosphate is richer in energy than simple glucose. The next reaction is a familiar one. The fructose-1-6-diphosphate is split by an enzyme, *aldolase,* into two trioses, 3-phosphoglyceraldehyde, which you will recognize as a major product of photosynthesis, and *dihydroxyacetone phosphate,* which can be converted into 3-phosphoglyceraldehyde by the enzyme *phosphotriose isomerase.* (See fig. 9.34.)

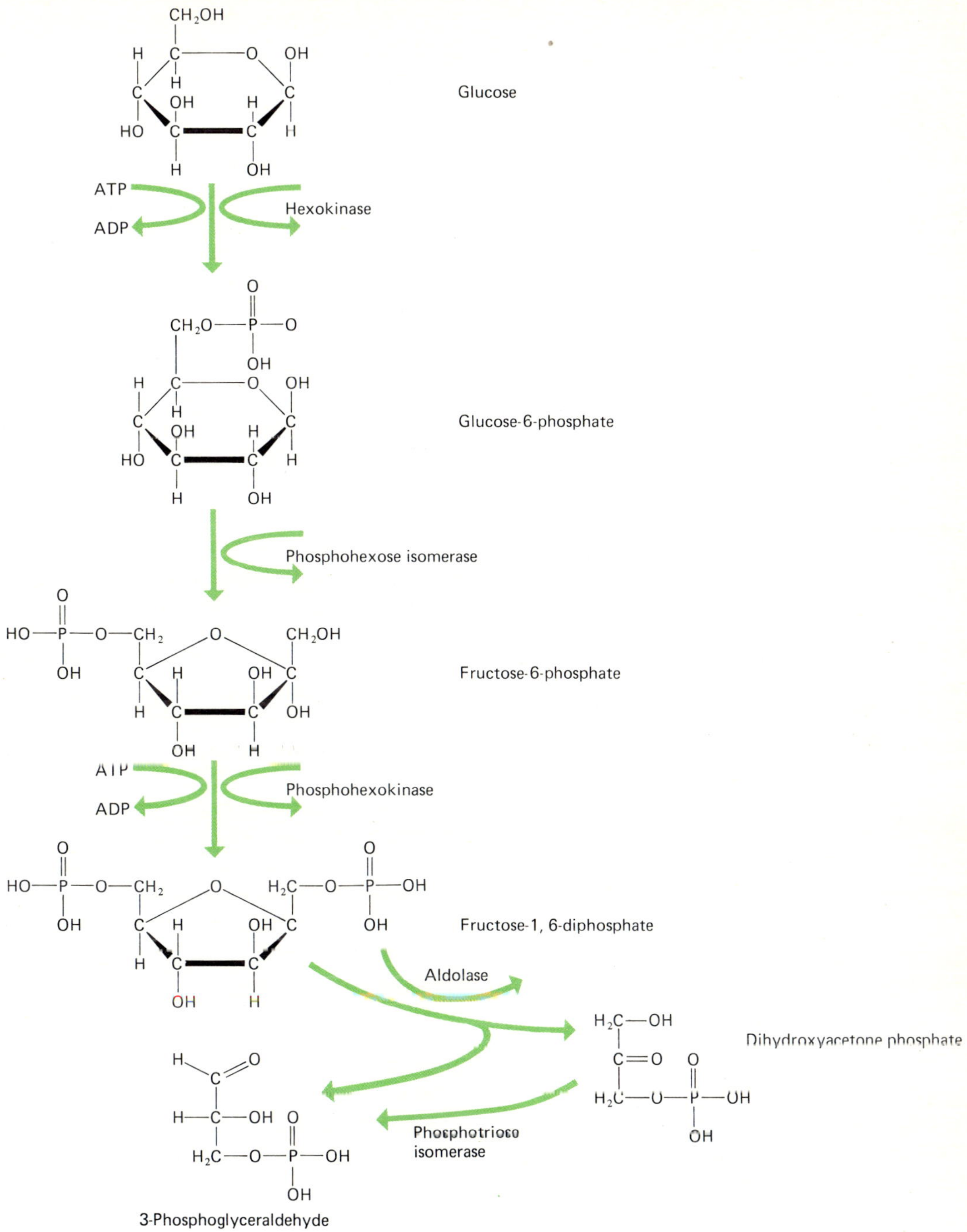

FIGURE 9.34 Early reactions of glycolysis.

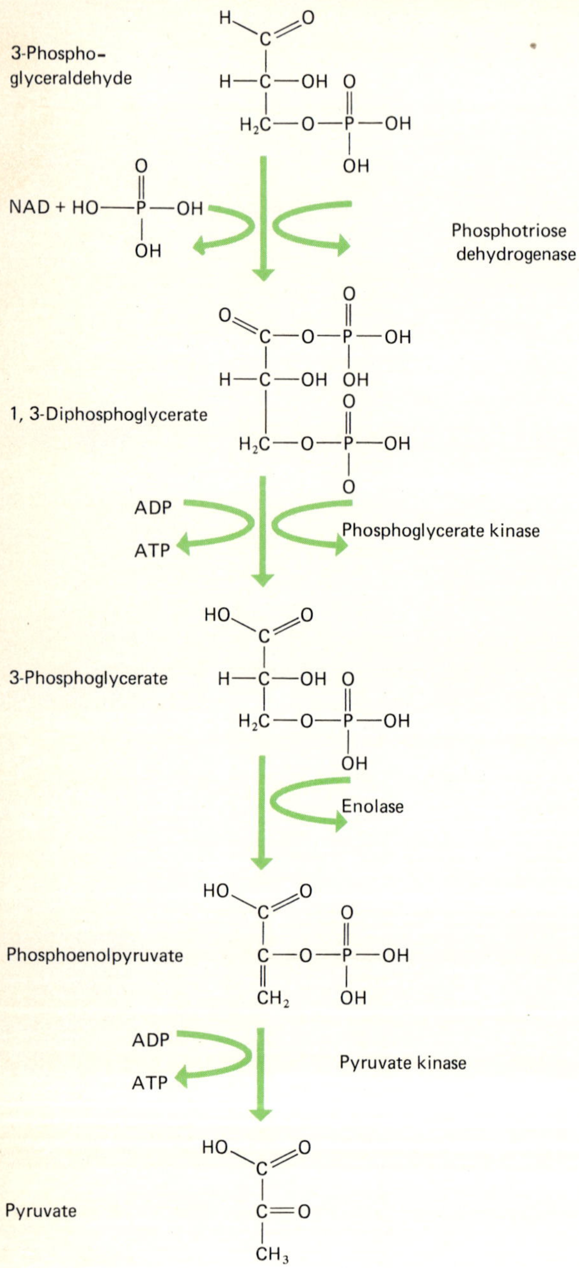

FIGURE 9.35 Subsequent reactions of glycolysis.

Next, another series of reactions takes place. Each molecule of 3-phosphoglyceraldehyde is phosphorylated; each now contains two phosphate groups and is called *1,3-diphosphoglycerate.* The phosphorylation requires that a hydrogen atom be removed from each, and this has the effect of converting the bonds of the 1,3-diphosphoglycerate into high-energy bonds. At this point in the sequence, there are two molecules of 1,3-diphosphoglycerate with one high-energy bond each, and the series has consumed two molecules of ATP. Each of these phosphorylated high-energy trioses, though, goes on through a short series of conversions which result in the synthesis of two ATP molecules for each 1,3-diphosphoglycerate. (See fig. 9.35.)

The end product of the reaction series is a three-carbon compound called *pyruvate*. To convert glucose to two molecules of 1,3-diphosphoglycerate takes two ATP molecules, but the conversion of 1,3-diphosphoglycerate to pyruvate produces four ATP molecules; thus, there is a net gain of two ATP's in the degradation of glucose to pyruvate.

This reaction series is anaerobic, and requires no oxygen. In fact, it can take place only in the absence of oxygen. (Recall that processes requiring the presence of oxygen are said to be aerobic.) Two things can occur in the continued absence of oxygen. First, pyruvate can react in an enzymatic sequence with $NADH_2$ to become ethyl alcohol. Should that occur, some of the oxygen in the pyruvate will contribute to the production of CO_2, leaving the alcohol as a two-carbon compound. (See also box 9.4.) The second thing that can happen to pyruvate is that it can react with $NADH_2$ in another way that does not produce CO_2, but does result in the generation of a three-carbon compound called *lactate*.

Glycolysis provides sufficient energy for the metabolic requirements of many lower organisms, such as yeast cells. Higher organisms, however, require far more absolute quantities of energy to sustain life than do yeast cells, and thus in

Muscles and Wine

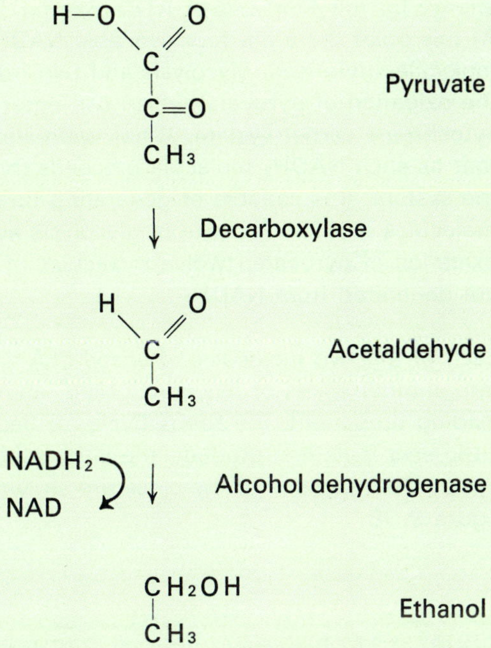

In the absence of oxygen, pyruvate undergoes the reactions shown; it is degraded to ethanol or to lactic acid. Ethanol is ethyl alcohol and is produced by yeast cells. Thus the addition of yeast to grape juice, which is rich in maltose, will produce wine, as long as oxygen is kept out of the container in which the reactions take place. The yeast, however, cannot tolerate an environment in which the concentration of alcohol exceeds about 12%; thus the maximum "proof" of wine is about 24. A vintner may control the proof by allowing the yeast to be exposed to air, thus killing it and keeping the potency of the wine down. Because the yeast has not reached the limit of its tolerance, such a wine would be sweeter than one in which the yeast died a natural death as a result of its own *exometabolites* (waste products of metabolism). A 24 proof wine that is sweet, of course, is one to which sugar or alcohol has been added after fermentation.

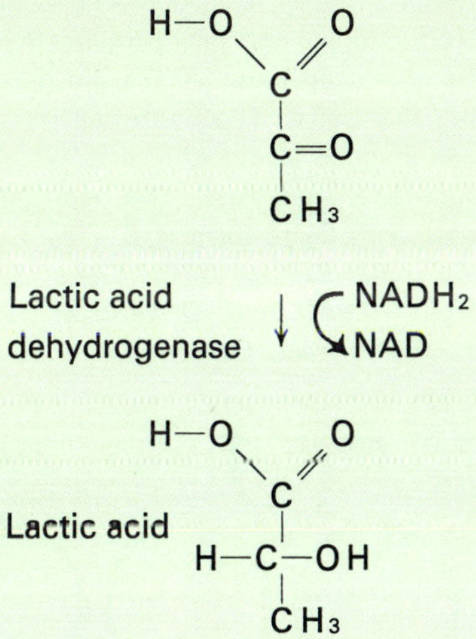

An equally interesting reaction series which may be more familiar to the reader occurs in muscles and produces lactic acid. If the oxygen supply of a muscle is reduced or stopped entirely, the muscle will continue to contract, using up its store of oxygen and energy-rich compounds. In the absence of oxygen, however, pyruvate will be converted to lactic acid will break down and the resulting pyru- cell will interfere with the other chemical reactions, and the muscle will "fatigue." Thus the feeling that we call muscle fatigue is explained in chemical terms as a buildup of lactic acid. When the oxygen supply is restored, the lactic acid will break down and the resulting pyruvate will go on to be oxidized in the series of reactions that we shall examine next.

BOX 9.4

higher organisms, glycolysis is followed by a series of reactions in which greater amounts of ATP are synthesized from pyruvate. These reactions require oxygen and thus are *obligately aerobic*.

The Krebs Tricarboxylic Acid Cycle

Glycolysis releases about 7% of the potential energy in a molecule of glucose. Under aerobic conditions, a great deal of the rest is released in two series of reactions beginning with pyruvate. The first series, which we shall not go into in detail here, produces CO_2 and an active form of acetic acid called *acetyl-CoA* (or acetyl *coenzyme A*, the structure of which is shown in figure 9.36). Hydrogen is removed in this reaction and reduces NAD to $NADH_2$. Since glycolysis breaks glucose into two molecules of pyruvate, two molecules each of acetyl-CoA and $NADH_2$ are produced.

The $NADH_2$ is used to reduce oxygen and form water, but as in photosynthesis, the electrons are carried through a cytochrome carrier system where their "downhill" progress provides the energy for the synthesis of ATP. (See fig. 9.37.) At this point there are four available $NADH_2$ molecules (two from glycolysis and two from the oxidation of pyruvate) which can enter the cytochrome carrier system. It has been shown that as each $NADH_2$ molecule proceeds through the system, it is capable of generating three molecules of ATP. Thus, from glycolysis and the oxidation of pyruvate, twelve molecules of ATP are generated from $NADH_2$.

Each of the two molecules of acetyl-CoA is fed into another series of reactions called, after its leading proponent, the Krebs Cycle, or because citric acid is its first product, the citric acid cycle. The reactions of the cycle are shown in figure 9.38.

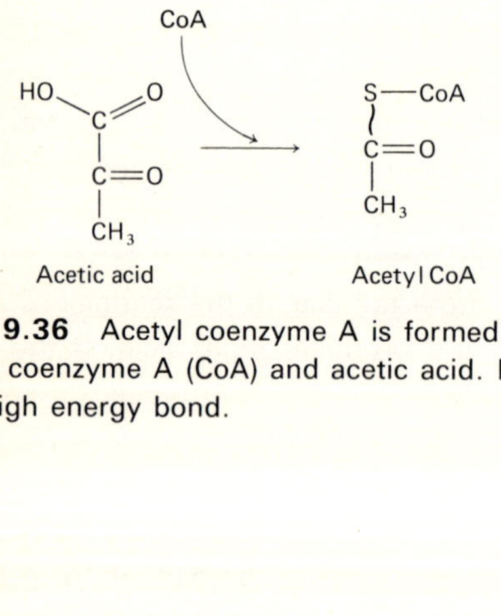

FIGURE 9.36 Acetyl coenzyme A is formed by reaction of coenzyme A (CoA) and acetic acid. It contains a high energy bond.

FIGURE 9.37 Respiratory chain. As $NADH_2$ is generated during glycolysis it is fed into a respiratory chain made up of coenzymes. It is not known with certainty which reactions produce ATP from ADP and P_i, but it is thought that at least one ATP is produced in the reduction of the flavoprotein. If reduced FAD enters the chain, only two ATP's are generated, thus supporting this belief. The hydrogen ion subsequently reduces a series of cytochromes, which when oxidized produce ATP from ADP. The hydrogen ion eventually reacts with O_2 to form H_2O. *(From James D. Watson,* Molecular Biology of the Gene, *Second Edition, copyright © 1970 by J. D. Watson; W. A. Benjamin, Inc., Menlo Park, California.)*

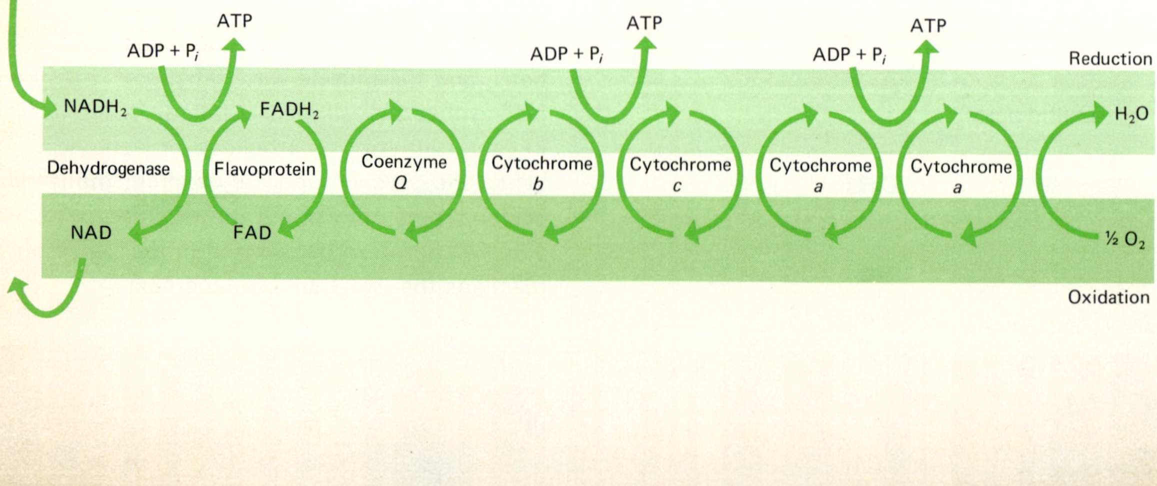

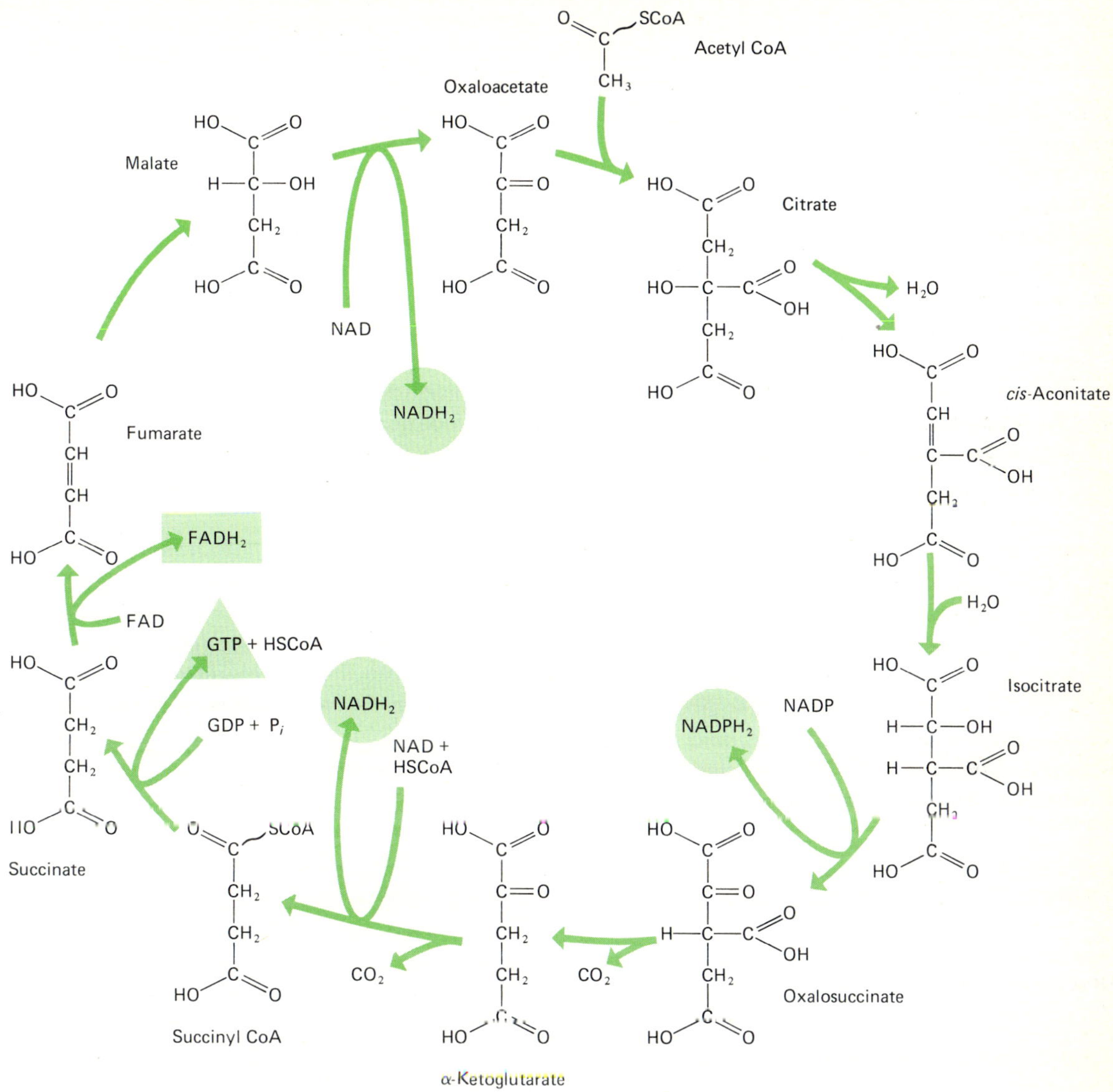

FIGURE 9.38 The Krebs, or citric acid, Cycle. This simplified diagram traces the conversion of one compound into several others. The cycle is usually thought to begin with the reaction of acetyl coenzyme A and oxaloacetate. Note that the coenzymes NADP and NAD are not the only coenzymes reduced in the sequence, and also that HSCoA, a product of the first reaction in the cycle, is used again in the conversion of α-ketoglutorate to succinyl CoA.

Notice that as each of the two molecules of acetyl-CoA passes through the cycle, two molecules of $NADH_2$ and one of NADPH are generated. These six molecules are fed into the cytochrome carrier system to produce more ATP, and since each generates three molecules of ATP, there are eighteen molecules of ATP generated by the $NADH_2$ and NADPH that result from the Krebs Cycle. Added to the earlier twelve, that makes thirty. Note also that each of the two acetyl-CoA molecules generates a compound called *guanosine triphosphate* (GTP), which is an equivalent of ATP. Also, as succinate is converted to fumarate, the coenzyme that is reduced is FAD, not NAD. $FADH_2$ enters the cytochrome carrier system at a later point than $NADH_2$ and only generates two molecules of ATP. Thus, from $FADH_2$, four molecules of ATP are generated in the carrier system. Added to the thirty ATP molecules accounted for above there are a) four from $FADH_2$ and b) two as GTP, for a total of thirty-six. If we now add the two molecules of ATP from glycolysis itself, we reach a total of 38 ATP molecules (including two GTP equivalents) from the complete oxidation of a molecule of glucose. Table 9.1 summarizes this accounting.

The oxidation of a molecule of glucose to produce thirty-eight ATP molecules means that less than 50% of the energy stored in glucose is recovered. The remainder is used to drive the reactions or is lost as heat or an unusable form of energy. It is estimated that steam engines are not much more than 10% efficient, and internal combustion engines about 25% efficient, at extracting the potential energy from their fuels. The efficiency of cells is truly remarkable.

TABLE 9.1 Generation of ATP-GTP in Glycolysis and Respiration

Reaction Series	*Products*			
	$NADH_2$	$FADH_2$	ATP	GTP
glucose → pyruvate	2		2	
Pyruvate → acetyl-CoA	2			
Krebs Cycle	6	2		2
	10	2	2	2
Cytochrome conversion				
$NADH_2$ (# of NADH × 3)			30	
$FADH_2$ (# of FADH × 2)			4	
			36 + 2 GTP	

Although we have concentrated on the oxidation of glucose as a source of usable energy, we should point out that lipids and proteins are important as well. Lipids are broken down to glycerol and fatty acids, and the glycerol is converted to 3-phosphoglyceraldehyde which enters the glycolytic pathway at the appropriate point. The fatty acids are then broken down into a number of two-carbon compounds that are converted to acetyl-CoA and enter the Krebs Cycle. Since fatty acids are highly reduced, they yield far more hydrogen for reduction of coenzymes, thus producing more energy per unit weight. The oxidation of one gram of fat produces from two to two and a half times as much energy as that of an equal weight of carbohydrate.

Proteins are hydrolyzed into amino acids, and the amino group is removed to prepare the compound for the respiratory process. Some are converted into pyruvate, some into acetyl-CoA, and some into the intermediate products of the Krebs Cycle. Equal amounts of protein and carbohydrate yield approximately equal amounts of energy.

Mitochondria, Chloroplasts, and Energy-Processing

Thus far, we have dealt with photosynthesis, glycolysis, and respiration as though they occurred independently of cells. Obviously, that is not so. For one thing, the processes are controlled by enzymes that carry out each reaction, and some of the products of the reactions are bound to enzymes. For another, the processes occur in very specific parts of the cell, and many of the enzymes involved are thought to be bound to the internal membranes of the cell and to transfer the reactants from enzyme to enzyme as a conveyor belt in a factory transfers partially assembled automobiles. As we mentioned in the section on the origin of life, the cellular structures that are the sites of photosynthesis are chloroplasts and, where respiration takes place, mitochondria. The enzymes that carry out glycolysis are found in the soluble part of the cell and are not associated with the cell membrane or with either of the organelles mentioned above.

Electron microscopic studies of mitochondria and chloroplasts have shown that each consists of an outer membrane and a highly folded internal membrane. (See fig. 9.39.) The cells of plants contain both chloroplasts and mitochondria, the former to synthesize carbohydrate from CO_2, H_2O, and light energy, and the latter to oxidize the glucose and synthesize ATP. Within the chloroplast and mitochondria, most of these conversions take place on the surface of the inner, folded membrane. The spaces between the membranes are *thought* to be filled with the enzymes necessary for these chemical reactions, as well as with some concerned with the synthesis of the membranes themselves. Thus, the intricate structure of the mitochondrion and chloroplast controls the organization of energy processing. The ordered structures are necessary for the ordered complex processes of photosynthesis and respiration.

FIGURE 9.39 (a) An electron photomicrograph of a mitochondrion showing the double outer membrane and the infolding to form the inner membranes where metabolic reactions occur. (b) An electron photomicrograph of a chloroplast, showing the inner folded membranes and the grana (also see fig. 9.42). *((a) Courtesy of Keith R. Porter, University of Colorado; (b) courtesy of Albert E. Vatter, University of Colorado Medical School.)*

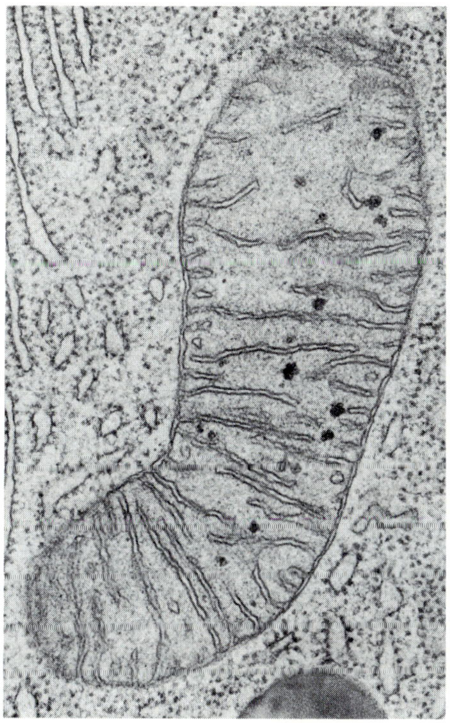

(a)

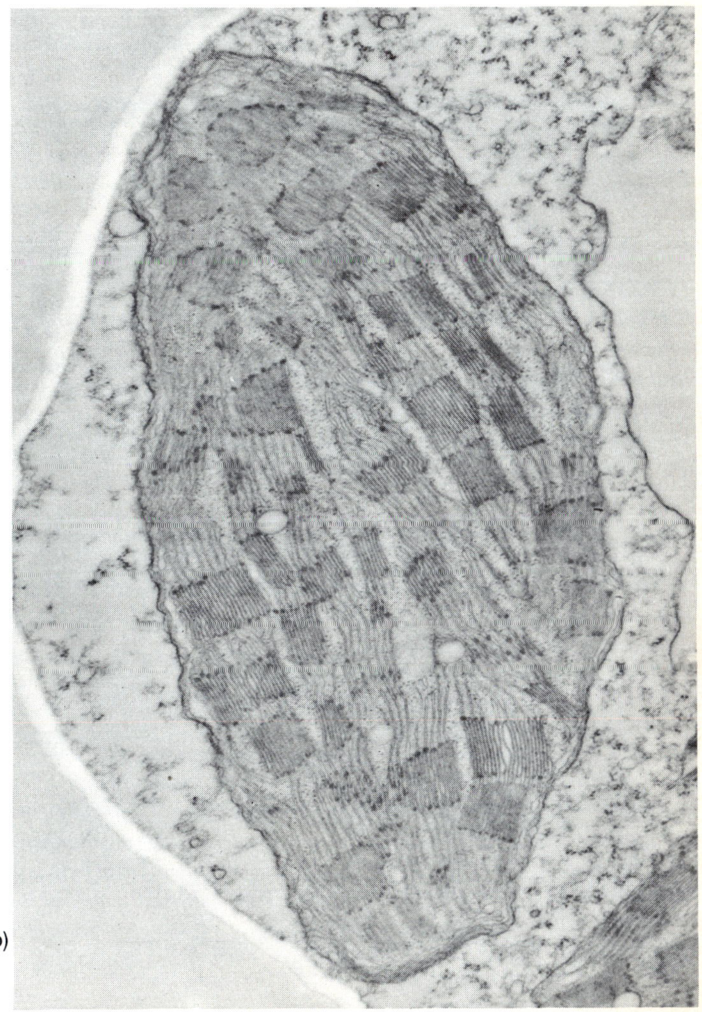

(b)

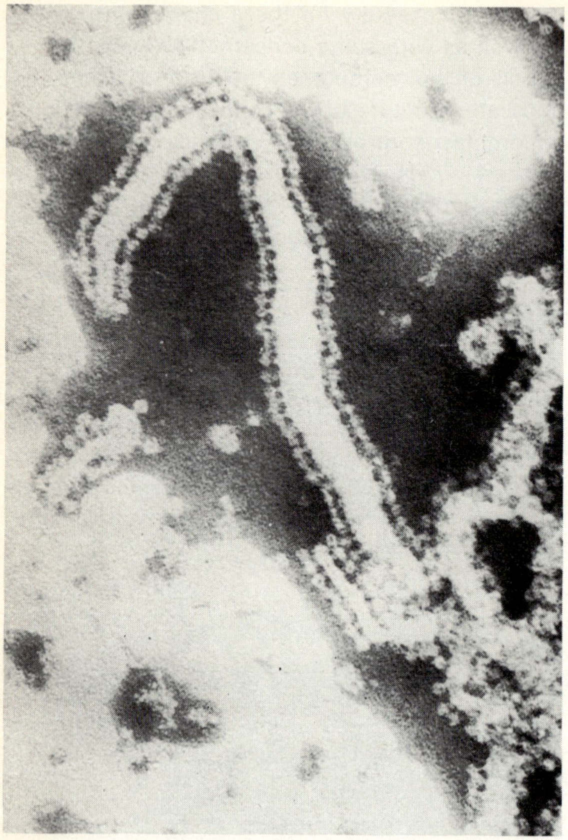

FIGURE 9.40 An electron photomicrograph of a section of the inner mitochondrial membrane. The attached small structures are where the biochemical processes of respiration and ATP formation are thought to take place. *(Courtesy of Britton Chance, University of Pennsylvania. Reprinted from* Science, *Vol. 142, 29 Nov., 1963, p. 1176. Copyright © 1963 by the American Association for the Advancement of Science.)*

Electron microscopic studies have revealed, on the inner surface of mitochondria, the presence of a large number of stalklike projections which terminate in a "head." (See fig. 9.40.) The stalk and head are thought to be hollow. It is also thought by some that the balance of hydrogen on either side of the inner mitochondrial membrane is the key to the synthesis of ATP, and that as hydrogen atoms flow through the inner membrane to the space between the inner and outer membranes, other hydrogen atoms flow back across the inner membrane through the stalk of the projection. When the hydrogen atoms reach the head, they lose some energy there as they react with ADP, inorganic phosphate, and an ATP synthesizing enzyme. The hydrogens then resume their flow down the energy gradient of the cytochrome carrier system. (See fig. 9.41.)

In the chloroplast, there appears to be an arrangement of membranes that resembles stacks of coins with interconnecting bridges. The bridges usually enter one stack at a point above or below that at which they left the other stack. The stacks, or *grana*, seem to contain the greatest concentration of chlorophyll and, thus, are believed to be the sites where the radiant energy of the sun is trapped. The bridges seem to serve as the entry points for NADP and ADP, and as exit points for NADPH and ATP as well as being the location of the greatest concentration of inorganic phosphate. (See fig. 9.42.)

There is still a great deal to be learned about the relationships between the structure and function of chloroplasts and mitochondria. They have been seen with the electron microscope, but exactly how they function to convert energy is not known in great detail.

In Chapter 6, we said that almost all of the DNA in the cell is concentrated in the nucleus, and that the development of the cell is chiefly controlled by nuclear DNA. There is a great deal of evidence to suggest that the chloroplast and mitochondrion are *not* under the complete control of the nucleus. One of the most impor-

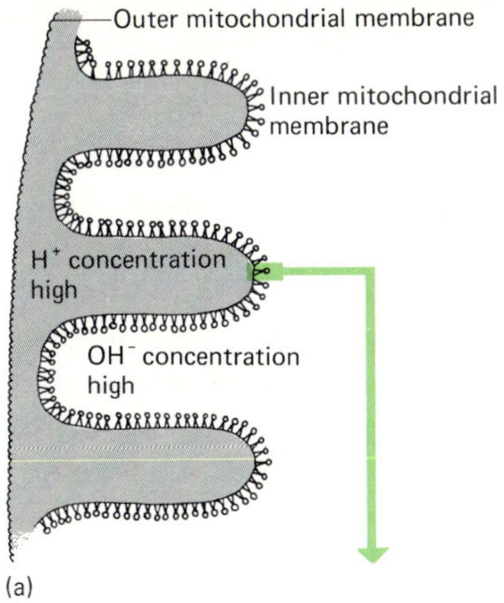

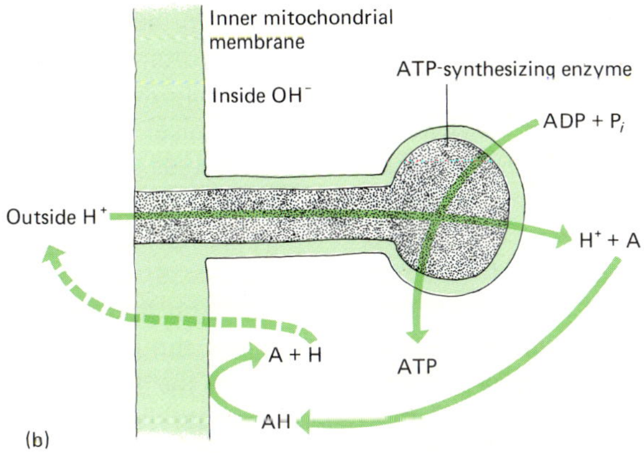

FIGURE 9.41 The flow of hydrogen in mitochondrial inner membrane subunits. Hydrogen ions that pass into the interior of the mitochondrion through the "stalks" are thought to be transported back through the membrane after bonding to an unknown compound, as above in the reaction $H^+ + A \rightarrow AH$. *(Figure 9 from "Energy Transduction in Cells" by Robert E. Beyer, in* Topics in the Study of Life, *the BIO Source Book. Copyright © 1971 by Harper & Row, Publishers, Inc.)*

tant clues that led to this conclusion was the discovery of a significant and constant quantity of DNA in mitochondria and chloroplasts. Such DNA has been found to differ significantly from that found in the nuclei of the cells from which the organelles were taken. Using ultraviolet light, investigators have been able to induce mutations in nuclear DNA that had no effect on the DNA in these organelles, and the reverse situation has been possible, though more difficult. The mitochondria and chloroplasts appear to be capable of synthesizing new DNA using their own as a template. Finally, the timing of DNA synthesis in the chloroplasts and mitochondria seems to be different from that of the cell as a whole. There is also evidence that these organelles can synthesize their own RNA, and chloroplasts and mitochondria have been found to contain their own ribosomes, although these are usually smaller than those transcribed from the nuclear DNA. There are also several

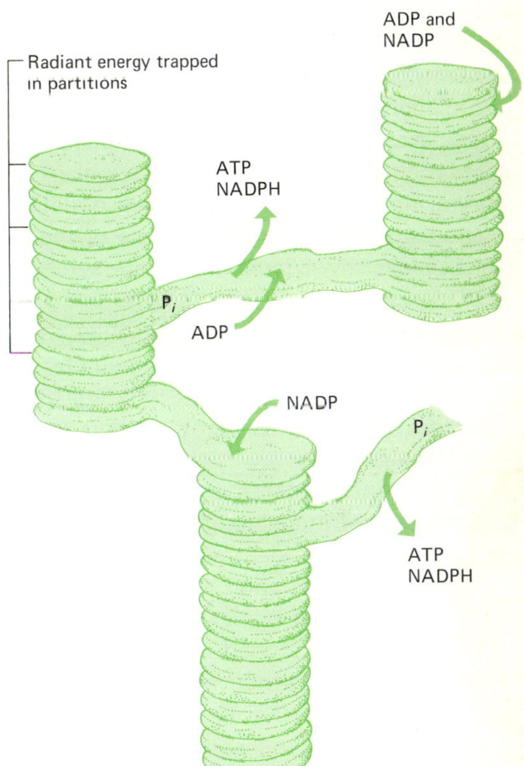

FIGURE 9.42 A three-dimensional representation of grana showing the flow of ATP and ADP and of NADP and NADPH. Inorganic phosphate seems to be concentrated within the interconnecting "bridges," and as coenzymes and ADP enter the bridges they react with P_i and other compounds in the grana to carry out the complex series of chemical reactions of photosynthesis.

differences in the ribosomes. Chemicals that inhibit the function of cytoplasmic ribosomes have no effect on mitochondrial or chloroplastic ribosomes, and inhibitors of the ribosomes of these organelles have no effect on cytoplasmic ribosomes.

The study of a microscopic alga, *Micromonas*, has shed some light on the reproduction of chloroplasts and mitochondria. This alga contains only one of each of these organelles. When the alga divides, the organelles also divide, indicating that the cell does not manufacture these organelles from cytoplasmic constituents. Evidence supporting this conclusion was gained in the study of another alga, *Nitella*. The chloroplasts of *Nitella* are stationary, anchored in rows to a layer of cytoplasm. Using time-lapse photography, one investigator made a film in which the chloroplasts of this organism were clearly seen to be *growing and dividing independently of the cell.* More sophisticated experiments by other investigators have shown that mitochondria, too, are capable of independent growth and division.

All the evidence, then, appears to point to the independence of chloroplasts and mitochondria as functional and genetic units. One piece of information, however, places important qualifications on their independence. Some of the proteins, especially cytochrome *c*, are synthesized in the cytoplasm of the cell and are probably coded by nuclear genes. Thus, while we may speak of the independence of mitochondria and chloroplasts, it appears that they are not completely free of nuclear control; their autonomy is only partial.

The evidence that mitochondria and chloroplasts are somewhat independent entities within the cell is often used to support the symbiotic theory of the origin of higher forms of life. It is persuasive evidence, indeed, but we must remember that other alternatives have been suggested. What is important for us to understand here is that these organelles are the sites at which life is sustained and perpetuated through the conversion and organization of energy to do work.

The work for which we have thus far accounted consists largely of the synthesis of ATP. But ATP is not, in itself, the end product of the processes of photosynthesis and respiration. Rather, ATP is the currency that lies intermediate between the input of energy, either as light or carbohydrate, and the real work of the organism, surviving and reproducing. As we saw in Chapter 5, to survive some organisms must secure food and avoid predators, and to reproduce they must find a mate. The means by which evolution has provided for survival and reproduction are diverse and often fascinating. But all species share essentially the same means of surviving in the biochemical sense. All carry on the same essential processes of biosynthesis, a much less visible, but incredibly important form of work. The assembly of a polysaccharide from molecules of glucose, of proteins and lipids from amino acids on the one hand and fatty acids on the other, the postulated separation of a DNA molecule, or the movement of RNA off the DNA template all consume ATP. The very membrane that encloses the cell is made up of proteins and phospholipids, the synthesis of which requires ATP. The energy required to maintain the structural integrity of the cell and, thus, of the tissues and organs that constitute the body of any plant or animal is supplied in the form of ATP. In other words, in order to keep the body together as a functioning unit requires an input of energy. In every cell where a great deal of energy-processing must take place, such as a muscle cell, it has been found that mitochondria exist in larger numbers than in cells where little energy-processing takes place, as in bone cells. However, except for very specialized cells such as red blood cells, no cell lacks the capacity to process energy. It is a measure of the unity of life that in cells taken from almost any source, the process is essentially the same.

Much of what we know about photosynthesis, glycolysis, and respiration came from the study of relatively few organisms, but there is a convincing amount of evidence to indicate that the processes are the same for almost all living forms. The differences lie in the different products synthesized and reflect the differences that exist between species and the differing ways in which those species respond to environmental stimuli. *Understanding this unity in diversity is a most important point, for it is shared by man with every other living thing on this planet.* Basically, all life has the same source of energy and all living forms process it in the same way.

Thus far, we have examined the biochemical nature of energy transfers, and we have seen some of the systems that have evolved in plants and animals for the gathering and processing of energy. As we indicated early in this chapter, energy drives all living systems, whether we consider the cell or the ecosystem. It is now time for us to turn our attention to the latter, for it is in the processing of energy at the levels of the biological community and the ecosystem that man has had his most clearly visible effects.

Selected Readings

*Baker, Jeffrey J. W., and Allen, Garland E. *Matter, Energy, and Life.* Reading, Mass.: Addison-Wesley Publishing Co., Inc., 1970. A fine brief treatment of thermodynamics, biochemistry, and living systems.

*Butler, Warren, and Roth, Thomas. *Chloroplasts and Their Development.* New York: Academic Press, Inc. This collection of very good electronmicrographs contains little text as such. Each of the electronmicrographs, however, is accompanied by extensive captions.

Giese, Arthur. *Cell Physiology.* Philadelphia: W. B. Saunders Company, 1968. A solid, standard text that outlines many of the experimental techniques used in the study of cellular metabolism.

Loewy, Ariel, and Siekevitz, Philip. *Cell Structure and Function.* New York: Holt, Rinehart and Winston, Inc., 1969. A good introduction to cell biology. This text first treats structure (the static aspect of the cell) and the function (the dynamic aspect).

*Pauling, Linus, and Hayward, Roger. *The Architecture of Molecules.* San Francisco: W. H. Freeman and Company, Publishers, 1964. The collaboration between a Nobel Prize-winning chemist, Pauling, and an artist has produced an imaginative survey of basic chemical principles with superior, full-color illustrations.

Tandler, Bernard, and Hoppel, Charles L. *Mitochondria.* New York: Academic Press, Inc., 1972. Like the Butler and Roth book, mentioned above, this is a collection of electronmicrographs accompanied by extensive captions.

Yudkin, Michael, and Offord, Robin. *Biochemistry.* Boston: Houghton Mifflin Company, 1975. A new text that concentrates on major topics.

Goodenough, Ursula, and Levine, R. P. "The Genetic Activity of Mitochondria and Chloroplasts." *Sci. Am.,* Vol. 223 (November, 1970), pp. 22–29. Offprint number 1203.

Kendrew, John C. "The Three-Dimensional Structure of a Protein Molecule." *Sci. Am.,* Vol. 205 (December, 1961), pp. 96–110. Offprint number 121.

Zuckerman, Sir Solly. "Hormones." *Sci. Am.,* Vol. 76 (March, 1957), pp. 76–84. Offprint number 1122.

*Paperback edition available.

energy in ecosystems

chapter ten We have attempted in the preceding chapters to emphasize how organisms interact with each other and with their environments. At times, it was necessary to concentrate on the internal workings of organisms and of cells, but only to understand better how an organism functions within the biosphere. It was necessary, for example, to understand cellular and molecular genetics to conceive of the intricacies of the process of evolution. Thus, we have spent time with the ideas of energy transport, photosynthesis, and respiration to set the stage for a discussion of how energy and materials are transferred and cycled between organisms in the environment.

The organisms that make up the biological communities of the world do so because they survive to reproduce. Thus, they must have adaptations enabling them to obtain and process the needed energy and materials. We have seen how natural selection has produced these mechanisms, how some organisms process energy and chemicals, and how different species obtain their food. Our understanding, however, is incomplete. An organism is only a small portion of a species population, as a species is only a small part of a community, and a community is only a small part of the biosphere. It is now necessary to put all of these elements together to form a picture of the whole.

Sunlight, acting on vegetation, is the initial source of energy that drives the functioning ecosystem. *(Courtesy of Pierre Berger from Photo Researchers, Inc.)*

Thus, we must look at three things that go on within a community: (1) the transfer of energy and materials, (2) the growth and interactions of species populations, and (3) the formation, fluctuation, and stability of communities and ecosystems.

The Role of Plants—Primary Productivity

There is a rich diversity of plants on this planet, including single-celled phytoplankton, grass, wheat, ferns, mosses, trees, from sequoias to oaks to palms, and even the green algal scum on the shores of a polluted lake. They have a wide variety of means for obtaining sunlight and inorganic nutrients for conversion into carbohydrates, lipids, and proteins to sustain life. As we have seen, though, they all share a basic biochemical mechanism for the conversion, and thus, we can look at their role in a community in terms of certain generalizations.

Incoming Solar Radiation

To make the first, and perhaps most important of the generalizations, we can consider almost any aggregation of plants from a field to a tropical rain forest. Light energy enters the community as ''white light,'' a combination of all the wavelengths in the visible spectrum. Some of it, as we have said, is reflected by the leaves, some heats the leaves, and some passes through or by the leaves to heat the ground or be reflected back into the atmosphere. Some is absorbed by the leaves. One of the first questions that we can ask is how much energy is available to the community? The energy that reaches the outer atmosphere amounts to 2 calories per square centimeter per minute (2 cal cm^{-2} min^{-1}) on a plane perpendicular to the

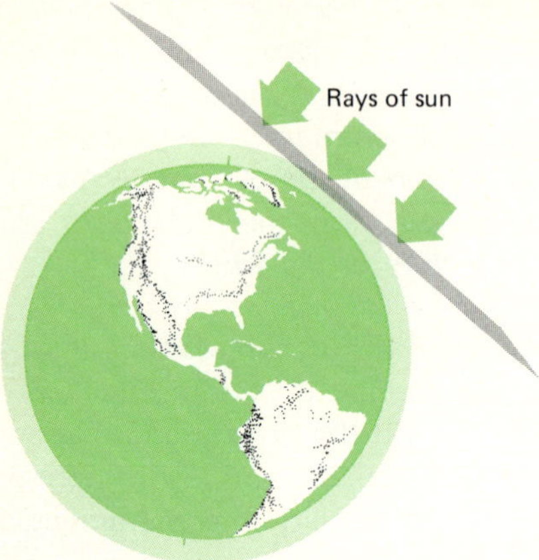

FIGURE 10.1 The solar constant is defined as the amount of radiant energy striking a plane perpendicular to the sun's rays at the outer limits of the atmosphere.

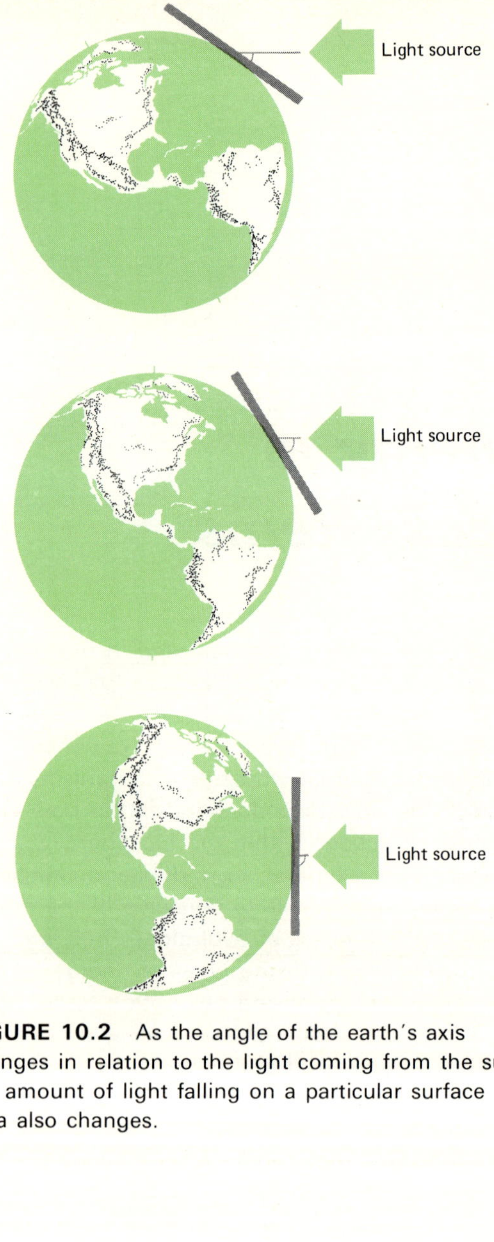

FIGURE 10.2 As the angle of the earth's axis changes in relation to the light coming from the sun, the amount of light falling on a particular surface area also changes.

sun's rays; this amount is called the *solar constant.* (See fig. 10.1.) Only about 67% of the solar constant reaches the surface of the earth, since some is reflected by the atmosphere and some is absorbed by and heats the atmosphere itself. Thus, at the surface of the earth on a clear day with the sun directly overhead, the maximum amount of radiant energy available from the sun is approximately 1.34 $cal/cm^2/min$. The amount of energy that reaches any particular location also varies, depending upon the *angle of incidence* of the light. Since the earth is not flat and the light is coming from a single source, some of it must pass through a greater amount of atmosphere and, thus, give up more energy before it reaches the surface. As the angle of the earth's axis changes throughout the year, those points on the surface where the angle of incidence is acute undergo an increase in the amount of light available as the angle becomes increasingly obtuse (see fig. 10.2). Similarly, the amount of light reaching those places where the angle is obtuse is less as the acuity of the angle increases. Thus the amount of energy available to plants depends in part upon the time of the year and the latitude at which the plants are growing. The effect of the angle of incidence is also seen in the

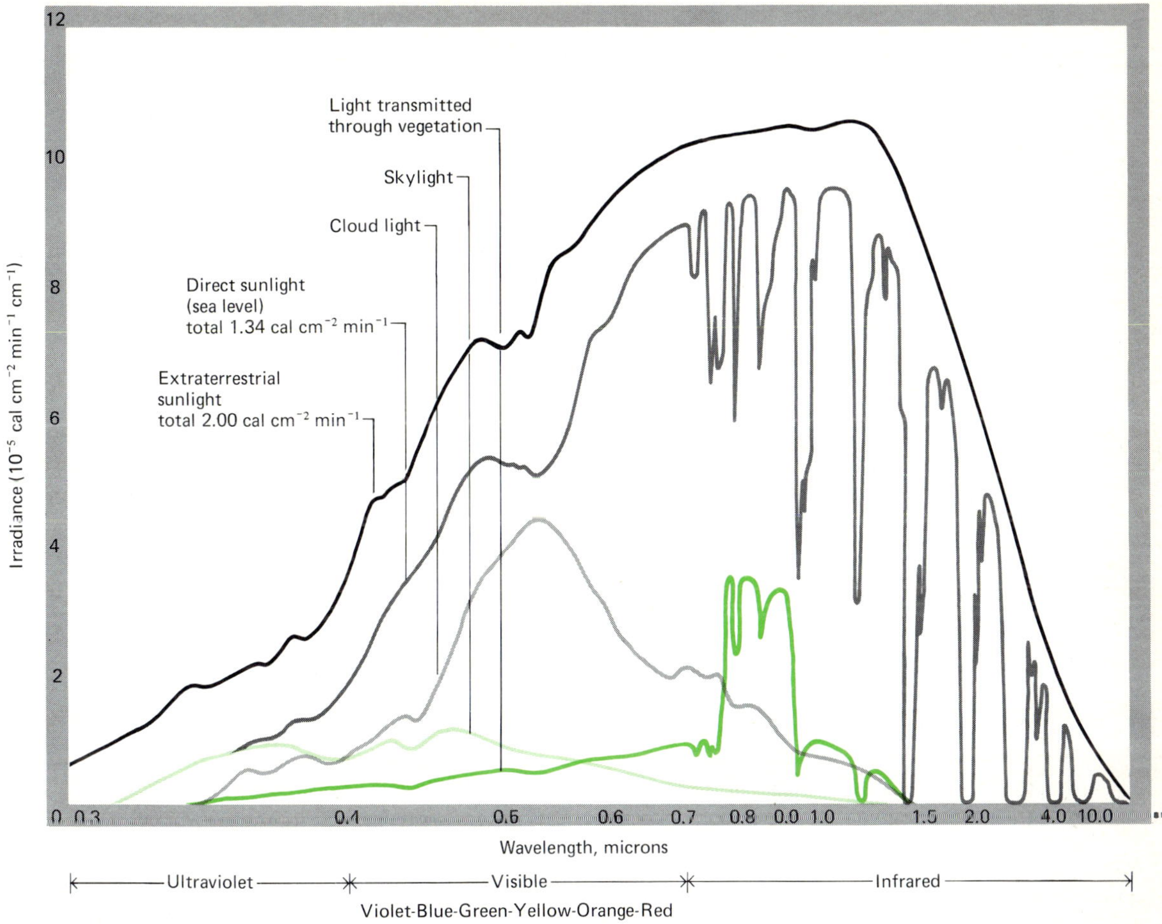

FIGURE 10.3 The distribution of solar radiation under different conditions. *(From E. P. Odum, 1971, after Gates, 1965.)*

length of time that plants are exposed to the light of the sun during the day. As the earth rotates, the angle of incidence is relatively acute in the early morning and grows more obtuse until noon, when it begins to grow acute again. Thus, the combination of the time of day, the time of year, the latitude at which plants are growing, and the region of the earth affect the amount of photosynthesis that may take place within a community.

In addition to those factors, we must consider climate as well. The amount of water vapor or cloud cover over a particular location affects the amount of energy available at the surface of the earth. And finally, vegetation itself reduces the amount of light available beneath it; thus, on the floor of a forest the amount of energy available to plants is further reduced. Figure 10.3 shows the distribution of solar radiation available at the earth's surface at noon on a clear day, on a cloudy day, and beneath vegetation.

We should also point out that plants living in water face similar constraints on the amount of energy available for photosynthesis. Those plants living nearest the surface receive the greatest amount of light and, of course, absorb some of it in particular wavelengths. The water itself also absorbs a great deal of light. Thus, in bodies of water, the availability of light energy

FIGURE 10.4 The shapes of tree crowns are probably in part due to adaptation to the angle of incidence of light at different latitudes.

diminishes as depth increases, and it diminishes more rapidly than in an equal distance in the atmosphere. Consequently, aquatic plant life is limited to the lighted zone of bodies of water. In the clearest ocean water, most active plant growth is limited to a depth not exceeding 30 meters. In some very turbid lakes and ponds, at depths of greater than half a meter there is not enough light to support photosynthesis.

Clearly, then, the amount of energy available at the surface of the earth is a highly variable factor, and 1.34 cal/cm^2/min is a *maximum* value, subject to significant decrease. Nevertheless, there are many adaptations in the plant world that have permitted plants to ''fit'' the constantly varying amount of available light. Adaptations to the angle of incidence of light can be seen in the shapes of crowns on trees. On palm trees, the leaves are concentrated at the top, since most of the available light is directly overhead. Deciduous trees found in the temperate zones of the world are adapted to the angle of incidence of light in those zones and, therefore, have a large mass of leaves extending relatively further down the trunk of the tree than the leaves of the palm. In the subarctic regions of the world, where the angle of incidence is more acute, tree crowns are more conical, with relatively less photosynthetic area being found at the top of the tree and far more on the sides. (See fig. 10.4.) The competition for light among plants has produced selection for height in some instances, and as we have seen, this has brought about the evolution of vascular systems in some plants. In phytoplankton, flotation mechanisms have evolved that enabled the species to remain close to the surface where the amount of available light is greatest. Other species of plants have adaptations that permit more efficient utilization of light available. Mosses and other shade-tolerant plants absorb light more efficiently. Vines and epiphytes ''climb'' walls or trees to get above vegetation canopies. Blades of grass are adapted to cover more area and thus expose more photosynthetic surface to light.

The Utilization of Light

Though many adaptations to light availability may be found in the plant kingdom, we may still ask whether those adaptations have allowed plants to utilize all of the available energy that enters the biosphere. The answer, as we know, is no. Over a year, under the most favorable conditions, only about 50% of the energy reaching the surface of the earth is absorbed by plants. Obviously, there are places where nearly all of the energy is lost to the biological community, such as deserts or cities. Under very good conditions, the amount absorbed may exceed 50%, but that is rare. If it occurs in a deciduous forest, for example, the loss of leaves in winter will bring the average to less than 50%. Most of the light absorbed, however, is of unsuitable wavelengths or reaches the leaf without activating photosynthetic pigments. This light only heats the leaf. A certain amount of the light energy that is absorbed by the plant serves to evaporate some of the water produced as a by-product of photosynthesis, which the plant gives off into the environment. This process (transpiration) accounts, in part, for the kind of differences that can be found between the desert climates where there are few plants and tropical rain forests where there is an abundance of plant life. It is the light absorbed by chlorophyll and the accessory pigments that makes energy available to the biological community. Though the other light energy that enters the biosphere can have important effects, it is the *energy that activates chlorophyll and accessory pigments in the plant to produce chemical energy* (called *gross primary productivity*) that is the driving force of all living systems.

Gross and Net Primary Productivity

As a percentage of the total energy available at the surface of the earth, gross productivity depends on a variety of conditions: The species present in a location, the time of year, the availability of nutrients and moisture, temperature, and so on. Under favorable conditions, gross productivity averages around 1% of the total energy reaching the biosphere, and since only about 50% of the total reaching the biosphere is absorbed by plants, gross productivity is about 2% of the energy absorbed.

Some of the energy is used for the plants' metabolic activities. Some goes into heat, and some is lost due to the inefficiencies of the process of photosynthesis itself. Some is used directly by the plant without ever being converted to a carbohydrate, and of the energy that does get converted to carbohydrates, some is used for the metabolism of the plant. Potatoes, for example, are energy stores for the plant, and the energy may later be used by the plant for other metabolic activities. Some of the gross production is used for growth, which is accomplished in multicellular plants by both the enlargement of cells and by the production of new cells.

The physical enlargement of the plant is a form of energy storage, whether it is in the form of high-energy carbohydrates, or cellulose, or even lipids, as in some algae. This stored, or ''excess,'' energy is known as *net primary productivity* and is *the energy that remains in some material form after the metabolic activities of the plant have been accounted for.* Net productivity is available either to the plant at some later time, for example, for seed production, or to other organisms such as herbivores and decomposers.

Thus, we can write a series of four word-equations to describe what happens to the radiant energy of the sun that reaches the biosphere:

(1) Energy (E) arriving at the biosphere minus (—) energy absorbed by atmosphere = energy available at surface.

(2) E available at surface — E reflected from surface — E passing by or through plants = E absorbed by plants.

(3) E absorbed by plants — E heating plants — E evaporating water in plants = gross productivity.

(4) Gross productivity — plant respiration = net productivity.

Though apparently ''lost'' to the biological community, the energy that arrives at the biosphere but does not go into gross productivity is not without its importance, since it largely determines the climate of the earth and becomes a part of the global energy budget. It is important to note that a part of the energy running the climate of the world does enter the plant community (Equation 3), and even though it does not contribute directly to productivity, it does have an indirect effect, since climate affects gross productivity.

We have said that as a percentage of the energy absorbed, gross productivity averages about 2% under favorable conditions. Some biologists estimate that as much as 50% of the gross productivity becomes net productivity under similarly favorable conditions. They are quick to point out, however, that such conditions are hard to maintain over a long period of time. Taking all of the energy "losses" into account, we may rewrite the photosynthetic equation and see how much solar energy actually reaching the biosphere becomes available to plants and then to other organisms:

$$nCO_2 + nH_2O + 1{,}300{,}000 \text{ kcal light} \xrightarrow{\text{chlorophyll}} 26{,}000 \text{ kcal of } (CH_2O)_n + nO_2 + nH_2O + 1{,}214{,}000 \text{ kcal heat}$$

Gross production (26,000 kcal) — respiration (50%) = 13,000 kcal net productivity available to plant at a later time or to plant consumer

The problem with such general equations is that they assume that all plants are equally efficient, that favorable conditions are prevalent, and that all gross productivity goes into some form of carbohydrate. It becomes frustrating to have to deal with "typical" plants and "average" efficiencies and productivity. Part of the reason for having to take this approach is that there are not enough data available on individual plant species, and even if there were, it would be difficult to put the data together in some meaningful way.

Measurements of Gross and Net Productivity

Very frequently in the past, plant production was measured in terms of weight or volume. Agriculturalists, after all, are interested in selling bushels or tons of plant material, and the energy value of their products is of little importance to them. Such measures are unsatisfactory and inconvenient, however, when dealing with energy. They are particularly inconvenient when dealing with a community or ecosystem in which there are highly varied plant species, each with its own characteristic energy value. Thus, the most meaningful measurements of productivity are made in terms of calories and kilocalories. One may then take samples of the various species in a community, establish the energy value of each species, and arrive at figures for the gross and net productivity in a community. The measurements of kcal in some weight or volume of plant material vary greatly from plant to plant and between parts of the same plant. Such variations are due to varying H_2O content, affecting weight, "air" content, affecting volume, and varying bond strengths of different chemicals, affecting energy storage. Nevertheless, such measurements allow us to examine the total gross productivity of the community in relation to the amount of solar energy available, to determine net productivity and how efficient both gross and net productivity may be, and to determine how much energy is available to the rest of the community.

Moreover, measurements in weight and volume are unsatisfactory, because they are usually expressed in static terms, such as 40 bushels of wheat per acre. Such measurements tell us nothing about the time required to synthesize all of the compounds that make up a bushel of wheat. Gross productivity and net productivity, it must be understood, are *rates of energy transfer.* The solar constant is expressed as a quantity of energy reaching the biosphere *within a specified time period.* Net productivity is the *rate at which energy is stored,* and varies with time. For example, gross productivity at night is zero, as the plant uses energy that was stored during the daylight hours, and in different parts of the world, gross productivity varies greatly from season to season.

Two measurements often used, *biomass* and *standing crop,* are sometimes confused with net productivity. Biomass is usually defined as the total weight of living organisms, and although there is no handy definition of standing crop, it means approximately the same thing as biomass. Although both may be expressed in cal or kcal, biomass and standing crop are static measures. *The rate at which biomass is in-*

creased is proportional to productivity, but the *total biomass or standing crop is rarely proportional to productivity.* Since biomass can be expressed in units of mass or energy, we may refer to it in grams or kcal. The caloric value of a gram of protein is different from that of a gram of carbohydrate, and both are different from the caloric value of a gram of lipid. Thus, when biomass is expressed in grams, care must be taken to determine the protein, carbohydrate, and lipid composition of the organisms under consideration.

It is difficult to specify upon what productivity depends. As we have seen, it depends upon light, the amount of chlorophyll in the units of photosynthesis, and the availability of several inorganic nutrients. We have mentioned the roles of climate and latitude, but it is interesting to note that productivity in the temperate zones may be almost as high as that in the tropics due to the very long summer days.

Because of the many factors that may affect productivity, it is often necessary to make measurements for each particular community or ecosystem. Since such measurements are often difficult to make with full confidence in the data, it should come as no surprise that much is still unknown about productivity in real communities. However, we should discuss the methods used to measure productivity.

Methods of Measuring Productivity

Since the key process in plant productivity is photosynthesis, productivity measurement must be based on the products and reactants involved. Water can be omitted from consideration, because there is so much of it in the environment that it would be impossible to detect change; that leaves CO_2, O_2, carbohydrates, and plant tissue.

Using carbohydrates and plant tissue as a means of measuring productivity is called the *harvest method* and is useful in determining *net* productivity. Like the other methods we shall examine, the harvest method has certain limitations. It is most useful in situations in which plants are not being consumed as they grow and in which plant growth continues during the period when measurements are being made. Thus the harvest method is useful in measuring the productivity of agricultural systems, especially if an attempt is made to reduce the consumption of plants by animals such as insects, grazers, or birds. The method consists essentially of weighing harvested material and computing its caloric value, thus converting biomass to kcal. It must be done over a period of time that includes several sample harvests rather than a single harvest at the end of a growing season. This is especially true if the system being measured is not a cultivated field, but a natural one containing several plant species that mature and die at differing times of year, such as grasses. One of the drawbacks of the harvest method is the necessity of estimating the consumption by the plants of stored energy (that is, plant respiration), and the consumption by animals of the plants.

Certain measurements can be made using CO_2 as an index of photosynthetic activity. One such *carbon dioxide method* is to enclose a community in a transparent chamber, measure the amount of CO_2 in the air pumped into the chamber, and measure the amount remaining in the outgoing air. The problems associated with this method are obvious. The chamber may act as a greenhouse, and this may alter the rate of photosynthesis, and there is a limit to the size of the community that can be so enclosed. Again, net community production is measured, and other measurements must be taken to arrive at a value for gross production.

One very sensitive, and somewhat more recently developed, method involves the use of radioactive tracer materials. Using a radioactive isotope of carbon (C^{14}), investigators have been able to determine production in aquatic plants. Phytoplankton or other small aquatic plants are trapped in bottles to which a known quantity of the radioactive carbon is added. After a time, the plants are removed and dried, and the amount of radioactive carbon taken up is measured. Certain corrections have to be made for nonphotosynthetic carbon uptake before the real rate of photosynthesis is determined, but then productivity can be calculated from the rate of C^{14} uptake.

A final method is measurement of *oxygen,* from which an investigator can derive a value for gross production. This method is widely used to measure productivity in marine and aquatic communities and consists of taking samples of water from varying depths, measuring the oxygen content of the water, placing the water in a dark bottle and a transparent bottle, and returning the bottles to the depths from which they were taken. Photosynthesis will continue in the light bottle, but since no light enters the dark bottle, all that occurs there is respiration. After a time, the two bottles are removed from the water, and their oxygen content is measured again. Since the amount of photosynthesis that has occurred is proportional to the oxygen produced, the increase in oxygen in the light bottle will indicate productivity. However, since plant respiration in the light bottle has consumed oxygen, the oxygen increase in the light bottle will indicate *net* productivity. The measurement of oxygen in the dark bottle, where only respiration occurred, will indicate how much respiration took place in the light bottle. Thus, the amount of gross production can be calculated by adding the quantity of oxygen consumed in the dark bottle to the quantity of oxygen produced in the light bottle. (See fig. 10.5.)

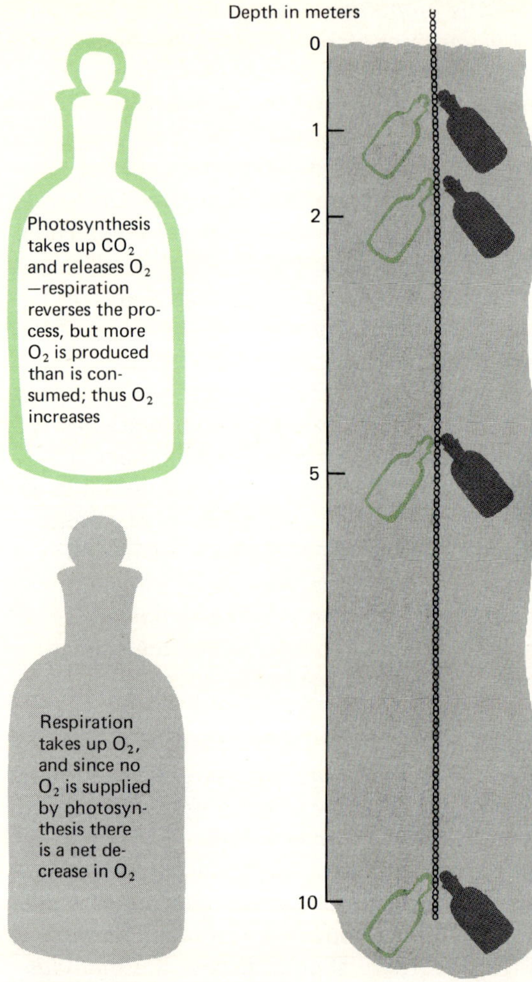

FIGURE 10.5 The rate of photosynthesis of the plankton community can be measured by measuring the differences in uptake and release of oxygen in transparent and darkened bottles.

Obviously, the same method cannot be used in all situations, although adaptations of the method involving the light and dark bottle have been used in terrestrial ecosystems. The enclosure method to measure CO_2 is a sort of ''light bottle'' on land, and a dark enclosure that is analogous to the dark bottle can be built to measure gross production on land. But in almost any system in which we would want to measure productivity, it is extremely difficult to exclude consumers whose respiration has a great effect on the measurements. Also, the amount of energy fixed is not in all cases proportional to the amounts of O_2 produced or CO_2 taken up. For example, some algae store lipids, not carbohydrates. As we have seen, carbohydrates contain more oxygen than lipids, and thus, in a community in which most of the energy storage is in the form of carbohydrates, less oxygen would be given off than in a community in which energy was stored in lipids. Using the oxygen method alone, it would appear that the carbohydrate-storing community would have a lower level of productivity than the lipid-storing community. Since lipids can contain approximately twice as much energy per unit of weight as carbohydrates, it is necessary to determine the biomass of an algal community, in addition to its oxygen production, if any comparative conclusions are to be drawn about its productivity *vis-à-vis* the carbohydrate-storing community.

Photosynthetic coefficients which make it possible to convert the materials measured to the amounts of energy fixed can be established and these coefficients, which vary slightly from species to species, must also be taken into account. Thus, to measure productivity for any community in a meaningful way, it is necessary to know the relationships between O_2, CO_2, the material produced (harvested), and any other relevant data. Unfortunately, the accumulation of such data requires a great deal of time and manpower, and at present we do not have complete and accurate data for *any* natural community. Thus, the data given in tables and figures in this book are only estimates.

The Implications of Productivity Measurements

There are many practical implications to the understanding of biological productivity. We will discuss some of the material factors that control productivity in the next chapter, and the importance of these implications will then become clearer. However, at this time, it is useful to consider some examples of how productivity affects the biosphere and what this means to man.

The productivity that we have been discussing, because it is the first input of energy into biological systems, is usually called *primary productivity,* and it is this input of energy that accounts for many of the differences between a tropical rain forest, a desert, a tundra, and a grassland. Understanding how this is controlled and the organisms responsible for it will do much to help man understand the management and use of major communities.

Equally important, the determination of productivity may reveal how much biological energy is available to man. A fertile river valley or a deciduous forest will support more people in more comfort than will a desert. Also, most of the world's fisheries are in fertile coastal waters where phytoplankton productivity is highest.

Excess primary productivity—that is, primary productivity that is not consumed by herbivores—can pose major problems. (See fig. 10.6.) The green scum of polluted lakes or bodies of water like Lake Erie, parts of San Francisco Bay, Jamaica Bay in New York, and innumerable small lakes is due to excess production. (See also Chapter 11.)

It is primary productivity that man controls and encourages by the practice of agriculture. Of the total ice-free surface of the earth, 24% is considered to be arable, though barely half of that is actually under cultivation. (The proportion that is not being farmed is considered by some to be marginally arable.) The practice of agriculture is one of the major ways in which man disrupts ecosystems. When we artificially select a new strain of corn or wheat, we are trying to increase primary productivity, although our interest is usually centered on productivity in one part of the plant. There are many implications to this, and although we will discuss

FIGURE 10.6 The masses of living, dying, and decomposing algae that develop when too many nutrients are added to a lake are unsightly, produce odor problems, and liberate materials that may damage other life in the lake. *(Courtesy of Grant Heilman.)*

them in more detail later, we can look at some of them here.

It is often argued that food supply is a limiting factor in the "third world." Consequently, many attempts have been made to increase productivity there. But such increases entail "costs." The first is the sacrifice of natural communities. The United States was once an undeveloped nation, and the cost of development included using the virgin forests and prairies to such an extent that there are very nearly no virgin forests or prairies left. The loss of a wild area, a truly natural community, is often an esthetic concern and is sometimes belittled as such, but we will have occasion to comment on that later. Some other costs of agriculture are more direct. The genetic alteration of many food plants to increase yield or nutritional quality has sometimes resulted in strains that are less efficient in other ways, like nutrient procurement. Thus, it has become necessary for us to supply the plant with manufactured fertilizers. To increase the amount of energy available to us, we must control the use of energy by other organisms. We must control or eliminate consumers such as insects with pesticides, build roads to transport the food from agricultural areas to population centers, and power the machines that have been designed to harvest and transport the food. Each of these activities requires the expenditure of fossil-fuel energy, and thus, each is a form of *energy subsidy.* The economic costs of these activities are included in the economic value of food. All too often, however, the energy costs of these activities are not computed when the energy value of the food is calculated. It has been suggested by some that we have passed the point at which we are expending greater amounts of energy to produce food than are present in the food being produced. The dollar costs alone of supplying these energy subsidies are often very high, and mean that if a presently underdeveloped nation wishes to increase its food production to keep pace with a growing population, it is faced with raising immense amounts of capital. And, ironically, the greater the investment, the smaller the return. In a nation such as the United States where agriculture is industrialized, the energy input per unit of area is ten times greater than the energy input for an equal area in Asia or Africa. However, the amount of food produced is only three times greater. Moreover, to achieve an optimal output, we often eliminate other plant species from the areas we wish to cultivate and rely on *monoculture,* or the cultivation of a single crop. While this does result in better yields for a single crop from a given amount of land, it is also subject to spectacular failures in the form of potato, rice, or wheat famines. Monoculture permits a single species of competitor to have a "field day" or a single abiotic factor such as drought to have an amplified effect.

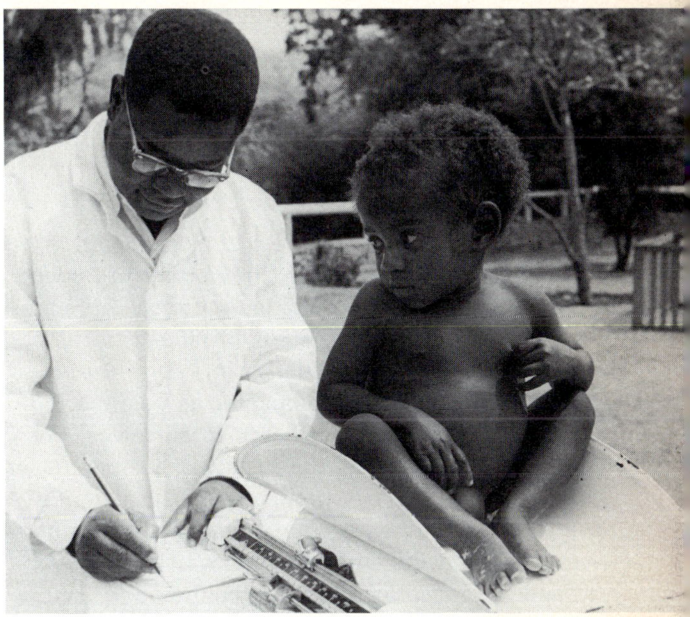

FIGURE 10.7 Kwashiorkor is caused by a lack of dietary protein. The most noticeable characteristic of the disease is a distended stomach, and the disease is usually most prevalent in infants. Sadly, many children do not receive the medical attention necessary to diagnose the disease, and even if they do the parents are often unable to provide the protein that is needed. *(Courtesy of World Health Organization.)*

An increase in carbohydrate production alone is not enough. Humans require carbohydrates, lipids, and proteins. One of the most common sicknesses in underdeveloped countries is *kwashiorkor,* which results from protein starvation.* Even a person well-supplied with calories in the form of carbohydrates can contract kwashiorkor if sufficient amounts of proteins are lacking. (See fig. 10.7.) (Although they may not have contracted kwashiorkor, some people in this country who have subscribed to strict Zen macrobiotic diets consisting solely of brown rice have suffered severely from protein deficiency, and at least one death has been reported.)

There are proteins in vegetables and grains, but these are not necessarily the proper proteins for human nutrition. Of the twenty amino acids required to assemble proteins, twelve can be synthesized from other constituents in the human body. The eight remaining amino acids are termed "essential" and must be a part of the human diet. They can be found in plants, but few, if any, plants contain all of the essential amino acids needed in sufficient amounts to sustain a human being. Corn, for example, is deficient in lysine. Thus, the most satisfactory source of the required proteins is meat, in which the essential amino acids are present in sufficient quantities. The consumption of meat, however, introduces another problem. The animals we eat must themselves eat plants, and thus, land must be allocated for protein production. Furthermore, since there is no way around the second law of thermodynamics, the animals

**Kwashiorkor* is a West African term meaning "the sickness that a baby develops when another baby is born." Kwashiorkor develops when a mother can no longer breast-feed an infant and the child must be weaned to a diet consisting almost exclusively of carbohydrates.

must consume more calories than they can pass on, and the land is therefore less productive. Indeed, in the future, we may have to eliminate the ''middleman'' (or animal), and subsist on plants. This is possible, but it is not convenient.

We will come back to these subjects again in much more detail. What is important to understand here is that agriculture requires very careful control of a wide variety of factors to achieve optimum production in a form that is usable but minimally damaging to neighboring communities, to the soil, and to ourselves. We know far too little of the variety of factors influenced by primary productivity. Our use of agriculture too often produces soil destruction and erosion and the subsequent loss of productive land. It is also an important contributing factor in water pollution. Worst of all, it contributes to overpopulation and to subsequent malnutrition and famine.

We know much about the biochemistry of photosynthesis, much about crop breeding and management, and much about how to get short-term agricultural gains. But we are grossly ignorant of the overall implications of our use of the products of photosynthesis. (See box 10.1.)

Higher Levels in the Food Chain

We have thus far largely avoided mentioning organisms other than plants, only because plants constitute the class of living things with the largest biomass and are the basis of all living systems; plants are the most important living things in the biosphere. Without animals, however, the world supply of CO_2 would eventually be exhausted, and huge quantities of ''waste'' organic materials from plants would accumulate. (The fact that the latter situation has, in fact, occurred is proven by the legacy of fossil fuels—coal, oil, and natural gas. These compounds have resulted from the deaths, millions of years ago, of plants, the remains of which were not completely decomposed by bacteria and fungi.) In the sense that we use the word here, *an animal is any organism that is a consumer of the energy stored in the organic compounds produced by plants;* that is, an animal is a consumer of O_2 and a producer of CO_2.

A better word for such an organism is *heterotroph.* Conversely, an organism that makes chemical energy available in the community is known as an *autotroph.* (Recall here our earlier discussion of the origin of life.) Most autotrophs are plants, although as we have seen, there are some bacteria that use light and H_2S instead of H_2O. There are also *chemotrophs,* which use the energy in the bonds of inorganic chemical compounds, and *myxotrophs* or organisms that can alternate between being autotrophic or heterotrophic as the occasion demands. Although bacteria and fungi may reproduce as plants, most of them gain energy as heterotrophs. They break down organic materials, usually from dead plants and animals, thereby producing CO_2 and other waste products and consuming O_2. As we shall see, bacteria and fungi may be *very* important in the energy economy of ecosystems.

What happens to the energy fixed by plants? We have already indicated this generally, but we should look at the process in more detail. Hypothetically, let us consider a cow in a field of grass. How much of the net productivity available to herbivores or heterotrophs enters the cow? To answer this question, consider the ''losses'' along the way. The cow does not eat the whole plant, but leaves the roots and the lower stems. A certain amount of material is lost due to ''sloppy'' feeding (the parts of the plant that are bitten off but never enter the mouth). Some material that may contain a great deal of energy passes through the cow undigested and is lost as feces. From the material that actually enters the cow's bloodstream, some energy is lost hydrolyzing starches or cellulose into soluble sugars, and proteins into usable polypeptides and amino acids. In a cow, some energy is consumed by bacteria that live

Energy and Efficiency

Although plants are relatively efficient at fixing solar energy and cells are relatively efficient at releasing the energy fixed by plants, we are less efficient at utilizing the energy available to us. We speak, of course, not of the biological energy in foods, but rather of fossil-fuel energy.

In the United States, currently the world leader in energy consumption, approximately one-quarter of all energy expended in 1970 was devoted to the transportation of goods and people, mostly the latter. Tables A and B reveal that most passenger miles are traveled in automobiles at a liberally estimated 32 passenger miles per gallon. Admittedly, the automobile provides us with a flexible and convenient transportation system, but its cost in energy-efficiency appears to be very high compared with rail or bus travel.

We also expend a great deal of energy in space heating; that is, heating our homes, schools, places of employment, and so on. In 1970, approximately 11% of the energy consumed in the United States was for residential space heating, and another 7% for commercial space heating. At the present, the use of electricity for space heating requires twice as much fuel as the direct combustion of oil or gas, but is largely based upon burning coal, a significantly more abundant fuel than oil or gas. Improvements in heating delivery systems are being developed that may make electrical heating as efficient as direct on-site combustion, though. No matter which form of space heating is used, however, measures can be taken to reduce its energy requirements substantially. Some of the effects of such measures are shown in table C.

TABLE A Passenger Traffic in the United States, 1970 (billions of passenger miles)

	Auto	*Air*	*Bus**	*Rail*	*Total*
Intercity					
miles	974.4	112.0	22.4	11.2	1120.0
%	87	10	2	1	100
Urban					
miles	688.7		21.3		710.0
%	97		3		100
Total					
miles	1663.1	112.0	43.7	11.2	1830.0
%	90.9	6.1	2.4	0.6	100.0

*Includes urban mass transit

(Adapted from Hirst and Moyer.)

TABLE B Fuel Consumption (passenger miles per gallon)

Automobile*	32
Conventional train	80–130
Electric Metroliner	50
Two-deck train	200
Ten-car subway train	75
Bus	100–125
Jet plane	22

*This rate of consumption assumes two persons per automobile traveling in a standard sedan. The average occupancy per auto is closer to 1.6 persons—and the passenger miles/gallon correspondingly lower at about 28 passenger miles/gallon.

(From Darmstadter.)

TABLE C Comparison Energy and Monetary Savings from Insulation of a New York Residence

	Baseline Gas	*Baseline* Electric	Gas	Electric
Wall Insulation (inches)	1⅞	1⅞	3½	3½
Ceiling Insulation (inches)	1⅞	1⅞	3½	6
Floor Insulation	No	No	Yes	Yes
Storm Windows	No	No	Yes	Yes
Savings ($/yr)	0	0	32	155
Energy Savings (%)	0	0	49	47

(Adapted from Hirst and Moyer.)

BOX 10.1

in its gut and hydrolyze cellulose. Though these microscopic organisms do make some energy available to the cow in the form of other compounds, they consume some energy themselves. Of the compounds that are finally delivered to the tissue in the cow, some is immediately metabolized and produces heat. Some, finally, is stored, but that means a conversion to new compounds and a consequent loss of energy as more heat. What is left becomes bone, connective tissue, fat, and meat.

It is difficult to calculate efficiency in the foregoing example. If the energy value of the grass eaten is compared to the energy assimilated, the efficiency will be very different from that calculated if the net productivity of plants is compared to the energy value of the tissue alone. The latter evaluates the losses incurred in feeding and the energy output that is of most interest to mankind. The amount of tissue produced, which we may call *net herbivore productivity,* stated as an efficiency, usually ranges from 1% to 5%.

Obviously, efficiencies depend on several things: the types of materials in the plants themselves, the feeding habits of the herbivore or heterotroph, the sizes of the populations of both consumers and producers, and so on. Thus, efficiency may vary greatly. On the whole, however, the upper limits of efficiency are calculated to be around 15%, and the lower limits, less than 0.1%, which may be the case when most of the plant is unused.

As an aside, we might point out that man often strives to optimize efficiency by feeding cattle (or other meat producers) materials that are easily converted to meat. Thus, grains such as corn are better than alfalfa. However, we must pay an economic and energy cost for this activity, since we are using only a small part of the plant's productivity.

There are probably no natural food chains as simple as our cow–grass–man example, so we should consider some of the conditions that prevail in the real world. First, there are many kinds of plants available to consumers, each having a different energy value and each being fed upon selectively by different organisms. Second, a cow, like any organism, has competitors that consume some of the net productivity available, typically including insects and birds. Third, herbivores are themselves eaten. Every food chain has a *top carnivore,* in many cases, man. Fourth, the "waste materials" of net productivity—roots or the leftovers of sloppy feeding—are also consumed by heterotrophs. This is usually done by bacteria or fungi, but some scavengers such as earthworms and soil insects are also involved. The efficiency of transfer from animal to animal is usually higher than that from plant to animal, ranging from 20% to as much as 30% due to the similarity in the chemicals involved. There are fewer inefficiencies involved in converting one protein into another than there are in converting carbohydrates to protein.

Thus, even the simplest communities are not really very simple. When we consider what appears superficially to be a simple community, such as a prairie, we often find that it is exceedingly complex. A prairie community may consist of a number of species of grasses and shrubs as producers, grazed upon by pronghorn antelopes, prairie dogs, gophers, bison, herbivorous birds, and insects. There are a number of carnivores, including coyotes, snakes, hundreds of species of carnivorous insects, and many species of carnivorous birds. There are also numerous, less visible bacteria and fungi that decompose dead organisms of any species.

Trophic Levels

How, then, can we generalize about such complex entities as communities? One way is to use a concept already introduced, but not yet specifically named, *trophic levels.* In any community, plants and other autotrophs form the first trophic level. Next come strict herbivores, or heterotrophs that consume only plant material. Organisms that feed on herbivores are called

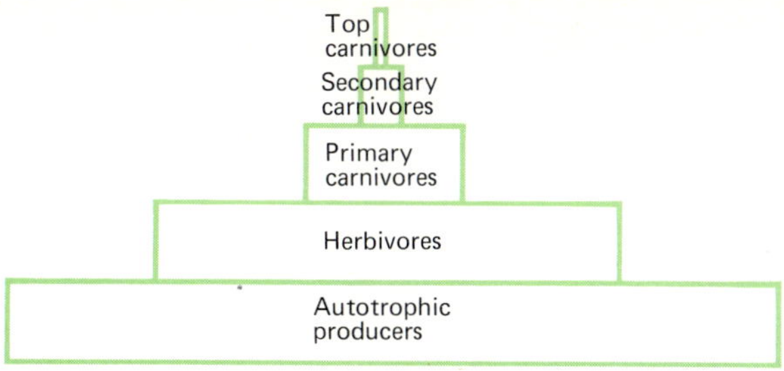

FIGURE 10.8 As energy produced by plants is passed from trophic level to trophic level less and less becomes available. A trophic energy pyramid is simply a diagram showing the relative rates of energy flow from trophic level to trophic level.

primary carnivores, carnivores that eat primary carnivores are called *secondary carnivores,* and so on. (See fig. 10.8.) Thus, we can envision a *series of conversions in which each lower stage provides an input to the next higher one.* The input can be measured either as an energy or biomass input, but if it is measured in terms of energy, there must be a loss of energy between each trophic level. Each trophic level in any community is made up of several species populations, and thus, the concept of trophic levels allows us to deal with all of the species in a complex community as coherent subunits of the community.

As usual, such generalizations have drawbacks. First, by lumping all of the organisms in one trophic level together, we must deal with average efficiencies for the whole level in that community; since species on the same trophic level may each have a different efficiency, such averages can be misleading. Furthermore, the efficiency of a species changes as the population of that species changes, and the population of a species in a community depends, in part, on the population of other species at the same or different trophic levels. To complicate matters further, some species operate on more than one trophic level. These omnivores (*omni* all) may act as herbivores at one time and primary or even secondary carnivores at another time, and the times need not be widely separated. Many birds and mammals fit into this category, and obviously man does, too. Omnivores can often be incorporated into trophic-level studies by observation of their general eating habits; that is, they can be assigned a place in the structure of the community on the basis of being, for example, 10% herbivores, 40% primary carnivores, and 50% secondary carnivores. As we indicated earlier, the idea of a food chain is something of an oversimplification. Food chains interlock to form food webs. We may find several organisms in a community feeding on a number of producers, and although a species might secure most of its energy from a single source, it is adapted to utilizing other sources as well.

In spite of the factors that complicate our understanding of communities, there is at least one factor that keeps the analysis from becoming hopelessly complex: Because the energy losses incurred at each transfer are so large, it is very rare to find a community with more than five trophic levels. And in spite of all of the qualifications and approximations, the system of trophic level analysis will still, in a general way, increase our understanding of the nature of biological communities, though it remains that anything we may learn is only a statistical approximation, and as such we are reminded to analyze the data with care.

Another major subdivision of the trophic structure of communities that we must consider, but that does not fit neatly into the categories we have so far assigned, consists of the decomposers. The "wastes" from a cow's feeding, as well as dead cows, plants, insects, and so forth, are consumed by decomposers. Though earthworms, soil insects, and carrion eaters also feed on dead organisms, the decomposers, or *saprotrophs* (*sapros* rotten or decayed) are the major consumers of such material. Their great importance lies in three key roles that they play. *First,* they clean up wastes. *Second,* they liberate CO_2, H_2O, and inorganic materials that would otherwise remain stored in dead tissue. *Third,* though they may be a small portion of

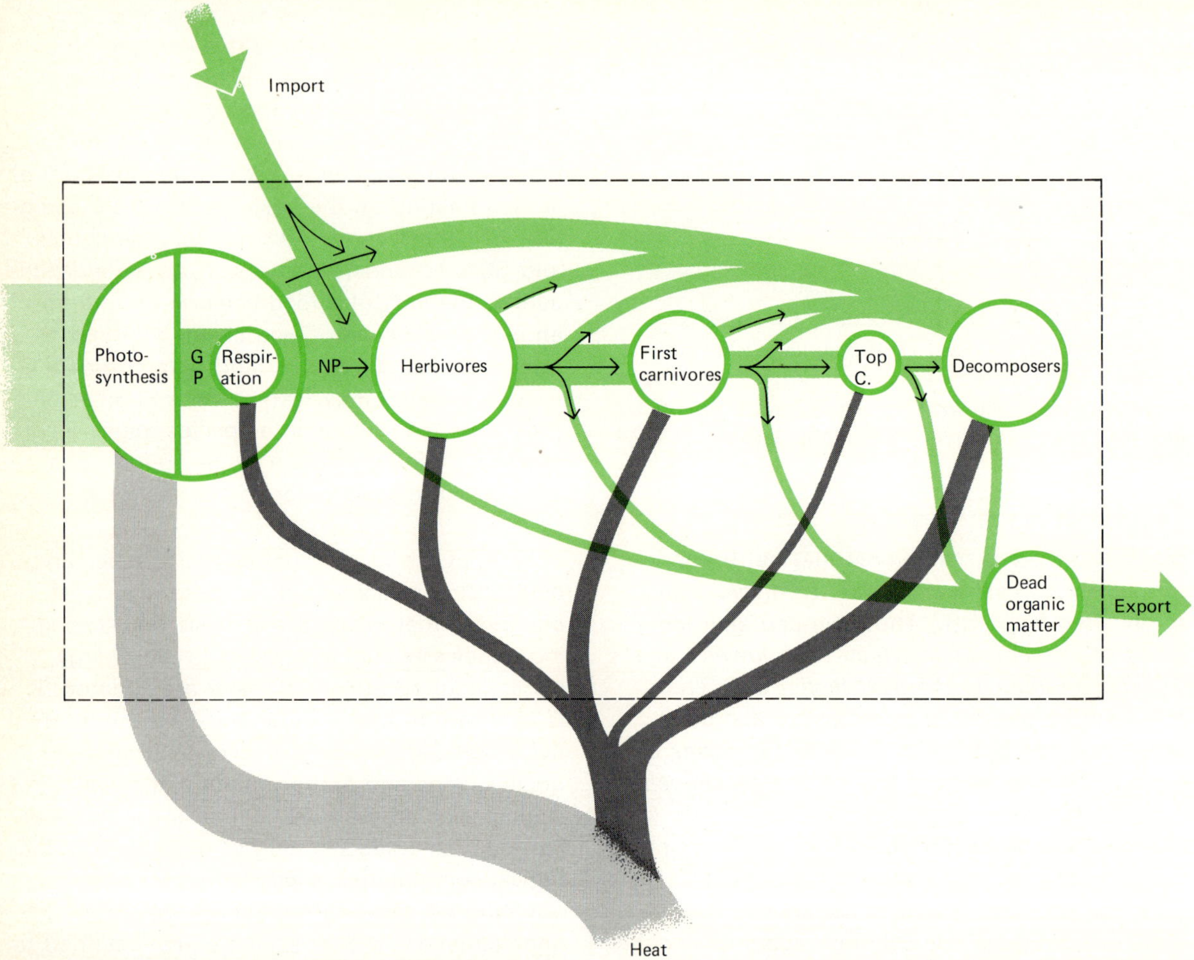

FIGURE 10.9 Energy enters the community mainly as sunlight to the plants, but there is also some import from outside. Some of the sunlight is lost directly as heat. Some is used in photosynthesis to give the gross production (GP), some of the gross production is used for plant respiration, which of course produces heat. The remainder is net production (NP), which becomes available to the animals in the community. Energy is then transferred from trophic level to trophic level with some being lost as heat from each trophic level and some being passed to the decomposers at each stage. Also, of course, some dead organic matter is exported from the system. The energy that goes to the decomposers from all of the other trophic levels is lost as heat, although some of the decomposers are also exported from the system as dead organic matter. *(Redrawn from H. T. Odum.)*

the biomass of the community, they very often carry on the major amount of the energy transfer that takes place (plants not included, of course).

The relationships between all of these kinds of organisms and energy flow in a community can best be seen in figure 10.9.

The food chains that we have thus far discussed, however briefly, are examples of *grazer* chains. There is another kind of food chain that is often overlooked, because the species involved are not easily seen. Much of the primary production of autotrophs is often not digested by heterotrophs, but broken down into small,

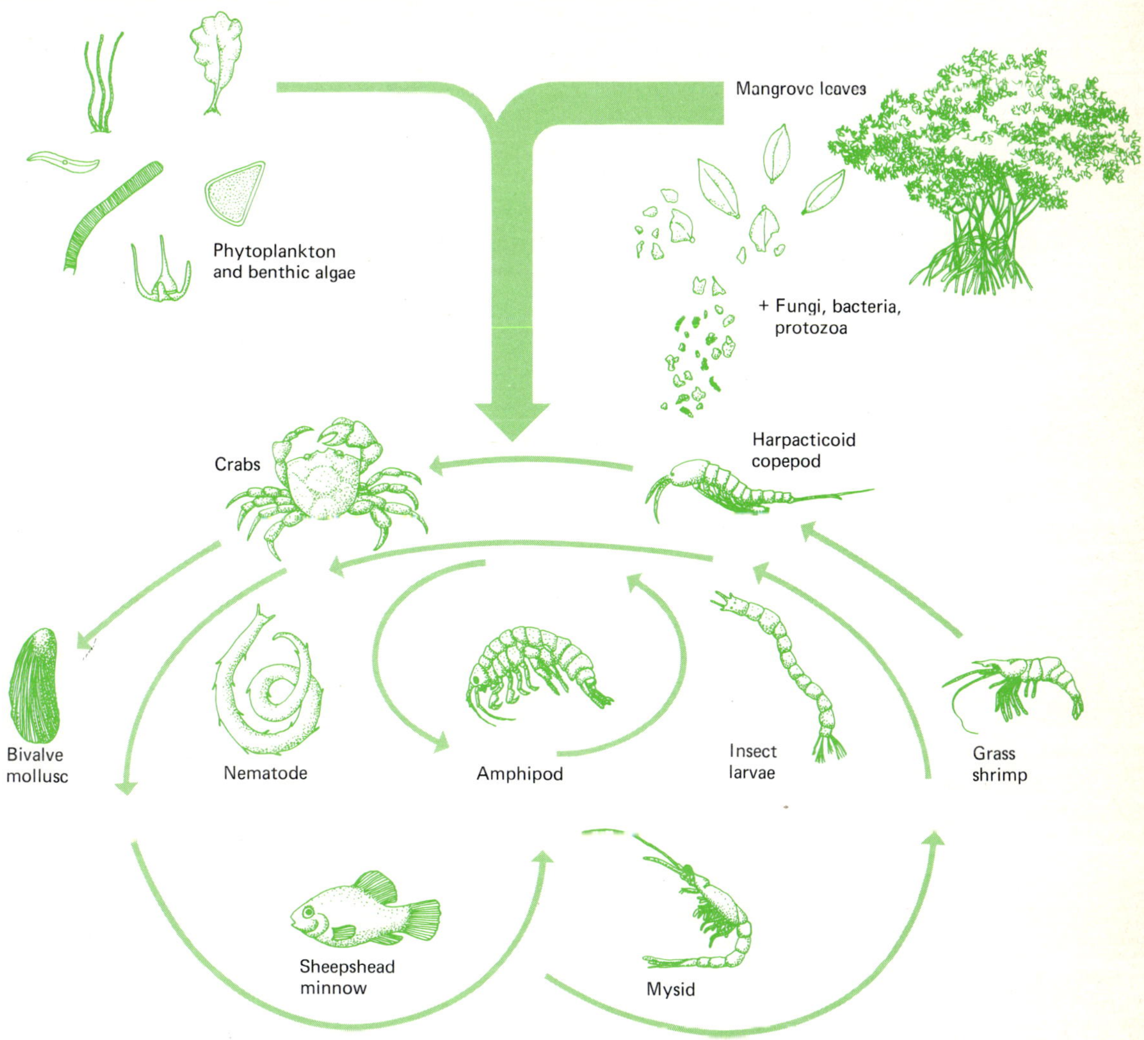

FIGURE 10.10 Most food chains are of course very complex. Also many, such as the food chain that depends upon the productivity of mangroves in southern Florida, are based on detritus, or finely divided organic material. In the mangrove ecosystem only a small amount of the energy comes from algae. The interrelationships between the higher trophic levels are very complex, and the energy flows shown here are only diagrammatic. *(Redrawn from W. E. Odum.)*

dead particles called *detritus* (L. worn away). Organisms that consume detritus, or *detritivores*, pass large amounts of the material through their bodies, some of which is then consumed by decomposers, but much of which is made available to other animals in the food chains that we have described. (See fig. 10.10.) Also, some small carnivores consume the detritivores, and so on up the food chain. Thus, detritus chains form the basis for many other food chains. The source of detritus is not by any means restricted to fecal material produced at other levels of the food chain; the best documented cases of detritus food chains begin with dead leaves from trees. A portion of the leaves may have been grazed by herbivorous insects, but that usually amounts to very little. When the leaf falls from the tree, it is consumed by detritivores such as certain bacteria, fungi, and protozoa. The resultant very small particles and the detritivores are consumed by various species of organisms depending upon the community. For example, if it is an aquatic community, various small filter-feeding shrimp, molluscs, and insect larvae consume the detritus, whereas on a forest floor, or even under a tree in a park, if the leaf litter is not removed, the small particles may be consumed by small soil-living insects such as springtails and mites or by various species of worms. In both cases, the detritivores are then consumed by organisms at the next trophic level. Obviously, it is difficult to make a fine distinction between detritus and grazer food chains, but since they form the basis of a number of food chains, some of which are of importance to man, and because they have so often been overlooked due to our interest in agricultural efficiency, we should understand their basic structure.

Trophic-Level Analysis

Most early ecological work failed to recognize the basic difference between standing crop or biomass and rates of production and transfer. One of the most common ways of presenting information about the community or ecosystem was to use *trophic pyramids.* (See fig. 10.11.)

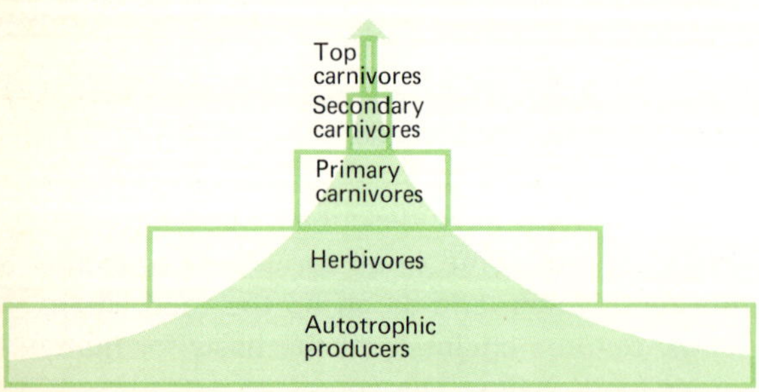

FIGURE 10.11 The energy available to organisms as it flows through the trophic pyramid always decreases from level to level.

As a rule of thumb, the construction of such pyramids was based on the idea that *the biomass of the organisms in each trophic level decreased as the separation of the trophic level from the photosynthetic activity of plants increased.* This is obviously true *if the pyramid is constructed in terms of energy transfer from level to level. It is not necessarily true, however, if biomass alone is considered.* If material is transferred very rapidly to herbivores, then the biomass of the first stage of the pyramid may be smaller than the next. (See fig. 10.12.) Such pyramids provide a useful tool for visualizing the *trophic structure* of a community by revealing a great deal about transfer rates. They are not, however, the only way to analyze a community, and they should not be understood as being a complete picture.

Another useful means of analyzing community structure is the *turnover,* which expresses the amount of energy that passes through the community compared to the amount of material in the community. The turnover in a community can be measured as turnover *rate* or turnover *time,* both of which are reciprocal. For example, if the photosynthetic producers in a pond are known to have a gross productivity of 5 grams (gm) per square meter (m^2) per day and a biomass of 10 gm/m^2, then the turnover rate for these organisms is 5/10 or 0.5 per day, and the turnover time is two days. In other words, the producers are replaced once every forty-eight hours. Terrestrial ecosystems are usually slower, and in a field with the same photosynthetic rate of 5 gm/m^2 per day, the biomass of the producers may be 500 gm/m^2. Thus, the turnover rate in the field is 5/500 or 0.01, and the turnover time is 100 days.

Thus, with a slow turnover time, the biomass of a community is ''doing very little,'' and we expect the trophic level pyramid to have a large base, with plants being the major organisms in the pyramid. Similarly, the pyramid that might be derived from a study of a pond having the sort of rapid turnover that was just shown might have a producer biomass smaller than

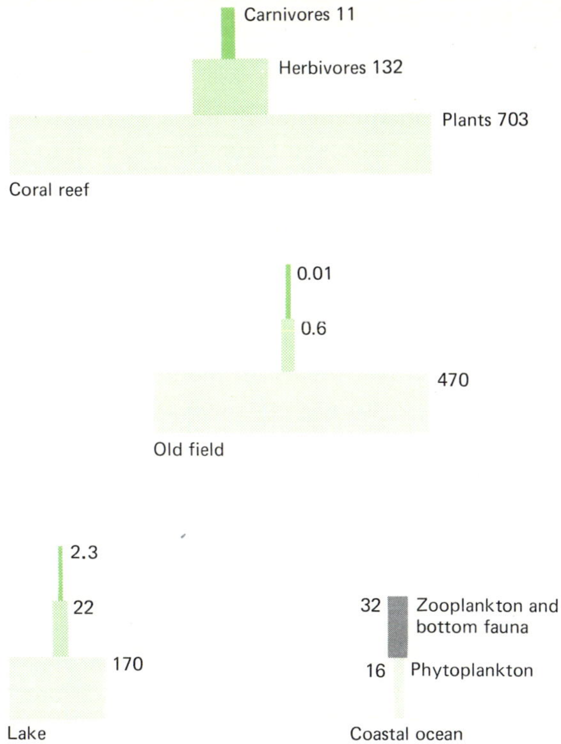

FIGURE 10.12 Biomass pyramids, like energy pyramids, usually decrease from bottom to top. This is not necessarily so, however, for a small plant biomass may support a larger biomass of animals if the productivity of the plants is used rapidly enough. "Typical" pyramids are found in the coral reef, old field, and lake ecosystems, but the coastal ocean ecosystem is frequently inverted as shown. All of the consumer trophic levels are combined in the coastal ocean example. The numbers give the grams of biomass per square meter in the community. *(Redrawn from E. P. Odum, 1971.)*

the herbivore biomass. Turnover can also be applied to higher trophic levels, in which case total assimilation is compared to total biomass.

Another useful measure is the photosynthesis/respiration ratio (P/R). If, in a whole community, the rate of photosynthesis is equal to the rate of respiration, then the community has achieved a steady state, and we can write the P/R ratio as $P/R = 1$. Such situations are not unknown in natural systems, though they are not very common. If P/R is greater than 1, biomass in the community is increasing; if P/R is less than 1, biomass is decreasing. These increases or decreases may be brought about because material is being imported or exported from the community, although an increase in biomass may be due to storage of the material produced in the community, as when a forest is growing in what was once a field. No matter what the status of a community, a knowledge of the P/R ratio can be an important tool in management of ecosystems. For example, it is to a farmer's advantage to insure that within the community he manages P greatly exceeds R in order for him to produce a crop. In contrast, if a sewage treatment plant introduces large quantities of organic material into a body of water, the effect may be the promotion of a high rate of respiration by decomposers and a situation in which P is far less than R; the P/R ratio can serve as a warning signal that the body of water into which the sewage discharge flows is becoming polluted. In a natural ecosystem, if P/R is greater than 1, it may also be considered a danger signal under certain circumstances, since it probably means that plants are accumulating. In a body of water, this is often a sign of *eutrophication* (which will be discussed in Chapter 11), and may also result from sewage discharges that carry high quantities of plant nutrients. Interestingly, treated sewage introduced into aquatic communities can sometimes increase the productivity of the community if it is introduced at a moderate and steady rate. The irregular dumping of large amounts, however, is more likely to destroy the community.

Clearly, there are a number of ways to analyze a community using the relationships between productivity and other phenomena. The foregoing is not a complete treatment, and since this is an active area of research, it is probable that other tools will be devised. No single measurement is sufficient for all communities, nor is any presently used measurement likely to be a final solution to the problems of understanding ecological interrelationships. But it is possible to analyze communities and ecosystems as whole units and to make generalized statements about them as whole units. It remains that each situation and ecosystem must be looked at separately and to do that the ecologist must remain, in part, a naturalist.

The Silver Springs Study

In order to make concrete some of the concepts and methods presented in this chapter, we will examine briefly a study in which many of them were used. Such a study was done during the 1950s by the ecologist Howard T. Odum, with the aid of a number of graduate students and some assistance from other professionals. Odum directed the analysis of a popular Florida tourist attraction, Silver Springs, and perhaps the key reason that his study is considered an important one is that it shows that such work can be done in spite of the complexity of natural communities.

The study area consisted of eighteen acres of the Springs, in which a large number of species were found. At the producer level, there were more than sixty species upon which more than fifty species of herbivores fed. There were more species of primary carnivores than either producers or herbivores, though as we will see, the biomass of primary carnivores was smaller than that of producers or herbivores. At the top carnivore level, six species were found. The decomposers were quite difficult to identify by species, but some twenty variants of species were found. The decomposer biomass was very small, but in terms of the energy flow in the community, decomposers played a very large role.

The task of identifying all of these species was by no means easy, and Odum had to turn to several persons to complete an accurate catalogue of species in the Springs. Observations also had to be made of the feeding habits of the various species to determine whether they remained at one trophic level or fed at different levels. Each species in a trophic level was obviously not equal to each of the other species in that level in terms of population or biomass or contribution to energy flow, and it was necessary to measure the species in the various levels in terms of their roles in the community. At the producer level, for example, it was found that twelve species were dominant, and these twelve constituted more than 95% of the biomass of the producer trophic level, thus contributing nearly all of the energy fixed in the community. The remaining species, more than fifty, constituted less than 5% of the biomass and made a negligible contribution to the energy flow of the community. Observation also revealed that the only changes that took place in the community's species populations were seasonal changes, and that over the time of the study (several years), there was no significant migration into or out of the community.

The physical components of the community environment were measured with great care, and because water temperature can affect the rate of photosynthesis, temperature readings were compared with measurements taken and published about a hundred years before Odum began his study. Because the springs boil up from the ground, the mineral content of the water was measured, and it was found that there were no significant variations in the amounts of minerals in the water during the time of the study. The temperature of the water also remained relatively constant throughout the year and throughout the whole area of the springs, and it did not appear to have fluctuated significantly over a period of many years. Dyes were injected into the water to determine the currents in the spring, and since the spring is a moving body of water and therefore changes rapidly, multiple samples and measurements were taken frequently by more than one investigator. Statistical comparisons were made of the measurements taken by the investigators when appropriate.

The first major conclusion the investigators reached was that the spring was a biostatic, chemostatic, thermostatic community. The *only* major variable was found to be day length. The stable nature of the community made it an ideal subject for study, since any changes that took place could be correlated with variations in light availability. Factors like changing species populations, the chemical contents of the water, and water temperature, all of which can affect productivity, remained constant and thus did not serve to increase the complexity of the interrelationships between the various components in the study. In short, the spring represented in a natural state the sort of controlled situation that an investigator strives for in the laboratory.

Measurements were taken to determine the amount of sunlight available at the surface under varying conditions of cloud cover and time of day, and it was found that trees growing around the springs diminished the amount of light available by 10% to 30%, depending upon the time of day and year. It was assumed that only 50% of the available light was in wavelengths suitable for photosynthesis, and that at an average depth of 1.8 meters, only 60% of the light penetrated the bottom of the springs.

The rate of photosynthesis was determined by measuring both O_2 production and CO_2 uptake. When the rate of photosynthesis was compared to the amount of useful light available to the plants, it was found that the efficiency of gross productivity was 5.3%, which as we have said, is a high level of efficiency.

The gross productivity of the dominant producers in the springs, algae and *Sagittaria lorata* (a bottom-growing higher plant) was measured in terms of dry biomass and was found to be 6390 $gm/m^2/yr$. Of this, 1900 $gm/m^2/yr$ was from *S. lorata*, and the standing crop of this plant was found to be 599 gm/m^2. Thus, the turnover rate of *S. lorata* was calculated at 1900/599, or 3.17 times per year, and the turnover time is a little less than four months. The algae, on the other hand, showed a productivity of 4490 $gm/m^2/yr$ and a standing crop of 177 gm/m^2. The turnover rate for algae, then, is 4490/177 or 25.4 per year, and the turnover time is a little more than two weeks. Thus, while the standing crop of algae measured at any time would make them appear to represent a less important resource in the community, their rapid turnover makes their contribution to the energy budget of the community of major importance.

The total biomass of all producers was measured at 809 gm/m^2, and the turnover time of the whole community was calculated to be about five months.

The next trophic level, herbivores, included more than fifty species, with about twenty constituting the bulk of the biomass for this level, which was measured at approximately 37 gm/m^2. Included among the species were turtles (biomass of 6.7 gm/m^2), mullet (5.6 gm/m^2), and a variety of snails, shrimp, midges, caddis flies, and other insects. In developing a biomass pyramid, it was necessary to include one species of fish, stumpknockers, in both the herbivore and carnivore trophic levels due to its being approximately 50% herbivore and 50% carnivore. Thus, 6.3 $gm/m^?$ of the herbivore biomass is stumpknocker biomass. The carnivore level in the biomass of the springs was measured at about 11 gm/m^2, including 6.3 gm/m^2 of stumpknocker. Catfish and sunfish together made up another 3.0 gm/m^2. Coelenterates, other invertebrates, and fish constituted the remaining biomass at this level, though altogether there were some seventy-four species of primary carnivores found in the springs. The top carnivores in the springs amounted to

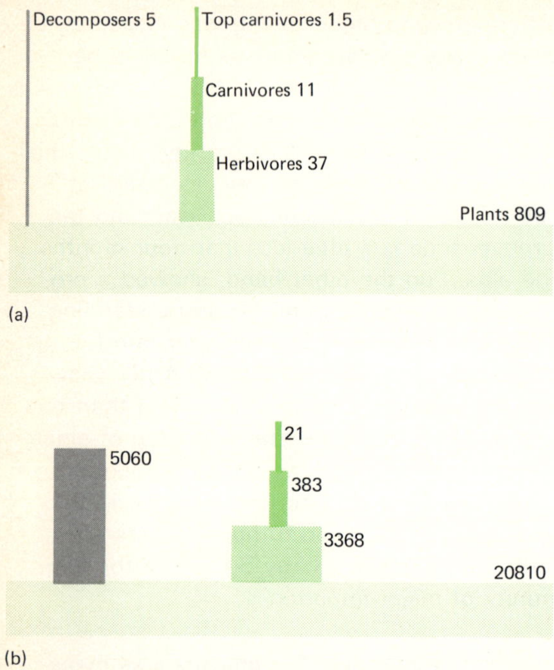

FIGURE 10.13 As in most ecosystems, the biomass and energy pyramids of Silver Springs are not the same. The relative amounts of herbivores and decomposers in the biomass pyramid (a) are less than they are in the energy pyramid (b). The numbers give the biomass in units of grams per square meter and the energy in kilocalories per square meter per year *(Redrawn: (a) from H. T. Odum; (b) from E. P. Odum, 1971.)*

roughly 1.5 gm/m^2, and while there were six species identified, two of them, gar and bass, constituted most of the biomass at this level. Decomposers amounted to 5 gm/m^2. Thus, the biomass pyramid for Silver Springs is constructed as in figure 10.13. As mentioned earlier, the trophic pyramids based on biomass are not necessarily indicative of the energy flow in the community, but by a variety of techniques biomass can be converted into energy units.

Using such techniques and conversion factors, Odum determined that gross primary production in the springs amounted to 20,810 kcal/m^2/yr, the greatest part of which was production by algae. The herbivore level can be converted to kcal, as can the other levels, including decomposers, and the ''shape'' of the pyramid when such conversions are made is significantly different from the shape of the biomass pyramid. The decomposers in the energy-flow pyramid constitute by far the greatest amount of energy used if plants are not included. The turnover rate for these organisms was measured and found to be greater than 2300, and their turnover time, then, is greater than *six times per day.*

The total energy budget for the community is as follows: Gross productivity is 20,810 kcal/m^2/yr, and net productivity, after deducting 11,977 kcal/m^2/yr for plant respiration, equals 8833 kcal/m^2/yr. Each year the net productivity of plants available to herbivores is supplemented by an estimated 486 kcal/m^2 of material imported into the community. This importation is in the form of bread thrown into the springs by tourists to feed the fish. When all other figures for respiration are taken into account, the total community respiration, including plants, equals 18,796 kcal/m^2/yr. When community respiration is deducted from gross production and imported energy,

$$21{,}296 \text{ kcal/m}^2\text{/yr} - 18{,}796 \text{ kcal/m}^2\text{/yr} = 2500 \text{ kcal/m}^2\text{/yr}$$

the investigators found that there was an excess of 2500 kcal/m^2/yr. The excess production, however, did not remain in the community as an accumulation of biomass, but rather was exported downstream. (In the adjacent community, of course, the excess becomes an import.) Thus, when all of the energy was accounted for, the P/R ratio was found to equal 1 (if export is included in R), and the community was described as having achieved a steady state.

There are two major points to be made about such studies. First, simple trophic-level analysis fails to convey the complexity of the interrelationships between species and groups of species and, thus, can yield somewhat misleading oversimplifications. (See fig. 10.14.) It should be clear, though, that analyzing communities and ecosystems is difficult. It is hard to collect

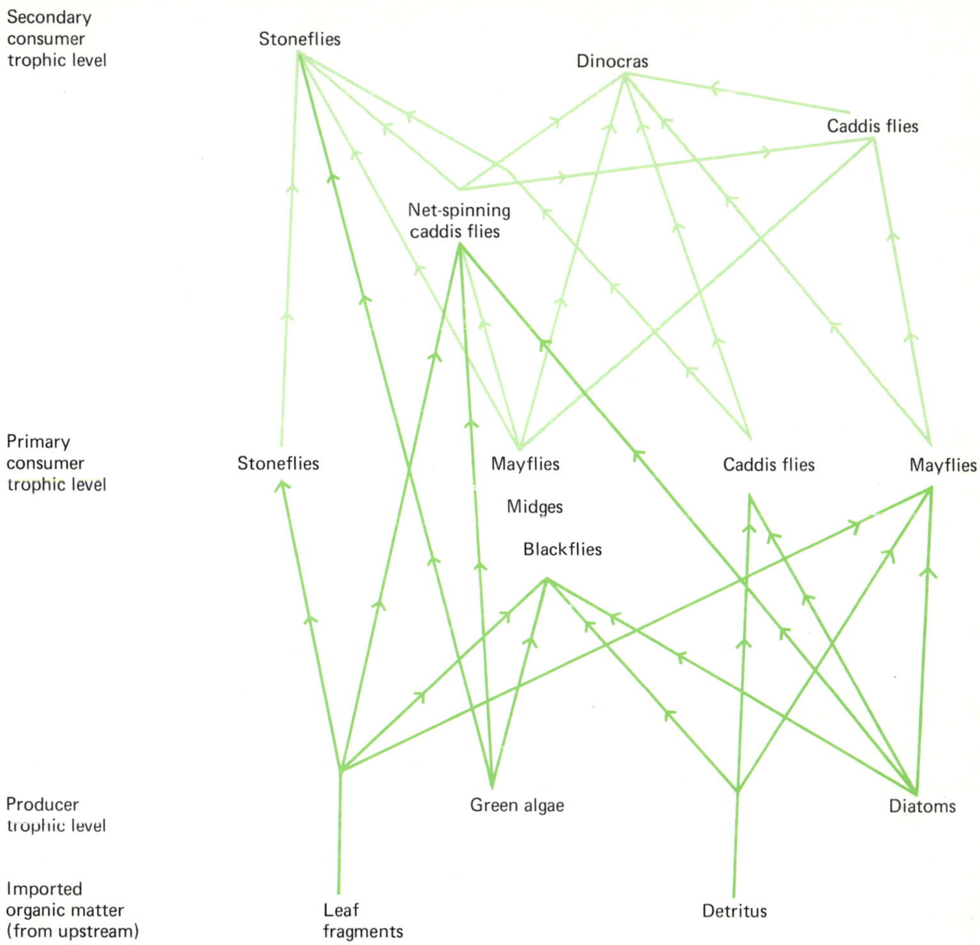

FIGURE 10.14 This diagram of feeding interrelationships between organisms in a Welsh stream illustrates the complexity of food webs. Note that most organisms rely upon leaf fragments, algae, and detritus, and that several organisms act both as herbivores and carnivores. Thus trophic relationships must be described as webs, and exact analysis of trophic energy transfer may be very complex. *(Redrawn from Jones, 1949.)*

sufficient data about the species involved and their feeding habits; such data accumulation requires intensive observation over a long period of time during which the conditions affecting the community or ecosystems can be measured. Once the data are accumulated, it is not always easy to know what to do with them. We are not yet certain what questions we must ask.

The second point to be made is that merely describing a situation is not enough. It is necessary to be able to *predict* what will happen as conditions change, especially when such changes may be induced by man. A great deal of ecological research is now being focused on the building of conceptual and mathematical models of communities and ecosystems. Block diagrams and visual ''compartment models'' such as we use in Chapter 11 (see pages 344–345) are the simplest forms of such models. Though they are useful for understanding the nature of communities, their predictive value is limited. If we can learn the rates of production and uptake of materials by species or groups of species with similar trophic interrelationships, the biomass of these groups or species, and the nature of their interactions at differing population levels, it should be possible to produce mathematical models to predict what will happen as conditions change. Such models would substitute equations for the arrows that show the transfers of energy or material. They use, among other techniques, the methods of systems analysis developed for communication studies and space science, and they rely heavily on the use of computers. It is beginning to appear that the complexities of biological communities are so great that the models necessary to accurately predict the future of a community or ecosystem will be even more complex than those necessary to guide a spaceship to the moon or design a nuclear reactor.

There are a number of efforts now underway to develop models for energy transport in ecosystems. One study is focusing on the great plains grassland biome, and another on the eastern deciduous forest. Other investigators are concerned with the prediction of the productivity of certain fisheries and the effects of organic pollution. How well these models will work remains to be seen.

The problems lie in constructing realistic food webs and collecting accurate data to check the validity of the models. The development of such models requires the skills of ecologists, biochemists, chemists, hydrologists, geologists, and mathematicians, and it involves some of the most fascinating interdisciplinary work ever undertaken by scientists. Eventually, we will be able to predict at least the general behavior of ecosystems and their responses to man's manipulation. That ability will make ecology one of the most useful and important of man's sciences. If the results of such large predictive models are reliable and are heeded by politicians, businessmen, and policy-makers, we may be able to more intelligently preserve, use, and enjoy our biological heritage.

In a few years, ecological research, and especially ecosystem models, may make the contents of present-day texts (including this one) obsolete; or at least we may learn that many of our concepts need refining. If well used, our present-day studies should have an effect on the health of the biosphere similar to that which the last fifty years of medical research has had on the health of man. *If well used.*

It should be clear by now that ecosystems are extremely complicated entities that function in ways determined by the long-term, complex evolution of a balanced community of organisms. They are ''run'' by energy, as is any other physico-chemical system, and their behavior may be measured, analyzed, and predicted. It should also be clear that we are an integral part of these systems and that we depend upon them for our survival.

Since we are not, at present, very well prepared to predict ecosystem function, we are equally ignorant of how best to manage these systems. Two major problems will be discussed in greater detail later. The first is that due to our lack of knowledge, it is difficult to make sound decisions. There are a series of questions for which we can find only incomplete and often equivocal answers. For example, what is the effect of irrigation on the long-term use of land for farming? What really are the effects of thermal pollution? How can we control insect pests? The list of partially answered questions is long. The second major problem that emerges from our imperfect understanding may be more difficult to cope with; we do not know well enough what we want. Are the Everglades worth saving, or could we use the land more economically for agricultural purposes? Should we worry about the recreational potential of Jamaica Bay, San

Francisco Bay, Chesapeake Bay, or other unique areas? How many people should the land support? Should people live in cities or "planned" suburban developments or on farms? Does an the industrial activity provides us with energy profit it eliminates a natural ecosystem? What if the industrial activity provides us with energy such as food or electrical power?

Such questions have social, political, and moral implications. Arriving at the answers will require that everyone have a better understanding of ecosystem function and especially a better understanding of the impact of our activities on the survival, stability, and existence of ecosystems.

We have got ahead of ourselves! We must look more closely at man's role in the natural world, but we must first understand more about the ways in which ecosystems function. No community can exist without an input of energy, but energy alone is useless unless it is converted into the chemicals that make up the organisms in communities. This means that the chemicals must be available to the community, and that this availability determines, in large part, how the ecosystem functions. We must examine, then, how the materials that enable organisms to use energy are cycled and transferred in the biosphere.

Selected Readings

*Odum, Eugene P. *Ecology.* New York: Holt, Rinehart and Winston, Inc., 1963. A brief survey of the field of ecology by the author of the standard text in the field.

Odum, Eugene P. *Fundamentals of Ecology.* Philadelphia: W. B. Saunders Company, 1971. Now in its third edition, this book was very important for turning ecology into a course of study. The chapter on ecosystem energetics is relevant to this chapter, and was written largely by Howard Odum, Eugene's brother.

*Odum, Howard T. *Environment, Power, and Society.* New York: John Wiley & Sons, Inc., 1971. In this largely speculative book, Odum attempts to show that the analysis of energy can be used to help answer certain legal and social, and even some religious questions.

*Phillipson, John. *Ecological Energetics.* New York: St. Martin's Press, Inc., 1966. A very brief, general treatment of the subject of this chapter.

Vollenweider, Richard A., ed. *Manual on Methods of Measuring Primary Productivity in Aquatic Environments.* Philadelphia: F. A. Davis Company, 1969. This is a more advanced treatment of the problem of measuring productivity. For the very interested student.

Woodwell, George M. *The Energy Cycle of the Biosphere. Sci. Am.,* Vol. 223 (September, 1970), pp. 64–74. Offprint number 1190.

*Paperback edition available.

The hydrologic cycle is largely responsible for the transport of materials through the nonliving portions of the ecosystem. The moisture from clouds is transported to the earth by rain, over and through the ground to rivers, and by the rivers to the sea. As it passes through the ground it is taken up by plants and transports the materials necessary for their growth. *(Courtesy of Alan Pitcairn from Grant Heilman.)*

materials in the biosphere

chapter eleven We have seen that the solar energy fixed by photosynthesis is the basis of ecosystem productivity, but we have also pointed out that although energy is the "driving force" in living systems, it is only one requirement. For plants to grow, solar energy must be converted to chemical energy and thus, a supply of matter must be available for the energy of the sun to work upon. And since populations of organisms have evolved in specific and differing environments, their adaptations have fitted them to the relative quantities of materials available in those environments. Thus, some species require more of some material than other species, and the amounts of materials available may largely determine which species can and cannot survive in a given environment.

Those materials necessary for the survival and growth of organisms are called *nutrients.* They include both energy-containing compounds, such as sugars, and inorganic materials, such as carbon dioxide, water, nitrate, phosphate, and ions of elements such as copper and iron. In this chapter, a somewhat arbitrary distinction will be made between materials that only provide energy, which we shall call *food,* and materials that provide a physical substance the organism incorporates into its tissues, such as nitrogen or vitamin B_{12}. This distinction is not always explicitly made by all biologists, but it is useful, and it clarifies the concept of nutrients. Of course, the distinction breaks down when we consider that all organic materials provide carbon and that some of the carbon is a nutrient that is incorporated into tissue. The presence or absence of inorganic and organic nutrients determines whether an organism can survive in an environment; thus, nutrients are, or can be, major factors of selection.

Micronutrients and Macronutrients

The subject of organic nutrients is complex because of the wide array of compounds that may be included in this category and the varying requirements of individual species. Organic nutrients include vitamins, certain amino acids, and for some plants and animals, other organic materials. The subject is made more difficult when we recognize that some areas of nutrition are very well known, while others are relatively poorly understood.

Consider vitamins, for example—a diverse group of organic compounds needed in varying, relatively small quantities by species that do not have the ability to synthesize them. Although a compound may be classified as a vitamin for one species, if it can be synthesized by the individuals of another species, it is not a required nutrient vitamin for the latter species. Man, for example, is incapable of synthesizing ascorbic acid (vitamin C) and therefore must obtain it in his diet. Many other mammals and many plants, however, are capable of synthesizing ascorbic acid, and it is not, therefore, a required vitamin for those species.

Since vitamins are often needed in such minute amounts, it is sometimes extremely difficult to prove whether a particular chemical compound is a vitamin. One way to attempt such a proof is to deprive an experimental organism of the compound and observe whether any metabolic effects occur as a result of the deficiency. The difficulty is in knowing with absolute certainty whether the compound has really been eliminated from the organism's diet. Purification techniques, however elaborate, are not perfect, and thus minute, but sufficient, quantities of the material may be unintentionally included in the diet of the experimental organism. Thus, generalizations about the dietary needs for certain vitamins are open to question. Moreover, it is by no means clear how nutritional requirements change under different conditions of age or health. Since the organisms that we depend upon for food and nutrients are composed of a wide variety of organic and inorganic compounds, it seems safe to say that a diet consisting of a carefully planned variety of plant and animal materials will supply a healthy person with sufficient amounts of vitamins to maintain health.

Our understanding of some inorganic nutrients is much more extensive. The most obvious inorganic nutrients are water, carbon dioxide, and oxygen. These compounds, being basic to all forms of life, are required in relatively large quantities and are therefore among the materials considered to be *macronutrients,* in contrast to vitamins which are considered as *micronutrients* (needed in measurable *trace* amounts). Among other important macronutrients are those elements needed for the synthesis of the major chemicals of life: nitrogen (N) for the synthesis of amino acids, certain phospholipids, DNA, and RNA; and phosphorus (P) for the synthesis of phospholipids, DNA, RNA, and ATP. Although nitrogen and phosphorus are the two main inorganic macronutrients needed by all organisms, there are several others, including sulfur (S), silicon (Si), calcium (Ca), potassium (K), sodium (Na), and chlorine (Cl). Not all of these elements are needed by all species under all conditions. We have already seen the roles that some of them play, such as that of (Na), (Cl), and (K) in the transmission of the nerve impulse or that of (S) in the structure of proteins and enzymes. Calcium is an important element in the structure of bones, teeth, egg shells, and shells of molluscs and other animals. Though these elements are of some importance, they are usually more readily available and needed in smaller quantities than nitrogen and phosphorus and are therefore not as universally important as (N) and (P). Altogether, there are slightly more than one hundred chemical elements known to man, but only about forty of them are presently known to be used in living systems. Interestingly, man makes use of far more than forty elements, but many are used for industrial rather than biological processes.

Our understanding of micronutrients in general is limited. We have learned much about some micronutrients, but much more remains to be studied. Among the more prominent micronutrients are iron (Fe), copper (Cu), aluminum (Al), magnesium (Mg), manganese (Mn), molybdenum (Mo), zinc (Zn), boron (B), vanadium (Vn), and cobalt (Co). Part of the problem involved in understanding the biological activity of these elements is associated with the problem of understanding vitamins. For example, (Co) is a part of vitamin B_{12}. Although needed only in trace amounts, these micronutrients are not uniformly distributed throughout the world, and they can, therefore, exert a limiting effect on the growth of species in places where they are unavailable or are available in quantities too small to promote much growth. In parts of southern Florida, for example, soil found to be rich in the organic nutrients needed by plants lacked copper and cobalt and was therefore unable to support the volume of crop production growers had expected. In another case, the addition of molybdenum to the water of a mountain lake in the amount of 100 parts of (Mo) per billion parts of water had a marked effect on the rate of photosynthesis in the lake. We have seen the role that magnesium plays in photosynthesis as a constituent of chlorophyll, and iron is a key constituent of hemoglobin.

We have already examined many of the constituent chemicals of organisms, and we have seen how they contribute to survival. It is not sufficient, however, to understand only what occurs within an organism if we are to understand biology. We must look at the global processes that determine the availability of these necessary chemicals, and we must examine how organisms are adapted for taking up these materials from their environment. We will then see that the biosphere may truly be said to regulate many of the chemical processes that occur in the atmosphere, waters, and rocks of the earth itself, and that the "nonliving" portions of the atmosphere, the geosphere and hydrosphere, are in an intimate balance with the living, dynamic communities of the earth.

Biogeochemical Cycles

All of the chemical elements that make up living things come from the earth, and thus, the chemistry of the earth, or geochemistry, is intimately associated with living systems. Obviously, not all of the elements found in rocks are used by living things. Silicon (Si), for example, is one of the most abundant elements found in the crust of the earth, but it is used extensively only by the diatoms, which construct their frustules from silicon dioxide (SiO_2). (See p. 81, Chapter 3.) The question of why some elements and not others with similar chemical properties came to be used in life processes is an interesting one, but it is not appropriate to the present discussion.

Elements not only come from the earth but return to it as well. The O_2 that a plant produces by splitting water is given off to the atmosphere and then taken up by animals. The atmosphere contains nitrogen in vast amounts, but not in a form usable by plants. It is taken up by some bacteria and blue-green algae, however, and converted to a form that can be used by plants for protein synthesis. Ultimately, the nitrogen is cycled through living systems and back into the atmosphere. This cycling of material from the earth through living systems and back to the earth is called *biogeochemical cycling.* Two basic kinds of biogeochemical cycles are recognized, *gaseous* and *sedimentary,* although as we shall see, the distinction is sometimes arbitrary. (See box 11.1.)

The Water Cycle

Of the materials that we will consider, water possibly plays more important roles than any other nutrient in the biosphere. *First,* the form and amount in which water is present largely determine what grows where. We differentiate between freshwater, marine, and terrestrial organisms, and between terrestrial organisms adapted to moist or dry environments. The diversity of available environments (such as jungles, deserts, tundra, grassland, and temperate forests) is caused primarily by the availability of water. *Second,* the distribution of water plays a major role in determining climates, in terms of both rainfall and temperature; and water in the atmosphere determines the meteorology of the air. *Third,* the nature of the earth's surface has been determined largely by the erosive qualities of moving water. Water has leveled mountain ranges and, as glaciers, produced major land forms. *Fourth,* water is a solvent for many chemicals, including *all* of the inorganic chemicals necessary for life. Thus, it is from water that the materials necessary for growth are taken up by the roots of plants or the cell membranes of phytoplankton. *Finally,* water forms one of the basic parts of many organic molecules and is, therefore, itself a major nutrient.

For our purposes, we will consider that the hydrologic cycle begins with the evaporation of sea water. Being basically an energy transformation, evaporation occurs as the radiant energy of the sun strikes the ocean surface, warming it and causing the molecules of water to move faster until some escape and form water vapor. The vapor rises into the atmosphere to form clouds, which may be moved by winds to other portions of the earth's surface. Usually, the water evaporated in one location returns to a nearby location as precipitation, but a sufficient amount resides in the atmosphere long enough to be moved to other parts of the world. When the water falls in the form of rain

Biogeochemical Cycling

Each of the cycles that we shall examine shares some basic properties with all of the others. In the most general terms, each may be said to consist of two components, a *reservoir* and an *exchange component.* The reservoirs are usually nonbiological and are characterized by a very slow turnover of the nutrient material. The oceans, the atmosphere, and deposits of rock are all reservoirs in which, as we shall see, nutrients are relatively unavailable. Although some quantity of nutrient is available from the reservoirs, the turnover times are immense. *All* the water on the planet, for example, eventually goes through the photosynthetic process, but at a rate estimated to be once every 2,000,000 years. Similarly, the oxygen that is produced from the splitting of water in photosynthesis cycles every 2000 years, and atmospheric carbon dioxide given off by the respiring cells of animals and plants cycles once every 300 years. Thus, under normal conditions, the reservoirs remain largely untapped and at any time very nearly all of the nutrient material in transit through the exchange component moves rapidly from organism to organism. The uptake from or introduction into the reservoirs, when it does occur, is relatively well balanced, a product of evolution. As we shall see, we have created a number of situations that cannot be classified as normal, and a great deal of the environmental consequences of human activities are the result of our intervention and disruption of nutrient cycles.

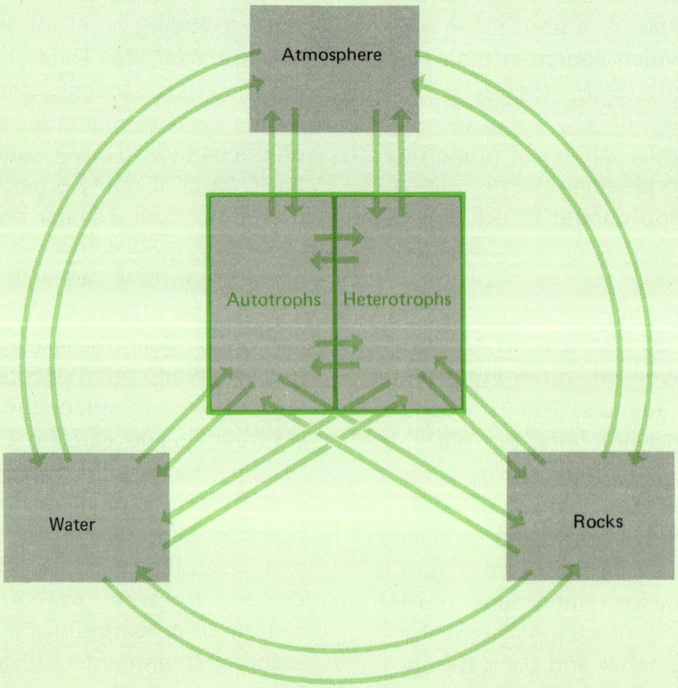

BOX 11.1

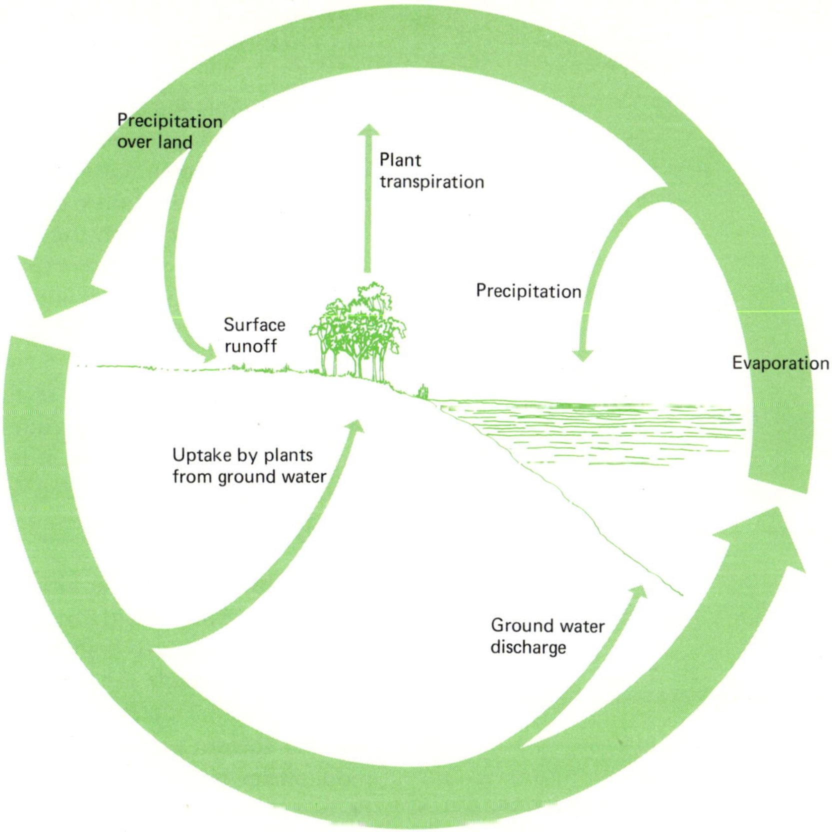

FIGURE 11.1 Obviously, the distribution of water and water vapor in the atmosphere, ground, rivers, lakes, and oceans is very important for the distribution of life. The distribution of water is controlled by the hydrologic cycle, which is in turn controlled by the distribution of the input of solar energy.

or snow, a number of things happen. Some small quantity re-evaporates before reaching the surface of the earth. Much of what does reach the surface runs off rapidly into streams, rivers, ponds, lakes, reservoirs, and oceans. The water that does not run off sinks below the surface and passes through rock and soil to become what we call *ground water.* Ground water may be absorbed by the roots of plants, or if it sinks below the root zone, it may slowly flow into lakes, ponds, streams, and rivers from which it is returned to the sea as fresh water or to the atmosphere by evaporation. (See fig. 11.1.)

Plant transpiration provides an extremely important source of atmospheric water vapor. It has been estimated that to grow one metric ton of live plant material requires approximately 100 metric tons of water. Thus, at least 99% of the material that enters the plant through the roots departs, mostly through the leaves. In terrestrial plants, of course, almost all of this water comes from ground water sources. If the one metric ton of plant material is analyzed, it will be found that only 250 kilograms (kg), or 25%, is dry weight. The remaining 75% (750 kg) of the live weight of the plant material is water in transit. Thus, of the 100 tons of water used to produce one ton of plants, only about 0.25% participates in photosynthesis. (See fig. 11.2.) The remaining 99.75% departs or is lost from the plant by way of transpiration. Much of the transpiration is brought about by the heat en-

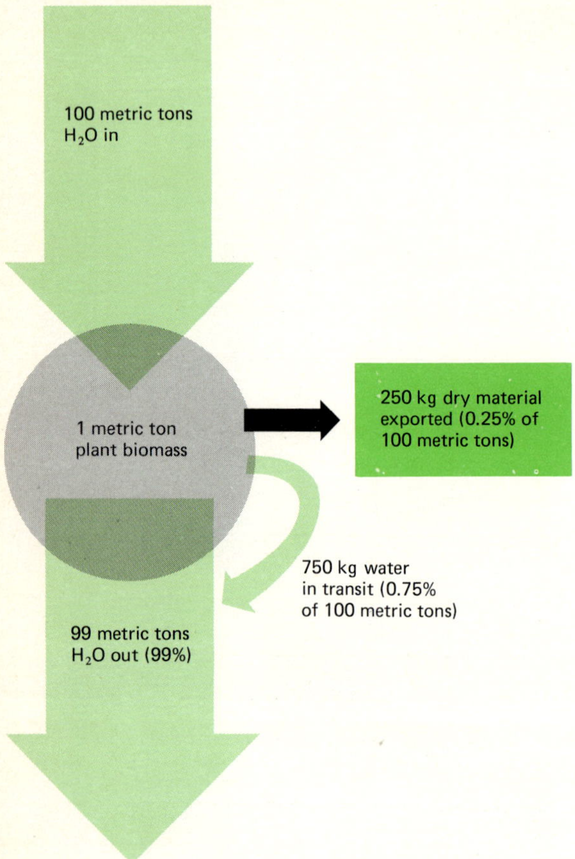

FIGURE 11.2 Most of the water that enters plants is used to transport nutrients and other materials through the plant biomass. Also, since so much water passes through the plants, the distribution of plants is very important in controlling the amount of water in the ground and in the atmosphere.

ergy that reaches the plant but does not enter into photosynthesis.

Thus, although the distribution of terrestrial plants is to a degree determined by the hydrologic cycle, the plants themselves help determine the climate in which they grow. Terrestrial plants also interact with the movement of water through the cycle by preventing erosion and the consequent loss of the ability of the land to retain precipitated water. Plants, then, play major roles in the biosphere, not only because they capture the radiant energy of the sun for conversion to chemical energy but also because they determine the nature of local climates and affect the shape of local land forms through their role in the hydrologic cycle.

Finally, the role of water as a solvent should be mentioned. Every inorganic nutrient required by plants and animals is transported in aqueous solution. Water reacts with other materials to form compounds that determine what can live where, and the solvent properties of water make it the medium for transport of materials within the bodies of organisms.

The Carbon Cycle

Next to water, carbon is the most important material that we will discuss. The reduction of carbon dioxide to carbohydrate provides almost all of the chemical energy on the planet, whether measured as the production of food plants or as fossil fuel. Besides its role in energy-bearing molecules, carbon is a key constituent of almost every chemical in living systems and is, therefore, a nutrient for every living thing. At the same time, certain compounds of carbon provide man with some of his most potent pesticides and herbicides. The role of carbon in living systems becomes geologically important as populations of organisms die, depositing their remains in such massive amounts that they become geological features. The Great Barrier Reef off Australia, the coral islands of the Pacific, the southern part of Florida, the Grand Bahama Banks, and the White Cliffs of Dover in England are but a few examples of geologically

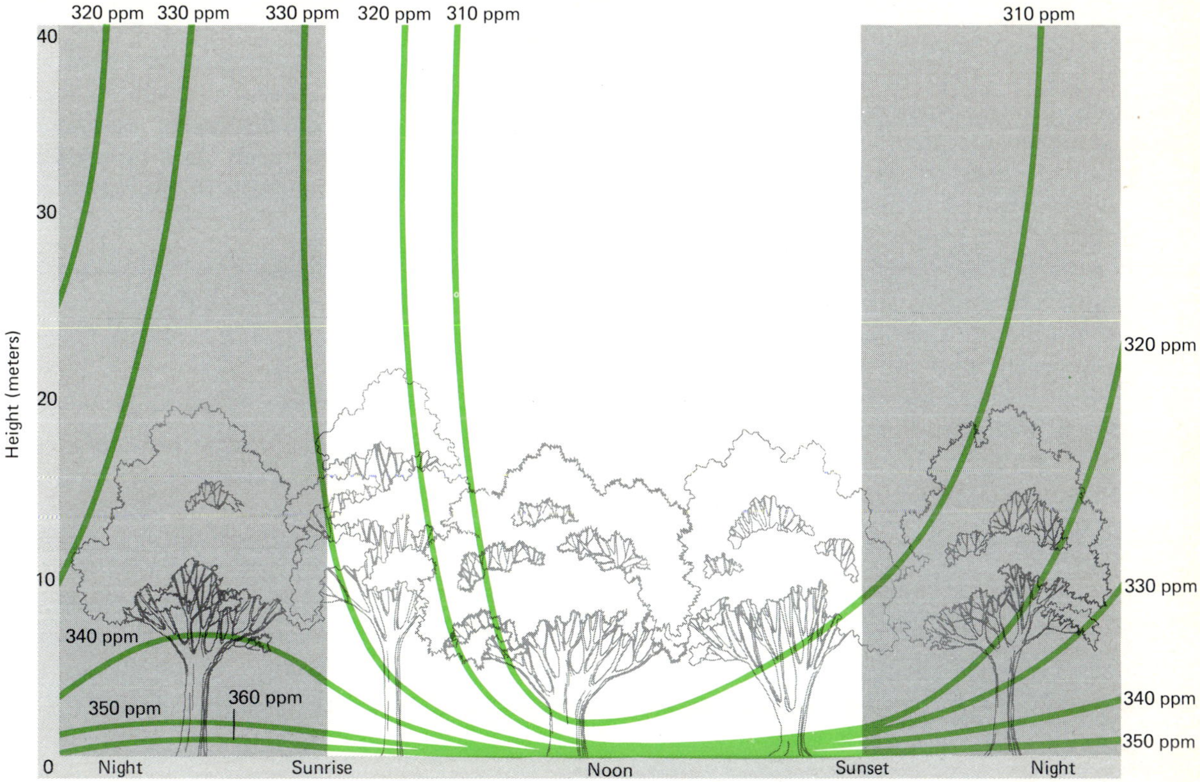

FIGURE 11.3 During the night, the respiration of a forest community releases carbon dioxide, increasing its concentration in the atmosphere, and during the day, the concentration decreases because of the processes of photosynthesis. *(From "The Carbon Cycle" by Bert Bolin. Copyright © September 1970 by Scientific American, Inc. All rights reserved.)*

important deposits of carbonate rock derived from living systems. Finally, CO_2 plays a role in the heat budget of the earth.

Most of the carbon that actually cycles in the biosphere is in the form of CO_2. As a constituent of the atmosphere, CO_2 amounts to approximately 0.03% by volume or 320 parts per million (ppm).* The distribution of CO_2 in the atmosphere is relatively uniform, though minor fluctuations occur in the atmosphere near plant communities. Plants take up CO_2 during the day while photosynthesis is going on, but at night, plant respiration increases the content of CO_2 in the atmosphere around the community. Perhaps the only place where the distribution of atmospheric CO_2 fluctuates greatly is in the canopy of very dense forests where the uptake is very rapid. (See fig. 11.3.)

*The entire weight of atmospheric CO_2 is estimated at 700 billion metric tons.

In both plants and animals, respiration produces CO_2 that is given off to the atmosphere. Thus, the CO_2 taken up from the atmosphere by photosynthetic plants is returned to it by plants, animals, and decomposers. (See fig. 11.4.) A great deal of carbon on the earth is bound up in living biomass and circulates rapidly between plants, animals, and the atmosphere. The next largest amount is tied up in rocks, which we shall briefly look at later. Finally, a portion is in solution in water, this portion tending to circulate more rapidly than the carbon bound up in rocks. Since much of the carbon that is present in rocks was deposited there through water, and since the amount of carbon in water may control the amount in the atmosphere and biosphere, we shall look next at the aqueous part of the carbon cycle.

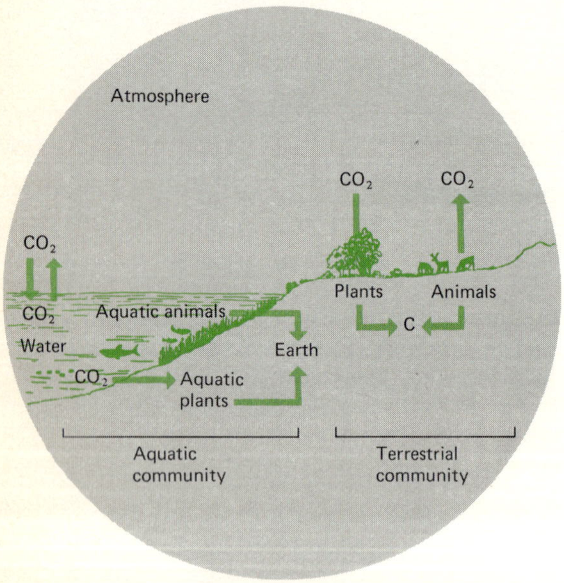

FIGURE 11.4 Carbon dioxide exchange obviously occurs between plants and animals, and between them and the atmosphere. Carbon dioxide also moves in and out of bodies of water, and in and out of the plants and animals in water. Some carbonate is precipitated from the water, sometimes through plants and animals, and becomes incorporated as carbonate or limestone in the rocks of the earth. Some forms of carbon also accumulate in the soil from the decomposed remains of plants and animals. Thus, the cycle of carbon and carbon dioxide takes place between the biosphere, the geosphere, and the atmosphere.

Before we can do so, however, it will be necessary to digress and consider one of the important properties of water itself. As it normally occurs, water is not simply H_2O, but rather contains H_2O and dissociated ions that are the products of the reaction

$$2\,H_2O \rightarrow H_3O^+ + OH^-$$

which is often written as

$$H_2O \rightarrow H^+ + OH^-$$

since H_3O^+ actually reacts as a free hydrogen ion (H^+).

Under normal conditions, pure water has a concentration of H_3O^+ of about 10^{-7} molecules for each molecule of H_2O (that is, one molecule of H_3O^+ for every 10 billion of H_2O). Since pure water is electrically neutral, it must also contain an approximately equal concentration, or 10^{-7}, of OH^-.

Consider what happens if an acid* is added:

$$HCl + H_2O \rightarrow H_3O^+ + Cl^-$$

On the other hand, if a base† is added, the reaction will be

$$NaOH + H_3O^+ + OH^- \rightarrow Na^+ + 2\,H_2O + OH^-$$

Thus, the addition of an acid raises the concentration of H_3O^+ ions, and a base lowers the concentration of H_3O^+ ions. Since both H^+ and H_3O^+ react as the same ions, we will, for convenience, hereafter use H^+ to represent both.

The conventional means of expressing the concentration of H^+ ions in an aqueous solution is as the negative logarithm of the concentration, which is called pH. Thus, in pure water, the concentration of H^+ is

$$pH = -\log 10^{-7} = 7$$

* An acid is any compound that can contribute H_3O^+ or H^+ to water and, thus, raise the concentration of these ions.
† A base is any compound that can contribute OH^- to water and, thus, raise the concentration of this ion.

TABLE 11.1 The Relationship Between pH and H^+ Concentration

	pH	H^+ or H_3O^+ Concentration	OH^- Concentration	Molecules of H^+ per Molecule of H_2O
	0	10^0	10^{-14}	1
	1	10^{-1}	10^{-13}	0.1
Very	2	10^{-2}	10^{-12}	0.01
Acid	3	10^{-3}	10^{-11}	0.001
	4	10^{-4}	10^{-10}	0.0001
Slightly	5	10^{-5}	10^{-9}	0.00001
Acid	6	10^{-6}	10^{-8}	0.000001
Neutral	7	10^{-7}	10^{-7}	0.0000001
Slightly	8	10^{-8}	10^{-6}	0.00000001
Basic	9	10^{-9}	10^{-5}	0.000000001
	10	10^{-10}	10^{-4}	0.0000000001
	11	10^{-11}	10^{-3}	0.00000000001
Very	12	10^{-12}	10^{-2}	0.000000000001
Basic	13	10^{-13}	10^{-1}	0.0000000000001
	14	10^{-14}	10^0	0.00000000000001

In an acid solution, the concentration of H^+ is greater than 10^{-7}, and therefore pH is less than 7; under basic conditions, the concentration of H^+ is lower, and therefore pH is greater than 7. (See table 11.1.)

Each aqueous solution has a pH value, and as we shall see, pH is always of importance for organisms in water. If the aqueous solution under consideration is not a body of water, but the solution in an organism, then pH is no less important. The pH of human blood, for example, is around 7.4. An increase or decrease of 0.4 pH units is usually fatal.

To return to the carbon cycle, consider what happens when CO_2 is introduced into water, either by the respiration of aquatic plants or animals or by diffusion from the atmosphere:

$$CO_2 + H_2O \leftrightarrows H_2CO_3$$

Carbonic acid is produced, and the water becomes more acidic. If a base is introduced next, say sodium hydroxide, the reaction

$$NaOH + H_2CO_3 \leftrightarrows H_2O + Na^+ + HCO_3^-$$

takes place, and the solution becomes more basic. As neutrality (pH = 7) is approached, more HCO_3^- ions, called *bicarbonate,* are produced.

As greater amounts of the base are added, more and more H_2CO_3 becomes HCO_3^- until nearly all of the CO_2 is HCO_3^- and pH approaches 7.

If more base is added, the following reaction will take place, forming carbonate:

$$NaOH + HCO_3^- \leftrightarrows CO_3^= + H_2O + Na^+$$

We can summarize the whole sequence of reactions as follows:

$$CO_2 + H_2O \underset{(1)}{\leftrightarrows} H_2CO_3 \underset{(2)}{\leftrightarrows} H^+ + HCO_3^- \underset{(3)}{\leftrightarrows} CO_3^= + 2\,H^+$$

The importance of these reactions clearly depends on the pH of the solution, and thus, at any time, water may be found to contain any of these forms of carbon in varying concentrations, depending on the pH.

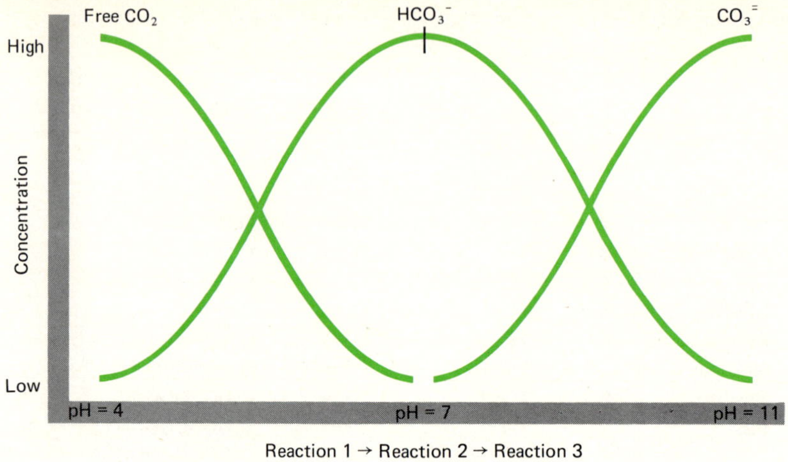

If CO_2 is added to water at pH 7 it adds H_2CO_3. This drives the equation to the right, liberating H^+ and making the pH lower. *Therefore, animal respiration makes water slightly more acidic.* If CO_2 is removed by photosynthesis, H_2CO_3 is removed, the reaction is driven to the left, H^+ is removed and the water becomes more basic. In ponds where the water is hard (that is, contains certain metals such as magnesium and calcium in aqueous solution) and photosynthesis raises the pH above 10, calcium carbonate ($CaCO_3$), a solid substance, may be formed (or *precipitated*).

The role of bicarbonate cannot be overstressed, since it is a *buffer*.* If H^+ is added to a body of water containing HCO_3^-, the reaction

$$H^+ + HCO_3^- \rightarrow H_2CO_3$$

takes place and thus prevents an increase in acidity. Conversely, the removal of H^+ produces the reaction

$$HCO_3^- \rightarrow H^+ + CO_3^=$$

and prevents the water from becoming too basic. Any naturally occurring body of water contains HCO_3^- in some amount. Thus, the pH of the water is held stable by the quantity of bicarbonate, and conditions of extreme acidity or basicity are rare. Under some abnormal conditions, the bicarbonate buffering system can be upset. Should the solution become very basic, very large quantities of calcium carbonate will be precipitated. The introduction of large amounts of acid, as from the draining of mines, can also upset the system. In either case, organisms adapted to the particular range of pH that is considered normal are driven out or killed. It is interesting that this same buffering system controls the pH of blood.

*In any solution, buffers prevent major shifts in pH from taking place.

In bodies of water, there is always a large pool of carbon in the form of HCO_3^-. Thus, even though CO_2 in nonionized form is not very soluble in water, the reactions shown above allow an immense amount of carbon to be dissolved in the world's water supply. This means that bodies of water form a *buffer pool* for atmospheric CO_2 and thus prevent very large amounts of CO_2 from accumulating in the atmosphere. For reasons that we shall examine soon, this interaction between naturally occurring bodies of water and CO_2 can also be of great importance in making CO_2 available for aquatic plants.

The amount of carbon stored in the world's water supply, however, is not as great as the amount stored in geological formations laid down by once-living organisms. Fossil fuels, of course, originated this way, but the amount of carbon in these fuels is smaller than the quantity that has been precipitated out of water by aquatic organisms as carbonate. Certain species of organisms—corals, clams, sponges, shellfish such as lobsters and crabs, and certain groups of microscopic zooplankton—are able to form external skeletons from calcium and carbonate through the precipitation of $CaCO_3$ from enzyme-mediated reactions. As individual organisms die, their skeletons accumulate, especially in tropical waters, and eventually cement together to form rocks. In some bodies of fresh water, photosynthesis can drive the pH high enough to precipitate calcium carbonate, (called *marl* in these situations) and in swampy areas, such as the Everglades, mats of blue-green algae create conditions of pH in the mats that may also precipitate layers of calcium carbonate. Calcium carbonate formation is now going on in the Bahama Banks, all coral reefs, and many ocean sediments. Most presently known deposits of limestone in the world are thought to be of biological origin, and a considerable portion of the basement rocks of England, France, southern Florida, the Shenandoah Valley, parts of Utah and Wyoming, and many other places in the world are also derived from biological processes. In many of these areas, microscopic fossils of aquatic organisms are extremely abundant. Thus, over geological time, enough limestone has been precipitated to create a major pool of inorganic carbon that may be eventually used for life processes.

Of considerable interest in any consideration of the carbon cycle is the result of the death of populations of organisms—either plants or animals—in environments where the action of decomposers has been ineffective. Under such conditions, over *very* long periods of geological time, the accumulated residues undergo the following reaction:

$$(CH_2O)_n \rightarrow (CH_2)_n + \tfrac{1}{2}\,(O_2)_n$$

Thus, hydrocarbons $(CH_2)_n$ such as petroleum or coal are formed. These pools of petroleum and coal have been exploited by man for a number of purposes. The energy-releasing reaction is:

$$(CH_2)_n + 3n\,O_2 \rightarrow 2n\,CO_2 + 2n\,H_2O + \text{heat}$$

Thus, the utilization of energy stored by photosynthesis in ages past is accomplished, but with the liberation into the atmosphere of large quantities of CO_2.

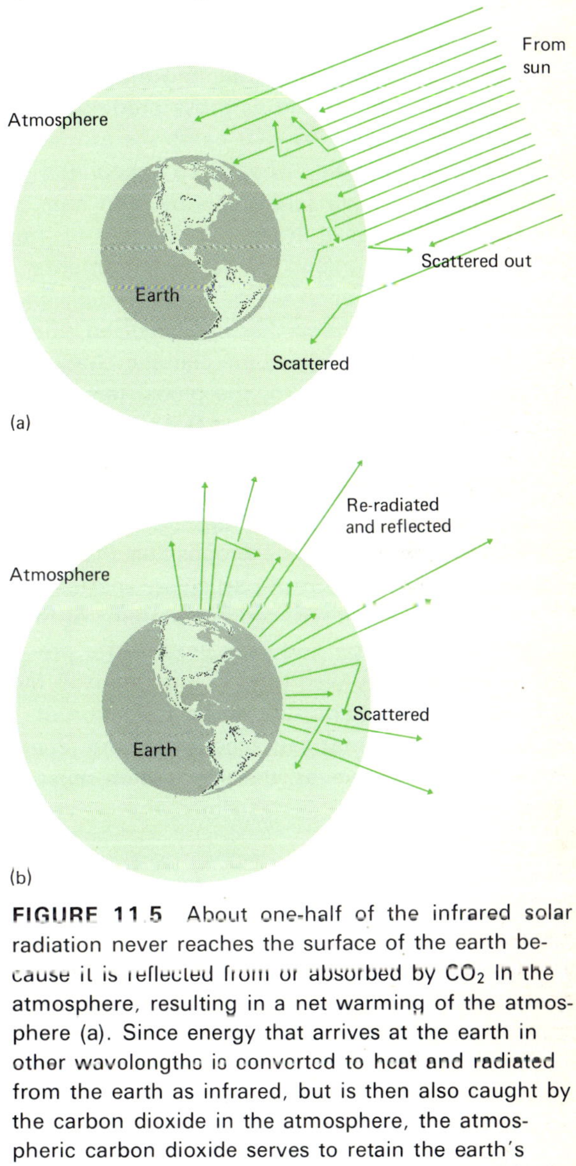

FIGURE 11.5 About one-half of the infrared solar radiation never reaches the surface of the earth because it is reflected from or absorbed by CO_2 in the atmosphere, resulting in a net warming of the atmosphere (a). Since energy that arrives at the earth in other wavelengths is converted to heat and radiated from the earth as infrared, but is then also caught by the carbon dioxide in the atmosphere, the atmospheric carbon dioxide serves to retain the earth's heat. It is speculated that an increase in atmospheric CO_2 will increase the proportion of the infrared that remains in the atmosphere, and thus will increase the heating of the atmosphere; this, however, has not been proved.

One phenomenon associated with the release of CO_2 deserves mention at this point—namely, the so-called "greenhouse" effect. (See fig. 11.5.) CO_2 in the atmosphere reflects about half of the incident infrared energy that reaches the earth. The other half of such energy is reradiated from the earth's surface, and most of the other energy that reaches the surface of the earth is also reradiated as infrared. (In both cases, of course, we are deducting the energy that enters the biosphere as gross productivity.) The energy reflected by CO_2 heats the atmosphere to some extent. It has been speculated that the utilization of fossil fuels by man has produced an increase in the CO_2 content of the atmosphere, and indeed, some measurements appear to confirm such an increase. If such an increase in atmospheric CO_2 has occurred, then we might expect to find some climatic changes. Between 1900 and 1945, the global temperature rose at an average rate of 0.8°C per 100 years, but since 1945 a cooling trend has set in. The proposition that the temperature increase was due to increased atmospheric CO_2 is pure speculation, as is the proposition that an increase in the amount of particulate matter in the atmosphere has reduced the amount of incoming solar energy and thus produced a precarious balance between the heating caused by increased atmospheric CO_2 and a reduction of incoming energy that can account for the recent cooling effect. However, the speculation *must* be tested.

The Oxygen Cycle

A third important material that moves through biological cycles is oxygen. We are most familiar with oxygen as molecular oxygen, O_2, a gas that constitutes approximately 21% of the atmosphere. Without oxygen, of course, heterotrophic organisms would be limited to glycolysis and thus would not be capable of processing the bulk of the energy stored in a molecule of carbohydrate, for example, nor would the carbon cycle be complete. Therefore, before complex multicellular organisms could evolve, a ready supply of oxygen had to be available. The atmosphere at the time of the formation of the earth about 4.5 billion years ago contained little or no molecular oxygen, as we have seen, but was made up largely of methane, carbon dioxide, and ammonia. The evolution of photosynthesis changed the atmosphere gradually but radically and thus made possible the evolution of a rich biosphere. The accumulation of molecular oxygen over the last 3.5 billion years is of such a magnitude that if all photosynthesis were to cease immediately, it would take more than one million years for heterotrophs to deplete the present supply. Indeed, if all photosynthesis were to cease immediately, the amount of oxygen available to heterotrophs would be of academic concern. They would all starve to death long before the oxygen content of the atmosphere was measurably diminished. Figure 11.6 shows the oxygen cycle.

In the recent past, in spite of the fact that the burning of fossil fuels has consumed an immense amount of oxygen, the atmospheric concentration of this gas appears to have remained virtually constant. Thus, the availability of oxygen is probably almost never a limiting factor in the growth of terrestrial animals. Although the concentration remains the same, at extreme altitudes above sea level the partial pressure of oxygen is low, because the total atmospheric pressure is low, and thus oxygen may be limiting to some animals, including man.

The amount of oxygen dissolved in water is not as constant as the amount contained in the atmosphere, and some waters may actually contain no oxygen (be *anoxic*). Depending on temperature, oxygen is soluble in water in the range of from 5 to 15 ppm. At low temperatures, water may contain oxygen in concentrations close to the maximum of 15 ppm. While this is a relatively low proportion, animals can still obtain sufficient amounts of oxygen to meet the demands of respiration. Adaptations such as gills, by presenting a large surface area over which water can pass, permit fish to extract oxygen from the water in the amounts required to carry on metabolism at a high rate. As long as there are plants in a body of water, there will

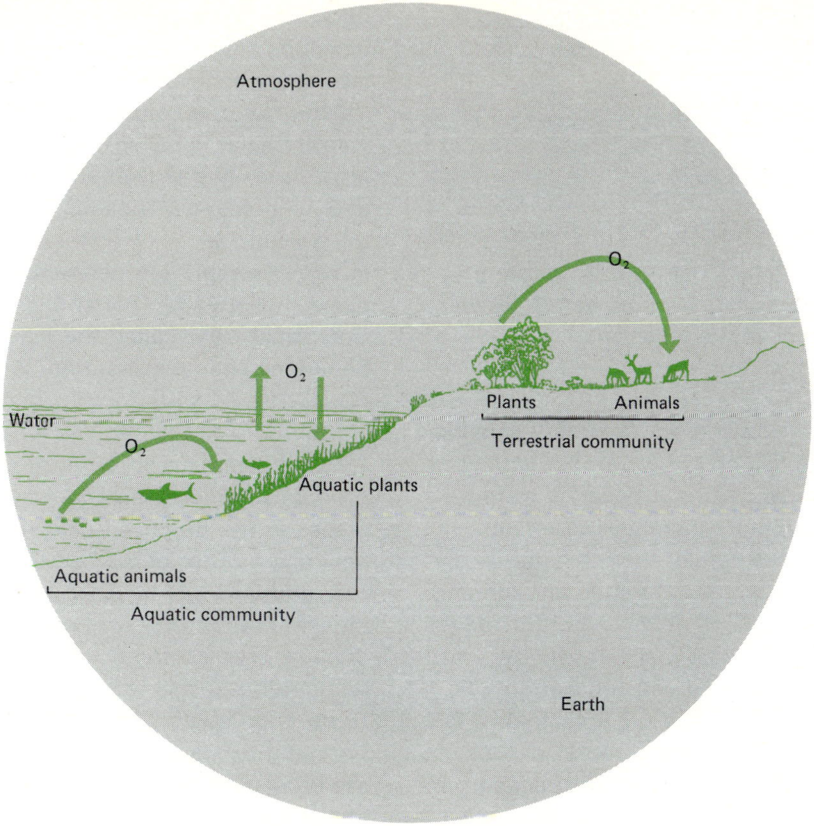

FIGURE 11.6 Oxygen travels in and out of the plants and animals of both the terrestrial and the aquatic environments, just as does carbon dioxide. This movement of oxygen through the oxygen cycle is of obvious importance in controlling the distribution of plants and animals.

be enough oxygen to support some animal life. In fact, water can become supersaturated with oxygen, and on a sunny day in summer, the excess oxygen may actually bubble out of the water. As we have seen, however, light (and thus photosynthesis) is limited to the upper regions of any body of water. What happens to heterotrophs below the lighted zone where the oxygen is being produced? Actually this is a complex question, because it depends on both the rate of heterotrophic oxygen uptake and the mixing and circulation of the water. This is an important problem, however, and since it illustrates some details that will be useful later, we shall examine it now.

In the oceans, there is *almost* always enough oxygen to support a small heterotrophic biomass with low rates of metabolism. Even in the very deep portions of the oceans (10,000 meters or more) where no light has ever penetrated, oxygen has been measured in amounts of more than 2 to 3 ppm. The oxygen reaches these great depths through the circulation of the water. Since all of the water in the ocean moves, at one time even the water in the deepest parts of the ocean was near enough to the surface to absorb oxygen, although in some parts of the ocean that might have been several thousand years ago. In some bays or fjords, however, water circulation may be reduced by a sill (see fig. 11.7), and under such conditions, the bottom waters may become anoxic. The bottom waters of the Black Sea, for example,

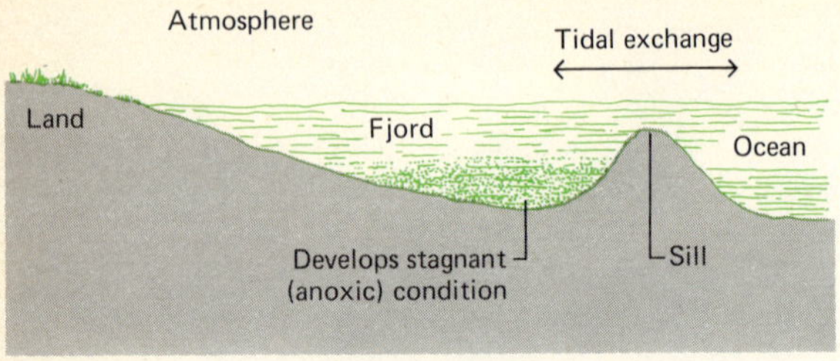

FIGURE 11.7 An underwater sill may prevent deep waters from fully circulating with the surface waters, and thus bring about anoxic conditions.

are completely anoxic and are thus only capable of supporting populations of anaerobic organisms. Waters at mid-depths may rarely become anoxic if the community metabolic rate is extremely high. On the whole, then, oxygen is distributed throughout bodies of water, though at decreasing concentrations as the depth of the water increases. The exact mechanisms by which this is accomplished concern patterns of circulation, temperature, and other factors. It is sufficient to understand that the great depths of oceans and deep lakes such as Lake Baikal in the Soviet Union (5314 feet, or approximately 1500 meters) are not completely lifeless, although the amount of life and the metabolic rates that can be supported may be limited.

Although ocean waters rarely become anoxic, the same cannot be said of lakes. Lakes provide, in fact, an interesting and instructive example of the cycling of oxygen and its relationship to productivity. Since understanding some aspects of lakes and ponds will be useful when we consider the cycling of nitrogen and phosphorus, and since they are seriously endangered parts of the biosphere, we will look briefly at some details of a typical lake. In the early spring, a lake in a temperate zone is found to be roughly the same temperature (*isothermal*) from top to bottom. The intensity of light is low, the day length is short, and thus phytoplankton producers are mixed evenly throughout. Therefore, at this time the lake has a low productivity. At some time from March to May, depending on the depth of the lake, the latitude, and climatic factors, a warmer layer of water develops near the surface. Since this water is less dense than the colder water below, it ''floats'' on the colder layer of water. The warmer layer is called the *epilimnion* (*epi* on + *limne* lake) and is isothermal down to a depth of 3 to 5 or more meters. Immediately below the epilimnion is a *thermocline,* a layer of water in which a rapid temperature change takes place. Immediately below the thermocline is an isothermal layer called the *hypolimnion* (*hypo* beneath). The lake therefore is *stratified* (see Fig. 11.8).

As the epilimnion is forming, phytoplankton productivity increases and may result in an increase in the dissolved oxygen content of the epilimnion. That portion of the lake where photosynthesis is taking place and where community respiration does not exceed community production is called the euphotic zone (*eu* true + *photo* light) and is largely in the epilimnion. Respiration continues in the hypolimnion at this time and increases as detritus drifts down from the productive epilimnion. Since the lake is stratified, oxygen production is cut off from the bottom where respiration takes place. In the epilimnion, oxygen may increase to the saturation point, and some is even given off to the atmosphere. In the hypolimnion, the only source of oxygen is from very slight mixing of water across the thermocline and some slight diffusion. Depending upon how much respiration takes place and the volume and temperature of the hypolimnion itself, the concentration of oxygen in the hypolimnion decreases. In the fall, as the surface waters cool, a mixing begins that ultimately results in an overturn of the lake waters. When the overturn takes place, the water from the stagnant hypolimnion often causes an odor near the lake.

To a great extent, the lake is a closed ecosystem and is thus a microcosm of oxygen cycling between the physical environment and living systems. Although there is much of intrinsic interest in the nature of the lake microcosm, it is especially important to us here, since it illus-

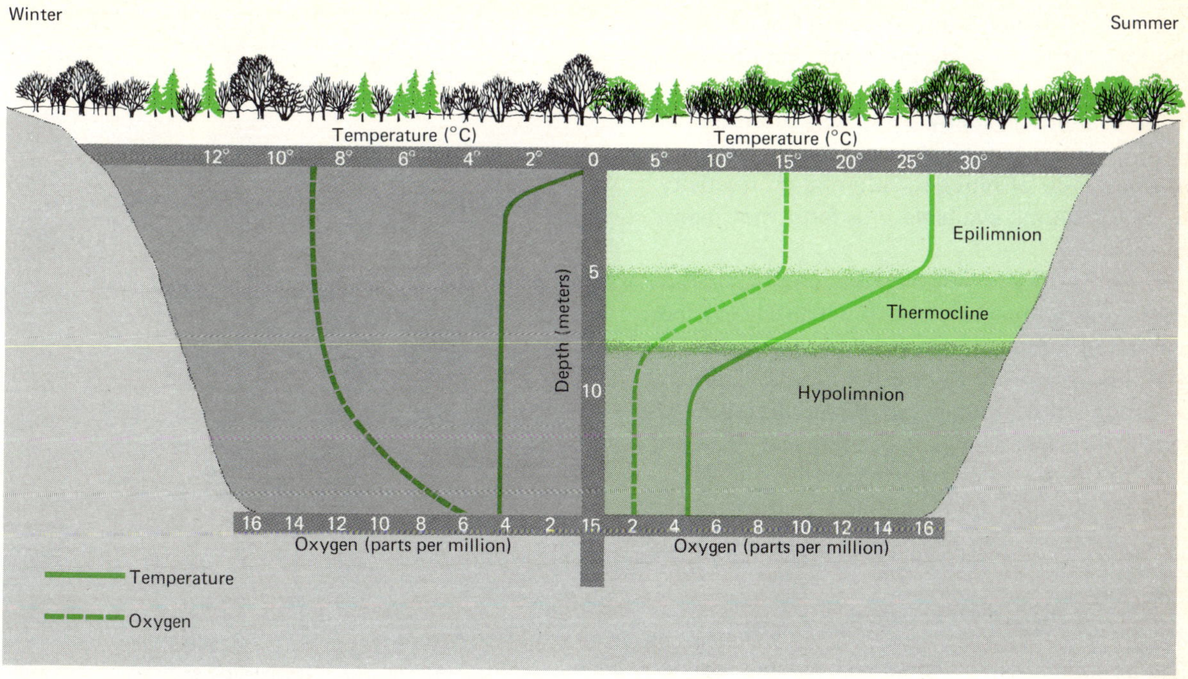

FIGURE 11.8 A lake is stratified in temperature during the summer, as shown on the right, and oxygen concentration becomes low in the hypolimnion. During the winter, as shown on the left, the oxygen concentrations are higher and the temperature of the lake is much more uniformly distributed from top to bottom. The temperature distribution is obviously determined by the seasonal change in climate, while the changes in oxygen concentration are due to the increased photosynthesis in the epilimnion and increased respiration in the hypolimnion that occur during the summer.

trates the closed nature of the oxygen cycle. The physical conditions we have described determine what lives in the lake and where in the lake the various organisms can live. They are presented not only because they provide a good example of cycling but also because they are intimately associated with some major ecological problems that we will soon discuss.

Oxygen and carbon are major nutrients and have important effects on the living and physical components of the biosphere; they are usually available in relatively abundant quantities. We will next look at two nutrients that are often far less abundant, and which, due to their relative scarcity, may exert a major control on the growth of species.

The Nitrogen Cycle

As an essential element in the structure of every amino acid, nitrogen is required by every living thing. As a constituent of phospholipids, nitrogen is a key element in the structure of cell membranes. Since nitrogen is a constituent chemical of the bases adenine, guanine, cytosine, thymine, and uracil, it plays a role in genetics and cellular metabolism. From a more applied point of view, nitrogen is the key nutrient upon which modern chemical fertilizers are built, and its use as a fertilizer has effected a revolution in agricultural productivity. Finally, we shall see that the availability and distribution of nitrogen have created some ecological problems of major proportions, problems that involve basic biological, economic, and social questions.

It would seem that since the atmosphere is approximately 78% nitrogen by volume, nitrogen would be a widely available and easily obtained nutrient. However, atmospheric nitrogen is in the form of N_2, a fairly inert gas. Plants require nitrogen in an oxidized form, usually nitrate (NO_3^-), although many can also use ammonia (NH_3^+). Thus, though there is an immense natural reservoir of nitrogen, it limits productivity by the amounts available in a form that plants can use.

There are two major ways in which N_2 can be converted into NO_3^-. First, it can be ionized in the atmosphere by electrical currents. The ions so produced eventually combine with water and reach the earth as a part of rainwater. The energy necessary to effect the ionization of N_2 may be obtained from lightning or from ionizing solar radiation. The first reaction that takes place is

$$2\,N_2 + 6\,H_2O \rightarrow 4\,NH_3^+ + 3\,O_2$$

The ammonia (NH_3^+) produced by this reaction may then react with hydroxyl ions in the atmospheric water vapor:

$$NH_3^+ + 3\,OH^- \rightarrow 3\,H_2O + NO_3^-$$

Thus, rainwater usually contains nitrogen in a form that is usable by plants. However, though rainwater can supply a significant amount of nitrate for plant productivity, the greatest proportion of the nitrogen that enters the plant comes from the activity of certain bacteria and algae that derive energy from splitting the N_2 bond.

Under aerobic conditions, bacteria of the genus *Azotobacter* convert N_2 to NH_3^+, and under anaerobic conditions, the same reaction is performed by bacteria of the genus *Clostridium.* These free-living soil bacteria are among the most abundant *nitrogen fixers,** although recent investigations have revealed that a host of less abundant, but equally important, microorganisms can fix nitrogen for uptake by plants. Certain genera of blue-green algae (*Anabaena, Nostoc,* and *Trichodesmium* among others) are also known to fix N_2, and they seem to be important sources of utilizable nitrogen in some marshes, lakes, and soil cover, and possibly in the ocean. One of the most important sources of fixed nitrogen are certain bacteria found living in the roots of plants. These bacteria of the genus *Rhizobium* can be found in soil, but they assume their greatest importance when they infect the roots of certain kinds of plants, principally legumes, such as clover, peas, and beans. The plant response to the infection is to isolate the bacteria in *root nodules* (see figs. 11.9 and 11.10). In the absence of the bacteria, the plant is not able to fix nitrogen, nor are the bacteria able to do so when not in association with the plant. Together, by some as yet unknown mechanism, the plant-bacteria complex is able to fix large amounts of nitrogen that the

FIGURE 11.9 The nearly spherical nitrogen-fixing nodules on the roots of a peanut. *(Courtesy of Hugh Spencer from National Audubon Society.)*

*To "fix" nitrogen means to chemically convert it from N_2 to some other form that plants can use, generally NH_3 (ammonia).

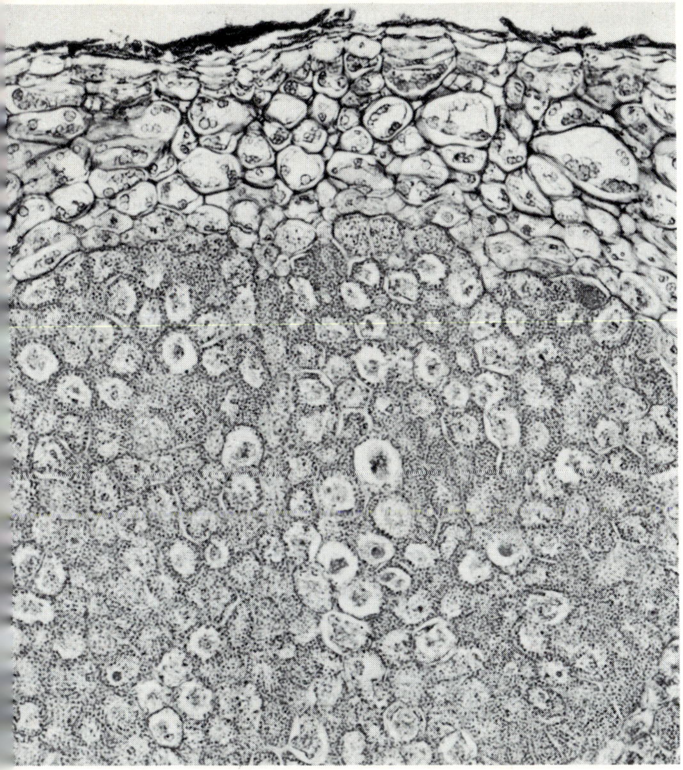

FIGURE 11.10 The very small dots on the interior of this cross section of a root nodule are the bacteria that result in the nitrogen fixation. *(Courtesy of Carolina Biological Supply Company.)*

plant can use for protein synthesis. The estimates of how much nitrogen can be fixed by the various kinds of bacteria vary, but apparently leguminous crops can fix as much as 480 kilograms of useful nitrogen per *hectare* (1 hectare = 2.47 acres), or around ten times as much as free-living soil bacteria. No matter what the source, once nitrogen is fixed in a useful form, it is taken up by plants for tissue production. Should the plant die before being eaten by a herbivore, decomposers break down the tissue and release nitrogen in the form of ammonia. The ammonia is taken up by another group of bacteria that oxidize it to nitrite (NO_2^-), and the nitrite is then oxidized by another group of bacteria to nitrate (NO_3^-) which can be utilized by plants. The diagram below shows the bacteria and the chemical reactants and products involved:

NH_3^+	Nitrosomonas →	NO_2^-

By the reaction:

$$NH_3^+ + 3\,OH^- \rightarrow H_2O + NO_2^- + 4H^+$$

and,

NO_2^-	Nitrobacter →	NO_3^-

By the reaction:

$$NO_2^- + H_2O \rightarrow NO_3^- + 2\,H^+$$

Not all of the ammonia participates in this system, however, since a small proportion is ''denitrified'' by certain bacteria that convert ammonia into either gaseous N_2 or nitrous oxide (N_2O), both of which are returned to the atmosphere. In the long run, the amount of nitrogen fixed by microorganisms must balance the amount of nitrogen returned to the atmosphere by denitrifying microorganisms. The rates of nitrogen fixation and denitrification are known to be relatively slow compared to the needs of plants, and most nitrogen is recycled through the biosphere.

It should be pointed out that the conversion of tissue from ammonia to nitrite to nitrate takes place whether the tissue is of plant or animal origin. However, once nitrogenous compounds are incorporated into the body of a heterotroph, there is another route by which it can return to the cycle before the organism dies and enters the decomposer trophic level. Most invertebrates and fish excrete ammonia as a result of the metabolism of amino acids or nucleotides. Birds and reptiles excrete uric acid,

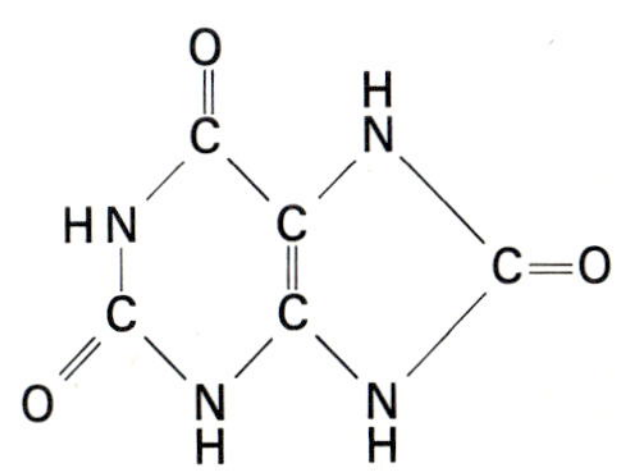

and many mammals, including man, excrete urea:

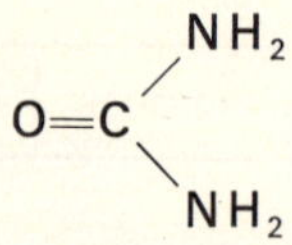

These compounds are included in urinary output, not defecation. Although feces contain protein and amino acids that can be converted to ammonia or nitrites by decomposers, most nitrogen that animals return directly to the cycle before death is in the form of urine.

Finally, most material in most natural communities is not broken down by animals and excreted. Rather, it enters detritus food chains and is then made available to decomposers that carry on the conversion to ammonia.

Thus, we can diagram the nitrogen cycle in terms of the compounds of nitrogen formed, and in terms of the movement of the compounds through the biosphere, as in figure 11.11.

The cycling of nitrogen is a common process everywhere—in water, in the soil of a forest floor, in a sewage treatment plant, or in any other place where proteinaceous material is released. It should also be pointed out that although this is a relatively well-understood cycle, it is not quite as simple as it appears here. There

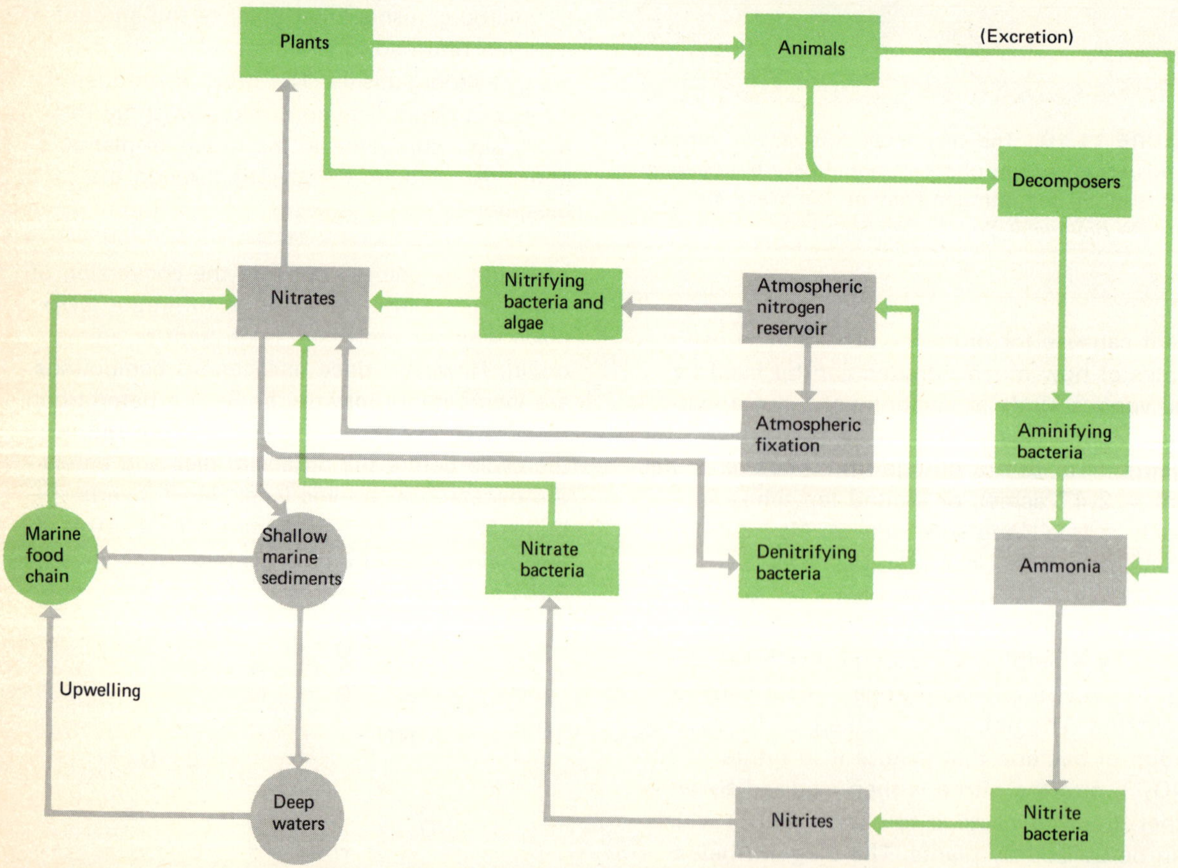

FIGURE 11.11 Nitrogen moves through the atmosphere, biosphere, and geosphere and through water just as does oxygen and other nutrients. The biological portions of the pathway are shown in color, and the marine parts of the cycle are shown as circles. *(Redrawn from E. P. Odum, 1971.)*

are a host of different bacteria that mediate the various conversions that take place, and the rates at which the conversions take place differ, sometimes radically, from place to place. How nitrate is transported in rivers, in ground water, and in the sea is a complex problem, and nitrogen fixation, an obviously important step in the cycle, is still poorly understood. As we shall soon see, the nitrogen cycle is of extreme importance to man and all of the biosphere, and we understand its complexities far too poorly.

The Phosphorus Cycle

The role of phosphorus in living systems, as we have shown, is critical. As a component of ATP, it is the chemical by which all energy conversions in living systems are carried out, and phosphorus forms a key compound in the structure of DNA. As a part of certain lipids, phosphorus is an indispensable component of the cell membrane. Thus, although it is generally needed in very small quantities, it can be a powerful factor in determining whether a species can survive and grow in a given place.

The major source of phosphorus for land plants is from rocks, and since different types of rock contain differing amounts of phosphorus, the availability of this nutrient to plants depends upon the kind of rock upon which the soil is based. As it is dissolved from rock or liberated from dead tissue by decomposers, phosphorus is usually in the form of phosphate ($PO_4^{\equiv}$), and, is usually taken up in this form by plants. The phosphate is carried in solution in ground water, and because it can enter into reactions with the soil itself, especially with clay particles, it is often removed from ground water by inorganic reactions. Thus, though it is often needed in relatively smaller amounts than nitrogen, in terrestrial and freshwater environments, phosphorus is often less readily available to plants. On the other hand, in oceans, where there is no soil to remove phosphate, nitrogen is usually in shorter supply. Thus, although its major source is rocks, not the atmosphere, and although it leaves the cycle as a sediment, not a gas, phosphorus often travels parallel to nitrogen in the biosphere (see fig. 11.12). Obviously these gen-

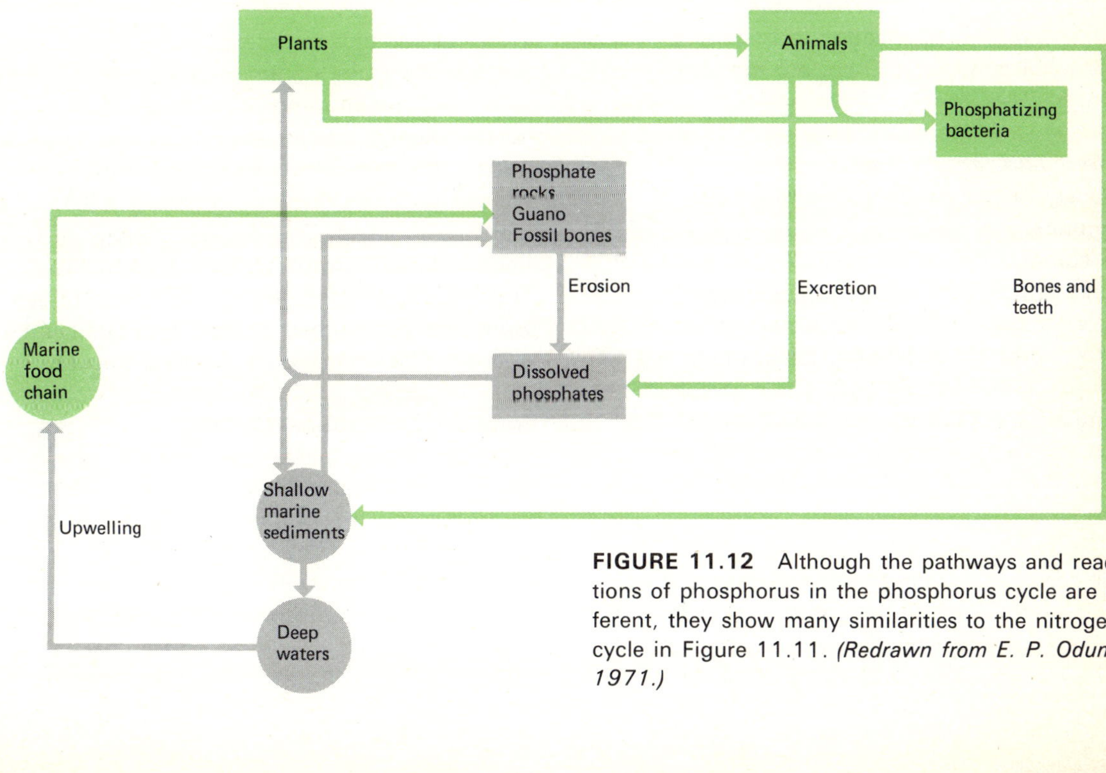

FIGURE 11.12 Although the pathways and reactions of phosphorus in the phosphorus cycle are different, they show many similarities to the nitrogen cycle in Figure 11.11. *(Redrawn from E. P. Odum, 1971.)*

eralizations are subject to a number of qualifications. The availability of phosphorus and nitrogen depends upon a variety of conditions such as the kind of soil, the activity of decomposers, the movement of ground water, and to a great extent, human activities.

Nutrient Regulation of Biological Communities

To understand how nutrients control the growth and development of ecological communities, we must understand one of the "rules" of ecology. First formulated around 1840 by Justus von Liebig, it has come to be called the "law of the minimum" (although it may not be considered a law in any strict sense). Basically, it states that *the rate of growth of a species population is controlled by the material present in the least quantity relative to the needs of the species.* Though this generalization has been raised to the nominal status of a law, it is really no more than a common sense observation and something of an oversimplification. Some investigators have interpreted it to include such things as temperature, soil type, and other factors, but we will restrict its application to materials. Under what conditions, then, are nutrients limiting?

Nutrient Limitation—Water

Obviously, water is often a limiting factor in communities. A major climatic determinant of large communities is the amount and timing of rainfall. In the same latitude or temperature zone, we can find both tropical rain forests and deserts. Parts of the Sahara, for example, are at about the same latitude as the rain forests of the Indian and Indochinese subcontinents (between 15° and 30° North). Rainfall is also a limiting factor over the interior portions of most continents.

As we have seen, marine and freshwater fishes possess adaptations to the osmotic properties of water. Other organisms are well-adapted to water supply by other physiological or anatomical means. Cactus plants, for example, are able to exploit an environment in which water is in extremely short supply. Some cacti have root systems that extend only a few inches into the ground but radiate from the plant as much as fifty feet, thus allowing the plant to absorb large amounts of water when infrequent rainfall occurs. Others send very deep tap roots into deeper, moister soil. Desert-dwelling vertebrates, such as the horned toad (actually a lizard), use very little environmental water, relying instead upon the water produced as a by-product of respiration. Another adaptation to scarce water resources is the excretion of uric acid, which is a solid and quite insoluble. Desert plants need water for photosynthesis, and some plant adaptations to desert conditions involve states of dormancy during periods of water shortage. When a rainfall occurs, the plants bloom profusely, reproduce, and then "shut down" as the water resources return to their normal, scarce levels. Few species can survive the rigorous conditions present in deserts, because most are unable to overcome the difficulty of evolving mechanisms to cope with the lack of water.

Thus, in addition to determining climate and the world's major communities, as we shall see later, and transporting other material as a "universal solvent," water is perhaps the most important limiting nutrient on the earth. Indeed, we can correctly think of irrigating a field as a kind of fertilizing. The vast technology of water impoundments in the western United States and the research on the desalinization of sea water are both aimed at providing sufficient quantities of a necessary, limiting nutrient.

Nutrient Limitation—Carbon

Although it is important geologically, chemically, and in the global radiation budget, carbon dioxide is much less important as a limiting nutrient. Because of the vast amounts of bicarbonate, carbon is probably rarely completely limiting in aquatic environments. In a greenhouse, CO_2 may reach a low level, and by adding it to the air in a greenhouse, production may be stimulated under ideal conditions, if nothing else is limiting. But such conditions are almost unknown in natural communities. It is possible to measure a decline in the CO_2 level of the atmosphere in the canopy of a rain forest, but it is not known that the levels ever become so low as to be limiting.

Nutrient Limitation—Oxygen

In terrestrial environments O_2 is rarely limiting, although at higher altitudes it might sometimes be thought of as a limiting factor due to low air pressure. In aquatic environments, however, the availability of O_2 can be a crucial limiting factor. We have seen that in the hypolimnion of a lake or pond the concentration of O_2 can fall to zero as the seasons change, and in many tropical lakes, the lack of seasonal change has brought about an unmixed condition in which the deeper portions of the lake are permanently anoxic, as also occurs in the Black Sea and some fjords.

Many aquatic organisms, of course, are adapted to low O_2 concentrations. Some, for example, have very efficient O_2 transport mechanisms, while many others have a very low metabolic rate. It depends on the species, of course, and generally, the less active species or species with more efficient O_2 transport mechanisms are adapted to very low concentrations of oxygen and, therefore, are most likely to be found in deep or polluted waters. Fishermen know well that active fish, such as trout, are usually found in more heavily oxygenated water.

Nutrient Limitation—Nitrogen and Phosphorus

After water, probably the most important limiting factors are nitrogen and phosphorus. Both putting manure on a field as a fertilizer and composting a garden are means of supplying nitrogen and phosphorus (as well as some other nutrients). In most cases, when we say that soil has been "worn out," we mean that nitrogen and phosphorus have been removed by plants and not returned to the soil by decomposers. This occurs when farmers remove a crop from the land, thereby removing the nitrogen and phosphorus taken up by the plants, before the nutrients can be returned by decomposers. As we shall see, the nutrients are rarely removed, and soil is rarely "worn out" under normal or wild conditions. We shall also see that with human intervention nutrients often reach decomposers at a point that may be very disadvantageous.

One of the most important agricultural "discoveries," dating back to Roman times, is that nitrogen and phosphorus must be replenished in the soil. The Romans did not understand that it was nitrogen or phosphorus that was being removed and replenished, but they were aware that in addition to manuring and crop rotation, the periodic planting and plowing under of leguminous crops had a beneficial effect on soil fertility.

In general, because it is bound by soil minerals, phosphorus is usually more limiting than nitrogen in terrestrial systems. Since the fresh water that feeds rivers, lakes, and ponds flows through the ground, phosphorus is usually also limiting in freshwater systems. Obviously, the amount of phosphorus available in lakes and rivers varies, depending upon the nature of the plant material growing in the soil through which the ground water flows. It also depends upon whether the vegetation is harvested or not. In oceans, phosphorus is not removed from the water by soil, and although there may be ten to twenty times as much nitrogen as phosphorus, the latter is usually not limiting.

Based on careful calculations of the chemical composition and energy value of protoplasm, we may write the photosynthetic equation in a way that takes into account much more than we have hitherto shown:

13,000,000 kcal of solar radiation + 106 CO_2 + 90 H_2O + 16 NO_3^- + 1 $PO_4^{\equiv}$ + (other mineral elements) $\longrightarrow$ 13,000 kcal of potential energy in 3528 grams of protoplasm consisting of 106 C, 180 H, 46 O,

16N, 1 P and including 815 g mineral ash + 154 O_2 + 1,287,000 kcal heat energy dispersed (99%)

This equation was calculated from values known for ocean plankton; since not all organisms have exactly the same chemical composition, they would not all have an identical energy value. As a statement of the efficiency of photosynthesis, the equation would be subject to considerable alterations that would vary from community to community and could only *very* roughly approximate terrestrial conditions. Nevertheless, it summarizes the productivity of ecosystems much more accurately than the equations we have thus far used. Its other value lies in that it shows us that photosynthesis is not simply the reduction of CO_2 to $(CH_2O)_n$; for production to proceed, other materials are needed in certain ratios according to the chemical composition of plants.

Nutrient Distribution in Communities

The problem of defining the boundaries of terrestrial and oceanic communities makes them difficult to study. Where, for example, does an ocean community begin? And even if it were easy to define the boundaries, the sheer size of ocean communities makes it difficult to comprehend all that is going on at a given time. Similar difficulties arise when dealing with a terrestrial community.

A lake or pond, however, presents investigators with fewer such problems. Boundaries are more easily defined, and sizes are frequently more manageable. This is not to say that lakes and ponds are divorced from the surrounding land. In fact, they are usually very sensitive to environmental changes that take place outside their boundaries. What makes lakes and ponds useful and interesting is that their characteristics are easily defined and measurable. Therefore, a lake or pond may provide investigators with a natural system on which hypotheses may be tested and from which generalizations can be drawn for further testing.

Nutrient Distribution in a Lake

Earlier we saw that oxygen distribution in a lake is related to the stratification of warm and cold waters and to photosynthesis, but we omitted one important detail: When plants start growing in the spring, they take up available nutrients; and as the summer progresses, the nutrients grow more scarce. The depletion of nutrients limits growth, and the spring bloom of phytoplankton slows in summer. Productivity may even fall to levels as low as those measured under winter conditions. This occurs in the epilimnion. In the hypolimnion, heterotrophs take up oxygen and release nutrients. Thus, if we look at a limiting nutrient in the lake over the year, we can see how its availability changes.

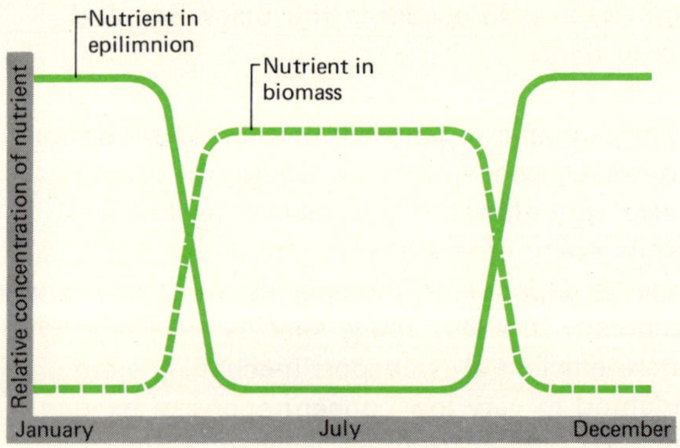

The diagram represents an ideal situation. In fact, things are rarely so well balanced. The basic point is that available limiting nutrients, while controlling the rates of growth of various components in the system, are regenerated and mixed throughout the community as physical conditions change. That is, *productivity is controlled by the availability of nutrients and, thus, by the seasons.*

Nutrient Distribution in the Sea

Having stated such a generalization from the nature of the lake, can it be applied to the ocean? Although the physical conditions in oceans are very different from those found in lakes, the basic principle is the same. Over a continental shelf, the situation is somewhat similar to that found in a lake. The depths are usually greater, but in temperate oceans, stratification takes place in the spring and summer with the same ultimate impact on plankton populations—that is, nutrients become less available as they are taken up by the population. Most plankton species have evolved life cycle patterns that permit maximum growth to take place in the spring. Thus, shallow areas of the ocean are very productive, since the nutrients are regenerated and remixed during the winter.

The tropical ocean presents particular problems for nutrient circulation. In tropical waters, there is a thermocline present all year long, which acts to prevent much nutrient material from reaching the water surface. Therefore, nutrients become almost unmeasurable at or near the surface. There is some productivity in tropical oceans, however, due to a high nutrient turnover. Thus, available nutrient material is used again and again, with only a slight replenishment from deep waters. Indeed, a limiting nutrient such as nitrogen is largely tied up in biomass. When it is released, it is very rapidly taken up again by other organisms before it can be measured in the water.

In the open ocean of temperate latitudes, an intermediate condition exists. There is permanent stratification of the water with a summer thermocline superimposed. Thus, although some mixing of nutrients to the surface during the winter produces a spring bloom, much of the nutrient is almost permanently lost in the deep waters and is not regenerated each year as it is over the continental shelf. Most of the nutrient materials in the ocean are in very deep waters and are only rapidly recycled under very special, but important, circumstances, in localities called *upwelling zones* (see fig. 11.13).

These upwellings are extremely productive. One such zone exists off the coast of Peru where the

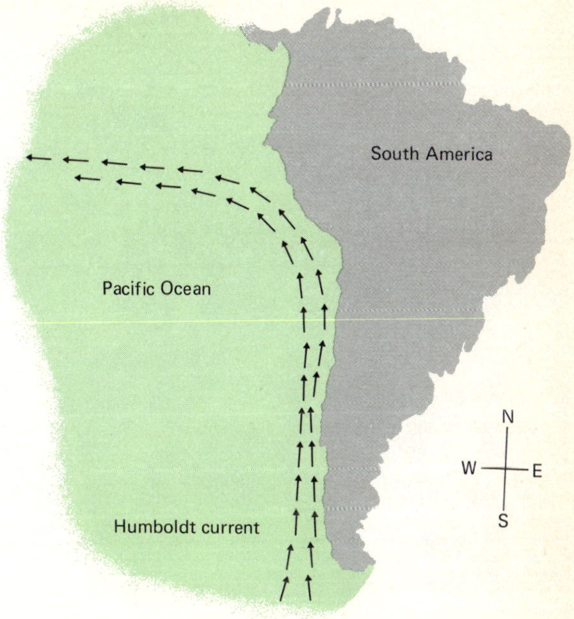

FIGURE 11.13 As the Humboldt Current passes to the north along the coast of South America, water is carried from below to the surface on the landward edge of the current, replenishing the nutrients. This upwelling results in a very high rate of productivity of the phytoplankton, creating one of the world's most productive fisheries.

Humboldt Current travels north along the coast and carries cold, deep waters to the surface. The current veers sharply westward at the equator, and the upwelling zone extends into the Pacific. Farther from the coast, however, the nutrients carried by these currents become increasingly scarce. The productivity of this upwelling is such that the Anchoveta fishery off the South American coast was for several years the most productive in the world. The ocean circulation occasionally changes for a few months, and when this happens, it has dire results. Production falls drastically, leading to immense levels of decomposition and even to the release of toxins into the air. In that part of South America the ocean food chain is a major determinant of life and living conditions. Much of the nitrate in the water is taken up by birds that feed on ocean fish, but not all of it is re-

FIGURE 11.14 It is obvious how a population of birds such as this can result in tremendous deposits of guano on the islands along the coast of Peru and Chile. The guano, of course, derives from the food the birds ate, which in turn came from the rich upwelled waters of the Humboldt Current. *(Courtesy of Bernard Silberstein from Monkmeyer Press.)*

turned to the ocean. A considerable proportion of it ends up high in the Andes, and an immense amount on the guano islands off the coast (see fig. 11.14). Over many centuries, birds nesting on these islands deposited vast amounts of nitrate in their droppings (*guano*) until some islands were almost completely white and could be mined for nitrate. For many years, until the early decades of the twentieth century, nitrate from these islands supplied a major proportion of the fertilizers used in North America and Europe. It was also used in less benign ways. Much of the nitrate from the bottom of the ocean that was deposited on these islands via the oceanic food chains of the Pacific was used for the very ''unpacific'' purpose of manufacturing munitions used in World War I.

Many people speak in glowing terms of some future day when we will ''farm the sea.'' We probably can gain large amounts of valuable protein from productive coastal and oceanic areas that are presently underexploited, but we will also probably not be able to do so for very long unless careful regulations are observed about the quantities and ages of fish taken. If all of the juvenile fish are caught, for example, where will future generations come from? Even assuming, however, that careful regulations are developed and that they are rigorously observed by all of the nations that will participate in the harvest, there are still several difficulties. First, nutrients are recycled to the surface only in certain localities, and most of the oceans are analogous to deserts. It may be possible to devise the technological means for bringing up nutrients from the deep ocean floor, but it will have to be recognized that to do so will require an energy subsidy of some sort. It will also have to be recognized that we do not yet know enough about deep ocean ecosystems to predict whether bringing up nutrients will have a beneficial or disastrous ecological effect. Second, in the sea, light is ultimately limiting, and the gross efficiency per square meter cannot be as high as on land, because the water itself absorbs a majority of the useful energy. Finally, if we eat fish, we will have to pay the cost of trophic energy conversion and loss. We may increase world fish production, but doing so will never completely solve the problems of feeding an expanding population.

Another highly productive marine environment deserves mention in this respect. *Estuaries* are shallow indentures along coasts where fresh water and the nutrients it contains meet the sea. For a variety of evolutionary reasons, estuaries and their bordering marshes and wetlands are the nursery grounds for a vast number of fish and invertebrates. The circulation of estuaries is typically such that nutrient-rich water flows landward near the bottom, mixes with the less saline surface water, and is "recycled" by the circulation patterns. The fresh water from the marshes and rivers contains abundant concentrations of detritus and thus is also a rich food source. The estuarine ecosystem, then, is highly productive and crucial to our marine fisheries. It is also among the most badly polluted of environments. More than 30% of the Eastern coastal wetlands in the United States have been destroyed in the last twenty years. How ironic it is that we describe a rosy future in which we will feed the world's hungry from the sea, while we destroy the basis for that hope by contaminating our most productive fishery resources!

Nutrient Distribution on Land

In terrestrial systems, humans have become a major factor in the cycling of limiting nutrients. Nitrogen and phosphorus are both used by mankind in immense quantities. We still mine most of the phosphate we use, and most of our nitrate is produced in factories where atmospheric nitrogen is electrically ionized. Thus, our supplies of nitrogen and phosphorus are subsidized through the use of fossil fuels, thereby allowing us to bypass naturally imposed scarcities. *For the present,* this subsidy appears to be more economical (financially) than reusing materials that contain nitrogen and phosphorus. The major use of nitrate and phosphate in terrestrial systems is to promote plant growth, especially in crops that we value highly. Today modern farmers only rarely spread manure on fields. Rather, manufactured fertilizers are used. There is nothing inherently wrong with this, since to a plant, nitrogen and phosphorus are useful regardless of the source. But there are other serious implications of the use of manufactured fertilizers.

Consider the NO_3^- in a field in Iowa. What happens to it? Before we began to farm the Great Plains, some of it went into ground water, but most of it was recycled in the community. Even after the introduction of farming, but before the advent of manufactured fertilizers, the input of nitrate was relatively low. Now, however, the common practice is to put more nitrate on the land than the plants can use, thus increasing the amount that goes into ground water and then into rivers, streams, and lakes. A large proportion of the crop that is grown on the land becomes food for livestock or humans. The waste plant material, however, is no longer used as mulch for the soil, but is usually disposed of in some body of water. Cow manure becomes stockyard or feedlot waste, which is also disposed of into some body of water, and the useless remains of slaughtered livestock (that is, that part of the animal for which there is no market) also goes into water. The butchered meat is shipped to a city where it ultimately passes through the human population. What is not absorbed also goes into water (see fig. 11.15).

Each time something goes into the water, it usually undergoes some kind of pretreatment, but in most cases, the treatment is marginal. Since approximately half of the land in the U.S.

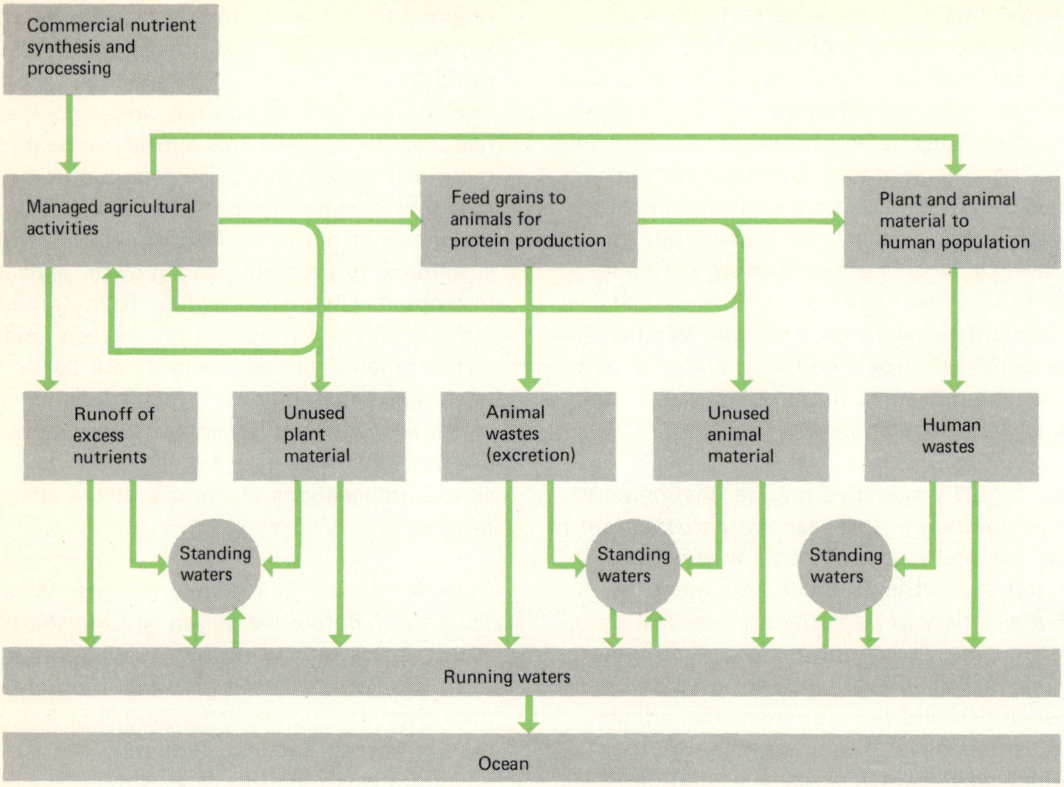

FIGURE 11.15 This diagram shows some of the ways that human activity influences the distribution and transport of nutrients, and why human activities have interfered with normal nutrient cycling.

is now in agricultural use (see table 11.2), humans have had a major disrupting effect on nutrient cycling in this country. In the progression of events just described, the nitrate put on an Iowa field has not remained in Iowa, where it might contribute to future crops. It has been transported from a location in the system where it could add to productivity that we value to another location in the system where it may promote productivity that we do not value. When industrial and household uses of nitrogen and phosphate are added to agricultural uses, it is clear that we have a problem of immense proportions. It seems ironic that our society is structured in such a way that it is less expensive (in financial terms) to discard nutrients into water supplies and use fossil-fuel energy sources to manufacture "new" nutrients than it is to transport the used, but still biologically active, nutrients back to the farm.

TABLE 11.2 U.S. Agricultural Land Use

Use	*Area (10^6 Acres)*	*Percentage of All Land in U.S.*
Cropland	387	17.1
Grassland pasture	547	24.1
Woodland pasture	82	3.6
Woodland not pasture	64	2.8
Farmstead, farm roads, etc.	30	1.3
	1110	48.9

All of the nutrients from agricultural, industrial, and household sources enter waste treatment disposal systems, and a brief consideration here of those systems can provide us with an important lesson in the effects of the availability of limiting nutrients. Consider the impact on a waterway of dumped, untreated sewage. This mixture of carbohydrates, proteins, and lipids is rich in the materials that supply energy and nutrients to populations of decomposers. Provided with an abundant supply of these materials, bacteria thrive, and their respiration removes oxygen from their environment. Therefore, downstream from the source of the sewage, a severe oxygen shortage may occur due to the respiration of decomposers. This oxygen depletion may kill off everything but bacteria. Thus, the water stops producing useful food. The measurement of the rate of this uptake of oxygen from the water, or *BOD* (biological oxygen demand), is an important means of monitoring water pollution. *BOD* has long been known to sanitary engineers, who try to minimize the effects of excess organic material by promoting oxidation under controlled conditions in sewage treatment plants. If sewage treatment is well-done, and very often it is not, then only dissolved materials will be released into the waterway. Though these materials can promote the removal of some oxygen from the stream, the overall effect is less drastic than that of the dumping of raw or partially treated sewage.

But the products of decomposition carried out in sewage treatment plants under controlled conditions, even with the complete removal of all organic compounds, still contain nitrate and phosphate and some trace elements. Since these materials offer no apparent hazards to health, and since they have no immediately obvious effects at the point of discharge, they are rarely removed. Moreover, and this is no small point, to remove the inorganic nutrients from sewage is an expensive activity. Only a few sewage treatment plants have been built to provide complete nutrient removal.

In freshwater systems, as we have mentioned, phosphate is usually limiting. Thus, downstream from the discharge, there is usually great algal growth. The water becomes green, the stream banks become slimy, the oxygen content rises, and when the algae begin to die, an oxygen depletion can be created as they are decomposed. The decomposition of dead algae, besides producing an unpleasant odor, can drive out the normal river flora and fauna that depend upon the oxygen. Rivers have an ability to recover, since nitrate and phosphate are taken up by plants, and oxygen can be replenished by the interaction of moving water and the atmosphere. Oxygen diffuses into moving water more rapidly than into still waters. As a flowing body of water, a river can recover if the nutrient input is stopped since the concentration of the nutrients will fall continually lower as new water is injected into the river from ground water sources. However, if the river encounters an impoundment, or reservoir, the flow is reduced, and the ability of the river to cleanse itself is impaired. It becomes like a stream that drains a fertilized field and ultimately feeds a pond. Once the nutrients get into a standing body of water, they continue to recycle.

Recall the relationships of nutrients, oxygen, and productivity in ponds as pointed out earlier. The addition of a nutrient brings about an increase in productivity in the epilimnion. That increases the organic material in the hypolimnion, and decomposers then remove oxygen and liberate the nutrients. This can easily cause the hypolimnion to become anoxic, result in fish kills, and cause an unpleasant odor when the pond turns over. If the fish in the pond are trout, it is not even necessary for completely anoxic conditions to exist to bring about a kill. Since trout require large amounts of oxygen and cold water, they usually remain in the hypolimnion during the summer months. An injection of phosphate into the epilimnion, by ultimately lowering the oxygen concentration in the hypolimnion, can kill off the trout population. Other fish might replace the trout, but the continued input of phosphate will bring the hypolimnion to complete anoxia and kill off all fish. Only anaerobic bacteria will be left. If the input of nutrients is stopped, the situation in the pond may, in some instances, be reversed. Since nutrients recycle in the pond, it is more usually an irreversible chain of events. Much research is underway to find ways of reversing such deteriorated situations.

Many reservoirs have been built, at considerable taxpayer expense, to supply drinking water. As standing bodies of water, they may undergo the same process as a pond, but the consequences are of more direct human concern. Irresponsible upstream waste disposal or the runoff of nutrients applied in excess or by accident to agricultural land destroys the quality of water in a reservoir and thus destroys the value of the investment.

Eutrophication

Lakes and ponds normally go through a process of aging, which will be described in more detail later. Basically, however, the lake fills with sediments, nutrients accumulate, and productivity increases. A bog mat forms around the water, eventually fills the lake, and a forest grows where a lake once was. Under normal conditions, this process takes from at least 1000 to 5000 years. Thus, populations in and around the lake ''adjust'' slowly to changing conditions. This process is called *eutrophication* (*eu* true + *trophe* food) and is a natural aging process. The increased productivity caused by pollution is in many ways similar to the increased productivity that occurs early in the eutrophication process, and because it is induced by human activities, it is called *cultural eutrophication.* Because of its accelerated rate, cultural eutrophication is far more drastic and destructive of natural communities. It denies them the time necessary to adjust. The excessive use of nutrients by man is one of the major causes of pollution in the world today, and in the United States, the major proportion of ponds and lakes are now in one stage or another of cultural eutrophication.

Nor are lakes and ponds the only bodies of water in this condition. San Francisco Bay, Jamaica Bay, the estuarine Potomac River, Great South Bay in Long Island, the Hudson River, and nearly every coastal area near a major American city is polluted, to some extent, by excess organic material and inorganic nutrients. Thus, ironically, although inorganic nutrients are necessary controlling factors in living systems, cultural and economic development has dislocated nutrient cycles with the result that nutrient excess has become one of the major sources of pollution.

Are there any solutions? We will discuss some possibilities later, but it will suffice here to say that it is not a simple problem yielding a simple solution. At one time, it was thought that the most efficient and least harmful means of disposing of the waste products of human populations was to dump them into the most convenient river. That has obviously failed, as have the solutions of dumping them into harbors and coastal waters. It seems that in the past, every ''simple'' answer has merely increased the complexity of the problem. Can we now use the deep oceans as dumping grounds? There is already an abundant supply of nutrients there, and frankly, we are not sure of what the long-term effects of deep ocean dumping might be. The sensible solution is to put the wastes back on land, but that raises other difficult questions that we shall deal with later.

Nitrates and phosphates are not the only ''problem'' materials to consider. We understand too little about the effects of organic micronutrients and other trace elements. It seems that their contribution is less important in controlling the productivity and biomass in a community, but could play a correspondingly large role in determining the species composition of a community. By favoring one species of plant in a community, an input of organic or trace micronutrients could lead to a population explosion of the favored species. Since plants compete for such things as light and space, the injection into a community of a material that limits the population of one plant species might permit it to outcompete other species. Thus, the entire balance of a community could be changed radically by an increase in the availability of a micronutrient, even if the dumping should take place on land. The possible effects of such dumping in the community can be amplified further if the plant species that "loses" the competition is replaced by a species that cannot be utilized by herbivores.

The Biological Effects of Nonnutrients

Besides nitrate, phosphate, and various micronutrients, we should also consider materials that travel through the hydrologic cycle and, although not nutrients, have biological effects. Two such materials have received much public attention lately—namely, DDT* and mercury (Hg). When taken up by insects, DDT, a synthetic organic compound, has a rapid toxic effect. When taken up by other organisms, it may also be toxic, but over long periods of time. It is not a naturally occurring compound, and since no microorganisms have evolved that are capable of readily breaking it down, it is not biodegradable. DDT is fat-soluble, and when taken up by an organism, it is absorbed into fatty tissue where it has a tendency to accumulate. It can enter a food chain at the producer level, and as the producers are fed upon by herbivores, and the herbivores by carnivores, the DDT can be found in ever increasing concentrations in tissues of organisms at higher trophic levels. In one area, for example, DDT or related compounds were measured in water at 0.005 ppm, and in marine phytoplankton at a level of 0.04 ppm. The top carnivores in the same area (both birds) were found to contain 22.8 and 26.4 ppm. The one with the higher concentration feeds on larger fish (see fig. 11.16).

DDT residues can now be found in very nearly all marine fish, and the harvest of California mackerel was halted by the Food and Drug Administration, because DDT levels in the processed fish were found to exceed federal guidelines. Since DDT is used for pest control in agriculture, it can be found in small amounts in vegetables. If a person were to eat ¼ pound of California mackerel daily as a part of a normal diet, the level of DDT in the fish (5 ppm) plus the levels now "normally" found in other foods would exceed the daily intake of DDT that the FDA considers safe for human consumption. Whether or not these guidelines are realistic has yet to be proved. It may be that humans are capable of ingesting larger concentrations of DDT without suffering any toxic effects whatever. On the other hand, the toxic effects may be cumulative over very long periods of time and may, therefore, only be significantly felt in the human population at a later time.

The biological effects of DDT in other species are varied and more readily detectable. In freshwater trout, concentrations of 5 ppm cause 100% failure in the development of young trout. Along the Texas coast, concentrations of 8 ppm have been found in the ripe eggs of sea trout, and such concentrations are suspected to be at the root of the decline of sea trout in Texas coastal waters from an average of 30 fish per acre in 1964 to 0.2 fish per acre in 1969. In recent years, no juvenile fish have been found in the areas of Texas coastal waters where there is known DDT contamination. Among some birds, and particularly those in coastal waters, there have been similar effects on reproductive and developmental processes, as DDT appears to interfere with calcium metabolism and, consequently, with the formation of egg shells. In 1969, in Southern California, only one brown pelican was known to have hatched out of about 1200 nesting attempts observed.

Mercury is used in many industrial processes, and it is also used as an agricultural pesticide for fungal control. Because of its great expense, users of mercury make an effort to recover it whenever technologically possible and economically feasible. Nevertheless, the maximum recovery rate is less than 50%. The remainder is dumped into the environment, where it forms methyl mercury ($HgCH_3$) and enters the biologi-

*DDT is the "shorthand" way of saying dichlorodiphenyltrichloroethane, a chlorinated hydrocarbon insecticide.

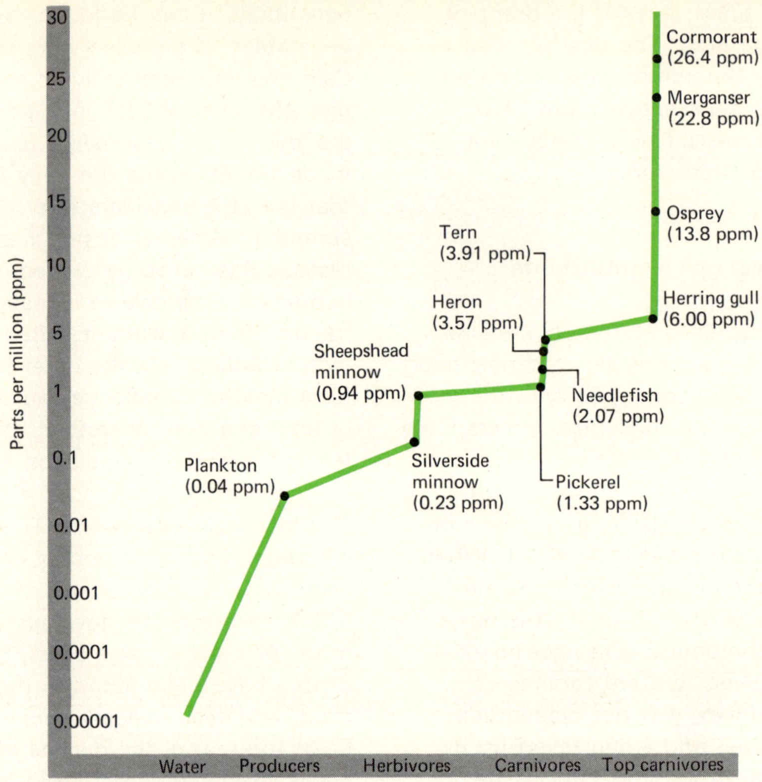

FIGURE 11.16 Pesticide concentration in a food chain. The graph shows the increasing concentration of the pesticides DDT, DDE, and DDD, all related toxic compounds. Note that the portion of the graph below 1 part per million is drawn on a logarithmic scale. The difference between intervals on the lower part of the graph is much greater than on the upper part of the graph; each number is ten times smaller than the next number above. If the lower portion of the graph had been drawn to the same scale we would not have been able to place the values for plankton and water on this page at all. To emphasize the difference, consider that the concentration in plankton is 800 times greater than the concentration in water, and the concentration in the cormorant, a fish-eating bird, is 660 times greater than that of plankton and 528,000 times greater than the concentration in water. *(Information from Woodwell* et al.*)*

cal community. Like some other metals (lead, chromium, cadmium, nickel, and arsenic), mercury is very toxic and in many instances has resulted in the poisoning of humans. In Japan, for example, 111 persons were poisoned, and 41 of them died as a result of eating fish and shellfish from Minimata Bay, which had been contaminated with mercury used in a nearby factory. Recently, one family in New Mexico was poisoned when they fed corn treated with a mercury fungicide to the family pig and then ate the pig. The corn had been treated for use as seed corn, not for human or animal consumption. Epidemics of mercury poisoning from treated seed have been reported in several Asian countries. Mercury is found in many fish, especially those in the higher trophic levels of food chains, such as tuna and swordfish. While there have been no well-documented cases of mercury poisoning from tuna and swordfish, mercury levels in some catches of these fish were recently found to be higher than deemed safe by the FDA, and the catches were confiscated. However, there are two unknowns about mercury that pose a problem with such actions: In the meat of tuna and swordfish, at what level, does mercury become toxic to humans, and how much of the mercury found is normal? There has been some recent evidence brought forth that tuna caught and canned many years ago contained mercury at levels close to those found in the confiscated catches. A great deal more work must be done on the effects of mercury and other toxic metals, especially on their effects in ''normal'' communities.

Both DDT and mercury provide cogent examples of two generally dangerous cases. First, since synthetic or manufactured compounds are not a normal part of biogeochemical systems, no natural populations of decomposers have been evolved that utilize them. Therefore, such compounds are not usually broken down rapidly into other less-harmful compounds. Sooner or later, of course, any compound will break down, but like DDT, some are so stable that they persist for many years. As long as they are being injected into the environment in quantities greater than can be broken down naturally, we are effectively stockpiling them. Besides DDT, there are a number of organic compounds that we do not normally monitor, and their effects on ecosystems remain largely unknown. Second, heavy metals, such as mercury, lead, and cadmium, are normally toxic, but they are also normally present in very small quantities. Some metals such as arsenic and zinc are even beneficial in trace amounts. We need to monitor the levels of these metals in biological communities and in the physical environment, as we need to monitor other uninvestigated materials. At the present, we are not doing so in any systematic way. It may well be that some of these materials have no deleterious effects at all, but we just do not know.

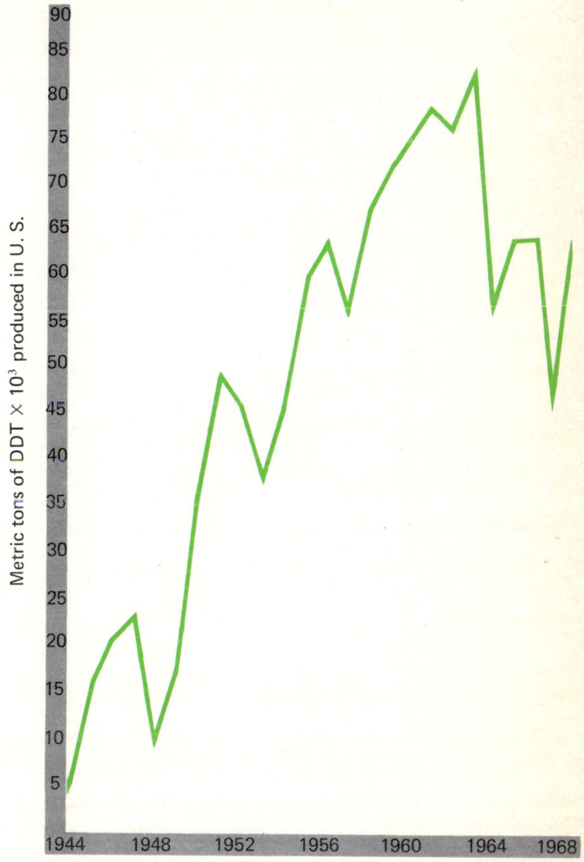

FIGURE 11.17 A graph of U.S. DDT production, 1944–68 *(From* Chemical Economics Handbook, *1969, Stanford Research Institute, Menlo Park, California.)*

We have frequently emphasized the slow pace of evolution, and that point is relevant here. Many small changes that take place in environments over long periods of time permit species to adapt and evolve, but when the changes are rapid, they produce drastic changes in communities. Adaptations could have arisen to degrade or metabolize DDT and heavy metals if such materials had been present for hundreds or thousands of years. And, of course, some adaptations to DDT have arisen in some insects in much less time than that. But the introduction of DDT into the environment only began in the mid-1940s. Between 1944 and 1968, more than 1,300,000 tons were produced in the United States alone (see fig. 11.17). Much of it was used here, and much was shipped out of the country for use around the globe. This sudden massive input has had widespread consequences that we will look at closely when we examine the structure and evolution of communities.

We do not intend to be prophets of doom. The many changes described here are taking place at this moment, and there may well be others taking place of which we are unaware. Many are the products of the technological society upon which we have become dependent, and many of the solutions may well be technological ones. The key problem that we face is our lack of fundamental knowledge about the nature of the biological world and our impact upon it. It is not necessarily prophesying doom to say that drastic, uncontrolled changes are poorly understood. What must be done? The answers to that question are as much political and philosophical as they are scientific. But before definitive answers can be given, or clear choices offered, a great deal more must be learned about the scientific aspects of the question.

By now it should be obvious why this book began by viewing biological systems as stable, evolved entities, why we paid so much attention to genetics and the process of selection, why we considered the physiology and adaptations of organisms so important, and why we presented energy transfer and productivity as a basic, all-important process. It should also be clear that even were we to consider biogeochemical cycles alone, human activities are intricately involved with the processes of the living world in ways that we understand too poorly or choose to ignore. If you doubted the relevance or importance of earlier sections of the book, perhaps you should review them and see whether, in the light of the material presented in this chapter, you still consider those earlier chapters to be unimportant. To be sure, there are sections that are digressive, but they digress to present more complete information, to cast interesting sidelights, to illustrate how scientific understanding has developed, or to provide material that will be of importance in the chapters to come.

The material in this chapter is of basic importance for understanding ecology, and some of the concepts are neither simple nor well-understood. A great deal of oversimplified and misleading information has been presented to the public in the last few years about many of these subjects, and many persons have tended to think they know a great deal about ecology. Professional ecologists rarely take that view, being only too painfully aware of their own ignorance and inability to provide reasonably simple, yet reliable, answers to complex and crucial questions. We may hope that simple and reliable answers may be found, but we do not yet know what they are. In some cases, the right questions have not even been asked. We will be better able to look at some of the questions and possible answers later. Thus far, we have treated organisms and ecosystems as though they were simply mechanisms for the processing of chemical energy. In some senses that idea is correct, and we understand the chemical aspects of ecosystems as well as any. But communities are made up of species populations, and populations are made up of individuals. Before we go further into the implications of biology and ecology, we must consider how populations develop and what controls them. We must also examine how species interact and come together to form the balanced communities of the biosphere.

Selected Readings

Bridges, E. M. *World Soils.* New York: Cambridge University Press, 1970. A somewhat technical treatment of soil structure and chemistry, but one of the few good books on a subject too often ignored.

Young, Keith. *Geology: The Paradox of Earth and Man.* Boston: Houghton Mifflin Company, 1974. A sound new text in introductory environmental geology.

*Scientific American, *The Biosphere.* San Francisco: W. H. Freeman and Company, Publishers, 1970. A reprint of the September, 1970, issue of *Scientific American,* containing articles on the biogeochemical cycles, as well as on other important topics.

United States Department of Agriculture, *Contours of Change.* Washington, D.C.: U.S. Government Printing Office, 1970. The USDA publishes yearbooks, each of which covers a specific topic. In the yearbook referenced, the general subject is "agricultural revolutions." Earlier yearbooks have dealt with soils, seeds, or other special topics. The publications can be useful sources of information.

Bormann, F. Herbert, and Likens, Gene E. "The Nutrient Cycles of an Ecosystem." *Sci. Am.,* Vol. 223 (October, 1970), pp. 92–101. Offprint number 1202.

Peakall, David B. "Pesticides and the Reproduction of Birds." *Sci. Am.,* Vol. 222 (April, 1970), pp. 72–78. Offprint number 1174.

Williams, Carroll M. "Third Generation Pesticides." *Sci. Am.,* Vol. 217 (July, 1967), pp. 13–17. Offprint number 1078.

Woodwell, George M. "Toxic Substances and Ecological Cycles." *Sci. Am.,* Vol. 216 (June, 1967), pp. 24–31. Offprint number 1066.

*Paperback edition available.

The dense aggregation of a species population at a nesting site often serves to regulate the size of the population by limiting the rate of reproduction. This is well exemplified by this penguin population. *(Courtesy of Ira Kirschenbaum.)*

the growth of populations

chapter twelve While it is the individual that carries the genes, and it is the individual that survives or fails to reproduce while processing energy and nutrients, we have seen that evolution operates at the level of the population. In the two preceding chapters, we focused on the transfer of energy and chemicals in ecosystems and the biosphere, and although we did not explicitly deal with populations, it should be clear that life is not merely a series of chemical reactions, but rather also a process controlled by the many individuals that make up diverse species populations. To fully understand the biology of communities and ecosystems and human interactions with them, we must now consider how populations grow and are controlled, as well as some of the implications of such growth.

The Growth of Populations

All organisms have the potential to produce more offspring in a generation than there are individuals in the parental generation. Oysters and many aquatic invertebrates, for example, produce millions of young per pair, and even species with low reproductive capacities, such as elephants or humans, are capable of producing more than two young per pair. Yet, in spite of this capacity, the populations of the earth usually remain fairly constant, or fluctuate about some constant or optimal number of individuals. Natural selection produces this relatively stable population size, and we have already seen how the process of evolution depends upon a reproductive excess.

Population Density

Any population in any habitat or ecosystem at any particular time may be described by a *population density.* This is one of the measurements most commonly made in an ecological study, and population density is defined and measured in two ways: (1) as a *crude density,* or the *number of individuals per total area* (or volume in aquatic systems); or (2) as *ecological density,* or the *number of individuals per habitat area,* where habitat area is defined as the area or volume that is actually available to a population. Sometimes these measurements are easily taken, as in the study of trees in a forest. In studies involving species that are highly mobile (such as birds) or difficult to see (such as fish, insects in leaf litter, or bacteria), or in which individuals are difficult to distinguish (such as insect pests), a number of specialized techniques have been developed to determine population density. The density of insect populations in soil samples, for example, may be measured by extracting the insects from the soil through the use of a Berlese funnel. (See fig. 12.1.) Plankton can be collected in silk or nylon nets that may range from 18 mesh to 200 mesh per inch. Some such nets may even be equipped with devices that measure the amount of water that passes through. Bacteria may be counted in water samples under a light microscope, or their density may be inferred from a variety of culture methods that will not be discussed here. Fishery scientists sometimes take test populations which, when carefully done, can tell them the density of the local populations. Obviously, any method which must rely on incomplete counts is subject to error, but modern methods

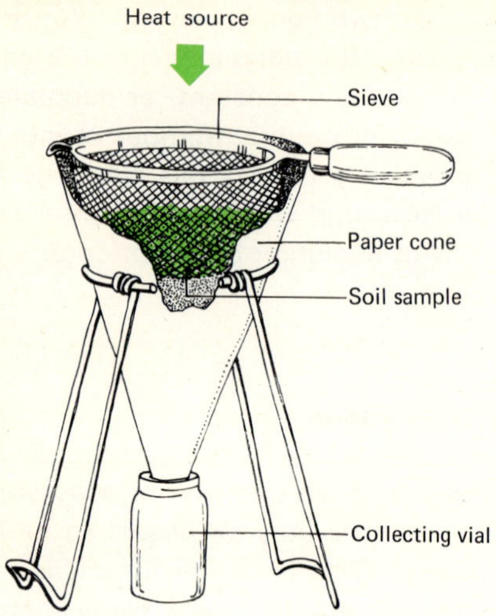

FIGURE 12.1 Heat will drive small arthropods from soil; thus an arrangement such as the Berlese funnel shown can be used to collect small soil arthropods.

of statistical analysis have allowed investigators to arrive at reasonably acceptable conclusions in many cases.

Intuitively, we can see that the density of a population is in some way regulated by the resources of the environment. We have already discussed some of the ways that limiting factors may control productivity and thereby control biomass and standing crop, both of which are measures related to population density, even though they do not take into account the numbers of individuals. In this and the next chapter, we shall look more closely at the factors regulating numerical population size.

The density of a population varies, obviously, with the species and the habitat. The population of oaks in a forest, for instance, will be vastly different from the populations of soil mites or deer in the same forest. Nevertheless, the population densities of these organisms are of great importance to the balance of the community. Anyone who has ever been concerned with the depredations of gypsy moths, boll weevils, mice, or lice recognizes that population density is of importance to us. As we shall see, it is often man who produces imbalances in the population densities of species that he considers pests.

Population Growth Curves

Although knowing the density of a population is useful, it is often of greater importance to learn whether the density is changing. Unfortunately, the language used to describe density is often very imprecise. Such terms as *rare, common,* or *abundant,* have often been used to describe a population. It is difficult to describe accurately the changes in density using such terms.

Density changes may be described in more precise terms and expressed in three ways: first, as a table of the numbers of individuals at specified intervals of time; second, as a graph of the table; or third, with an equation, this being the most precise and succinct method. All three means shall be used in this chapter. We advise caution at this time, however. While the *methods* of describing population growth are largely valid, the *study* of population growth is still very much a theoretical one. Many of the generalizations that we shall present are supported by good evidence, and by a great deal of observation of some species. Many more observations remain to be made to validate some of the models that shall be presented. The models, moreover, have some limitations, as will be pointed out. Finally, we should also point out here that what follows will give only a superficial understanding of the highly mathematical topic of population biology.

We have all seen examples of rapidly growing populations. Consider a bacterium introduced into a new culture medium, weeds colonizing a plowed field, or cockroaches in a new building. Each of these situations will produce a rapid and abundant growth in the numbers of individuals.

We can generalize about what happens to a species introduced into a new and favorable environment. Consider an actively dividing bacterium. For the moment, we will ignore death as a factor in the growth of populations. Every 20 to 30 minutes, the bacterial population will double: 2, 4, 8, 16, 32, 64, 128, 256, 512, and so on. Indeed, after *n* generations, the number of individuals in the population will be 2 , or two multiplied by itself *n* times. Clearly, such growth cannot continue indefinitely, since eventually the population would be equal to the number of particles in the universe and would be expanding outwardly at the speed of light. Death will occur, and sooner or later a balance of some sort will be achieved. We can make our understanding of population growth more concrete if we consider some simple mathematics. Changes in the density of a population obviously require a passage of time, and we may begin by considering an interval of time, t_0 to t_1. Since this is an interval, some change in time is dictated, and we may symbolize the change by the notation Δt.

$$\underbrace{t_0 \longrightarrow t_1}_{\Delta t}$$

The symbol Δ (delta) denotes "a change in." Over that interval of time, some change will presumably occur in the *number* (that is, density) of individuals in the population, namely *N*, designated as ΔN. Thus, we may calculate the *average rate of change* in the population as $\Delta N/\Delta t$. Suppose, however, that we are not interested in an average rate, but wish to measure the change that takes place in a population over a smaller interval of time. We could simply measure smaller intervals and let Δt become infinitesimally small, thus approaching zero. We would then be measuring not the average rate of change, but rather the *instantaneous rate of change*—the change that occurs at a particular instant of time. In order to distinguish between the average and instantaneous rates, we alter our statement $\Delta N/\Delta t$ to dN/dt.* (See fig. 12.2.)

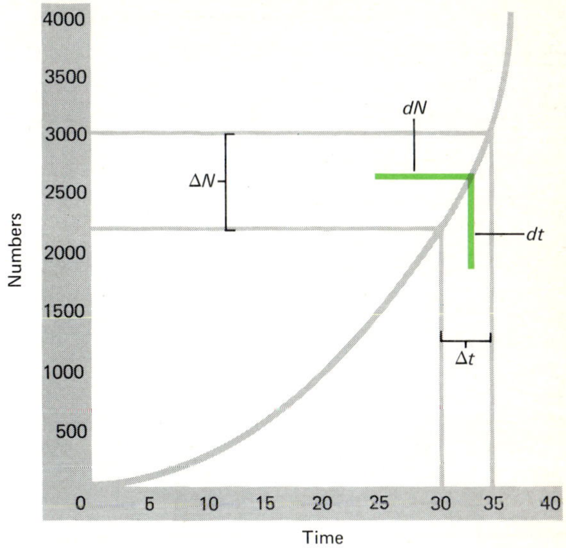

FIGURE 12.2 A diagram of ΔN over ΔT and dN over dT, showing the differences between the mean and instantaneous rates of change.

The rate of change in a population is, of course, the result of both the difference between births and deaths of individuals and the rate of migration. However, we will consider only closed populations, ones in which there is no migration. Under certain conditions, this is a realistic way of looking at the population, although at other times it is not. For example, were we to examine the world human population, migration would not be a factor, but if we concentrated our attention on the population of the United States or California, we would have to consider migration if we wished to make any projections of the size of future populations.

Any population has a birth rate, referred to as its *natality*, which is defined as the *number of individuals born per unit of time per individual.* Natality is symbolized as *b* (for birth). Of course, the population also has a death rate, or *mortality,* which is defined as *the number of individuals that die per unit of time per individual* and is symbolized as *d*.† For example, consider

*In the notation of calculus, *dN* indicates changes in *N* of an infinitesimally small nature.

†Not to be confused with the *d* in dN/dt.

a population of 5000 individuals in which 650 individuals were born and 350 died in one month (mo). Natality would be calculated as

$$\frac{650}{5000} = 0.13 \text{ per individual/mo}$$

and mortality would be

$$\frac{350}{5000} = 0.07 \text{ per individual/mo}$$

or simply 0.13 and 0.07 when one month is specified.

By subtracting mortality from natality ($b - d$), we may find the *rate* at which individuals will be added to a population per individual. To learn the rate at which individuals will be added to the whole population, we merely find the product of the rate at which individuals are added per individual and the number of individuals present or $(b - d)N$. This will give us the rate of change in the population, and thus

$$\frac{dN}{dt} = (b - d)N$$

Let us say that in the above example, the 5000 individuals were a valid sample of a population of 26,500 individuals and that ($b - d$) is the same in the whole population as in the sample. Then the rate of growth is

$$\frac{dN}{dt} = (0.13 - 0.07)\ 26{,}500$$
$$= 1590 \text{ individuals per mo}$$

In any population, the values of natality and mortality depend on several factors, including population density (N). For the time being, however, let us consider a population living under conditions that are ideal for the species—that is, under *optimal* conditions. In this case, then, the only limitation on b would be the physiological capacity of the species to reproduce, and the only lower limit on d would be the physiological capacity for longevity. Thus, b would be at a *maximum,* and d at a *minimum.* It should be pointed out that such conditions rarely, if ever, exist and that no one has ever been able to define an exact value for any species' physiological limits of reproduction and longevity. However, if we make the assumption that such conditions do exist, then we may speak about an *intrinsic rate of increase, r,* and we can say that under optimal conditions,

$$\frac{dN}{dt} = rN$$

This is sometimes called the *biotic potential* of the species. If we wished to know, however, the number of individuals that will be present in a population at the end of some interval, we would have to perform an additional step using the methods of calculus (which we will not go into here) to "solve" $dN/dt = rN$. The result of the solution is the equation

$$N_t = N_0 e^{rt}$$

Both $dN/dt = rN$ and $N_t = N_0e^{rt}$ are called logarithmic or exponential equations of growth. N_t is the size of the population after time t; N_0 is the initial population size; r is the intrinsic rate of increase; and e is a constant, the base of natural logarithms, equal to 2.71828... . When plotted graphically, this equation produces a J-shaped curve, as in figure 12.3.

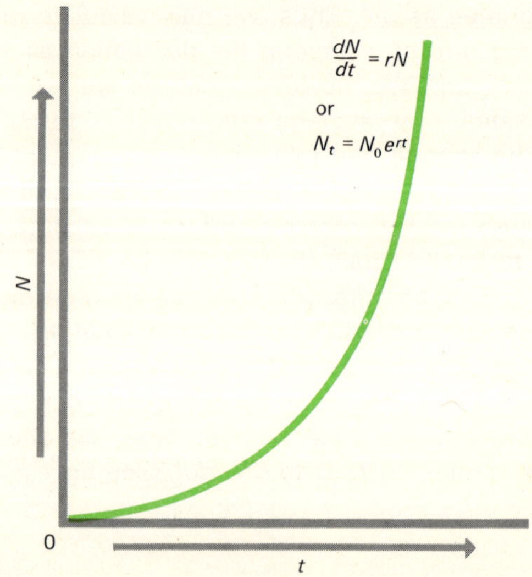

FIGURE 12.3 A J-shaped curve of population growth, or an unlimited growth curve.

Note that equal intervals of time produce unequal increases in the numbers of individuals. The growth in N accelerates, although r is constant.

Clearly, such growth cannot continue for very long; something must intervene to limit it. By our definition of the conditions under which such growth would occur, we effectively ruled out selection. Since r, the intrinsic rate of increase, assumes an optimal environment, when conditions are less than optimal, selection takes place and is reflected in the movement of b and d away from their maximal and minimal values. That is, the birth rate may go down, the death rate may go up, or as is most likely, both will change.

Natality and Mortality

Obviously, b and d are very basic ideas in both ecology and evolution, and we should briefly consider what determines the values of natality and mortality under less than optimal conditions. Certainly, the determining factors differ from one species to another, but in general terms, we can cite some of the factors that operate to limit b and d. Natality may depend on the frequency with which a species breeds, whether once in a lifetime, as in certain species of insects, or continually, as in the human species. The length of the gestation period can affect natality, as may the number of offspring produced per breeding. Population density may also affect natality, since populations existing at a very low density may have fewer contacts with potential breeding partners and, thus, may have a lower natality. Other factors might include weather conditions and how well-fed the members of the parental generation are.

Some of the same factors may affect mortality, but we can also add such things as disease, the availability of food and nutrients, predation, parasitism, competition (as we shall see in Chapter 13), and random events. Selection is usually more clearly measured by its impact on mortality, and in fact, we could describe mortality as the numerical product of selection. Nevertheless, both natality and mortality are products of the evolved adaptations of a species. Whether we cite feeding mechanisms, protective coloration, escape behavior, the seed dispersal of plants, family rearing, or egg deposition of animals, each is an adaptation with selective value for raising natality or lowering mortality.

Carrying Capacity and Environmental Resistance

We have said that the sizes of most populations remain fairly constant; that is, population size tends to achieve a steady state. In order for such a condition to exist, natality and mortality must be approximately equal, so that $b - d = 0$. Thus, dN/dt will also equal zero, and no increase will occur. As before, we can express this idea mathematically, although the expression is perhaps oversimplified and overgeneralized.

For any species in an environment, we can think of a density at which the population will become stable, which is called the *carrying capacity* of the environment, or K. As the number of individuals in the population approaches K, the rate of increase, $(b - d)$, must approach zero. We may modify the dN/dt form of the exponential equation to write

$$\frac{dN}{dt} = rN\left(\frac{K-N}{K}\right)$$

Consider the population example we used above. If the population contains 26,500 individuals, N, and the carrying capacity is 30,000, with an r of 0.06 individuals per individual per month (that is, $b = 0.13$ and $d = 0.07$), then

$$\frac{dN}{dt} = 0.06 \times 26{,}500\left(\frac{30{,}000 - 26{,}500}{30{,}000}\right)$$
$$= 1410 \times 0.1167 = 164$$

Thus, the rate of growth is 164 individuals per month.

Since $(K - N)$ is less than K, then $(K - N)/K$ must be less than one, and dN/dt will be reduced from the maximum rate of increase, rN. Moreover, as N approaches K, $(K - N)/K$ will approach zero, and the population size will reach a maximum at K. If a population is not growing, then $dN/dt = 0$, and therefore

$$rN\left(\frac{K-N}{K}\right)=0$$

This can be so only if one of the terms [r, N, or $(K - N)/K$] is zero. Since the intrinsic rate of increase, r, has been defined as maximum b and minimum d, then r cannot be zero. If $N = 0$, then the example is absurd, since no population exists to grow. If $K = N$, however, $K - N = 0$, and therefore $(K - N)/K = 0$ also. As N approaches K, the curve that would be produced on a graph becomes S-shaped, as in figure 12.4, and the population size will "level out." We speak of $(K - N)/K$ as *environmental resistance,* and the equation

$$dN/dt = rN\,[(K - N)/K]$$

as the *logistic* equation of growth.

FIGURE 12.4 An S-shaped population growth curve, showing the limitation of growth brought about by environmental resistance. The observed growth of a yeast culture is compared with the theoretical curve of limited growth and the J-shaped curve for unlimited growth.

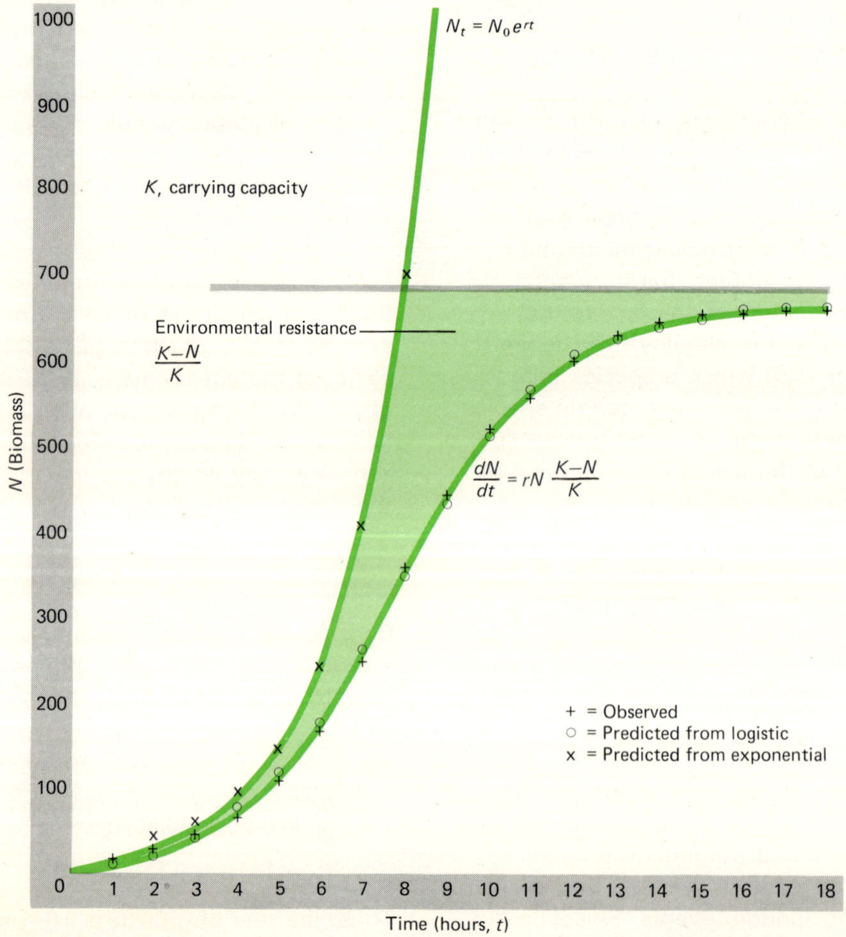

Since r is the maximum potential of a species when conditions are optimal, the actually realized value of $(b - d)$ is usually less than r. We may say, then, that

$$(b - d)\,N = rN\left(\frac{K - N}{K}\right)$$

and therefore

$$(b - d) = r\left(\frac{K - N}{K}\right)$$

The rate at which a population grows is equal to the intrinsic rate of increase times the environmental resistance. In other words, the maximum size of a population is regulated by the impact of various factors on the mortality and natality of the population.

A number of experiments with simple organisms in simple environments have shown that the logistic equation approximates the actual observed growth in the populations. Table 12.1 shows the results of calculations of population sizes as predicted by the exponential equation, as predicted by the logistic equation, and as actually observed.

Although the results from the logistic equation are very close to the observed values, the "fit" is not perfect. The discrepancies are, to some degree, indicative of the limitations of the equation. For example, the equation assumes that all individuals in the population are identical; that all are equally capable of producing young, that all have the same abilities to utilize the resources in the environment, and that selective pressures are the same on all genotypes in the population. We know that each of these assumptions is invalid. The equation also assumes that K is constant, and that may not always be so. One other limitation is worth mentioning at this point. According to the logistic equation, growth will be maximized when density is near zero. Some populations may reach a minimum density below which extinction becomes highly probable as the likelihood of locating a mate falls off. The Asian rhinoceros, *Dicerorhinus sumatrensis*, is believed to fall into this category. This animal ranges widely throughout Southeast Asia, and in recent years, it has undergone a severe population decline. Presently, it is estimated that there are not many more than fifty *D. sumatrensis* left. Their breeding season is short, and since they are solitary animals, they must wander in search of mates. Their chances of finding them are now considered remote.

TABLE 12.1 Growth of Yeast in Culture

Time (In Hours)	*Observed Biomass*	$\frac{dN}{dt} = rN\left(\frac{K-N}{K}\right)$	$N_t = N_0e^{rt}$
0	9.6	9.9	9.6
1	18.3	16.8	17.0
2	29.0	28.2	28.0
3	47.2	46.7	48.0
4	71.1	76.0	82.0
5	119.1	120.1	139.0
6	174.6	181.9	238.0
7	257.3	260.3	408.0
8	350.7	348.2	694.0
9	441.0	433.9	1238.0
10	513.3	506.9	2042.0
11	559.7	562.3	3504.0
12	594.8	600.8	5904.0
13	629.4	625.8	
14	640.8	641.5	
15	651.1	651.0	
16	655.9	656.7	
17	659.6	660.7	
18	661.8	662.1	14,765.0

(From Eugene P. Odum, *Fundamentals of Ecology,* 3d ed. (Philadelphia: W. B. Saunders Company, 1971), p. 187.)

Population Fluctuations

Complex organisms in complex environments are affected by a number of factors that regulate population growth and maximum size. If a population consists of individuals of a species having a complicated life cycle or a long period of individual development, there may be a delay in the response of natality as density approaches the carrying capacity. Such a delay may cause the population to grow past, or "overshoot," K. Thus, for a time, N will be greater than K, and $(K - N)$ will have a negative value, as will $[(K - N)/K]$. The population may then decline until it reaches K, or it may even fall below K. Once it falls below K, the population may again begin to grow. Thus, it may oscillate around K, as shown in figure 12.5.

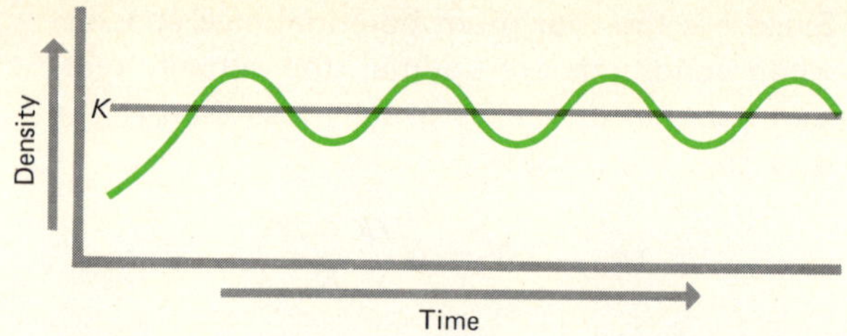

FIGURE 12.5 Due to predator-prey interactions and sometimes other factors, populations may increase and decrease cyclically.

Under nearly all realistic conditions, the oscillations will stabilize, and the population will reach an equilibrium with respect to K, as is shown in figure 12.6. Such stable oscillations are more likely to be measured in laboratory populations under very controlled conditions than under the conditions that exist in the real world.

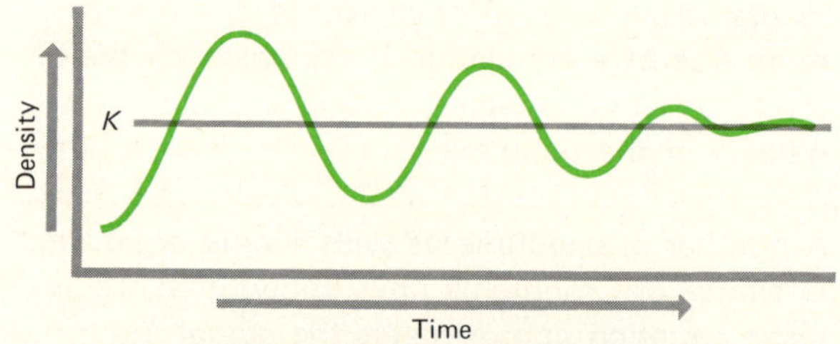

FIGURE 12.6 Sometimes the fluctuations of populations may be damped out to an equilibrium.

On the other hand, the oscillations may become more extreme, and the population may "crash," as shown in figure 12.7.

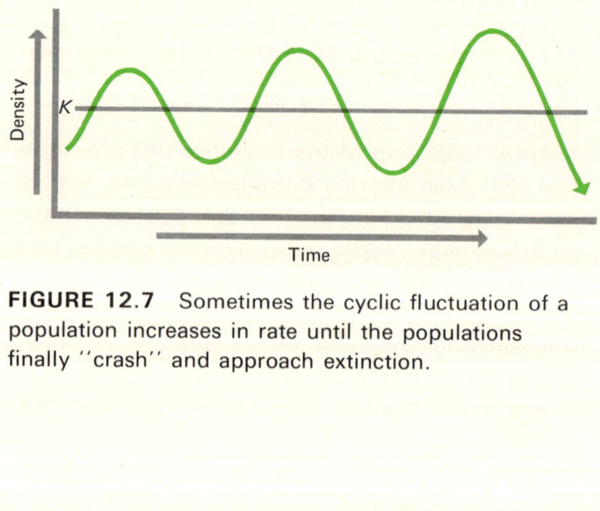

FIGURE 12.7 Sometimes the cyclic fluctuation of a population increases in rate until the populations finally "crash" and approach extinction.

Which of the possible situations develops depends entirely on the species' behavior and the environment. Clearly, seasonal variations, climatic changes, migrations, predators, and competitors will also influence what happens. We shall examine the effects of the latter two more closely in Chapter 13. Precisely what will happen to any population depends on so many variables existing both within and outside the population that it is almost impossible to predict the outcome of population growth. What can be said with some certainty, however, is that a population that is growing exponentially must encounter environmental resistance.

Implications of Population Growth and Control

The preceding ideas about population growth and control apply to all organisms, although there are many factors that may modify the basic equations that we have used. The concepts of growth and stability are also associated with the processing of energy and nutrients, which was discussed in the preceding two chapters.

Limiting Factors

An increase in population size almost invariably means an increase in biomass and, therefore, requires an increase in the energy available to the growing population. Such an increase in energy may be obtained at the expense of other species in the community, or it may mean an increase in community productivity. A growing population will require, of course, larger absolute quantities of nutrients, and thus, population growth also implies increased nutrient uptake. More importantly, however, population growth implies that previously defined limiting factors are no longer limiting. This may mean that nutrients have been available in the community, but for some reason have not been exploited, or it may also mean that some nonmaterial limiting factor has been at work. Thus, temperature, salinity, soil texture, altitude, predator populations, or disease may have been limiting and may have ceased being so. In some cases, an import of nutrient materials will promote population growth. A culturally eutrophic pond, for example, represents an explosive growth of algal populations as the level of a limiting nutrient such as nitrate in the pond is raised. On the other hand, the growth of a population of deer may occur if hunting is severely restricted. The efficiency with which certain populations of bacteria can take up and utilize nutrients may be correlated with the temperature of the environment. Thus, the thermal ''pollution'' of a body of water may promote growth in such populations. At the same time, of course, it may bring about a decline in the population of many species of fish. Although other factors can operate, population growth is usually limited by the availability of food and nutrients. When other factors do operate, they frequently have implications for energy and nutrient transfer in a community.

Density Dependence and Independence

The idea of material and energy being limiting factors is closely associated with density dependence and independence in the control of population size. If a population's growth rate or carrying capacity depends upon its size, the population is said to be *density dependent.* Thus, natality and mortality will change as the size of the population changes, and the carrying capacity will depend on the population density. Density dependent effects may be produced by the availability of food and nutrients, but other factors may also be directly observed. For example, some territorial animals and birds will not reproduce if the territory available per breeding pair is too small.* Among certain species, the response to population density can be quite striking. Wild rabbits in New Zealand, for example, terminate approximately half of all pregnancies when population density is high. Normally, these rabbits gestate for eighteen to thirty days, but female rabbits can reabsorb the fetuses of unborn rabbits as much as twenty days after pregnancy begins. Thus, the mother loses very little of the energy rich substances of which the fetuses are made. The survival value of such restraint is obvious, but it presents biologists with something of a puzzle. It serves to prevent overpopulation and a possible ''crash,'' but the altruistic organism that keeps the population density down by not breeding is selected against; its genes do not remain in the gene pool. How such behavior can persist in a population remains an unsolved problem of evolutionary biology. What is more, such altruism is not uncommon. Density-dependent effects may result from the accumulation of waste products or exometabolites to toxic levels. We cited an instance of this earlier in our example of yeast in winemaking (p. 29). Other density-dependent effects can be seen in predator-prey relationships and competitive situations, which we shall examine in detail in Chapter 13.

Some of the implications of density-dependent effects for humans are obvious; although we are on shaky ground if we try to draw too many in-

*A territory is the definable area that an individual or breeding pair will defend against intrusion by other individuals, usually males of the same species.

ferences for humans from observations of other organisms, for we are not a very typical species. Nevertheless, some interesting, and open, questions may be posed here. What are the effects of increased density in the human population? Under crowded, primitive conditions, such as those that existed in medieval Europe or now exist in Calcutta, disease may run rampant through the population and increase *d*. On the other hand, primitive conditions do not exist in most of Manhattan or Tokyo and although density is very high in both places, disease seems controllable. There seem to be behavioral responses to density in the human population, however, and some positive correlations appear to exist between violent crimes (homicide, rape, and assault) and high density. There also appears to be some correlation between density and mental pathology. However, truly definitive comparative studies of such relationships in different cultural contexts have yet to be made, and at the present we can do little more than speculate. Even problems that would seem relatively easy to solve are often more complex than they would at first glance appear to be. For example, our output of toxic substances into the environment seems to have increased as population has increased, but the output may not be purely the result of population growth. There is a positive correlation between the level of industrialization and the output of such materials, and thus, economics and technology may contribute at least as much to the problem as sheer population density. As we have said, of course, technology may well provide many of the answers to this problem.

It is harder to generalize about density-*independent* factors as regulators of population size. Most often cited as an example is climate. At the edges of a species' range, climate may often exert a powerful influence on a population. Thus, a hard winter may kill off overwintering insect eggs and thus produce a density-independent effect. The populations of many seasonal insects are limited by weather. In some parts of the country, Long Island, for example, the density of deer populations is controlled by a winter kill; poor nutrition due to overpopulation is involved as well, and to a great degree, the populations are density dependent. Indeed, as one may anticipate, density dependence and independence are not easily separable. In fact, the whole question has been a bone of contention among ecologists for a long time, probably due to a tendency to oversimplify situations. One long-term study of snails of the genus *Acmaea* revealed that the populations were, for the most part, controlled by density-dependent factors, but that after a severe frost, mortality rose as the portions of the rocks the snails lived on crumbled. Thus, both density-dependent and density-independent factors controlled population size. In one species of Australian insect, *Cardiaspina albitextura,* both types of effects were found. This insect lives on eucalyptus trees, and populations are small relative to the amount of food and living space available. One of the factors that maintains small population size is the normally prevailing weather. However, population size is also controlled by density-independent predation and parasitism on the nymphal stage of life, which is an unusual circumstance, since predation and parasitism are usually density dependent. In the adult stages, *C. albitextura* populations are regulated by density-dependent predation by birds. Occasional explosions occur in populations of *C. albitextura*, and when they do, the insects strip the eucalyptus trees of leaves and use all of the favorable locations for laying eggs. Since the insects have a much higher intrinsic rate of increase, *r*, than their predators, they increase much more rapidly and are controlled by the density-dependent effects of food shortage and lack of egg-laying sites. Soon, however, predator and parasite populations begin to "catch up" and produce additional density-dependent effects. In short, to assign either density-dependent or density-independent factors the sole role in regulating the size of a population is an oversimplification. Both may operate on a population at differing times, or even at the same time, and the observation that a density-dependent factor is operating should not rule out the possibility that a density-independent factor is operating as well.

Survival Strategies—*r* and *K*

Another way of looking at the control of a species population is in terms of its evolved strategy for survival. We hasten to add that such "strategies" do not imply conscious planning. Under given sets of environmental conditions, some species are adapted to undergo rapid population growth. Such species have a high intrinsic rate of increase and are thus termed *r strategists.* They are usually small organisms and efficient reproducers, and they are often characteristically found in the early stages of community development or in unstable or fluctuating environments. An *r* strategist is adapted to finding new habitats quickly, exploiting resources before other species can move in, and dispersing rapidly to find new habitats as the old ones begin to become unfavorable. Although many of the dispersing individuals do not survive to find new habitats, enough do survive to found new populations and keep the species going.

C. albitextura are *r* strategists, as are another species of insects, *Thrips imaginis.* Thrips are small flying insects that require pollen in their diets and thus must live in the flowers of a variety of plants. As a habitat, of course, flowers are ephemeral, and as they become sparse the thrips population must disperse. One investigation of thrips populations in a particular rose garden revealed that in the spring, when the flowers were abundant, each rose was populated by an average of 190 thrips. During the summer, however, the thrips populations fell to an average of around 5 per rose. Since each rose contained enough pollen for the thrips, the decline was not strictly caused by a shortage of food. Rather, the population fell to low levels because flowers became more scarce and therefore harder to locate. Those thrips that could locate flowers provided a base for population expansion when favorable conditions returned.

Other examples of *r* strategists include species of phytoplankton that undergo irregular "blooms" and disease bacteria in a weakened host.

Some species may maintain stable populations over long periods of time. These species are *K* strategists, since they tend to exist for long periods at or near the carrying capacity. *K* strategists tend to be larger animals or plants that live for a long time and produce few young, and they are characteristic of older, more stable communities. Populations of *K* strategists are most frequently disturbed by human activities, since they depend on undisturbed "virgin" or "natural" communities. Paradoxically, *K* strategists are often species highly valued by man for esthetic or recreational purposes. Shade trees, game animals, and many species we keep in zoos are *K* strategists.

Many species designated as "endangered" are *K* strategists. For example, estimates of the North American caribou population in 1907 were as high as 30 million. In the mid-1950s, the population was estimated at approximately 725,000, and it is now estimated to be about half of that. Caribou once inhabited the forests of the northern United States, but they are now restricted to remote parts of Canada. The decline is partially attributed to heavy hunting of the animals, but of greater probable importance was the loss of habitat due to lumbering and fire. The cutover areas were also invaded by populations of white-tailed deer, which compete with caribou for food. Another factor was the introduction in the late nineteenth century of the Eurasian reindeer into the habitat of the caribou. Although only 1300 reindeer were introduced, they underwent a rapid population growth and further encroached upon the caribou's habitat.

Among other examples of *K* strategists that are disturbed by human activities are species of whales, grizzly bears, the grey wolf, and mountain lions.

We shall return to this subject later when we consider the stability of communities, but it is worth pointing out here that *K* strategists are adapted to survive without any help from man, and yet these are the species that conservation departments and foresters most often seek to manipulate by stocking trout ponds, setting hunting seasons and bag limits, or regulating forests by selective tree cutting. There is a certain irony involved in "helping" nature whether it needs it or not.

We should also point out that *r* strategists, such as cockroaches, flies, mosquitoes, lawn weeds, nuisance algae, and trash fish such as ling, are the species that we try to manipulate with the application of poisons. Frequently, the result of such manipulation is merely the creation of another set of instabilities that can promote precisely the kind of growth we were trying to inhibit, sometimes in other species, sometimes in the species at which the application was "aimed." Modern insecticides are broad-spectrum chemicals that kill many species of insects and other animals. Since many insects are preyed upon by other insects and birds, the insecticide kills the predator and thus releases the prey population from a controlling factor. If the predator plays a key role in limiting the prey species, the surviving members of the prey population may make a spectacular comeback or may even exceed the density existing at the time of the first application. In one case, the use of the pesticide Azodrin* was shown to actually increase the cotton bollworms in California fields by having a more lethal effect on the bollworm predators than on the target population itself. We should add that the manufacturer, even in the light of this information, did not withdraw its product from the market. On the contrary, it was vigorously promoted as a cure-all for insect control in cotton production. The European red mite has become a major pest in apple orchards as a consequence of the use of DDT to control the codling moth which preyed upon the mites. In this sense, the effect of an insecticide is to simplify ecosystems and communities. As we shall discuss in greater detail later, such simplification is characterized by instability, which is one of the characteristics of communities that favor *r* strategists.

*An organophosphate insecticide that inhibits the activity of AChE (see p. 145) and produces, in acute cases, hyperactivity of the nervous system leading to death.

It is possible to carry the idea of *r* and *K* strategies too far. No species is exclusively one or the other, and like most concepts in a science too poorly understood, this one is oversimplified. Consider a pine tree, for example. Since it is often the first species of tree to colonize an abandoned field, it may be called an *r* strategist. Yet in some climates and soils, pines become the dominant species and may be called *K* strategists. Certain social insects illustrate the problem involved in designating a species as either an *r* or *K* strategist. The founding of a new colony is a risky venture. It has been shown, for example, that in the social termite, *Neotermes tectonae,* a colony with fewer than 50 to 75 individuals is more likely to be destroyed by ants than a larger one. Since colonies of social insects are usually founded by a single queen, a premium is placed upon the rate of increase in new colonies, and thus, they are *r* strategists. A number of observations about social insects support this generalization. First, the early broods in a new colony develop faster than later broods and are made up of smaller individuals. Thus, the population grows quickly. Next, the eggs laid by the queen in a new colony are more freely fed to the larvae of the early broods than to later broods. Finally, the workers of a new colony are less aggressive and therefore less likely to sacrifice themselves to predators than are workers of the mature colony, but their natural lifespans are shorter. Since workers do not reproduce, such altruistic behavior has no genetic consequences for the population. As the colony approaches the carrying capacity, all of these conditions appear to be removed, and thus, the colony may then be said to be a *K* strategist.

As with density-dependent and -independent effects, we cannot use the concept of *r* and *K* strategies without careful qualifications and reservations.

Population Explosions

Some of the most spectacular examples of population growth on record have occurred when some agent, usually man, has introduced a species into a new area. Earlier, we mentioned the Australian rabbit population, and had we wished to emphasize population growth alone, we could just as easily have considered the English Sparrow, the Starling, Dutch Elm disease, the Gypsy Moth (see fig. 12.8), the Japanese Beetle, the Fire Ant, or a host of agricultural insect pests. Such introductions of species sometimes have fascinating histories. They often have resulted from well-intentioned purposes. More importantly, almost all well-documented cases show a population growth that resembles, at least initially, the unlimited case $dN/dt = rN$, although the populations ultimately stabilized. Usually, these populations have exploited the productivity and nutrients of a community and have displaced other, less harmful species. By reproducing rapidly, they have also had some drastic effects on their new food sources or hosts. The United States Department of Agriculture has quarantine laws that are designed to prevent these ''outbreaks,'' but they often meet with little success. Pesticides are one means of control, but of course, the effects of pesticides are not limited to the pests. The best controls may be via predators or competitors, but as we shall see in Chapter 13, such controls are often difficult to develop and implement.

The growth of a population of organisms that cause human disease is of more direct and unhappy consequence to many persons. In the fourteenth century, the Black Death, a form of bubonic plague, rampaged through the human population of Europe and resulted in the death of an estimated 25% of the people. Actual figures are very difficult to come by for population data are scarce, but it appears that the

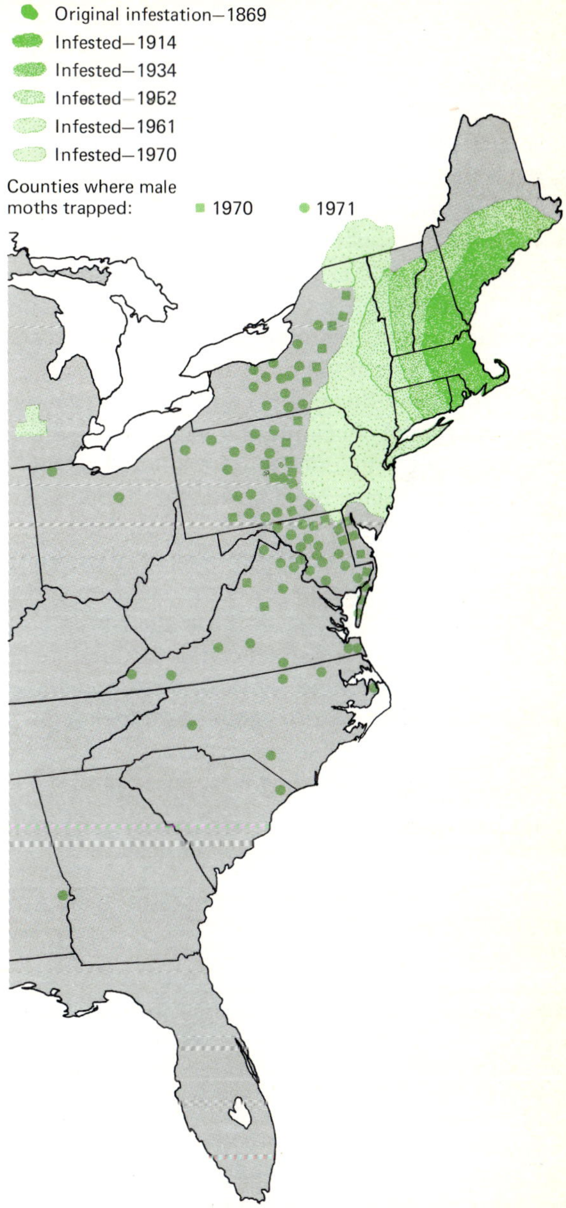

FIGURE 12.8 After the gypsy moth was originally introduced in the United States in 1869, it underwent a population explosion, rapidly infesting areas throughout the Northeast. During 1970 and 1971 it spread further west and south, and the pest is still not under control. This is a typical example of a ''population explosion'' due to introduction of an organism into a new area.

Black Death killed approximately 18 to 20 million persons. Thus, as the population of plague organisms rose, the population of human hosts fell. In the early days of American history, smallpox epidemics were frequent occurrences about which little could be done. Though it may seem that present public health measures have largely brought such outbreaks under control, in many parts of the world where such services are unavailable or insufficient for the needs of the population, the potentiality of widespread infection remains. Public health problems are compounded by the existence of modern means of transportation. Recently, for example, a strain of cholera thought to be restricted to the Southeastern Pacific has moved from there through Indochina, India, West Africa, East Africa, and the Mediterranean area. It was then carried by jet to Sweden by a woman who contracted the disease while on a vacation in Majorca. As the disease spread through Africa in 1970–72, an estimated 20,000 persons died. Cholera is not necessarily fatal if prompt medical attention is available; in some parts of the world, however, public health services are limited or widely scattered, and in the absence of treatment, the usual result is death.

Other population explosions are often the direct consequence of human activities. As we mentioned, monoculture may produce an unlimited resource for some herbivores such as cotton boll weevils, locusts, grasshoppers, white pine blister rust (a fungus), or the corn blight that this country experienced in the summer of 1970. In attempting to control these populations with insecticides or fungicides, we are also faced with the problems of controlling our control mechanisms.

Wildlife Management

We are also concerned with the maintenance of certain wild populations from which he derives pleasure or sustenance. This subject will be discussed in detail in Chapter 13, but since it is a matter of controlling the carrying capacity and $(b - d)$, a brief treatment seems pertinent here. Usually, for resource crops such as fish or lumber, the object is to maintain *maximum sustainable yield.* It can be shown that if a population behaves ideally according to the logistic equation, the maximum sustainable yield will be obtained if the population is harvested in such a way as to maintain it at half the carrying capacity—that is, $K/2$. That is not always so, however, and in the case of fisheries, a great deal of effort has been expended in determining what will not only produce the maximum yield but also prevent a population "crash." To prevent the overexploitation of fishery resources, certain international agreements have been developed involving both the sizes of harvests and the age of fish taken. Though some of these agreements have been tried and were successful, some have not been so successful. One failure can be documented by an account of the whaling industry.

Concerned with the declining production of world whale fisheries, those nations involved in whaling set up the International Whaling Commission (IWC) soon after World War II. The commission was given responsibility for controlling the depletion of whale stocks but lacked real power to insure that the signatory countries would comply with their directives. One of the commission's most important early decisions was to set harvest quotas. Such quotas were set not on the basis of the numbers of individuals of each species that could be taken, but on the basis of "blue whale units" (bwu). A bwu was defined as one blue whale, two fin whales, two and a half humpback whales, or six sei whales. Since it was easiest to fulfill the quota by taking blue whales, this species became the most sought after. Around 1950, the harvest of the blue whale began a sharp decline. In 1960, the IWC appointed a committee of biologists to investigate the future of Antarctic whale fisheries. After three years during which harvests continued to decline, the biologists reported back to the IWC on the overexploitation of whale stocks. They urged the IWC to abandon the bwu system and to assign harvest quotas by species. They also recommended that the IWC immediately order a halt to the harvest of blues and humpbacks and reduce the harvest of fins.

The biologists estimated that if such actions were not taken, the whaling industry would be faced with continuously declining stocks of whales. They predicted that in 1963–64 only 8500 blues and 14,000 fins would be harvested, and that the harvest on fins would amount to about three times the maximum sustainable yield. Their predictions were surprisingly accurate, as 8429 blues and 13,870 fins were taken. The biologists continued to make recommendations about the sizes of catches permissible, and the countries involved in the Antarctic whaling industry continued to ignore them; the limits set by the IWC were usually about twice the biologists' recommendations. Predictably, the catch continued to decline. Of the four nations involved—Japan, Norway, the Netherlands, and Russia—only two are still in the business of hunting whales. The Netherlands sold its fleet to Japan in 1965, and Norway was forced out of the industry in 1968.

For some species, of course, there is a minimum density. Should the population fall below this theoretical minimum, the likelihood of mating falls off, and the population becomes extinct. Fortunately, it seems that the exploitation of a population usually becomes uneconomical before the population reaches its minimum density. Thus, fishery stocks may have a chance to recover, since the fishermen go bankrupt before the species become extinct. Whales present a slightly different problem, since even rare species, such as sperm whales, are shot during the hunt for the more common species. A danger also arises among species that aggregate for mating, because it is possible for a whole population to be wiped out in a single catch. If an animal's skin or feathers have a high market value, its value will increase as the stocks of the animal decrease, and commercial hunting may drive it near extinction. The danger to the species is amplified by sport hunters, since the "value" of a trophy depends upon the rarity of the animal. Thus, sportsmen will continue to hunt long after commercial exploiters have stopped.

The main point is that the whole concept of game, fish, or forest management is "unnatural." Man, in taking the yield of these resources for his growing population, of necessity deprives other species of the resources. This is so whether we hunt the animals intentionally or merely occupy their habitat. In short, we are a competitor, and as such disturb the stability of ecosystems. Occasionally, the effect is not serious, because man produced the excess in the first place. Deer often thrive on cornfields, to the chagrin of farmers, who then propose extended hunting seasons and a relaxation of regulations to allow the shooting of deer of both sexes and all ages. Usually man's disturbing effects can be withstood more easily by herbivores, omnivores, and plants. Larger carnivores, however, animals with a very low intrinsic rate of increase, suffer most. One important conclusion to be drawn from all the implications of population growth is that organisms have evolved niches which allow stability for K strategists and occasional bursts of growth for r strategists. Anything that we do with technology will alter some habitat and thus produce some change in the populations of organisms there. This should not be construed as a blanket condemnation of technology, but it is an argument in favor of careful, intelligent planning of technological activities.

Another important implication of population growth, and its control, has already been discussed in Chapter 7. In any environment, a species population has a characteristic r and K, both of which are determined by its evolution and are thus products of its niche. Mortality and natality imply that some organisms survive to reproduce and some do not, or that within the population some have a higher birth rate than others. What really matters, of course, is survival through the reproductive period. It is not necessary to go into population genetics any further, but one should be reminded that the whole question of an evolved niche and speciation is intimately associated with population growth and control. The existence of mortality implies selection. That selection may be stabilizing, as in most "natural" communities, but when populations are undergoing rapid growth it can be directional. Thus, we have come full circle to our original point, the excess young produced by a population permit evolution to occur by providing grist for the mill of selection. It should be clear that evolution is *the* basic process for ecology.

We must make one final point about selection and mortality. When we "manage" a population, whether wildlife or domesticated, we alter mortality. That is selection, and it is rarely random. When we manipulate the resources of a species by managing other species, developing land, introducing competitors, or putting out toxins, we change both natality and mortality. Again, that is selection. Consider that the populations of some species of whales are now so small that if they ever recover, the amount of genetic variability in the population may be seriously limited. Those species that we have artificially selected, such as corn, potatoes, and rice, may be approaching a loss of genetic variability, and some attempts are now being made to collect and stock seeds from plants considered to be in danger of such a loss. We worry a great deal about pollution, the overexploitation of resources, and driving species to extinction. Perhaps we should also consider these slower, more subtle effects as well.

The Human Population

Census-taking techniques for the human population vary, and in some cases no reliable data exist. Even in nations where very careful measurements are made, as in the United States, certain anomalies are known to creep in, and these discrepancies cast some doubt on our ability to ascertain the size of our own population. The 1960 U.S. census, for example, reported that there were 1670 fourteen-year-old widowers in the population. Nevertheless, good estimates are available that indicate that approximately 3.8 billion humans inhabit the planet at the present time. Estimates of the size of the human population in the past are shown in table 12.2. Estimates of the world human population made in 1963 by the United Nations Statistical Office projected the population to the year 2000, the lower estimate was 5.4 billion people, and the medium and high estimates were 6.1 billion and 6.9 billion. Since the estimates were made, the world population appears to have grown at a rate between the medium and high projections.

TABLE 12.2 Human Population Size

(A.D.) Year	*Estimated Population*	*Estimated Density/km^2*
1650	545,000,000	3.7
1750	728,000,000	4.9
1800	906,000,000	6.2
1850	1,130,000,000	7.7
1900	1,610,000,000	11.0
1950	2,495,000,000	16.4
1972	3,780,000,000	25.7

Doubling Times

In order to dramatize the growth in the human population, it is common to speak of it in terms of *doubling times*, which is defined as *the time required for the population to double its size.* In this connection, the growth rates of populations are usually referred to in terms of births and deaths per thousand individuals. By using growth rates per thousand, it is easy to express growth in terms of percentages. For example, the world birth rate in 1968 averaged 34 per 1000, and the death rate averaged 14 per 1000. Thus, $(b - d)$ was 20 per 1000, or 2%. At a 2% annual increase, the world human population will double in 35 years. Should the rate of growth reach 3%, the doubling time will fall to 24 years; at 4%, it will be 17 years. In the 200 years between 1650 and 1850, the human population doubled. By 1930, or 80 years later, it had doubled again, and by 1975, only 45 years later, it will have doubled again. At its present rate of growth, it will double in just under 35 years. A word of caution is necessary in respect to doubling times. Because the doubling time for the human population has consistently fallen, there is a temptation to assume that it will continue to decrease. Obviously, a bottom limit to doubling times is dictated by the ability of the species to reproduce. It would be absurd to think that the doubling time could ever reach zero, since that would require each female child to give birth to another child almost at the time of her own birth.

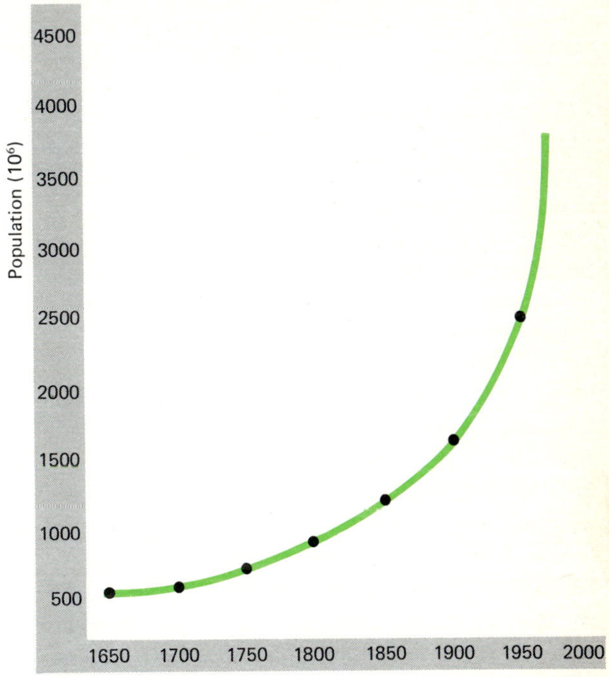

FIGURE 12.9 Obviously the human population is falling into a typical pattern for J-shaped growth. Although the rates of increase are not equal for all nations or all social groups within nations, on a worldwide basis it is obvious that a carrying capacity must eventually be reached. Yet, the rate of growth shows no sign of tapering off to a carrying capacity. This has led to many alarmist predictions of famine, disease, social unrest, and catastrophic sociological events. It is discussed later in the book; we do not know what will eventually bring about an approach to a carrying capacity. We can only hope that the human population will stabilize in a manner that minimizes human suffering. The resources available to man are obviously finite, and it is obvious also that a J-shaped population growth curve leads to an infinite population.

A graphical representation of the growth of the human population since 1650 reveals a curve such as would be produced if the population were growing according to the equation $dN/dt = rN$. (See fig. 12.9.) Like any other species in the biosphere, man is born and dies, and the equations of population growth and regulation apply no less to him than to other species. Since humans possess all of the attributes of a *K* strategist (large size, high resource requirements, small litter size, and a long period to sexual maturity) but is undergoing the unrestrained growth characteristic of an *r* strategist in an unlimited environment, we must ask what determines human carrying capacity. In part, this is a sociological question, but sociology and ecology are so tightly interwoven in the area of human population growth that we cannot separate them.

Consider human natality first. In industrialized countries, human natality may be determined by a combination of sociological, economic, and personal factors that dictate or influence decisions on family size. Chemical and mechanical birth control devices are now so widely available that only religious attitudes and personal viewpoints prevent their use. That is a development of the twentieth century, however. In less well developed nations, as in times past, if choice was a factor, birth control was achieved at the cost of suppressing the innate urge for sexual enjoyment. To control natality the means exist, or can be made available at a low cost. (See box 12.1.)

We may also ask what controls mortality in a human population. In the past, mortality was almost solely determined by a combination of resource limitation (usually famine), disease (which goes hand in hand with famine), and a phenomenon unique to man, warfare. Geological disasters, such as volcanic eruptions, tidal waves, and earthquakes, were probably the only density-independent factors, and they played only minor roles. In our efforts to reduce mortality, we have been fantastically successful. The effects of disease and famine have been greatly suppressed. The mortality associated with warfare may have increased slightly, although this is difficult to determine, since in past centuries, the effects of pestilence, famine, and warfare were often difficult to separate. While some causes of mortality have declined markedly, others, such as degenerative diseases (heart disease, cancer, and hypertension), have increased. On the whole, however, mortality in the human population has declined over the past century.

There is no reason to think that our carrying capacity is unlimited. In fact, it *cannot* be. What, then, *is* our carrying capacity? Whatever it was a few centuries ago, it seems to have changed. We can deduce that from the fact that our population is no longer constant. As an animal, man is unique in several ways. First, we *may* determine our carrying capacity. It is theoretically possible for us to determine what the optimal carrying capacity should be and to adjust natality and mortality accordingly. Whether this can actually be done is a social and political question. What the human carrying capacity *should* be is a complex question involving economics, cultural attitudes, and a host of other factors that may not be measured quantitatively. In many ways, the implications of the question are frightening.

Clearly, it is undesirable to limit population by increasing mortality. The only option available, then, is to decrease natality. This may seem to be an oversimplification, but it is not. Decreasing natality seems to be the only desirable way to limit the human population. Yet this may be the most complex and potentially disruptive "simple answer" ever faced by mankind, since it is by no means clear exactly *how* it can be implemented.

There *is* one other option. Do nothing. By taking this option, resource limitation or perhaps human behavior through social pathology will increase mortality and limit natality until $(b - d)$ is less than 1. It does not seem necessary to go into detail about what might occur if we choose not to choose. We have had recent examples in Biafra and Bangladesh, and human history is studded with similar events.

There is a philosophical question embedded in the problem of human population control. Is it "better" to allow complete freedom and the social disorder and suffering that such freedom might entail, or is some sacrifice of freedom in matters of reproductive behavior more desirable? It seems relevant to point out that at present we do not have complete freedom in many aspects of our private and public lives. Under most circumstances, for example, we are not free to commit homicide or other antisocial acts. Similarly, whether we like it or not, some forms of sexual behavior are legally forbidden even between mutually consenting adults. Such abridgments of freedom have been determined by our culture, and we might ask if reproductive freedom can or should be similarly limited.

The final point that we should consider is selection. A decrease in natality or an increase in mortality, unless birth and death are completely random, implies that selection is occurring. We have mentioned that selection can be strongest in a stable population. A voluntary decrease in natality or an imposed limitation on reproduction in one segment of a population would be genetically selective against the traits of that segment. The effect of the decrease would mean a genetic change in the population. The

Contraception

Until the mid-1950s persons who wished to practice birth control were limited to contraceptive devices that were essentially the same as those in use for nearly 200 years, or to methods involving self denial. With the testing, in 1956, of an orally administered chemical contraceptive, a "second generation" of birth control measures was ushered in. "The Pill" is now the most widely used and effective contraceptive available. The pill, however, has several side effects and disadvantages, not the least of which is that it is available only for use by women. This places most of the burden of choice of contraception upon women, and it should be obvious that women alone should not have to suffer the burden of choice or the side effects of a contraceptive method.

All attempts to develop an oral contraceptive for males have thus far met with failure. In some cases the failure has been due to the (unrealistic) conditions under which the drugs work, such as a pill that is effective only if the user totally abstains from alcoholic beverages, or because the sexual desire of the user declines significantly, in which case the contraceptive effect of the drug seems somewhat redundant. As a consequence, most males either must continue to use condoms or to submit to a surgical procedure known as *vasectomy.* There are, however, some problems associated with vasectomies. Since the operation is irreversible in at least 50% of the cases, it may amount to permanent sterilization. Nevertheless, an estimated one million males had vasectomies performed in 1972, an increase of nearly 250,000 over 1971. Recent research on vasectomized rodents has indicated that other problems may arise. In some of the experimental animals cysts have developed in the vas deferens, while in some the size of the testes has become reduced. Exactly what these effects mean with regard to humans, if indeed they mean anything at all, is still an open question, but the researchers advise that caution be exercised.

Perhaps the development of a "third generation" of oral contraceptives is in the making, based upon the immune response. In one series of recent experiments, investigators were able to significantly reduce conception among female mice by injecting them with antibodies. The isolation of an enzyme, lactate dehydrogenase, from the testes of male mice was followed by the injection of the enzyme into rabbits. In the bloodstream of the rabbits the lactate dehydrogenase acted as an antigen, and the antibody produced was then injected into female mice, where it reduced conception by 60%. The injection of the antibody into male rabbits seems also to have interfered with sperm production, and thus promoted sterility. Since lactate dehydrogenase has been detected in several animals including man, the development of such antibodies may hold some promise for the development of oral contraceptives that can be used by both males and females.

BOX 12.1

racial implications are obvious. The only way to prevent such selection would be by a completely random lottery. *Or do we want selection?* We might ask the question this way: Is any political system in existence in the world today capable of implementing a program of controlled genetics (*eugenics*) without initiating a justifiable political revolution among those against whom selection will be directed? Such questions can be addressed by biologists as citizens, but the questions cannot be answered unless *every* citizen has some knowledge of the biological implications of the problem.

The Age Structure of Populations

Examining the numbers of individuals in a population is not the only means of analyzing it. As we saw in Chapter 7, populations may be analyzed in terms of their genetic composition, and as we shall see in Chapter 13, they may be analyzed in terms of their distribution, which has a great deal to do with the behavioral patterns that contribute to survival. For our present discussion, however, it is important to look at the ways in which populations may differ in terms of the ages of their members. Understanding the age structure of a population may tell us something not only about its potential for growth but also about the strategies that lead to the survival of the population. While similar to r and K strategies, the strategies we refer to here are slightly different.

Survivorship Curves

Consider the problem faced by the actuarial staff of a life insurance company. Their job is to predict the life expectancy of a potential customer at the time a policy is purchased. Obviously, they cannot conduct exhaustive physical examinations of each prospective buyer; all they ask is whether he or she is in reasonably good health. If so, then the probability that the individual will die in the near future is no greater than the probability that anyone else of the same age will do so. What actuaries do, then, is to examine massive quantities of statistics on the age composition of a human population and the distribution of the ages of death, and from this information they predict the probability of any individual's survival. The same approach may be used in analyzing a population of organisms.

Consider table 12.3. The key information for our purposes in such a table is the survivorship data. If we include all of the members of the population at the instant of birth, survivorship must, by definition, be 100%. If we consider the entire lifespan, survivorship must, also by definition, be 0%. Thus, we can draw a pair of axes representing the percentage of lifespan and the percentage surviving. The curve that we would derive for any species would be a *survivorship curve.*

Of special interest in such exercises are the different shapes of curves derived for different species. (See fig. 12.10.) Basically, there are three types of curves. One shows low initial survival, but high survival of those individuals that live beyond a certain age or proportion of their possible lifespans (curve A). Another shows high initial survivorship with most of the population surviving to an age near the maximum for the species (curve B). The last is a curve that shows a gradual, more or less linear decrease of survivorship with age (curve C).

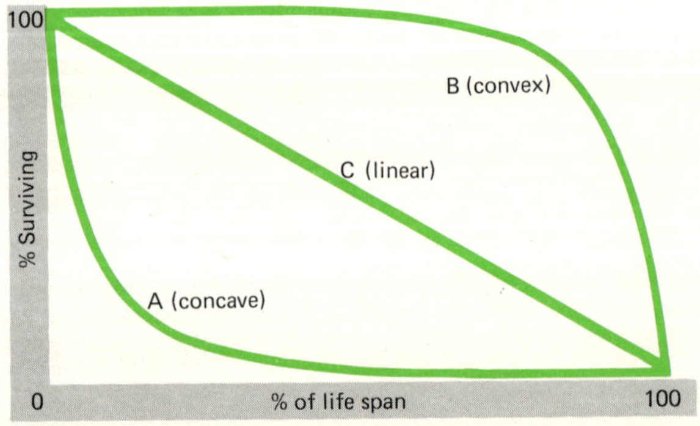

FIGURE 12.10 Survivorship curves show the different reproductive strategies adopted by different organisms.

TABLE 12.3 Life Table for Dall Mountain Sheep (*Ovis d. dalli*) Based on Known Age at Death of 608 Sheep

Age (In Years)	*Number Surviving at Beginning of Interval (Out of 1000 Born)*	*Number Dying During Interval (Out of 1000 Born)*	*Mortality Rate per 1000 Alive at Beginning of Interval*
0.0– 0.5	1000	54	54.0
0.5– 1.0	946	145	153.0
1.0– 2.0	801	12	15.0
2.0– 3.0	789	13	16.5
3.0– 4.0	776	12	15.5
4.0– 5.0	764	30	39.3
5.0– 6.0	734	46	62.6
6.0– 7.0	688	48	69.9
7.0– 8.0	640	69	108.0
8.0– 9.0	571	132	231.0
9.0–10.0	439	187	426.0
10.0–11.0	252	156	619.0
11.0–12.0	96	90	937.0
12.0–13.0	6	3	500.0
13.0–14.0	3	3	1000.0

(Adapted from E. S. Deevey, Jr., ''Life Tables for Natural Populations of Animals,'' *Quart. Rev. Biol.* 22 (1947): 283–314.)

The first type of curve (curve A) is typical of species that produce a large number of young ''expecting'' a high level of mortality in the early stages of development. Many marine invertebrates, such as oysters and barnacles, show such survivorship curves, as do many species of insects. To some extent, a concave survivorship curve is the strategy of an *r* strategist, since it implies a very large intrinsic rate of increase if conditions are optimal. The curve also implies almost no parental care of the young, and thus, it may be expected of simpler organisms with little or no behavioral adaptations for care of the young or for internal development of the young.

The convex curve (curve B) is the opposite of curve A. Curve B implies that some mechanism exists for insuring a high survivorship among the young. This curve is typical of the human species and of other ''higher'' mammals. The mechanism involved is the birth of a fairly well developed infant that is given prolonged postnatal care. The curve indicates (1) that the species has a low intrinsic rate of increase, (2) that the rate is almost realized, (3) that most individuals reach reproductive maturity, and (4) that the offspring must reproduce if the species is to avoid extinction. Thus, convex curves usually indicate *K* strategists. The curve is typical of species in which a transfer of behavioral information, or learning, takes place during the postnatal period. Such learning is a characteristic of primates, and as we shall see in Chapter 16, it sets the stage for the development of culture.

The linear curve (curve C) implies that death is random at any age. In reality, very few species are known to produce such a curve. One exception is the hydra, a freshwater coelenterate.

A curve that is characteristic of species with poor initial survivorship but wherein individuals

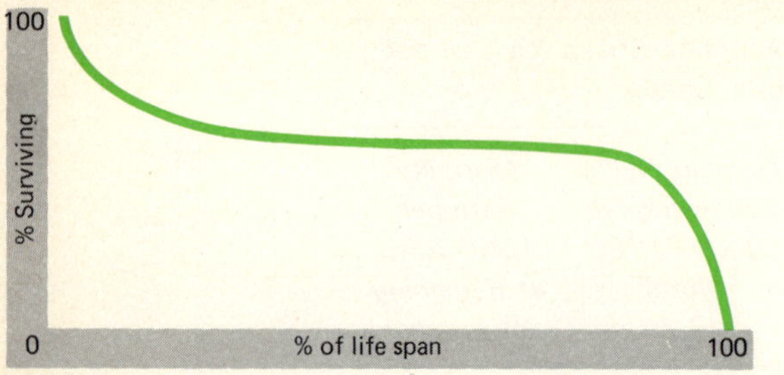

FIGURE 12.11 A survivorship curve showing high initial mortality but a very low mortality during most of the rest of the lifespan.

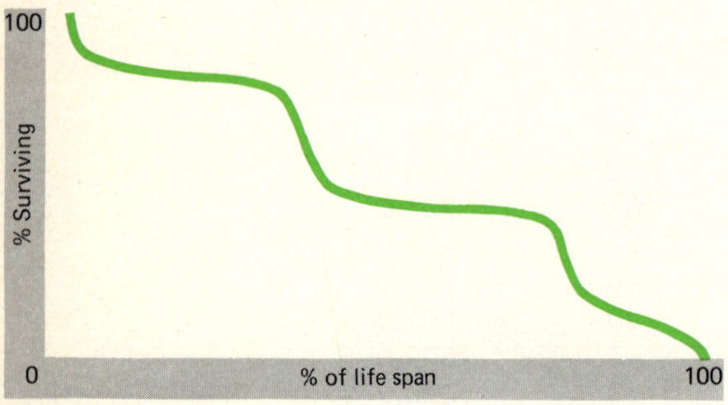

FIGURE 12.12 Insects have survivorship curves similar to this, because there is a high mortality at the time of each molt.

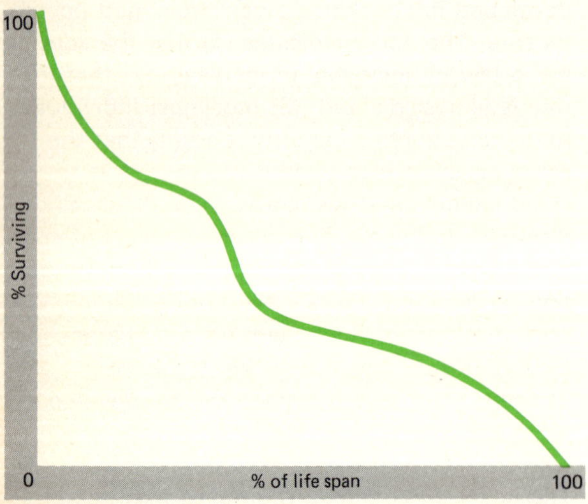

FIGURE 12.13 The survivorship curve of a "primitive" human population reflects a high mortality in early years, survival through prereproductive years, and a high mortality associated with the period of reproduction.

that reach the postjuvenile stages live out most of their possible lifespans is shown in figure 12.11.

This is called a *sigmoid curve* and seems typical of organisms such as rabbits, reptiles, rodents, and birds. The young are born fairly well developed but receive relatively less parental care than organisms described by convex curves. Learning does not play a very large role in such species, but the lack of learning is compensated for by a genetically determined stereotyped behavior, by some parental care, and by a large litter or clutch size.

Finally, there is the curve produced by species with complex life cycles. (See fig. 12.12.) Insects that molt in order to grow may be described by such a curve, since each molt exposes the individual to a higher than normal probability of death.

A survivorship curve is in part the product of the adaptations of a species, but it is also in part a product of its environment. If, for example, we were to measure the mortality in a population of mosquitoes, we would be likely to find a concave curve under normal conditions. However, a similar population of mosquitoes in an optimum environment would probably show a more convex curve. Thus, care must be taken to interpret the shape of the curve correctly, since the conditions under which the measurements are taken may alter the shape. Over long periods of time, the curve characteristic of a species might change. At one time, for example, the human population was probably described by a curve intermediate between the convex curve and the linear curve. (See fig. 12.13.)

The number of children a human female is physiologically capable of bearing has not changed for perhaps tens of thousands of years. Through her entire reproductive life, a woman can bear between twenty and thirty children. Twenty thousand years ago, women probably did bear a large number, although probably not the maximum. Infant mortality was high, however, and so was mortality among children. As humans developed the capacity to control their environment, the shape of the human survivorship curve changed and became more convex. The discovery of agriculture provided a predictable, if not always entirely reliable, abundance of food. This further contributed to human control of the environment and further altered the shape of the curve. More recent developments in the practice of medicine have served to reduce mortality in all age classes of the human population, thereby contributing to the present shape of the human survivorship curve. Thus, with no significant changes in our biological adaptations, we have brought about a major revolution in our capacity to survive. We have accomplished this feat by radically altering the environmental component. Whether, in the very long run, we can continue to survive is still an open question.

Age Distribution

Another important kind of information about a population is its *age distribution*, or the *relative proportions of the population in each age class.* Organized graphically into *age pyramids,* such data can be very useful in the analysis of a population. We shall consider some hypothetical age distributions, how they relate to survivorship curves, and some of the alternative interpretations that may be drawn from the pyramids.

The pyramid in figure 12.14(a) represents a population in which a large proportion of the individuals are relatively young. We would expect that the population would produce a concave survivorship curve. However, if the curve is convex, we can expect the population to undergo rapid, even explosive, growth. In the latter case, those individuals represented by the base of the pyramid will soon begin reproducing, and since they represent a major proportion of the population, they will produce a proportionally large number of offspring.

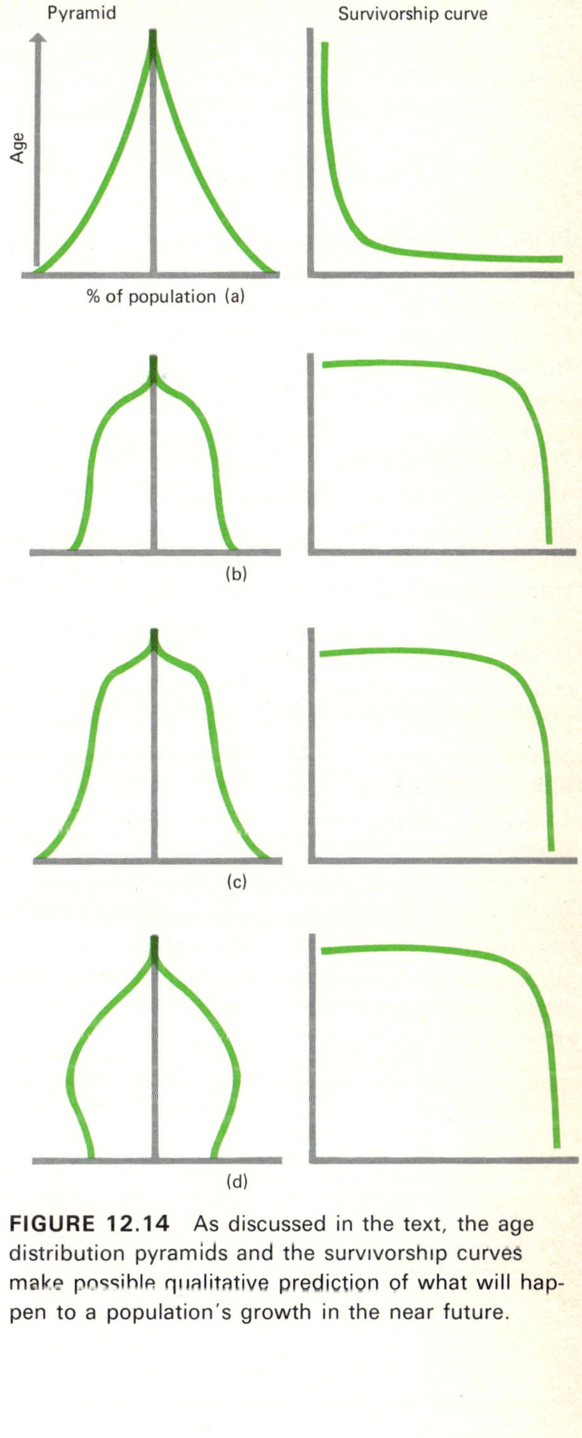

FIGURE 12.14 As discussed in the text, the age distribution pyramids and the survivorship curves make possible qualitative prediction of what will happen to a population's growth in the near future.

The pyramid in figure 12.14(b) represents a stable population if the survivorship curve is convex, as shown. If the curve is concave or linear, a population decline can be expected, as mortality will be high in the younger age classes at the base.

If the combination shown in figure 12.14(c) is found, we can expect the population to grow, but less rapidly than in the first situation we described. Natality will be higher than mortality, because something is acting to raise natality or lower mortality in the early stages of life.

The pyramid and curve shown in figure 12.14(d) indicate a declining population. The greatest proportion of the population is in either the late reproductive or early postreproductive stages of life.

In interpreting age pyramids, it is important to consider the proportion of individuals in the reproductive stages of life, since it will to some extent determine natality. A human population that is structured as in pyramid A in figure 12.15, in light of the human survivorship curve, can be expected to grow because of the large proportion of potential reproducers it contains. Should the age structure appear as in B, we may expect a decline in the population, since those in the postreproductive stages represent a large proportion of the population and may be expected to die soon.

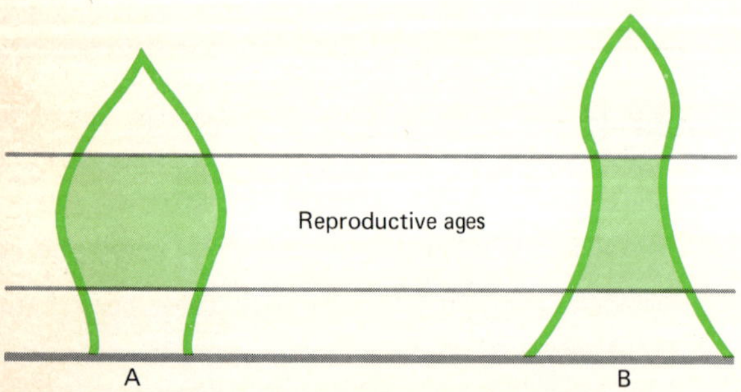

FIGURE 12.15 The proportion of population of reproductive age is very important in determining whether the population will increase or decrease.

The basis for the predictions that we have made from the structures of age distributions should be clear: If the proportion of individuals in the reproductive or prereproductive stages of life is large, a great number of young will be produced in the future. The *proportions* may not change in the future, but the *number* of individuals will increase. It is important to understand that although the age structure alone may be interesting, we can draw conclusions from it only when we take into account the survivorship curves of the species.

Conclusion

The problems associated with finding food and nutrients, taking them into the body, and utilizing them to grow and reproduce have been solved in varying ways by species' evolved adaptations. Moreover, the precise mechanisms of genetics have provided species with the means to insure their survival by passing on the evolved adaptations. It should be clear that these adaptations play an important role in determining the growth rates and sizes of populations.

It is not sufficient to consider species populations as isolated entities, however. We must examine how each population fits in and contributes its niche to a community. This means that we must consider first the interactions between species populations and then the dynamics of communities.

Selected Readings

Ehrlich, Paul R. and Anne. *Population, Resources, Environment.* San Francisco: W. H. Freeman and Company, Publishers, 1972. A widely used and well documented discussion of the human implications of the "population explosion."

*Hardin, Garrett. ed. *Population, Evolution, and Birth Control.* San Francisco: W. H. Freeman and Company, Publishers, 1969. A broad ranging collection of pieces on these three topics. Some of the sources are surprising, as are some of the statements.

*Lack, David. *Natural Regulation of Animal Numbers.* New York: Oxford Book Co., Inc., 1954. A classic treatise on population control.

*McLaren, Ian. ed. *Natural Regulation of Animal Populations.* New York: Lieber-Atherton, Inc., 1971. A collection of viewpoints, some conflicting with others, on the problem of population regulation.

Slobodkin, Lawrence. *Growth and Regulation of Animal Populations.* New York: Holt, Rinehart and Winston, Inc., 1961. Although a small book, all the basically important topics are covered, often better than anywhere else, and always rigorously

Cited earlier, but relevant to this chapter, are Wilson and Bossert, *A Primer of Population Biology* and MacArthur and Connell, *The Biology of Populations.*

Deevey, Edward S. "The Human Population." *Sci. Am.,* Vol. 203 (September, 1960), pp. 194–198. Offprint number 608.

Huxley, Julian. "World Population." *Sci. Am.,* Vol. 194 (March, 1956), pp. 64–74. Offprint number 616.

Pequegnat, Willis. "Whales, Plankton, and Man." *Sci. Am.,* Vol. 198 (January, 1958), pp. 84–86. Offprint number 853.

*Paperback edition available.

Cattle egrets follow the large herbivores through the African grasslands, feeding on the insects that the herbivores disturb. The birds, in turn, act as an advance warning system for the herbivores, often sensing and reacting to the presence of predators before the larger animals perceive them. *(Courtesy of Ira Kirschenbaum.)*

population interactions

chapter thirteen No species lives in total isolation, and therefore the growth of one species' population has an effect on the growth of other species' populations. Thus, the evolution of one species affects the evolution of species with which it interacts, and the species coevolve. The existence of trophic levels implies the existence of predator-prey relationships, which in turn exert selective pressure on all species involved. Moreover, the existence within a single trophic level of more than one species almost inevitably leads to competition between species for energy and nutrients. (Of course, some competition for resources such as space may also occur between species in different trophic levels.) Such competition is selective. Therefore, before we can discuss the ecology of whole communities in any detail, we must first consider the kinds of interactions that occur between species.

Two species that have never come in contact with each other have no evolved interactive behavior or adaptations, and yet if they should come in contact, they will probably interact in some direct or indirect manner. If they exist in the same community for long periods of time, it is likely that interactive behavior and adaptations will evolve. We will consider some of this highly evolved interactive behavior, and we will look at why interactions between species that have not coevolved is often disruptive.

The possible kinds of interactions between species are summarized in table 13.1.

It should be apparent from our previous discussion that competition and predation are the most common, and probably the most important, interactions. We shall briefly defer our discussion of these interactions to look at some of the more unusual and interesting kinds of interactions. We should first point out, however, that any classification of interactions is relative; that is, the advantages or disadvantages felt by any interacting species may be severe or minor, or they may be very clearly defined or very general. For example, since all plants produce oxygen and all animals consume it, we may say that any animal interacts with any plant. Such generalizations, however, are so broad as to be meaningless. What we shall discuss are interactions of a significant nature, although that is begging the question, because the definition of significance is relative, and the judgment of significance is subjective.

Furthermore, all organisms in a community are interconnected by trophic relationships, however indirect, and thus, they all interact. We shall discuss these "large-scale" relationships when we consider communities as a whole. In this chapter, we will discuss the interactions between pairs of species as though they were the only interactions that occur. The reader should keep in mind that a species may be observed to interact with many other species in a community, and that a species may be designated as a predator in one instance, while under other circumstances, it may be classified as prey, a competitor, or a commensal. Often within the context of a whole community, the role of a species may not be very simply defined, and in this chapter we have for the most part ignored the community context in order to deal more easily with the phenomena under consideration.

TABLE 13.1 Interactions Between Species

	Species Effect		
Interaction	*1*	*2*	*Description*
Neutralism	0	0	Neither species affects the other
Competition	—	—	Each species disadvantages the other, either directly or indirectly
Amensalism	—	0	One species disadvantaged; other species not affected
Parasitism	+	—	One species advantaged; other species disadvantaged
Predation	+	—	Same effect as parasitism
Commensalism	+	0	One species advantaged; other species not affected
Protocooperation	+	+	Both species benefit
Mutualism	+	+	Both species benefit

+ Indicates an enhancement of population survival or growth.

0 Indicates that no significant interaction takes place.

— Indicates that the population growth or survival is inhibited.

NOTE: the term *protocooperation* is preferable to *cooperation,* since it does not imply any conscious choice.

Henceforth, we shall frequently refer to this scheme of classifying interactions. When doing so, we shall use the terms *zero effect, minus effect,* and *plus effect* to indicate the effects under discussion.

(After Eugene P. Odum, *Fundamentals of Ecology,* 3d ed. (Philadelphia: W. B. Saunders Company, 1971), p. 211.)

Neutralism and Amensalism

As we suggested, two species in a single community are probably never completely independent, and thus, zero-effect relationships are impossible. Even if we refer only to direct, significant interactions, zero-effect relationships are still extremely rare.

Neutralism (0/0) can only exist if two species share no common resource or have evolved so as to ''partition'' the resource between them in such a way that they absolutely avoid competition. Neutralism can most easily occur between two species that are on different trophic levels and have no feeding relationships between them.

Superficially, it may appear that many species in a community are neutralistic with respect to many other species. Consider Spanish moss, which is not really a true moss. This epiphyte grows on trees in the southeastern United States. Since it derives its nutrients from the air and is fed upon by different species from, say, a ground moss, one might think that there would be little interaction between an epiphyte and a ground moss. But consider that the Spanish moss creates shade and thus competes for light. Is the relationship one of amensalism (–/0), with the epiphyte gaining no benefit and the ground moss being at a slight disadvantage? If the ground moss has a low tolerance for light, then the epiphyte gains no benefit from the relationship and the ground moss does. Could it be commensalism (+/0)? Perhaps. If the tree on which the epiphyte grows gains some benefit from the ground moss, then the epiphyte probably benefits indirectly. The point of the example, of course, is not to say something about the interaction of epiphyte and moss, but rather to indicate the care that must be taken when deciding what relationship exists between two species.

Neutralism may be approached when two competing species undergo selection that minimizes competition. Thus, closely related species of birds will often evolve to different nesting sites or food preferences. Complete neutralism, however, is probably never obtained. The two species simply come to share the resources sufficiently to permit each to survive. Neutralism may also be approached by species with very different habitats and needs. Thus, a seed-eating bird probably has little effect on a lichen growing on a rock on a forest floor.

We can reason that neutralism is very unlikely in the case of newly introduced species, since such species often produce drastic readjustments in a community and are likely to affect nearly all other species in some way. We shall look at some human effects in this regard later.

Amensalism (−/0) is a very unlikely relationship in a truly balanced community, since the zero effect species will exert strong selective pressure on the negative-effect species to adapt or be eliminated. Clearly, amensalism merges with competition. In the Spanish moss example, if the ground moss benefited, a grass would have been at a disadvantage. Amensalism does exist, however, and if grass were to grow under a tree with epiphytes, amensalism might be said to describe the relationship between the grass and the epiphyte.

The two most frequent occurrences of amensalism are (1) cases involving competition in which one species so strongly outcompetes the other that the first is not affected, and (2) those cases in which one species produces a chemical substance that is disadvantageous to another species. The epiphyte-grass situation is an example of the first type of amensal relationship, as would be instances of one species living on the extremity of its range. We mentioned an example of the second type earlier regarding the California chaparral and herbs. Another example is found in species of algae that produce "red tides" in many tropical waters. When such algae bloom, they produce a toxin that may result in massive fish kills.

We might also mention that because of the waste products we create, we have often appeared to have an amensal relationship with other species. However, as the effects of these waste products have become clearer, it has become apparent that the relationship is not truly an amensal one. It may be characterized as a (−/−) interaction, and yet it cannot be called competition, since the negative effect on man is not produced by the other species. This is a new type of interaction, and one which for obvious reasons could have deleterious evolutionary effects on the species involved.

Parasitism and Other Interactions in Which at Least One Species Benefits

Parasitism (+/−), commensalism (+/0), protocooperation (+/+), and mutualism (+/+) are relationships in which evolution has usually led to adaptations whereby at least one species benefits. Before discussing these interactions, we should point out three things. First, we shall avoid using the word *symbiosis* (literally, "living together"). As some biologists have used the term, it has been restricted to describing mutualism. However, it has also been used, not strictly incorrectly, to describe parasitism, commensalism, and protocooperation as well. Thus, the term has too many meanings to be very useful. Second, these relationships have classically been thought of as occurring between very closely associated species, such as intestinal flora and ruminants, or species in host-parasite relationships. Actually, however, the species may be more widely separated, as in the Spanish moss example. Third, parasitism and predation have ill-defined boundaries. Though it is possible to determine whether one population is predatory or parasitic on another, it is not always easy. Distinctions are sometimes made on the basis of size. Predators are usually larger than their prey, and parasites are usually smaller than their hosts. Size considerations can be misleading, however, since some predators may be smaller than their prey. Usually, parasites are dependent on their hosts, whereas predators may depend on several species. But consider mosquitoes. They are not dependent upon man, and yet they are considered a human parasite. Another criterion used is that predators entirely consume their prey, but there are species of insects considered to be parasites that also consume their hosts entirely. The point, of course, is that using a criterion to determine whether the relationship between populations is one of predator-prey or parasite-host can be misleading. When it is not an obvious case of one or the other, more observation of the interaction is usually necessary. Clearly, our terminology may occasionally break down, but we should not be too concerned about the name of a relationship as long as the nature and effects of the interaction are clearly understood.

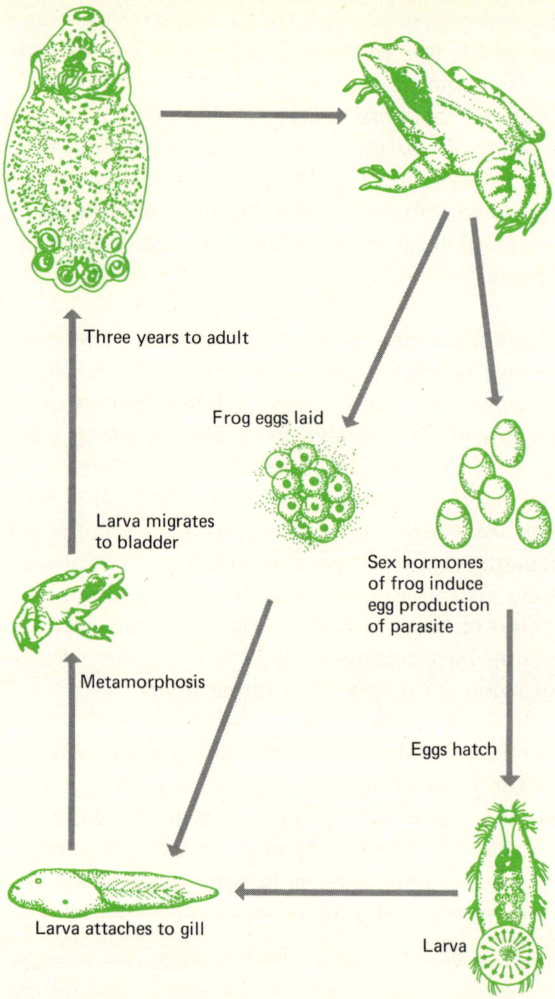

FIGURE 13.1 The life cycle of the trematode parasite of a frog. The adult parasite lives in the frog's bladder and its eggs are released with the urine. The larva develops and infects the frog tadpole. During metamorphosis of the tadpole to a frog the larva migrates to the bladder and matures to form a new adult parasite. At this point the cycle starts again. Many parasite life cycles are much more complex. *(After Barnes.)*

Parasitism

Since it illustrates the extent to which organisms may evolve specializations, parasitism, from an evolutionary point of view, is one of the most interesting interactions between species. Most parasites are *host-specific*; they live on only one species of host and cannot survive on another. Basically, we may divide parasites into two classes: *ectoparasites* (*ecto* outer), such as fleas, lice and many bacterial and fungal infections; and *endoparasites* (*endo* inner), such as intestinal nematodes, flatworms, and bacterial parasites. Endoparasites are usually the most highly evolved, since they are adapted to a very specialized environment. This kind of specialization may be somewhat disadvantageous, however, since it presents problems to the parasite species in getting from one host to another. As an evolutionary solution to this problem, many endoparasites are adapted to two hosts. One of the hosts is called the *primary host* and is the one in which the parasite matures and reproduces. The other is called a *vector* (or *secondary host*) and transports the immature stages of the parasite from one primary host to another. You may recall our discussion of the Myxoma virus in the Australian rabbit population. (See fig. 13.1 for a parasite life cycle.)

The human parasites that cause the extremely debilitating disease *schistosomiasis* illustrate the complexity a parasite's life cycle may reach. The disease is caused by at least three species of blood flukes of the family Schistosomatidae. *Schistosoma haematobium* is found in most of Egypt, in Iraq, parts of Israel and Jordan, and Portugal; *S. mansoni* is found in Egypt north of Cairo, in Africa, and in the West Indies and parts of South America (possibly as a result of the slave trade); and *S. japonicum* is found in Japan, Taiwan, China, the Philippines, and Malaysia. The life cycle of each species differs slightly, but in general, the adult schistosome lives in the blood vessels near the intestines. When ready to reproduce, it migrates to the intestines and lays its eggs in the capillaries of the intestinal wall. The eggs are deposited in such quantities that they cause the capillaries to rupture and expel the eggs into the intestinal tract itself, whence they reach the outside of the body in the feces. If the feces are not diluted with water, the eggs will die in a few days. If the feces are diluted but the water is treated in a sewage plant, the cycle will also end. However, in many of the countries where the schistosome is common, human wastes are not well-treated or are used as fertilizers (particularly in rice paddies). If the eggs reach a body of water, whether it be a rice paddy, an irrigation ditch, or a river, they will hatch in a short time (from a few minutes to about sixteen hours). The tiny, ciliated larvae that hatch can live in the water for up to twenty-four hours. They swim about until they encounter a snail of a certain species. The larvae bore into the snail

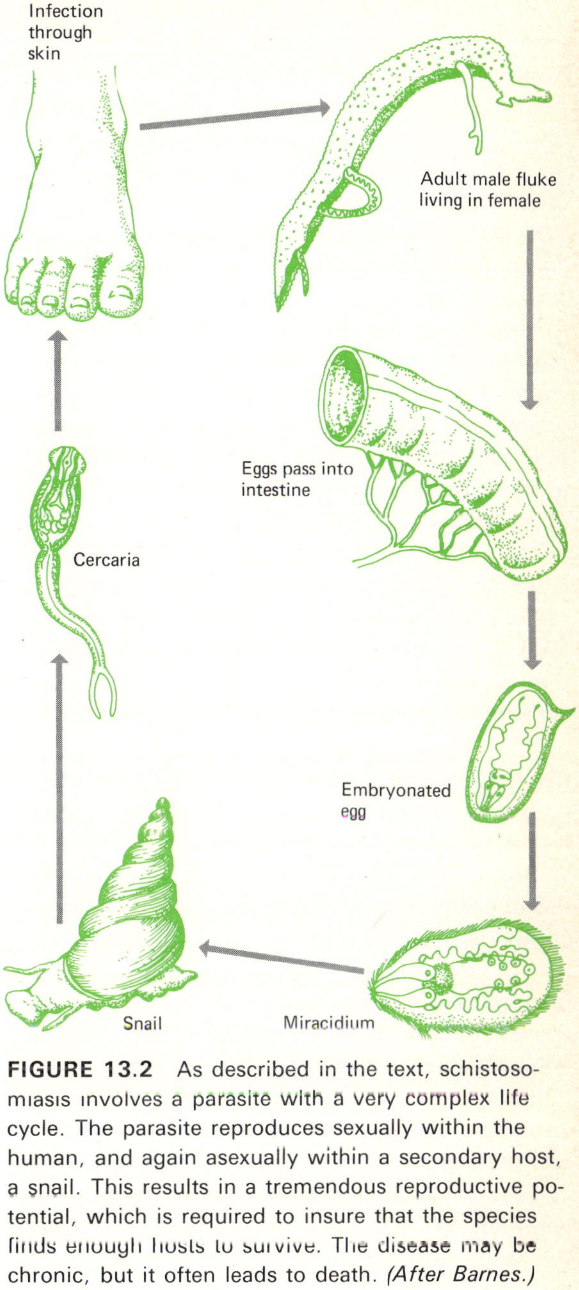

FIGURE 13.2 As described in the text, schistosomiasis involves a parasite with a very complex life cycle. The parasite reproduces sexually within the human, and again asexually within a secondary host, a snail. This results in a tremendous reproductive potential, which is required to insure that the species finds enough hosts to survive. The disease may be chronic, but it often leads to death. *(After Barnes.)*

and make their way to a digestive gland in the innermost part of it. Once in the snail, they reproduce asexually and are capable of enormous population explosions. One species of snail harboring *S. mansoni* was observed to discharge an average of 3500 schistosomes daily, and the offspring of one larva was found to exceed 200,000. The immature schistosomes are discharged from the snails in "puffs," and the new individuals enter the water, where they can survive for two to three days. Should they come into contact with a human being while in the water, they attach themselves to the skin and digest their way into the blood stream. They are then carried through the heart, lungs, and liver to the intestines, where they remain to mature and begin the cycle anew (see fig. 13.2). While in the heart, lungs, and liver, they sometimes lay eggs which can accumulate and produce tumors. The disease, which is not itself usually fatal, frequently so weakens the human host that he may die of some other cause normally considered insignificant.

Recently, the incidence of schistosomiasis in Egypt appears to have risen sharply as a result of the construction of the Aswan High Dam. Until the dam was completed, the irrigation of the upper Nile Valley depended solely on the annual flooding of the river. The snail vector for the disease in Egypt, *Bulinus truncatus,* persisted in small populations, though the periodic dry spells killed off many snails. Now, under conditions of constant irrigation, the snails have a more favorable environment, and so too do the parasites. Unfortunately, not enough careful records were kept of the earlier incidence of the disease in Egypt, and thus, it is not certain whether the recent "outbreak" is real or merely the result of better recording procedures.

Many people do not realize just how common parasites actually are. Of the 40,000 animal species that are found in Germany, fully 10,000 are parasites on the other 30,000. In wild animal populations, it is rare to find an individual that is not parasitized by protozoa, flatworms, or roundworms. Nearly all gray squirrels, for example, are infected with chronic blood protozoa. Wildlife parasitologists expect that every specimen they examine will have at least one infection, if not more.

The point is that a parasite is a highly adapted, specialized organism. We may ask why wildlife, chronically infected humans, and domesticated animals are not more debilitated by such infections. The answer, as we pointed out in Chapter 8, is simple: Only a very poorly adapted parasite kills its host, for in doing so, it kills itself. Some eggs may be laid or cysts produced before the host's death, but not nearly as many as would be produced if the host were alive. Only one human parasite, the virus that causes rabies, invariably causes death of the host, and the rabies virus is not an exclusively human parasite. Nevertheless, the adaptation of parasite to host does not mean that the host is perfectly healthy. Parasites such as hookworm and pinworm do not kill their human hosts, but they can lower the resistance of the host to other diseases that can cause death. On the other hand, there are human parasites such as the protozoan *Giardia lamblia* that are sometimes not even treated when found, because their effect is often so mild. Should a parasite infect a host that is already debilitated from some other disease, death may ensue.

Why, then, are many human deaths attributed to parasites? Often, some of the worst parasites infecting human populations are really the parasites of domesticated animals. *Trichinosis* is endemic in pigs and rats, roundworms (*Ascaris lumbricoides*) are parasites of pigs, and most flukes and tapeworms were originally parasites of other domestic animals. This is yet another example of the secondary effects of civilization. In some cases, man becomes an "accidental host" for parasites that can cause diseases. Tularemia, plague, endemic typhus, and Rocky Mountain spotted fever are all primarily diseases of rodents, carried by insect vectors such as fleas and ticks.

There are two important points to be made about "well-adapted parasites." The first is that *as ecosystems evolve and develop, mutually disadvantageous interactions between species tend to be minimized, and positive interactions tend to grow more predominant.* Thus, the survival of interacting species is enhanced. The second point is that *new associations between interacting species tend to be more severe and negative than the interactions between species that have been associated with each other for a long time.* An important corollary is that: *because man often produces new associations of species, man's activities are often disruptive of natural communities.*

Commensalism

The next three kinds of interactions that we shall examine, commensalism (+ / 0), protocooperation (+ / +), and mutualism (+ / +) represent an order of increasing adaptation and are sometimes hard to tell apart.

Perhaps the most common form of commensalism is that in which one species provides a habitat space for another. A robin nesting in a tree and a lichen growing on the bark are both commensal with the tree. Commensal interactions are often far more specialized than these examples. A remora (fig. 13.3) is a fish that has a highly specialized suction device at the top of its head that allows it to attach itself to a shark and ''hitch a ride.'' The remora's body is very streamlined and thus minimizes the ''drag'' that the shark encounters while swimming with these fish attached to its body. As the shark feeds on its kill, the remora scavenges small bits of food that the shark cannot eat anyway. Another commensal on the shark is the pilot fish, which often rides the bow wave in front of the shark's nose. (See fig. 13.4.)

Before going on, it is appropriate to introduce some more terminology. The interactions that we have been discussing may fall into two categories: *obligate* and *facultative.* An obligate interaction indicates that the plus (+) species cannot survive without its partner. Facultative interactions indicate that the plus species can survive on other species in the absence of its

FIGURE 13.3 A shark with remoras attached. *(Courtesy of William Stephens from Tom Stack & Associates.)*

FIGURE 13.4 The pilot fish often ''cruises'' in front of or alongside a shark, and benefits by gathering bits of food which the shark does not eat. *(Courtesy of Al Bloom from Monkmeyer Press.)*

partner. The remora is an obligate commensal, whereas the pilot fish is facultative. Some breeds of dogs are obligate, and "hot house" flowers tend to be obligate. Man, of course, is the only species that has deliberately produced obligate commensals.

Mutualism and Protocooperation

Commensalism (as well as protocooperation and mutualism) may arise from parasitism. Man's intestinal flora were probably at one time parasitic. Thus, as parasites become better adapted, they may become commensals or mutualists.

The obligate or facultative quality of a relationship between species distinguishes between protocooperation and mutualism. Both are highly evolved interactions, but mutualistic species are much more specialized. Good examples of both kinds of interactions may be found among different species of bees. The common honeybee is not dependent upon any particular species of flower as a source of food, and many flowers pollinated by bees are also pollinated by other insects and birds. Certain species of bees, however, have become extremely specialized. Among species of tropical bees, each species is attracted to a single, separate species of orchid. Thus, although there are several species of bees and several species of orchids, each bee species is dependent upon only one kind of orchid for its food, and each orchid species is dependent upon only one species of bee for its pollination. Many of the orchid species produce a single chemical that bees of several species can detect. However, they also produce other chemicals that repel all but one species of bee. The advantages of such interaction should be clear. For the flower, it prevents cross fertilization and wasted reproductive "effort," and for the bee, it prevents competition. This is another example of niche specialization. There are also, however, some disadvantages. Should the local population of one of the species become extinct, as might happen to the orchid population if the land were cleared, then the other population would probably also become extinct.

Mutualism is less frequently found than protocooperation, because it requires extreme specialization. Nevertheless, it is not uncommon, since it provides the advantages of survival security and reduced competition.

Certain species of termites, as we mentioned earlier, are unable to hydrolyze the bonds that hold glucose units together to form cellulose. Therefore, these termites must live mutualistically with certain protozoa that are found in the gut. These protozoa can hydrolyze such bonds and thereby provide food for themselves and the termites. Ruminants, such as cows, sheep, goats, camels, and giraffes, also live mutualistically with a variety of microorganisms. The food supply of the ruminant is basically grass and other fodder, which is so high in cellulose content that the ruminant, like the termite, cannot digest it. The ruminant takes the material into its body and deposits it with a large amount of saliva into the first part of its "stomach," the *rumen*, where a host of microbial populations begin to break it down. The microbial population in the rumen of a cow has been estimated to be as high as 10^{10} cells per milliliter. The bacteria and protozoa attack the ingested material and digest it. The simple fatty acids that are produced are absorbed through the lining of the rumen, and the gases produced (carbon dioxide and methane) are expelled by frequent belching. In effect, the rumen becomes a fermentation vat. The population of microorganisms grows rapidly, and then, along with some of the undigested plant material, a large portion of the population of microorganisms is passed into the rest of the digestive tract, where the ruminant secretes enzymes that break down some of the plant material and much of the microbial population. Thus, the ruminant derives its protein indirectly by digesting the organisms that have digested the plant material and incorporated it into their bodies.

We are leading up to a point that will become apparent when we discuss communities but bears mentioning here. Niche specialization prevents excess competition and provides survival security *if* no drastic changes take place. Such specialization leads to more complex relationships between species and to more complex communities, but it requires a great deal of time and a great deal of evolution. Therefore, it is found only in the oldest and most stable communities. In the next chapter, we shall examine the implications that this has for man's planning.

We have discussed the more unusual, although important, interactions between species first in order to (1) show the extent to which interacting species are interdependent, and (2) point out that negative interactions are minimized over evolutionary time. We also wanted to emphasize how elaborate some of the adaptations for interspecific interactions may become.

We have left for last the two most important types of interaction, competition and predation.

Competition

As shown in table 13.1, competition is classified as an interaction in which each species is adversely affected by the other. We may further categorize competition into two forms: first, competition in which two species interfere with each other but do not compete for some single resource; and second, resource competition in which both species require a single resource, whether a food, nutrient, or habitat space.

The first type, *interference competition*, most frequently occurs either when each species produces some material that places the other at a disadvantage or when the behavior of each species is a disadvantage to the other. Thus, if two species of fish are too crowded in a pond, their waste products may interfere with each other. Or, if two species of frogs have very similar mating calls, much reproductive effort may be wasted. This sort of behavioral interference is extremely rare, however, for the obvious reason that it would be strongly selected against.

Resource competition is much more common, since some resources are almost always shared by species in the same trophic level. If one of those resources approaches a minimum limit for either species, then that species will be at a disadvantage if the other species uses the resources. Almost all of the trees in a forest compete for light, space, and nutrients. Filter-feeding herbivorous zooplankton compete for phytoplankton food. Deer compete with some leaf-eating insects, and many insect species compete with each other. Birds compete for nesting sites, and territorial animals compete for territory.

We have already seen many of the evolutionary effects of competition. The purpose of our treatment here is to make some precise generalizations about the competition. Recall the logistic equation from Chapter 12:

$$\frac{dN}{dt} = rN\left(\frac{K-N}{K}\right)$$

or, by simple algebra,

$$\frac{dN}{dt} = N\left(\frac{rK}{K} - \frac{rN}{K}\right) = N\left(r - N\frac{r}{K}\right)$$

For the most part, environmental resistance was discussed as a factor or factors outside the population. Competition within the population, or *intraspecific competition*, is included implicitly in the statement $(K - N)/K$. The addition of any individual to the population will increase N and thus bring the value of N closer to K, as the subtraction of an individual will reduce the value of N. Since these additions and subtractions are the result of births and deaths (we are still ignoring migration), and since $b - d = (K - N)/K$, we can express a relationship between r and K: r/K. We may define r/K as *the amount by which the birth of one individual decreases the rate at which the population grows.* (Examine the previous equation to see why this is so.) If r/K reduces the rate of increase by some amount for each individual born, then the births of N individuals will reduce the rate by $N(r/K)$. Thus,

$$r - N\left(\frac{r}{K}\right)$$

is the total amount by which the rate of increase of the population may be reduced by intraspecific competition, and therefore the growth of the population is expressed by the equation

$$\frac{dN}{dt} = N\left[r - N\left(\frac{r}{K}\right)\right] \quad \text{as above}$$

Now, let us suppose that a second species (species 2) inhibits the growth of this species (which we now call species 1), and that the effect of species 2 on the population of species 1 is similar to the effect of intraspecific competition, except in terms of the amount. That is, r/K is the effect of one individual of species 1, and we multiply by a constant, α (alpha), to find the effect of an individual of species 2; thus,

$$\alpha\left(\frac{r}{K}\right)N_2 \qquad (N_2 = \text{density of species 2})$$

Writing the whole equation for species 1, taking into account the inhibiting effect of species 2, we see that

$$\frac{dN_1}{dt} = N_1\left[r - \left(\frac{r}{K}\right)N_1 - \alpha\left(\frac{r}{K}\right)N_2\right]$$

This is the basic form of the Lotka-Volterra equations, named after the two mathematicians who independently formulated them in 1925 and 1926. To define the effects of species 2 on species 1, we use the subscript 1 to refer to the N and r of species 1, and α to represent the *competition coefficient* of species 2 on species 1. Similarly, to determine the effects of species 1 on species 2, we use the subscript 2 to refer to the N and r of species 2, and β (beta) to represent the competition coefficient of species 1 on species 2. Thus, the Lotka-Volterra equations for interspecific competition are

$$\frac{dN_1}{dt} = N_1\left[r_1 - \left(\frac{r_1}{K_1}\right)N_1 - \alpha\left(\frac{r_1}{K_1}\right)N_2\right]$$

$$\frac{dN_2}{dt} = N_2\left[r_2 - \left(\frac{r_2}{K_2}\right)N_2 - \beta\left(\frac{r_2}{K_2}\right)N_1\right]$$

It is not important that these equations be memorized, but it *is* important that the principle behind them be understood. These equations have been used to derive several conclusions about competitive behavior between populations, but before we examine these conclusions in detail, we should mention that other equations have recently been proposed that may predict other competitive behavior.

It is difficult to test the Lotka-Volterra equations in the field, and thus, much of the experimentation that has been performed to validate them has been done in the laboratory. Two particular sets of experiments are of interest. The first were performed by Thomas Park of the University of Chicago, who was working with his students and associates on competition between two species of flour beetle, *Tribolium castaneum* and *Tribolium confusum* (see fig. 13.5). These two species have very similar habitat requirements, and in both cases, a simple medium consisting of flour or wheat bran provides both their habitat and food. The periodic addition of fresh flour to a jar containing either species can keep the population going for an indefinite time. In Park's experiments, each species was introduced into the medium under varying conditions of temperature and humidity. Under similar conditions, each species was found to be able to survive. However, when both species were present, one species was

FIGURE 13.5 *Tribolium confusum,* a flour beetle. This and a very similar species, *T. castaneum,* were the subjects of the very interesting competition experiments described in the text. *(Courtesy of Grant Heilman.)*

TABLE 13.3 Results of Competition Between Two Species of Tribolium

Culture Climate	*% of Times Competition Won by:*	
	T. confusum	*T. castaneum*
Hot-wet	0	100
Warm-wet	14	86
Cool-wet	69	31
Hot-dry	90	10
Warm-dry	87	13
Cool-dry	100	0

Hot = 34°C Wet = 70% relative humidity
Warm = 29°C Dry = 30% relative humidity
Cool = 24°C

NOTE: Twenty to thirty replications of each climate culture were examined.

(From Thomas Park, "Experimental Studies of Interspecific Competition," *Physiol. Zool.*, 27: 177–238 (1954).)

always found to eliminate the other. The conditions and the results of this experiment are summarized in table 13.3. An examination of the data shows that *T. castaneum* invariably eliminates *T. confusum* under hot-wet conditions, and *T. confusum* eliminates *T. castaneum* under cool-dry conditions. Under conditions intermediate between these extremes, the competitive abilities of the two species tended to vary, although it is clear that differing climatic regimes strongly affected the outcome of competition. When the beetles were placed in single-species cultures under differing climatic conditions, it was found that there was a high correlation between the rate of increase, *r*, and the ability of the species to win the competition under the same climatic conditions in two-species cultures. However, it was also found that if the *r* of *T. confusum* and *T. castaneum* differed only moderately under the same conditions in single-species cultures, the species with the higher *r* did not always outcompete the other under the same conditions in two-species cultures. Regarding other factors affecting the outcome of competition, it was found that the presence of a virus in one of the populations affected its ability to compete, and that genetically different strains within the population might differ significantly in competitive situations.

The other experiments that we should look at deal with competition between species of clover, and were performed by J. L. Harper of the University College of North Wales. Coincidentally, the genus of the two species of plants experimented with was *Trifolium;* the species were *T. repens* and *T. fragiferum*. It was found that under identical environmental conditions of light, temperature, soil, water availability, and so forth, both species grow equally well. *T. repens* grows more rapidly and develops a full growth of leaves sooner, but *T. fragiferum* grew taller and thus was able to obtain sufficient light for growth, especially after *T. repens* had stopped growing. The density of both plants in mixed stands was approximately equal to the density of either plant in unmixed stands. Thus, each species inhibited the growth of the other, but because of morphological differences and different times of maximum growth, they could coexist in mixed stands.

Thus, there is experimental evidence for the Lotka-Volterra equations. Regardless of whether or not the equations are absolutely correct, they at least allow us to intuitively see how competition occurs.

One of the most controversial ideas in ecology is that of competitive exclusion. The *competitive exclusion principle* states that *two species with identical niches cannot coexist in the same habitat without one of them eventually being eliminated.* One of the species is likely to be less efficient at exploiting some resource for which the two must compete, and no matter how small the difference in the ability to exploit the resource may be, the less efficient species will eventually suffer local extinction. The Lotka-Volterra equations can be used to show that unless α is much less than K_1/K_2, and β much less than K_2/K_1, one of the species will be eliminated. However, it is very unlikely that the constants would be so low.

The competitive exclusion principle seems to hold true in communities under conditions of equilibrium, and in changing communities, observations may reveal that although two species are present and competing for the same resources, one of the species is on the way to local extinction. Thus, the principle appears to operate in unbalanced conditions as well. Nevertheless, there is conflicting evidence, both from the laboratory and from the field. For example, many species of phytoplankton apparently have the same niches and compete for the same nutrients, and yet they appear to be able to coexist over long periods of time. We do not have the time or space here to examine all of the evidence for and against the competitive exclusion principle, and thus, only a cursory treatment is given. Part of the problem with the principle is that it does not take into account several salient factors. First, species are not always made up of genetically identical individuals that produce copies of themselves. As we have seen, the Hardy-Weinberg conditions of equilibrium rarely, if ever, exist in nature, and thus, the frequencies of various alleles in a population constantly undergo change. Indeed, adaptations to the conditions of the environment, including the presence of competitors, alter the gene frequencies of competing species. Second, the environment itself is not a homogeneous factor; it undergoes change and evolution that may affect competing species and alter the outcome of competition. Third, the validity of the principle may depend strongly upon the concept of the niche. Whether two species *can* have *identical* niches is still open to question.

Until now, our definition of the niche has been useful, but not very rigorous. It is time to introduce a more complete definition in order to understand competition and its implications more clearly. Suppose the requirements of a particular species could be shown on a graph with two axes—that is, a two-coordinate system. Let one axis represent some temperature range, and the other a range of relative humidity. After observing and measuring a species' tolerances for temperature and humidity, we could plot an

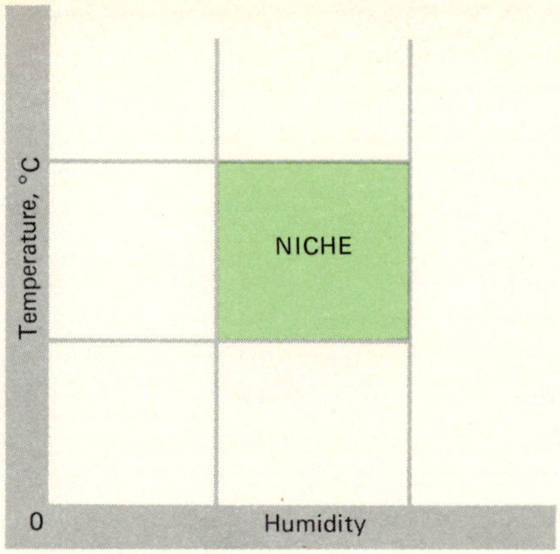

area on the graph that would show its niche in terms of temperature and humidity. We could add another coordinate representing sunlight. By projecting the areas on the graph, we could draw a volume that represented the niche in three dimensions. Although it could not actually exist in "real" space, we could add fourth, fifth, sixth, up to *n*th coordinates and define a *multidimensional niche* for the species. A space or volume of more than three dimensions is called a *hyperspace* or *hypervolume,* and thus, we could consider the niche of a species to be

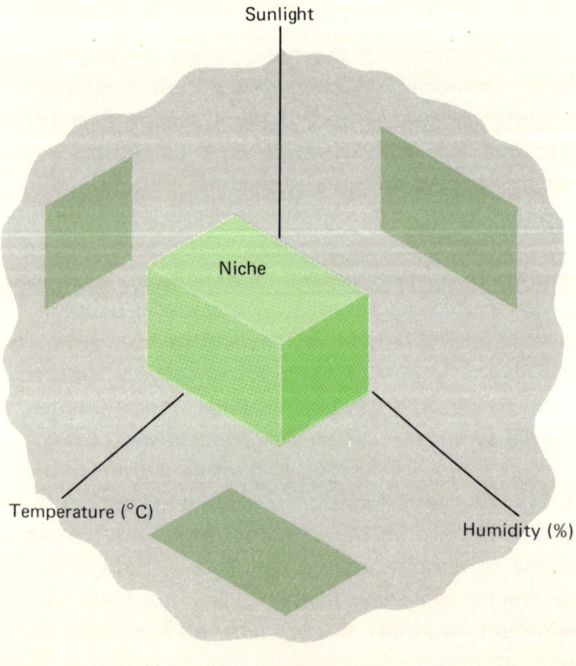

a hypervolume defined or limited by the upper and lowermost bounds of the species requirements. To accurately define the niche hypervolume, we could include, indeed if we are to accurately describe the species we *must* include, many factors such as food size, nutrients, temperature, water availability, sunlight, lipid and protein requirements, spatial requirements, energy requirements, pH, and a host of others.

The volume so described would not be unchangeable. Evolution would alter the bounds over a long period of time, and the ''shape'' of the volume would therefore change.

One other point should be made about the concept of the niche hypervolume. To describe the upper and lower limits of any of the requirements of a species implies that all individuals of the species can exist equally well within that range. That is not so, however, and we must define an optimal set of conditions within the boundaries of the hypervolume under which the species best lives. Outside the optimal hypervolume, as the maximal and minimal limits are approached, the likelihood of survival falls off. It must also be recognized that the individuals in any population vary, and that some individuals may be more efficient at surviving nearer the maximal or minimal bounds than others.

Now consider two species whose niche hypervolumes overlap. (See fig. 13.6.) Any habitat, at one point in time and in real space, may fall within a niche volume. For example, at some specific location in a pond, some particular combination of NO_3^-, temperature, food size, vitamin B_{12}, pH, pressure, light, and many other factors will exist. If that location falls within the overlapping volumes, the two species will compete and one should win (assuming the validity of the competitive exclusion principle). If the location falls within the volume of only species one, then species two can live there only temporarily, if at all. If the location is outside both hypervolumes, of course, neither species can live there. Now let us assume that in some

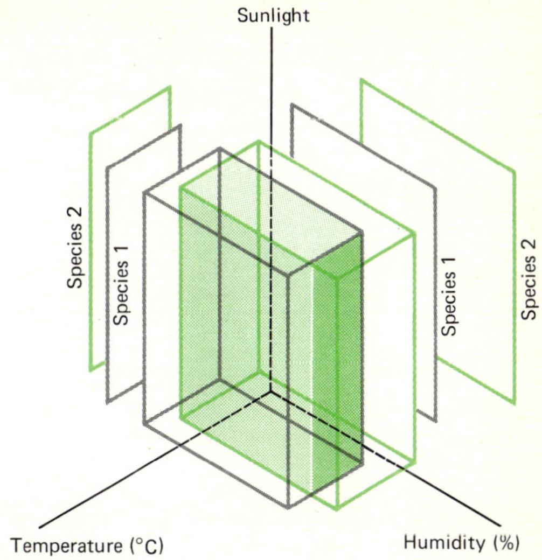

FIGURE 13.6 The niche hypervolumes of two species may overlap. In the overlapping volumes the species compete.

pond all things are equal except temperature—that factors such as nutrient availability, light, and pH, among other things, are not factors that limit either species. Then suppose that in this pond, the temperature from the surface down to 2 meters is only in the niche volume of species 1, that the temperature from 2 to 4 meters is in the niche volumes of both species, and that the temperature from 4 meters down is only in the volume of species 2. Competition will occur only in the portion of the pond from 2 to 4 meters in depth, and one species will probably be eliminated from this portion of the pond. (See the illustration on the following page.)

We have come to a point made implicitly earlier and which we should now make explicit. *If two species compete, then it is to their mutual advantage if their niches evolve so as to avoid competition*. Therefore, selective pressure is exerted for what is sometimes called *character displacement* or *niche displacement.* Recall the discussion of new species being introduced into an area in Chapter 7. Niche displacement may be expected to cause differences in a species' behavior, food preferences, and nesting site preferences, among other things, and to result,

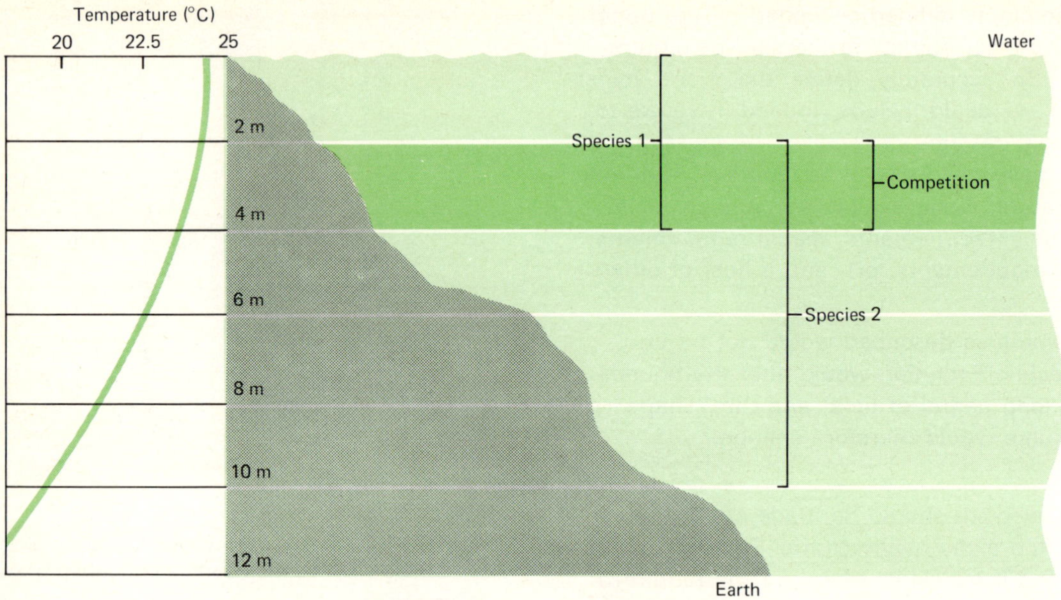

ultimately, in increasing specialization by the species. As we shall see in the next chapter, such *increasing specialization has a tendency to make communities evolve to greater and greater diversity,* a point that we shall see to be of great importance to man.

We will discuss the question of what the human niche might be in Chapter 16, but until then you might think about the question. It provides some interesting insights.

It should be apparent that competition is one of the most important factors in the process of evolution by natural selection. It is usually, for obvious reasons, strongest between species at the same trophic levels. The other major type of interaction, predation, occurs between species at different trophic levels and is responsible for the transfer of energy and nutrients.

Feeding Interactions

As we mentioned in our discussion of parasitism, it is often difficult to distinguish between parasitism and predation. Usually, common sense observations provide the most reliable guides. Predators usually kill their prey and rely on the whole prey population rather than an individual for their food supply. Herbivores, except for plankton feeders, are usually exceptions: a single herbivorous insect may live its entire life on one plant. Predators are usually free-living organisms; they move around independently of the prey population. Finally, predators are usually not as specialized in their food requirements as parasites.

Nevertheless, predators and prey exert strong selective pressures on each other. It is to the predator's advantage to become more efficient at capturing prey, and to the prey's advantage to become more efficient at eluding predators. There are, of course, limits to the advantages to be gained by such efficiency. If a predator population becomes 100% efficient at capturing prey, it may cause the extinction of its primary food supply and thus endanger its own survival. Should the prey population increase its elusive ability to 100% efficiency, it may begin to grow beyond the ability of the environment to sustain it; that is, it may exceed its carrying capacity and suffer a crash.

Predator-prey interactions offer some of the most striking examples of adaptation, usually in the prey population. Kenneth Roeder of Tufts University has studied the means by which moths avoid capture by bats. The bat emits a high frequency ultrasonic signal that can "bounce off" a flying moth and be detected by the bat in flight. This form of *echolocation* (see fig. 13.7) has been adapted to by certain species of moths through the evolution of a *tympanic organ* that is able to detect sound in the frequency range that the bat emits. At distances of more than 10 feet, bats seem unable to detect the echoes from objects the size of a moth, but moths can detect the sound of a bat from as much as 100 feet. Even though the bat can fly much faster than the moth, this seems to provide the moth with sufficient time to get out of the path of an incoming bat by merely turning in a direction away from the source of the sound. However, bats do not fly about in perfectly predictable paths. Rather, they "cruise" in search of prey, and thus the bat and moth often randomly encounter each other at distances too close for the moth to escape by instinctively turning away from the source of the sound. The bat can then easily overtake and eat the moth. At very short distances, the moth must take more rapid evasive action. To test the moths' reactions when the sound was detected at close range, Roeder set up a loudspeaker that emitted sounds in the same frequency range as a bat would emit. He also set up a camera near the loudspeaker and illuminated the area. When a moth flew into camera range the shutter was opened, triggering a timing device that released a series of batlike sounds from the loudspeaker precisely one second later. The moths' responses were at first confusing. Some folded their wings and fell directly toward the ground, some went into a "power dive," and others went into a series of looping twists. (See fig. 13.8.) In all cases, however, two elements were present: (1) the moths flew or moved toward the ground, and (2) they showed no orientation toward the source of the sound. Roeder admits that he was confused by the moths' reactions until he realized that they were also probably confusing to the bats trying to locate the moths with their sonar. The very unpredictability of the moths' behavior appears to be the key to their survival.

FIGURE 13.7 A bat emits pulses of high frequency sound. These are reflected from objects in the path of the bat, including potential prey such as this moth, and the reflected pulses of sound are received by the bat. When the bat hears these reflected sound pulses, it can determine both the distance and location of the moth. The photo shown above is a simulated diagram of echolocation. *(Courtesy of Frederic A. Webster.)*

When predators hunt visually, the prey populations may evolve adaptations of several sorts. Some species evolve protective coloration that makes them indistinguishable from their surroundings. It should be pointed out that such coloration may also be used by predators to conceal themselves from their prey. A leopard's spots, for example, make the leopard very difficult to see as it lies in wait on the branch of a

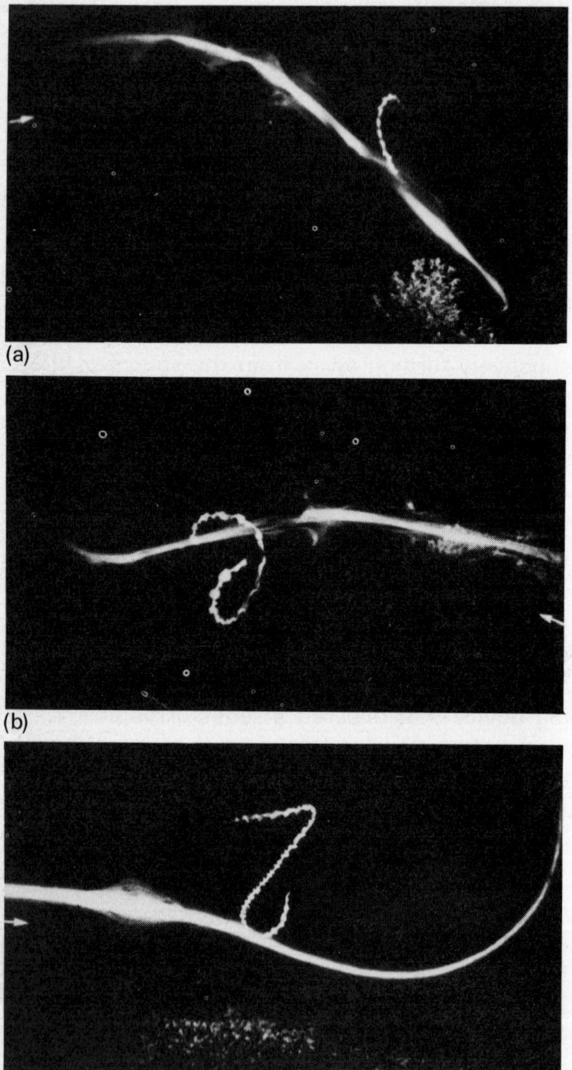

(a)

(b)

(c)

FIGURE 13.8 When a moth first hears the ultrasonic pulses of a bat, it initiates various evasive movements. These are sometimes successful, and sometimes not. These "streak photographs," in which the broader streak is the bat and the smaller streak is the moth, show successful and unsuccessful attempts at capture. The arrows indicate the direction of approach of the bat: (a) shows a successful capture in which the moth dove but the bat dove at the same time, (b) shows a capture in which the bat predicted remarkably well the movement of the moth and succeeded in capturing it, while (c) shows a successful evasion by the moth. *(Courtesy of Frederic A. Webster.)*

tree. Many insects have a taste that will make predators nauseous or even poison them. Since it would do the species little good to be poisonous or ill-tasting but not recallable by the predator species, such insects are often vividly colored or distinctively marked. This warning coloration prevents the predator from eating such an insect on the next encounter. The black and yellow banding of the bumblebee warns the hungry frog or bird that it will be stung if it attempts to eat the bee.

Such defenses are sometimes adopted by species unrelated to the unpalatable animals or insects. Such instances of *mimicry* are not uncommon, and they appear to provide palatable species with good protection from predation. The classic case of mimicry concerns the monarch butterfly, *Danaus plexippus,* and a completely unrelated species of another family, the viceroy, *Limenitis archippus.* (See fig. 13.9.) The case was a controversial one for a long time, as some biologists insisted that both monarch and viceroy butterflies were palatable to birds. Recent work by Jane Brower of Amherst College, however, has gone a long way to work out the controversy. The monarch's palatability depends upon its food source in the larval stage. Some populations of monarchs feed on milkweed plants containing variable levels of a poison that in birds will cause vomiting, and, if it reaches the heart, will cause death. Other populations feed on other plants. Thus, there are both palatable and unpalatable monarch populations. The majority of monarch populations, however, contain both palatable and unpalatable individuals.

Brower brought several wild bluejays into her laboratory at Amherst and attempted to feed them on monarchs. In most cases, the birds refused to eat them. If left to go hungry, however, most of the birds were induced to eat, and they were fed palatable individuals. Subsequently, the jays ate monarchs daily. Then, some of the jays were fed monarchs raised on poisonous milkweed. Within a few minutes, the birds began to vomit and continued to retch for

(a) (b)

FIGURE 13.9 The viceroy butterfly (b) almost perfectly mimics the appearance of the monarch butterfly (a). *((a) Courtesy of Hal Harrison from Grant Heilman; (b) courtesy of Grant Heilman.)*

FIGURE 13.10 A bluejay vomiting after feeding on a monarch butterfly which had eaten poisonous milkweed. *(Courtesy of Lincoln P. Brower, Amherst College.)*

around half an hour. (See fig. 13.10.) After one experience with a poisonous monarch, the birds rejected them on sight for several days. In some cases, twenty-four hours after eating one, the mere sight of a monarch made some birds retch.

Some monarchs were raised on a species of milkweed that is particularly poisonous, and these butterflies were known to contain more than four times the lethal dose of the poison for a jay. Yet, none of the jays ever died from eating a monarch. The dosage sufficient to produce vomiting is much lower than the lethal dose, and the vomiting is a physiological response that rids the birds of the poison before it can kill them. It would seem at first glance that it would be more advantageous to the prey population to be poisonous enough to kill the predator, since that would then rid the population of a threat to its survival. However, as we pointed out, negative interactions tend to be minimized in favor of interactions that enhance survival. Jays are aggressive territorial birds, and if one jay abandons a territory, another will move in. If the first jay in the territory can be "educated" to avoid monarchs, then the monarch population sacrifices fewer individuals to maintain an educated jay population than it would have to sacrifice if it killed each jay that ate a monarch.

The advantage for a mimic is clear. As birds learn to avoid the monarch, the predator pressure on the viceroy population is reduced as well. Most of the butterflies that are closely related to the viceroy are brown, black, or blue, with a white band across their wings. It has been postulated that point mutations arose in the viceroy species, and that the mutations gradually accumulated until the viceroy evolved to its present form. The mutants, of course, were conferred with selective advantages, and some produced more offspring, thus perpetuating the mutations. Gradually, the proportional numbers of offspring of mutant forms increased in succeeding generations until they outnumbered the offspring of nonmutant parents. It has been argued that the mutant forms were more conspicuous than the nonmutants and thus were more strongly selected. However, Brower has shown that many birds are easily conditioned to avoid a certain stimulus and will then tend to avoid all stimuli that are similar, even though they may be different from the original stimulus. This is called *stimulus generalization,* and when conditions are right, this phenomenon will cause birds to reject many potential prey species that are only vaguely similar. Thus, an advantage can be gained by a mimetic species that bears only a very small resemblance to the model species. We should point out that although Brower's evidence and reasoning seem to offer an explanation for the rise of mimicry among insects that are preyed upon by birds, there are still many biologists who dissent from this view. Brower's hypothesis needs more testing before it can be accepted as a complete explanation of the evolution of mimicry. (See box 13.1.)

It is clearly not to the predator's advantage to exhaust the supply of its prey, as it is not to the advantage of a parasite to kill its host. Thus, a "strategy" usually evolves whereby the predator does not kill in excess, but takes only what it can use. Usually, the organisms easiest to take are the weakest or most ill-adapted individuals, and thus predation serves to increase the vigor of the prey species. Wolves, for example, cull out older, weaker, or sick individuals from caribou herds. Caribou calves that wander from, or fail to keep up with, the herd are also preyed upon by wolves. Similarly, a herbivore rarely kills its food supply or grazes the land sufficiently to bring about soil erosion, man's domestic herbivores being the frequent exception. Only rarely do insects badly damage or destroy vegetation unless the environment is disturbed or the insect is a newly introduced species. The gypsy moth is an example of the latter case, and the boll weevil would not be a pest if there were no cotton monoculture. Parenthetically, most plant species have evolved in such a way that they do not cause devastating destruction of the soil or excessive nutrient re-

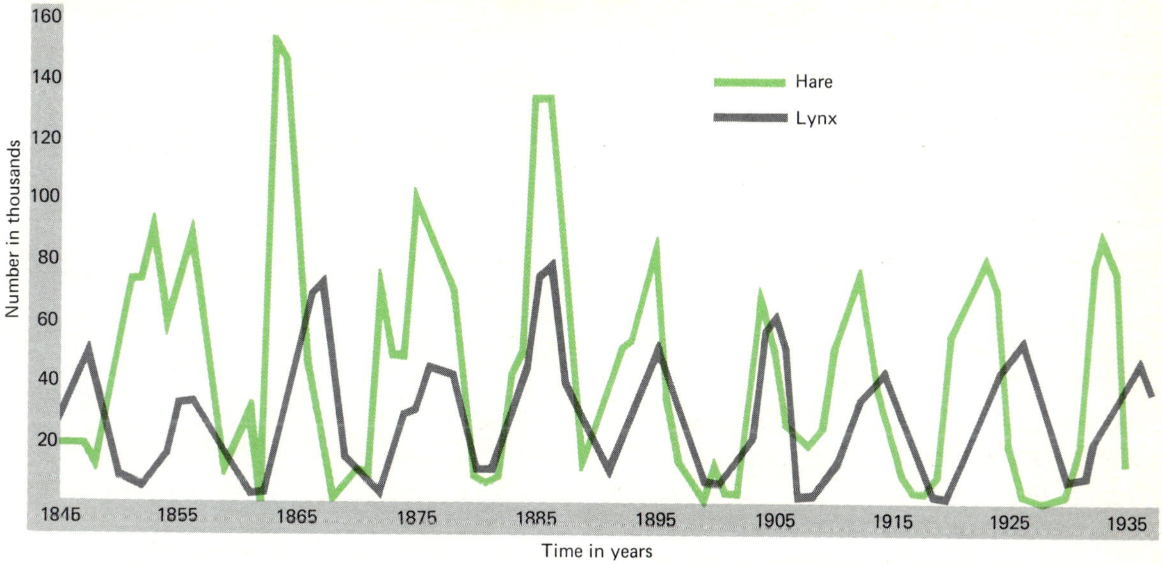

FIGURE 13.11 The oscillations of the Canadian hare and lynx populations are a classic example of the cyclic fluctuations found in some predator-prey interactions. *(After E. P. Odum, 1971, redrawn from MacLulich, 1937.)*

moval. Again, we emphasize that negative interactions are minimized when species have evolved together or have coexisted for a long time.

Predators and prey do mutually ''adjust'' their population sizes. We could write more equations that predict the population growth rates of interacting predator-prey species, but we shall not do so here. They are developed by reasoning similar to the Lotka-Volterra equations already presented. What we shall do is present some of the important and interesting points made by such equations

The simplest equations predict that as predators increase, the prey population will decrease until the predators begin to decrease because of starvation (or migration into a habitat where prey are more abundant). This will allow the prey population to begin increasing again and lead to an increase in the predator population. Thus, the predator and prey populations may oscillate. A number of examples of such oscillations have been observed in both the field and the laboratory. The classic case of such oscillations can be seen in figure 13.11, which shows the variations in abundance of lynx and snowshoe hares in Canada from 1845 to 1935. A major problem with this example is that it is based not on observations of the lynx and hare populations themselves, but rather on populations approximated by the number of pelts taken by trappers and received by the Hudson Bay Company. It is known that although the lynx preys upon the hare, hare populations may also show significant oscillations in areas where no lynx populations are present. Similar oscillations have been seen in the populations of North American lemmings and their chief predators, arctic foxes and snowy owls.

However, such oscillations usually become ''damped out'' as shown in figure 12.6, and occur only in simple communities. Predators are usually not restricted to single species of prey, and when a decline occurs in the population of one prey species, the predator preys more heavily on secondary prey species. If the predator population remains stable—that is, if it does not increase or decrease significantly—the primary prey species can ''make a comeback,'' and predation will once again increase on the primary

Insects and Flowers

Not all forms of mimicry involve species that are members of the same order, as in the case of monarchs and viceroys. Some of the most striking examples may be found among species that are not even members of the same kingdom!

Several species of orchids, for example, so closely resemble insects that they have been given descriptive names that incorporate the names of the model insects. Thus, *Ophrys apifera* and *O. bombyliflora* resemble bees, while *O. aranifera* appears to be a spider. Although such species were known for many years, it was not until 1961 that the interactions between the flowers and the insects were carefully observed and the relationships between them worked out satisfactorily. Such observations revealed that male insects regularly attempt to copulate with the flowers, and while doing so either receive or deposit pollen sacs, thus cross pollinating the plants. Since the reproductive behavior of insects often requires the reception of visual, tactile, and chemical stimuli, and since the flowers present only a partial stimulus, such copulation is not known to be consummated. Thus far, the most careful investigations have failed to show that the insects ejaculate sperm on the plants, and so the reproductive effort of the insect is not "wasted." In short, the plant species benefits from its interaction with the insect species, the insect species is not significantly disadvantaged, and the relationship may be described as commensalism.

Ophrys bombyliflora (Courtesy of Milton J. Terrell.)

BOX 13.1

The petal-like appendages of the *Hymenopus coronatus* nymph (above) enable it to mimic closely the orchids among which it lives (below). *(Photos courtesy of Edward S. Ross, California Academy of Sciences.)*

The resemblance of insects to plants, or plant parts, is more commonly known than the resemblance of plants to insects. Some of the most familiar examples include insects that appear to be bits of bark or thorns. One family of insects, the Mantidae, includes a number of cryptically colored species, such as the familiar "praying mantis," as well as some which are almost indistinguishable from the flowers among which they live.

Hymenopus coronatus is a mantid that is found in Malaysia. The pink form of this species, pictured here, lurks in orchids and other pink forest flowers. Small insects which land on this mantid in search of nectar are usually ignored, but when larger ones land their search is quickly ended.

prey species until an equilibrium is reached. If predator populations achieve equilibrium, so may prey populations.

Mankind as an Interacting Species

As we have seen, the interactions between species are highly evolved and specialized. Negative interactions tend to be minimized by evolution, and populations tend toward stability. However, man has often brought about such rapid changes in communities and ecosystems that other species have not had enough time to adjust to them. Such disruptive human activities are common. We shall look at man's effects in terms of the niche and evolution in a later chapter. We shall also look at effects on communities in the next chapter. It is appropriate here, however, to look briefly at man's historical role as a predator and competitor and to outline the very important changes that have taken place in that role. Our discussion is brief not because the topic deserves only brief treatment, but because we shall return to it in greater detail when we consider the evolution of man in Chapter 16.

The study of human evolution is very controversial. The available evidence is sparse and subject to conflicting interpretations. Thus, the reader should bear in mind that much of what follows is somewhat speculative. Some of what we know about early man is based on our knowledge of contemporary primitive populations, such as African Bushmen or Australian aborigines. In some instances, our information is purely speculative; we are making educated guesses about what life might have been like on the basis of what we know about life in other populations. The reader ought not to be misled by the apparently uncomplicated story we present here. Man's evolutionary past remains shrouded in mists.

(a)

Let us begin by considering a primitive human population. As examples we could cite the !Kung Bushmen of Africa's Kalahari Desert or the aborigines of Australia's Gibson Desert. (See fig. 13.12.) In either case, we would be dealing with a population that is considered to be representative of man as he is *believed* to have lived 20,000 or more years ago. In anthropological terms, these populations are hunter-gatherers. They depend for their sustenance on the edible roots, berries, and fruits they can gather and on the few animals they can kill with primitive weapons. Obviously, the population density of early man, like that of Bushmen or aborigines, must have remained low, in part, because of severely limited food resources. The population densities of Bushmen and aborigines are low, and we can make an interesting comparison between the two. The !Kung recognize 54 species of edible animals and 85 species of edible plants. On the average, a !Kung group consists of 25 to 30 individuals. In contrast, the aborigines of the Gibson Desert recognize only 47 species of edible animals, including termites and lizards, and only 38 species of edible plants. Owing to their relatively impoverished environment, the aborigine populations are much smaller than the !Kung populations, sometimes being less than 10 individuals. What sort of interactions might such populations have with other species?

(b)

First, early man was probably parasitized, perhaps to the extent that wild animals are today, and suffered from a variety of chronic diseases. He probably had some camp followers, such as dogs, that played a protocooperative role by cleaning up man's refuse. The followers may also have unintentionally warned man of danger and they may possibly have helped man hunt other animals. Other camp followers were probably also present but acted more like parasites. Rodents and some insects probably followed human populations about, deriving benefit by living on man's refuse. But these organisms may also have been disease vectors and thus had negative effects.

Like any omnivore, man was a predator and competitor. He competed with species that ate the same plants he ate and with scavengers such as hyenas, jackals, and wild pigs for the remains of kills made by large predators. He may also have competed with some top carnivores for the large animals he hunted. Nevertheless, although man obviously must have had some effects on other species, human ecological relationships and selective pressures remained stable, and the population probably remained low for hundreds of thousands of years. On the whole, man achieved a balance with other species. But there was an immense difference between humans and the species with which they

FIGURE 13.12 Hunting and gathering are the ways most primitive populations obtain food, as do, for example, (a) the bushmen of Africa's Kalahari Desert or (b) the aborigines of Australia's Gibson Desert. *((a) Courtesy of Constance Stuart from Black Star; (b) photo by Dr. R. A. Gould, courtesy of the American Museum of Natural History.)*

interacted. Humans survived not because of any morphological adaptation (although they did undergo some important morphological changes), but rather because of the ability to reason. We shall examine this ability and the role that it played in the survival of the human population in a later chapter.

In spite of the balance man was able to achieve with other species, it is hard for us to say that life was "enjoyable." From our vantage point, such a life hardly sounds appealing. Disease was endemic, although probably rarely epidemic. Few individuals lived past forty, and on the average, a human lifespan was around eighteen years.

We shall consider later what might have happened to the human population to make it change its hunting-gathering way of life, but such a change did take place. Man became an organized agricultural nomad, moving from place to place, raising a crop or two for a few seasons and then moving on. He was still a predator and a competitor (as we are today), and for the same reasons as when he was a hunter-gatherer. But man developed new relationships with domestic plants and animals, and through them with other species.

To practice agriculture, man must clear land, and in doing so acts as no other species does; man becomes a herbivore that does not utilize the energy in the biomass removed from production. Certainly, some of the plant material taken from the land may have been used as fuel for fires, but even that use of energy is radically different from biological consumption, since energy is ultimately denied to decomposers. Much of the material that man removed to cultivate the land was recycled for decomposers, but in a pathway radically different from that found in a wild community. As an agricultural nomad, man was now in the position of completely changing or destroying communities. Whatever species of plants remained in the community after man had finished clearing it had to compete with man's crops for nutrients. Thus, as crops competed with wild organisms, man indirectly changed the course of evolution. But the wild plants were not the only species affected by man's crops. Those organisms that fed upon the crops were conferred a selective advantage. They competed with man and thus became "pests."

Now consider man's crops and domesticated animals themselves. It was probably at the beginning of the agricultural nomadic stage that man began to practice artificial selection. Man's interrelationships with crops and domesticated animals, such as swine, goats, and cattle, clearly became protocooperative. As man became more dependent on them, and they on him, the relationship became mutualistic. It is certainly mutualistic today, since without such crops and animals, mortality in the human population would be enormous. Indeed, without the human input of fertilizers, pesticides, water, and energy in planting, weeding, and so forth, the mortality among crop and domestic animal species would be similarly great. Through artificial selection, man directly changed the course of evolution. Moreover, the practice of artificial selection has probably had unmeasurable evolutionary effects on the populations of wild species that competed with, or preyed upon, the species that man selected. We do not know what the effects of such pressures might have been, for we have not been able to look very closely at how man's domestication of certain species has affected their competitors. More research is needed in this area.

What about parasitism in early agricultural human populations? In the process of domesticating wild animal species, man probably came into contact with new, less well adapted, parasites. Thus, mortality from parasitism probably increased, although in time, evolution probably caused man and these new parasites to adapt to each other.

The shift from a hunting-gathering way of life

to an agricultural one probably was accompanied by an increase in the human population as the carrying capacity of the environment was expanded. The shift also made it possible for man to settle in villages. Twenty-five thousand years ago, well before the advent of agriculture, the population of the earth is estimated to have been approximately 3.3 million persons, or 0.04 persons per square kilometer. Fifteen thousand years later, although the population had risen to 5.3 million, the density remained approximately the same. (We would have to extend the expression of density to several decimal places to see a significant change.) Eight to ten thousand years ago, around the time that agriculture was established and the earliest human villages began to appear, the total population rose to about 86 million and the density rose to 1.0 persons/km^2.

We could (and shall later) ask if man's niche had changed, and whether, in an ecological sense, we are still the same species.

In less time than any similar change could have taken place if natural selection and evolution alone had been at work, man's niche and its impact on natural communities changed. Something about human behavior had a unique impact on man's evolution. Most primitive agricultural societies were nomadic, in part because they exhausted the nutrients in the soil and destroyed the ability of communities to support human populations. But man was able to adopt new behavioral patterns that led to the next step in evolution, the management of agriculture and urbanization.

Before we can discuss human effects on biological communities, however, we must first examine the nature of such communities.

Selected Readings

Chandler, Asa, and Read, C. P. *Introduction to Parasitology.* Philadelphia: W. B. Saunders Company, 1961. Now in its tenth edition, this classic text remains one of the best overall treatments of parasite-host interactions available.

*Klopfer, Peter. *Behavioral Aspects of Ecology.* Englewood Cliffs, N.J.: Prentice-Hall, Inc., 1972. An excellent treatment of population interactions from the behavioral point of view.

*Mowat, Farley. *Never Cry Wolf.* Boston: Little, Brown and Company, 1963. A delightful and often amusing tale of a naturalist's observations of wolves in their natural habitat. In spite of its entertaining nature, the book goes a long way toward destroying our stereotypic beliefs about the behavior of the wolf.

Schaller, George B. *The Serengeti Lion.* Chicago: University of Chicago Press, 1972. A truly first-rate study of the behavior and ecology of predators and their prey. As the title indicates, the lion is emphasized, but other predators are also treated. The cheetah, the leopard, and the wild dog each receive a separate chapter.

*Wickler, Wolfgang. *Mimicry in Plants and Animals.* New York: McGraw-Hill Book Company, 1968. A concise and yet thorough treatment of a subject that is more widespread and diverse than is generally realized.

*Zinsser, Hans. *Rats, Lice, and History.* Boston: Little, Brown and Company, 1935. A fascinating and easy to read history of typhus and typhoid fever.

Ehrlich, Paul R., and Raven, Peter. "Butterflies and Plants." *Sci. Am.,* Vol. 216 (June, 1967), pp. 104–111. Offprint number 1076.

Roeder, Kenneth. "Moths and Ultrasound." *Sci. Am.,* Vol. 212 (April, 1965), pp. 94–102. Offprint number 1009.

*Paperback edition available.

FIGURE 14.1 This portion of Long Island (photographed from an altitude of more than ten miles) shows the mosaic pattern of biological communities and some of man's effects on those communities: The South Shore and Fire Island are at the bottom of the picture; the farmlands and resort communities of the North Shore are at the top. (a) is an undeveloped coastal marsh, little affected by the activities of man; (b) is an area of marsh which has been filled with material dredged from the bottom, and thus de-

communities

chapter fourteen Species interact in a variety of ways, depending upon their niches and adaptations, and because they interact, they continuously evolve. But species rarely interact with just one other species. Rather, they interact with many others and with the chemical and physical factors in their environments. The multitude of different species populations of an ecosystem make up a *community.*

Consider the high-altitude aerial photograph shown in figure 14.1. The landscape we see in that photograph is made up of a varied mosaic of biological communities, characterized usually by their dominant vegetation. Coastal marshes, plowed land, dune vegetation, farmed fields, and pine and hardwood forests are all apparent, and each constitutes a distinct ecosystem and community. Each area contains lesser communities within it, and each is a part of an even larger community. We have seen how individual species evolve mechanisms that insure their survival, how they interact with the environment to survive, and how energy and materials cycle through communities. Now we must look at the communities themselves, for these most dominant features of the landscape are in many ways of great importance to the human species. Many human activities exert profound influences on the distribution and character of communities, and in turn, the human species is dependent on these large biological systems.

There are some fundamental, but very difficult, questions to answer concerning the nature of communities. For example, can the boundaries of a community be specified? Is there some minimal physical size or number of species in a community? What determines the species composition of a community? Do communities change? To answer these rather difficult questions let us consider a pond. It is an ecosystem containing a definable community of aquatic organisms. But consider a rotting log on the bottom of the pond. It is also an ecosystem and has its own community of organisms. A field surrounding the pond, of course, is also an ecosystem containing a community, but does it include the pond? The field and pond are parts of a watershed that is itself an ecosystem. Even larger units exist, called *biomes*, each consisting of a major set of species distributed over a large geographical area. A biome may cover a major portion of a continent and is considered to be the largest definable community. We shall discuss biomes in more detail in a later chapter.

stroyed; (c) is an area of coastal dunes, some of which have been developed for residential purposes — because the dunes frequently move, this is a hazardous place for building; (d) is a patchwork of mixed pine and oak scrub forest, protected because it is in a park; (e) is an area that was once similarly covered with pine and oak forest but is now farmland devoted to a wide variety of crops. Similarly (f) was also once forest or marshland, but is now almost entirely devoted to urban and suburban housing. In some areas, such as (g), dwellings have been built around the small ponds and lakes, causing problems of eutrophication. Other construction such as airports (h) and golf courses (i) vary in their effects on the natural communities. The nutrient waste from the residential areas drain into lakes and into the bay along the South Shore, causing extreme algal growth problems. (Scale: one inch = two miles.)

Obviously, however, each of the examples represents a very different system in which very different processes must operate. Thus, generalizations are difficult to make.

Clearly, the definition of a particular community or ecosystem depends on the interests, background, and viewpoint of the observer. It is usually convenient if some kind of easily recognizable or measurable boundaries can be defined, as is possible for a watershed or pond, but sometimes it is necessary to define rather arbitrary boundaries for a community or ecosystem. For example, in the ocean, the continuity of the water obliterates sharp boundaries, thus making communities difficult to define. On land, a grove of hardwoods, a stand of pines, and a plowed field, all adjacent, might be considered three distinct communities. Yet, they would share some species that migrate between them, and as we shall see, the field might eventually grow into a grove of pines, and the pines might eventually be replaced by hardwoods. Thus, although the boundaries of a community may be distinct in space, they may be indistinct in time.

As we pointed out earlier, the evolution of the component species in a community produces change and evolution of the community itself. We will consider some of the implications of that long-term evolution later. First, however, we must consider how communities may change, or "evolve," in another, much more rapid way. With time, the species composition of a community changes as species make way for, and are replaced by, other species. An understanding of this process, called *community succession*, will tell us much about the other processes that take place in the community.

Community Succession

Communities usually change and develop in an orderly way and seem to follow a pattern that is fairly similar in many different environments. When a plowed field is abandoned, it will soon become grown over with grasses. In time, the grasses may be replaced by shrubs, the shrubs by pines, and the pines, in one or two human generations, by hardwoods. In most cases, a set of definably different stages is passed through and leads to a community that reaches a *climax.* The whole set of stages is called the *sere*, and the process is called *seral succession.* The *climax* is usually a mature, stable, and balanced community. A major change in a community may produce readjustments of all the species populations in the community. Thus, logging, defoliation, or fire produces changes not only in the species composition of plants in a community but also in the populations of microorganisms, insects, birds, and other animals as well. After such a change, the system can usually return to a climax; the system, it would appear, can "heal" itself. Of course, if an irreversible change in soil, climate, or geomorphology takes place, the system cannot return to the same climax. Let us look now at a classic example of succession as it occurs under natural conditions.

The Lake Michigan Dunes

For the last several thousand years, along the southern and eastern shores of Lake Michigan the action of wind and water has deposited sand in the form of dunes. (See fig. 14.2.) Soon after their deposition, the dunes are colonized by growths of dune grass, which acts to stabilize the sand formation by preventing wind erosion. Once stabilized, the dune is able to support various "weedy" growths and, in time, stands of trees. The death and decomposition of these plant populations gradually transform the dry sand into a soil that is rich in humùs and is capable of retaining more moisture. As larger plants appear, the dunes are further stabilized, the soil becomes increasingly enriched and moist, and the way is paved for the climax. On the dunes, the climax is dominated by beech and maple trees, and the physical environment is measurably cooler and more moist than in the early stages of succession. Overall, the time required for the whole process, from dune formation to climax, is estimated to be approximately 1000 years. (See fig. 14.3.)

FIGURE 14.2 A sand dune near the shore of Lake Michigan, with grass as the earliest colonizing plant. *(Courtesy of the State of Michigan, Department of Natural Resources.)*

FIGURE 14.3 Primary succession is very well illustrated by the progression of vegetation communities on the Lake Michigan dunes. *(After E. P. Odum, 1971.)*

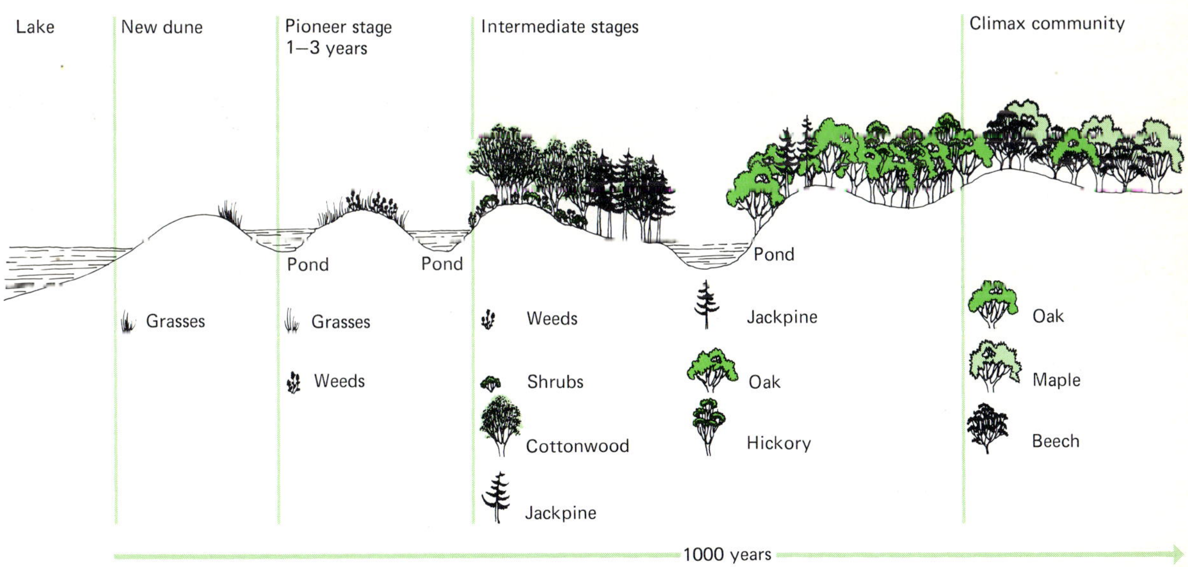

The Lake Michigan dunes provide an example of *primary succession*—that is, the succession of a community in a newly produced habitat. Another example of primary succession might be a volcanic island that has recently risen from the sea. Such instances of primary succession, however, are rare. Of greater importance is the case in which a community undergoes great change, usually as a result of human activity, and then returns to the climax condition. Our next example, therefore, shall show this process, known as *secondary succession.*

Succession in the Southeastern United States

The abandonment of agricultural land in the upland Piedmont region of the southeastern section of United States provides an example of how secondary succession proceeds. The initial conditions are very different from those cited in the dunes example. The climate is different, the substrate is soil rather than sand, and the species involved are different. Nevertheless, some basic similarities can be seen between the two forms of succession.

The bare field is colonized in the first year by a few grasses and various species of weeds. These rapidly growing, easily seeded species (usually *r* strategists) are outcompeted for light and nutrients, and within a year or two, they are replaced by grasses. Soon species of perennial, woody shrubs begin to appear (see fig. 14.4), and the grass-shrub community persists for approximately 20 years. Pine and cedar seedlings begin to take root well before the grass-shrub community is 20 years old, but because these species grow more slowly, it takes more time for them to compete effectively with the grass and shrub species present. About 20 years after the field was abandoned, the pines assume a position of dominance in the community. It should be pointed out that the 20-year period is approximate, since the actual time required for the pines to achieve dominance depends on a variety of factors. The pines remain the dominant species in the community for a period of 75 to 80 years. However, early in the development of the pine forest, oak and other deciduous seedlings take root. Since oaks

FIGURE 14.4 Perhaps the best studied and best understood successional pattern is that leading to the deciduous forest climax in the southeastern Piedmont region of the United States. *(After E. P. Odum, 1971.)*

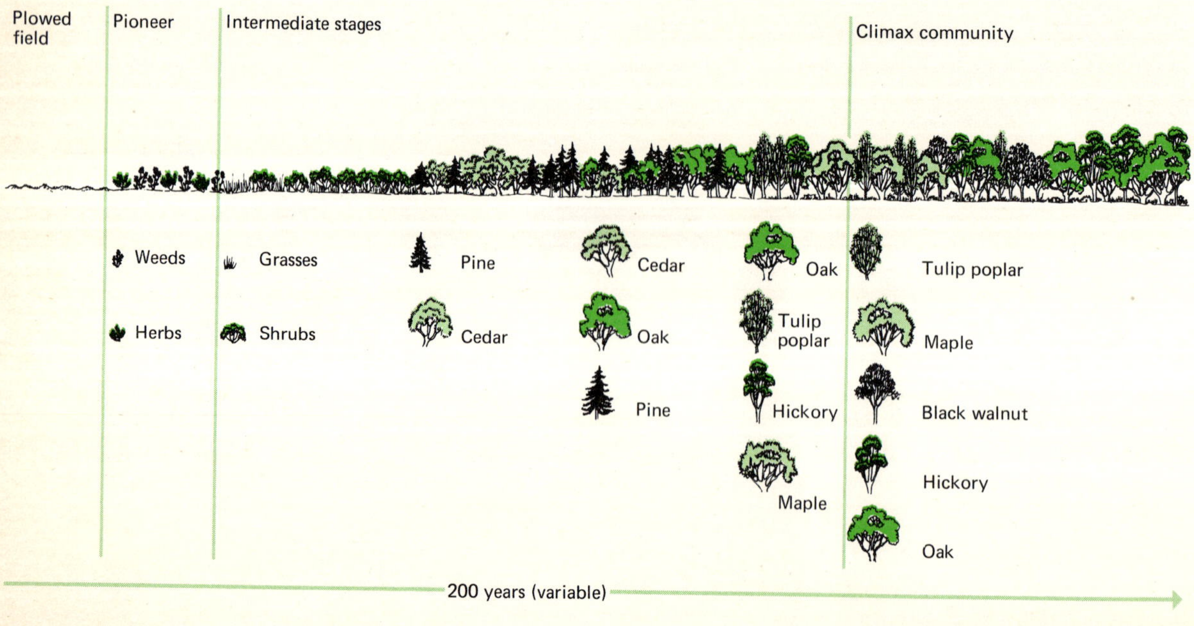

FIGURE 14.5 Hardwoods grow in the understory of a pine forest, eventually converting it to another type of community. *(Courtesy of Harriet S. Gallant, Science Photo/Graphics Ltd.)*

grow more slowly, they form an understory of hardwoods among the pines. (See fig. 14.5.) As the oaks grow, they outcompete the pines for moisture, nutrients, and light, and approximately 100 years after the abandonment of the field, they begin to achieve dominance over the pine. Hickory seedlings begin to appear in the late stages of the pine forest. Soon the oaks and hickories completely replace the pines. The culmination of the sere is normally an oak-hickory climax, which is usually firmly established approximately 150 to 200 years after abandonment of the plowed field.

Succession in Aquatic Systems

Succession also operates in aquatic systems such as ponds and lakes. Imagine a lake formed by the withdrawal of a glacier. Initially, such lakes usually have sandy bottoms formed by the immense sheet of ice. The water is cold, and the lake is fed by a cold, low nutrient stream. Gradually, silt and nutrients from the surrounding land accumulate in the lake and provide resources for pioneer populations of phytoplankton and bottom-growing vegetation. Productivity increases as the nutrient concentrations in the lake rise, and organic materials also begin to accumulate. As we have seen, the nutrients recycle in the lake, although some are lost from the productive zone as they are incorporated with the organic material into bottom sediments or are removed from the lake by an outflowing stream. In the early stages of succession, the lake might support a population of trout or similar fish in the clear, cold water of the deep hypolimnion. Nutrients continue to accumulate, providing a base that supports larger populations of phytoplankton and other algae, and these larger populations cause the water to become somewhat turbid. The increasing accumulation of nutrients promotes greater levels of productivity, and thus, as the lake grows shallower and sediments accumulate on the bottom, productivity rises. Cold water species, such as trout, are replaced by bass, sunfish, or pickerel, and in the summer months, the water is greenish. Rooted vegetation, such as lily pads, begin to grow up from the bottom in the shallower

FIGURE 14.6 As pond succession continues, rooted aquatic plants grow near the edge. Eventually a bog mat encroaches upon the surface. *(Courtesy of Thomas H. Brooks from Stock, Boston.)*

portions of the lake. The sandy, rocky shores begin to be covered with plants growing out from the edge of the lake, and thus, a bog mat begins to form. (See fig. 14.6.) The pond is by now very shallow. Depending on a variety of factors, such as the initial depth of the lake, level of nutrient input, water temperature, and geographic location of the lake, the process takes an estimated 1000 to 4000 years.

The bog mat continues to accumulate and grow toward the center of the lake. Soon nutrients begin to become bound up in undecomposed organic matter. As the organic materials dissolve and decompose, some of them form humic acid, a by-product of decomposition that is resistant to further action by decomposers. Such materials accumulate in the water, turning it yellowish brown. Productivity declines because nutrients are bound up by stored organics. Sooner or later, unless something interferes with the process, the bog mat covers the entire lake and slowly consolidates sufficiently to support trees. A forest is formed over the bog, and ultimately, the forest progresses to the same seral climax as a plowed field in the same geographic area.

Generally speaking, the lake goes through three stages before bog formation: It is first *oligotrophic,* then *eutrophic,* and then *dystrophic.* (See fig. 14.7.)

Marine scientists and oceanographers have not extensively studied succession in the ocean. Oceans are generally considered to be in a steady state of development, having been biologically and chemically stable for many thousands of years. However, some investigators are now turning their attention to the problems of succession in the ocean, since the ocean's stability seems to be threatened by man's input of nutrients and toxins.

The General Features of Succession

As we can see from the examples given, communities and ecosystems develop from relatively simple to more complex forms. In the early

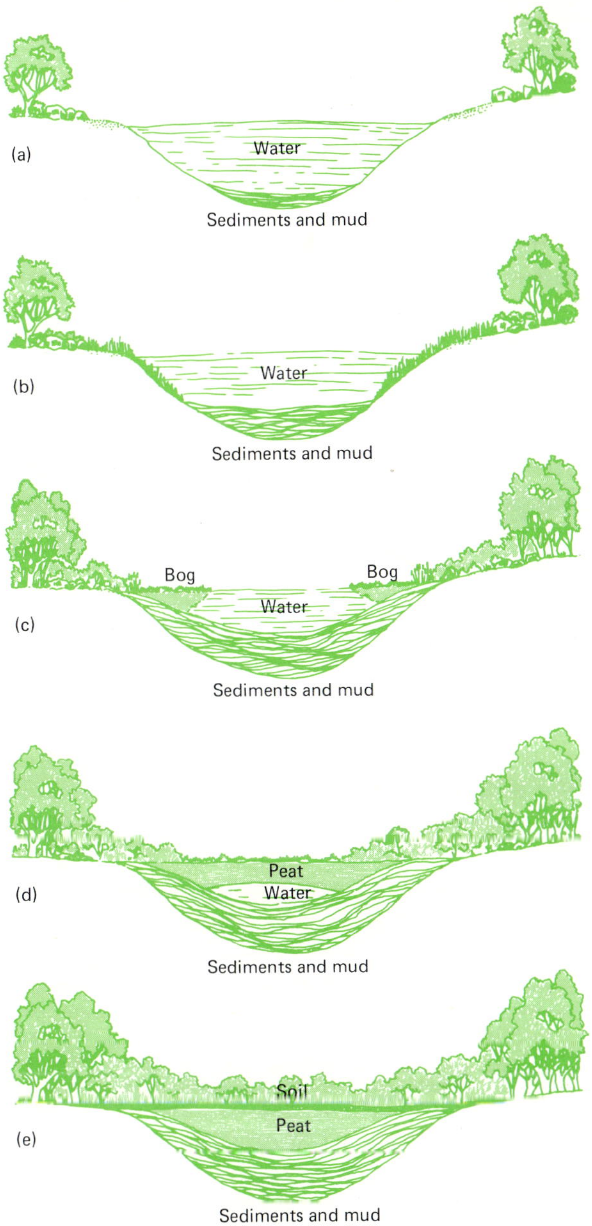

FIGURE 14.7 As described in the text, a lake goes through an oligotrophic (a), eutrophic (b), and dystrophic (c) stages before becoming a bog (d) and eventually a forest (e).

stages of development, a community is characterized by relatively few species subject to a great deal of change. As the community grows older, however, the number of species increases, and change is slower and has less of an impact on the community as a whole. Associated with the simplicity of the new community is the relative dominance of the physical factors of the environment. As the community develops, species diversity increases, and the community becomes more complex and biologically dominated. Thus, although physical factors still play an important role in the climax, living things play a proportionally greater role in controlling the processes of the system.

We should now introduce some new terminology related to the idea just mentioned. In examining community succession, we can distinguish between *allogenic changes,* or *changes brought about by the action of influences outside the community,* and *autogenic changes,* or *changes brought about by the community itself.* Allogenic factors include, among other things, temperature, water availability, geological and meteorological events, nutrient input, and human activities such as agriculture, logging, city-building, and general economic development. Autogenic changes may include changes in the soil substrate brought about by plant and animal decomposition, changes in local climate due to forest cover, or purely biological changes, such as the introduction of a population of owls made possible by the availability of food resources and nesting sites in trees.

Allogenic changes tend to be more frequently observed in the early stages of community development. In the Lake Michigan dunes, for example, the populations of grasses on recently deposited dunes are often buried by wind-borne sand. This is not to say that autogenic changes are rare in the early stages of succession. One investigator has shown that in the upland Piedmont region, one of the species of weeds that colonizes plowed fields produces organic matter and detritus that inhibits its own growth in the second year, thus paving the way for its own replacement by grasses. Such phenomena are believed to be widespread.

The particular combination of allogenic and autogenic factors at work in a specific community determines the rate of succession and the species composition of the climax. Since some of the factors, particularly allogenic ones, are difficult to predict, it becomes equally difficult to predict the rate and outcome of succession except in the most general terms. Nevertheless, an understanding of the development and succession of communities and ecosystems will provide us with some general principles that will be important in our understanding of man's role in the biological world.

Although succession has been studied for some time (the Lake Michigan dunes were first examined carefully in the late nineteenth century), many of our generalizations are the result of more recent investigations. Many of the following ideas about communities and the process of succession have been formulated only within the last fifteen or twenty years and, thus, are subject to much more rigorous examination and proof than they have so far received. Many exceptions can be found to many of the statements that will be made in the following sections.

Let us look first at the role of energy and nutrients in community succession.

Community Succession—Productivity and Nutrients

Consider a community at climax. Community respiration must be nearly equal to gross productivity, and thus, $P/R = 1$. If energy is imported into the community, as in the earlier example of Florida's Silver Springs, the input will be balanced by the export of energy. At climax, then, there will be little or no accumulation of biomass.

Now consider a community in the early stages of development. A field, two years after abandonment, can serve as an example. Biomass would be accumulating, production would be greater than respiration, and the P/R ratio would be greater than 1. We could expect net community productivity to rise rapidly as the field was colonized by plants and then to begin to decline as herbivores moved into the com-

TABLE 14.1 Stages of Community Development and Succession

All in kcal/m²/yr	*Early* Pine Plantation (England)	*Medium* Oak-Pine Forest (New York)	*Climax* Rain Forest (Puerto Rico)
Gross Production	12,200	11,500	45,000
Plant Respiration	4,700	6,400	32,000
Net Production	7,500	5,000	13,000
Animal Respiration	4,600	3,000	13,000
Net Community Production	2,900	2,000	0
% Gross Prod Converted to (NCP/GPP) Biomass	23.77	17.40	0
Accumulation of Biomass	Yes	Yes	No
P/R Ratio	12.2/9.3 = 1.31	11.5/9.4 = 1.22	45.0/45.0 = 1

(Adapted from Eugene P. Odum, *Fundamentals of Ecology,* 3d ed. (Philadelphia: W. B. Saunders Company, 1971), p. 46.)

munity in greater numbers and began utilizing more and more of the productivity of the plants. The total biomass of the community would increase in the initial stages of development, as would leaf litter and organic matter in the soil. Thus, the ratio of productivity to biomass (*P/B*) would initially be high. Then, as detritivores, decomposers, and herbivores moved into the community, the *P/B* ratio would decline.

Some actual measurements of productivity and respiration are shown in table 14.1.

Thus, we may see from table 14.1 that in the early stages of community development a significant proportion of the energy in the community is used to produce biomass. In the later stages of development, a decline occurs in the proportion of energy that goes into biomass production. At climax, most of the energy in the community is utilized in plant and animal respiration.

Therefore, it appears that the climax reaches an equilibrium with regard to energy flow. All of the energy that enters is utilized by the community. Almost no energy is stored in excess biomass or exported from the community, and the only losses of energy are those that can normally be expected from plant and animal metabolism. At climax, the community is in a state of dynamic equilibrium.

The idea that a climax reaches an energy equilibrium has been borne out in some laboratory experiments. Such experiments usually are performed on *microecosystems*—that is, on very small, and theoretically representative, samples of systems taken from nature. In one such experiment, samples taken from a pond were cultured in large flasks. The samples contained phytoplankton and various small heterotrophic organisms. The flasks were plugged with cotton to permit gas exchange, and light energy entered the system through the glass. It was found that such systems maintained themselves at an equilibrium for indefinite periods of time. Thus, they could be thought of as miniature climax ecosystems.

Succession was found to occur when samples from the "climax" flasks were injected into fresh media in smaller flasks. Initially, daytime gross production was found to exceed nighttime respiration, allowing biomass to accumulate. Thus, *P/R* was greater than 1. Productivity and respiration rose for about thirty days, and subsequently, both rates began to decline. After sixty to eighty days, total daily productivity and respiration became approximately equal. Therefore, *P/R* equaled 1, and biomass no longer accumulated. (See fig. 14.8.) Biomass production was measured in terms of grams of carbon produced. These measurements were compared to the quantities of carbon in the system in order to determine the relationship between productivity and community development. It was found that in the initial stages of succession, 1 gram of carbon was produced per day for every 20 grams of carbon supported by the productivity. As the system approached its climax, the ratio fell dramatically, as only 1 gram of carbon was produced for every 100 grams of carbon supported. Thus, it was concluded that at climax, larger populations of organisms could be supported by overall lower rates of productivity and respiration.

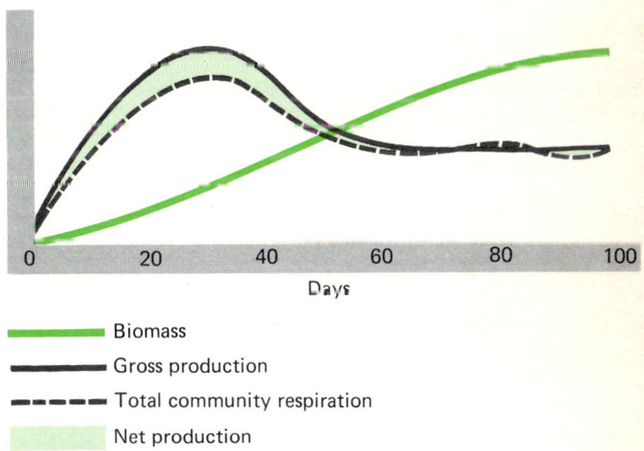

FIGURE 14.8 As described in the text, the changes in community productivity, respiration, and biomass of laboratory microecosystems follow a very regular pattern. *(From E. P. Odum, 1969.)*

It is much more difficult to obtain measurements of succession in natural systems, such as forests, than in laboratory microecosystems. The rate of change is much slower, and measurements must be made over several generations. The complexity of the system is far greater, and therefore, many more variables exist that can obscure the nature of the relationships between species and their physical and chemical environments. As a means of overcoming the problems associated with the long time spans involved in studying forest ecosystems, some students have developed computer simulations of ecosystem development. These simulations are based on mathematical models such as we mentioned in Chapter 10. The models seem to indicate that the relationships between productivity, respiration, and biomass accumulation in successional forests proceed in a way similar to those found in laboratory microecosystems (although it is difficult for the researcher to avoid programming his expectations into the model). (See fig. 14.9.) Initially, *P/R* is greater than 1, and biomass accumulates. As the climax is approached, productivity declines, the *P/R* ratio approaches 1, and the accumulation of biomass ceases. Intuitively, we could expect other systems to behave in the same way.

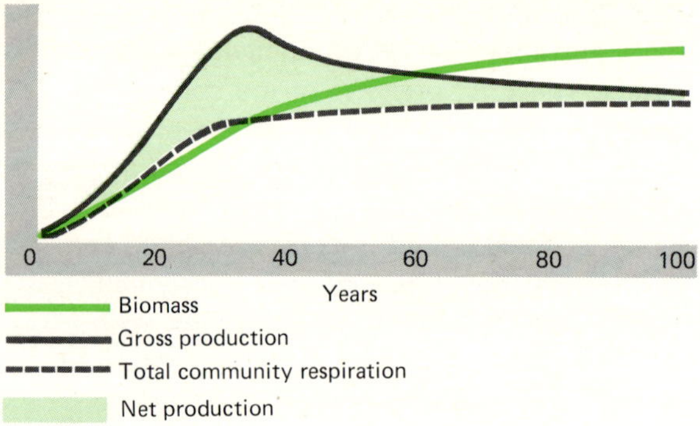

FIGURE 14.9 The relationships between productivity, respiration, and biomass that are predicted by mathematical models of natural communities are similar to the relationships observed in laboratory microecosystems. We must keep in mind, of course, that the models are theoretical and may not represent what happens in the "real world." *(From E. P. Odum, 1969.)*

We should point out again that biomass is a measure of population density that does not take into account the numbers of individuals. The accumulation of biomass in the early stages of community succession is analogous to the growth of populations as described by the exponential equations of growth. As the carrying capacity is approached, the growth levels off in the manner described by the logistic equation. The curves representing biomass in figures 14.8 and 14.9 resemble the growth of many species populations.

As we pointed out at the beginning of Chapter 11, energy alone does not produce growth in individuals or populations. Nutrient materials are essential for growth, and we shall now consider the role of nutrient cycling in community succession. Initially, of course, most of the nutrients in an ecosystem are in the soil. However, as plant populations grow, more and more nutrients become bound up in biomass. Thus, the main nutrient pool shifts from the physical to the biological component of the ecosystem. This means that there is proportionately less nutrient material available to producers, and that the efficiency of nutrient use must increase. If the efficiency of nutrient use cannot or does not increase, productivity must decrease since the nutrients necessary for growth are not available to producers. The shift in the nutrient pool, then, may be one of the reasons for the decline of productivity in the later stages of succession. The shift in nutrient availability may by itself be an important cause of succession.

In the early stages of community succession, the species present are usually made up of relatively small individuals having simple life cycles and rapid rates of reproduction (*r* strategists). As the populations grow and the nutrients in the community become bound up in biomass, larger individuals of the same species (or species made up of larger individuals) are conferred a competitive advantage, since they frequently have a greater capacity to store nutrients. They also usually live longer and can survive periods during which nutrients are not readily available. It is speculated that such species may be adapted to exploiting periodic releases of nutrients and are thus better able to survive and grow than are species that require a steady supply of some nutrient.

In the early stages of community development, nutrients are more likely to be removed from the community by allogenic processes. The import or export of nutrients can have a greater effect on the rate of development at this time than at later times. In later stages, nutrient turnover in the community may be more rapid, but the cycling is closed; therefore, allogenic processes may not have the same effect in the community as at earlier stages, since nutrients are cycled from organism to organism.

By briefly contrasting productivity and nutrient cycling in early and climax stages of community development, we can see a clear tendency toward the achievement of equilibrium. In the initial stages of development, the energy that enters the community produces a measurable quantity of excess biomass. As the community develops, however, the P/R ratio approaches 1, and biomass ceases to accumulate. The energy available to the community is equal to the energy consumed by autotrophic and heterotrophic respiration. Nutrients, at first bound up in the physical component of the system, shift to the biological component, and nutrient cycles close as the flow of nutrients is increasingly dominated by biological interactions. Thus, the effects on the community of allogenic factors are reduced.

An increase in biomass in an ecosystem may reflect the rapid growth of a single species population. However, even the most casual examination of a climax will reveal that it consists of a relatively large number of different species. We shall consider this aspect of community development next.

Distribution of Population Numbers —Species Diversity

During succession, the changes in productivity, respiration, and biomass are intimately related to changes in the food chains and feeding interactions in the community. Initially, the only food in a community is living plant material. The immigration of herbivorous insects follows, and the deaths of both plants and insects pave the way for the introduction of detritus feeders and decomposers. Herbivorous animals also move into the community, and these are followed, of course, by carnivores. In short, as a community grows older, it provides a wider variety of resources that may be exploited by an increasing number of species.

The increase in species diversity at the producer level may even affect the nature of the physical environment. As the number of plant species increases, plant transpiration may alter the *microclimate.* The microclimate is *the climate in the immediate vicinity of an organism or object.* Microclimates bear definite relationships to the general climate but may vary considerably from it.

In addition to altering the microclimate, plant species in a new community provide a habitat for other organisms in the form of food resources, nesting sites, and shade. These habitat effects increase the potential diversity of the community, allowing species with specialized niches, including specialized plants, to migrate into the community. The interactions among the species in the community grow more complex, and simple food chains are replaced by more complex food webs. Once complex webs are established, it has been found that much of the energy in the community is processed in detritus food chains. One study of a mature forest community, for example, showed that less than 10% of the net annual production was consumed in the living state. The remaining 90% became detritus and was utilized by the more poorly understood detritivores and decomposers.

As the community progresses through the various stages of succession, detritus and decomposer food chains become increasingly important and thus may be said to be characteristic of more mature communities.

Thus far, we have discussed species diversity as if it were simply a measure of the number of different species present in a community. We should point out, however, that diversity consists of more than just the number of species present in a community; it also involves the *relative numbers of individuals of each species* in the community. Imagine two communities, each containing ten species. Were we to consider the number of species alone, the two communities would be equally diverse. Let us say, though, that 90% of all the individuals in community A are of one species, and that in community B each species contains an equal number of individuals. If we were to take random samples of ten individuals from each of these two communities, it is probable that *on the average* the sample from community A would contain nine individuals of one species and only one of another species. The samples from community B would *on the average* contain an individual of each species in the community. Thus, community A is considered less diverse than community B.

Ecologists have recently paid much attention to species diversity, and a number of quantitative indices have been proposed which would allow numerical comparisons to be made between ecosystems. We need not go into the details of these indices, but we should point out that they *may* eventually become useful tools for examining the health of ecosystems. Most of the proposed measures indicate that diversity increases as a community matures. There are usually more species in mature communities than in young communities, and there is less of a tendency in mature communities for one or two species to greatly outnumber others. However, even in mature communities, there are usually major species present that contain greater numbers or biomass than the other species. In contrast to communities in mature stages, pioneer communities tend to have fewer species present, and of those, a very few often have very large populations.

Niche Specialization in Community Succession

As the species diversity of a community increases, the variety of available habitats increases. Consider, for example, the oak-hickory climax mentioned earlier. In such a community, it is possible to see a number of strata, each of which provides a unique set of conditions that favor different species. (See fig. 14.10.) At the ground level, there is a stratum of leaf litter where detritivores and decomposers thrive. Growing in and on the litter are mosses, ferns, a variety of short herbs, and some taller ones, all providing food resources, cover, and nesting sites for small, ground-living animals such as voles. Above the taller herbs are found a variety of shrubs and some small trees that may provide nesting sites and food for small birds and food resources for browsers, such as deer. The next level is formed by an understory of small trees that may also provide food for deer and nesting space for some species of birds. The larger trees may be divided into two or more strata consisting of the lower or upper branches where different birds and small mammals such as squirrels may find nesting sites and food resources.

Each species of plant, of course, has its own characteristic transpiration rate, and thus, the microclimates, and particularly the relative humidity, at each stratum may differ somewhat. The particular density of shrubs at any location in the forest will depend on the amount of cover, and therefore available light, at the understory level. The density of shrubs and small trees will in turn determine the light available at the leaf-litter level.

Thus, in a climax, a wide variety of habitats are available and may be occupied by species hav-

FIGURE 14.10 The vertical structure of a climax forest community provides habitats for a wide variety of plants and animals, and thus encourages a relatively high diversity of species.

ing quite specialized niches. The variety in the food supplies, microclimates, nutrient availability, and light availability, to mention only some of a species' requirements, is produced primarily by the community itself—that is, by autogenic processes. As a community develops, the increasing variety of habitats leads to colonization by organisms with more specialized niches.

One study of insects in an oak-hickory climax showed that of 240 species collected, 181 species (approximately 75%) were found in one stratum. Of the remaining 59 species, 32 (13%) were found in two strata, and 19 (8%) in three strata. Only 4% could be found in all strata of the community. A study of a coniferous forest in New England showed that three species of warblers of the same family could be found at distinctly different levels in the crowns of trees. Magnolia warblers (*Dendroica magnolia*) occupied the lower levels of the trees, black-throated green warblers (*D. virens*) occupied the middle levels, and blackburnian warblers (*D. fusca*) occupied the upper levels.

The increase in niche specialization is accompanied by another adaptation mentioned earlier—the shift in a community from populations of *r* strategists to populations of *K* strategists. In the initial stages of community development, *r* strategists can move in and undergo the rapid population growth that is characteristic of them. As the community develops, however, the conditions that favor *r* strategists are removed. The habitat is no longer free of competitors. The increasing variety of the developing community favors populations that are adapted to slower growth and continued existence at, or near, their carrying capacities, or *K* strategists. As must be clear from the previous chapter, a climax is usually characterized by populations that have a high degree of positive interactions, such as mutualism, protocooperation, and commensalism. Negative interactions, such as competition, parasitism, and predation, have "evened out" and produced stable populations.

To summarize community succession as thus far presented, we may think of a mature ecosystem as one in which (1) more highly specialized organisms are found, (2) the different species are well-adapted to each other, (3) there is an optimal utilization of productivity, (4) there is a minimal waste of yield, and (5) most of the materials in the community are circulated through the life cycles of the species present.

Community Succession and Homeostasis

As a community develops, its species composition changes and "adjusts," ultimately producing a stable, diverse climax. The resulting mature system is relatively unchanging, except for very long-term evolutionary changes, and the population interactions, nutrient cycles, energy budget, and other processes in the community approach a steady state. A comparison may be made between the balance achieved in such systems and the steady states achieved in organisms through the mediation of homeostatic mechanisms such as those described in Chapter 5. (See pp. 163–164.) As in the organism, should something disturb the balance within the community, homeostatic mechanisms will return the system to a balanced condition. Obviously, such mechanisms cannot be simple, and they are of the utmost importance in the human management of natural communities. We should point out here that homeostasis *does not imply an absolutely unchanging quality,* but rather means that as changes occur in the conditions of a community, the response of the community enables it to restore the balanced condition and thus to survive. As the term *homeostasis* is usually used, it also implies that definable limits exist within which a community may change and survive, but that beyond those limits, no response by the community can restore it to a steady state.

Thus, just as the metabolism of an individual organism is controlled by homeostatic mechanisms, so too are the processes that take place within a community. In the complex food webs that characterize climax, if a population of one organism declines, its predators may shift to another food resource, and the prey population may then begin to grow again. Or, short-term changes in the general climate of a community can produce changes in the community microclimates, as for example, during periods of drought. But since the microclimate of a community is in part the result of autogenic factors, adjustments may be made to restore equilibrium. As we pointed out in Chapter 11, when photosynthesis is taking place at a high rate, transpiration is also; thus, during sunny, dry weather, plants may maintain the relative humidity within a community. In a climax, then, the community operates in such a way as to stabilize not only the populations of species, but also the physical environment.

Obviously there are limits beyond which the homeostatic mechanisms of a community will fail to restore equilibrium. Approximately thirty years ago, a severe windstorm felled a large proportion of Engelmann spruce trees in a forested area of the Colorado Rockies. Within the community lived a bark beetle that laid its eggs in the bark of older trees, thus killing off the older trees and making room for younger growths of spruce. The beetle population itself was kept in check by the woodpecker population and thus did not reach such proportions that the beetles would attack younger trees. After the storm, however, the beetle population rose rapidly, since the woodpeckers could not reach them in the tangle of branches among the fallen trees. The beetle population grew so large that they were able to attack even the younger trees and kill them. The effect of the beetles was even more devastating owing to the fact that they carried a fungal disease that also attacked the trees. Since the woodpecker's rate of increase was slow relative to the beetle's, its control was ineffective. By the time the woodpecker population could "catch up," it was too late. In short, the environmental conditions changed so radically in so brief a time that the homeostatic mechanisms maintaining the equilibrium of the community failed.

The implication of homeostasis in the climax is great stability. We may explain this stability through the use of an analogy. Imagine a ball at the top of a hill. If we were to push the ball slightly, it would roll down the hill, but it would not return. If the ball were at rest at the bottom of a steep-sided valley and we were to push equally hard, it would come back to rest quickly with few oscillations. If the valley were shallow, a ball pushed with the same force would also come to rest at the bottom, but the magnitude of the oscillations would be greater. The first instance, the ball on the hill, is an example of an unstable system; the ball cannot return to its original state. The second and third instances are examples of stable systems; the ball returns to its original state. The steep-sided valley may be thought of as a more stable system than the shallow valley, because the original state is restored with smaller oscillations. Examples of oscillating stable systems are common in nature. The beating of a heart or the alternating expansion and contraction of the lungs are stable oscillations. The whole physiology of organisms is a stable oscillating system, and when disease occurs, it produces instability by greatly increasing the magnitude of the oscillation. In the biology of communities, a plowed field is an unstable system, one that undergoes continuous change (succession) to produce a stable climax community. Evolution nearly always acts to produce homeostatic, stable, dynamic systems.

Thus, we have seen that the system produced by the process of succession is a diverse, stable climax that through homeostatic mechanisms achieves a condition of dynamic equilibrium similar to the dynamic stability of individual organisms. We will now look more closely at the types of climax that are produced.

The Distribution of Communities at Climax

Thus far, we have discussed only a few examples of succession. Although it is clear that the process of succession is very common, perhaps universal, it ought also to be clear that the climax that results is not the same in every environment. The nature of the particular climax that will develop in any area depends on a variety of geological and climatic factors. It would seem to be possible to define major climatic climaxes typical of very large areas, often known as *biomes*, and to define local variations of the climax within these large communities that are caused by local variations in environmental conditions.

In our discussions of succession in the Lake Michigan dunes and the upland Piedmont region, we described two different climaxes: beech-maple and oak-hickory. When we discussed the succession of a pond, we also noted that once the bog mat was consolidated, the area would go to a climax typical of the particular geographic region in which the pond was located. If we ask, however, what the particular climax for a given geographical region is, we are again faced with a difficult question to answer. The theory that all successional stages ultimately result in a stable, diverse community can be supported by observation, but there are two ways to interpret such observations. The first interpretation is that within any given geographical region, all communities develop toward the same climax, although the rates of succession may

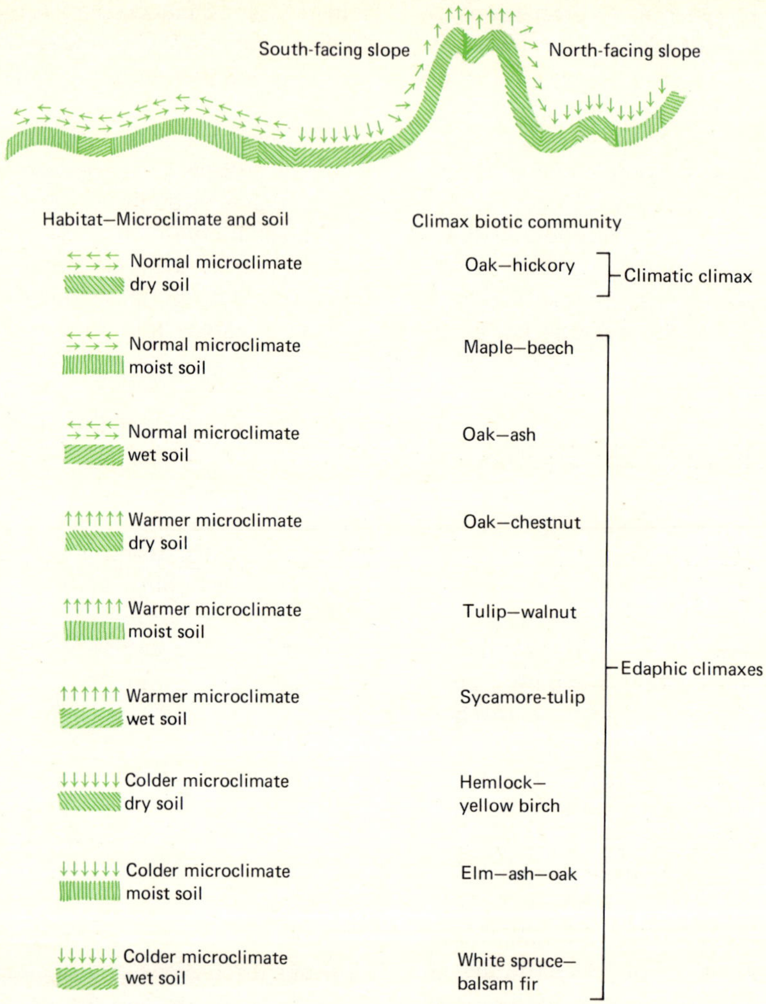

FIGURE 14.11 A variety of edaphic climaxes with differing dominant vegetation may develop, depending upon soil type and microclimate. These examples are for the communities of southern Ontario. *(After E. P. Odum, 1971.)*

vary greatly and thus produce a region that appears to be made up of many climaxes. The other interpretation is that the physical conditions within geographic regions vary so much that we cannot expect the same type of climax community to result throughout the entire area. It is unlikely that the controversy will be resolved within our lifetime, since the time required by a community to reach its climax is usually measured in human generations.

A compromise has been suggested by some ecologists. They propose that there are *climatic* and *edaphic* climaxes. The climatic climax is *a community that is in equilibrium with the gen-*

eral climate of the area in which it is found. An edaphic climax is defined as a *community that is modified by local conditions of the substrate or possibly by the local climate.* Indeed, under some conditions, the climate in a particular locality may deviate significantly from the general climate of a region. Mountain tops, for example, are usually cooler and more humid than the lower lying areas in the same region. The utility of this compromise lies in providing us with a constant, the climatic climax, against which we may measure certain variables, the edaphic climaxes. It also provides us with a basis for gauging the possible effects of human intervention in natural ecosystems. If we adopt this point of view about natural systems, it becomes clear that the loss of a wilderness area is not just an esthetic concern, but a scientific concern as well, since we lose a resource from which we could learn how better to manage those ecosystems upon which our survival may directly depend.

Two brief examples of climatic and edaphic climaxes will clarify the concept. In southern Ontario, on well-drained but moist soil substrates with normal microclimates, the community most usually found is dominated by beech and maple trees. Thus, the beech-maple community is considered to be the climatic climax for this region. Other climaxes are found, however, that are dominated by other associations and can be correlated with differences in the soil substrate and the conditions of the local microclimate. For example, when the soil substrate is dry and the microclimate normal, the usual association is an oak-hickory community such as those found in the southeastern United States. Figure 14.11 shows the other edaphic climaxes and the soil moisture and microclimatic conditions that produce them.

According to the theory that all communities in the same general region will ultimately develop to the same climax, the oak-hickory communities found in southern Ontario will ultimately produce sufficient organic matter to raise the moisture capacity of the soil, and then they will succeed to beech-maple associations. However, according to the compromise theory, other factors may prevent that from occurring. If the oak-hickory community, or any other edaphic climax is subjected to environmental extremes, it is unlikely that the succession will proceed. If, for example, the edaphic climax is based on the side of a steep hill or ravine where water runoff is rapid, the oak-hickory community may never be able to produce enough organic matter to alter the moisture retention capacity of the soil. Thus, the succession to a beech-maple community is blocked. From this point of view, considering man's life span and the probable time required by communities in extreme situations to reach a climatic climax it is more practical to consider edaphic climaxes as stable communities which will probably not be altered until changes occur in the regional climate or the geology of the substrate.

Another example of climatic and edaphic climaxes can be seen in the pygmy forests and redwood communities of the Mendocino area of California. A transection of the California coastline (see fig. 14.12) shows that in this region of northern California, large communities of redwoods can be found side by side with forests of stunted coniferous trees. These pygmy forests are the result of the presence of a layer of impenetrable soil, called *hardpan*, that lies about 18 inches below the soil surface. The hardpan layer prevents the development of the extensive root systems needed by large trees and restricts the movement of water and nutrients. The soil above the impervious hardpan is low in nutrients and extremely acid (pH 2.8 to 3.9). Thus, the type of community that can develop under these conditions is quite different from the sort that can grow in adjacent areas that lack the hardpan layer.

The distribution of climatic and edaphic climaxes may give a region the appearance of a "mosaic" made up of a number of very different communities, produced by differences in the soil, which are in turn produced by the uneven distribution of geological features. (See

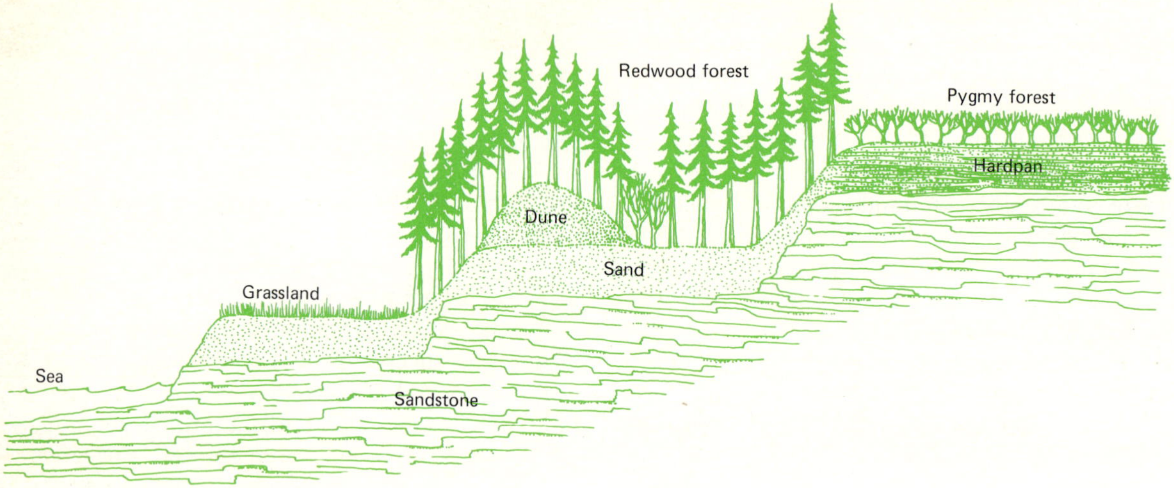

FIGURE 14.12 The redwood and pygmy forest communities of the California coastal area form another good example of edaphic climaxes. *(From E. P. Odum, 1971, after Jenny* et al.*)*

FIGURE 14.13 Although much of the mosaic of communities shown here is due to farming, many of the communities are also edaphic, caused by differences in geology and soil moisture. The bogs, such as in the lower left, and the differences in hardwood forest represent edaphic climaxes. *(Courtesy of Grant Heilman.)*

fig. 14.13.) Although the mosaic quality of community distribution may be due to variations in geology and other local conditions, and may be produced by the juxtaposition of climatic and edaphic climaxes, we must be careful to distinguish these communities from slowly changing subclimaxes. Under many conditions, the oak-hickory forest may be slow in going to the beech-maple climax, or the pine forest that may follow a forest fire may last a very long time before being replaced by a deciduous climax.

Moreover, many subclimaxes may be maintained by man or by recurrent natural events. Foresters often cut pine woods whenever a cash crop of lumber is available, thus artificially maintaining the pine subclimax and also, of course, its potential for economic yield. Light grazing by cattle, sheep, or goats, occasionally combined with management practices such as controlled burning, may maintain a community in a state resembling an "old field" community. Most biological communities in the temperate zones of the world have been so strongly influenced by man that few areas of real climax still exist.

Communities resembling climaxes in their longevity but differing from the regional climatic climax may also be maintained by periodic events. The pine forests covering much of the southeastern coastal plain of the U.S. are maintained by periodic fires. Although such fires occasionally occur naturally, they are usually set by man, and they are often deliberately set by foresters to maintain maximum lumber yields. Crown fires are usually destructive of the entire community, but light surface fires burn off competitors to the pines and hasten the breakdown of bodies of plants, thus aiding microbial decomposition and the cycling of nutrients in the community. (See fig. 14.14.) The pine and scrub oak communities of Cape Cod seem to be maintained by periodic fires and severe storms. The beautiful heaths and moors covering most of Highland Scotland were once forested and are now, in part, maintained by sheep grazing

FIGURE 14.14 In order to maintain productive and economically valuable stands of pine, foresters in the southeastern United States usually practice controlled burning of the undergrowth. This artificially duplicates what may have been a frequent natural occurrence in the area since pre-Columbian times. *(Courtesy of the United States Forest Service.)*

and burning. Such an ''artificial'' climax is called a *disclimax*, and to the casual observer it may easily be confused with an edaphic climax.

Thus, we can see that the communities of any region may be maintained by a variety of factors and may or may not be approaching the regional climatic climax. Nevertheless, all of these communities share many of the characteristics outlined earlier, and all of them are composed of organisms evolved to a particular situation.

Predicting what will occur in the process of succession is extremely difficult. One of the problems that arises in making predictions is that our theory of ecosystem development is based on relatively short-term observations of systems that undergo change only over very long time spans. Where historical information about communities is available, it is often unreliable or incomplete. The rates of succession in most communities, as we said earlier, may be measured in human generations, and the lack of reliable data from the past makes it difficult to predict the future.

The Evolution of Ecosystems

In Chapter 7 we discussed in detail the evolution of individual species, and we saw that as the frequencies of alleles in the gene pool change, the niches of the species change. We also saw that speciation is frequently the result of geographic isolation. In the last two chapters, however, we have implied that few selective factors act on one species alone. Similarly, geographic isolation will usually isolate more than one population. Thus, several species usually evolve together in response to the environmental changes affecting them. They also coevolve by interacting with each other. The mutual adaptations of parasite and host to minimize mutually disadvantageous effects is an example of the latter. Since there are many environments on earth that are climatically and geographically similar, many similar climaxes have evolved.

One can walk inland from a beach on the western coast of Australia and see a range of communities containing species that appear similar to those that can be seen while walking inland from a beach on the eastern or western coasts of the United States. Yet, the apparently similar species are totally unrelated. Grassland communities, deciduous forest communities, and desert communities have *independently* evolved on both continents. The species are different, and the original immigrants were different, but convergent evolution has occurred not only in individual species but in whole ecosystems as well. Similarly, many islands, such as the Galapagos, have evolved complete communities through coevolution of the individual species.

The ways in which species can ''solve'' environmental problems are limited, and thus convergent evolution produces similar species from totally different stocks. These species fill very similar niches in the communities of the biosphere. Consider some of the species found in Australia, for example. There is fragmentary evidence indicating that the first marsupials to colonize the Australian continent were mouselike and opossumlike ancestors of the more than one hundred species of marsupials found there today. Subsequent to the colonization by marsupials, the continent was isolated from the rest of the world, and except in South America and Australia, most species of marsupials became extinct around 25 million years ago. In Australia, however, they thrived, and at about the time that the marsupials in most other parts of the world became extinct, the Australian marsupials split into two lines—carnivores and herbivores (see fig. 14.15). Subsequently, a number of marsupial lines have emerged, and thus, there are marsupial analogues of mice, moles, cats, and wolves. Although they share the common names of some placental mammals, these marsupials are not at all closely related to their placental namesakes. In fact, the marsupial mouse differs in more than its reproductive system, since unlike the ordinary mouse, it is a predatory carnivore. Since the niches of most of the marsupials are similar to those of many pla-

(a)

FIGURE 14.15 Australian marsupials. Unlike the ordinary mouse, the marsupial mouse (b) is a carnivore; and, unlike the placental bear, the Koala "bear" (a) is strictly a herbivore. *((a) Courtesy of Scott Ransom from Photo Researchers, Inc.; (b) courtesy of Tom McHugh from Photo Researchers, Inc.)*

(b)

centals, the communities as a whole must also be similar. The important point is not that evolution has produced superficial similarities between unrelated species, but rather that great similarity has been produced between very different communities made up of unrelated species. The similarities that are of greatest importance to man may be seen by comparing communities in terms of stability, diversity, *P/R* and *P/B* relationships, nutrient cycling, homeostasis, and so forth. These community similarities are often also associated with the degree of stress influencing the community. Thus, we must examine the effect of such stress in some contrasting communities. That will allow us to say something about the way in which we should view our use of different major communities.

Stability, Stress, and Climax Ecosystems

We have seen that there are regular variations in *P/R, P/B*, diversity, niche specialization, and stability associated with the process of succession. These are summarized in table 14.2. We have also seen that there are a wide variety of climaxes such as biomes, edaphic climaxes, disclimaxes, and intermediate successional stages. We are now in a position to compare some of these in order to see what generalizations may be made regarding the differences between them. As we shall see, such a comparison may provide us with some important lessons for our use of biological resources.

We should also emphasize again that many of the generalizations that we have made are

Table 14.2 Trends to Be Expected in the Development of Ecosystems

Ecosystem Attributes	*Immature Stages*	*Mature Stages*
1. *P/R* ratio	Greater or less than 1	Approaches 1
2. Gross production/biomass (*P/B* ratio)	High	Low
3. Net community production	High	Low
4. Food chains	Linear	Weblike
5. Total organic matter	Small	Large
6. Inorganic nutrients	Extrabiotic	Intrabiotic
7. Diversity of species	Low	High
8. Diversity of habitat	Low	High
9. Niche specialization	Broad	Narrow
10. Mineral cycles	Open	Closed
11. Nutrient exchange rate between organisms and environment	Rapid	Slow
12. Role of detritus in nutrient regeneration	Unimportant	Important
13. Growth strategy	*r* strategists	*K* strategists
14. Internal symbiosis	Undeveloped	Developed
15. Nutrient conservation	Poor	Good
16. Stability	Poor	Good

neither rigorous nor well-proven. Just as the evolutionary process can only be inferred from observation, long-term community processes can only be inferred from measurements and observation. Community processes are not amenable to laboratory analysis, as is possible with the hypotheses of the physiologist or biochemist. Thus, much of what we have said in this chapter is the result of inference, and the generalizations and hypotheses must be confirmed and tested by further observation. Many may have to be changed. The following generalizations about climaxes are particularly subject to the previously mentioned limitations. Nevertheless, we may learn a great deal from such generalizations about how we should deal with climaxes.

Let us compare two sets of contrasting climaxes, the Arctic tundra and the tropical rain forest, and the marine intertidal community and the deep sea benthos. We shall see that for all their striking differences, the tundra and intertidal communities share a number of characteristics, as do the rain forest and deep sea benthos.

The tundra is a huge region encompassing the northernmost parts of the North American and Eurasian continents. In all, the tundra encompasses around 3 million square miles, or approximately 5% of the land surface of the earth. The conditions there are so harsh that no trees can develop, and species diversity is close to that of an "old field" community. (See fig. 14.16.) In winter, the sun does not rise, and in summer, it never falls below the horizon. Thus, the tundra is subjected to long periods of darkness and light. Winter temperatures are extremely low, sometimes falling to −57°C (−61°F), and summer temperatures are sometimes as high as 32°C (90°F). On the average, rainfall in the tundra amounts to 8 inches per year. Beneath the soil surface is a layer of *permafrost*, or permanently frozen soil, which may rise to within 4 cm of the surface or be as deep as one meter. In summer, the soil above the permafrost thaws. If the land is on an incline, the sodden, thawed soil above the permafrost may tend to slide. Such soil movements may only amount to 2 or 3 inches per year, but they are sufficient to interfere with root development. Thus, in the tundra, the development of deep rooted plants is prevented by both the impervious permafrost and the instability of the soil. The permafrost also keeps water from seeping deeply into the ground, and this, along with the very low evaporative rate of water in the cold air, makes the tundra, in spite of a very low annual rainfall, a very wet environment. Pockets of ice under the soil may also develop, push up the soil, and promote erosion by winds.

There are relatively few species of plants on the tundra. These plants are fed upon by a few species of birds, insects, and rodents. The herbivores are fed upon by carnivores such as snowy owls, foxes, and gyrfalcons (gull-like birds). During the summer, the tundra is also populated by migrating herds of caribou, by wolves, and by a number of species of birds that nest in the tundra and fly south in winter.

The populations undergo cyclic fluctuations of significant proportions. For example, the populations of lemmings, a species of rodent, grow exponentially for about four years and then undergo rapid decline. The population growth of the lemmings causes them to overgraze the land, even occasionally to the point of promoting erosion. When food supplies are exhausted, the lemmings starve in huge numbers and undergo chaotic mass movements that may or may not be migrations. Predation is extremely heavy as the populations of lemmings grow. Owls and other predatory birds may be seen flying over the mass of rodents and eating their fill, while on the ground, the populations are followed by foxes and wolves.

The growth of the lemming population promotes growth in the populations of its predators, chiefly the arctic fox and snowy owl. After the lemming population crashes, the fox and owl populations do too. In some instances, the lack of food has caused snowy owls to expand

FIGURE 14.16 The polygonal patterns in this photograph of the Alaskan tundra are caused by movement of the permafrost. The hillock (called a pingo) probably has an ice core. *(Courtesy of Steve McCutcheon, Alaska Pictorial Service.)*

their range into central and northern parts of the United States.

Thus, the tundra presents a situation of great physically, as opposed to biologically, controlled environmental stress. Much of the stress is produced by the extreme fluctuation in environmental conditions. This, then, results in a situation in which few species may survive, and those few that do survive usually have large populations. Productivity, *P/R,* amd *P/B* vary widely over the year and over the course of years, and populations are often unstable. Homeostasis, although effective in the long run, is not effective enough to prevent these population fluctuations.

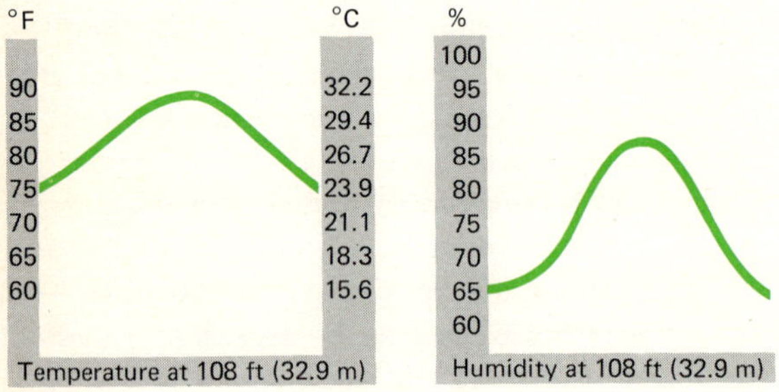

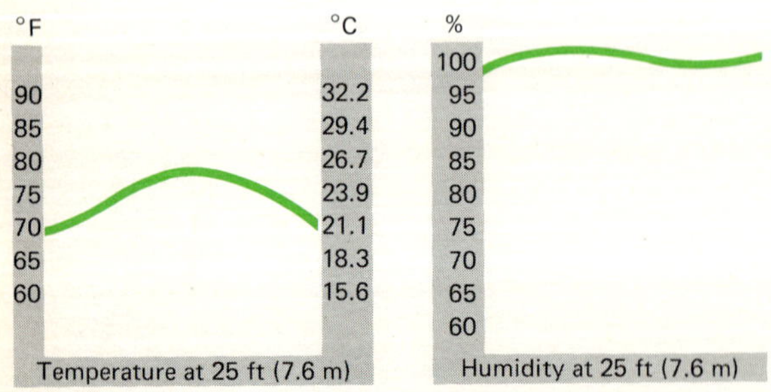

FIGURE 14.17 The tropical rain forest provides extreme control of the microclimate within the forest community. Daily temperature and humidity variations are almost entirely determined by the vegetation.

The tropical rain forest provides a striking contrast to the tundra. Mean annual temperatures in the region of the Southeast Asian rain forests are near 27°C (80°F), and compared to the tundra, there is little fluctuation from the mean. Although rainfall may vary from one year to another, there is no dry season as such; there are only humid and more humid times of year. In short, climatic fluctuation is nearly insignificant. Day length is also approximately the same throughout the year. Microclimates in the tropical rain forest may vary considerably, however. Measurements made in the rain forest of Sarawak indicate that at 25 feet above the ground, relative humidity remains constant at nearly 100%. Relative humidity in the canopy of the same forest, 108 feet above the ground, may fluctuate from 65% to 90% during the same time period. Temperatures may fluctuate at the 25-foot level from the low 70s to the high 70s, but at the 108-foot level, they vary from the mid-70s to the high 80s. Nevertheless, these differences may be largely attributed to the action of the community itself, not to general conditions of the environment. (See fig. 14.17.)

Species diversity in the tropical rain forest is extremely high. In the rain forests of Malaysia, for example, the different species of trees in two 4-acre plots were counted and were found to number 197 and 227. Sumatra and Borneo are estimated to contain approximately 3000 species of trees having trunks ⅓ meter or more in diameter. The number of species of smaller trees, shrubs, and plants may never be known.

Thus, the tropical rain forest is quite a contrast to the tundra. In the rain forest, the environment is uniform, and a tremendous diversity of organisms is able to survive. The forest produces a variety of habitats, thus allowing niche specialization and a further increase in diversity. There is much less fluctuation in productivity, respiration, and biomass, and *P*/*R* approaches one. Nutrient cycling in the rain forest is very efficient, in contrast to that in the tundra, with a rapid turnover of most of the nutrients. Most of the nutrients in the tropical rain forest are bound up in biomass. There is little population fluctuation, and homeostatic mechanisms are very effective. In many ways, the tropical rain forest is the epitome of the stable climax community.

FIGURE 14.18 The intertidal zone, such as on the coast of Maine, is an area of extreme environmental variation. To some extent the large marine algae protect the smaller organisms from the environmental fluctuation. *(Courtesy of Roy A. Gallant, Science Photo/Graphics Ltd.)*

Roughly analogous to the tundra is the temperate intertidal zone, or that portion of the shoreline along the oceans that lies between the highest and lowest points of tidal inundation. (See fig. 14.18.) Obviously, the physical environment in the intertidal zone undergoes great fluctuations. When heavy rains fall, the community is washed with fresh water, and within hours, it may be inundated with salt water. During dry periods, the community may go from complete inundation to levels of relative humidity as low as 20%. On a spring day, extreme temperature variations are not unusual, and throughout the spring, extremes of temperature are the rule. Along the Pacific coast of

North America, temperature variations are not as extreme as on the Atlantic coast, but conditions of humidity are as variable.

The diversity of species in the intertidal community may approach the low levels of the tundra. Only a few species of algae, snails, clams, mussels, and other invertebrates may survive. Most of these possess adaptations to minimize the stressful variations of the physical environment. There is some specialization in the intertidal community, but for the most part, it is specialization to withstand stress and does not allow high diversity. Population fluctuations may be extreme. Whole beds of mussels and algae may be removed by winter storms. Barnacle populations become overcrowded and are often decimated. The ratios of P/B and P/R are high, because there is sufficient nutrient material and light, but wave erosion and wind remove a great deal of the production from the community. The populations of the intertidal community are often sources of detrital material for deeper ocean communities. A major difference between the intertidal zone and the tundra is that food is imported into the intertidal zone by the flooding tide. Nevertheless, the intertidal community and the tundra share many of the aspects of an immature ecosystem.

The deep sea benthos is a contrast to the intertidal zone. Environmental change in the benthic zone of the ocean below about 1000 meters depth is very small. Temperature variations rarely exceed 1 °C per year, and the salinity of the water and the water pressure are constant. (Pressures at different depths vary, of course, but they are constant at any particular depth.) There is darkness at all times, and the food in the community enters from above, usually as a fine ''rain'' of detritus. Recent investigations have revealed that on a transection from Cape Cod to Bermuda, from the intertidal zone to depths of 2000 meters, the indices of species diversity show a continuous increase. Though we do not know as yet what the niches of most of the deep sea benthic species are, it is speculated that they are specialists. The populations of the oceanic benthic community apparently remain stable, and although evidence is scanty, it is thought that homeostasis is high. Ratios of P/R and P/B are meaningless, of course, since no light energy enters the community. Nevertheless, the benthic community resembles the rain forest in terms of high diversity, low stress, and little environmental fluctuation.

Many other community comparisons might have been made. For example, we could have compared the intertidal zone to the coral reef, since the reef, like the rain forest, is a diverse, biologically dominated community in which populations are stable, many habitats exist, and niche specialization is prominent. Or we might have compared a desert to the rain forest, since the desert is a low-diversity, stressed community in which specialization tends to be a response to the rigorous physical environment.

Obviously, there are many intermediate stages between the extremes that we have presented, several of which are more familiar to the reader than the Arctic tundra or a tropical rain forest. We shall examine these intermediate stages in the next chapter.

A dichotomy appears to emerge from our observations. Some of the mature communities seem to be more similar to the early stages of community succession, while others fulfill the conditions described for climax communities. Is there any apparent cause for this? Why do some mature climaxes, such as the desert, the tundra, or the intertidal zone, have low diversity, physical control, large population fluctuations, fluctuations in P/R and P/B, and low homeostasis, while other climax communities, such as the rain forest, the benthos, and the coral reef, exhibit the opposite characteristics? And why are other communities intermediate?

It is fairly well accepted that communities in more stressed environments resemble early successional stages, but why this is so, and what is meant by stress are not as obvious. Consider again a plowed field or a burned forest. A wide variety of species exist that will start and maintain the sere to a climax. Many more species are adapted to the variety of conditions in less stressed environments, but few are able to maintain populations in environments with low rainfall, low temperatures, or other extreme conditions. If the species must also be adapted to a wide range of environmental conditions such as exist in the tundra or intertidal zone, still fewer will succeed.

The question arises as to why there are not more species adapted to very stressed environments. Every extreme environment, from caves to hot springs to the deep ocean, is occupied, and evolution has produced a tremendous number of adaptations. In an extreme environment, however, it is probable that few genetic combinations will provide for survival. If environmental conditions fluctuate over a wide range, the probability must be low that adaptations can arise that will permit survival under all conditions.

A few species, but very few indeed, have evolved that can survive under extreme environmental conditions or under conditions that fluctuate greatly. We can speculate that stressed environments will be colonized eventually, and that a very slow but continuing evolution will take place to stabilize them. Many ecologists disagree, however, and the question really remains unanswered.

The Effects of Mankind on Communities

There are probably no communities on earth that have escaped the direct and catastrophic, or indirect and subtle, effects of human activities. What can we learn about the effects of human activities from the perspective of the present chapter?

Consider two of the generalizations that we have made about communities. First, the destruction of whole biological communities or the availability of completely new environments sets in motion a process of succession from immature, low diversity, unstable, inefficient communities to mature, stable, diverse, and efficient ones. Second, drastic environmental fluctuations produce stress that lowers the diversity, increases instability, and lowers the efficiency of communities. Climaxes under stressful conditions may resemble immature communities.

Many, perhaps most, human activities produce these conditions. We have already discussed the effects of plowing and burning, which are activities upon which man is dependent. The application of pesticides, heavy fishing pressure, selective forest cutting and clear cutting, pollution of rivers, lakes, and estuaries, and many other human activities produce stress that lowers community stability and diversity and brings about the other effects discussed earlier. Many of our activities, whether they are essential to our survival or not, produce community changes that prevent homeostasis and balance and lower the efficiency of energy transfer and nutrient turnover. Only rarely do we consider the effects of our activities from the viewpoint of community and ecosystem ecology.

Many of our agricultural activities deliberately utilize the results of inefficiency and instability caused by environmental stress. Recall that P/R and P/B ratios are highest in early successional stages and in stressed communities, and that high P/R and P/B ratios result in the accumulation of biomass. Humans harvest the accumulation of biomass and export it from the community, thus maintaining the system in a low diversity, low stability stage. This accounts for the yields of grazed rangeland, selectively cut forests, controlled clam and oyster beds, and so forth. The most extreme conditions occur when we practice crop monoculture, in which there is no diversity and no long-term increase in biomass, and all stability is maintained by our agricultural practices. We have also artificially selected strains of domestic species having the highest P/R and P/B ratios in order to increase yield. Nutrient cycling is almost nonexistent, since crops are harvested, and nutrients must be artificially replaced.

The lower diversity and stability that are necessary for agricultural yield produce secondary problems. Outbreaks of insect pests, weeds, crop blight, and parasites are typical of the instability of immature or stressed communities. These outbreaks must be controlled if high agricultural yields are to be maintained, and thus herbicides and pesticides are applied. But, as we have seen, these chemical agents have well-known secondary effects on other communities, and although they enable us to maintain stability in farmed regions, they increase instability elsewhere. We are faced, then, with several dilemmas. How do we maintain both high *P/R* and *P/B* ratios and satisfactory stability? And how do we maintain stability without the undesirable effects of pesticides and herbicides? How do we feed an increasing population while maintaining "reasonable" biological stability? Once again we have to say that there are no easy answers to such questions. We must grow crops, and we must harvest certain biological resources. We cannot completely eliminate the effects of lowered diversity and instability. At best, we may be able to minimize such effects whenever the demands for human survival dictate our engagement in stress-producing activities. We may be able to prevent unnecessary disruptions of natural communities in some instances.

Many of the methods used to prevent the effects of instability have worked. But some of these methods, such as biological insect control and certain soil conservation and wildlife management practices, have usually been directed at one species or resource and have ignored community effects. Thus, pesticides may kill pests, but they may also kill other species that are necessary for the stable functioning of the community. Or they may remove predators or competitors and thus allow the development of outbreaks of other pests. They may also remove food resources of desirable species such as birds, thereby "starving out" the normal pest-controlling predator. The proper fertilization of a field will increase its agricultural yield, but it may also cause eutrophication in adjacent bodies of water. Selective forest cutting may maintain a disclimax community at the highest lumber yield, but it may also irreversibly change the nature of the soil. The problem is, therefore, that although we can recognize the effects of specialized control procedures, we usually do not understand community processes well enough to suggest better alternatives. It is possible that some of the computer simulation models now being developed will allow us to manage communities in a better way.

At present, we are unable to avoid many of the negative effects of community instability. However, many of the disruptions caused by man are avoidable, since they are often traceable to ignorance, carelessness, or irresponsibility. We have discussed the effects of cultural eutrophication brought about by the careless disposal of nutrient wastes. In the time it took to write the last few pages of this chapter some 3000 acres of forest in Big Sur, California, were destroyed by a carelessly set campfire. The military use of defoliants, the discharge of heated effluents from power generating stations, "spilled" pesticides, oil pollution of oceans and estuaries, the smog that is killing vegetation in the Los Angeles basin and impairing human health in Boston, New York, Chicago, and other American cities, the building of housing developments on flood plains and steep hillsides, the increased salinity of groundwater resources caused by highway snow removal programs, and many other activities of man that are generally classified as "polluting" are common, avoidable causes of community and ecosystem stress and instability. In some instances, the effects are subtle, as in increased groundwater salinity, and sometimes they are obvious, as in forest fires and military defoliation. Some of these activities are unavoidable; some, such as our use of forest resources, are even necessary. Many are avoidable, although such avoidance may be expensive.

If the effects of our activities are unknown, then we should determine how potentially dangerous or disruptive they might be. If the effects are known to be disruptive, and if we judge the expense of eliminating or minimizing those effects to be too great to pay, then we are acting irresponsibly if we persist in engaging in the activity. As a nation, we spend tremendous amounts of money, talent, and human energy on defense, space, and highway programs and comparatively little on overcoming our lack of knowledge concerning the most common features of the earth we live upon.

If we are to manage our biological resources intelligently, the immediate tasks are neither simple nor quick. We must identify which of our activities may have stressful effects; we must study the activities carefully; we must test the expected effects in controlled situations; we must monitor the quantities of potential pollutants in the environment; and we must decide on the degrees of change due to stress that we would find acceptable. The achievement of these goals will demand a great deal of careful ecological research and will require an informed public that is capable of making sound decisions.

We may conclude that our effects on most of the communities of the earth have been dramatic. Sometimes the changes that man has made have been controlled and desirable. In many cases, however, they have been unexpected and undesirable. Nearly all of the changes have had effects on biological communities that operate more or less on the principles brought out in this chapter. We must develop means of predicting the changes that might result from our activities and determining which changes are tolerable and should be encouraged and which are unacceptable and should be discouraged. Such decisions will affect us all and will have to be made by well informed citizens. The alternative is ignorance, irresponsibility, and environmental deterioration.

We have made many generalizations in this chapter that apply to a wide variety of situations. Obviously, we must look more closely at the ecology of some natural communities and the effects that we have had on those communities. In the next chapter, we will examine in more detail some of the major communities, or biomes.

Selected Readings

*Hazen, William, ed. *Readings in Population and Community Ecology.* Philadelphia: W. B. Saunders Company, 1970. A collection of articles from the original literature. Several of the topics covered are relevant to Chapter 10.

Teal, John and Mildred. *Life and Death of the Salt Marsh.* Boston: Little, Brown and Company, 1969. A superbly written account of succession in an eastern coastal ecosystem.

*Whittaker, Robert H. *Communities and Ecosystems.* New York: The Macmillan Company, 1970. A brief, somewhat rigorous treatment of community ecology with an emphasis on plants.

Deevey, Edward S. ''Bogs.'' *Sci. Am.,* Vol. 199 (October, 1958), pp. 114–116. Offprint number 840.

*Paperback edition available.

biomes — the major biological communities of the world

chapter fifteen In the early chapters of this book, we discussed the factors that control, regulate, and determine the functioning of organisms and many of the principles we discussed were sufficiently general to apply to most organisms in most environments. In the last four chapters, we have described the properties of ecosystems, and although some of the principles we discussed apply only to specific locations, many can be applied to a majority of ecosystems and communities. The readers of this book probably have had in mind specific communities or ecosystems with which they are especially familiar and about which they may wish to learn more. Clearly, it is impractical for us to deal with that many communities. Instead, we shall discuss the major communities of the world, and even then our discussion must be superficial. Nevertheless, we shall see that these major biological communities, or *biomes*, are very important to man. Furthermore, understanding them will help the reader to understand more completely the communities which are already familiar.

The ecology and function of the taiga is determined largely by harsh winters and large amounts of snowfall. *(Steve McCutcheon, Alaska Pictorial Service.)*

Aside from practicality, there is another reason for considering biomes rather than smaller, more specific communities. We can no more afford to be isolationists in our understanding of biological communities than we can be isolationists in our political involvements. For good or ill, American economic interests, technological capabilities, and resource use have global implications. The United States, in conducting a war in Southeast Asia, developing oil resources in the arctic tundra, dredging for manganese ore on the deep ocean bottom, or mining copper in the Rocky Mountains, has by technological means influenced biological communities on an unprecedented scale. Our economic aid programs have fostered the development of monoculture in the rain forests of South America and Asia, and DDT, manufactured in the United States and used for control of malaria elsewhere, has had secondary effects in many countries. If we are to understand the roles that we have played or might play in the world, then we must look beyond the boundaries of our own nation and understand all of the kinds of communities in the biosphere.

Major Terrestrial Biomes

A biome is generally characterized by its dominant climax plant community, the locations and nature of which are determined by climate, and by the adaptations of the plants and animals to the climate. As we have already seen, however, a variety of subclimaxes, edaphic climaxes, and man-maintained communities are usually considered to be part of a biome. A swamp or river plain in a temperate deciduous biome may not support the oaks and maples that are characteristic of the biome, but the swamp or plain is still a part of the biome. Nevertheless, there are often so many exceptions within a biome that generalizations become difficult to make. For example, tundralike communities may be found on or near the summits of mountains throughout the world, even in equatorial Africa and South America. Or, human use may make the biome type hard to recognize, as when the clearing of grazing land in a forest biome produces what appears to be a typical grassland community. In spite of these difficulties, the biome concept is useful and provides a focal point from which to examine many of the implications of human activity.

What we must do, then, is examine some of the major biomes of the world in order to see some of their ''natural'' characteristics, the uses to which they have been put by man, and the changes that have followed human intervention and utilization. We shall learn from such an examination that human activities have done a great deal to modify many of the biomes, and that the biomes themselves form a logical basis for major land-use and development plans.

Obviously, if we are to consider biological communities and resource use on a global basis, we must have some sort of organizational framework. Figure 15.1 shows one way of organizing biomes on the basis of temperature and rainfall. As with most generalizations, there are a number of exceptions, but the temperature-rainfall scheme is useful, if somewhat oversimplified.

Although mean annual temperature and rainfall are the main determinants of biomes, the geomorphology and latitude of a particular location, in that they affect climate, also determine what biome type may be found. Another very important factor is the timing of rainfall and temperature variation. Seasonal variations produce

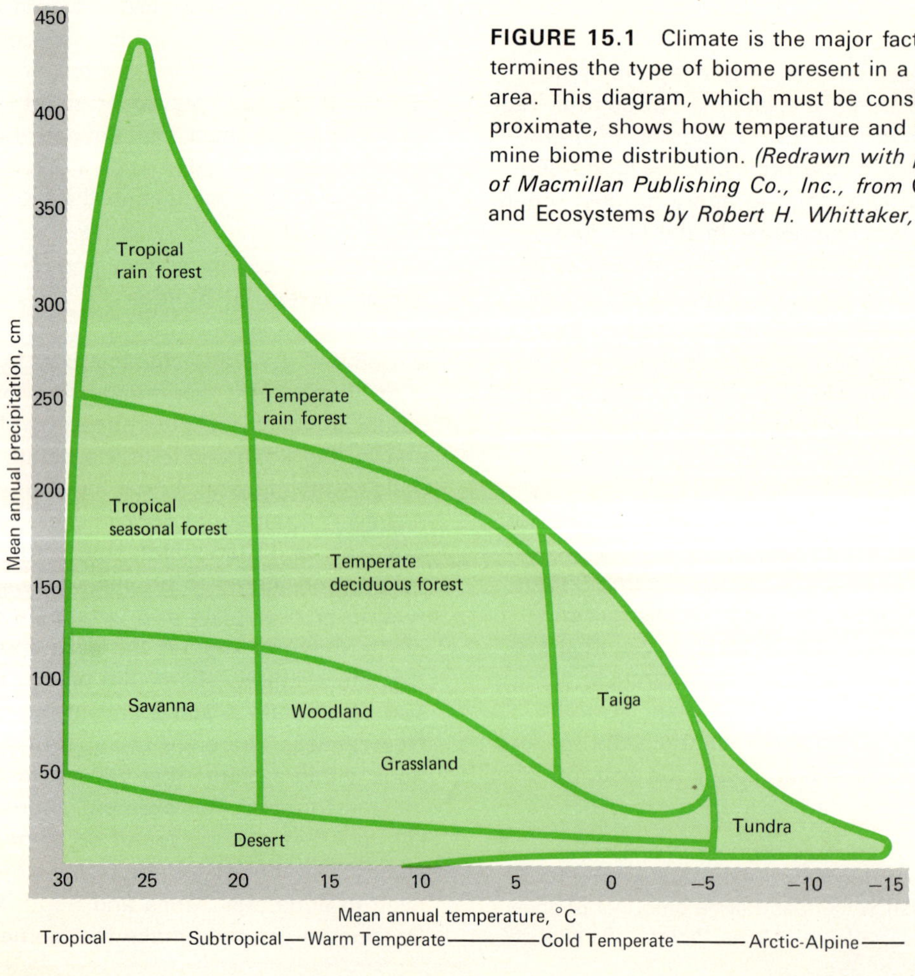

FIGURE 15.1 Climate is the major factor that determines the type of biome present in a particular area. This diagram, which must be considered approximate, shows how temperature and rainfall determine biome distribution. *(Redrawn with permission of Macmillan Publishing Co., Inc., from* Communities and Ecosystems *by Robert H. Whittaker, 1970.)*

stress factors that we discussed in the last chapter and are probably as important as annual means of temperature and rainfall.

The four biomes that are most widely distributed and cover the greatest proportion of land in the world, the tundra, the taiga, the temperate deciduous forest, and the temperate grassland, are all characterized by mean annual temperatures of less than 15°–20°C (59°–68°F). Notice in Figure 15.2 that most of the land mass of North America, Europe, and Asia is occupied by these biomes. In figure 15.2, it can also be seen that the tundra and taiga are distributed in bands running in an east-west direction, the tundra lying north of the taiga. Lying south of the taiga is an interrupted band of grassland and temperate deciduous forest biomes, the grasslands usually being located in the interiors, and the forests being located on the edges of continents. The reasons for this distribution pattern are obvious. The tundra is characteristic of the arctic regions, where precipitation is scant. As we have seen, the short growing season and the extremes of temperature in the tundra produce low, somewhat marshy vegetation. The taiga (which takes its name from the Russian word for "primeval forest"), or northern coniferous forest biome, is characterized by trees such as spruce and fir, and has mean averages of temperature and rainfall that are somewhat higher than the tundra. (Actually, most of the precipitation in the taiga is snowfall.) South of the taiga, the mean

FIGURE 15.2 As described in the text, most of the biomes occur in quite regular patterns, although there are areas of considerable complexity. *(After E. P. Odum, 1971.)*

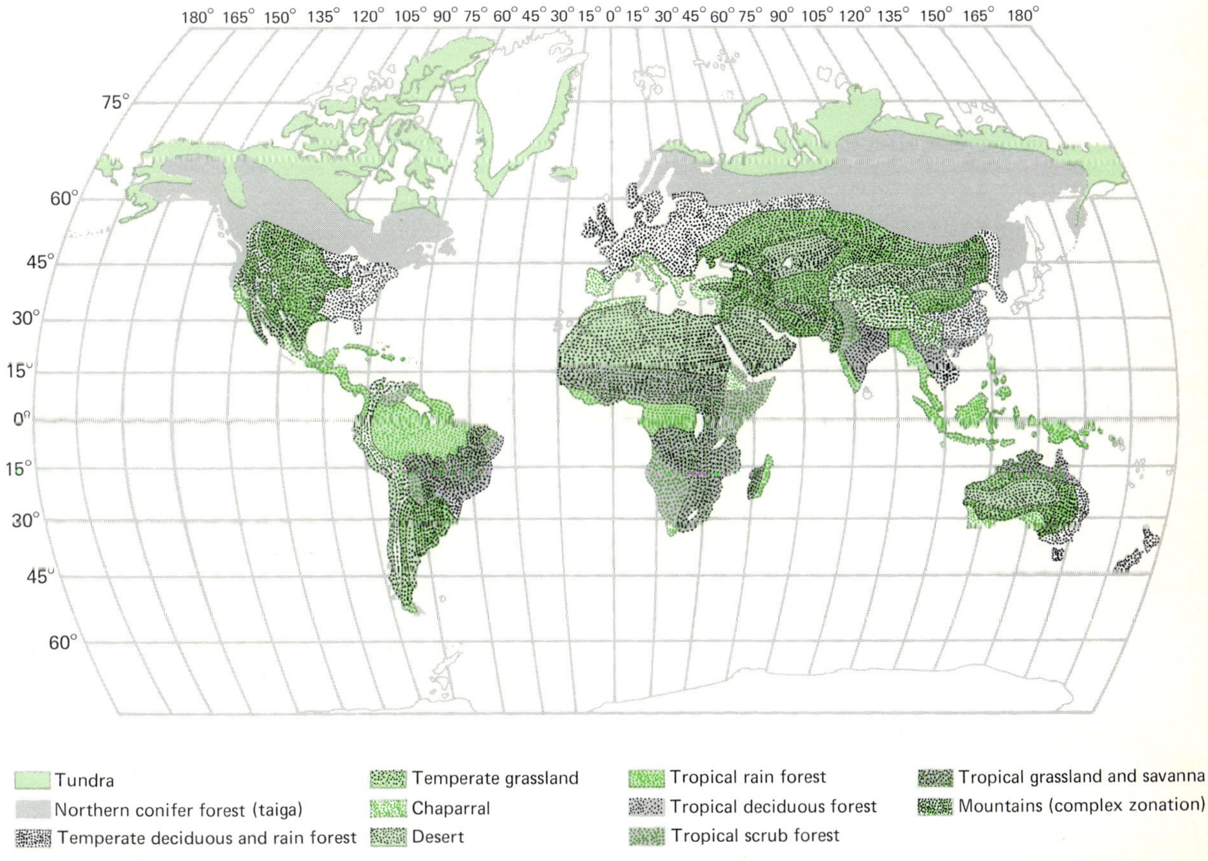

annual temperatures of the grassland and temperate deciduous biomes are approximately the same, but mean annual precipitation differs significantly. The prevailing winds near the edges of the continents tend to carry moisture-laden air in from the oceans, and most of this moisture is precipitated as rain near the coasts and where mountains are found. The greater rainfall can support the growth of larger plants, such as trees, and thus, the temperate deciduous forest biome is located along the coasts. The interior of a continent receives less rainfall, and thus, only smaller plants like grasses can grow there.

Before considering specific biomes in detail and examining their uses by man, we should point out that although we shall be discussing each biome as though it were a uniform community, each is in fact a varied aggregation of smaller communities. We should also point out that the plant and animal species characteristic of a biome type in one location of the world differ from those that characterize the same biome type in another part of the world. The niches that these differing species occupy are frequently similar, however, For example, in the grasslands of Eurasia and North America may be found unrelated species of grazers. In the Eurasian steppes the predominant grazers are wild horses and asses, while in North America the same ecological role is filled by bison and antelope. (The horse was extinct in the New World when it was introduced by the Spanish conquistadores in the sixteenth century.) We should also point out that the boundaries that exist between biomes are usually not sharply defined. Rather, one biome gradually gives way to another over some distance. These gradual *ecotones* are characterized by changing patterns of species dominance from place to place and by an overlap of animal species.

The Tundra

A permafrost layer, low grasses and shrubs, lichens, and cycles of herbivores and predators are all characteristics of the tundra, as we have already discussed. (See fig. 15.3.) As we have seen, the tundra is a classic example of a high stress, low diversity system. Fortunately, since the community is very susceptible to damage, the tundra has not yet been subject to much human use. Its low diversity is an indication that the tundra is an easily perturbed system, and the low rates of decomposition characteristic of the tundra would allow the accumulation of materials frequently associated with human use, such as hydrocarbons, wastes, and insecticides.

Many difficulties stand in the way of developing the tundra for human use. Constructing buildings in permafrost is difficult, since the soil is unstable, and building foundations are subject to sinking or fracturing. Moreover, although the tundra is marshy in many places, for most of the year the only means of obtaining water is by thawing ice and snow. Since the tundra is isolated from population centers, transportation is costly, and all buildings are expensive to construct and maintain. Since it is also isolated from agricultural areas, food is expensive. Sea routes are less expensive but are closed for a portion of the year, and the building of overland transportation networks is difficult and expensive because of the permafrost. The recent discovery of mineral resources, especially petroleum, and the growing scarcity of such resources in other parts of the world may accelerate the development of the tundra.

Exactly what effects such development will have, both on the tundra and adjacent biomes, is currently a matter of sometimes heated debate. For instance, the discovery of oil on Alaska's North Slope prompted proposals to build a pipeline from the site of the discovery to a town in southern Alaska where the harbor is open all year round and from which oil could be transported by tanker to refineries in the contiguous United States. In order to keep the oil flowing in the pipeline, it would have to be heated to between 71° and 82°C (160°–180°F). In those areas where the pipeline would be underground, it would probably melt the permafrost. The effects of such local melt-

FIGURE 15.3 In southern portions of the tundra biome, caribou may be important grazers. *(Courtesy of Leonard Lee Rue III from National Audubon Society.)*

ing of the permafrost are still largely unknown. Another problem of development is that during the summer in the tundra, mosquitos are abundant pests. What effects will insecticides have on other populations?

It is not likely that we will be able to develop tundra biomes without having some ill effects on the system. The important question is whether we are willing to spend money on the research necessary to plan intelligently enough to minimize the damage. The controversy surrounding the building of the Trans-Alaska pipe line has prompted some valuable research, but a great deal more remains to be done.

The Taiga

Like the tundra, the taiga is a relatively unexploited biome, although some timbering has taken place in parts of Canada, Scandinavia, and Russia. As in the tundra, transportation is a problem, but the suspected presence of rich mineral deposits will encourage the rapid construction of transportation networks. Species diversity in the taiga is low. A few shrubs grow under the trees, but the canopy and the low angle of the sun create a great deal of shade and thus prevent a lush undergrowth from developing. Not many other characteristics of the tundra are found in the taiga, however. Although it is probably easily disturbed, changes in the taiga are probably reversible. However, the soil of the taiga is created by the slow decomposition of leaf litter on the forest floor, and the removal of the upper soil layer would probably have drastic effects. The effects of lumbering, however, may be reversible. Since we know relatively little about the possible effects of development on the taiga, it is probably fortunate that we have not exploited it to any great extent.

The coniferous trees of the taiga are adapted to the cold conditions by several mechanisms. The pitch in the trees acts as an antifreeze, and the needles of the trees resist freezing. Thus, the trees that characterize the taiga biome are able to maintain some photosynthesis throughout the year. The regular falling of leaves occurs, although never all at once as in the deciduous forest biome, and the leaf fall in the taiga supports a number of species of soil invertebrates. The cones that are characteristic of conifers provide a food resource for small mammals such as ground squirrels. Most of the species of large animals that characterize the taiga, such as moose and caribou do not depend upon it for their sole source of food, but rather migrate between the taiga and the deciduous biome.

FIGURE 15.4 The temperate deciduous forest biome, such as in the mountainous Appalachian region of the United States, is characterized by immense areas of magnificent and economically important hardwood forests. *(Courtesy of National Park Service, United States Department of the Interior.)*

Other species in the tundra, as we have seen in the case of the snowy owl, may migrate into the taiga and even into the deciduous forests of North America.

The taiga and the tundra, then, remain relatively undisturbed by human activities, and we need to learn a great deal more about them before we exploit their resources. The next two important biomes, the deciduous forest and the grassland, are the most intensively utilized and consistently disturbed biomes in the world, for it was in the grassland biome that man developed agriculture and in the deciduous forest biome that the industrial revolution occurred.

The Deciduous Forest Biome

In the moist seasonal environment of the deciduous forest biome (see fig. 15.4), the adaptations of trees provide for growth during the summer months when solar energy input is high. During the winter months, the leaves fall from the trees, and there is therefore no photosynthesis; however, since most of the water leaves the exposed portion of the tree, there is no freezing either. A much greater diversity of plants and animals can be found in the deciduous forest biome than in the taiga, and an undergrowth of shrubs and herbs develops during the portion of the year when the canopy is sparse. (See fig. 15.5.) The fallen leaves, seeds, and nuts support a greater diversity of animals, including both insects and vertebrates, and especially birds and mammals. Most readers are familiar with this type of environment, since it constitutes a substantial proportion of the populated areas of the United States, Canada, and Europe.

However, it must be pointed out that there are almost no true climax deciduous forests remaining in the world. Most of the forest that once

FIGURE 15.5 A variety of vegetation grows on the forest floor. It is provided with light that filters through the canopy, especially in the spring and fall. *(Courtesy of Hugh Rogers from Monkmeyer Press.)*

covered Europe was laid waste during the Iron Age and even earlier when vast expanses were cut over or burned down to make room for agriculture. Some sizable remnants remained until the beginning of the industrialization of Europe, but they too disappeared to make way for a burgeoning urban population and to provide fuel for growing industrial enterprises. At present, about all that remains in Europe are managed forests that, however pleasant, are not true climax communities.

A good example of how such depredations on natural communities occur can be found in Scotland. The beautiful Scottish moors are not natural communities at all. (See fig. 15.6.) During the late Middle Ages, English ironmongers began to move to Scotland, because the deciduous forests of England were being depleted to make room for agriculture and sheep grazing and because the tax situation in Scotland was more favorable to a profitable business. This resulted in the massive cutting of the deciduous forests of Scotland to manufacture charcoal for the smelting of iron. Moreover, for a long time, the political relations between Scotland and England were stormy, as were Scotland's internal politics. During the many small wars that

FIGURE 15.6 At one time the land on these Scottish mountains was completely covered with forest. *(Courtesy of W. B. Finch from Stock, Boston.)*

broke out between the two nations and between rival factions in Scotland, much forested land was burned over. The subsequent soil changes, increases in soil acidity (lower soil pH), and the low moorland vegetation maintained by heavy sheep-grazing have prevented the recovery of the original community. Absentee landlords, living primarily in England and Ireland, also maintained a disclimax community by burning over the land so that they could hunt red deer and grouse. Although there is now an active reforestation program underway in Scotland, the new forests consist of conifers, not the broad leafed deciduous trees that originally populated the nation. Thus, the development of an iron industry, the growth of a sheep-grazing economy, and the practice of repeatedly burning the moors have maintained a completely artificial disclimax community over nearly all of Scotland.

The deciduous forests of the New World have not been as seriously disturbed, but there are few remaining areas of virgin forest. The image of the early American pioneer as a settler striving to clear the land and make room for his fields of wheat, corn, and tobacco is basically accurate. Nearly all of the fertile, fairly level land along the eastern seaboard was cleared for farming during the period from the seventeenth to the nineteenth century, while areas that were too rocky or infertile for farming were actively logged. The changes made in the continent were frequently necessary, although many of the means that were used were crude. What is sad is that so few areas remain of virgin communities. When the forests were cleared, the animal populations that inhabited them, including wolves, mountain lions, bison, and moose, nearly or completely disappeared as well.

In Eurasia, to the southeast of the deciduous forest biome, and in America, to the west of the deciduous forest biomes, lie two of the great grassland biomes of the world. It was in the grassland biome that agriculture began, and we shall now turn our attention to what may be some of the most devastated land in the world.

The Grassland Biome

Most of the principal food plants that man has domesticated and that now provide the greatest proportion of human nutrition have been derived by the artificial selection of grasses. Moreover, most of our domesticated animals, especially those upon which we depend for food, were originally inhabitants of grassland biomes. To this day they are grazers.

The grassland condition is apparently normally maintained by a combination of low rainfall, fire, and grazing. The fire, of course, burns off the shoots of trees before they can develop and outcompete the grasses for light and space. In the moister locations of the grassland biome, as in the eastern Great Plains of North America, for example, the loss of grazers and the prevention of fire have allowed the grassland to grow over into deciduous forests. The grass species found in this biome are highly varied. In North America, for example, there are about fifteen species of grasses that are considered to predominate in various parts of the biome and many more that are also present in smaller populations. Many of the grasses put down large root systems, some penetrating as deeply as 2m (6 feet), and the roots of many species play a very important role in forming the sod upon which the biome is located.

Many of the animal species of the grassland biome are burrowers, as we have said, but there are many large mammals as well. The grassland biome is well suited, of course, to supporting large herds of migratory grazers, the most familiar species of which, in North America, are pronghorn antelope and bison. It is thought that the migratory behavior of many grassland herbivores is intricately associated with the growing seasons of the grass species and serves to prevent overgrazing.

This biome, of course, is also well suited to grazing agriculture and is subject to overgrazing and consequent damage to the soil. When it is too intensively utilized, as it is when domesticated herds are introduced and prevented from

FIGURE 15.7 Poor agricultural practices can completely destroy a community, as in the Oklahoma dustbowl. Although some grass regrows, the prairie will never return. *(Courtesy of Charles Wunder from Monkmeyer Press.)*

migrating, the soil in this biome is deprived of the stabilizing factor provided by extensive root systems and is easily eroded. In the moister locations of the biome, the plow can equally effectively disrupt the soil, and although this may lead to a few years of high crop production, the soil can be easily exhausted of its nutrients and water-retaining humus. In both cases, the change is usually irreversible, as the dust bowl of Oklahoma attests (see fig. 15.7). With the exception of a few carefully maintained preserves, there is practically no real prairie left in the United States. It seems safe to say that the prairie will never return, since the changes that have been brought about in the rich organic soil are almost irreversible.

We have thus far considered briefly four biome types that make up most of the land mass of America and Eurasia, as well as sizable proportions of Australia and South America. Two of the biome types, temperate deciduous forests and grassland, have been subjected to heavy human use and are now hard to find in their natural states. We must now examine the remaining major biomes—the very wet and very dry ones. As it happens, most of these are found in tropical and subtropical environments. The scheme shown in figure 15.1 breaks down somewhat in considering these biomes, since *when* moisture is available is often as important as *how much* is available. These biomes are highly varied and difficult to classify, and we can consider only a few of them.

The Tropical Savanna

It is difficult to draw precise lines between the temperate grassland, the tropical savanna, and the temperate and tropical shrubland and desert. Perhaps the most important of these biomes, and the one most endangered at

FIGURE 15.8 Grasses, with interspersed low trees and herds of large grazers, are characteristic of the African savanna biome. *(Courtesy of Ernst Haas from Magnum Photos.)*

present, is the savanna. (See fig. 15.8.) As shown in figure 15.2, the savanna covers a major part of Africa and is also found in South America and Australia. Savanna biomes are characterized by plains of grasses interspersed with trees or clumps of trees. The plant species of the savanna vary widely and are adapted to seasonal rainfalls alternating with very dry conditions. During the dry periods, fires are periodic occurrences to which the plant species are adapted. The vast savanna regions of the world support a diverse and spectacular assortment of mammals, such as the many herbivores and predators of the African Veldt or the marsupials of Australia. The savanna biome, like the grassland, is maintained by a combination of fire and grazing, but the wet and dry seasons are also important.

The productivity of the savanna may be increased by irrigation, and the land is suitable for cattle; however, many problems surround the development of this biome. If irrigation is introduced, permanent changes may result, as the plants, adapted to periods of drought, may not be able to survive. Overall productivity in the savanna is low because of the dry seasons, and therefore the biome could not support large sedentary herds of intensively grazing cattle unless the plants that characterize the savanna were supported through extensive irrigation. Without irrigation the present plant species would be destroyed by heavy grazing in the dry season. Introducing herds of cattle which would be prevented from migrating is only one of the problems, since the cattle themselves would likely be susceptible to a variety of diseases and parasites which the native species are adapted to already.

Although the savanna may appear promising for human development, it is a low diversity system containing some highly specialized species. As such, it is very sensitive to change and may very easily be destroyed. Unfortunately, the growing nations of Africa and South America, in which savanna biomes are located, need the productivity of the land for the well-being of their people. One may argue that the population growth of these nations should be slowed down, but that evades the problems of the immediate future and the problems that must be faced by the people who are alive today and must be fed. Achieving a stable population is a long-term goal that must be pursued. In the interim, however, we must recognize the need for taking actions that will enhance the quality of life for the present world population.

Experiments are now underway on the domestication of animal species that are native to the savanna biome, and some ecologists think that it may be possible to harvest wild populations of such animals as wildebeest or antelope on a sustained-yield basis. In this way, it may be possible to avoid destruction of the savanna biome, while at the same time providing for the nutritional needs of people. If the savanna can be preserved by such programs, we may all learn how we could have kept some of our natural ecosystems intact while intelligently utilizing others.

The Chaparral Biome

A typical shrubland, or chaparral, develops in Mediterranean climates. The Mediterranean climate is somewhat cooler, and rainfall is more limited (though seasonal) than in the climates of the savanna biomes. As the name of its climate suggests, the chaparral biome is found in the nations surrounding the Mediterranean Sea, but it is also found in parts of the southern Pacific coast of North America and in parts of Australia. The dominant plant species in a chaparral biome vary from area to area, as do the species of animals.

In the Mediterranean basin, the chaparral biome is the result of thousands of years of grazing, lumbering, and agriculture. Portions of what is now modern Greece were once covered with stands of oak and cedar, but as civilization progressed, the trees were cut for a variety of purposes such as shipbuilding, fuel, and home construction, and their regrowth was prevented by grazing herds of domesticated goats. Exposed to the effects of wind and rain, the soil in this hilly region gradually eroded until it was no longer able to support the growth of large plants.

The chaparral of southern California (see fig. 15.9) may be viewed as a fire-maintained disclimax. Periodic fires sweep through the biome, burning out the shrubs and also preventing the

FIGURE 15.9 Chaparral in the Cleveland National Forest of California. *(Courtesy of United States Forest Service.)*

growth of trees. Many of the shrub species are adapted to the fires, and the germination of some species is even stimulated by them. After a fire, the seeds of many chaparral species germinate with the first rain, thus preventing the possible erosion of the soil.

The chaparral biome is a relatively low diversity system, but it is fairly stable. Irrigation, fire prevention, or intensive agriculture may, however, bring about irreversible effects, since the soil is relatively loose and easily disrupted. Such changes have already occurred in the Mediterranean, although the California chaparral and the ''mellec scrub'' of Australia remain fairly undisturbed.

The Desert Biome

Although there are a variety of other dry communities besides the chaparral, most are less important in their distribution, and they intergrade with the desert biome. Deserts are characterized by their low, variable rainfall. There are only two places in the world where there is almost no rainfall at all, one being in the interior of the Sahara in North Africa, where less than 2.5 centimeters (1 inch) falls annually, and the other being in the Atacama Desert of northern Chile-Peru, where annual rainfall is less than 1.25 centimeters (0.5 inches). The image of the great sand dunes of the desert is somewhat exaggerated, since less than 10% of the Sahara is dunes. Perhaps a more accurate idea of the desert can be gained from, of all places, the American western movie. Sagebrush and cactus are characteristic of the sort of vegetation usually found in deserts. Desert species of plants are adapted either to retaining water or to making use of the water available from sporadic rainfalls during short periods of growth

FIGURE 15.10 The hot American desert is characterized by cactus and low shrubs such as creosote bushes. *(Courtesy of Allan D. Cruickshank from National Audubon Society.)*

FIGURE 15.11 Sagebrush characterizes the cold American desert. Most of the rain (and snow) falls in the mountains. *(Courtesy of Richard Weymouth Brooks.)*

and reproduction and relatively longer periods of dormancy. Such *xerophytic* (Gr *xeros* dry) plants have evolved from a number of families but have converged substantially and often appear very similar on all continents.

Ecologically, deserts can be characterized as hot or cool. In North America, the former are dominated by cactus and creosote bushes, and the latter by sagebrush. (See figs. 15.10 and 15.11.) The animals of the desert are adapted to retaining the water produced by metabolism, and some are so efficient at using metabolic water that they need never drink at all.

It is difficult to conceive of a desert being damaged by human activity, but any traveler in the deserts of Africa and the Middle East soon has such notions dispelled. The number of irrigation projects from ancient times is astonishing. Why attempts to turn the desert into productive land failed is not always known, but one probable cause is common and still present in modern agricultural practices. When water is put on the soil and then evaporated, it leaves behind small quantities of mineral salts that can accumulate and ultimately destroy the productivity and utility of the soil. The history of the Tigris-Euphrates valley is an example of the misuse of irrigation. Periodic floods of the river valley were channeled into irrigation projects from the time of the ancient Babylonians. Although the flood waters provided the land with nutrients and moisture, the gradual buildup of salts made the soil unfit for agriculture. The failure of a series of civilizations in the Tigris-Euphrates river valley and other parts of the Middle East is thought to have been partially caused by the salinization of the soil. The same problem appears to be occurring in the southwestern United States, where one of the major costs of irrigation is incurred in flushing the land to remove accumulated salts. Without great expense, the increased concentration of salts is apparently irreversible.

We are apparently capable of both destroying deserts and creating them. Much of the desert region in present-day Arabian and Middle Eastern countries was at one time forested. In fact, the last trees were removed as recently as the time of the Crusades. It appears that the growth of agriculture in that part of the world and the agricultural methods practiced led to an immense loss of topsoil and nutrients in the region. As a consequence, the vegetation died off, and the soil lost its ability to retain water. Thus, the land shifted from forest or shrubland to desert in many areas, and the shift may have produced regional climatic changes. Exactly what happened is not very well understood, but we do know that many of the civilizations in the region failed and that some of the Crusades failed, and we suspect that both failures may have been partially produced by the devastation of the biological communities of the Middle East.

The one remaining terrestrial biome that we will consider is the opposite of the desert.

The Tropical Rain Forest

As we saw in the last chapter, the tropical rain forest is the most diverse, complex, and stable terrestrial community. (See fig. 15.12.) To a greater extent than any other terrestrial biome, the tropical rain forest forms its own environment. The characteristic vertical stratification of the plant species in the tropical rain forest, the rapid turnover of nutrients, most of which are bound up in the community biomass, and the highly diverse populations of species with narrow niches are all indications of the stability of the biome. Under normal conditions, there are few perturbations in species populations, and the recovery from minor perturbations is rapid. The complete removal of the forest will usually result in recovery to very lush vegetation, but not to rain forest conditions. The actual recovery to a climax state may take a very long time, but precisely how long it will take and through

FIGURE 15.12 The tropical rain forest forms a dense, vast expanse of forest cover over many of the tropical portions of the world. Pictured here is a stretch of rain forest in Brazil. *(Courtesy of Carl Frank from Photo Researchers, Inc.)*

what stages the recovery will pass are not yet well understood.

The practice or attempted practice of agriculture in areas from which rain forests have been cleared has often resulted in the permanent destruction of the soil by a process known as *laterization.* The soils in many tropical rain forests are of a type called *laterite* and are characterized by a very thin surface layer of nutrient-bearing soil which also contains high concentrations of iron and aluminum. Removal of the vegetation cover may cause the nutrients and other salts to be washed out of the soil by rains, leaving the iron and aluminum behind. Upon exposure to oxygen and direct sunlight, a series of chemical reactions takes place within a short period of time, hardening the soil into a rocklike substance. The term *laterite,* in fact, is from the Latin word for brick, *later.* Thus, the combined nutrient loss and soil transformation inhibit both the practice of agriculture and the regrowth of the forest.

Human use of the tropical rain forests of the world may be a very important consideration in our future. We do not yet know how the tropical forests of Vietnam and Cambodia will recover from the ecological destruction wrought upon them by the dropping of millions of tons of explosives and the massive spraying of defoliants and herbicides to facilitate military reconnaissance. There are perhaps wider implications to the question of the adoption of agriculture in nations having extensive rain forests. Vast areas of Brazil have been cleared for agriculture or city building. The expected population of Brazil (200 million persons by the year 2000) must be fed and housed. There remains, however, a great need to preserve some areas of forest, and we need to know more about the expected stages of recovery and the recovery rates. If all of the rain forest is removed from some major region, for whatever reason, we would expect some definite changes to occur in evaporation patterns. What effect such changes might have on the overall climate of the region and the world is impossible to determine, but we can expect that some climatic changes would be brought about by such activity.

Clearly, to achieve the best quality of life, we must take biological communities into consideration in planning our activities. One of the key points of this book is that we do not yet know enough about such communities to plan effectively. Nevertheless, we hear a great deal about national land-use planning. Those plans that are made *too seldom* take into account the nature of biological communities or consider the factors that regulate communities. An abundance of sociological, economic, and demographic factors may be considered, as indeed they must be, but even when understood the biological factors are often overlooked or ignored. We have seen some of the results of such oversights in the Tigris-Euphrates river basin, the Highlands of Scotland, and the dust bowls of Oklahoma.

The problems faced by planners are relevant not only at the local level, as in urban-suburban planning, but also at the national and international levels. Some areas within the United States are better suited for agriculture, while others are better suited for urbanization, mineral exploitation, or resort development. Likewise, some nations are better suited biologically and geologically for some purposes than for others. The Los Angeles basin, for example, contains some highly fertile soil, but it is now so heavily paved over that it can probably never be used for agricultural purposes. The social problems that would be attendant upon an attempt to reconvert the region to agriculture would be tremendous. The distribution of biomes suggests how development should or should not proceed if the interests of the people and the biological environment are to be best served. Unfortunately, this has yet to be recognized by some of those individuals responsible for planning and resource use.

We should point out that some of the sections in this chapter dealing with the human use of biological resources are debatable and reflect only the opinion of the authors. We hope, however, that the reader will agree with us on two points: (1) that we need more information about the functioning and analysis of biological communities, and (2) that the scientists who understand the nature of the functioning of communities should play a greater role in planning and decision-making on the local, state, national, and international levels. Although editorializing may be considered inappropriate in textbooks, we feel very strongly that if the counsels of ecology continue to be ignored, we will place ourselves and the rest of the world in needless jeopardy.

Communities in the Aquatic Environment

We have already discussed the properties of water and many of the adaptations to aquatic environments, and we have also discussed some of the correlations between nutrient cycling, productivity, and transfer processes in bodies of standing fresh water. What remains is to survey some of the general communities in standing and running fresh waters and in the oceans.

Lakes and Standing Waters

We should mention that apart from the generalizations made earlier (see pages 340–341), lakes may vary widely. The variations that exist may be due to climate, morphology, or chemistry.

Climate affects the stratification characteristics of a lake or pond. In the far north, a lake may never get warm enough to develop stratification. In other climates, a lake may stratify in winter, under the ice, as well as in the summer. As we pointed out earlier, some tropical lakes may be permanently stratified. Obviously, stratification affects productivity by affecting nutrient cycles, but productivity may also be affected by the "flushing" of a lake, which depends on rainfall, and by basin hydrology. Thus, a spring-fed lake will have a different level of productivity from a lake fed by streams; and of course, differing temperatures will allow for different species in the aquatic community.

The morphology of a lake affects the mixing of its waters and thus nutrient cycling. In a deep lake, there will probably be a deep hypolimnion which may remain cold all summer and contain a large nutrient and oxygen reserve; thus, there is less likelihood that the lake will become anoxic. If the lake is shallow, the thermocline may reach the bottom of the lake, allow nutrient recycling to occur all summer long, and therefore, increase the productivity of the lake. Anoxia cannot occur, and therefore there will be no death of hypolimnion-dwelling organisms. Morphology also influences shoreline communities and affects how such communities may dominate the lake. Thus, in a shallow lake, a large shallow limnetic zone exists, and the major proportion of the productivity of the lake is accounted for by benthic plants.

The chemistry of the water may influence many aspects of life in the lake. Which species can live in the lake depends upon the pH of the water, its hardness, and of course, the presence of chemical nutrients such as nitrate and phosphate. The chemistry may also affect the mixing of the waters. In a few lakes, the deeper waters are denser and more saline and contain more dissolved chemicals. This results in a permanent stratification of the lake and usually in a permanently anoxic hypolimnion.

In most areas of the world, the esthetic and recreational values of lakes are very important, and these values contribute to the importance of the communities of the lake. When we consider the esthetic and recreational uses of a lake, then, we must also consider the use of the surrounding land, since some of the factors that affect the lake are found in the terrestrial biome of which the lake is a part. The runoff of nutrients from the land is an obvious factor and depends, in part, upon human activities in the drainage basin of the lake and upon upstream land use. If the land is forested and unpopulated (by humans), then there will be no problems with sewage, and nutrient cycling in the lake will be closed. If the lake is in an agricultural region, however, farming practices and soil conservation measures will dramatically affect the nature of the lake. The excessive use of fertilizers can contribute to cultural eutrophication. Inefficient soil conservation practices can lead to the loss of topsoil from the land and a consequent buildup of silt in the lake. If the water becomes too muddy, productivity will decline, and the benthic community will be stifled. If the lake fills in as a result of silting, its recreational value is destroyed. Proper resource planning, then, means that downstream effects on standing waters must be examined, and the limnologist becomes an important member of the planning team.

Rivers and Streams

In most cases, the communities in rivers and streams depend upon the particular and frequently unique combination of watershed morphology, total flow, and gradient of the river, as well as upon the activities that take place within the river basin. Whereas lakes are studied by dividing the lake into vertical zones, the communities of rivers can be described and distinguished by dividing the river into *near-source, middle,* and *coastal plain* segments.

The near-source segment of a river is usually characterized by a steep gradient, low volume, and rapid flow. (See fig. 15.13.) This produces a rocky bottom, and most species present are adapted for attachment to the substrate or for strong swimming. This segment is usually low in nutrients, and the rapid flow prevents much plant biomass from accumulating. Thus, most biological energy in the community comes from outside the stream, partially in the form of leaf litter and soil detritus. The species diversity in the near-source segment is usually low, and the river is therefore particularly sensitive to contamination in this area. Acid mine drainage usually has its most devastating effects here. The

FIGURE 15.13 The uppermost reaches of a river are characterized by boulders and a rapid flow of water. The organisms require very specialized adaptations to live in this harsh environment. *(Courtesy of Jeff Albertson from Stock, Boston.)*

oxygen concentration of the water in the near-source segment of a river may be high, because rapids and riffles produce continual reaeration of the water. Many land-use plans ignore the upper reaches of flowing waters, although the high esthetic value of such sections can have a positive effect on land prices.

The intermediate range of a river is usually characterized by a higher volume of flow, a shallower gradient, a well-developed flood plain, and a sandy bottom. (See fig. 15.14.) During summer months in the temperate zone, the producers include a thin mat of algae on the bottom. In quieter sections of the stream, the producers may also include some rooted aquatic vegetation. Insect larvae may be found burrowing in the stream bottom, and in areas of heavy leaf and organic matter accumulation, large populations of other invertebrates may be found. Most energy in the stream still comes from outside, although if nutrient pollution occurs, the river may undergo some of the symptoms of eutrophication as productivity rises. This may bring about a downstream decrease in oxygen as producers decompose and cause unpleasant odors, and this results in a loss of esthetic value. The river may also, of course, be severely damaged in the intermediate segment by sewage pollution, causing oxygen removal and anoxia.

FIGURE 15.14 Meanders, eroded flood plains, and a moderate gradient are characteristic of the middle reaches of a river. This is an aerial view of part of the Shenandoah River near Harper's Ferry, West Virginia. *(Courtesy of Blue Ridge Aerial Surveys.)*

The lower reaches are characterized by the meandering flow of the river, oxbow lakes, and large flood plains. (See fig. 15.15.) The bottom is silt or mud, and a very different fauna is usually present. The turbidity of the water may be so great that there is not sufficient light penetration to support photosynthesis on the river bottom, and due to the slowly moving current, phytoplankton productivity increases in importance. Usually, in this segment, more energy is produced in the stream than is imported from the outside. By the time the river reaches a coastal plain, it carries everything from the major drainage basins upstream. The lower reaches of many of the great rivers—the Mississippi, the Hudson, the Cuyahoga, the Merrimack, the Potomac, the Sacramento, the Thames, the Danube, the Volga, and the Seine—are among the most damaged and polluted rivers in the world. The damage frequently begins long before it becomes obvious to the casual observer, for even upstream agriculture contributes silt that destroys the benthic communities. Each of the rivers mentioned above has played an important role in the development of cities and industries, and each has become a sink for industrial and urban wastes.

There is an important point to be made here about the health of flowing waters. River communities may well serve as the "pulse" of aquatic and terrestrial communities in the upstream portions of their drainage basins. The silt produced by careless agricultural practices, nutrient input from excess fertilizers, and heavy metals and wastes from industrial plants will all, sooner or later, show up in some portion of the river communities. Changes in productivity may signal the beginning of eutrophication and indicate that the agricultural practices must be reviewed. A decrease in species diversity may indicate pollution due to poor land-use practices. The presence of heavy metals and pesticides in the higher trophic levels of the community food chain indicates other misuses of resources. Thus, an understanding of river ecology and the development of adequate methods of monitoring rivers could well become the best means we have of identifying the misuse of terrestrial ecosystems and the misallocation of resources.

Communities in the Ocean

Although lakes and rivers are very important, both as indicators of the health of terrestrial ecosystems and for economic, recreational, and esthetic reasons, the oceans, if only by virtue of their immense sizes, are among the most important ecosystems in the world. While many freshwater and marine species share many characteristic adaptations determined by the nature of water itself, differences in total area, volume, depth, and salinity make ocean communities very different from freshwater ones.

In the ocean, the circulation of water and the distribution of dissolved gases and chemicals are critical to the existence and distribution of communities. Changes in atmospheric pressure, density, and temperature create winds which

FIGURE 15.15 Oxbows frequently become separated from a river and form lakes, such as the one in this photograph of the Minnesota lake country. *(Courtesy of Grant Heilman.)*

act upon the surface of the water. These winds, along with changing density of the water due to salinity and temperature changes, affect the movement of the water, as does the rotation of the earth. In the northern hemisphere, the currents produced by these factors generally move in a clockwise direction, while in the southern hemisphere, the motion is counterclockwise. (See fig. 15.16.) In addition to surface currents, of course, there are also deep currents which serve to move the water both horizontally and vertically. One such vertical movement, the Peruvian upwelling, was discussed in Chapter 11. In Antarctica and in some other places of the world, ''downwellings'' (usually called convergence zones) occur, carrying surface waters to great depths, where they circulate for hundreds or perhaps thousands of years.

The circulation of ocean waters has profound effects on marine communities, not only because it transports organisms but also because it transports oxygen and nutrients. The deep waters of the ocean, as we have seen, are a nutrient sink supplied by the decomposition of organisms from all levels of the food chain, as well as by the excretion of wastes. These nutrient supplies are carried to surface waters either by vertical mixing or by upwelling. In the upwelling zone off the coast of Peru, for example, nutrients derived from antarctic organisms are carried north and are augmented by materials that have sunk from surface waters along the way.

Oxygen is almost never limiting in the ocean, except very rarely at mid-depths, in the deepest trenches or in some fjords and the Black Sea, as we have seen. On rare occasions, at depths of 200 or 300 meters below very productive areas of the ocean, oxygen may become scarce as large quantities of plankton die and are decomposed as they settle from above. Oxygen enters the ocean at the surface and is carried to deeper waters, where amounts sufficient to support organisms may remain for several hundred years.

Light is of very great importance in the ocean, as it determines the depth at which photosynthesis may be carried on. Although the surface of the ocean reflects some light, the amount reflected is not significant. The water itself absorbs much more light than is reflected, and the amount available for photosynthesis drops off sharply within a relatively short distance. (See table 15.1.)

Two phenomena related to light in the ocean deserve special mention. First, *bioluminescence,* the light produced by organisms through enzymatic reactions, is of great importance in the ocean. Nearly every phylum contains some species that are bioluminescent. The light produced usually is blue green in color, the same color as the maximum transparency of sea water. The organs in which the light is produced are highly diverse, and the light apparently serves many purposes, including predator avoidance, sexual recognition, and prey attraction.

Another light-related phenomenon is *diurnal vertical migration.* Many larger planktonic species of animals live well below the surface during the daylight hours but rise to the surface during the hours of darkness. The adaptive significance

TABLE 15.1 Penetration of Light in Ocean Waters of Varying Clarity

(Measured in luxes*)

Depth (m)	*Clearest Ocean Water*	*Average Ocean Water*	*Average Coastal Water*	*Turbid Coastal Water*
0	120,000	120,000	120,000	120,000
10	65,000	42,000	6,800	2,600
20	42,000	20,000	380	70
50	14,000	1,800	—	—
100	2,500	33	—	—

*(A lux is the equivalent of approximately 0.1 foot candles.)

(Adapted from Harold Sverdrup, Martin Johnson, and Richard Fleming, *The Oceans, Their Physics, Chemistry, and General Biology* (Englewood Cliffs, N.J.: Prentice-Hall, Inc., 1942), p. 776.

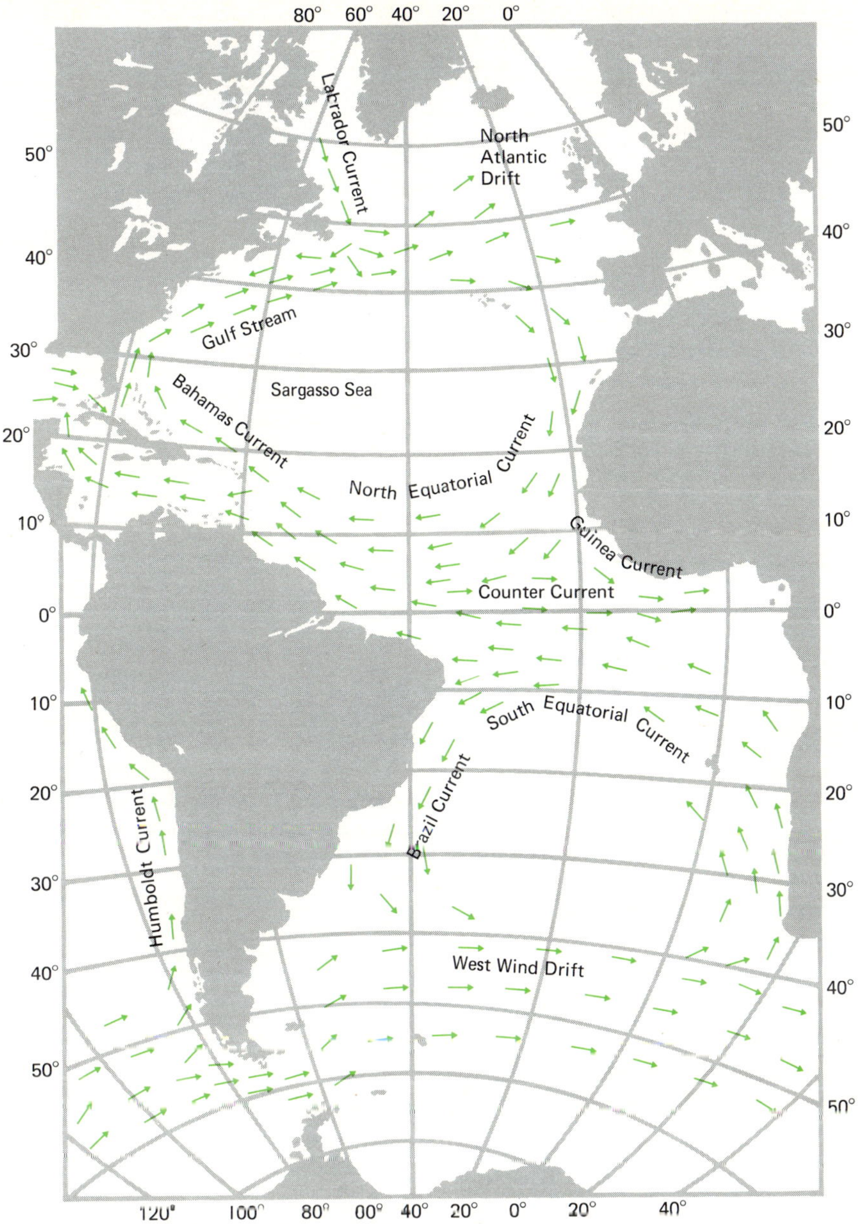

FIGURE 15.16 The currents in the north and south Atlantic illustrate well the clockwise and counter-clockwise patterns of circulation. *(Redrawn from* The Sea and Its Mysteries *by John S. Colman, G. Bell & Sons, Ltd., London, 1950.)*

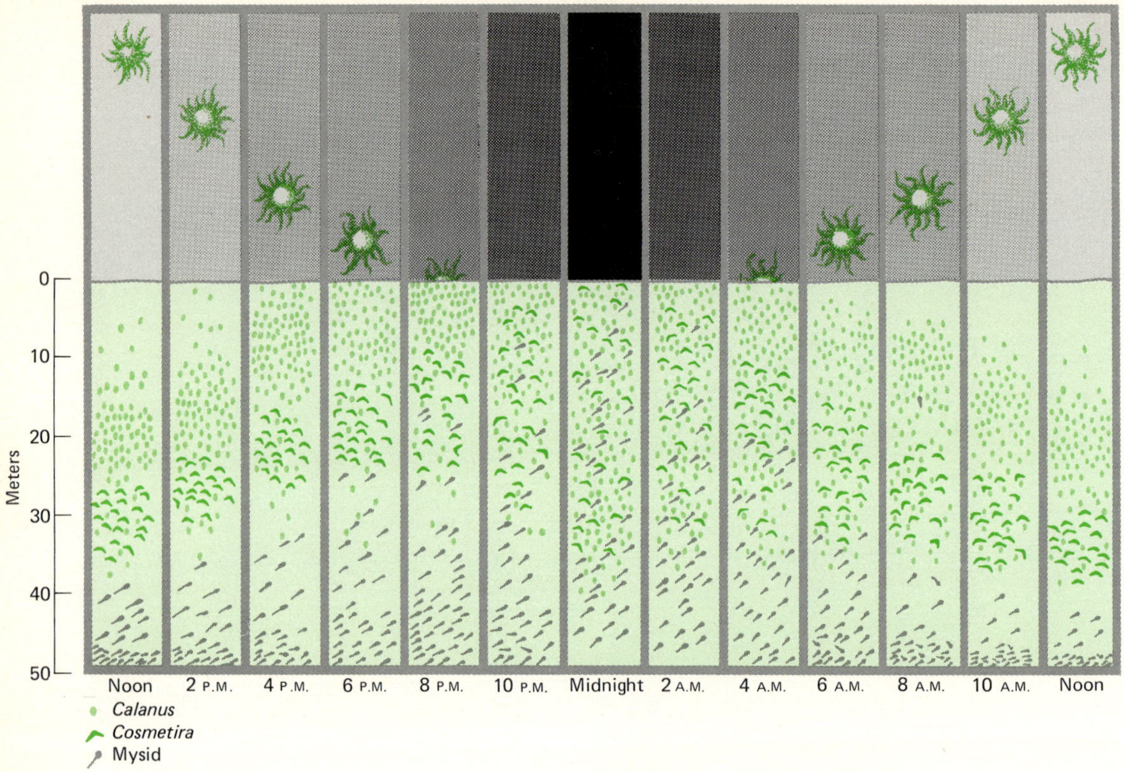

FIGURE 15.17 Many pelagic marine animals undergo daily vertical migrations, often in complex patterns. This serves to concentrate the populations at certain depths during the day and at other depths at night, and sometimes to disperse the populations. The migrations of the three crustacea shown here are typical. *(From E. P. Odum, 1971, after Russell and Yonge.)*

of these migrations is not fully known, but it is thought to be associated with predator avoidance during the day and feeding at night. (See fig. 15.17.)

A variety of physical conditions in the ocean determine where organisms may live and, thus, produce a variety of communities. However, the transport of organisms, materials, and gases in the ocean prevents very distinct boundaries from being created between communities. Vertically, of course, a boundary does exist between the lighted zone of water where photosynthesis takes place and deeper waters where photosynthesis cannot occur. Another vertical boundary exists at depths of between 500 and 1000 meters where major changes in fauna occur, and is probably associated with species' ability to see. At greater depths, most boundaries in the ocean are very indistinct. We know very little about very deep ocean communities, however, and because of the rate at which new species are being discovered, it is thought that a large proportion of the species that live in the very deep ocean have not even been seen and described. The deep ocean organisms that have been brought to the surface for study could not be kept alive long enough for detailed physiological and biochemical studies.

As we emphasized earlier, we must distinguish between benthic and pelagic communities. (See fig. 15.18.) Most of the comments made previously in this section refer to pelagic communities, but many apply to benthic communities as well. Changes that occur in benthic communi-

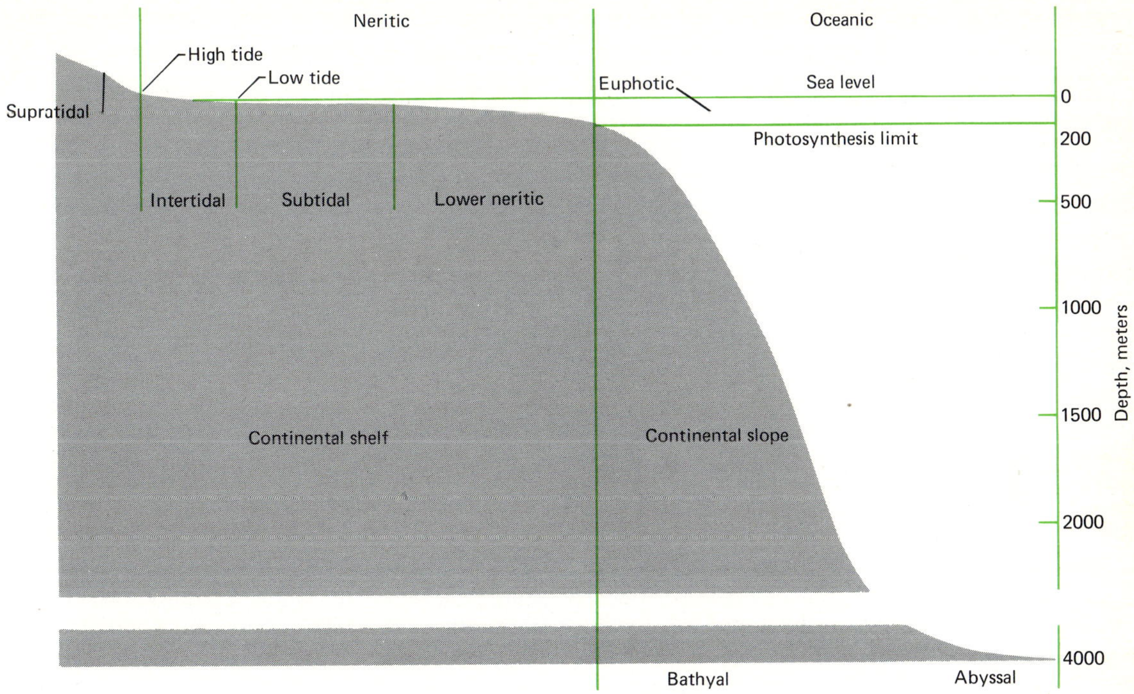

FIGURE 15.18 The communities of the ocean floor are divided into several zones. The intertidal is the region between the low and high tide marks; the subtidal is the region immediately below, often having extensive growths of seaweeds; while the lower sublittoral is the rest of the continental shelf out to the edge of the continental slope. Regions of intermediate depth down to 2000 meters are called bathyal while the abyssal communities exist over most of the ocean floor to depths of 4000 meters. Although not illustrated, deeper waters such as in ocean trenches are characterized by what are called hadal communities. The pelagic region is divided into the neritic zone, or the zone over the continental shelf, and the oceanic zone, or the zone over the continental slope and deep ocean regions. The euphotic zone is the zone within which photosynthesis occurs. *(Redrawn from E. P. Odum, 1971, after Hedgpeth.)*

ties frequently depend upon the depth of the water. In the relatively shallow waters over the continental shelf, the benthic community is usually very rich—that is, densely settled. In such areas, nutrients are relatively well mixed throughout the water, and the productive zone of the water is fairly near the benthic community. At depths deeper than about 200 meters, however, the density of the benthic community declines. The benthic communities of the continental slopes, or from about 200 meters to 1000 or 2000 meters, are less dense owing to the increased distance from the productive zone and lower energy availability. On the level plains that constitute most of the ocean floors, at depths averaging around 4000 meters, benthic communities live under relatively uniform conditions, burrowing in the soft muds and ooze that cover most of the ocean floors. Although not densely populated, the communities of the very deep ocean are large and appear to be highly diverse. The very deepest parts of the oceans—trenches that may reach depths of nearly 11,000 meters—support communities about which we know very, very little.

For the most part, while vertical community changes occur over tens or hundreds of meters, horizontal changes in communities are gradual, occurring over tens or hundreds of kilometers. The plankton species of the Atlantic and the Pacific, for example, differ greatly, and some even greater differences can be found in the communities in tropical waters and those in arctic and antarctic ones. Also, over a continental shelf, the communities may be very different from open ocean communities. These differences are due in part to the productivity and nutrient availability in shallow waters but are also partially explained by differing coastal currents, lower salinity, and other effects of land drainage.

Perhaps the best way of differentiating surface plankton communities is on the basis of productivity. Most subtropical and tropical open oceans are very low in productivity. Productivity is greater at higher latitudes, and it is highest at the coasts. Why is this so?

In tropical waters, the permanent stratification holds nutrients in the deep water, and thus, little nutrient material is available to the productive zone except by the very slow mixing of water through the thermocline. Most productivity is thought to be due to nutrient cycling in the mixed zone. But although tropical oceans are low in productivity, the conditions there are fairly stable, and therefore the oceans exhibit a high diversity and low biomass. The tropical ocean is characterized by a wide variety of specialized plankton and nekton, but it contains few commercially important species save tuna, which travel through the tropical oceans. Most other species in the tropical and subtropical ocean are too difficult to harvest, and since productivity is low, such difficulty is probably good in that the stock cannot be depleted rapidly.

Temperate regions of the ocean show a greater seasonal variation in productivity similar to that found in a temperate lake. Stratification occurs in the temperate ocean, and as the incident radiation increases during the spring months, the nutrients available from the deep waters are taken up and produce a spring bloom. As the nutrients are taken up by phytoplankton populations, productivity declines in the summer. There are more fisheries in temperate oceanic waters, although not as many as in coastal waters.

Coastal waters are highly productive for several reasons. There is some runoff of nutrients from the land, but not a sufficient amount to account for the greater productivity of coastal waters. More important is the mixing over continental shelves and the upwelling of deeper nutrient-rich water. During the winter months, the water over a temperate continental shelf is usually approximately the same temperature from top to bottom, as is a freshwater lake at the same time of year. Nutrients are evenly distributed throughout the water, and when spring comes, they are taken up by phytoplankton populations, causing a decline in productivity in the summer. As the fall arrives and the upper layers of water cool, nutrients mix, and the mixing begins anew. Coastal upwellings result from strong currents along coasts.

We have briefly indicated here, as well as in other chapters, that humans use the oceans for a variety of purposes, but the question remains of whether we have a very large effect on marine communities. In fact, we do not know, for our knowledge of ocean communities is very limited. We have already documented some of the deleterious effects of man's utilization of marine resources, as in the case of the antarctic whaling industry, and there are other disturbing signs that our interactions with marine communities have been less than mutually beneficial. At present, there is a thin film of oil over virtually *all* ocean surfaces. Exactly what effect such oil will have on marine communities is not yet known. Petroleum is known to be poisonous to many marine species, but the threshold concentrations required for poisonous effects remain unknown. Paradoxically, some of the detergents used to clean up large oil spills have caused more damage to marine organisms in the intertidal and shallow waters than the oil itself. Apparently, the safest and most effective means of removing crude petroleum from the sea is still the very old fashioned method of spreading chopped straw on the oil and then gathering up the straw after the oil has been absorbed by it. Oil spills such as that which occurred when the tanker *Torrey Canyon* went aground, and major seepages such as those which occurred in the Santa Barbara Channel are easily identifiable sources of petroleum pollution of the ocean. The danger of such occurrences increases as offshore drilling operations become more widespread and as oil tankers grow larger. At the present time, there are a few 327,000-ton "supertankers" in operation, some 500,000 ton tankers being planned or constructed, and some 800,000-ton vessels being projected. Although these sources could present major problems should an accident occur, it is estimated that spills and major seepages are the sources of only a small proportion of the oil that threatens marine communities. Approximately 90% of the oil introduced into the ocean, or an estimated 2.1 million metric tons per year, is thought to come from the normal operation of oil tankers, other ships, and oil refineries and the disposal of used lubricants and automotive and industrial oils. Another possible source is the fallout of airborne hydrocarbon residues emitted by motor vehicles and industrial sources. Such hydrocarbons are currently being emitted into the atmosphere at the rate of some 90 million metric tons a year, some fraction of which becomes caught in rain. To the extent that the rain carries these materials to ocean surfaces, they increase the base load of oil pollution. If only 2% of the residues emitted enter the ocean, then the annual input of petroleum products may be doubled.

Oil, of course, is not the only material used by man that may represent a danger to marine communities. As we have shown, DDT and other similar pesticides are known to have entered ocean food chains. In connection with this, a disastrous combination of effects may be in the making. Measurements of the natural oil slick on Florida's Biscayne Bay showed that the concentration of dieldrin, a chlorinated hydrocarbon insecticide similar to DDT, in the uppermost one millimeter of water was ten thousand times higher than in the underlying water. (Recall that DDT and similar compounds are soluble in fats and oils.) It is known that the animal larvae and marine phytoplankton and zooplankton spend a portion of the night hours very near the surface and may concentrate these insecticides in their tissue. We have already mentioned the effects that such concentrations may have on other species in the food chain. What this example points out, however, is that the introduction of two or more substances, in this case petroleum and an insecticide, may have a combined effect that is greater than the sum of their separate effects. Among the other potentially damaging materials we have deposited in the ocean are artificially produced radioactive isotopes and "surplus" toxic gases, originally manufactured for military purposes. What effects these substances might have, should they ever escape from the containers in which they have been sealed, are not known. It has been proposed that the sludge remaining after sewage treatment be dumped in the deep ocean, but our knowledge of deep ocean communities is so sketchy that there is no way to estimate the effects of the sludge on ocean communities.

Coastal Communities

Although they represent only a small proportion of the entire ocean ecosystem, coastal communities are a unique and uniquely endangered resource. There are a number of types of coastal communities, including mangrove, coral reef, rocky shoreline, and beach communities, but we shall concentrate here on estuarine and coastal wetland communities, since these are presently in the most precarious position.

Several features of estuaries make them very important. They are highly productive, as we have already said, and this productivity contributes to their importance for coastal fisheries and underlies their importance as breeding grounds for many species of open ocean fish. They are also a widely used recreational resource. At the same time, estuaries are the sites of many coastal cities and receive most land-based waste products.

The circulation in an estuary is unique because of the combination of tidal inflow from the ocean and the net outflow of rivers. As the tides rise and fall, the waters of estuaries move back and forth, which tends to mix the saline and fresh waters. The difference in the densities of saline and fresh water, however, prevents complete mixing. The less saline river water moves out to sea on top of the incoming sea water, and a salinity gradient is established that tends to prevent mixing. Thus, we can characterize an estuary as having saltier bottom water intergrading with less salty surface water, the whole system being sloshed back and forth by tides. Since the saline sea water moves inland, the *net* transport at the bottom, averaged over several tides, is shoreward, whereas it is seaward at the surface. Thus, sediments, detritus, and other materials, as they settle to the bottom of the estuary, tend to be transported in an inland direction. The net result is that nutrients carried in from the deep ocean waters and nutrients carried to the ocean by the river are trapped in the estuary to a great degree and make the estuary a highly productive ecosystem. (See fig. 15.19.)

FIGURE 15.19 The net transport of water at the surface of an estuary is seaward. On the other hand, saltier water moves landward along the bottom, mixing with the fresher water from above, Thus, the surface water in an estuary becomes saltier as one approaches the ocean and at any point estuarine water is saltier at the bottom than it is at the surface. *(After E. P. Odum, 1971.)*

As a nutrient trap, the estuary is an ideal location for the survival of larval forms, and it is not surprising that many open ocean species of fish spawn in estuaries. The land-based input of detritus partially accounts for the richness of shellfish found in estuaries, since nearly all shellfish species are filter feeders.

The productivity of the estuary is the result of phytoplankton, benthic algae, and rooted plant populations. Although these provide food resources for the larvae and juvenile populations of many ocean species that spawn in the estuary, they are also fed upon by species that live only in estuarine ecosystems. These obligate estuarine species are adapted to the tidal and seasonal variations of estuaries.

While the circulation patterns described generally hold for all estuaries, there are large differences between specific examples that we might cite. In the Mississippi River Delta, for instance, the volume of fresh water is so great that the river water overflows the salt water of the Gulf of Mexico, and a "plume" of less saline water is formed that extends a considerable distance to sea. Where the tidal inflow and freshwater outflow are more nearly equal, as in the Chesapeake Bay, a distinct salinity gradient is formed, and salt and fresh water stratify within the estuary. When the action of tides is strong and there is little freshwater outflow, as in San Francisco Bay, the water is well mixed and approaches the salinity of the ocean.
A variety of other factors such as the force of prevailing winds, the nature and slope of the bottom, the amount and timing of rainfall, and the rate of evaporation also affect circulation and salinity.

Unfortunately, however, the same mechanism that serves to trap nutrients and detritus in an estuary also serves to trap contaminants such as heavy metals, pesticides, pathogenic bacteria, and toxins. The introduction of excess nutrients can bring about estuarine eutrophication that is similar to that of freshwater lakes. It is not yet well known what effects such pollution will have on estuaries. Nor can we predict the levels that pollutants might reach with the further growth of cities and the increasing concentration of the human population in and around cities.

The characteristic net upstream transport of materials along the bottoms of estuaries accounts for another of their unique features: They act as sediment traps. Thus, most estuaries are gradually filling in, and the edges are frequently made up of intertidal marshes. These marshes, in turn, tend to trap nutrients and produce great amounts of detritus on which downstream populations may feed. Marshes, however, may more easily be filled in by man than the estuary itself, or navigation channels may be cut through them. Thus, these important communities are rapidly being destroyed. During the 1950s, the rate at which coastal marshes on the eastern seaboard of the United States were filled in approached an average of 2 acres per hour. Such economic development can have little long-term benefit.

Allied to estuaries, although shallower and lacking the characteristic salinity gradient, are coastal lagoons. In many areas along coastlines, the action of tides and coastal currents has deposited more material along the shore than has been carried away; this has resulted in the formation of long shore islands. Behind these islands are bodies of water characterized by their high productivity, delicate food chains, and extensive marshlands. Like estuarine systems, these lagoons are often nursery grounds for many marine fish and shrimp and are also of primary recreational importance. Seeking economic gain, developers have frequently filled in the marshlands and dredged the lagoons to allow shipping enterprises to use them. Waste disposal systems serving the developed areas, cities, and agricultural land tend to increase nutrients in the lagoons and to bring about eutrophication, which decreases land values and the output of coastal fisheries.

Perhaps the most striking example of this sort of deterioration, but by no means the only one, is the South Shore of Long Island in New York. Figures 15.20 through 15.22 show some typical locations in Long Island. The dredging of Jamaica Bay has removed most of the shallow bottom that is normally highly productive of shellfish. The material dredged up, called *dredge spoil*, has been placed on wetlands. Other means of filling in such areas have removed more wetlands and greatly decreased the value of the land. In Hempstead Bay, to the east of Jamaica Bay, some dredging and filling have occurred, but the major damage has been caused by massive growths of thick, matted algae. It is speculated that the algal growths have been promoted by nutrient input from the developed land around the bay. As the algae decompose in their own lower layer, oxygen is removed from the sediments, bringing about the death of benthic animals and plants. Further to the east (see fig. 15.22), agricultural wastes have caused blooms of undesirable phytoplankton species that do not serve as food for other members of the community and outcompete the food species.

These coastal communities and others like them have borne the brunt of a great deal of economic development and are in danger of being completely destroyed. The irony of this destruction, if it occurs, is that it will provide no long-term human benefit—no benefit, that is, except perhaps in making waste disposal a bit more convenient and providing for the economic gain of a few developers. It is possible to justify the destruction of a forest community to bring agriculturally needed lands into production, but the destruction of a biological resource simply as an unintentional by-product of urbanization seems much less excusable and much more avoidable.

FIGURE 15.20 In the western part of Long Island, near New York City, the land has undergone extensive industrial and urban development. Material dredged from the bays and inlets to allow navigation has been piled on marshes, sometimes simply remaining as waste, as in the middle of the photograph; sometimes it is used in construction of residential areas, as in the background.

FIGURE 15.21 To the east of the area shown in figure 15.20, in Hempstead Bay, there is a tremendous influx of nutrients from nearby residential land. This results in development of vast, thick, dark beds of algae which decompose, disturbing other marine life and causing odor and other esthetic problems.

A strong effort is being made by conservation groups to preserve our estuarine and marshland resources. We can only hope that such effort can overcome the political pressures of economic special-interest groups.

The marine environment supports a variety of communities and ecosystems about which we know too little. Although we are aware that many coastal areas are endangered from uncontrolled pollution, as are many lakes and rivers, we still know too little about the ecology of such systems and less about the ocean ecosystem. Only by acquiring more knowledge can we enable ourselves to make the right decisions about the management, utilization, and preservation of these resources.

In this chapter, we have all too cursorily examined a variety of complex communities and ecosystems and our effects upon them. Most such communities are obviously older than *Homo sapiens.* We have seen that humans have had an impact upon the communities of the world, but we may say the same thing of any species with a rapidly growing population. However, the development of agriculture, the consequent growth in the human population, and the recent growth of industrialism have accelerated our impact on the communities of the world. If we are to survive as a species, we must, of course, utilize natural communities. But if we depend on those communities for our survival, then we shall have to learn how to use them wisely; how to insure the survival of the communities themselves.

FIGURE 15.22 The coastal embayments on the eastern portion of Long Island have been less disturbed than on the western part. As shown here the sand flats have little excess vegetation growing on them, except for some dark areas of eelgrass, which is typical of a "healthy" environment.

Thus far, we have looked at the evolution and adaptation of organisms within the natural environment. We have seen how their evolution and adaptation are mediated by genetics and expressed in the physiology and biochemistry of the organism. We have looked at the communities of the world and at populations, and we have briefly explored the effects of mankind on those communities. It now remains for us to examine the evolutionary history of mankind, the interaction of man and natural communities, and the means by which we might gain and use the knowledge that will allow us to live wisely within the biosphere.

Selected Readings

Bardach, John. *Harvest of the Sea.* New York: Harper & Row, Publishers, 1968. An examination of marine communities and what we may do to utilize them intelligently.

Darlington, Philip J. *Zoogeography.* New York: John Wiley & Sons, Inc., 1957. A comprehensive survey of animal distribution.

Kendeigh, S. Charles. *Animal Ecology.* Englewood Cliffs, N.J.: Prentice-Hall, Inc., 1961. A classic text which concentrates largely on North American species.

Neill, Wilfred T. *The Geography of Life.* New York: Columbia University Press, 1969. A well-balanced and interesting treatment of biogeography with a global scope.

Raymont, J. E. *Plankton and Productivity in the Ocean.* Elmsford, N.Y.: Pergamon Press, Inc., 1963. A useful advanced text on a specialized topic of biogeography, in its third edition.

Ruttner, Franz. *Fundamentals of Limnology.* Toronto: University of Toronto Press, 1963. A comprehensive but somewhat dated treatise.

Shelford, Victor E. *The Ecology of North America.* Urbana, Ill.: University of Illinois Press, 1963. A comprehensive account of the communities of North America as they are thought to have existed between the years 1500–1600.

Southward, A. J. *Life on the Seashore.* Cambridge: Harvard University Press, 1967. A brief survey of environments, adaptations, and the methods of study of intertidal and estuarine ecosystems.

Lowdermilk, Walter C. "The Reclamation of a Man-Made Desert." *Sci. Am.,* Vol. 202 (March, 1960), pp. 54–63. Offprint number 1152.

Love, R. Merton. "The Rangelands of the Western U.S." *Sci. Am.,* Vol. 222 (February, 1970), pp. 88–94. Offprint number 1169.

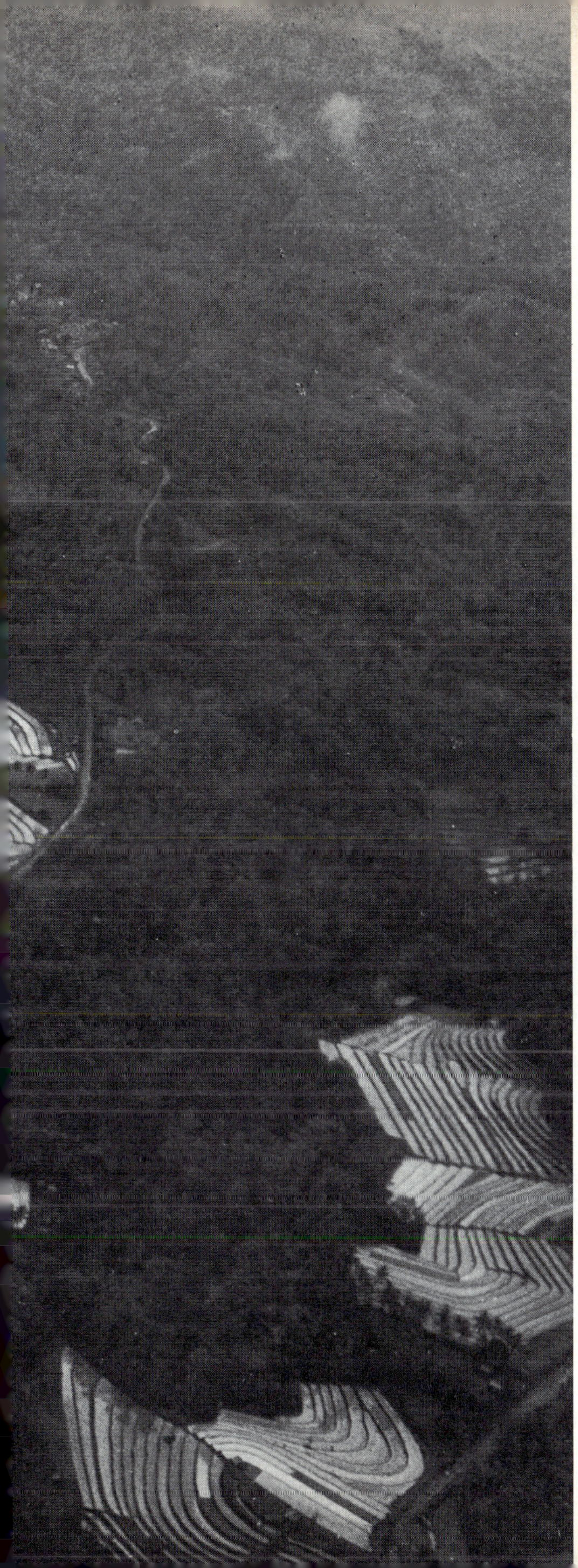

part four
evolution and ecological effects of man

The use of tools has been critical in determining human culture, evolution, and use of the biosphere. *(Courtesy of Muriel V. Williams.)*

the evolution of man

chapter sixteen We can safely say that man is the dominant factor in many communities, having the ability to change all of them. We also know that humans have very little ability to accurately predict the effects that their actions might produce on communities. Thus, we are in a delicately precarious situation that makes research in ecosystem biology crucially important. The purpose of this and the following chapter is to examine the roots of our situation—that is, to understand how we progressed from being members of a few communities to being the dominant force in so many, to learn how our use of communities has changed over time, and to see if there are any general statements that we can make about our future in the natural world.

The first subject that we must examine, then, is the nature of the human ecological niche and how modern society and technology fit into that niche. Therefore, we must consider the evolution of man from the perspective of his adaptations to survive, just as we would examine the evolution of any other animal species. This means that we shall have to consider primitive man in nature and the inherited biological and behavioral abilities that enabled him to survive.

It is appropriate to emphasize here that man *is* an animal, that we evolved from other animals. It is equally important to emphasize that man possesses many abilities that are not shared by other animals. This is not to say that man is better than other species; we are simply differently adapted. There are many species that are more successful than man, and there are species with longer evolutionary histories and larger numbers. The unique abilities of *Homo sapiens* are adaptations of which man can be proud (recognizing that pride itself is a uniquely human characteristic), but at the same time, man must maintain an attitude of humility toward the rest of nature's species. From the evolutionary perspective of man as an animal we shall go on to examine the history of man as an active ecological agent.

We should point out from the start that much of what follows is by necessity oversimplified, and that much of it will fall into the realm of hypothesis.

The Unique Characteristics of Man

First, man is bipedal, walking on two legs in an upright position. Of course, bipedalism is not a uniquely human characteristic, since ostriches, kangaroos, and a number of other animals also walk upright on two legs. Second, man has grasping hands with control of the individual digits. This enables us to hold tools and make artifacts with those tools. Associated with having grasping hands is the fact that man is the one of very few animals that carries food to his mouth rather than putting his mouth in the food. But again, man is not the only animal that has grasping hands and makes and uses tools. The woodpecker finch of the Galapagos Islands uses tools but does not make them. (See fig. 16.1.) Chimpanzees use tools, and they even make them. Chimps have been observed choosing a twig, cleaning the bark and leaves from it, and carrying it from one termite nest to another, where they insert the twig into the nests, withdrawing it with termites attached. They then carry the twig to their mouths to eat the termites. Third, man has binocular vision and can therefore see in three dimensions. Our eyes are located in the head in such a way that the visual fields of each eye overlap, enabling us to view the same object from two slightly different perspectives and to gauge very accurately its location relative to other objects, including ourselves. Like the other characteristics of man, binocular vision is not uniquely human. Most members of the primate order have binocular vision, as do some other animals.

FIGURE 16.1 The woodpecker finch of the Galapagos Islands uses cactus spine to pry insects from wood, and thus it is a tool user. Most nonhuman animals that use tools, however, do so through instinct and do not manufacture the tools themselves. *(Courtesy of Dr. I. Eibl-Eibesfeldt.)*

Thus far, it would seem that what we have cited are the similarities of man and certain other species. The important point about these characteristics is that they are *all* present in the human species. Bipedalism has freed man of the necessity of walking on his knuckles and has enabled him to use his hands to manipulate objects, to look at his work, and to become a superb toolmaker.

But man does not make tools instinctively, as does the Galapagos woodpecker finch. Our toolmaking is a complex, *learned* behavior that requires us to communicate information about how to make things and to use the things that are made. If the tools made and used are to contribute to his survival, then man must also be aware of, and be able to communicate, information about his environment; he must know what a chipped stone, a bone scraper, or a steering wheel is used for.

Thus, the fourth of man's unique abilities is language. Many other animals can communicate information about their environment, although most animal communication is restricted to information about the immediate physical or emotional state of the communicator. But to use tools to *control* the environment requires a special sort of ability—the ability to understand and to communicate, at a level of some complexity, the abstract idea of causes producing effects. It may be speculated that the basic syntactical structure of sentences is of considerable significance—that is, that the subject-verb-object sentence structure reflects an awareness of causality that may be very important in man's evolutionary history. Although it has been claimed that the development of language required a particular sort of evolution of human mouth parts, such structures are not wholly necessary for the development of language. Mynas and parrots can readily reproduce human sounds and human sentences without the structures that are used by humans. This is not to say that the evolution of language in the human species could have been accomplished without the particular structures that allowed the species to produce a special pattern of sounds that we call human speech, but it does imply that such evolutionary developments were of secondary importance. The crucial evolutionary developments that permitted speech occurred in the human brain.

The fifth, and, perhaps the only, unique human characteristic, then, is a brain that is structured and organized in such a way as to allow abstract thinking. Such an organ seems to have evolved only in the human species, and although we know more about it now than we did, say, a hundred years ago, we are still uninformed as to the real nature of the structure of the human brain. Because brains cannot be fossilized, we are even less well informed about the evolution of the brain. There appears to be some correlation between the ratio of cranium and face sizes and the ability to think abstractly, but even this generalization is subject to some serious qualification, as we shall see. In any event, the very complex abilities to think abstractly and to communicate ideas evolved in the human species. Such abilities are critical to, and required for, learning, using what can be learned, and communicating to others what has been learned.

Communication, of course, implies the existence of both a sender and a receiver of information, thus bringing us to the sixth human characteristic, group behavior and social structure. Again, neither group behavior nor social structure is uniquely human. Schools of fish behave as a group, and individual behaviors may be observed within, and relating to, the school. Some animals form societies only at particular times, usually for breeding, as with seals; but some species, such as sticklebacks, form societies only during non breeding periods. In other species, the social structure changes throughout the year. In caribou, for example, herds may be observed at any time of the year, but careful observation will reveal that during the breeding season, a "herd" is usually made up of a single male and a harem of cows, while in the nonbreeding season, the herds are made up of groups of adult males and separate groups of cows and calves. Although most species of mammals are solitary—that is, nonsocial—man and most other higher primates are highly social. The existence of such social groups provides the basis for what in man has come to be the communication of ideas.

Social structures are required for group behavior and thus impose a set of learned behaviors regarding survival within the group. Social structures also provide the context within which such

FIGURE 16.2 The interaction within groups of primates is thought to be a precursor of culture. Although the apparent immediate purpose of grooming behavior is cleanliness, the physical contact involved in grooming activities of primates, such as the baboons shown here, serves to regulate group behavior. *(Courtesy of Masud Quraishy from Bruce Coleman, Inc.)*

learned behavior can be transmitted. Admittedly, such transmission of behavior does not require the existence of a social structure of very complex, or even very large, groups. It can and does occur at a very simple level and in very small groups. The family unit, of course, is relevant to our discussion here. Recall that one of the characteristics of mammals is the care of the young. Thus, mammals usually give birth to few offspring, but there is a great likelihood of infant survival. Since the young learn from their parents, the stage is set for learning from the social group as well, if the young are a part of such groups. They, in turn, pass on what they learn to their offspring. *The transmission of information that is learned rather than genetically determined* is part of the definition of *culture.* Thus, we see that social structures leading to group behavior and the transfer of learned behavior from generation to generation set the stage for the evolution of *culture*, a term usually restricted to describing human activity. Chimpanzees, for example, transmit learned behavior from one generation to another, but we do not refer, except very loosely, to the ''culture'' of chimpanzees. Because culture refers to humans, we may say that it is dependent upon the transmission of learned, abstract information from one generation to the next.

Culture contributes to the survival of the individual and, through the individual, to the survival of the group. In a very real sense, cultures survive through the selection of the fittest, but cultural adaptations that allow survival usually do not arise before selective pressures are exerted by the environment. Thus, the ''evolution'' of culture does not generally operate by the same rules as the evolution of a species. As we discussed earlier (see pages 107, 110, chapter 4), culture is analogous to the inheritance of acquired characteristics (although acquired characteristics cannot actually be inherited).

Thus, biology, anthropology, and sociology come together when we consider the evolution of man. Obviously, all three are related to the ability of man to survive, and evolutionary and environmental biologists have much to say that is of interest to the sociologist, anthropologist, and psychologist. Since man is ecologically the most important single species on earth at the present, environmental and evolutionary biologists may also learn a great deal from an understanding of sociology, anthropology, and psychology.

Before we go on to discuss the details of human evolution, we should first consider some of the disadvantages of the anatomical adaptations that we have thus far presented. It is often said that man is weak and vulnerable when compared to other animal species. Many of our weaknesses have resulted from some of our anatomical adaptations.

As man evolved from other primates, the development of bipedalism required major anatomical changes. Gorillas and chimpanzees can maintain themselves in an upright position for a short time, but to do so is tiring to them because of the structure of their pelvic bones and knees and the development of the gluteus maximus. (See fig. 16.3.) In man, a shorter pelvic bone with a relatively broader ilium, a more well developed gluteus maximus, and locking knees allow him to stand and walk erectly without excessively tiring himself. (See fig. 16.4.) The drawbacks, however, are that man, having a greater proportion of the weight of the body centered over the feet, is subject to flat feet, since the arch of the foot is a feeble support; he is also subject to trick knees, broken knee

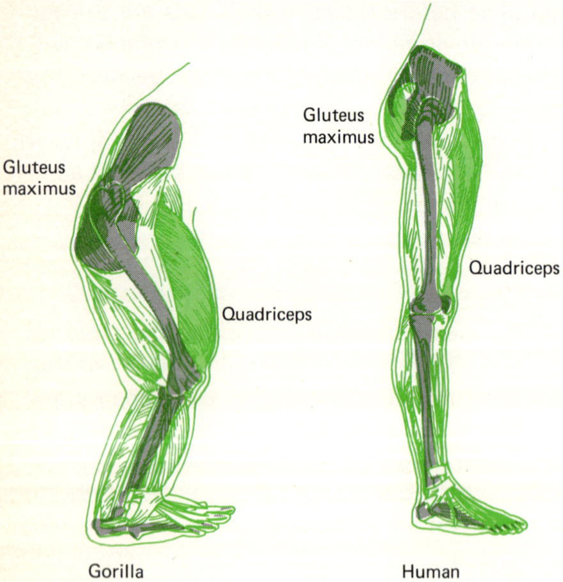

FIGURE 16.3 The poorly developed gluteus maximus muscle of a gorilla makes upright posture difficult and does not allow for efficient bipedal walking. In humans, the gluteus maximus pulls the torso upright and in conjunction with the quadriceps muscle makes possible the swinging motion of the legs. In addition, the weight is distributed vertically on the knee, and the knee itself locks in position so as to hold the legs straight and the body upright.

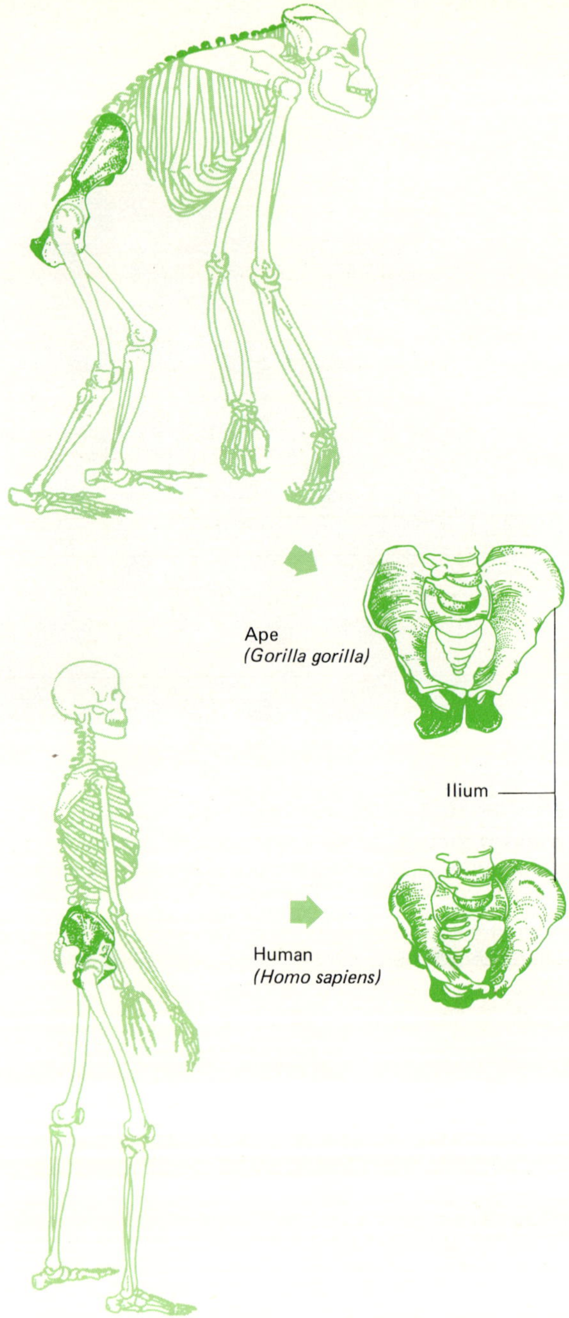

FIGURE 16.4 Like the musculature and legs, the human ilium is better constructed to allow upright posture than that of apes.

caps, various joint diseases, to dislocation of the hip joint, as well as being restricted to a slow mode of locomotion, and to an upper limit of the size of an infant's head due to the size of the female pelvis.

The evolution of bipedalism also brought about changes in the vertebral column. In most vertebrates, the vertebral column forms an arch from which ribs and abdomen hang, thus forming a sort of bag in which the major organs are suspended and cushioned from shock. (See fig. 16.5.) In man, the vertebral column assumes an S-shape, and the internal organs are not cushioned in a hanging sac, but rather rest upon the bony pelvis. A variety of problems arise as a consequence, including abdominal hernias, diaphragmatic hernias, and the parabolic spread of middle-age. Upright posture also exposes the internal organs to wounds that would not be so easily inflicted if man were a quadruped. In the female, the fetus and womb rest on the bony pelvis and are not well cushioned against shock.

The effects of bipedalism on the back are also pronounced. Strained back muscles are common, as are various deformations, and it is easy to rupture an intervertebral disc. In an upright position, the hydraulic pressure produced by a sneeze is tremendous and even can result in such ruptures, more commonly called ''slipped discs.''

The assumption of an upright stance brought about changes in the way the head was supported. The head of a quadruped is attached to the vertebral column in a nearly horizontal position and is supported by very strong neck muscles. In man, the head rests atop the vertebral column, and most of the musculature of the neck is devoted to moving the head, not supporting it. Neck injuries are common, and the skull is very vulnerable in its exposed position.

Brain growth, so closely associated with the evolution of man, occurred within specific

FIGURE 16.5 Forces on the human skeleton are nearly all vertical, but the vertebral column forms an S-shape, which does not provide the archlike support that is possible in a quadruped. The gravity forces on a quadruped's vertebral column are also vertical, but the musculature of the abdomen serves to provide much better ''architectural'' support.

portions of the brain. Those parts of the brain involved in smell and hearing assumed a relatively less important role than those portions of the brain concerned with the control of the hands and abstract thinking. The increased size of the brain was accompanied by a reduction in the thickness of the brain case, and thus, we have well-developed brains in fragile skulls that are in exposed positions.

A host of other human problems may be associated with bipedalism and increased brain size. Sinus infections and diseases of the sinus cavities, problems with the palate, and many dental problems are among the ills to which man is subjected as an evolutionary consequence of the adaptations that have enabled him to reach his present evolutionary stage.

Clearly, the anatomical adaptations of man have been, and still are, largely counterselective. Yet man has been a successful species, and it is equally clear that the reason for this success must lie in other adaptations. To understand the advantages conferred on the human species from these other adaptations, let us review the evolution of man from our primate ancestors.

Apes and Man

Man is a primate, sharing the characteristics of prosimians, monkeys, and apes. These characteristics have provided the potential, the raw material, for the unique capabilities of man.

Since primates are also mammals, the tendency of mammals to care for their young is shared by all primates, even though primates branched off early from other mammals and followed a distinct evolutionary path. Over some span of evolutionary history, a group of mammals, probably insectivores, adapted to life in the trees and gave rise to the primate line. This event undoubtedly took a long time, but it must have occurred relatively early in the evolution of mammals, since some primitive primates living today do not possess very good thermoregulatory mechanisms. Their anatomy is relatively primitive, and they appear closely related to insectivores. The contemporary primates most closely related to ancestors of the primate line are prosimians, and although they once were widespread, they are now largely confined to the island of Madagascar (now the Republic of Malagasy), off the eastern coast of Africa. All of the twenty species of prosimians on Madagascar are lemurs, which are agile, arboreal animals possessing long limbs and grasping digits. Like other arboreal primates, lemurs have binocular vision, which enables them accurately to gauge distances. Thus, the lemurs possess two very important characteristics shared by man. (See fig. 16.6.)

However, lemurs exhibit few of the behavioral adaptations that are normally associated with man. In particular, lemurs are not especially so-

FIGURE 16.6 The mongoose lemur of Madagascar is one of the many prosimians remaining on that island. Its binocular vision helps in locomotion in trees. *(Courtesy of Arthur W. Ambler from National Audubon Society.)*

FIGURE 16.7 Because of their arboreal existence, most primates, including the gibbon, have the ability to rotate their arms over the head and swing from object to object. Humans may use this ability only for occasional calisthenics, but we do have it. Obviously, most other quadrupeds do not.

cial animals, and, except for one species, group behavior is not well developed. Monkeys are much more closely related to man than are prosimians. There are more anatomical similarities between monkeys and man than between prosimians and man, and many monkeys, especially Old World monkeys, are highly social. We have already examined the social structures of the baboons, a species of Old World monkeys. A variety of evidence indicates, however, that man is even more closely related to apes than to monkeys. Among other things, men and apes have brachiating arms—that is, arms that can move in an arc (see fig. 16.7)—and in both men and apes, the snout is very flat, allowing for a large degree of overlap in the visual fields of the eyes. This overlap allows three-dimensional vision at very close distances and thus the manipulation of objects in the grasping digits. Other evidence of relationships between man and apes has resulted from biochemical studies of protein structure. By a technique called *electrophoresis*, certain constituents of blood plasma can be electrically separated and deposited on a treated surface in a pattern that is species-specific. Electrophoretic separations of human, chimpanzee, gorilla, and orangutan plasmas reveal patterns that are very similar and indicate that some biochemical relationships exist between the apes and man.

The ancestry of apelike man is still vague. There are large and significant gaps in the fossil record, and even in the instances when the record is relatively well defined, it is still pitifully sparse. Most of the fossil evidence that is available is fragmentary and consists largely of teeth and jawbones, some skulls, pelvises, and thigh bones. Since teeth have been among the most abundant fossil finds, a great deal of attention has centered on them. One of the things that investigators noticed rather early was that nearly all fossil and contemporary prosimians have at least thirty-four teeth, while nearly all fossil and contemporary monkeys, apes, and men have no more than thirty-two teeth. The number of teeth is only one useful bit of information enabling us to differentiate the remains of prosimians from those of primates. Another key bit of evidence can be found in the shape of the molars, or grinding teeth. All animals that grind their food (horses, cattle, monkeys, or humans) have teeth that are corrugated on top. The arrangement of the corrugations, or *cusps*, is quite revealing. Monkeys have four-cusp molars, but the teeth of apes and men have five cusps. Thus, on the basis of the number of teeth and the arrangement of cusps, anthropologists can determine whether the fossil under examination is from a prosimian, a monkey, or an ape or man. Unfortunately, from these two bits of information alone, one cannot finally determine whether the specimen is from an ape or a man. Another way of looking at fossil teeth and jaws provides the necessary information. In all four genera of apes, the teeth are set in a parallel fashion, and the roof of the mouth, or *palate*, is flat. The human palate is arched, and the teeth are arranged in a flaring jaw that is wider at the rear of the mouth than at the front.

In the 1930s, an anthropologist digging in India found a number of fossils that, from the evidence supplied by teeth and jaws, were classified as monkeys or apes. But one differed from the others. Found in the Siwalik Hills, this fossilized upper jaw showed teeth and a palate characteristic of humans. Its discoverer, G. E. Lewis of Yale University, called it *Ramapithecus* and announced that it was the forerunner of man, an announcement that most paleontologists held to be somewhat premature. For one thing, a lower jaw showing the same characteristics was already in the collection of one museum under the name *Bramapithecus,* and although both dated from the Miocene (a period that began 25 million years ago and lasted some 12 million years), it was still possible that other fossils might turn up that would be more conclusive.

Approximately a quarter of a century later, while digging in the Olduvai Gorge in present-day Tanzania, Louis S. B. Leakey turned up an intact upper jaw so strongly resembling *Ramapithecus* that he concluded that they were of the same genus. In fact, if they were placed together, they formed an almost complete, and unmistakably hominoid (manlike), upper jaw. They were of nearly the same age, approximately 14 million years old. Very importantly, all of the teeth were of the same approximate size. Among apes, the front teeth, incisors and canines, are usually larger than the molars, thus allowing them to tear off the leaves and forage that constitute a large part of their diet.

The lower jaw of *Bramapithecus* still posed something of a problem, but it appears to have been solved when someone suggested that the two might be of the same genus. The two jaws were fitted together, and it became clear that they were from the same creature. Thus, *Bramapithecus* was dropped from the scheme of human evolution.

Although the evidence continues to mount that *Ramapithecus* is an ancestor of man, the first unequivocal hominid is *Australopithecus*. (Note that the term hominoid is defined as ''manlike.'' Gorillas, chimpanzees, and *Ramapithecus* are hominoids. The term *hominid* refers to species that are direct ancestors of man and to species of man.) *Ramapithecus* appears in the fossil record near the end of the Miocene (some 14 million years ago). The next geological epoch, the Pliocene, lasted 10 million years and is distressingly, and as yet inexplicably, devoid of fossil evidence of the evolution of man. Whether *Ramapithecus* evolved into a true hominid remains utterly conjectural.

The next link in the chain of human evolution was discovered almost accidentally in 1924. A professor of anatomy at a South African university was sent some fossil-bearing rock from a quarry in Tuang, South Africa. Most of the fossils in the rock were of monkeys, but one appeared to differ significantly from the rest. He slowly chipped away the surrounding rock and found the face and most of the skull of what was apparently a five- to six-year-old child. He called the skull *Australopithecus africanus* (South African ape) and claimed that it was a hominid. With but a single skull as evidence, the claim was greeted skeptically. Most anthropologists believed the fossil represented a hominoid somewhere between chimpanzees and gorillas. The difficulty with this hypothesis was that it failed to explain how the skull had got so far south. The period during which the "Tuang baby" lived, the late Pliocene, was characterized by dry weather and the gradual shrinking of the forests. By the late Pliocene, the portion of South Africa where the Tuang baby lived was relatively treeless, and if it were an ape, then it must not have been arboreal, which is unlikely. The discoverer of the Tuang baby pointed out one particular feature of the skull that indicated it was not an ape. At the point of the skull where the spinal cord joins the brain, is an aperture called the *foramen magnum*. In the Tuang baby skull, this aperture is located in such a position as to suggest that the skull rested directly on top of the vertebral column rather than nearly horizontal to the vertebral column, as in chimps and gorillas. The Tuang baby, it appeared, was bipedal.

In the 1930s, another paleontologist, digging at sites not far from the one where the Tuang baby had been found, uncovered several nearly complete adult skulls and some partial pelvises that clearly indicated their owners had been bipedal. This find strengthened the claim that *Australopithecus* was a hominid, but it also revealed that there may have been two different australopithecines in South Africa.

In the meantime, after nearly six years of chipping, the upper and lower jaws of the Tuang baby were freed from the stone in which they had been embedded. The teeth that were revealed were basically the same as those of a human child. The incisors and canines were relatively small, unlike the same teeth in an ape, and the jaws were lighter and shorter than an ape's. The skull lacked the prominent ridges of bone to which the heavy muscles characteristic of the massive jaws of a vegetarian are usually attached. In short, all of the evidence available indicated that *Australopithecus* was a hominid.

Man the Toolmaker

At some time in the 9 to 10 million years that separate *Ramapithecus* and Australopithecus, the latter became a gound-dwelling species. The reason for this major evolutionary step seems fairly obvious: The australopithecines, or their ancestors, adapted to a new way of life in order to exploit the food resources of the ground. Like other biological adaptations, the move from the trees to the ground was not a consciously made choice, but was determined by the nature of the environment. Even today, apes spend a portion of their time on the ground, and what is hypothesized to have happened during the Pliocene is that some species of apes simply began to spend more time on the ground than others. These apes had a slightly greater capacity to exploit the food resources of the ground than did others. As we saw in the chapter on population genetics and evolution, if the selective pressures favoring an adaptation to the food resources of the ground were minute, then the transition could have occurred easily over the time span we are considering. Once on the ground, the hands of early man were free to be used for carrying things (food to the mouth, for example), and true bipedalism could begin.

Chimpanzees and gorillas spend a portion of their time on the ground, though, and we might ask why bipedalism has not evolved in these species. It has usually been hypothesized that bipedalism preceded tool use, but a strong case has been made by a University of California anthropologist, Sherwood Washburn, that tool use preceded and favored the evolution of bipedalism.

Biochemically and behaviorally, chimpanzees are the closest relatives to man. They use tools for a variety of purposes, including food gathering, defense, and even personal cleanliness. Chimps in the wild have been observed using sticks to retrieve termites from nests and to probe dead logs for the presence of insect larvae. They have also been observed chewing leaves to make them into "sponges" to soak up water that the chimp cannot reach with its lips, to clean out the brain case of an animal the chimp has killed and eaten, to clean the anal area after an attack of diarrhea, or to dab at wounds. They use sticks as levers to lift large bees' nests and also as weapons, and they have been observed to throw stones at baboons and humans. Perhaps early man was no better than a chimpanzee at throwing stones. But early man probably had the ability to recognize the potential of a thrown stone as a weapon and came to depend on them more than chimps do in hunting small animals. This would have led, according to Washburn's hypothesis, to selection for greater tool use and therefore for bipedalism to completely free the hands for such activities. Such selection would obviously have also favored the evolution of the brain and therefore changes in the size and organization of the hominid brain. Thus, the three developments—tool use, bipedalism, and increased brain capacity—reinforce each other in a positive feedback system and become inextricably intertwined in the evolution of man from his primate ancestors.

Unfortunately, in the early years of our understanding of *Australopithecus,* no definitive proof existed that it was a tool user. The lack of proof was not of great interest, since the question under debate was whether *Australopithecus* was even a man. Complicating the subject further was the existence of two somewhat different australopithecines. One of these, called *Paranthropus* by its discoverer but now known as *Australopithecus robustus,* was larger than *Australopithecus africanus* and was at first thought to be more recent than *A. africanus.* But *A. robustus* exhibited several distinctly more primitive characteristics than *A. africanus.* Its jaws were more massive, its teeth larger, and it possessed a bony ridge on the top of its skull to which the heavy jaw muscles of an apelike vegetarian would have been attached. It became clear by the mid-1950s that *A. robustus* and *A. africanus* were probably different species, and it also became apparent that they had existed side by side for a long time. Equally important, it appeared that *A. robustus* had evolved as a large hominid rather early and remained fundamentally unchanged during its evolutionary lifetime, whereas *A. africanus* had evolved in a smaller, more primitive form and had grown and changed considerably over evolutionary time.

Another theory concerning *A. africanus* and *A. robustus* is that they were members of the same species. According to this theory, *A. robustus* was the male, and *A. africanus* the female. Because of the great differences in the skull, it is not likely that they were of the same species, but this theory cannot be dismissed too lightly.

In the early 1950s, Louis and Mary Leakey discovered the first of a number of hominid fossils at Olduvai Gorge. When this fossil, which was dated at over 2 million years, was studied, it was found to be an *A. robustus.* Later digging at Olduvai revealed a number of *A. africanus* fossils and a variety of chipped tools. It has been hypothesized that the tools were made by

A. africanus, not *A. robustus,* and the hypothesis is supported by evidence from other sites, for although *A. robustus* can be found in a number of places and over a long time span, it has not yet been shown that it can be unequivocally associated with tools. If it was not a tool user, then some of the mystery surrounding the species can be explained. It did not evolve because it did not use tools. Thus one of the key elements of the feedback system was missing, and there was no selective pressure exerted for the use of tools. *A. robustus* would have been subjected to stabilizing selection, but not to directional selection. It is generally agreed that bipedal, tool using *A. africanus* is the direct ancestor of man, and that man evolved in Africa.

We have repeatedly emphasized the slow pace of evolution and the difficulty associated with objectively separating species over time. It is by no means certain that the most recent members of *A. africanus* are not, or do not at least deserve the label of, *Homo.* By whatever name we wish to call them, hominids continued to evolve and make evolutionary advances along the same lines laid down by the brain-tool use-bipedalism mechanism. They also moved out of Africa and radiated to other parts of the world, evolving into a new genus and species, *Homo erectus.*

This next step in human evolution is represented by a fossil skull first discovered in the 1890s. Called Java man after the place where it was found, this skull shows some major evolutionary advances over *Australopithecus,* but because so little was known of the geology of Southeast Asia, it could not be dated very accurately. Later, in 1927, excavations were begun in Northern China in a hill laced with limestone caverns, called Choukoutien. After some two years of digging, a skull was revealed, and the next four months were spent carefully removing it from the limestone. When the skull was free, casts were made of it, and then the skull was compared to Java man. Apparently they were of the same species. They had heavy ridges over the eyes, a low, sloping forehead, and thick skull bones. The Chinese skull, called Peking man, was much more complete, and measurements of the cranial capacity revealed that the brain of this individual had been approximately 1000 cubic centimeters, substantially greater than any ape's. The leg bones of *Australopithecus* that have been uncovered have indicated that although it was an efficient running biped, it walked with something of a waddling, slow gait. The leg bones discovered in the caves at Choukoutien and in later Javanese excavations are almost indistinguishable from those of a modern man. They are longer and straighter than those of *Australopithecus* and suggest that Java and Peking man were better walkers. Given the size of the brain and the other anatomical details known about them, Java man and Peking man were designated as *Homo erectus,* the first true humans.

The caves in China where *H. erectus* was found have revealed a great deal more about his life. A variety of animal bones have been dug up, almost three-fourths of them the bones of deer, not a cave-dwelling species by any means. It appears that *H. erectus* was an efficient hunter. It is also apparent that the meat that *H. erectus* ate was frequently cooked, since a large number of the bones are charred. The earliest signs of the use of fire come from the caves at Choukoutien and are dated at around 400,000 years ago. (See fig. 16.8.)

In addition to the bones of deer, there are bones of rhinoceroses, elephants, horses, camels, and a variety of small game. It is unlikely that the large game was brought down by a single hunter, and thus, it is thought that cooperative hunts must have taken place. Certainly, no solitary man dragged the carcass of a slain rhinoceros anywhere. The important point about such evidence is that successfully hunting such game demanded some skill in communication to organize and conduct the hunt. Such communicative skills were made possible by the continuing evolution of a large brain. The concep-

tual abilities of *H. erectus* were probably almost nil when viewed in terms of modern man, but they were sufficient to allow him to survive and to radiate from Africa through the Middle East to Southeast Asia and China.

Let us speculate on the kind of culture that might have existed. We can readily see the niche that was evolving. Man was clearly an important predator and hunter-gatherer who survived because of intelligence, society, and the use of tools and fire. Humans lived in groups that may have been large enough to be called tribes, and they lived in caves. Whether they had begun to erect other living structures is unknown. Man was a migratory species, but developed the habit of returning to a home base. We will consider the ecological implications of such behavior later, but here we should point out that *H. erectus* was probably an integral part of his ecological community. He was an important part, especially because of the use of fire, but he was, nonetheless, merely a part of the community. At this stage in evolution, man had not yet achieved dominance over any community. Evolution was proceeding, perhaps even accelerating, and all of the elements necessary for a revolution in the human role in biological communities were present.

The evolution of *H. erectus* from *A. africanus* cannot be deduced from the evidence unearthed in Java or China, although there is some tentative evidence that *Australopithecus* had radiated to Java. That they were directly related was assumed by most anthropologists, but only recently has evidence linking them been found. Once again, Louis Leakey, while

FIGURE 16.8 Anthropological excavations require patient, tedious work by people who are dedicated to the careful unraveling of the human past. In an excavation such as this one at Choukoutien, the use of spoons, trowels, and fine brushes is more appropriate for the movement of the soil and rock than are other means of excavation. The markings on the side of the excavation serve to locate artifacts and other remains. *(Courtesy of the American Museum of Natural History.)*

digging at Olduvai Gorge, found fossil fragments of a well-developed *H. erectus*, and even more recently, in strata between the sites where *Australopithecus* and *Homo* were found, he found what appear to be transitional forms between the two.

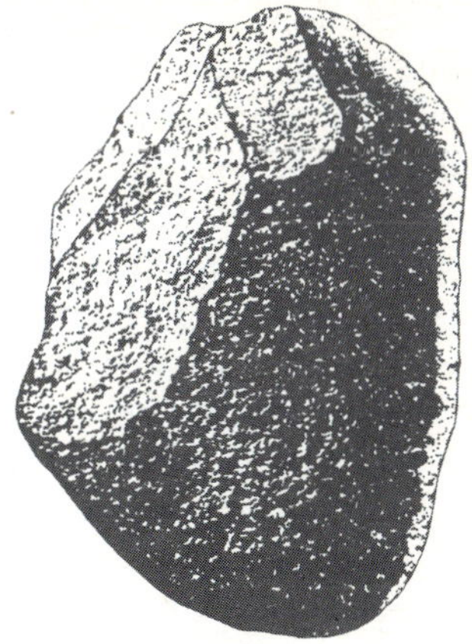

FIGURE 16.9 The earliest human tools were simple chipped pebbles.

Cultural Evidence of Human Evolution

Fossilized bones and teeth tell us only so much about early man, and to learn more we have to turn to the evidence of culture. Obviously, if culture contributed significantly to survival, then we may be able to learn more about early man's niche from a study of culture.

The earliest cultural evidence of man, of course, are his tools. In the fossil beds at Olduvai Gorge, Louis and Mary Leakey discovered so many tools that they gave a name to the culture that they represented, *Oldowan.* Generally, any tool more than 500,000 years old is considered an Oldowan tool, even though it may have been found outside of Olduvai Gorge. The very earliest of these tools are hardly recognizable as such. They are little more than pebbles or small stones with a few chips knocked off one side to provide a very crude cutting edge. (See fig. 16.9.) Such primitive tools persisted through the evolution of *Australopithecus* and into the time when *H. erectus* appeared and are found in Africa, Asia, and Europe. Some small changes took place in Oldowan tools over the 1.5 million years that they were in use, but the changes were usually insignificant. Their makers chipped away larger portions of the original stone, but the tool remained basically a hand-held chopping tool.

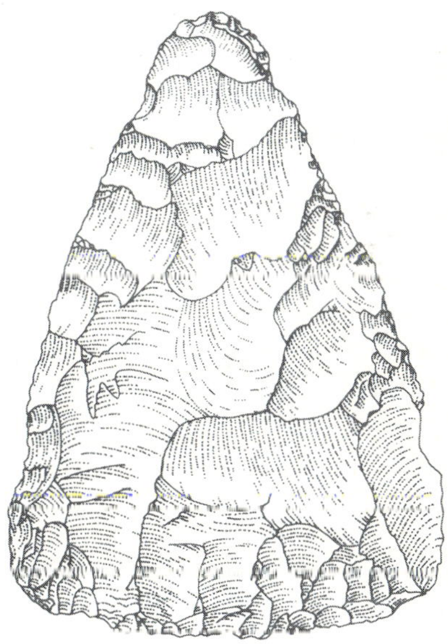

FIGURE 16.10 Obviously, considerable skill is required to chip and form an Acheulian tool such as this cleaver.

Then, about 500,000 years ago, toolmaking underwent a major change as *H. erectus* adopted a more refined toolmaking technique. (See fig. 16.10.) The new tool industry is called Acheulian, after a small town in France, St. Acheul, and came to an end only 75,000 years ago. It is characterized by a relatively large diversity of tools, all of which are thoroughly worked and finely formed. In fact, some are so well made that people today can duplicate them without spending a considerable amount of time practicing the techniques.

Acheulian tools fall into two categories: core tools and flake tools. As their names imply, core tools were produced by chipping away at a stone until the core was shaped into the desired tool, and flake tools were those made from the stone flakes produced in the chipping process. At first, the flake tools were quite primitive, but they reached a state of development that was impressive and required a well-developed intelligence. The original flake tools were simply the chips that remained after producing a core tool, but as time went on, the flakes themselves received more attention from the toolmakers, as they began to work them with great care, producing a flake of some size and then working the edges until they had produced a bladelike flake with two sharp sides. A certain amount of what can be called primitive "mass production" was developed as early toolmakers prepared cores from which usable flakes could be chipped off in quantity. A stone of suitable material was formed into a conical core and then placed on a surface with the point down. By striking around the edge of the upper surface of the cone, a toolmaker could chip off one flake tool after another, the flakes becoming progressively smaller until the core was thrown away. (See fig. 16.11.)

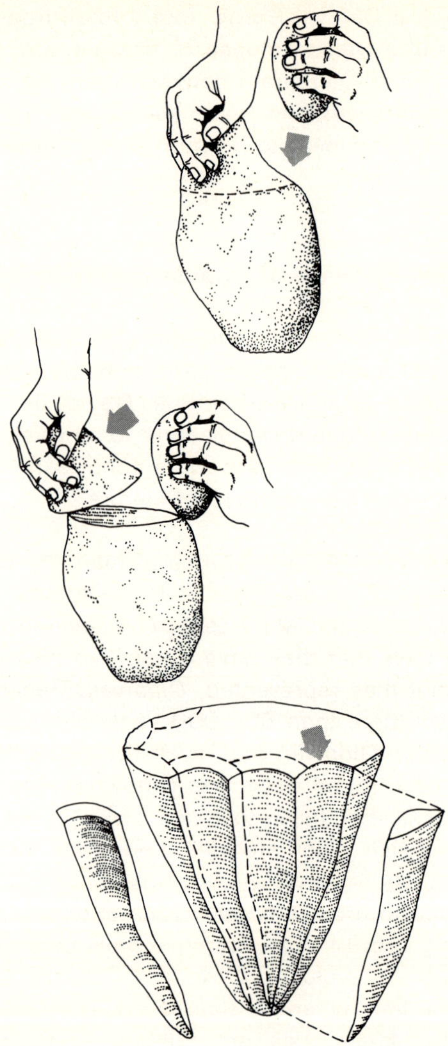

FIGURE 16.11 As shown here and as described in the text, the production of stone tools requires considerable ingenuity and skill.

Acheulian toolmakers expanded their abilities to make tools as they adopted different sorts of striking tools. The original means of producing a stone chopper was to strike the intended tool stone with another stone. Such hammers produced large chips and irregular edges. At some time, it was learned that wood, bone, or antler hammers could produce smaller, more controllable chips and thus more sophisticated tools.

The Acheulian tool industry was begun by *H. erectus,* but before the industry died out, its founder disappeared. The first appearance of *H. erectus* in the fossil record is dated at 500,000 years ago, and the last appearance at around 300,000 years ago. *H. erectus'* successor, Neanderthal man, appears in the fossil record around 110,000 years ago, leaving us with a 190,000 year gap in the evolution of man. Two interesting exceptions to this gap exist. They are both skulls found in Europe in the 1930s, one in Steinheim, Germany and the other in Swanscombe, England. The German skull is the better preserved of the two, consisting of the entire skull and face, but lacking a lower jaw. The Swanscombe find consists only of bones from the top and back of the head. The Swanscombe skull was, in fact, a very disturbing find to anthropologists, since the size and shape of the bones are very similar to the size and shape of the same bones in *H. sapiens,* and yet Swanscombe Man was found in association with Acheulian tools in a stratum that has been reliably dated at an age of between 200,000 and 300,000 years. How could a modern man have evolved so rapidly? When the Swanscombe skull was compared with the Steinheim find, however, certain similarities became apparent. The back of the Steinheim skull is similar enough to that of the Swanscombe skull for both to be classified as normal variations of the same species, but the face of the Steinheim skull, with its heavy ridges over the eyes and relatively low and sloping forehead, suggests its kinship with *H. erectus.* Although it is somewhat speculative, it is thought that the Swanscombe and Steinheim skulls indicate that the back of the human head was evolving more rapidly than the facial bones.

Thus, modern man was emerging. The Acheulian industry developed even more sophisticated techniques and tools, including tools such as burins, which could be used to make needles from bone and antlers, denticulate tools, which could be used for sharpening sticks to a point, and boring tools, which could be used to punch holes in animal skins to be made into clothing. The development of an extensive tool kit followed and allowed man to live in and exploit a variety of environments ranging from tropical and neotropical ones to colder northern climates, where he may even have constructed some crude dwellings. Such a species of man may have been nearly as intelligent as we are now, and hand in hand with the evolution of his culture went a variety of biological changes.

The Rise of Modern Man

The successor of *H. erectus* appears clearly in the fossil record around 110,000 years ago and is called *H. sapiens neanderthalis.* The classic Neanderthal was a short, powerfully built hominid who was probably slightly bow legged. He was very heavily muscled and probably extremely strong. The face of a Neanderthal is likely to be familiar to most readers. A large bony ridge over the eyes gave an appearance that is described most widely and aptly as "beetle-browed." The skull accommodated a brain that was on the average as large as that of a modern man, but was structured differently. The crown of the head was lower and flatter than modern man's, but was longer and rounder at the sides and back.

For some years the only fossil specimens of Neanderthal man were found in Europe, and they all appeared to be relatively the same. Other fossils of other men were found, and it appeared from the fossil record that Neanderthal man disappeared abruptly around 35,000 years ago. It was assumed that the more recent forms of man had evolved separately from Neanderthal and had eliminated him from Europe. As a consequence, it was assumed that Neanderthal man was an evolutionary offshoot of *H. erectus,* and that the true line of human evolution from *H. erectus* lay elsewhere. Then in 1932, a fossil skeleton of a woman was found in Palestine at Tabūn. In most respects she was a classic Neanderthal, but the brow ridges were less pronounced than could be accounted for by normal variations, and the back of the skull was more rounded than that of any Neanderthal hitherto found. At about the same time at Shkūl, a nearby location, some ten skeletons were found in a rock shelter. These skeletons exhibited a startling variety. Some were almost classically Neanderthalian, but some others could hardly be classified as Neanderthals. Their limbs were longer and straighter, and their skulls higher, than Neanderthals', and they had more pronounced chins and smaller faces. Some twenty-five years later, in 1957, a complete skeleton of a Neanderthal hunter was found in a cave in northern Iraq at a place called Shanidar. Like the Tabūn woman, his eye ridges were less prominent than they should have been, and his face looked vaguely modern. The skeletons of five more adults and one infant were also dug up at Shanidar, all showing similarly modern facial characteristics. The specimens from Shanidar appeared to be related to the Tabūn woman and have been considered to be members of the same variant tradition of Neanderthal, but not as clear a departure as the Shkūl specimens. From these skeletons, it appears that Neanderthal man in the Middle East was evolving into a more modern man; these skeletons represent a transitional stage between Neanderthal and the man that replaced him in Europe, Cro-Magnon.

It is appropriate here to recall our earlier discussion of population genetics and evolution. We saw there that a species may contain several geographic variations within its range, but that speciation usually occurred only when there was no gene flow between the demes of the species. The fossil records of Neanderthal man and his successor, Cro-Magnon, seem to suggest that they were variations from place to place and from time to time. The ability of man to survive in a variety of environments, as we have seen from the distribution of *H. erectus* and possibly even *A. africanus,* prevents the isolation necessary to produce complete speciation. There was always some contact and, therefore, some cultural interchange and gene flow. Thus, although there were differences from place to place, they may be seen as forming a cline that exhibited "advances" in culture, brain size, and morphology throughout the world. Some general adaptations conferred selective advantages in all environments, although there were some strictly local variations. The classic Neanderthal of western Europe, for example, may be considered to have been one such local variant. Man is certainly now, and for a great proportion of his evolutionary history may have been, a single species. Geographic subspecies have arisen, but although they may even have been temporarily isolated, the isolation was not sufficiently prolonged to produce speciation. Moreover, human subspecies have exhibited only slight modifications of the general plan and are usually well adapted to the environmental conditions that they have lived in.

Most of what we know of Neanderthal's successor comes from fossils found largely in Europe. First discovered by highway builders who were widening a local road near a cave called *Cro-Magnon* (meaning, in the local dialect, "great big"), these skeletons are difficult to distinguish from those of many humans living today. Cro-Magnons were a taller people than Neanderthals and had noses with high bridges and prominent chins. Contemporary peoples who probably bear the greatest resemblance to Cro-Magnon man are the Scandinavians, Irish, and some Anglo-Saxons. We know, and can know, nothing about the skin, eye, and hair color of Cro-Magnon, nor about the texture of their hair or the shape of their eyes, all characteristics too frequently used to emphasize the differences that exist between humans. Cro-Magnon man possessed a brain that, on the average, was about the same size as that of a contemporary man. The eye ridges were often slightly prominent, but within the range of current human variation. Dressed in the clothing of today, he would probably not draw a second look from anyone.

Cro-Magnon man is notable, however, not merely in the physical differences from his predecessor but also in the use that he made of his intelligence. He adopted some of the Neanderthal tool kit, but made a number of improvements in it, particularly in the realm of bone tools. Much more impressive, however, is another of the cultural achievements of Cro-Magnon man—namely, *art*. Scattered throughout most Cro-Magnon fossil sites are an abundance of paintings, carvings, and sculpture that suggest several things about his life that we can only very tentatively guess at among his forebears. The art of Cro-Magnon man seems to have had magical import for him. Cave paintings, which abound in parts of France, were rarely executed in the caves that were inhabited by Cro-Magnon peoples. To the contrary, most are found in caves that bear few signs of habitation, and often the paintings are located in places that are difficult and sometimes even dangerous to reach. Thus, it would appear that they were not for public esthetic enjoyment. The paintings are predominantly of the animals that were most abundant and that Cro-Magnon man hunted. (See fig. 16.12.) Many of the paintings show the animals with spears in them, or trapped in pits, or otherwise hunted. It is thought that the artists were practicing sympathetic magic; that is, they might have believed that by drawing what they wished would happen, they could somehow make it occur or at least increase the likelihood that it would occur. Conspicuous by its absence is the figure of Cro-Magnon man himself in his art. When man does appear, he is usually drawn as a stick figure, in contrast to the rich detail and color of the animals depicted. This is thought to indicate the magical qualities of the paintings, for if man believed that by drawing the animals he could somehow influence the actions of unknown forces, then by leaving himself out of the picture he would not draw the attention of those mysterious forces to his presence.

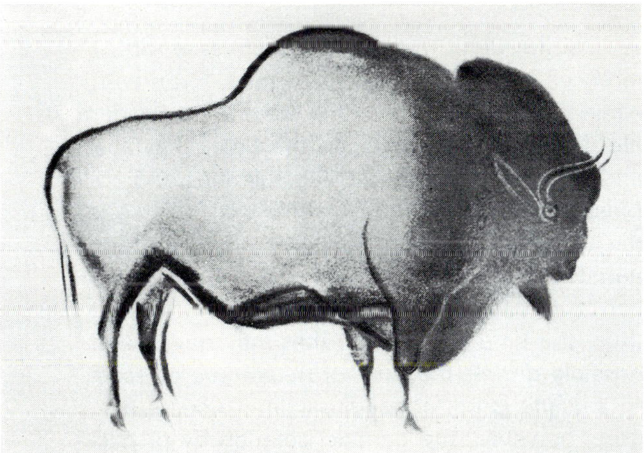

FIGURE 16.12 The creative skill required for the production of cave paintings is evident in this reproduction of a painting of a European bison from a French cave. It is ironic that the hunting of these animals by the Cro-Magnon peoples nearly resulted in their extinction. It is surprising that the bison was not domesticated while many of the other animals portrayed in cave paintings were. *(Courtesy of the American Museum of Natural History.)*

One important exception to the rule of not depicting the human form in Cro-Magnon art is the existence of a number of small statues of females. They are all highly stylized figures, carved from bone or stone, or modeled in clay. The emphasis in the figures is generally on the breasts, abdomen, and hips, the extremities being disproportionately small. These statuettes are thought to be fertility symbols, but they also are speculated to reflect the importance of the role of Cro-Magnon woman during certain periods. At times, the weather in Europe was bitterly cold, and during such times, women remained in the dwelling places and played a marginal role in food-gathering. Their role perhaps became as subordinate as that of a contemporary Eskimo woman. There are very few statuettes from such periods of cold weather. During times that were warmer, although not as warm as the present, women engaged in food-gathering to a greater extent, gathering root vegetables, fruits, and berries in season. The males were engaged in hunting animal food, and the women are thought to have planned and organized, stored and rationed out the food for the winter. During such periods, the artists may have reflected more on the importance of women to their survival and on the questions of birth and fertility, and such thoughts may have given rise to the figurines that are abundant from such periods.

Cro-Magnon peoples took great care with their dead, and a number of graves have been found in which the skeletons bear distinct traces of having been treated after death. One practice was to sprinkle the body with red ochre, perhaps to give it a more lifelike appearance. Hunters were sometimes buried with their spears and bones of animals or mammoth tusks. There are other indications that Cro-Magnon peoples believed in the existence of a life after death.

A consideration of Cro-Magnon man brings us to the edge of recorded history, to the use of metal for tools, and to the discovery of agriculture. These are the topics of our next chapter, but before we go on to them, we shall turn our attention to a subject germane to this chapter and to our lives today, the concept of race in the human species.

The Concept of Human Races

Depending on the criteria used, the races of man have been variously estimated to number from as few as four to as many as thirty-six. The criteria themselves have often been subject to a considerable amount of disagreement, and only relatively recently have anthropologists accepted a biological definition of race, or at least a definition grounded in population biology. A *race,* as defined by one student of evolution, Theodosius Dobzhansky, is *a population that differs in the frequencies of some genes.* One of the advantages of such a definition is that it makes no reference to a phenotypically "normal" individual and thus does not indulge in physical or behavioral racial stereotypes. It also recognizes that even within the gene pool of the race, there is room for individual variation. Another advantage is that the definition recognizes that changes may occur in the race under question as gene frequencies change. Table 16.1 shows the differing frequencies of certain genetically determined blood groups in differing human populations.

TABLE 16.1 Blood Type Frequencies in Selected Human Populations

	Frequencies			
Blood Type	Navajo	Eskimo	English	Aborigine
A	0.013	0.333	0.250	0.216
B	0.000	0.027	0.050	0.023
O	0.987	0.642	0.692	0.766
M	0.917	0.913	0.524	0.176
N	0.083	0.087	0.476	0.824

NOTE: If the frequencies for M and N in the Navajo and Eskimo populations are compared it would appear that they were the same population, but that when the frequencies of the genes for A, B, O groups are compared we can see clear difference for the populations. A similar situation, in reverse, exists with the frequencies in the English and aborigine populations. These differences point up the necessity for examining more than one sort of datum when trying to estimate the differences that exist between populations. The above data were taken from small samples in some instances, and are subject to an unknown amount of sampling error.

(From I. Michael Lerner, *Heredity, Evolution, and Society* (San Francisco: W. H. Freeman and Company, Publishers, 1968), p. 224.)

We cannot deny that there are physical differences between the races of man. Even within some geographical localities there are considerable differences, as between the Scandinavians and Spaniards in Europe, the Pygmies and Masai in Africa, or the Koreans and Chinese in the Orient. But all of these differences are gradual. They are accompanied by a gene flow that has been accelerated by Neolithic culture, and thus, the kind of isolation that may have been possible during the time of *H. erectus* or early *H. sapiens* is no longer possible. To some degree, however, the genetic isolation by physical factors of the environment may have been replaced by some forms of cultural isolation that only recently have begun to break down. The Indian caste system, for example, is thought to have been instituted around 1500 B.C. by light-skinned invaders of India who, after conquering the darker, aboriginal population, set up a rigidly enforced social code that prohibited the intermarriage of light-skinned and dark-skinned peoples. Although some gene exchange undoubtedly took place during the almost 3500 years that the system persisted, this form of cultural isolation may have been nearly as effective as physical isolation. A great deal of study remains to be done, however, to determine the true genetic effects of the system. In China, a rigidly enforced cultural isolation of the mandarins from the rest of the population had what were probably only marginal genetic effects, since the mandarin class was open to the general population through competitive examination.

Attempts are frequently made to draw inferences about the superiority of a particular race on the basis of its involvement in the development of a superior culture. Any anthropologist working with so called "primitive" peoples and cultures immediately recognizes the questionability of such inferences. One can, for example, show that Bushmen of South Africa are better adapted to the kind of life they lead than many Europeans to the lives they lead. Consider the Bushman as a hunter. When he leaves his base camp to bring back food, he may be gone for several days at a time. He may track an animal for many miles, often at a trot. By the time he catches up with his prey, he may have been on the hunt for many hours and many miles. He is probably fatigued, but when he spots his prey he becomes excited, and as a physiological response to the excitement, his body provides a dose of epinephrine, a hormone that operates to release glycogen, cholesterol, and fatty acids from the liver. It also acts to increase the flow of blood to the heart, lungs, central nervous system, and limbs, while reducing the flow to the abdominal organs. The response permits a man to temporarily overcome the effects of fatigue, move more rapidly, and respond to stimuli more alertly. Thus physiologically prepared, the Bushman can overtake and kill his prey. If the prey is a large animal, his arrow or spear may only wound it, and he may have to follow the animal for several more miles until the hunt is over. Epinephrine allows Bushmen to survive, and the increases in blood sugar and lipids are worked off by the strenuous exercise that follows their buildup in the bloodstream.

In contrast, consider a more enlightened, "culturally superior," modern man in American society. Under conditions of stress, his body responds in the same way as the Bushman's. Some stressful situations might easily be imagined; the chagrin of finding oneself stranded in a traffic jam when one must be at a particular place at a particular time is but one of many possible examples. In many situations of stress, a "civilized" human being does not have recourse to the strenuous physical activity necessary to utilize the materials that have entered his bloodstream as a response to the stress. To the extent that any of these materials contributes to ill health (and cholesterol has been implicated in heart disease), our culture is killing us, and we can hardly claim that it is superior to that of the Bushman.

The differences between the cultures of Bushmen and Europeans are of too recent an origin to have been accompanied by any but the most superficial and slight physical and behavioral differences, however, and the differences that do exist can probably be shown to be adaptive.

The attitudes that we often call "racism" are the result of attempts to see human variations as absolute differences between human populations or to attach value judgments to variations in the human population. As we have seen, the variations that may be found in the human species form gradual clines. And as we have just seen in the Bushman example, the idea that some cultures are superior to others is highly questionable.

Much of what is called racism is based on *typological* thinking. The assumption is made that some invariable norm or standard exists to describe *all* the members of a race. One form of such thinking was expressed in the ideas held by Nazi theorists. Nazi typological racism characterized Jews or Slavs as "cunning" or "stupid" or "weak," and Aryans as "strong" or "destined to rule." In each case, these subjective behavioral or cultural characteristics were identified with physical characteristics. Nor have Nazis been the only practitioners of typological racism. We have all heard comments like "Orientals are inscrutable," "Negroes are musical," or "the Irish are drinkers," "this race is lazy" or "that race is strong." It is an easy step from making such comments to asserting that some race is "superior," or another is "inferior." The first obvious fallacy of typological racism is the assumption that all members of a particular race are identical. We have seen again and again that all animal species are composed of highly varied individuals. The argument is also fallacious in its assertion that there is a perfect correlation between certain characteristics, that individuals with a particular eye or skin color will have certain behavioral characteristics.

There may be heritable behavioral differences between the races, but if so, they have not been dissociated from differences ascribable to cultural backgrounds. This has become an acute problem in the measurement of intelligence. The tests that have been used to gauge intelligence have often been highly subjective, and some do little more than measure the cultural identity of the person being tested with the persons who designed the test. Any objective evaluation of intelligence derived from such tests is almost incidental to what has been tested and is largely suspect. Even some of the more "culture free" tests are significantly subjective. There is no question that intelligence is an inherited capacity, but when considering any individual, there is also an important and significant environmental component at work in the expression of intelligence. The relative roles of heredity and environment in the determination of the intelligence of any person is a matter that is far from settled, and the two components appear to be inseparable at this time. The controversies that have arisen about the interpretation of intelligence tests have all too often also included a certain amount of typological thinking. The problem, in part, is that no one has yet been able to gain the objectivity necessary to view different races clearly enough to judge what differences are inherited and how significant they may be.

Most racism in the past has taken the form of typological thinking that we have reviewed above, but a subtler sort of typological racism is also possible. This form of racism is expressed in the tendency of some "liberal" thinkers to deny any possibility of variation in the human species, whether it is of genetic or cultural origin, or whether it occurs within one population or between two different ones. This position has often been adopted by persons arguing against the more obvious form of typological thinking. The form that this subtle racism often takes is that of forcing a minority group to conform to the attitudes of the majority, thereby artificially promoting homogeneity where little or none exists. No individual should be forced to conform to a racial stereotype, either by being segregated into a minority group or by being "fitted in" to some majority pattern.

At this point we should reiterate one of the most important points made in our earlier discussion of variation within a species. The variations that exist in the gene pool of a species contribute to the overall ability of the species to survive. The variations that exist in the human gene pool are, in this context, both beneficial and desirable.

The story of man is not yet complete. Two cultural developments transformed man from a passive to an active ecological agent. *H. erectus*, through the use of tools and fire, and probably dwelling places, modified the immediate environment and was able to radiate into a tremendous number of physical and climatic environments. He remained, however, a "well-adjusted" member of the community. But the use of fire initiated the independence of man from biological energy sources, and the development of agriculture permitted not only the control of the immediate environment but also the manipulation of entire communities. Further refinements in the use and exploitation of energy resources gave rise to modern technology. All of these changes occurred within an amazingly short time and have produced the effects that are part of the major point of this book and the subject of the next chapter.

Selected Readings

*Alland, Alexander. *The Human Imperative.* New York: Columbia University Press, 1972. A consideration of the possible evolutionary sources of human aggression. The role of culture is so important in human evolution, Alland argues, that human aggression cannot be explained in the terms used for other animal species.

*Coon, Carleton. *The Hunting Peoples.* Boston: Little, Brown and Company, 1971. A fine survey of the human populations that, until relatively recently, led the life of the hunter-gatherers.

*Dobzhansky, Theodosius. *Mankind Evolving.* New York: Yale University Press, 1962. Dobzhansky reviews the evidence of human evolution, and considers man's present condition and possible future evolution.

Kummer, Hans. *Primate Societies.* Chicago: Aldine-Atherton, Inc., 1971. An introductory overview of primate behavior which emphasizes the ecological context.

van Lawick-Goodall, Jane. *In the Shadow of Man.* Boston: Houghton Mifflin Company, 1972. The record, told in nontechnical terms, of Miss Goodall's ten years studying a troop of chimpanzees in the wild. A fine and sensitive book.

Lenneberg, Eric H. *The Biological Foundations of Language.* New York: John Wiley & Sons, Inc., 1967. A survey of the evidence for, and an attempt at the development of, a biological theory of the evolution of human language.

*Pfeiffer, John. *The Emergence of Man.* New York: Harper & Row, Publishers, 1972. Drawing together the evidence of human evolution from the fossil record and from primate behavior, Pfeiffer has produced a sound and absorbing book.

Binford, Sally R. and Lewis R. "Stone Tools and Human Behavior." *Sci. Am.,* Vol. 220 (April, 1969), pp. 70–2. Offprint number 643.

Bodmer, Walter F. and Cavalli-Sforza, Luigi Luca. "Intelligence and Race." *Sci. Am.,* Vol. 223 (October, 1970), pp. 19–29. Offprint number 1199.

Clark, J. Desmond. "Early Man in Africa." *Sci. Am.,* Vol. 199 (July, 1958), pp. 76–78. Offprint number 820.

Deevey, Edward S. "Radiocarbon Dating." *Sci. Am.,* Vol. 186 (Feb., 1952), pp. 24–28. Offprint number 811.

Glass, H. Bentley. "The Genetics of the Dunkers." *Sci. Am.,* Vol. 189 (August, 1953), pp. 76–81. Offprint number 1062.

Washburn, Sherwood L. "Tools and Human Evolution." *Sci. Am.,* Vol. 203 (September, 1960), pp. 62–75. Offprint number 601.

*Paperback edition available.

the ecological niche of modern man

chapter seventeen We have seen that each species has a niche that is the product of its evolution and that is transmitted genetically from generation to generation. We have also thus far considered man from an anatomical, physiological, and evolutionary point of view—that is, as a species of animal. It is therefore logical to ask "What is the niche of modern *Homo sapiens*?" In attempting to answer this question, we will be led to some interesting speculations about the place of man in nature and about man's past, present, and future relationships with the biosphere.

We should state at the outset that this chapter is meant to be speculative. We present here our point of view, and we present it to stimulate thought on the part of the reader. Although facts are dealt with in this chapter—facts that deserve the attention of everyone—what we propose here is not as a whole to be taken as absolute truth. It is to be understood as a series of points of view, and it must be understood

The burning of photosynthetic products has drastically modified the earth's biological communities. *(Courtesy of Jacques Jangoux.)*

that there are different viewpoints. We shall take a historical perspective on the activities of man as they have affected biological communities. Such a perspective, we believe, will aid in developing the questions relevant to man's niche.

Man as a Hunter-Gatherer

Largely as a consequence of bipedalism, man is a physically weak and ill-adapted organism. Yet it was the potential for bipedalism that allowed man to use tools, encouraged the capacity for abstract thought and language, and gave rise to culture. Thus, by means of a unique combination of adaptations, man survived in spite of his weakness and vulnerability. What, then, was the ecological niche of early species of *Homo*? To put the question differently, could the niche of early *Homo* be defined in the same terms as those of other animals?

Most young mammals are reared in a protected environment, learning from their parents, from other adults, and from experience. Thus, both learned and inherited behaviors are important factors in the survival of mammals. In primitive man, the learned behavior became more important than the inherited behavior. The advantage so gained was essential to survival, for learning is essential to the development of language, toolmaking, the use of fire, and the other accoutrements of human culture. It was man's learned abilities that allowed him to overcome his innate weaknesses.

Early man was a tribal organism and depended for his food on the biological community in which he lived. As a hunter-gatherer, he was, so to speak, at the mercy of the elements. To this day, there are tribal societies of hunter-gatherers, living much the same way that early man is strongly suspected to have lived. Probably all ancient tribal hunter-gatherers were restricted to living in fairly warm climates with moderate rainfall, to traveling about on foot, and to using simple weapons and tools. Such men

(a)

(b)

FIGURE 17.1 The construction of dwelling places, such as a Masai hut (a) and an Eskimo igloo (b), to protect people from the weather was one of the first steps by which mankind became independent of the immediate environment and escaped the environmentally dominated niche. *((a) Courtesy of Owen Franken from Stock, Boston; (b) courtesy of Steve McCutcheon, Alaska Pictorial Service.)*

were able to survive, as their modern counterparts survive, through a detailed knowledge of their environment, through a thorough knowledge of the location and harvesting times of plants and the behavior of animals. They learned where and how to get food, and they passed that information on to subsequent generations, often via religion or superstition. Insofar as it instilled a belief in what may or may not be done or eaten, religion may be speculated to have had survival value.

Early man's knowledge, gained by experience, was transmitted by the mechanism of culture, but early man still had a definable niche and still had to live within certain environmental constraints determined by his genetic inheritance. At the same time, he had developed a rudimentary ability to change the environment, and thus, the limits of environmental constraint were undoubtedly wide. As man's ability developed, a point was reached at which human culture and learned skills enabled him to extend even further his normal, biologically controlled range. Even before the discovery (or invention) of agriculture, hunter-gatherers were living in every conceivable environment on the earth, from tropical forests and savannas to the arctic tundra. Biological evolution in the human species continued, of course. Human mortality was not random, and thus, the biological characteristics of Eskimos and Masai differ, in part, because of differential survival.

It appears, however, that biological evolution took place at a much slower rate than the other changes that enabled man to survive. The unique set of abilities that man possessed were the product of natural selection, and those abilities enabled him to deliberately or indirectly affect every environment on the earth. For the first time in the history of life, an organism was able to survive by escaping the restraints imposed by its heredity. The ability to think abstractly, to learn, to develop language, and to transmit such abstract learning from generation to generation has enabled man to transmit knowledge and information for survival to subsequent generations outside the information-transfer system of the genetic code. Thus, the abilities and adaptations of man (as determined by culture) could change much more rapidly than those of other animals which were dependent upon the genetic transfer of abilities and the random processes of evolution. The use of rational thought produced further changes in human ability, and the rate at which such changes could take place was not limited by the same factors that limit biological evolution.

For the first time, an organism arose with the capacity to consciously bring about changes in where and how it might live. In a sense, the ecological niche, or at least the niche as determined by heredity, no longer bound man to a particular way of living.

But no animal is ever entirely freed from environmental constraints, and as an animal, primitive man was (as modern man still is) ultimately dependent upon the natural world. In spite of his ability to think abstractly, he still was not able to perceive the consequences of many of his actions. The rational powers of man were limited, and many of the abstract ideas he held were the product of his culture, the vehicle whereby he received the information he needed for survival.

As a hunter-gatherer, man progressed slowly. Clearly, he had the ability to do more than merely passively accept what the community around him provided, and he used that ability to some extent, thus expanding it. But it is at the stage beyond hunting and gathering that man becomes an extremely important ecological agent. Man remained a hunter-gatherer for hundreds of thousands of years, and his ability to go beyond his inherited niche developed slowly. Subsequently, he reached the point at which a cultural development occurred that gave him far greater independence from natural communities than had hitherto been possible and also meant that the concept of the niche no longer applied to him.

Man and Agriculture

We do not know precisely when agriculture was invented or what the first stages in the development of agriculture were. We do know that it was the next, and probably the most important, point in the development of further "advances" in man's control of his environment. The domestication of plants and animals certainly occurred over and over again in both the Old and New Worlds.

Wheat and barley, two staple sources of carbohydrate energy for man, were among the first plants cultivated and appear to have been first domesticated in the Middle East and Turkey. (See fig. 17.2.) Cereals were cultivated at least by 14,000 B.P. ("before the present"). In some locations, the two major progenitors of modern wheat, einkorn and emmer, grew in river valleys as abundantly as in cultivated fields. One theory holds that semimigratory tribes of men learned to harvest these plants and, recognizing the function and value of the seeds, began to carry seeds with them. Apparently, barley was similarly domesticated from several species of the genus *Hordeum*. Other early domesticated grains were millets and sorghums, and these appear to have been domesticated in several places and from several genera. Rice, which is now the major food crop of the world population, was apparently first domesticated in the Orient, probably in what is now India, before 5000 B.P. The domesticators of rice had probably succeeded in domesticating other grains earlier, for the culture of rice is a more complex process than the culture of many other crops. Rye and oats are both hardy crops that are thought to have been domesticated from weedy contaminants found in wheat fields. They may have been cultivated with wheat originally, but the recognition of their hardiness led to their cultivation on land that was not suitable for wheat. All of these domesticates were first cultivated in the Old World.

In the New World (North, Central, and South

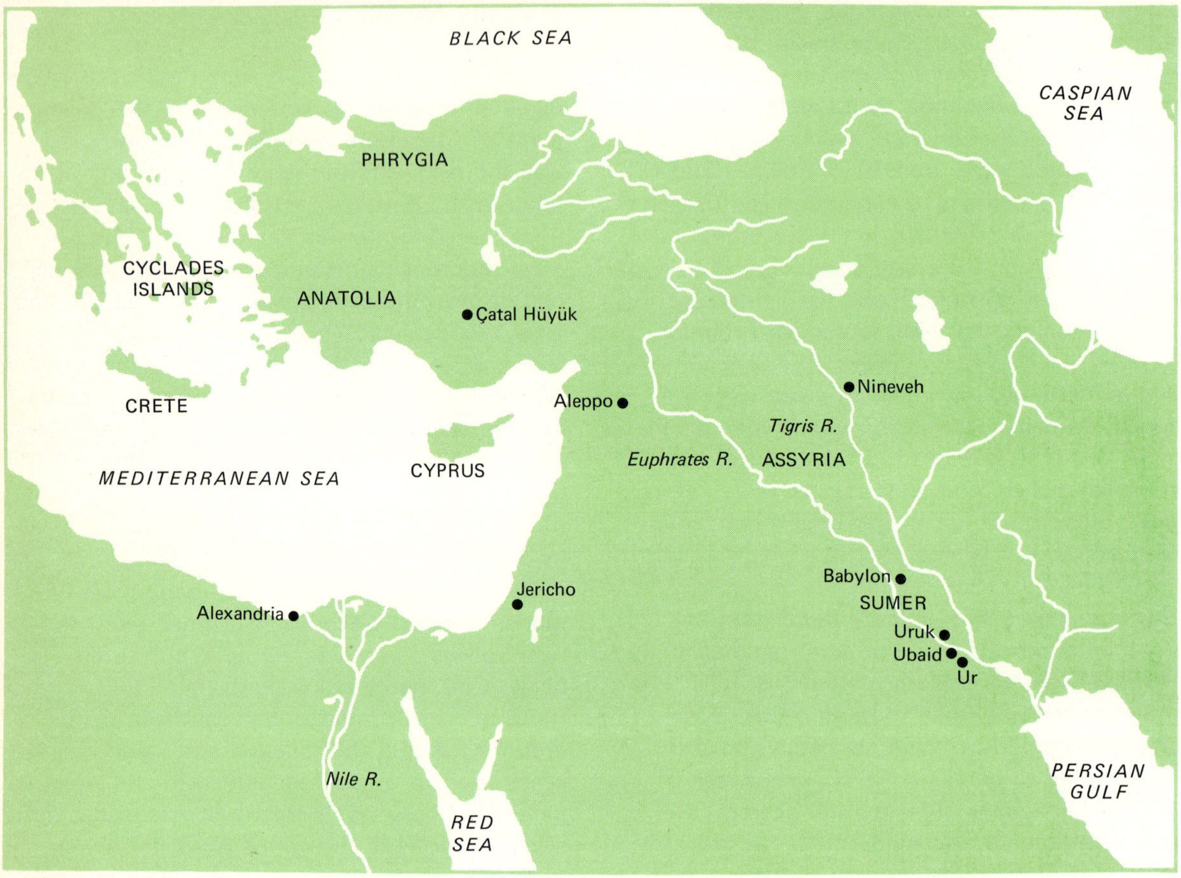

FIGURE 17.2 A map showing some of the important sites in the Near East that have been excavated and where discoveries regarding the introduction and early practice of agriculture have been made.

America), tomatoes, potatoes, Indian corn, pumpkins, summer squash, and peanuts were domesticated. Some New World crops pose something of a puzzle, though. Both cotton and gourds were domesticated not only in the New World but also in the Old World. It is probable that the regional domestications were independent events, but some have argued that this provides evidence of cross-cultural exchange between the Old and New Worlds that considerably predates the sixteenth-century voyages of discovery.

The domestication of animals also probably took place several times, perhaps at around the same general time as the domestication of plants, although not necessarily by the same people. Goats and sheep were probably first herded in or near the mountains of Turkey and the Middle East. Cattle were probably kept by early nomadic tribes living in the steppes of Russia and Asia. In the southern part of central Asia, the yak and the water buffalo were herded in areas where the herding of cattle was difficult, and these animals still fill the same roles in the life of man that cattle fill elsewhere. The original range of pigs, like that of cattle, extended from the European Atlantic to the Asian Pacific coasts, and pigs may have been domesticated several times across the range, but chiefly in the Middle East. One of the earliest beasts of burden is thought to have been the familiar two-humped Bactrian camel, which was probably first herded in northern India. In the

Middle East, the one-humped dromedary was an early domesticate and a beast of burden. The ass was probably one of the earliest African domesticates. The horse was probably domesticated some time later, as it requires more sophistication to use. Dogs, derived from the same stock as wolves, were very early domesticates, although little can be said about where and when they were first domesticated. Cats came later and appear to have been domesticated first in Egypt, where they were useful for killing pests. They were later taken into the home and ultimately deified.

Therefore, most of the plants and animals that are today considered domesticated were domesticated in prehistoric times. Indeed, the activity itself signaled a complex change in the relationships between man and the rest of the living world and formed the basis upon which rests the social, political, and economic institutions of civilization. Hunter-gatherer populations were necessarily small migratory groups that effected few changes in the surrounding communities. Not so agriculturalists. Where soils were somewhat fertile, agricultural groups settled for one or two generations, organized a village-type community, and possibly devoted some time to the pursuit of manufacturing and the arts. In areas of greater soil fertility, as in the Nile, Indus, and Tigris-Euphrates river valleys, the potential existed for the development of small cities dependent upon the continued productivity of the surrounding land. Of course, by contemporary standards, the small cities of prehistoric times would be considered villages. But, although they lived off the land as did hunter-gatherers, the populations of these cities did so by first intentionally altering it. To cultivate domesticated plants, they removed native communities. Such changes in the patterns of distribution and numbers of human populations had profound effects. In the 5000 years preceding the present era, the majority of Old World temperate ecosystems were drastically changed.

FIGURE 17.3 Although we think of the domestication of animals as something that occurred in prehistoric time, the process still goes on among primitive peoples. This bird has become the pet of a Venezuelan Joti Indian girl, but it will probably eventually serve also for food. The domestication of animals is natural to mankind. *(Courtesy of Jacques Jangoux.)*

These changes were accompanied, of course, by correspondingly important changes in human needs and populations.

Before 11,000 B.P., a neolithic culture settled at Jericho, built huts, conducted burial ceremonies, and probably cultivated cereal crops. By 9800 B.P., Jericho was an 8-acre town of approximately 3000 persons. Within a moat that had been dug 10 feet deep and 30 feet wide in solid bedrock, they had built a defensive wall nearly 20 feet high. The wall indicates that its builders may have been active traders who were fearful of attack from other peoples. They also built the earliest known colonnaded temple and may also have constructed granaries to store surplus food. Since they remained sedentary for so long, it is speculated that they practiced crop rotation. They may also have herded domesticated animals, but they left no evidence of it.

A major neolithic settlement to the north of Jericho in present day Turkey tells us a little more about early cities in the Middle East. Çatal Hüyük has been dated at around 8500 B.P. and is twice the size of Jericho. Food particles found in pottery at Çatal Hüyük indicate that the people cultivated both wheat and barley. They also ate peas, apples, lentils, almonds, acorns, fish, fowl, and eggs. They depended to some extent on hunting deer and boar for meat, but they may have domesticated sheep, since they made woolen textiles.

The living sites at Çatal Hüyük are arranged around a series of shrines, reflecting the central role of religion in everyday life, as well as the dependence of the people on agriculture. Their religion appears to have been a fertility cult, which in one form or another is often characteristic of early agriculturalists who prayed for the fertility of the land.

Between 7000 and 5000 B.P., a succession of cultures flourished in the Middle East, each of which is now known by the name of the site that has been excavated. In the Tigris-Euphrates valley, several important developments took place during this period. At Ubaid, excavations have revealed the existence of irrigation canals and examples of the oldest known sailing boats. The importance of the canals is obvious; the sailing boats may indicate that trade and commerce had begun as a major activity. *In nearby Uruk were found the first signs of writing. This writing dates from about 5550* B.P. *and marks the beginning of history.* As importantly, writing enabled man to communicate his ideas and his learning with greater accuracy than was possible when he had to rely solely on memory. As a consequence, the impact of culture was heightened.

At about the same time as writing was invented, so was the plow.

The biological communities of the region, it is safe to assume, were completely changed in this period. The presence of fertile soils, the periodic flooding of river valleys, which replenished the soil and soil nutrients, and the practice of irrigation were all crucial factors in the development of permanent human settlements. As we have seen above, the settlements had profound effects on the land and productivity of the Middle East earlier. These settlements in time developed into the first civilizations.

The agricultural way of life spread slowly from the Middle East into Europe. But in Europe, the problems faced by agriculturalists were different, and the soil could not be treated in the same way. Europe was largely forested, and in order to farm, man first had to clear the trees from the land. Thus, slash-and-burn agriculture was begun. (See fig. 17.4.) An area was cut over, or the trees burned off, and a settlement founded. The removal of the plant community produced some immediate erosion of the soil, but more importantly, as crops were removed, so were nutrients. Without the natural irrigation provided by a river containing silt and nutrients, the soil degenerated, sometimes rapidly. Once the soil lost its productivity, the settlement was abandoned. Nearly all of the temperate decid-

(a)

(b)

FIGURE 17.4 Slash-and-burn agriculture, although not now practiced in most localities, is still used in many parts of South America. The forest is first cleared (a) and then a variety of crops are grown. Corn, gourds, and bananas are being cultivated in the garden shown in (b). The people in this region of Venezuela are practicing a form of agriculture similar to that of the earliest farmers. *(Courtesy of Jacques Jangoux.)*

uous forest east of the Ural Mountains was affected by slash-and-burn agriculture in the period prior to 5000 B.P., and the results were devastating to natural communities. Lakes were filled in with silt, animal populations became extinct, and erosion changed the very terrain in parts of the region.

The Ecology of Civilizations and Empires

The periodic inundation of the major river valleys replenished soils and soil nutrients and enabled man to build permanent cities, develop civilizations and empires, and grow in numbers. Thus, a variety of civilizations flourished in Mesopotamia, Egypt, and elsewhere after 7000 B.P. These civilizations developed industries and engaged in trade with each other and with less advanced societies. They spread their ideas, their culture, and learning through commerce; they also spread the practice of agriculture and its biological effects.

Continued irrigation, however, has the effect of producing accumulations of salts on the land and thereby destroying its productivity. In areas where the annual rainfall is evenly distributed, the accumulation of salts is not a great problem, since the rainfall flushes the land; but in these areas, soil erosion and the loss of nutrients are a problem. Another "revolution" enabled man to overcome this problem to a degree: He began to spread manure on the land, rotate his crops, and grow some crops that were to be plowed back under. These practices made it possible to develop cities and empires outside the major river valleys.

The civilizations of Greece and then of Rome were able to rise. Although the agricultural practices of ancient Greece wrought great changes on the landscape in that country, the legacy of Rome was even more far reaching. The Romans developed a system of huge estates manned by slaves and peasants. Throughout Europe and the other parts of the Roman

Empire, these estates formed the foundation upon which the Empire was supported. The tribes to the north of Rome still practiced slash-and-burn agriculture, but many adopted the Roman system and settled into a sedentary agricultural life. The decline of the Roman Empire carried many Roman institutions with it, but their agricultural system was retained and gave rise to the medieval European manor system. (See fig. 17.5.)

It would be helpful if we could provide more details of the changes that took place in the natural communities of the Old World as agriculture developed and spread. Sadly, the data are lacking. It is known that most of Europe in the early Stone Age was forested, as was eastern North America in the early seventeenth century,

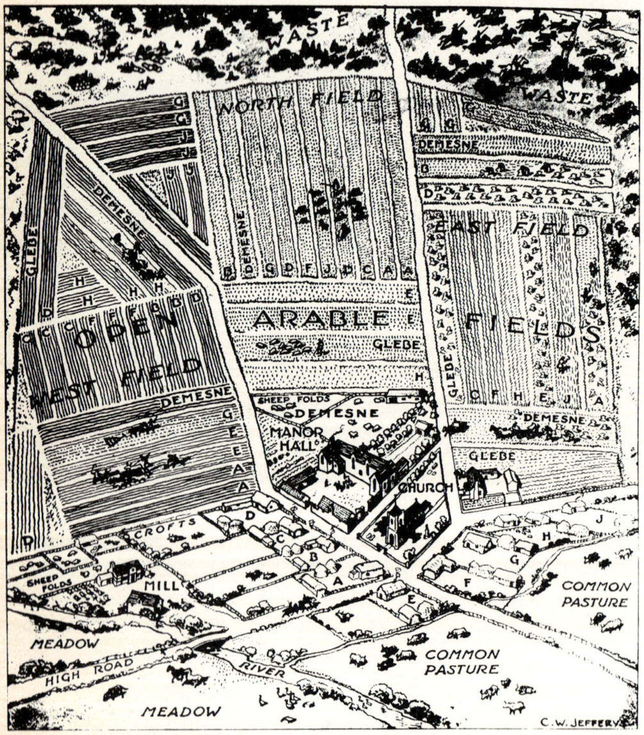

FIGURE 17.5 The organization of farming in this fourteenth-century English manor was derived directly from the Roman estate system, with a central defensive dwelling for the landowner, a small town and farmhouses, and surrounding fields. *(Courtesy of Historical Pictures Service, Chicago.)*

and that most of the Middle East was grassland, not the near desert it is now. The changes that occurred were immense and were brought about without the intervention of modern technology. The ecological "crises" brought about by human activity are not new.

Before the changes became evident, however, the development of agriculture initiated a period of growth in the human population as the carrying capacity expanded. Consider a preagricultural tribe of hunter-gatherers for a moment. There is some evidence that early man, as a hunter-gatherer, may have practiced infanticide. Without cultivation, food suitable for infants is often very hard to find, and nomadic women were probably forced to breast-feed their children for three to four years (as contemporary hunter-gatherer women still do). Having to carry two children while breast-feeding both may impede the search for food resources. The solutions to this problem include abortion, taboos against pregnancy during lactation, and infanticide. Data gathered recently from aboriginal hunter-gatherers indicate that in the not too distant past, they killed between 15% and 50% of their infants. As agricultural villages arose, however, the food supply grew, and the domestication of sheep, goats, and cows provided alternatives to the milk of a mother. Thus released from the constraints of breast-feeding for extended periods, the women could raise more children, and taboos and harsh practices such as infanticide and abortion were relaxed or done away with. Infant mortality fell, of course, as did other mortality rates. The food supply was regular and predictable, if not always abundant. The sick or lame did not have to be left behind as the tribe moved on to find food. They could be fed until they were well, or in small cities, they might turn to pursuits other than food-gathering. Natality also rose as parents grew healthier.

Thus, as agriculture spread and agricultural practices improved, the carrying capacity of man expanded, allowing large civilizations to rise and the human population to grow.

The growing population led to an increase in the demand for food, however, and therefore to the utilization of greater amounts of land. Ever larger cities had to be built, and yet more land had to be put under cultivation. The growth of cities, civilizations, and empires led to the use of land at its maximum capacity. At the same time, the growing cities provided ideal conditions for disease organisms. The rich of Rome moved away from the city in the summer to avoid the malarial swamps of the Tiber, and examinations of Egyptian sarcophagi indicate a high incidence of schistosomiasis. Traders and armies spread disease and parasites, and organized warfare killed people and crops, either directly or indirectly. Typhus and typhoid fever accompanied all armies, and the literature of the ancients is filled with references to famine. The populations of many areas stabilized, while others declined, and the ''golden'' ages of most civilizations produced populations barely able to be supported by the land, societies based on slavery and massive peasant classes, and massive outbreaks of famine and warfare. Man was approaching his carrying capacity and was beginning to encounter environmental resistance.

The carrying capacity of man took on a new connotation. It was no longer the product of the natural community, but the product of the cultural and agricultural systems that man had devised. In more primitive peoples, there were no parallels to epidemic disease. Rather, disease was confined to a small, somewhat isolated population. Food shortages among primitive peoples affected only a few families who could survive by gathering some other food. Having come to depend on monoculture, whole cities and nations starved. In many ways, the diversity of the biological community had protected preagricultural populations. That diversity had been done away with at the founding of civilization.

We have, in the span of a few pages, considered a great deal of history. In the light of this history, we should now pose again the question raised at the beginning of this chapter. What is the hereditary niche of man? We have answered the question for the hunter-gatherer; he lived as a part of the community and was limited by his food supply and the vagaries of the community he depended upon so intimately. The development of agriculture altered man's relationship to the communities around him. He no longer lived as a part of a community; he produced a low diversity, crop-dominated community to support him. This brought about drastic changes such as the destruction of the soil through nutrient depletion and salinization, the removal of forests by migratory slash-and-burn agriculturalists, and the growth of cities in river valleys.

Population increases followed, and as agriculture improved, man approached a new carrying capacity that was in large part determined by the use of nutrient cycles. The Roman Empire depended largely on the estate system of agricultural production and on the commerce in cultural and agricultural products which supported cities as the centers of empire. These cities, however, disrupted nutrient cycling, and downstream aquatic communities suffered as the land gradually lost much of its productivity.

To a great degree, then, human culture supplanted the hereditary niche. No longer dependent for survival on the evolved role within communities, man came to use the information-transfer processes of culture and learning rather than the information transfer provided by his genes to survive. Although man still survived in numbers that were determined by the food he could produce within the constraints of his agricultural, commercial, and political systems, certain aspects of culture, such as the transmission of disease by armies and traders or warfare itself, limited his population. *The idea of a hereditary biological niche does not seem to apply well to man.*

With the fall of Rome and the decline in commerce and stability of government, many other changes in man's use of resources occurred. But the stage was being set for another series of changes that were to have effects nearly as profound as the initial development of agriculture. Moreover, the rate at which these changes occurred was greater than the rate of past changes. We shall examine some of the factors that led to the industrialization of Europe, for like the use of agriculture, this phenomenon of Western culture has influenced every aspect of life on earth.

Nutrient Cycling and the Rise of Civilization

Growing populations and expanding agricultural activities were not without their effects on nutrient cycles. River valley agriculture or agriculture stimulated by irrigation allowed a flow-through cycling with seasonal removal and replenishment of necessary nutrients such as phosphate and nitrate.

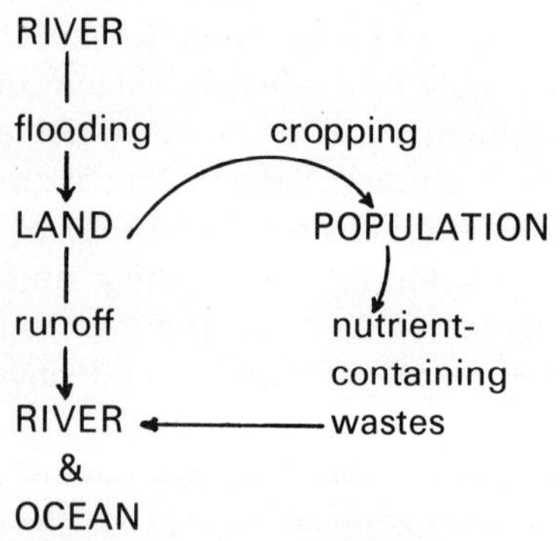

The practice of slash-and-burn agriculture lacked this sort of replenishment.

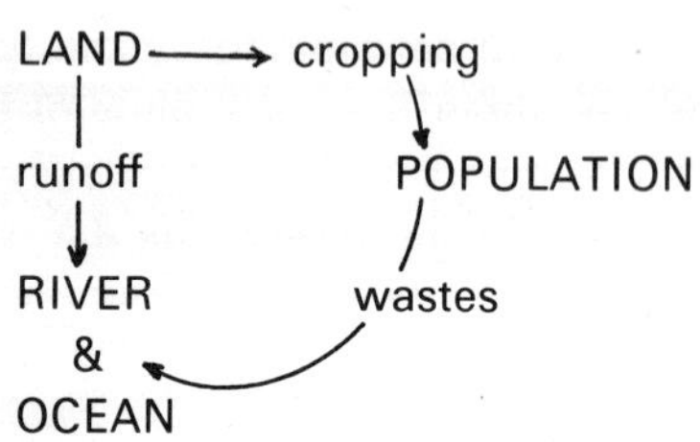

Obviously, supporting long-term productivity was not possible. The use of fertilizers, chiefly manure, altered the cycling somewhat and made more extended use possible.

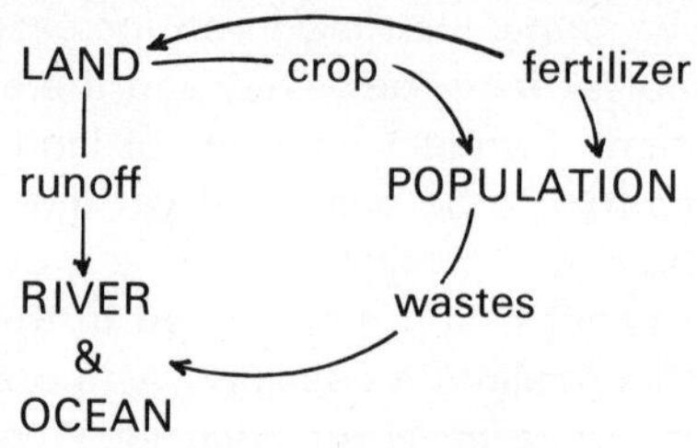

The development of cities brought about further changes in nutrient cycles. While recycling is possible in a small agricultural village, on an estate, or on a medieval manor, it is not economical in large cities. Thus, the countryside surrounding the large city becomes depleted of nutrients, and water pollution can begin to take place. The stench of the Tiber below Rome was legendary.

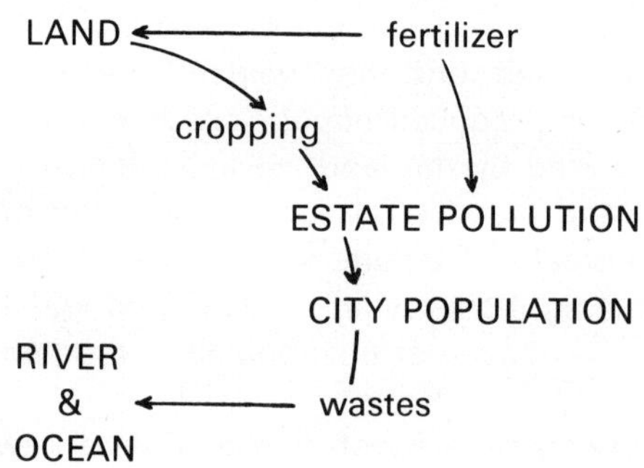

A limit was thus imposed on the number of people a city could contain, the limit being determined by nutrient cycling (and by transportation of agricultural goods). It is important to point out that man had now altered natural communities not only deliberately but also inadvertently by his waste disposal systems.

Ecology of Europe to the Beginning of the Industrial Era

The decline of Rome and the disintegration of its empire were accompanied by major changes in commerce, the arts, religion, and law, but not in agriculture. Agricultural practices, crops and farming communities continued much as they were. Europe and most of Asia prospered agriculturally, supporting and being supported

FIGURE 17.6 Improved techniques of agriculture, particularly the introduction of a suitable plow, revolutionized farming in northern Europe. *(Courtesy of Historical Pictures Service, Chicago.)*

by a large peasantry and limited, though important, commerce. In Asia and the Byzantine Roman Empire, ''high'' culture continued, and a number of factors more properly discussed in history books eventually led to a resurgence of European commerce. On the Asian continent, the human population suffered the vagaries of periodic famine and disease, while in the Middle East, the soil that gave birth to agriculture gradually deteriorated, and the human population declined. In Europe, the manor system continued.

After about A.D. 1000, commerce with the Orient, the Crusades, the opening of river traffic, improvements in maritime shipping, and the stabilization and growing centralization of governments were among the many factors leading gradually to a decline of the manor system and to the slow growth of large, trade-based cities. These cities were all situated either on rivers or coasts because of the importance of water-based transportation. Once again, the patterns of nutrient cycling, the use of land, and the sizes of human populations were to change. The changes occurred very gradually, of course, although much less gradually than the changes discussed above.

One medieval farming practice illustrates how agricultural innovation provided some impetus to the growth of cities. In Roman times, the horse was rarely used for plowing, since the only means of harnessing a horse to a plow was to tie a rope around its neck. A heavy load would thus choke the animal. Moreover, the hoof of the horse was relatively soft and easily injured in the broken, rocky sod turned over by the plow. Also in the damp soil of northern Europe, the hoof of a horse was subject to rot. As a consequence, oxen were the predominantly used draft animals in Europe. The use of oxen continued well into the medieval period. Around the seventh or eighth century, a heavy, wheeled plow was invented. This plow was equipped with a curved metal plate that broke, lifted, turned, and pulverized the heavy soil characteristic of northern Europe. (See fig. 17.6.) Such a plow was far too heavy to be pulled by horses, and teams of eight oxen were widely used. During the reign of Charlemagne, the horse collar was invented and came into use. But the horse did not immediately replace the ox at the plow. Around the fourteenth century, the iron horseshoe was invented, enabling man to use horses as draft animals.

Replacing the team of oxen with a team of horses increased agricultural productivity. A horse eats more than an ox, does not live as long, and when it grows too old to work, cannot be fattened for food. But a horse is faster than an ox and thus can do more work in the same amount of time. The horse is not only faster but also stronger; it has greater endurance than an ox and can work two or three hours more each day. The peasant could consequentially put more land under the plow, and in some areas of Europe, land came to be measured in units based on how much land could be plowed in one day by two oxen or one horse. Thus, with less human labor, greater agricultural productivity was achieved, and food surpluses gave impetus to population growth. At the same time, the cultivation of legumes was increased, improving the diet of the peasant and the fertility of the land.

The Renaissance was presaged by the improved use of wind and water energy in processing cereal, further improvements of the plow, the development of canals and shipping, and improved road building. All of these activities stimulated the economy and were closely associated with the use of energy. To perform work, man no longer was solely dependent upon the energy he gained by eating. As we shall see, this has been of major importance.

But medieval Europe was essentially an agrarian society. Manors were self-sufficient entities, dependent on the outside world only for luxuries. The excess food supplies that they had started to produce were shipped to cities and larger towns. The surpluses allowed a degree of stability to be achieved as long as trading relations remained stable. The state of biological communities depended on the economy, and archaeological investigations in England have suggested that patterns of land use varied from time to time. Thus, forests were cleared and then allowed to grow again, towns appeared and disappeared, and the species that were favored by man or flourished in the patchwork of communities encouraged by man became dominant. Among the species so favored were the hare, fox, grouse, and hedgehog.

The patterns of nutrient cycling resembled those described above for cities:

FIELDS ⇅ (and FIELDS ⇄ MANOR)
PEASANT FARM → MANOR → TOWN OR CITY → RIVER & OCEAN

Some soils deteriorated due to the export of nutrients and were left to lie fallow until they could recover their fertility, thus dictating changes in land use patterns. In the cities, rivers and canals frequently became little more than open sewers, and aquatic communities deteriorated.

The human population grew steadily under these conditions. At the time of the Roman Empire, the population of Europe had probably reached a plateau that remained relatively stable until about the tenth century and then began to grow slowly in the next three centuries. The urbanization of Europe led, however, to an overload in the system. Warfare and epidemic diseases reduced the populations of the cities and of the peasantry tilling the land. Thus, a sort of unstable carrying capacity was reached. The bubonic plague (accompanied and furthered by warfare and armies), fire, and famine swept back and forth through Europe beginning in the fourteenth century. A form of bubonic plague called the Black Death, originating in the Orient and carried to the west (probably on Genoese ships) broke out in 1348, and in the course of a few years, it is estimated to have wiped out between a quarter and a third of the population of Europe and Byzantium. In some places, its effects were even more devastating; the city of Florence is thought to have lost nearly two-thirds of its population to the Black Death.

The population of Europe remained relatively stable from the middle of the fourteenth century to the end of the sixteenth century. To be sure, in some localities, it rose, but in others it fell. Germany suffered a major population decline during the Thirty Years' War (1618–48); France suffered declines during the War of the Spanish Succession (1701–14) and the Religious Wars (1562–98); and Russia, after the death of Ivan the Terrible, suffered a decline during the Times of Trouble (1530–84). In England, the population grew, but slowly. Nor were the direct effects of war the only causes of population fluctuations. Armies indirectly caused population reverses by destroying crops, but poor weather also caused crop failures and famines, and diseases took their toll as well.

Thus, a complex set of interactions involving agriculture, nutrient use, population, economics, politics, religion, and social structure determined the effects of man on biological communities. We must now consider another factor, enegy use, for another revolution was in the making. Developments in the availability of energy were to bring about drastic changes in the life of man and the biological communities upon which he was dependent.

Energy Use and the Rise of Industrialism

Thus far, our discussion of man's effects on biological communities has centered on the impact of agricultural practices and the modification of nutrient cycling. Clearly, human affairs were intimately related to the human use of biological resources and communities, and during the period we have reviewed, man was almost entirely dependent upon the energy produced by

FIGURE 17.7 The worst effects of the bubonic plague were felt in the city of Florence, where the mass burial shown here occurred. The plague was most devastating in the cities of Europe, and although at the time many sought mystical or religious explanations for the occurrence of the catastrophe, we now know that it was caused by ignorance of city planning, hygiene, and sanitation. *(Courtesy of The Bettmann Archive, Inc.)*

photosynthesis. Photosynthesis provided man with food and with heat for homes. Man had also begun to exploit other energy sources, primarily wind and water. The next major change in the relationship between man and biological communities hinged on changes in the pattern of energy use.

In order to gain some perspective on the changes that were to come, let us briefly review some of the salient features of man's use of energy before the rise of industrial Europe. With one exception, the hunter-gatherers' only consumption of energy was the eating of plants and animals by the members of the tribe. The exception was the energy that they used when burning wood for heat or cooking. Hunting, harvesting, and transportation were all performed by the expenditure of human energy derived from the food that man ate. Thus, primitive man was limited by the energy that he was able to gather in his hands.

The domestication of animals allowed man to exploit the energy resources of animals and allowed him to engage in trade, transport agricultural goods to markets, transport luxury goods, and extend culture overland at an accelerated pace. The use of draft animals in agriculture for plowing, hauling crops, and grinding grains also enabled man to exploit hitherto unavailable energy resources that were critical in the development of more extensive agriculture and in the growth of towns and cities.

During the Middle Ages, man began to tap the energy of wind and water currents, energy derived directly from the heating of the atmosphere and water by the sun. Windmills and waterwheels came into limited use in processing crops and aided in increasing agricultural production, for some land that had been devoted to providing forage for draft animals could be turned to producing food for the human population. (See fig. 17.8.)

In spite of all of these changes, though, as late as the middle of the eighteenth century, man

(a)

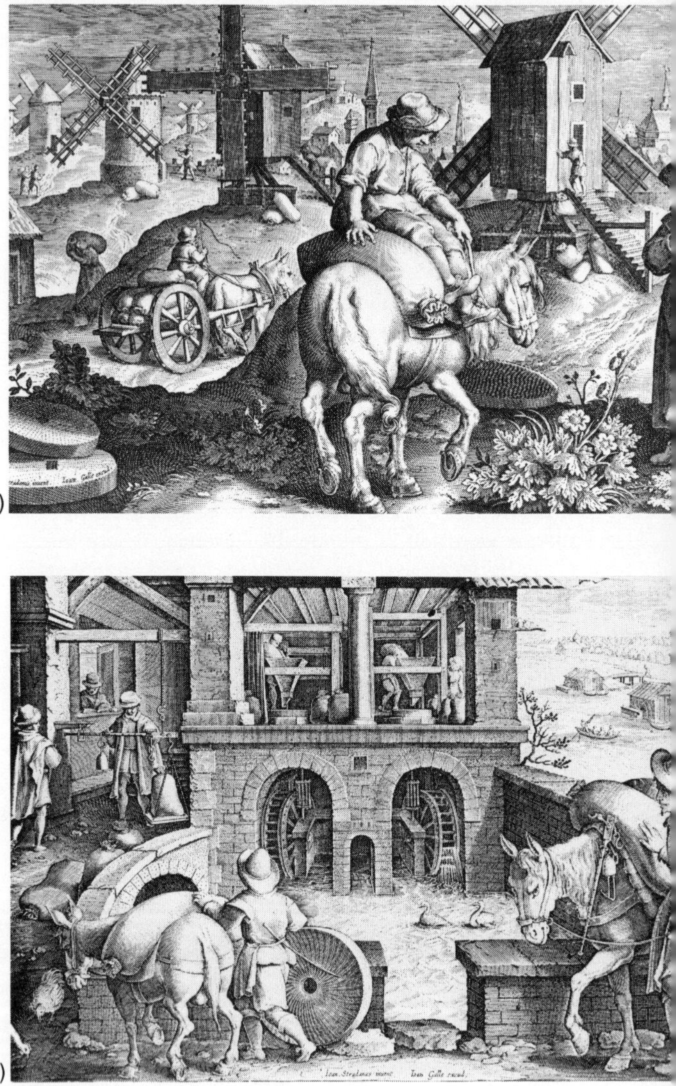

(b)

FIGURE 17.8 The use of wind (a) and water (b) as sources of energy was introduced during the Middle Ages, and for the first time mankind had available a source of energy that was not directly derived from the processes of photosynthesis. *(Courtesy of The Bettmann Archive, Inc.)*

was still primarily dependent upon but a few sources of usable energy—photosynthesis, wind, water, and draft animals. The photosynthetic production of carbohydrates and plant tissue was still the chief source of energy, whether the products were burned for warmth, converted to meat by domesticated animals, or directly consumed as grains and vegetables.

Beginning in the mid-eighteenth century and extending through to the twentieth century, major changes took place in man's exploitation of energy resources. The changes that occurred were bound up with a number of cultural factors and thoroughly altered patterns of land use in Europe and North America, and ultimately around the world.

At first, industrialism was a phenomenon largely confined to England, although its effects were felt throughout the world. One of the key factors in the rise of industrialism was the existence of a market, a population sufficiently large to spur technological innovation, and England was undergoing a population boom. In 1707, the unification of England and Scotland into a single political entity created a free trade area that made Britain the nation with the largest internal market in Europe. Between 1750 and 1832, the population of England and Wales doubled. The growth of industrialism also requires concentrations of labor that do not exist in an agrarian society, but the patterns of land use in England created cities having just the concentrations needed to man the factories that would spring up. English landlords, seeing greater profits in using their land to graze sheep for wool than in renting fields for cultivation, began placing their properties in pasturage. In the meantime, the English yeoman farmers were able to produce sufficient food to feed a growing urban population.

A variety of technological advances took place in the textile industry, among them the substitution of water power for human energy in spinning yarn and weaving it into cloth. The English textile industry, in fact, provides an instructive insight into the agricultural and social implications of changing patterns of energy use. Early in the eighteenth century, cotton was introduced into England as a competitor to wool. At first, all cotton was imported from India, already woven and printed. Under pressure from the wool interests, Parliament passed a law restricting the importation of printed cotton goods. This induced the fledgling cotton industry to seek efficient, economical ways of producing cotton in England. Among the innovations were water-powered spinning machines that allowed those who spun yarn to keep up with those who wove cloth. Previously, it had taken five persons spinning to keep one person weaving. The new technology of spinning was so successful that weavers could not keep up at first. Mechanized looms were slow to take hold, but by the 1820s, such looms outnumbered hand looms. As a result, Indian agricultural cotton production could not keep pace with the English demand. The invention of the cotton gin in America in 1793 had made it possible for the English to use American cotton, and ever-larger areas of land in the American south were devoted to the cultivation of cotton. Between 1793 and 1813, American cotton production rose from 1.5 to 85 million pounds per year. Accompanied by the increase was a large increase in the importation of slaves into North America, and the forced deportation of Cherokee Indians to land in the west. Thus, by changing the source of energy to power spinning machines, land use patterns in America and India were altered.

But the growth of the English textile industry cannot be attributed solely to the introduction of water-powered spinning machines. The looms that were developed required greater inputs of energy than could be supplied by water alone, and the growth of the textile industry coincided with another invention, the steam engine. At first, the steam engine was only used to pump water out of mines, but by the end of the eighteenth century, a variety of improvements in the engine allowed it to be used in other ways. Water-powered spinning machines and looms

FIGURE 17.9 This advertisement for a portable steam engine made in 1798 illustrates well the difference between biological and mechanical sources of energy. *(Courtesy of the Science Museum, South Kensington, England.)*

were quickly replaced by steam-powered models. By the end of the first third of the nineteenth century, steam engines were in use in printing plants, flour mills, breweries, lumber mills, and a host of other industries. The biological energy of man and draft animals and the solar energy in wind and moving water were about to be replaced by the energy in fossil fuels.

The steam engine also had an immense impact on transportation. The impact was slow to be felt, however, for the development of a rail-based transportation system required large amounts of iron. At first, the smelting of iron ore required large amounts of charcoal. England, its forests depleted, imported charcoal from nations such as Sweden and Russia where the timber supply was more plentiful. Soon, however, further technological advances made the use of coal practical in smelting iron, and the railroads were able to develop rapidly. Any agricultural land overlying a coal deposit sud-

FIGURE 17.10 The railroads in the United States expanded greatly between 1840 and 1860.

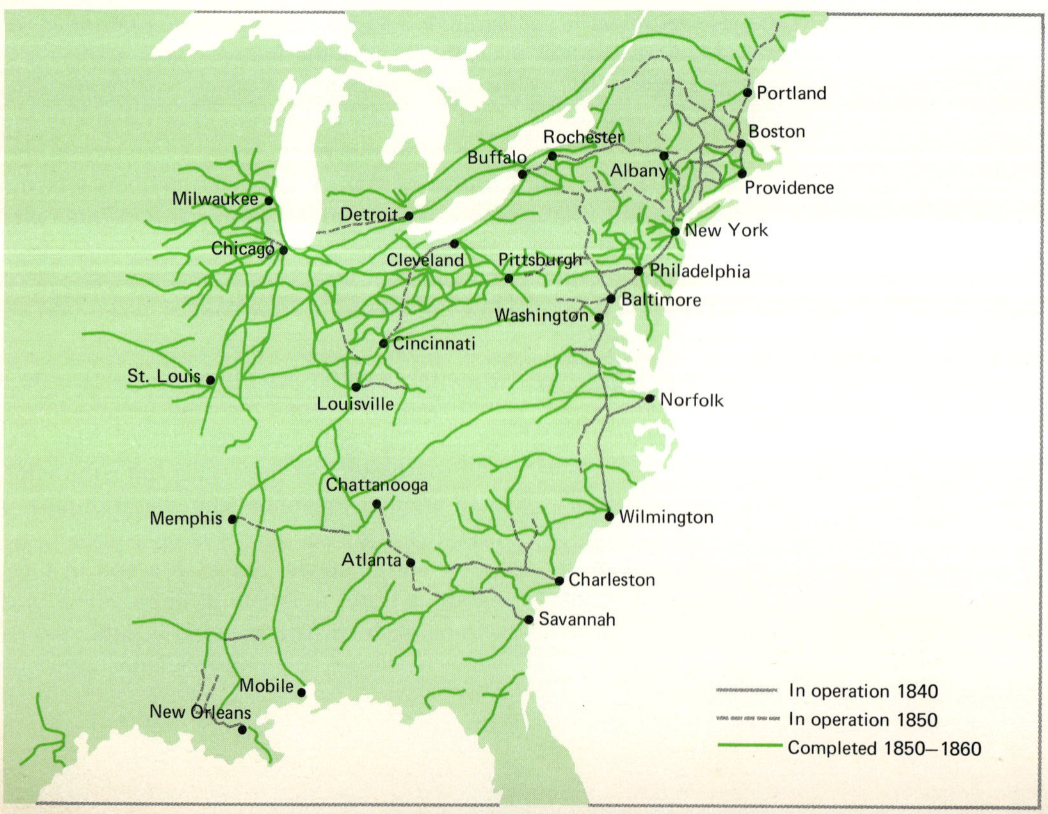

denly became an industrial asset. As a transportation system, the railroads were initially a locally useful device for moving food from agricultural areas to cities and for moving manufactured goods from the cities to agricultural areas. In 1848, there were about 6000 miles of railroad in America, most of which were east of the Appalachians. During the 1850s, the railroads extended into the American Middle West. By 1860, there were 30,000 miles of track, and agricultural products from the American grassland biome were being shipped to the eastern seaboard in forty-eight hours, a trip that had taken two to three weeks. (See fig. 17.10.) During the same period, American agriculture became mechanized. By 1860, for example, 80,000 horse-drawn mechanical reapers had been sold, and the combination of mechanized farming and an extensive rail system made American agriculture a world force. (See fig. 17.11.)

After the Civil War, oceangoing steamers replaced the sail-powered "clippers," and the cost of shipping wheat across the Atlantic fell by 75%. In spite of a number of restrictive laws designed to protect it, English agriculture collapsed in the 1870s. It was simply more economical to buy grain that had been shipped more than 4000 miles by train and steamship than to devote the dwindling supply of English land to the production of foodstuffs. Other countries also became food exporters. Denmark, for example, switched to dairy products—milk, eggs, butter, and cheese—and began to ship them to the increasingly industrialized nations of Europe.

The opening of the American Middle West to agricultural exploitation had effects on the grassland biome, which have not been subtle and have already been discussed to some extent. In Iowa and Kansas, Illinois and Minnesota, natural communities were replaced by monocultures of wheat and corn. At the same time, the grassland biome became populated by large herds of domesticated livestock produced by the artificial selection of wild Texas longhorns and English Hereford cattle. By 1880, for example,

FIGURE 17.11 The use of steam-powered farm machinery revolutionized agricultural practices in the United States, causing exploitation of the prairie and its eventual destruction as a natural biological community. *(Courtesy of the Library of Congress.)*

there were an estimated 4.5 million head of cattle grazing the range from Kansas to Montana and the Rockies. These developments drove out the species native to the biome. The buffalo were eliminated, and nearly made extinct in a veritable orgy of senseless killing, and many other species were systematically killed, because they were considered predators or pests or both. The practice of agriculture, which resulted in the erosion of topsoil, increased the silt load of North American rivers, as it had the rivers in many other regions of the world.

The advent of steam engines using fossil fuel made these developments possible. Before the development of the railroads, shipping grain or beef to large eastern markets was a long and expensive process. The rise of the railroad dramatically lowered the cost and shortened the trip. Even before the Civil War, more than two-thirds of the meat eaten in New York, for example, was produced west of the Appalachians. The development of fossil fuel resources (coal and later oil and natural gas) has allowed agriculture to achieve undreamed of levels of productivity both by opening up new agricultural regions and by supplying the motive power for agricultural machinery. But consider the effect that the industrialization of agriculture has had on nutrient cycling. Waste materials, rich in nutrients, now become concentrated at particular points in the processing pathway.

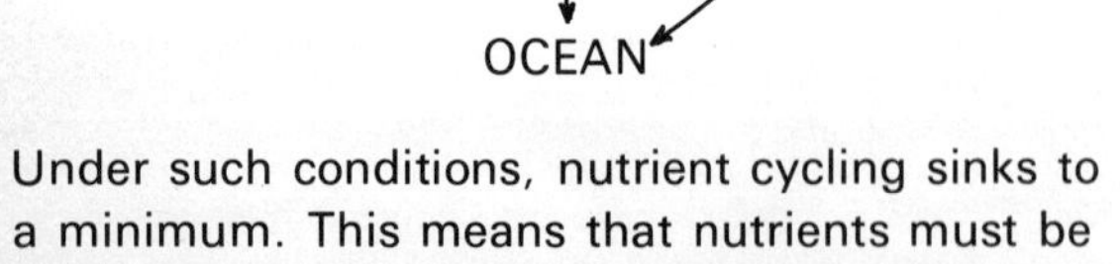

Under such conditions, nutrient cycling sinks to a minimum. This means that nutrients must be replenished if the soil is to retain its productivity. Around 1840, Justus Von Liebig, whose "law of the minimum" was introduced earlier in this book, published a book on plant nutrition, the principles of which became the basis for the chemical fertilizer industry. Initially, chemical fertilizers such as nitrate were mined from the guano islands off the coast of South America. But as other techniques became available, the nutrient supply came to be manufactured or processed in factories and required the input of fossil fuel energy. Such fertilizers not only maintained the productivity of the soil, they increased it, and this led many of the users of fertilizers to increase the amounts applied, often to levels well beyond the point of diminishing returns.

The development of a low-cost internal combustion engine in the late nineteenth and early twentieth centuries further accelerated the industrialization of agriculture in the U.S. and abroad, as the gasoline engine replaced the horse and the steam engine in the reapers and other agricultural machines. Beginning in the middle of the nineteenth century in America and earlier in parts of Europe, fewer and fewer people were required to produce greater amounts of food. By 1920, nearly half of the human population in the United States lived in urban areas. The revolution in the use of energy resources had brought about equally revolutionary changes in the sizes and distribution of the human population.

Prognosis

At first, the headlong rush to the cities produced unprecedented slums and contributed to some of the social upheaval of the nineteenth century. Slums, of course, had always been a part of city life, but the life of the nineteenth-century slum dweller was particularly unpleasant, especially noticeable, and given the social temperament of the age, quite distressing to some who did not live there. Man and his technology were thought to be but a step or so below the angels, heading toward complete control of the natural world; the underfed and the exploited who lived in the slums of the cities of the world were indications that something might not be quite right. A number of attempts were made to ameliorate such conditions. Settlement houses patterned after those that had been founded in England were established in major cities. Social legislation was passed, and labor organizations were founded to improve the lot of the workingman. Many of these attempts achieved limited success, but as we know all too well, they did not eliminate the slums from American cities.

Developments in medical care and public health also served to improve the life of man. Medical institutes, hospitals, and public health agencies and programs all contributed to better health, and mortality in the industrialized nations declined. Malaria, smallpox, cholera, typhus, bubonic plague, yellow fever, and tuberculosis were brought under control beginning in the late nineteenth century. Infant mortality decreased with improved maternal and infant care. After World Wars I and II, the benefits of industrialized agriculture and the medical advances of the industrialized nations were extended to the nonindustrial nations of the world. The population, which had begun to grow at the beginning of the eighteenth century, was increasing even more rapidly by the mid-nineteenth century. By the beginning of the twentieth century, it had begun to soar.

Undoubtedly, the welfare of man has improved. Few, if any, of Malthus' dire predictions have been fulfilled, for industry and agriculture have managed to keep well ahead of the burgeoning population. However, the general welfare of the population has been better provided for in some nations than in others. In fact, some startling discrepancies exist and, sadly, appear to be growing greater! The rise in agricultural output, the improvement of medical care, and the production of luxury goods have not been without their costs, however. Many biological resources have been uncontrollably exploited and destroyed; species of animals have unnecessarily become extinct; soil erosion in some portions of the world has been major; the disruption of nu-

(a)

(b)

FIGURE 17.12 (a) These squatters' huts in Central Park, New York City, in 1869 illustrate the effect of mass immigration in urban areas. (b) Similar conditions are now present in many cities throughout the world, such as in Caracas, Venezuela. *((a) Courtesy of Historical Pictures Service, Chicago; (b) courtesy of Owen Franken from Stock, Boston.)*

trient cycles has contributed to the deterioration of the world water supply; and the machines that the global transportation system has come to depend upon have contaminated the atmosphere. Many of the natural phenomena that are thought to enhance the quality of life have been injured, if not destroyed.

The increasing affluence of the United States has produced a situation in which the economic values placed on natural communities are beyond the ability of many ordinary citizens to pay, while they would be utterly out of the reach of most of the people of the world, many of whom are still peasants. Many Americans hike, boat, fish, or travel and demand places for these activities. Yet the preservation of suitable communities for such leisure use requires that lumbering be curtailed, strip mining be reduced, agricultural methods be improved, and waste disposal technology be improved and implemented. All of this requires new technology and use of energy and thus requires additional resource use. It costs money as well, and therefore increases the economic value of the land.

The use of energy, land, and nutrient materials by a species no longer limited by a hereditary niche has led to an entirely new biological phenomenon: A single species is the dominant force in nearly every biological community on earth. Intelligence, toolmaking, culture, and technology have created a huge human population that either deliberately or accidentally has placed its mark on the entire biosphere. Man is no longer limited by the biological materials that he can gather in his hands. The implications are awesome. We live in the midst of the most rapid change in the biosphere that the world has ever known. Where is it to lead?

It would be instructive to review again what has happened in the past when human culture has allowed man to reach his carrying capacity. Time and again throughout history, the imposition of a natural limitation on man has been accompanied by social and cultural upheaval. The failure of the land has been accompanied by warfare, pestilence, and famine; the latter in part the result of the failure of the land, but also often the result of warfare. Religious and political institutions have been overthrown, quietly on occasion, but often violently. The causes and effects of these phenomena are extremely difficult to sort out, even more difficult to prove. From the viewpoint of a political or economic historian, matters of political or economic interests come to the fore in any analysis, and questions of food supply, land use, and the health of biological communities assume secondary importance, if any importance at all. In the face of scanty historical data, how does one determine whether enough food was in fact available for the population of some region? We lack accurate, basic information about population sizes, agricultural production, and distribution of food supplies for the better part of human history. *It is plausible to hypothesize that biological phenomena may have been the root causes of the social upheaval that has characterized human history.* The carrying capacities of biological communities, as affected by cultural, economic, and political factors, may well have determined both the timing and the effects of most of the social unrest in history, from the rise of civilizations, through the Fall of Rome, the misery of the late Middle Ages, the Industrial Revolution, and into the present century.

The present situation? We are now in a period of rapid global population growth, accompanied by apparent increases in the use of nonrenewable resources. The use of atomic energy, the automobile, a superlative highway system, the "green revolution," major finds of undersea mineral wealth, the possibilities of using direct solar energy, the use of computers to manage our affairs and coordinate our technology, breakthroughs in health care—almost all are developments of the past thirty years, or merely one four-hundredth of the time that has elapsed since the first cereal grains were domesticated. It may be that these developments alone are the equivalent of all technological developments that occurred in the preceding 12,000 years.

It has been argued on the one hand that our current growth and the corresponding use of resources may continue unabated on the expectation that technological breakthroughs such as have occurred in the recent past will continue to occur. On the other hand, many have predicted the occurrence of resource limitation, famine, and social breakdown some time in the next few decades. Basically, the problem is whether the carrying capacity produced by technology can keep pace with the increase in population. Technological advances all seem to have occurred in steps or spurts of innovation. The human population, if unchecked, will grow continuously and exponentially. The time between the steps of technological advance have, over the past two centuries, certainly seemed to grow smaller. Technological growth has come to seem continuous, but such continuity may be illusory. We simply do not know whether technological growth will, or even can, allow us to produce greater quantities of food, produce or convert energy cheaply and efficiently, or substitute abundant resources for needed ones that are growing increasingly scarce.

Some economists think that because technological solutions have always arisen in the recent past, they will continue to appear in the future. This view assumes that history always repeats itself, yet its proponents sometimes ignore or dismiss the other implication of that viewpoint. If history is likely to repeat itself, then humanity is likely, sooner or later, to have to endure the sort of deprivation and misery that the populations of the Middle East, Rome, and fourteenth-century Europe experienced when technological innovation was unable to keep pace with population growth. These economists have dismissed Malthus, primarily because many of the predictions he made have not *yet* been realized. But they *were* realized many times in history. Risking redundancy, let us say again that we simply do not know whether further technological advances will allow us to sustain for very long the world population at its current rate of growth.

However, we *do* know that there must be some limitation to the population of man that the world can support. We also know that because of international commerce, problems associated with approaching the carrying capacity of man will not, as they have in the past, be limited to one region or nation. It is the opinion of the authors that within 50 to 200 years, the society of man may undergo a period of major upheaval similar to those in the past, but that its intensity may be far greater and its effects global. How, where, or when it might begin we cannot know, but unlimited growth can never continue indefinitely.

Perhaps a more optimistic point of view is possible. As we have seen, man may exercise some rational control over his culture. Perhaps we should ask not when we shall pass into chaos, but how we may use our culture and technology to produce the best quality of life for man. What is the optimum world population, and what technology is needed to support it? Can chaos be prevented and a "high quality" of life be produced for everyone? Certainly it should be within our technological grasp.

If our efforts to prevent chaos do not succeed, it will be because the evolved abilities of man are ultimately maladaptive; that is, they will not allow us to overcome biological restraints until it is too late.

In the minds of some, statements of opinion are outside the prerogative of the authors of a biology textbook. Nevertheless, some admittedly subjective observations are in order concerning how man may consciously limit the use of resources and stabilize the population in order to maximize the quality of human life.

Considering the size of the present world population and the size that it is likely to reach even if extreme population control were to be *immediately* achieved, we must, if we are to maintain the present quality of life, continue to depend upon technology. If we are to improve the conditions of a significant proportion of the world population, we shall probably have to increase our resource use. This is not to say that technological advance itself ought to be the goal of society. Man is a tool user, but he does not use tools for the sake of using them; he uses tools to benefit himself. Without humanely intelligent, humanly oriented technological advances, we will continue growing unchecked, depleting and misusing the resources available to us. By no means will the challenges of science and technology be diminished. We think, in fact, that they are likely to become greater.

At present, most of our technological use of science has been an outgrowth of our understanding of physics and, secondarily, chemistry. Agriculture, medicine, and, to the greatest degree, ecology and environmental sciences are in relatively primitive stages of development. We cannot predict with a high degree of confidence the effect of population growth and technological development on the use of our biological resources. An immense effort in basic biological research will be necessary if we are to intelligently regulate our society. We were able to send men to the moon in an astonishingly short time after we had made the commitment to do so, but the achievement was based on an understanding of physics and chemistry that had taken us several generations to accumulate. With our present knowledge of biological sciences, we are by no means in a position to intelligently and effectively regulate our nonphysically based relationships with the world.

Regulation of our relationships with the world, regulation of the size of the human population, regulation of our use of resources—regulation invariably seems to involve the loss of choices available to individuals and groups of individuals, presumably for "the common good." Thus we enter the realm of the political scientist, the historian, and the sociologist, to name but a few. Is it possible to regulate people without depriving them of some freedom? The actions of people are seemingly regulated most of the time by law and custom, by the shared ethical values of a social group, or if you prefer, by the decree of God. Whether it is possible to use predictions of what may be best for the stability of population and the quality of life in order to regulate the actions of the members of society is a question that scientists alone cannot answer. Three possible alternatives suggest themselves: *First*, we can do nothing and endure a repetition of the misery and suffering that have occurred before: *Second*, we can turn over the decisions to some "authority" that presumably knows best: *Third*, we can involve the whole of society in making voluntary choices. If we choose the third alternative, then we shall have to rely on scientists for information, but we shall also have to evaluate that information before acting upon it. In spite of all the calls for scientists to regulate the technological applications of their work, we believe that scientists *must not* force their own social and ethical decisions upon mankind. This is not to be construed as a proscription of the scientist's right as a citizen to make social and ethical decisions. That right belongs to all citizens. But the responsibility for social decisions rests with society. Insofar as citizens must make decisions based on scientific information, they must possess sufficient knowledge of science to make them intelligently. The need for both information and intelligent decisions regarding the control of population and the likely results of population growth and technological innovations is critical.

Benjamin Franklin, at the conclusion of the Constitutional Convention in 1787, was asked whether the government was to be a monarchy or a republic. Franklin answered, "A republic, madam, if *you* can keep it." (Our emphasis.) The implications of Franklin's remark are clear to us in a political sense, but they have particular cogency regarding the roles of the scientist and the citizen in a free society. We believe that the Jeffersonian ideal, shared by Franklin, of an educated, intelligent electorate possessing a social conscience informed by rational decisions is necessary for the creation of the kind of "high quality" life we are capable of leading, even within the limits of our present scientific knowledge and technological understanding. Any number of predictions, inventions, scientific laws, or new resources are of no purpose if their implications for the quality of life must be maintained by a tyrant, however benevolent. Each person must come to his own decision, and then the society must reach a decision so as to preserve the freedom of individuals. The keystone of the process is education, for without it, individuals cannot judge the options, if indeed they can find out what the options are.

The process is a tedious one and sometimes takes a long time. As a consequence, some people counsel revolution, either of the "right" or of the "left." But we do not believe that there is time for the anarchy that generally initially follows revolutions. Education that increases the power of individuals—that is, increases their power as a thinking, informed electorate, free to exercise those qualities of man that have evolved over millions of years of natural selection—will produce results more rapidly. Man's mind, tools, language, and culture have set him apart as a member of the biological communities of the world. He has control, and to use that control in ways that will enhance his own life and the lives of other species requires, indeed demands, that *each* person acquire knowledge. Perhaps this is a naïve view. Perhaps. But we believe that the only other choices are tyranny or chaos.

Selected Readings

Bell, Daniel, ed., *Toward the Year 2000.* Boston: Houghton Mifflin Company, 1968. A collection of articles from *Daedalus* magazine which deal with society and environments in the future.

Boyden, S. V., ed., *The Impact of Civilization on the Biology of Man.* Toronto: University of Toronto Press, 1970. Papers from a symposium held in Australia. Some very interesting and provocative hypotheses are developed.

Dale, Tom, and Carter, Vernon. *Topsoil and Civilization.* Norman, Okla.: University of Oklahoma Press, 1955. A fascinating survey of agricultural history dealing largely with Europe and the Middle East.

Snow, C. P. *The State of Siege.* New York: Charles Scribner's Sons, 1969. A cogently argued alarmist view, written by a scientist-novelist. Snow predicts that man will exceed his carrying capacity before this century is ended.

Thomas, William L., ed., *Man's Role in Changing the Face of the Earth.* Chicago: University of Chicago Press, 1956. A large and provocative book that is now somewhat dated, but nevertheless bears reading.

Turnbull, Colin. *The Mountain People.* New York: Simon & Schuster, Inc., 1972. An anthropologist's record of the life of the Ik, an African mountain tribe forced to live under considerable environmental stress. The effects on their society are chilling.

Braidwood, Robert J. "The Agricultural Revolution." *Sci. Am.,* Vol. 203 (September, 1960), pp. 130–134. Offprint number 605.

Boerma, Addeke J. "A World Agricultural Plan." *Sci. Am.,* Vol. 223 (August, 1970), pp. 54–63. Offprint number 1186.

Iversen, Johannes. "Forest Clearance in the Stone Age." *Sci. Am.,* Vol. 194 (March, 1956), pp. 36–41. Offprint number 1151.

Kelly, Clarence F. "Mechanical Harvesting." *Sci. Am.,* Vol. 217 (August, 1967), pp. 50–59. Offprint number 329.

Langer, William L. "The Black Death." *Sci. Am.,* Vol. 210 (February, 1964), pp. 114–118. Offprint number 619.

glossary, credits, and index

glossary

abiotic not including living organisms, their effects, or products.

absolute refractory period a stage in the conduction of a nerve impulse during which part of the membrane of a neuron is impermeable to sodium ions.

acid a solution in which pH is less than 7; a substance which contributes or liberates H^+ ions in a solution.

acoelomate (see *coelom*).

activation energy some quantity of energy required to initiate a chemical reaction.

active site a location on the surface of an enzyme that reacts with a substrate to bring about a chemical reaction.

active transport a process, requiring the expenditure of energy, whereby dissolved material is moved across a cell membrane.

actomyosin a contractile molecule found in muscle cells and made up of the proteins actin and myosin.

adaptation an evolved characteristic of an organism that enables it to survive or adapts it to its niche.

adaptive radiation the divergence of a single species or phyletic group of species into unoccupied niches.

adenine a purine base found in DNA and ATP as well as in other important biological molecules.

aerobic containing molecular oxygen (O_2). Usually refers to the condition of an environment or to an organism that requires molecular oxygen for metabolic processes.

algae the lower plants. Plants without highly evolved reproductive organs or vascular tissue. Usually live in water or damp places.

allele one of two or more alternative forms of a gene.

allogenic change change in a community brought about by influences from outside the community.

allopatric refers to populations of the same species which live in different and not overlapping portions of the species' range.

amensalism an interaction between two species whereby one is disadvantaged but the other is not affected.

amino acids the molecules of which a protein is constructed: there are several types; they are characterized by a carboxyl group at one location, an amine group at the opposite end, and a radical that identifies the specific amino acid.

anaerobic devoid of molecular oxygen, as an anaerobic environment. Also refers to an organism capable of living in such an environment.

angiosperm seed plants in which the embryo is enclosed by a seed case. Includes the broad leafed plants.

anion (see *ion*).

anoxic without molecular oxygen.

antibody a protein that is synthesized as a response to an antigen and that serves to remove or otherwise incapacitate the antigen.

anticodon a sequence of three bases on the tRNA molecule which interacts with the codon of an mRNA molecule to attach an amino acid to a growing polypeptide.

antigen a molecule, usually a protein, that can induce the synthesis of an antibody.

ATP (adenosine triphosphate) a compound consisting of the nucleoside adenosine and three molecules of phosphate. Two of the bonds, those between the phosphates, are "high energy" bonds.

autogenic change change in a community brought about by factors in the community itself.

autosome a chromosome that does not determine the sex of an organism.

autotroph an organism that is capable of utilizing the energy of light to synthesize organic molecules and provide its metabolic energy. Includes the green plants and some bacteria.

axon a long extension of the cell body of a neuron which conducts the nerve impulse to the next nerve cell or to an effector.

bacteriophage a virus that infects a bacterium, injects its own genetic material, and utilizes the metabolic apparatus of the bacterium to produce more bacteriophages.

base a chemical that raises the pH of a solution by contributing OH^- ions. Also, in genetics, the chemicals adenine, guanine, cytosine, thymine, and uracil that form molecules of DNA and RNA.

base-pairing the bonding within DNA and RNA molecules, by weak hydrogen interactions, of adenine to thymine and cytosine to guanine.

benthic refers to the bottom substrate of any body of water as well as to those organisms that inhabit the benthic environment.

bilateral symmetry the property whereby an object can be divided so that each half is the mirror image of the other half.

biogeochemical cycle the process whereby chemical elements are transported from the nonliving environment into living organisms and then back into the nonliving world; for example, the nitrogen, carbon, phosphorus, and hydrologic cycles.

biomass the total living material, measured in terms of dry weight, energy, or some other unit, in a specified area or volume.

biosphere the portion of the earth and its atmosphere capable of supporting life.

biotic potential the inherent capacity of a species to reproduce and increase in numbers. (Compare *intrinsic rate of growth*.)

blastula a hollow, fluid-filled ball of cells formed during mitotic division of the zygote.

blue-green algae the Cyanophyta—a group of photosynthetic procaryotic organisms closely allied to the bacteria.

BOD biological oxygen demand. The oxygen concentration decrease in a water sample over some specified period of time, usually five days, caused by bacterial metabolism. Used for evaluating the oxygen-removing potential of sewage.

buffer a substance or compound which prevents large changes in the pH of a solution.

calorie the quantity of energy required to raise the temperature of one gram of water at 15°C one degree C. A kilocalorie or Calorie is the energy required for the same temperature increase in one kilogram of water.

Calvin Cycle a metabolic pathway in photosynthetic organisms in which 1-5 ribulose diphosphate is converted into carbohydrate.

capillary the smallest type of vessel in the circulatory system of an animal.

carbohydrate a molecule containing carbon, hydrogen, and oxygen in the proportion $(CH_2O)_n$, where n is some number greater than 2.

carnivore an animal which kills and feeds on other animals.

carrying capacity the maximum population density of a species that can be supported in some specified area.

catalyst a substance that participates in and accelerates a chemical reaction of other substances, but without undergoing any change itself.

cation (see *ion*).

centromere a structure on the chromosome where the chromatids are attached.

cerebellum a portion of the vertebrate brain, evolved from the hindbrain, which controls many muscular functions as well as balance.

cerebrum a portion of the vertebrate brain that has evolved from the forebrain and in lower forms is largely specialized to serve olfactory functions. In humans it is the largest part of the brain and the center of conscious thought.

chemotroph an organism that obtains energy from chemical reactions solely using inorganic materials. Includes some bacteria.

chlorophyll the chemical pigment which absorbs light during photosynthesis and initiates the photosynthetic fixation of energy.

chloroplast in eucaryotic plants, the cellular organelle where the reactions of photosynthesis take place—the plastid.

chromosome the cellular structure which appears during cell division and which contains the genes. It consists, usually, of two chromatids attached at the centromere.

cilium a contractile filamentous organelle on the outside of some cells. It is usually shorter than a flagellum and is used for locomotion or transport of fluids.

cleavage the stage of cell division in which the cell membrane forms separating daughter cells. Often refers to the initial division of the zygote or to the early stages of development of an embryo.

climatic climax a climax that is in equilibrium with the general climate of the area within which it is found.

climax the last stage in community succession (sere), which is perpetuated under the local climatic and edaphic conditions.

codon a triplet sequence of bases on the mRNA molecule which carries the information required to add a specific amino acid to a polypeptide.

coefficient of selection a quantitative value assigned to the relative fitness of a specific genotype which indicates the intensity of selective pressures operating on that genotype.

coelom the body cavity between the digestive tract and body wall in all advanced vertebrates and invertebrates. An animal lacking a coelom is referred to as being *acoelomate,* while one possessing a coelom is said to be *coelomate.*

coelomate (see *coelom*).

coenzymes molecules which interact with enzymes to give the latter the capacity to catalyze certain reactions.

commensalism interaction between two species in which one species benefits and the other is not affected.

community an integrated group or association of species in a particular locality.

competition interaction between two species in which both species are disadvantaged, often through seeking the same resource. There is also intraspecific competition—competition between individuals in a species population—which serves to decrease the rate of population growth.

condensation reaction a reaction between two molecules which results in the synthesis of a third molecule and one molecule of water. Also called a dehydration reaction.

cones the photoreceptors of the eye that sense color and high intensity light and also provide good image resolution.

crop a storage organ for food which allows an organism to practice discontinuous feeding. Also, of course, agricultural produce.

crossing over the breaking and rejoining of chromosomes during meiosis in such a way that certain genes on the chromosomes exchange places.

cytosine a pyrimidine base, important in the structure of DNA and RNA.

dark reactions in the process of photosynthesis, a series of reactions that do not require the input of light as activation energy.

daughter cells the two cells that arise by the mitotic division of a single cell.

deme a specific and usually local population of some species and its gene pool.

demersal refers to organisms dwelling near

and dependent on the bottom of a body of water, although not attached to or living in the bottom. Cod, catfish, and suckers are demersal fish.

denaturation the loss of specific structure in a protein brought about by the breaking of weak hydrogen bonds. Denaturation is often irreversible and results in the loss of biological activity.

dendrite the location on a nerve cell, usually at the terminus of a process, where nerve impulses are received.

density dependent refers to factors that control the density of a population and whose intensities depend on the density.

density independent refers to factors that control the density of a population but that are not influenced by the density.

detritivore an organism that feeds on detritus.

detritus in biology, particulate material resulting from the breakdown of organisms. Other meanings in geology: see dictionary.

dicotyledons seed plants which have two cotyledons or seed leaves and nonparallel veins in the leaves. Example, most deciduous trees.

differentiation the process, during the development of an embryo, by which particular genetically identical cells or groups of cells develop into specialized cells, organs, and tissues.

diffusion the movement of dissolved materials across a cell membrane without the expenditure of energy by the cell.

diploid the condition of a cell containing two sets of chromosomes. (Compare *haploid.*)

disclimax a climax, not the climatic climax, maintained by man or human activities.

DNA deoxyribonucleic acid, the chemical of the gene.

dorsal the ''upper'' surface of a bilaterally symmetric animal. The surface away from the ground.

doubling time the time required for a doubling of population density.

dystrophic a brown-water lake or pond, usually of low productivity and surrounded by a bog mat. Usually the last aquatic stage in lake succession.

echolocation navigation by means of emitting sound, usually high frequency, and detecting the location of objects through the echoes created by the sound.

ecosystem the biological community together with the associated abiotic environment.

ecotone the boundary region between two communities.

edaphic climax a climax with modifications of the climatic climax caused by local conditions.

effector a cell, tissue, or organ, which upon stimulation acts to bring about a particular effect or result, e.g., a muscle.

electron a subatomic particle that is negatively charged and orbits the positively charged nucleus of the atom.

embryo an organism in the early stages of development of the life cycle.

endplate the portion of a muscle cell where a motor neuron synapses with it.

energy the capacity to do work.

enzyme a catalytic protein.

epilimnion the upper isothermal layer of a lake or pond above the thermocline.

estuary a partially enclosed body of coastal water, usually with a salinity gradient caused by fresh water input.

eucaryote a cell or organism that possesses a membrane-bound nucleus containing the largest part of the genetic material of the cell. Adj., *eucaryotic.*

eutrophic refers to highly productive lakes and ponds (or occasionally other communities), frequently with oxygen depletion in the hypolimnion. In a strict sense, brought about by natural lake succession. Cultural eutrophication is caused by excess inorganic nutrients from waste disposal.

eutrophication the ''aging'' process of a lake

whereby nutrient content and production increase. Eventually the lake fills in with a bog and then forest. Cultural eutrophication refers to a similar process with increased plant growth caused by input of waste nutrients.

external fertilization fertilization occurring outside the body of an organism.

F_1 in the notation of genetics, "first filial" generation. Denotes the offspring in the first generation after mating in the parental generation.

facultative refers to the ability of an organism to live under a variety of conditions, such as in either aerobic or anaerobic environments. For example, a facultative anaerobe. (Compare *obligate*.)

feces solid wastes produced in the digestive system of an animal.

ferredoxin a protein that is believed to accept electrons from chlorophyll *a* in the initial reaction of photosynthesis.

filter feeder an animal which obtains its food (usually detritus or plankton) by filtering it from the medium. For example, sponges and copepods.

firing threshold a characteristic of a nerve cell referring to the quantity or strength of stimuli required to induce the neuron to conduct an impulse.

flagellum a whiplike organelle extending from a cell. Usually used for locomotion in the protozoa.

fossil fuels hydrocarbon energy sources produced in the earth, under pressure and heat, from the partially decomposed remains of organisms.

fovea a region of the vertebrate eye on a direct line with the center of the lens, where image resolution is greatest, and where the density of cones is very high, and of rods, very low.

gamete the haploid reproductive cell, a sperm or ovum, for example, which fuses with another gamete to form the diploid zygote.

ganglion an aggregation of nerve cells that carries out integrative functions.

gastrovascular cavity an internal space, found in simple organisms, which serves to digest food and to circulate the products of digestion.

gastrula a stage in the development of an embryo—following the blastula stage—during which dividing cells begin to invaginate, or fold inward.

gene the unit of heredity. A sequence of nucleotides within a molecule of DNA from which a molecule of mRNA can be synthesized.

gene frequency the proportion of an allele in a population relative to the proportions of other alleles.

gene pool the entire genetic makeup of a population, taking into account the gene frequencies of all alleles.

germ cell (see *gamete*).

glycolysis the energy-liberating process in respiration whereby a carbohydrate is converted to pyruvate.

grana internal structural components of the chloroplast.

gross productivity the rate of total energy or carbon assimilation by an organism, population, or community, including that used for metabolic processes.

guanine a purine base. An important structural component of DNA.

gut the digestive tract.

gymnosperm seed plant in which the embryo is not contained by a seed case. Includes the conifers.

habitat all the environmental conditions in a place occupied by an organism, population, species, or community.

Hardy-Weinberg Law a statement that gene frequencies will not change from one generation to another unless migration, mutation, selection, and nonrandom mating are introduced.

hemoglobin an iron-containing protein which makes up the largest part of the red blood cells and which is active in the exchange of carbon dioxide and oxygen in vertebrates.

herbivore an animal that eats plants.

heterotroph an organism that uses organic material synthesized by other organisms as its source of energy. Includes the animals, fungi, and most bacteria.

heterozygous possessing two unlike alleles for a particular characteristic. (Compare *homozygous.*)

homeostasis the maintenance of a steady state within the body of an organism or within a community. Implies a range of fluctuations that are within the ability of an organism to control.

homeotherm an animal that maintains a fairly constant body temperature.

homozygous possessing two like alleles for a particular characteristic. (Compare *heterozygous.*)

hydrolysis the splitting of bonds between portions of a molecule with the addition of a water molecule (as H^+ and OH^-) to the products of the reactions.

hypolimnion the deeper isothermal waters of a lake below the thermocline.

hypothalamus a portion of the vertebrate brain, evolved from the forebrain, where the controls over emotions and drives such as sex, hunger, thirst, and pain are centered.

internal fertilization fertilization occurring within the body cavity of an organism.

intrinsic rate of growth the theoretical maximum rate, per individual, at which a population may grow.

in vitro literally ''in glass,'' referring to observations of processes carried on in culture vessels or in the laboratory.

in vivo literally ''in life,'' referring to observations of processes occurring in nature or in whole organisms.

involuntary muscle tissue, usually composed of smooth muscle cells, not directly under the control of the central nervous system. For example, intestinal muscle or uterine muscle.

ion an atom or molecule possessing extra electrons or lacking a full number of electrons, and thus having a net negative or net positive electrical charge. A negatively charged ion is called an *anion,* and a positively charged ion is called a *cation.*

irritability a characteristic of protoplasm, defined by the reception, transmission, and response to a stimulus.

K strategist a species whose populations tend to remain near the carrying capacity, and whose population densities do not fluctuate greatly.

kilocalorie (see *calorie*).

kinetic energy the energy of matter that is in motion.

Krebs Cycle a cyclical series of chemical reactions in the respiratory process whereby energy is derived from chemicals.

larva an early, free-living stage in the life cycle of an animal that differs, often markedly, from the adult stage.

light reactions photosynthetic reactions initiated by light. (Compare *dark reactions.*)

lipid a chemical consisting usually of glycerol (a three-carbon compound) and three fatty-acid chains of varying lengths. Phospholipids contain a nitrogenous, phosphorus-containing compound in place of one fatty acid. Steroids are also classed as lipids.

locus the location of a gene on a chromosome map.

linkage groups groups of genes which segregate together.

logistic growth population growth which exhibits a slow-growing lag phase, a rapid rate of increase, and then a slow or nongrowing phase as carrying capacity is reached.

medulla a portion of the vertebrate brain evolved from the hindbrain, which controls and coordinates certain unconscious functions.

meiosis the cell division process whereby diploid cells are reduced to haploid cells, usually gametes. (Compare *mitosis.*)

membrane potential the electrical potential of a membrane caused by the unequal distribution of anions and cations across the membrane. Measured in millivolts.

metabolism the process of utilization of energy by plants and animals.

microclimate climate in the immediate vicinity of an organism or object.

microecosystem a small, fairly self-contained laboratory ecosystem.

micronutrient a nutrient material required by an organism in trace quantities.

microvilli the cylindrical outgrowths of cells in the villi—the projections lining the small intestine.

mimicry the resemblance of one species (the mimic) to another (the model), usually unrelated, species. Mimic species often gain some selective advantage from their appearance.

mitochondrion the cellular organelle where the oxidative reactions of respiration are carried on.

mitosis the division process by which two daughter cells arise from the parent cell. Each daughter cell has the same number of chromosomes as the cell from which it is descended. (Compare *meiosis.*)

molecular weight the sum of the atomic weights of all of the atoms in a molecule.

molt a process, exemplified best in the life cycle of an insect, during which a new exoskeleton grows within the old one which is then shed by the organism.

monocotyledons seed plants with one cotyledon or seed leaf and parallel veins in the leaves. Includes the grasses.

mortality the rate of death per individual in a population.

motor neuron a neuron that innervates an effector.

motor unit a complex of a motor neuron and the muscle fibers which are innervated by the neuron. Thus a single nerve impulse can effect the contraction of several fibers.

mRNA messenger RNA, transcribed on the DNA template, and which in turn translates its encoded message on the ribosome.

muscle the aggregation of muscle cells which are bound within a membrane of connective tissue and which are attached to the skeleton by tendons. The tissue of which muscle and other contractile organs are composed.

mutation a change in genetic makeup. A mutation may involve as little as a single base in DNA, or as many genes as there are in an entire set of chromosomes, as in polyploidy.

mutualism an interaction between two species in which both species benefit.

natality the rate of birth per individual in a population.

negative feedback the output of a system which when fed back into the system serves to inhibit the rate of output.

nekton the strong-swimming animals.

nematocysts independent stinging effectors, nearly always found in the tentacles of coelenterates, which discharge threadlike venom-containing barbs that in turn penetrate and poison prey.

nerve impulse a nerve signal; the result of a stimulus conducted along a neuron.

net productivity the rate of assimilation of energy or carbon by a species, trophic level, or community, less the loss by metabolism.

neuron a cell in the nervous system specialized for the receipt, transmission, or response to a stimulus.

neutralism interaction between two species in which neither species is affected.

niche the role of an organism in the environment; its activities, adaptations, and relationships to the environment.

nitrogen fixation the conversion of molecular nitrogen into biologically usable forms.

notochord a gelatinous supporting rod lying below the dorsal nerve cord in embryonic and primitive adult chordates. The characteristic of that phylum.

nuclear membrane the porous membrane surrounding the nuclear region of a cell, wherein the chromosomes are found.

nucleotide a molecule consisting of a sugar (either ribose or deoxyribose), a phosphate, and a purine or pyrimidine base.

nucleus the cellular organelle, surrounded by a membrane, which contains the hereditary material, nuclear DNA.

nutrient any material used for purposes of growth or metabolism. Often restricted to the inorganic nutrients used by plants, that is, materials such as nitrate, phosphate, and trace elements.

obligate refers to an organism that requires a specific condition or mode of life; for example, an obligate aerobe, an obligate anaerobe, or an obligate parasite. (Compare *facultative.*)

oligotrophic refers to a body of water containing few nutrients, with low productivity and clear water.

operon the adjacent structural genes coding for the transcription of a single mRNA molecule. The activity of the operon is controlled by the regulator gene and operator.

organelle literally "little organ." A discrete structural entity within the cell specialized to carry out some particular function.

ostium an aperture in a blood vessel through which blood may be forced out.

oviparous refers to an organism which releases unfertilized eggs for external fertilization.

oxidation the removal of electrons from an atom or molecule, often with the addition of an oxygen ion or the removal of a hydrogen ion.

oxyhemoglobin hemoglobin with which molecular oxygen has loosely bonded. The form of hemoglobin that transports oxygen through the vertebrate body.

parasitism interaction between two species in which one benefits and the other is disadvantaged, but in which the disadvantaged species is usually not killed.

parthenogenesis the development of an organism from an unfertilized gamete.

pelagic refers to open water, without reference to the shore or bottom.

peptide bond a bond formed between the carboxyl of one amino acid and the amino group of another. The bond between amino acids in a protein.

pH the negative logarithm of the hydrogen ion concentration. Relates to the acidity or alkalinity of an aqueous solution.

phenotype the physically manifest characteristics of an organism which result from the interactions of its genotype and its environment.

phloem vascular tissues of a plant which conduct water and materials away from the sites of photosynthesis.

phosphorylation the addition of a molecule of phosphate to some other molecule, often forming a high-energy bond.

photosynthesis the synthesis of carbohydrates from CO_2 and H_2O, in which light is used as a source of energy.

phytoplankton planktonic plants.

placenta a complex of maternal and embryonic tissue formed in the uterus and serving as an exchange organ for nutrients and wastes.

plankton aquatic organisms, usually small, which drift or swim weakly and whose horizontal distribution depends on water currents.

planula the early larva of a sponge.

pleiotropism the expression of a single gene in more than one phenotypic characteristic.

poikilothermic refers to animals that do not maintain constant body temperature and which have temperatures the same as their surroundings.

polymorphism the existence within a population of discontinuous phenotypic variations.

polypeptide a protein, a chain of amino acids joined by peptide bonds.

population density number of individuals of a species population per unit area or volume.

potential energy energy that has been

''stored,'' in the bonds of a molecule for example, and that can be used to do work.

primary host the host of the adult (sexually reproductive) stage of a parasite.

primary productivity the rate of productivity of plants in a community.

primary succession seral succession from a completely new environment. (Compare *secondary succession.*)

protein any of a class of large biological molecules made up of amino acids.

pupa an apparently dormant stage in the life cycle of an insect during which the adult body is formed by the extensive reorganization of existing tissue.

r strategist a species whose population grows rapidly under optimal conditions but which subsists at low numbers at most times. An opportunistic organism.

radial symmetry the property of being symmetrical about a center line; for example, a doughnut or a jellyfish has radial symmetry.

radula a rasping organ, used for feeding, near the mouth of certain snails.

random mating mating in which an organism has an equal chance of mating with any other organism of the same species and of the opposite sex.

receptor a neuron specialized for the reception of a stimulus. Photoreceptors are sensitive to light, chemoreceptors to the properties of molecules, and mechanoreceptors to movement, pressure, or shear.

reduction the addition of electrons to an atom or molecule, usually by addition of a hydrogen ion or removal of an oxygen ion.

relative fitness a statistical value assigned to an allele in a population indicating the contribution to survival made by that allele within the context of the genetic background in which it is found.

relative refractory period an interval during the conduction of the nerve impulse when the membrane of the neuron is permeable to sodium ions; a larger stimulus must be presented to reach firing threshold than is necessary when the membrane is at ''rest.''

respiration the conversion of food to energy. Also the process of breathing.

retina the layer of photoreceptors in the eye.

ribose a monosaccharide containing five carbon atoms; thus a pentose.

ribosomes organelles, consisting largely of ribosomal RNA, where mRNA interacts with tRNA to synthesize proteins.

RNA ribonucleic acid, consisting of ribose, phosphate, and the bases cytosine, adenine, guanine, and uracil.

rods photoreceptors sensitive to low intensity light.

saccharides sugars. A single saccharide is a monosaccharide; two sugars bonded together form disaccharides. Many monosaccharides bonded together form a polysaccharide, such as starch.

salinity the quantity of dissolved salts in water, usually expressed as parts per thousand ($^0/_{00}$) by weight. Salinity of oceanic seawater ranges from 30 to 38 $^0/_{00}$.

saprotroph an organism that obtains its sustenance from decaying materials.

secondary host the host of the immature, nonsexually reproducing stage of a parasite.

secondary succession seral succession from a modified community, such as a cleared forest. (Compare *primary succession.*)

sediment feeders organisms which feed by passing bottom sediments through the gut, digesting the organic material present.

selection pressure the impact of selection upon a particular genotype in a population.

sere the series of stages or communities that follow one another in the process of community succession.

sessile an organism attached to an object, living in one place; for example, barnacles, oysters.

sex linkage refers to the occurrence of other genes on the chromosome carrying sex information.

standing crop the biomass of a species at some particular time.

stomata microscopic openings in leaves permitting gas exchange with the atmosphere.

substrate in biochemistry, any molecule which is acted upon by an enzyme. Also, a surface upon which an organism rests or across which it moves.

succession the natural orderly replacement of one community by another ideally leading to the climax. (Compare *sere, climax.*)

survivorship curve a graph showing the percentage of individuals of a population born surviving to different percentages of total possible lifespan.

symbiosis literally means a living together. A close association between two or more organisms of different species. Includes parasitism, commensalism, and mutualism.

synapse the region at which two neurons interact.

synaptic cleft the space between two neurons across which a nerve impulse is passed.

territory in biology, the definable area that an individual or breeding pair will defend against other individuals, usually males of the same species.

test cross a genetic technique in which an organism of unknown genotype is crossed with one of a known genotype to determine the genetic constitution of the former.

thallus the body of a multicellular alga.

thermocline the depth of maximum rate of change in temperature in a body of water.

thermoregulation the regulation of body temperature by an animal.

thymine a pyrimidine base found in DNA.

transcription the assembly of an RNA molecule on the template provided by DNA.

translation the assembly of a protein from information coded upon messenger RNA. Translation occurs at the ribosome.

transmitter substance a compound, such as acetylcholine, which passes across the synaptic cleft and initiates a nerve impulse in an adjacent neuron, thereby transmitting the impulse from one neuron to the next.

transpiration the loss of water through the surfaces of a plant, especially the leaf surfaces.

trophic level the position of an organism in a food chain: producer, herbivore, primary carnivore, secondary carnivore, etc.

turnover time in ecology, the time required to completely change and replenish the specified energy or material in a unit such as a community, population, or trophic level.

turnover rate in ecology, the rate at which material or energy in a population, trophic level, community, or other unit is removed and replenished.

trace element in biology, an element needed in extremely small amounts.

uracil a pyrimidine found in RNA.

vascular tissue in higher plants the tissues that conduct fluids within the plant. The xylem and phloem.

vector in biology, an organism, usually a secondary host, which serves to transmit a disease or parasite.

ventral the lower surface of a bilaterally symmetric animal. The surface nearest the ground.

vivaparous refers to an organism which bears living young.

voluntary muscle muscle tissue that is under the control of the central nervous system.

wild type the ''normal'' form of a gene as opposed to its mutant form.

xylem the vascular tissue of a plant which carries water and dissolved materials away from the roots.

zygote the fertilized egg.

credits

Figure 2.3 Rerendered from Biological Sciences Curriculum Study, *Biological Science: Molecules to Man,* rev. ed., Houghton Mifflin Company, Boston, 1968.

Figure 3.11(a), (c) From *Invertebrate Zoology,* 2d ed., by Robert D. Barnes. Copyright 1968 by W. B. Saunders Co. After *Protozoology* by R. P. Hall. Prentice-Hall, 1953.

Figure 3.11(b) From *Invertebrate Zoology,* 2d ed., by Robert D. Barnes. Copyright 1968 by W. B. Saunders Co.

Figure 3.12(a) From *Invertebrate Zoology,* 2d ed., by Robert D. Barnes. Copyright 1968 by W. B. Saunders Co.

Figure 3.12(b) From *Invertebrate Zoology,* 2d ed., by Robert D. Barnes. Copyright 1968 by W. B. Saunders Co. Reprinted with permission from John O. Corliss, *The Ciliated Protozoa,* 1961, Pergamon Press, Inc. After "Contribution à la connaissance des Infusoires planktoniques," by E. Fauré-Fremiet. Bulletin Biologique de la France et de la Belgique, 1924.

Figure 3.12(c) From *Invertebrate Zoology,* 2d ed., by Robert D. Barnes. Copyright 1968 by W. B. Saunders Co. After *General Protozoology* by V. A. Dogiel, revised by Poljanskij and Chejsin, 2d ed., 1965. The Clarendon Press, Oxford.

Figure 3.13(a) From *Invertebrate Zoology,* Second Edition, by Paul A. Meglitsch. Copyright © 1967, 1972 by Oxford University Press, Inc. Reprinted by permission.

Figure 3.13(b) From *Invertebrate Zoology,* 2d ed., by Robert D. Barnes. Copyright 1968 by W. B. Saunders Co.

Figure 3.13(c) From *Invertebrate Zoology,* 2d ed., by Robert D. Barnes. Copyright 1968 by W. B. Saunders Co. After "On the biology of *Calanus finmarchicus.* IX. Feeding and digestion in the young stages," by S. M. Marshall and A. P. Orr, in the *Journal of the Marine Biological Association of the United Kingdom,* Plymouth, 35:587–603 (1956). By permission of the Cambridge University Press.

Figure 3.14 (coiled cell) From *Invertebrate Zoology,* 2d ed., by Robert D. Barnes. Copyright 1968 by W. B. Saunders Co.

Figure 3.16(a) From *Invertebrate Zoology,* 2d ed., by Robert D. Barnes. Copyright 1968 by W. B. Saunders Co. After *Animals Without Backbones,* 2d ed., 1948, by Ralph Buchsbaum. Published by the University of Chicago Press. Copyright 1938 and 1948 by the University of Chicago. All rights reserved.

Figure 3.16(b) From *Invertebrate Zoology,* 2d ed., by Robert D. Barnes. Copyright 1968 by W. B. Saunders Co.

Box 3.2(a)–(d) From *Invertebrate Zoology,* 2d ed., by Robert D. Barnes. Copyright 1968 by W. B. Saunders Co. (d) After "On the feeding mechanism of the copepods, *Calanus finmarchicus* and *Diaptomus gracillus,*" by H. G. Cannons. *Journal of Experimental Biology,* Vol. 6 (1929). Reprinted by permission of Cambridge University Press.

Box 3.3(a) (larval and adult) From *Invertebrate Zoology,* 2d ed., by Robert D. Barnes. Copyright 1968 by W. B. Saunders Co.

Box 3.3(b) (larval) From *Invertebrate Zoology,* 2d ed., by Robert D. Barnes. Copyright 1968 by W. B. Saunders Co. (adult) From *Invertebrate Zoology,* 2d ed., by Robert D. Barnes. Copyright 1968 by W. B. Saunders Co. After "Sexual Phases in *Crepidula,*" by W. R. Coe, in *Journal of Experimental Zoology,* Vol. 72, No. 3, 1936, pp. 455–477.

Box 3.3(c) (adult) From *Invertebrate Zoology,* 2d ed., by Robert D. Barnes. Copyright 1968 by W. B. Saunders Co. (larval) From "Metamorphosis in Crustaceans," by John D. Costlow, Jr., in *Metamorphosis, A Problem in Developmental Biology,* edited by William Etkin and Lawrence I.

Gilbert. Used by permission of Plenum Publishing Corporation.

Figure 3.23 From *Invertebrate Zoology,* 2d ed., by Robert D. Barnes. Copyright 1968 by W. B. Saunders Co.

Figure 3.24 From *Invertebrate Zoology,* 2d ed., by Robert D. Barnes. Copyright 1968 by W. B. Saunders Co.

Figure 3.25(b) From *Invertebrate Zoology,* 2d ed., by Robert D. Barnes. Copyright 1968 by W. B. Saunders Co. After "Evolution of the Annelida, Onychophora and Arthropoda," by Snodgrass. Smithsonian Miscellaneous Collection, Vol. 97, No. 6, 1938.

Figure 3.25(c) From *Invertebrate Zoology,* 2d ed., by Robert D. Barnes. Copyright 1968 by W. B. Saunders Co.

Figure 3.29 From *Invertebrate Zoology,* 2d ed., by Robert D. Barnes. Copyright 1968 by W. B. Saunders Co.

Figure 3.33(a) From *Invertebrate Zoology,* 2d ed., by Paul A. Meglitsch. Copyright © 1967, 1972 by Oxford University Press, Inc. Reprinted by permission.

Figure 3.33(b) From *Invertebrate Zoology,* 2d ed., by Robert D. Barnes. Copyright 1968 by W. B. Saunders Co.

Figure 3.37 From *Vergleichende Anatomie der Wirbeltiere* by J. E. W. Ihle, P. N. van Kampen, H. F. Nierstrasz, and J. Versluys. Heidelberg: Springer-Verlag, 1927. Used by permission.

Figure 3.41 From *The Algae: A Review* by G. W. Prescott. Houghton Mifflin Company, Boston, 1968.

Figure 3.44 From *The Algae: A Review* by G. W. Prescott. Houghton Mifflin Company, Boston, 1968.

Figure 4.15 From *The Vertebrate Body, A Shorter Version of the Second Edition,* by Alfred S. Romer. Copyright 1956 by W. B. Saunders Company.

Box 4.1 From *The Vertebrate Body, A Shorter Version of the Second Edition,* by Alfred S. Romer. Copyright 1956 by W. B. Saunders Company.

Figure 4.19(a) From *The Vertebrate Body, A Shorter Version of the Second Edition,* by Alfred S. Romer. Copyright 1956 by W. B. Saunders Company.

Figure 4.22 From Biological Sciences Curriculum Study, *Biological Science: Molecules to Man,* 3d ed., Houghton Mifflin Company, Boston, 1973.

Figure 4.23 From Biological Sciences Curriculum Study, *Biological Science: Molecules to Man,* 3d ed., Houghton Mifflin Company, Boston, 1973.

Box 4.4 (seeds) From Biological Sciences Curriculum Study, *Biological Science: Molecules to Man,* 3d ed., Houghton Mifflin Company, Boston, 1973.

Figures 5.1–5.5 From Biological Sciences Curriculum Study, *Biological Science: Molecules to Man,* 3d ed., Houghton Mifflin Company, Boston, 1973.

Box 5.1(c) From *Visual Perception* by Tom N. Cornsweet. Copyright 1970 by Academic Press, Inc. Use by permission. After *Vision and the Eye,* 2d ed., by M. H. Pirenne, Associated Book Publishers, London, 1967.

Box 5.1(d) From "The Problem of Visual Excitation," by George Wald, Paul K. Brown, and Ian R. Gibbons, in *Journal of the Optical Society of America,* Vol. 53, No. 1, pp. 20–35, January 1963. Copyright by the Optical Society of America. After *Visual Perception* by Tom N. Cornsweet. Copyright 1970 by Academic Press, Inc. Used by permission.

Figure 5.14 (man's brain) From Biological Sciences Curriculum Study, *Biological Science: Molecules to Man,* rev. ed., Houghton Mifflin Company, Boston, 1968.

Figures 5.14–5.16 (nonhuman brains) From Biological Sciences Curriculum Study, *Biological Science: Molecules to Man,* 3d ed., Houghton Mifflin Company, Boston, 1973.

Figure 5.18 From Biological Sciences Curriculum Study, *Biological Science: Molecules to Man,* 3d ed., Houghton Mifflin Company, Boston, 1973.

Figure 5.23 From Biological Sciences Curriculum Study, *Biological Science: Molecules to Man,* 3d ed., Houghton Mifflin Company, Boston, 1973.

Box 5.4 From Biological Sciences Curriculum Study, *Biological Science: Molecules to Man,* 3d ed., Houghton Mifflin Company, Boston, 1973.

Figure 6.22 From Helena Curtis, *Biology,* Worth Publishers, Inc., New York, 1968. After *Fundamentals of Comparative Embryology of the Vertebrates,* rev. ed., by Alfred F. Huettner. Macmillan Publishing Company, 1949. Used by permission.

Box 6.4(a), (b) From *Patterns and Principles of Animal Development,* by John W. Saunders, Jr. Macmillan Publishing Company, 1970. Copyright © John W. Saunders, Jr., 1970. After

Comparative Embryology of the Vertebrates, by O. E. Nelsen. McGraw-Hill Book Company, 1953.

Box 6.4(c), (d), (h) From Helena Curtis, *Biology,* Worth Publishers, Inc., New York, 1968. After *Fundamentals of Comparative Embryology of the Vertebrates,* rev. ed., by Alfred F. Huettner. Macmillan Publishing Company, 1949. Used by permission.

Box 6.4(e)–(g) From *Patterns and Principles of Animal Development,* by John W. Saunders, Jr. Macmillan Publishing Company, 1970. Copyright © John W. Saunders, Jr., 1970.

Figure 6.23 From *An Introduction to Embryology,* 2d ed., by B. I. Balinsky. W. B. Saunders Company, 1965. After *Embryonic Development and Induction* by Hans Spemann. Yale University Press, 1938.

Figure 7.2 From Biological Sciences Curriculum Study, *Biological Science: Molecules to Man,* 3d ed., Houghton Mifflin Company, Boston, 1973.

Box 8.1 After J. Clausen et al., *Experimental Studies on the Nature of Species.* Washington: Carnegie Institute of Washington, Pub. 520, 1940.

Figure 8.7 From *Adaptation,* 2d ed., by Bruce Wallace and Adrian M. Srb. Copyright 1964. By permission of Prentice-Hall, Inc., Englewood Cliffs, N.J.

Figure 8.8 From *Populations, Species, and Evolution,* by Ernst Mayr. Copyright 1970 by the Belknap Press of Harvard University Press.

Figure 8.9 From *Biology of Populations* by R. H. MacArthur and J. H. Connell. Copyright 1966 by John Wiley and Sons, Inc. After J. H. Connell in *Ecology,* Vol. 142, 1961. By permission of Duke University Press.

Figure 10.3 From *Fundamentals of Ecology,* 3d ed., by E. P. Odum. Copyright 1971 by W. B. Saunders Company. After "A. Heat, Radiant and Sensible, Chapter 1—Radiant Energy, Its Receipt and Disposal," by D. M. Gates, in *Meteorological Monographs,* Vol. 6, No. 28, July 1965, the American Meteorological Society.

Box 10.1 Tables A, C From "Efficiency of Energy Use in the United States," by E. Hirst and J. C. Moyer, *Science,* Vol. 179, pp. 1299–1304, 30 March 1973. Copyright 1973 by the American Association for the Advancement of Science. Table B From Joel Darmstadter et al., *Energy in the World Economy: A Statistical Review of Trends in Output, Trade and Consumption since 1925.* Published for Resources for the Future by Johns Hopkins University Press, 1971.

Figure 10.9 From "Energy Flow of the Silver Springs Community," by H. T. Odum, in "Trophic Structure and Productivity of Silver Springs, Florida," *Ecological Monographs,* 27:55–112, 1957. Used by permission of Duke University Press.

Figure 10.10 From *Pathways of Energy Flow in a South Florida Estuary,* by W. E. Odum, Ph.D. dissertation, University of Miami, 1970.

Figure 10.12 From *Fundamentals of Ecology,* 3d ed., by E. P. Odum. Copyright 1971 by W. B. Saunders Company.

Figure 10.13(a) From "Energy Flow of the Silver Springs Community," by H. T. Odum, in "Trophic Structure and Productivity of Silver Springs, Florida," *Ecological Monographs,* 27:55–112, 1957. Used by permission of Duke University Press.

Figure 10.13(b) From *Fundamentals of Ecology,* 3d ed., by E. P. Odum. Copyright 1971 by W. B. Saunders Company.

Figure 10.14 From "A Further Ecological Study of a Calcareous Stream in the 'Black Mountain' District of South Wales," by J. R. Erichsen Jones. *Journal of Animal Ecology,* 18:142–159 (1949).

Figures 11.11–11.12 From *Fundamentals of Ecology,* 3d ed., by E. P. Odum. Copyright 1971 by W. B. Saunders Company.

Figure 11.16 From "DDT Residues in an East Coast Estuary: A Case of Biological Concentration of a Persistent Insecticide," by George M. Woodwell, Charles F. Wurster, Jr., and Peter A. Isaacson, in *Science,* Vol. 156, pp. 821–824, 12 May 1967.

Figures 13.1–13.2 From *Invertebrate Zoology,* 2d ed., by Robert D. Barnes. Copyright 1968 by W. B. Saunders Company.

Figure 13.11 From *Fundamentals of Ecology,* 3d ed., by E. P. Odum. Copyright 1971 by W. B. Saunders Company. After "Fluctuations in the numbers of the varying hare (*Lepus Americanus*)," by D. A. MacLulich. University of Toronto Studies. Biological Series. No. 43 (1937).

Figures 14.3–14.4 From *Fundamentals of Ecology,* 3d ed., by E. P. Odum. Copyright 1971 by W. B. Saunders Company.

Figures 14.8–14.9 From "The Strategy of Ecosystem Development," by E. P. Odum, *Science,* Vol. 164, 18 April 1969, pp. 262–270.

Figure 14.11 From *Fundamentals of Ecology,* 3d ed., by E. P. Odum. Copyright 1971 by W. B. Saunders Company.

Figure 14.12 From *Fundamentals of Ecology,* 3d ed., by E. P. Odum. Copyright 1971 by W. B. Saunders Company. After "The Pygmy Forest-Podsol Ecosystem and Its Dune Associates of the Mendocino Coast," by H. Jenny, R. J. Arkley, and A. M. Schultz, in *Madroño*, Vol. 20 (1969).

Figure 15.2 From *Fundamentals of Ecology,* 3d ed., by E. P. Odum. Copyright 1971 by W. B. Saunders Company.

Figure 15.17 From *Fundamentals of Ecology,* 3d ed., by E. P. Odum. Copyright 1971 by W. B. Saunders Company. After F. S. Russell, in *The Seas* by Russell and Yonge, Frederick Warne & Co., Ltd., 1963.

Figure 15.18 From *Fundamentals of Ecology,* 3d ed., by E. P. Odum. Copyright 1971 by W. B. Saunders Company. After Hedgpeth, Joel W. "The Classification of Estuarine and Brackish Waters and the Hydrographic Climate," in "Report of the Committee on a Treatise on Marine Ecology and Paleoecology 1950–1951," No. 11, Washington, D.C., National Research Council, December, 1951, pp. 49–56.

Figure 15.19 From *Fundamentals of Ecology,* 3d ed., by E. P. Odum. Copyright 1971 by W. B. Saunders Company.

index

B

C

F

G

H

N

O

P

T

X

Y

Z

ABCDEFGHIJ–H–7987654